Lecture Notes in Networks and Systems 678

The series "Lecture Notes in Networks and Systems" publishes the latest developments in Networks and Systems—quickly, informally and with high quality. Original research reported in proceedings and post-proceedings represents the core of LNNS.

Volumes published in LNNS embrace all aspects and subfields of, as well as new challenges in, Networks and Systems.

The series contains proceedings and edited volumes in systems and networks, spanning the areas of Cyber-Physical Systems, Autonomous Systems, Sensor Networks, Control Systems, Energy Systems, Automotive Systems, Biological Systems, Vehicular Networking and Connected Vehicles, Aerospace Systems, Automation, Manufacturing, Smart Grids, Nonlinear Systems, Power Systems, Robotics, Social Systems, Economic Systems and other. Of particular value to both the contributors and the readership are the short publication timeframe and the world-wide distribution and exposure which enable both a wide and rapid dissemination of research output.

The series covers the theory, applications, and perspectives on the state of the art and future developments relevant to systems and networks, decision making, control, complex processes and related areas, as embedded in the fields of interdisciplinary and applied sciences, engineering, computer science, physics, economics, social, and life sciences, as well as the paradigms and methodologies behind them.

Indexed by SCOPUS, INSPEC, WTI Frankfurt eG, zbMATH, SCImago.

All books published in the series are submitted for consideration in Web of Science.

For proposals from Asia please contact Aninda Bose (aninda.bose@springer.com).

Marcelo V. Garcia · Carlos Gordón-Gallegos
Editors

CSEI: International Conference on Computer Science, Electronics and Industrial Engineering (CSEI)

Advances and Applications in Computer Science, Electronics and Industrial Engineering. Proceedings of the Conference on Computer Science, Electronics and Industrial Engineering (CSEI 2022)

Springer

Editors
Marcelo V. Garcia
Universidad del País Vasco
Bilbao, Spain

Carlos Gordón-Gallegos
FISEI
Universidad Técnica de Ambato
Ambato, Ecuador

ISSN 2367-3370 ISSN 2367-3389 (electronic)
Lecture Notes in Networks and Systems
ISBN 978-3-031-30591-7 ISBN 978-3-031-30592-4 (eBook)
https://doi.org/10.1007/978-3-031-30592-4

This Springer imprint is published by the registered company Springer Nature Switzerland AG
The registered company address is: Gewerbestrasse 11, 6330 Cham, Switzerland

Preface

This book contains the best papers accepted for presentation and discussion at the 2022 International Conference on Computer Science, Electronics, and Industrial Engineering (CSEI 2022). This conference had the support of the Technical University of Ambato and Springer, Lecture Notes in Networks and Systems. It was held at Ambato, Tungurahua, Ecuador, on November 07–11, 2022.

CSEI 2022 is an international scientific conference especially dedicated to Computer Science, Electronics, and Industrial Engineering and its solutions. The main objective of the CSEI 2022 conference is to provide a forum for researchers, developers, and practitioners to review and discuss the most recent trends in the Computer Science, Electronics, and Industrial Engineering areas.

The International Conference on Computer Science, Electronics, and Industrial Engineering 2022 is built on the successes of the First CSEI 2019, Second CSEI 2020, and Third CSEI 2021 held in Ambato, Ecuador. The program committee of CSEI 2022 was composed of a multidisciplinary group of more than 100 experts in different areas of Information Systems, Electronics, and Industrial Engineering. They have had the responsibility for evaluating, in a double-blind review process, the papers received for each of the main themes proposed for the conference: (1) Computer Science, (2) Electronics, and (3) Industrial Engineering.

CSEI 2022 was honored to invite 6 experts to give the keynote speeches: Dr. Ruxandra Stoean from the University of Craiova, Rumania; Dr. Carlos Cano from the University of Almeria, Spain; Dr. Carlos Rovetto from Technological University of Panama, Panama; Dr. Daniel Jerez from Technical University of Ambato, Ecuador; Dr. Pedro Escudero from Indoamerica University, Ecuador; and Dr. Carlos Peñafiel from National University of Chimborazo, Ecuador.

The International Conference on Computer Science, Electronics, and Industrial Engineering 2022 was a successful, fruitful online academic event and received about 80 contributions from 10 countries around the world. From the exhaustive and exigent review process, only 44 best papers were accepted for presentation and discussion at the conference. The accepted papers are published by Springer in the series "Lecture Notes in Networks and Systems (LNNS)" in one volume and will be submitted for indexing by ISI, EI Compendex, Scopus, DBLP, and/or Google Scholar, among others. The program included keynote speeches, oral presentations, and discussions covering a wide range of subjects.

The conference was an unforgettable experience thanks to the organizers, the authors for their valuable contributions, and the attendees for their active participation. Besides, it is important to express appreciation to the reviewers, who provided constructive criticism and stimulating comments and suggestions to improve the author's contributions. Also, we are grateful to the conference chair, co-chairs and host, and all members of the technical program committee for their active contribution to the execution of the CSEI

2022. Finally, we thank the internationally renowned scientists who acted as keynote speakers at the CSEI 2022 conference.

November 2022

Marcelo V. Garcia
Carlos Gordon

Organization

General Chair

Marcelo Vladimir García Sánchez	Technical University of Ambato, Ecuador

Co-chairs

Carlos Nuñez	Technical University of Ambato, Ecuador
Carlos Gordón	Technical University of Ambato, Ecuador

Organizing Committee

Pilar Urrutia	Technical University of Ambato, Ecuador
Carlos Sánchez	Technical University of Ambato, Ecuador
Clay Aldás	Technical University of Ambato, Ecuador
Christian Mariño	Technical University of Ambato, Ecuador
Geovanni Brito	Technical University of Ambato, Ecuador
Julio Balarezo	Technical University of Ambato, Ecuador
Anita Larrea	Technical University of Ambato, Ecuador
Mauricio Carranza	Technical University of Ambato, Ecuador
Mario García	Technical University of Ambato, Ecuador
Fernando Urrutia	Technical University of Ambato, Ecuador
Ismael Ortiz	Technical University of Ambato, Ecuador
Daysi Ortiz	Technical University of Ambato, Ecuador
Israel Naranjo	Technical University of Ambato, Ecuador
David Torres	Technical University of Ambato, Ecuador
Marlon Santamaría	Technical University of Ambato, Ecuador
Andrea Sánchez	Technical University of Ambato, Ecuador
Sandra Carrillo	Technical University of Ambato, Ecuador
Santiago Jara	Technical University of Ambato, Ecuador
Santiago Altamirano	Technical University of Ambato, Ecuador
Franklin Tigre	Technical University of Ambato, Ecuador
César Rosero	Technical University of Ambato, Ecuador

Reviewer Committee

Josef Tomas	Aalen University of Applied Sciences, Germany
Cristina Paez Quinde	Instituto Superior Tecnológico España, Ecuador
Franklin Tchakounté	University of Ngaoundéré, Cameroon
Qinghe Zheng	Shandong Management University, China
Anand Nayyar	Duy Tan University, Vietnam
Vu Kien Phuc	University of Economics Ho Chi Minh City (UEH), Vietnam
Jose E. Naranjo	Universidad Técnica de Ambato, Ecuador
K. Selvakumar	Anna University, India
Özgür Tonkal	Samsun University, Turkey
Sajag Chauhan	CSEI, India
Gandhiya Vendhan	Bharathiar University, India
Mohamed M. F. Darwish	Department of Electrical Engineering, Faculty of Engineering at Shoubra, Benha University, Egypt
Predrag Dašić	Academy of Professional Studies Šumadija—Department of Trstenik, Serbia
Pablo Bengoa Ganado	Tecnalia Research & Innovation, Bilbao, Spain
Mónica Aresta	University of Aveiro, Portugal
Krzysztof Ejsmont	Warsaw University of Technology, Faculty of Mechanical and Industrial Engineering, Poland
Resmi N. G.	Muthoot Institute of Technology and Science, India
M. Mujiya Ulkhaq	Diponegoro University, Indonesia
Aitziber Mancisidor Barinagarrementeria	University of the Basque Country (UPV/EHU), Bilbao, Spain
Ivan Pires	Instituto de Telecomunicações, Universidade da Beira Interior, Covilhã, Portugal
Ratko Pilipović	University of Ljubljana, Faculty of Computer Science, Slovenia
Sergio Luján Mora	Universidad de Alicante, Spain
Asier Salazar Ramirez	Universidad del Pais Vasco/Euskal Herriko Unibertsitatea (UPV-EHU), Bilbao, Spain
Sebastian Saniuk	Department of Engineering Management and Logistic Systems, Faculty of Economics and Management, University of Zielona Góra, Poland
Vitor Santos	Universidade Nova de Lisboa—NOVA IMS (Nova Information Management School), Portugal
Giulia Cisotto	University of Milano-Bicocca, Italy

About the Speakers

Deep Learning for Computational Support in the Restoration of Cultural Heritage Artifacts

Ruxandra Stoean

Is Associate Professor at the University of Craiova, Romania, and Principal Investigator at the Romanian Institute of Science and Technology, Cluj-Napoca, Romania. She holds a Ph.D. in computer science, focused on optimization through evolutionary computation. Her currents research interests involve the development of deep learning models for images and signals, with applications in medicine, engineering, and cultural heritage.

Multi-feed Antenna Designed by Characteristic Modes Theory

Carlos Cano Domingo

Graduated in telecommunication. He received two master's degrees in telecommunication and mobile systems from the University of Malaga, Málaga, Spain. He is currently pursuing a Ph.D. degree at the University of Almeria, Spain. His main research interests include electromagnetic propagation, signal analysis, embedded systems, wireless synchronization, and machine learning. Recent advances in the Ph.D. topic have been motivated by the international collaboration between several research institutions, mainly formed by the University of Almería, the University of Málaga and Craiova, where he had been researching through 4-month Ph.D. stays

Approach to Optimize the Production and Demand of Electrical Energy through Systems of Inequalities and a Regression Model

Carlos A. Rovetto

Ph.D. in Computer Systems Engineering, University of Zaragoza, Spain; master's in Computer Systems Engineering, University of Zaragoza, Spain; Bachelor of Systems and Computing Engineering, Technological University of Panama Specialization in Higher Education, Universidad del Istmo, Panama; Bachelor of Programming Technology and Systems Analysis, Technological University of Panama; and Engineering Technician with Specialization in Programming and Systems Analysis, Technological University of Panama

Agile Computing

Daniel Sebastian Erez Mayorga

Currently dedicated to the analysis and implementation of information systems, management of open, democratic projects with autonomy and responsibility, promoting a proactive and effective attitude of the actors, and optimizing results-oriented processes that generate value for organizations. I am also ready to contribute my knowledge in the field of IT Audit and Control.

Sensors and Microfluidics, from Biomedical Applications to Industry

Pedro-Fernando Escudero-Villa

Ph.D. in Electronic and Telecommunications Engineering, Autonomous University of Barcelona, Biomedical Applications Group (GAB, IMB-CNM), Spain; master's in Biomedical Engineering, Polytechnic University of Madrid, Spain, work and research area: bioinstrumentation and nanotechnology used in optical hyperthermia tests with gold nanoparticles; master's in Electrical, Electronic, and Automatic Engineering, Carlos III University of Madrid, Spain, work and research area: optoacoustic spectroscopy used to characterize contrast elements in tomography, gadolinium nanoparticles, and carbon nanotubes; Electronics and Computer Engineer, Chimborazo Higher Polytechnic School, Riobamba, Ecuador.

Multi-feed Antenna Designed by Characteristic Modes Theory

Carlos Ramiro Peñafiel-Ojeda

Engineer in Electronics and Telecommunications at the National University of Chimborazo, master's in Telecommunications Engineer, and Ph.D. in Telecommunications Engineering at the Polytechnic University of Valencia, Spain. He also participated in the "Messageri della Conoscenza" Project, on the subject of sensor networks to command drones, managing to win a research grant at the Université de Technologie de Compiègne, Compiegne, France. He collaborated with the Spanish company Netllar, in the development of microwave networks for the generation of shaped beams, and collaborates with the company Huawei based in Finland, for the design of MIMO antennas that can be integrated into mobile devices. He participated as Researcher in the Antennas and Electromagnetic Group of the Queen Mary University of London and Professor at the Salesian Polytechnic University of Quito and the National University of Chimborazo.

Contents

Artificial Intelligence

Electronics Engineering

Industrial Engineering

About the Editors

Marcelo V. Garcia, Ph.D. (**Scopus ID:** 56597269000, **Web of Science Researcher ID:** J-3195-2019) He studied electronics and instrumentation engineering at the University of the Armed Forces-ESPE. In 2013, he obtained his master's degree in Control, Automation, and Robotics Engineering, and in 2018, he obtained his doctorate at the University of the Basque Country (UPV/EHU). His studies were carried out thanks to a grant from the Ecuadorian government. From 2008 to 2013, he worked as Engineer in different companies in the area of oil and gas in Ecuador such as Schlumberger, Petrobras, and Petroamazonas EP. His research interest is focused on the design of next-generation architectures based on Industry 4.0 in various domains such as automation and smart manufacturing.

Carlos Gordon Gallegos, Ph.D. (**Scopus ID:** 56419359600) He received his Electronic engineering and the M.Sc. degree in Networks and Telecommunications from the Technical University of Ambato (UTA) Ecuador in 2005 and 2010, respectively. He joined the Department of Optoelectronics and Laser Technology, Universidad Carlos III de Madrid, UC3M, Madrid, Spain in 2012, and he received Ph.D. degree in Electrical, Electronics and Automation Engineering in 2016, under the supervision of Prof. Guillermo Carpintero del Barrio. He has written more than sixty SCOPUS scientific papers. His current research interests include photonic integrated circuits in InP for the Millimeter-wave & Terahertz signal generation for high data rate wireless communication systems, semiconductor short-pulse laser systems, Artificial Intelligence in Robotics, and Renewable Energy.

Computer Science

QR Codes as a Strategy in Political Marketing 2.0

Leonardo Ballesteros-López[1(✉)], Carlos Mejía-Vayas[1], Sonia Armas-Arias[2], and Carla-S. Castro-Altamirano[2]

[1] Facultad de Ciencias Administrativas, Grupo de investigación Marketing C.S., Universidad Técnica de Ambato, Ambato, Ecuador
{ca.guerrero,carlosvmejia}@uta.edu.ec

[2] Facultad de Ciencias Humanas y de la Educación, Grupo de investigación Marketing C.S., Universidad Técnica de Ambato, Ambato, Ecuador
{sp.armas,ccasrto4835}@uta.edu.ec

Abstract. QR codes as a strategy in political marketing 2.0 is becoming more noticeable its impact on digital communication; it has become more entered to the different uses that can be given to this technology, projecting in the use of innovative strategies within the political 2.0. The purpose of this research is to determine the acceptance of QR codes as a strategy in political marketing 2.0 through the acceptability survey, these questions were developed in the political 2.0 context related to digital media focused on QR codes. The methodology applied in this research is of exploratory and experimental type, in this research the quantitative approach was applied through a survey of the Evaluation Model of Communication 2.0, which develops a qualitative analysis, in the same way the emotions that were transmitted in the campaigns directed to the voting population of the Technical University of Ambato were analyzed and quantified somatically. Within the results it is evident that political communication in social networks is not strategic, there is an excess of information that saturates the social network and lacks content of value to the citizen. Finally, it is concluded that political communication 2.0 is incipient in Ecuador, Facebook is used as a traditional communication channel, with rhetorical content, uninteresting, but loaded with positive emotionality to try to connect with the electorate, therefore the creation of QR codes as a new strategy is one of the most successful ways in political campaign.

Keywords: ICT · QR Codes · Marketing · Policy 2.0 · Marketing Strategies

1 Introduction

For [1] political marketing emerges in the mid-twentieth century, being this that goes from assumptions that goes into the psychology that goes hand in hand

Supported by organization x.

M. V. Garcia and C. Gordón-Gallegos (Eds.): CSEI 2022, LNNS 678, pp. 3–13, 2023.
https://doi.org/10.1007/978-3-031-30592-4_1

with advertising, this goes back to 1952 where the candidate Dwight Eisenhower was one of the first to resort to advertising agencies of that time with a campaign aimed at the public on television.

According to [2] QR codes are generally considered a key tool in political campaigns, since their objective is recruitment, considered as effective due to the immediate access to digital content, experts in acceptance campaigns agree and suggest that QR codes be taken into account as a media tool.

Similarly, [3] One of the aspects to consider is that it is European countries as it is in Spain has been increasingly taken into account that QR codes is a tool that is ready for use in political campaigns, For the young generation voter in what refers QR codes is returned a better acceptance, which means a better distribution of the budget, ensuring better performance in campaigns with more voters.

With reference to [4] QR codes in metrics is an application to be taken into account for candidates in general that implement the instrument of information collection that is going to be the survey with the application of QR codes is in practice an easy way to perform online which can be done without any problem anywhere, considering putting them in strategic places where people usually feel bored which happens in scenarios such as bus stops, waiting lines and even in the subway.

As a result [5] these codes are a fundamental tool for marketers in the digital area, this tool involves a large number of applications, the two-dimensional codes have evolved for good that today are no longer just a simple black code if not now you can give way to customization, including the logo and colors of the company, which contains information such as a Vcard cvon url, text links or SMS.

The purpose of the project with the present theme is to be applied to the current and future political campaigns, we must consider that every day there are technologies that are applicable in relation to the political world that goes hand in hand with marketing, considering that nowadays mobile devices have become more accessible to most of the population in different ages.

The research aims to provide the population with a useful solution for people who want to enter or are already in the political world or for people who want to learn more about the different options of political campaigns, obtaining a broader picture and giving the option to correctly choose the person or group of people who will represent them with the use of digital technologies such as two-dimensional codes.

The constant increase in technologies for communication, becoming increasingly accessible to most of the general public, which currently have a smartphone that have a camera are able to recognize codes called BIDI, with an application designed for that purpose, thus being intended as digital tools in political campaigns, political candidates every day are opting for this type of technology as a means of communication to reach people with campaign information easily and efficiently [3].

This type of codes has several advantages for immediate access to data and information quickly and accurately, with its information storage system as links that lead directly to the website of the electoral party or simply social networks of the political candidate or election campaigns that provide more in-depth information about the candidate or the political party itself.

QR codes have become a resource of utmost importance in advertising campaigns, opening more and more in the public that still has some misgivings when using this technology in advertising campaigns, The use of these technologies is becoming more and more popular among the public that is still a little suspicious at the moment of using this technology, giving way to the possibility of companies to have an identification seal, as well as in each political candidate, in addition, this type of codes offer the user the option of being personalized, with the integration of logos, color modification and even shaping the candidate's face [6].

Political marketing has gained strength in recent elections which is a good way for a political party to take strength in the candidacies, it is increasingly clear the picture in the use of social networks and websites are the most efficient means of communication and more support today, giving way to the QR codes as an essential tool link between the user and the information, as regards political marketing, which every day is gaining more strength, so now is reflected in the political environment [5].

The motive of this research is to publicize the degree of acceptance of QR codes in the population, specifically in the university population, the purpose and motive of this research is to publicize the advantages that will give this type of digital tools as a link of communication of information in a competent political campaign, in addition to demonstrating the obvious competitive advantages over other political candidates, whose budget is higher but opt for campaigns with traditional or obsolete tools for today.

2 State of the Art

The quick response code known as QR codes is a two-dimensional code which has similarity to what is known as a bar code with the function of encoding a message with limited length, if a QR code is a two-dimensional bar code, which its default form consists of black and white boxes that take the name of modules capable of encoding information such as the URL or simply a text message which to access this information requires the use of a cell phone, if the two-dimensional codes include different applications such as payment of digital services, in advertising for its ease of creating and accessing your information when scanned [7].

The QR codes the author [4] states that they mainly work in identification processes as well as in the collection of products, avoiding human errors, in simplification they are quick response codes, objectively designed to be used by smartphones, a QR code is identified by black modules which are superimposed on a white background, capable of storing text, links or other data.

The advantages and facilities of QR codes are many, it stands out mainly so that people interested in such information can access by simply scanning the two-dimensional code on the cell phone, giving the opportunity to save the code and read it later in more convenient places from anywhere they want, either by the time they have at that moment or prefer at any other time or place [8].

QR codes are two-dimensional quick response codes with coded information of great magnitude compared to traditional barcodes used in supermarkets, they are codes used to obtain information quickly, codes that are increasingly entering establishments or companies to be part of the process as they are in payment of products, services and to exchange customer and company information.

According to the author, he mentions that in political campaigns the benefits are clear at the time of the campaigns to attract the audience. Effectiveness in attracting more people compared to traditional media.

Easy and fast updating of data and information. Reduced costs compared to traditional forms of information massification.

It is known that technology advances by leaps and bounds so the implementation of QR codes as a strategy is essential for virtual resources either multimedia, impact the public in order to attract attention, it is worth mentioning that QR codes are considered a technological tool that allows the user to access information quickly and easily. In addition, QR codes is a free digital technology, so it gave way too many applications in marketing and different areas of daily life of people [9,19].

2.1 QR Codes in Digital Marketing

According to the author the quick response codes over the years until today have been an important tool that goes hand in hand with marketing, this is due to the high popularity of mobile devices, the codes with its technology capable of storing information at low costs are mainly used in digital marketing. QR codes is a tool of digital marketing becoming a necessary point to consider achieving the objectives in terms of obtaining data that is given by the interaction of users such as, for example, with scanning the code with promotions on downloading free products or discounts on purchases.

Political marketing [10] mentions that it goes from communication that has electoral purposes, which even reaches the management and use of research in order to generate strategies that allow in the design of products, and services, in addition to improvements in politics in electoral campaigns that will have an impact during the period of governance, a clear difference that has with companies whose objective is to maximize profitability with a greater market share, instead in politics seeks a population in order to get the largest number of voters through a message through strategies that ensure the presidency.

The practice of political marketing [11] states that it lies in strategies that are projected to communication, whose function is to inform and interpret political data that seeks to deepen the reality of cultural and popular, its main objective is to capture attention, generate confidence in large masses, a change in political marketing is to generate cyber candidates, also seeks to focus on generating ties with millennials.

According to the author [12] defends that political marketing involves analyzing the electoral market, since with this information it will be possible to give a better understanding that voters have sufficient motivation to give the vote or accept as a representative to a political candidate, everything related to digital technology provides for the collection of information, with the use of devices technologies, social networks and the internet in general, so mentions the brand as a key strategic part of political marketing, since it will be the identity that will give the political candidate [13,16,17].

2.2 Success Factors in Political Marketing

The author [14] states that in recent elections the best candidate does not win, the one with the best marketing strategies wins, past elections are a sign of this where the best marketing managed to position the president, demonstrating the power that marketing has in something as important as the presidential elections of a country here it is clearly applicable that not the best product is the best positioned in the market, nor the best candidate wins the presidential elections.

One of the success factors in recent political elections was the application of social networks, cases such as Barak Obama, remaining for history with the use of digital media strategies focused on political marketing, customizing websites, Mauricio Macri in Argentina managing to capture voters mostly undecided through social networks like Twitter and Facebook, taking initiative in these digital media to create a fast and efficient way to answer questions, clarifying the decision in citizens at the time to say the vote [15,18].

According to the author, mobile marketing [8] considers that it is the most appropriate tool to reach billions of users, especially focused on the new generations, since the cell phone has become so basic in their daily lives, with information available to them always, with facilities to enter digital platforms such as social networks, entertainment, in addition to stay connected with friends, family and even make new friends.

3 Methodology

In this research the exploratory type was applied, since this type of research is given in order to obtain the approach to the subject to which it has been raised to be studied and approached, in order to obtain the maximum information concerning the subject, giving a broader picture on the subject, it is the beginning of the research, the collection of initial information to give the step to more exhaustive research.

In addition, experimentation was applied since this type of research had the purpose of verifying the experimental variable not tested, in order to test the study variables, taking into account the conditions to control the increase or decrease of those variables and the effect to be achieved, in order to verify the hypothesis.

The instrument used in this research was the instrument applied for data collection, which consisted of 3 sociodemographic questions, 10 questionnaire questions comprising 7 Likert scale questions and 6 multiple choice questions, for a total of 13 questions in the questionnaire.

The Kolmogorov-Smirnov test allowed identifying the distribution of data within the experimentation, therefore allowing the acceptance or rejection of the hypothesis established for the research; having a p-value of less than 0.05 in the most representative questions that were selected for the calculation of the statistics; the null hypothesis is rejected and the acceptance of the alternative hypothesis is considered, which corroborates that QR Codes contribute as a strategy in political marketing 2.0 (Fig. 1).

		Use QR codes to stay informed about electoral proposals, political candidate profiles or simply keep abreast of political campaigns.	Occasions makes use of QR codes	Devices used most frequently when using the Internet	Most frequently used social network
N		376	376	376	376
Normal parameters[a, b]	Media	3,4149	2,2128	1,5745	1,8218
	Standard desviation	,76816	,64700	1,01445	,92817
More extreme differences	Absolute	,256	,461	,438	,269
	Positive	,256	,461	,438	,269
	Negative	-,226	-,342	-,286	-,188
Kolmogorov-Smirnov's Z		4,963	8,945	8,489	5,225
Sig. asintót. (bilateral)		,000	,000	,000	,000

a. The contrast distribution is the Normal distribution.
b. They have been calculated from the data

Fig. 1. Kolmogorov-Smirnov one same test

For the development of the proposal, the creation of QR codes in political campaigns focused on politics 2.0 is taken into account; the use of innovative technology that attracts the voting population is promoted.

Figure 2 shows the development of QR codes by means of the QR Monkey tool; a free platform where all kinds of information is personalized, in order to obtain additional information and motivation in the voter (Fig. 3).

Political communication 2.0 is determined by the interaction and relationship established through QR codes between political candidates and voters, this relationship is based on a communication that develops in equal spaces, therefore, it should not be traditional or unidirectional, in which the candidate speaks from a virtual platform and the citizen listens, on the contrary, the citizen has a voice, and the candidate establishes spaces for dialogue with them.

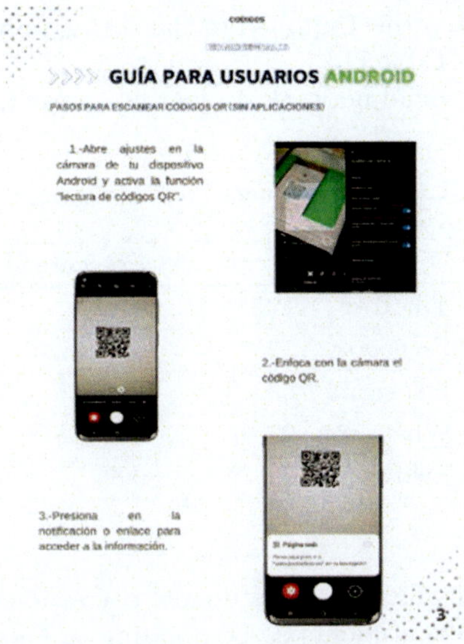

Fig. 2. Development QR code

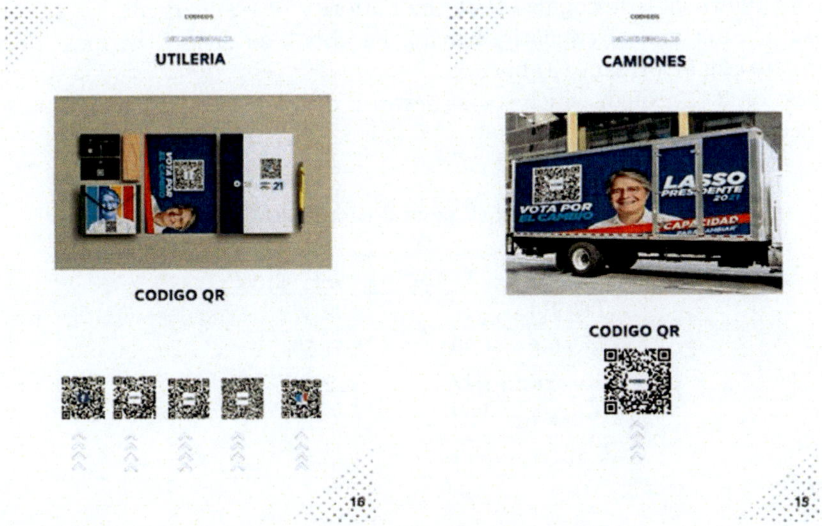

Fig. 3. Applications QR code

4 Results

The results of the 376 people surveyed, with a total of 16 questions consisting of 3 sociodemographic questions and 13 questions from the questionnaire referring to the topic of the research project applied to the voting population of the Technical University of Ambato (Table 1).

Question 1. Which social network do you use most frequently?

Table 1. Frequency social network

Options	Frequency	Percentage
Tik-Tok	172	45,7%
Facebook	125	33,5%
Instagram	53	14,1%
WhatsApp	23	6,1%
Twitter	2	0,5%

45.74% which is equivalent to 172 students corresponding to TikTok, 33.51% which is equivalent to 126 students corresponding to Facebook, 14.10% which is equivalent to 53 students corresponding to Instagram, 6.12% which is equivalent to 23 students corresponding to WhatsApp and 0.53% which is equivalent to 2 students correspond to Twitter which formed the total of the surveyed sample. This allows us to recognize that the majority of respondents use the social network TikTok more frequently, leaving Facebook as one of the most popular social networks at present (Table 2).

Question 2. Through which social network did you stay informed during the 2021 election campaigns?

Table 2. Electoral campaigns

Options	Frequency	Percentage
Tik-Tok	110	29,3%
Facebook	205	54,5%
Instagram	47	12,5%
WhatsApp	2	0,2%
Twitter	12	3,2%

From a total of 376 students corresponding to 100% among the sample, 54.52% which is equivalent to 205 students correspond to Facebook, 29.26% which is equivalent to 110 students correspond to TikTok, 12.50% which is equivalent to 47 students correspond to Instagram, 3.19% which is equivalent to 12 students correspond to Twitter and 0.53% which is equivalent to 2 students correspond to WhatsApp which formed the total of the surveyed sample. This allows us to recognize that most respondents were informed through the social network Facebook, with the social network tiktok gaining strength (Table 3).

Question 9. How likely are you to use QR codes to stay informed about electoral proposals, profiles of political candidates or simply to keep abreast of political campaigns?

Table 3. QR code usage

Options	Frequency	Percentage
Not likely	6	1,6%
Unlikely	24	6,4%
Neither probable nor improbable	117	47,1%
Likely	146	38,8%
Very likely	23	6,1%

Out of a total of 376 students corresponding to 100% among the sample, 1.60% which is equivalent to 67 students who are not at all likely to use QR codes to keep themselves informed about political campaigns, 6.38% which is equivalent to 24 students who are unlikely to use QR codes to keep themselves informed about political campaigns, 47,07% which is equivalent to 177 students who are neither likely nor unlikely to use QR codes to keep themselves informed about political campaigns, 38.83% which is equivalent to 146 students who are likely to use QR codes to keep themselves informed about political campaigns and 6.12% which is equivalent to 23 students who use QR codes to keep themselves informed about political campaigns that formed of the total sample surveyed.

This allows us to recognize that the majority of respondents have a decision that they are neither likely nor unlikely to make use of QR codes to keep informed during political campaigns.

5 Discussion

The survey yielded favorable results for the implementation of QR codes in political campaigns because several respondents agreed that QR codes, also noted the familiarity with the two- dimensional codes thanks to the digital messaging platform that is WhatsApp, since the use of QR codes is needed to enter from other devices on the website.

The application of the survey also helped to demonstrate that social networks and mobile devices go hand in hand, showing that they are currently the main elements to keep informed and intercommunicated with news and even when informing political campaigns with proposals and political profiles, being so that social networks are present in daily life, as well as mobile devices.

The contents they disseminate, as proven in the research, are thought in the interests of the candidate, not in the needs of the audience, for that reason, the participation and engagement of voters are low. It cannot be a coincidence that the four candidates with the highest popular acceptance and voting are those who obtain the best score in the communication process, this is mainly because they manage to build a community, maintain a 2.0 attitude, publish frequently in their social accounts, although the level of response and interaction is low.

Acknowledgment. Thanks to the Technical University of Ambato, to the Directorate of Research and Development (DIDE acronym in Spanish) for supporting the research group: Marketing C.S. and the project Aplicación del marketing digital como herramienta de transformación en la política 2.0 dentro de la provincia de Tungurahua: predicción y toma de decisiones mediante web semántica.

References

1. Peltokorpi, J., Isojärvi, L., Häkkinen, K., Niemi, E.: QR code-based material flow monitoring in a subcontractor manufacturer network. Procedia Manufact. **55**, 110–115 (2021)
2. Mathivanan, P., Ganesh, B.: QR code based color image stego-crypto technique using dynamic bit replacement and logistic map. Optik **225**, 165838 (2021)
3. Hill, G.N., Whitty, M.: Embedding metadata in images at time of capture using physical Quick Response (QR) codes. Inf. Process. Manag. **58**(3), 102504 (2021)
4. Peng-Cheng, H., Chin-Chen, C., Yung-Hui, L., Yanjun, K.: Enhanced (n, n)-threshold QR code secret sharing scheme based on error correction mechanism. J. Inf. Secur. Appl. **58**, 102719 (2021)
5. Ajeet Masih, E.: Feasibility of using QR code for registration & evaluation of training and its ability to increase response rate - the learners' perception. Nurse Educ. Today **111**, 105305 (2022)
6. Fu, Z., Fang, L., Huang, H., Yu, B.: Distributed three-level QR codes based on visual cryptography scheme. J. Vis. Commun. Image Represent. **87**, 103567 (2022)
7. Donaldson, A.: Digital from farm to fork: infrastructures of quality and control in food supply chains. J. Rural. Stud. **91**, 228–235 (2022)
8. Davidson, S.: The world wants to reopen: will vaccine passes be the key? Biom. Technol. Today **2021**(6), 5–7 (2021)
9. Reddy-Kummitha, R.: Smart technologies for fighting pandemics: the techno- and human- driven approaches in controlling the virus transmission. Gov. Inf. Q. **37**(3), 101481 (2020)
10. Ketron, S., Kwaramba, S., Miranda, W.: The "company politics" of social stances: how conservative vs. liberal consumers respond to corporate political stance-taking. J. Bus. Res. **146**, 354–362 (2022)
11. D'Attoma, I., Ieva, M.: The role of marketing strategies in achieving the environmental benefits of innovation. J. Clean. Prod. **342**(15), 130957 (2022)
12. Atkinson, A., Meadows, B., Emslie, C., Lyons, A., Sumnall, H.: 'Pretty in Pink' and 'Girl Power': an analysis of the targeting and representation of women in alcohol brand marketing on Facebook and Instagram. Int. J. Drug Policy **101**, 103547 (2022)
13. Zheng, W., Hwee-Ang, S., Singh, K.: The interface of market and nonmarket strategies: political ties and strategic competitive actions. J. World Bus. **57**(4), 101345 (2022)
14. Duarte, L.O., Vasques, R.A., Fonseca Filho, H., Baruque-Ramos, J., Nakano, D.: From fashion to farm: green marketing innovation strategies in the Brazilian organic cotton ecosystem. J. Clean. Prod. **360**, 132196 (2022)

15. Gao, Q., Zhang, Z., Li, Z., Xhao, X.: Strategic green marketing and cross-border merger and acquisition completion: the role of corporate social responsibility and green patent development. J. Clean. Prod. **343**, 130961 (2022)
16. Flores, E., Cumbajin, M., Sanchez, P.: Design of a synchronous generator of permanent magnets of radial flux for a pico-hydropower station. In: García, M.V., Fernández-Peña, F., Gordón-Gallegos, C. (eds.) Advances and Applications in Computer Science, Electronics and Industrial Engineering. AISC, vol. 1307, pp. 135–151. Springer, Singapore (2021). https://doi.org/10.1007/978-981-33-4565-2_9
17. Gualpa, T., Ayala, P., Caceres, J., Llango, E., Garcia, M.: Smart IoT watering platform based on orchestration: a case study. In: Garcia, M.V., Fernández-Peña, F., Gordón-Gallegos, C. (eds.) CSEI 2021. LNCS, vol. 433, pp. 191–204. Springer, Cham (2022). https://doi.org/10.1007/978-3-030-97719-1_11
18. Quintero Lorza, D.P., Duque Méndez, N.D., Gómez Soto, J.A.: GLORIA: a genetic algorithms approach to tetris. In: Nummenmaa, J., Pérez-González, F., Domenech-Lega, B., Vaunat, J., Oscar Fernández-Peña, F. (eds.) CSEI 2019. AISC, vol. 1078, pp. 111–126. Springer, Cham (2020). https://doi.org/10.1007/978-3-030-33614-1_8
19. Villacis-Copo, P.M., Saltos, L.F., Ponce-Sanchez, Y.E., Naranjo-Robalino, A., Garcia, M.V.: Comparative study of the level of bullying in students from a public and private institution; [estudio comparativo del nivel de acoso escolar en estudiantes de una institución pública y privada]. RISTI - Revista Iberica de Sistemas e Tecnologias de Informacao **2021**(E43), 56–68 (2021)

Methodology for Cyber Threat Intelligence with Sensor Integration

João-Alberto Pincovscy[1(✉)] and João-José Costa-Gondim[1,2]

[1] Post Graduation in Electrical Engineering (PPEE), Department of Electrical Engineering, University of Brasília (UnB), Brasília, DF 70910-900, Brazil
joao.pincovscy@aluno.unb.br, gondim@unb.br

[2] Department of Computer Science (CIC), University of Brasilia (UnB), Brasilia, DF 70910-900, Brazil

Abstract. Identifying attacks on computer networks is a complex task, given the huge number of machines, data diversity, and a large volume of data. Cyber Threat Intelligence consists of collecting, classifying, enriching, classifying data, and producing knowledge about threats in network defense systems. In this scenario, we find network Intrusion Detection Systems that specifically analyze network traffic and through signatures detect anomalies, generating records for system operators. The purpose of this work is to present a methodology to generate knowledge about Threat Intelligence, from the records of network sensors, collecting Threat or Compromise Indicators and enriching them to feed Threat Intelligence Sharing Platforms. Our methodology speeds up the decision-making process as it incorporates an up-to-date, public repository of signatures already in the collector, eliminating the threat identification phase in an additional step. For the demonstration and evaluation of the methodology, a proof of concept was carried out that covered the entire threat identification cycle.

Keywords: Threat Intelligence · Intrusion Detection · Anomaly Analysis · Threat Indicators

1 Introduction

Most publicized reports of known intrusions involve those that occur over networks, so if a computer is connected to a network, it is more susceptible to being hacked [21]. Especially in recent years with the development of digital communication technology and the increase in teleworking across the planet [13].

The Internet became popular if a large-scale growth of the Internet of Things (IoT), which in recent years has published a growing industry in clouds, smart, and an industry 4.0, also increasing the attack surface diversity of them [1]. In the sphere of Digital State Intelligence, the technological evolution of networks has enabled the evolution of threats, marked by the advent of Advanced Persistent

The authors are grateful for the support of ABIN TED 08/2019.

M. V. Garcia and C. Gordón-Gallegos (Eds.): CSEI 2022, LNNS 678, pp. 14–28, 2023.
https://doi.org/10.1007/978-3-031-30592-4_2

Threats (APT), with advanced, stealthy, continuous, and long-term attacks on specific target networks [37].

In this scenario, any device that connects to the Internet can potentially be a vector of invasion or target, with thousands of devices generating a huge volume of information about their connectivity, which may require Cyber Threat Intelligence (CTI) services, [9] to analyze and filter the data identifying possible attacks, whose main objective is to support organizations in understanding the risks and known threats, APT and unknown threats called zero-day [30,37].

Existing intelligence systems lack mechanisms for collecting and preliminary classification of information [18], and CTI sensors that are commonly part of network firewall systems are used [6].

As we will see below, despite recent advances in the collection, analysis, and storage of incident indicators used in CTI [1,2], the solutions adopted to support collection are not optimized for identification and correlation with Threat Indicators, being Indicators of Compromise (IoC) or Indicators of Attack (IoA) [33]. In addition, IoCs need additional information to be more easily evaluated and categorized in Threat Intelligence Sharing Platforms (TISP) [29].

This article proposes the integration of the Intrusion Detection System (IDS) or Intrusion Prevention System (IPS) [23] for collection using signatures in the first stage and the behavior of applications in honeypots [11], generating records of possible attacks or compromises. Then the enrichment of the data from the records through the collection of relevant complementary information. Its main contribution is to present a methodology to generate knowledge about Threat Intelligence, from the records of a network sensor, together with a proof of concept.

This article is organized into sections. In Sect. 2, definitions and related works are presented, followed by Sect. 3, which presents the methodology. In Sect. 4 the proof of concept is presented together with the results and their discussion. The conclusions are presented in Sect. 5.

2 Definitions and Related Works

Some relevant definitions along with related work are presented in the sequel.

2.1 Threat Intelligence Generation Process

Threat information is any information that helps the organization protect itself from a threat or detect the activities of an attacker. Security teams need a high degree of maturity to be able to interpret technical data from collections, organize them into information and correlate this information-producing CTI [20]. Thus, the threat intelligence generation process can be described as [7]: Collection, Processing, Analysis, Deploy, and Dissemination.

2.2 Data Model

In the search for the most appropriate data model, we found an important study that fits perfectly into our methodology, called the 5W3H method (What, Who, Why, When, Where, How, How much, and How long) [7]. This method subsidizes decision-making regarding the choice of data to be enriched. It answers the questions presented in Table 1.

Table 1. Description of the 5W3H method [7].

Question	Description
What	Directly describes the topic being addressed
Where	Specifies geographic references to the topic
When	Specifies relevant time frames to the topic like date and time
Who	Associates the topic with an entity capable of executing it
Why	Describes possible motivations for the occurrence of the topic
How	Describes the main characteristics and mechanisms of the topic
How much	Refers to the costs and impacts generated by the topic
How long	Description of the topic's effectiveness in terms of time

This is a method commonly applied as a management tool used in the strategic planning of companies. It is originally known as 5W2H (what, who, why, when, where, how, how much). However, it is applied in different areas to objectively evaluate a certain element [5,15].

The 5W3H method proved to be interesting because, in addition to dealing with the record in all its dimensions, it also deals with persistence (how long). Thus, this method allows the complete characterization of a threat [7].

In the suitability of this work, "What" is used to define the element under analysis. In CTI, it can be translated as the threat classification. You can create several classification parameters from types of threats to groups of signatures used in the collection. Then we have the "Where" that can characterize the origin. The "When" provides the moment of registration, characterized by the date and time of the event. "How" provides the method or Tactics, Techniques, and Procedures (TTPs) used by the threat. In all evidence of a threat or incident, it is essential to attribute the action to an author, characterized by the "Who". For a more assertive attribution, it is important to seek to define the "Why", contextualizing the scenario through the motivations of the event. Another question to be answered is the intensity of the event, answered by "How much". And finally, the durability of the event was answered by "How long". This questioning of the method is especially important for CTI when it comes to identifying APT-type threats.

2.3 Criteria for Data Enrichment

To define the criteria for data enrichment, the usability and relevance of the aggregated information were taken into account. The choice of data to be enriched is strategic to support real-time decision-making, use network defense rules and policies, or strengthen further investigations. Thus, a data model was created, prioritizing the source IP address of the flow, the DNS Domain, or the e-mail address, in that order.

Automated Enrichment. When it comes to Threat Indicators, there are basic data to be evaluated such as IP addresses, domain names, domain servers, malware behavioral elements, and email headers, among others. Indicators contain one or more elements that contextualize the threat. The context can include "When" timestamps, how long they were active, "How long", an indication of the severity of the incident, as well as information about the mechanism and dynamics of an attack (e.g. infection process, spread, and action) of malware. But to add intelligence to the event, we need to add information about TTPs [24] to enable the identification of "Who" threat actors and possible ongoing campaigns [29,36]. For all this information to be produced, the data enrichment process is necessary before the analysts evaluate the Threat Indicators, and it must be done in the most automated way possible.

When dealing with the topic of Cyber Threats, automation is essential due to the evolution in the number and sophistication of threats, the diversity of services to be protected, and the lack of specialized resources in the area of cybersecurity. We can list the basic needs in cybersecurity about automation [27,36]:

- Detection of Threat Indicators,
- Enrichment of information about threats,
- Detection and prevention of security incidents,
- Triage in the treatment of incidents in terms of their severity, and
- Information sharing control (5W3H).

Data enrichment can be used to support sophisticated malware detection frame-works that use Machine Learning (ML) for a signature generation [12]. There are situations where, after the performance of the ML process, the information is still not enough for the elaboration of signatures. Then, through the enrichment process, external data is aggregated to support the analysis process for creating the signatures.

Several data sources can be used to enrich Threat Indicators, such as social networks [17], Domain Name System (DNS) records, identification of routing prefixes that identify an autonomous system (AS), and hashes in malware analysis repositories, among others.

2.4 Related Works

In the researched texts, we found several proposals for recent solutions using IDS to protect the network infrastructure in different application scenarios [4,10,28]. We also found the use of machine learning and artificial intelligence in security analysis in network flows [32] and data enrichment. However, the proposed methodology is to explore flexibility in the use of incident detection in various scenarios, combining the use of signatures with a method of enriching the data of the record produced and subsequent storage in TISPs.

Table 2 was constructed through searches in academic articles databases and selecting the most relevant ones relating to the term Suricata and the term MISP.

Table 2. Summarizes the most relevant works

Article	Summary
[19]	FISHY uses IDS collection and a system for identification, categorization, classification, and enrichment of IoCs using ML for IoT systems.
[22]	ECAD introduces a new concept to integrate ML and analytical tools into a real-time intrusion detection and prevention solution.
[16]	They propose a system called INTIME, an integrated framework based on Machine Learning and Deep Learning.
[26]	The work proposes the identification of IoCs using a Convolutional Neural Network. The proposal calls iGen.
[14]	CyTIME is a framework for managing CTI data and collecting data from external repositories through TAXII interfaces. Automatically generate security rules.
[35]	The solution called APIRO consists of an API-specific word embedding model and a Convolutional Neural Network (CNN) model.

3 Proposed Methodology

The proposal is the implementation of sensors to collect anomalies using pre-selected signatures, aligned with the security policy and the organization's business strategy. Knowing the traffic pattern on the network, we can choose signatures and identify anomalies. Thus, all generated records are Threat Indicators. In addition to the use of traditional sensors that use signatures, the possible use of honeypots is also proposed, with services related to those of the organization installed for real-time recording of threats to possible vulnerabilities. Figure 1 presents the Proposed Methodology.

Some log data coming from the sensors are complemented through automated enrichment processing, in support of the Computer Security Incident Response Team (CSIRT) [11]. This processing aggregates other information for better investigation of the anomaly. As an example, we can mention the IP address (Internet Protocol) of origin "Who". Enrichment of this data may result in a Fully Qualified Domain Name (FQDN) through automated access to Domain Name System (DNS) databases stored on root name servers.

The FQDN can immediately identify the country of origin and associated autonomous system, speeding up investigation or decision-making in the event

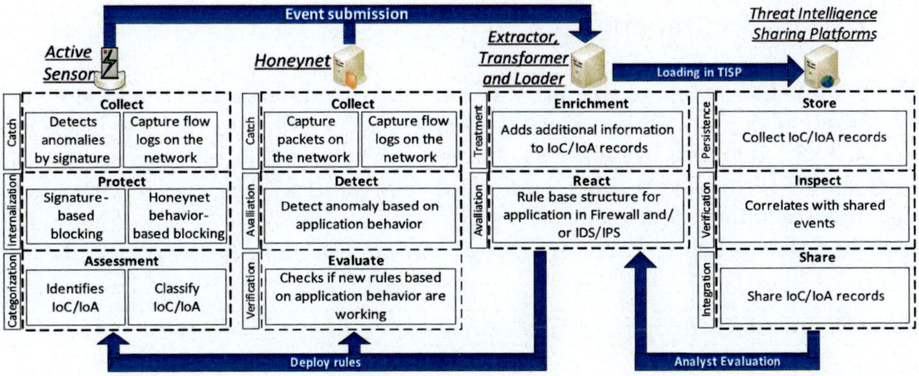

Fig. 1. Proposed Methodology.

of attacks. In the evaluation of attacks with specific TTPs to the organization's services, a more detailed investigation must be carried out by the CSIRT being subsidized by the enrichment of the data. After enrichment, the record is uploaded to the TISP to verify other occurrences involving the record, and two situations are possible: if there are already reported events, it is a threat confirmation and the cybersecurity team starts to follow and monitor the event as an attack, and if necessary, adjust existing rules; otherwise it could be a false positive or a Zero-Day attack attempt. Then, the TISP records are monitored, waiting for possible other related events. In the case of identifying an attack, we can apply the existing response protocols. The flow of analysis of the records generated by the sensors (IDS/IPS, Honeypots, or Firewall) must follow the following flowchart (see Fig. 2):

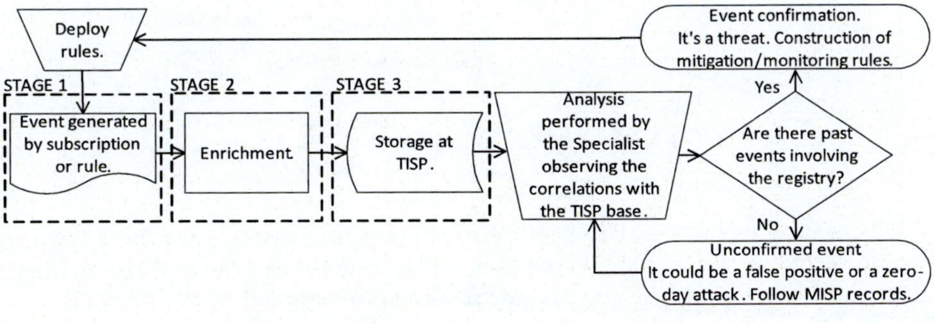

Fig. 2. Proposed Threat Management Flow.

In a final step, all records must be stored on TISPs to enable storage with event tracking by observing new correlations obtained from CTI information shares.

4 Proof of Concept

The intended objective of this proof of concept is to build a functional model of the methodology and demonstrate the activities according to the flow described in Fig. 3. An IDS/IPS system was chosen for the assembly of the sensor, using open source technologies (open source) of IDS (or IPS), associated with mechanisms for anomaly detection using signatures and analysis of flows previously described that provide minimally categorized and organized threat records.

4.1 Proof of Concept Architecture

For effective performance, we configure the IDS sensor to collect traffic on a switch between the external router and the firewall, as shown in Fig. 3. Thus, the collection takes place before any filter or intervention. However, in this position the IDS sensor would be very vulnerable to attacks, so we configured the sensor to only perform traffic collection in promiscuous mode, in which case it does not respond to connections. We put a web service machine as a mirror of the signature repository protected by the firewall, updating the signatures daily. So the sensor looks for these signatures on the intranet and updates its context.

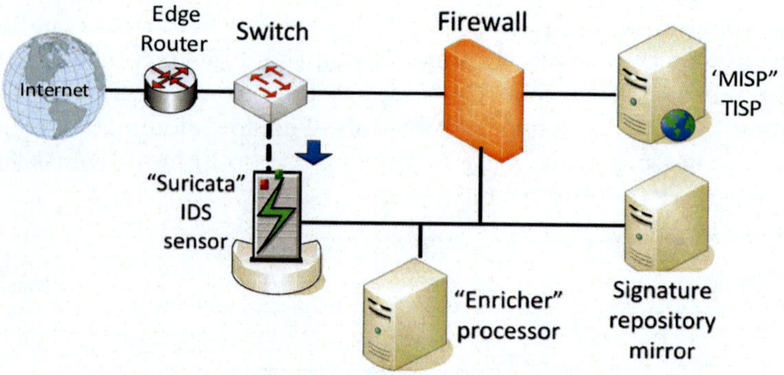

Fig. 3. Proof of concept architecture.

The sensor sends threat indicators to a log storage system on the intranet, also protected by the firewall, where they are enriched. Then the already enriched threat indicators go directly to the MISP or can be applied to the firewall.

The sensor chosen for the collection is the Suricata IDS [25]. Suricata can perform all traffic collection or just subscription-based logging. Taking into account the agility and optimization of the agility-oriented information, we will only focus on our class registrations in subscriptions, which we already provide a prequalification of the events. Thus, the policy is already considered with only the records of the activity classified as normal and security adherent.

In this proposal, we take into account the existing signature base in the Emerging Threats repository https://rules.emergingthreats.net/. In this case, signatures are made available by type, already classified and categorized by Threat Indicator. The time is updated every day at 22:00 UTC, with the information identified with a time stamp in the Universal Sortable Date "Z" format and with the registration made and extended the YYY-MM-DDTHH:MM:SSZ [8].

Sensor: Suricata and Emerging Threats. The Emerging Threats signature repository is fully compliant and configurable on IDS Suricata [31], featuring a set of rules pre-classified into threat groups, All Knowingly Suspicious, Potentially Malicious, or Hostile Activities by the cybersecurity community. Thus, there is a daily download mechanism of this Threat Intelligence base in the form of alert rules and the rules are applied as signatures [3].

When configuring an IDS, just activate and update these rules, as there is only interest in detecting malicious activities. The result is the generation of a record base of events that matches the rules. The choice of rule classes that will be chosen to become blocking depends on the information and communications security policy adopted by the organization.

Data Enricher: Enricher. A mechanism for data enrichment is a program with the function of adding information to the data and the records, facilitating the interpretation and analysis of the information. In the proof of concept presented, the Enricher [34], was used, but any system that searches repositories for complementary information relevant to the Threat Indicators could be used. The option for Enricher was due to its simplicity in the implementation of the framework assembled for the proof of concept.

Enricher searches Suricata records in JSON format for one of three types of initial search parameters: IP address, DNS domain, or email address. Thus, the program follows the following steps: it searches for one of the parameters of the target and will carry out an initial validation; once the validation is done, the program will make the first connection with the TISP to verify if the parameter informed is in the base; if there is no event, the program will create a new event with the parameter informed and send it to TISP, retrieving the eventid for future treatment; if the event already exists, TISP will return the available information to the tool, as well as the eventid of the event; the tool will then start a search based on the integrated tools for information about the selected target, where such a search will take place in a crossed way, and, once the IP of a Domain is obtained, it will trigger the search for information about that IP, and vice versa back; once the search is completed, the program will make a new connection with the TISP, updating the event information with the data found; the data will be made available on the platform in the form of proposed attributes, leaving the analyst responsible for analyzing which information is or is not relevant.

TISP: MISP. The last element configured is the MISP as a TISP. Suricata software and MISP were found in current academic documents, which is the reason for the choice [16,19,22]. Thus, the proof of concept meets all the elements proposed in the methodology, including the collection, evaluation, enrichment, structuring, and sharing of Threat Indicators.

4.2 Results

After implementing the proof of concept in the laboratory, several records were obtained in JSON format, but with many records of tracking DNS communications and monitoring alerts to control TLS and HTTP connections, as shown in Fig. 4. However, they all came mixed with the real ones signatures with alerts of possible threats, as we will see below:

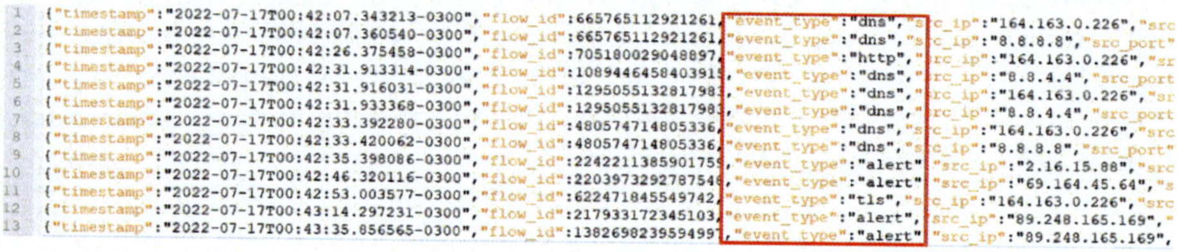

```
{"timestamp":"2022-07-17T00:42:07.343213-0300","flow_id":665765112921261,"event_type":"dns","src_ip":"164.163.0.226","src
{"timestamp":"2022-07-17T00:42:07.360540-0300","flow_id":665765112921261,"event_type":"dns","src_ip":"8.8.8.8","src_port"
{"timestamp":"2022-07-17T00:42:26.375458-0300","flow_id":705180029048897,"event_type":"http","src_ip":"164.163.0.226","sr
{"timestamp":"2022-07-17T00:42:31.913314-0300","flow_id":108944645840391,"event_type":"dns","src_ip":"8.8.4.4","src_port
{"timestamp":"2022-07-17T00:42:31.916031-0300","flow_id":129505513281798,"event_type":"dns","src_ip":"164.163.0.226","sr
{"timestamp":"2022-07-17T00:42:31.933368-0300","flow_id":129505513281798,"event_type":"dns","src_ip":"8.8.4.4","src_port
{"timestamp":"2022-07-17T00:42:33.392280-0300","flow_id":480574714805336,"event_type":"dns","src_ip":"164.163.0.226","src
{"timestamp":"2022-07-17T00:42:33.420062-0300","flow_id":480574714805336,"event_type":"dns","src_ip":"8.8.8.8","src_port"
{"timestamp":"2022-07-17T00:42:35.398086-0300","flow_id":224221138590175,"event_type":"alert","src_ip":"2.16.15.88","src
{"timestamp":"2022-07-17T00:42:46.320116-0300","flow_id":220397329278754,"event_type":"alert","src_ip":"69.164.45.64","s
{"timestamp":"2022-07-17T00:42:53.003577-0300","flow_id":622471845549742,"event_type":"tls","src_ip":"164.163.0.226","src
{"timestamp":"2022-07-17T00:43:14.297231-0300","flow_id":217933172345103,"event_type":"alert","src_ip":"89.248.165.169","
{"timestamp":"2022-07-17T00:43:35.856565-0300","flow_id":138269823959499,"event_type":"alert","src_ip":"89.248.165.169",
```


Fig. 4. Record generated by the IDS.

For a better analysis of the threats, filtering was carried out in the logs, looking for only the entries with alerts generated by the signatures, and among these only the important information, such as timestamp, src-ip, signature, and category, as shown in Fig. 5. For better observation of the alerts, have been transformed into the following log format:

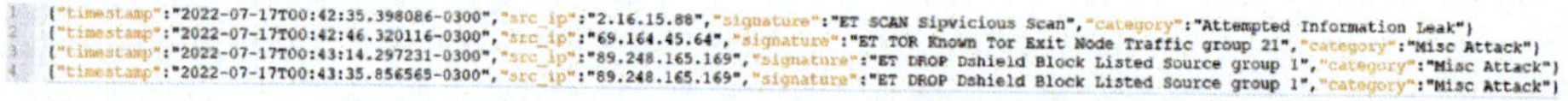

```
{"timestamp":"2022-07-17T00:42:35.398086-0300","src_ip":"2.16.15.88","signature":"ET SCAN Sipvicious Scan","category":"Attempted Information Leak"}
{"timestamp":"2022-07-17T00:42:46.320116-0300","src_ip":"69.164.45.64","signature":"ET TOR Known Tor Exit Node Traffic group 21","category":"Misc Attack"}
{"timestamp":"2022-07-17T00:43:14.297231-0300","src_ip":"89.248.165.169","signature":"ET DROP Dshield Block Listed Source group 1","category":"Misc Attack"}
{"timestamp":"2022-07-17T00:43:35.856565-0300","src_ip":"89.248.165.169","signature":"ET DROP Dshield Block Listed Source group 1","category":"Misc Attack"}
```


Fig. 5. Record generated by the IDS.

Then, the analysis of the records was performed, classifying them by signature and by IP using a python3 script developed specifically for this purpose, as shown in Fig. 6. Thus, alerts with more entries in the registry were obtained, which were then submitted to enrichment.

After loading the MISP, the following day it was observed that an IP address was correlated with two other events already reported by other organizations that perform record sharing with our CTIR, as shown in Fig. 7. In the figure, data identifies the system in production.

```
In [13]: #Monta um novo dataframe apenas com o IP e a assinatura tratada e remove as duplicadas
         df_ip = df_all_files[["src_ip","signature_parsed"]].drop_duplicates(ignore_index=True)

         print(df_ip)

                  src_ip    signature_parsed
         0      2.16.15.88  ET SCAN Sipvicious
         1    69.164.45.64        ET TOR Known
         2  89.248.165.169     ET DROP Dshield

In [14]: #Conta a quantidade de vezes que o ip aparece para cada assinatura tratada
         df_count = df_ip.groupby(['src_ip'])['src_ip'].count()
         print(df_count)

         src_ip
         2.16.15.88        1
         69.164.45.64      1
         89.248.165.169    1
         Name: src_ip, dtype: int64
```

Fig. 6. Record generated by the IDS.

	Published	Creator org	Owner org	ID	Clusters	Tags	#Attr.	#Corr.	Creator user	Date	Info ↓	Distribution	Actions
☐	✖	CTIR-	CTIR-	—96926		type:OSINT	1		joao@ .br	2022-07-20	Analyzer	Organisation	
☐	✖	CTIR-	CTIR-	—96928		type:OSINT	1		joao@ .br	2022-07-20	Analyzer	Organisation	
☐	✖	CTIR-	CTIR-	—96930		type:OSINT	1	2	joao@ .br	2022-07-20	Analyzer	Organisation	

Powered by MISP 2.4.151 Operated by CTIR - 2022-07-21 17:38:31

Fig. 7. Record generated by the IDS.

4.3 Proposed Methodology Versus Related Works

We identified specific characteristics of an Intrusion Detection solution, taking into account IoC and IoA, in addition to the possibility of generating rules for detection in Firewall and IDS. Another important characteristic is the solution's application scenario, being important in the scalability and flexibility of use in different network scenarios. Table 3 compares those techniques with the proposed methodology.

An interesting architecture found in the research was the FISHY presented in the article [19]. As an architecture based exclusively on ML performance in registry bases, the proposal does not include the possibility of detecting attacks in real-time and does not use IDS signature bases for detection. In addition, it does not include the generation of rules for application in Firewall Systems existing in the networks and does not have integration with TISPs.

Another proposal found was the ECAD presented in the article [22]. As an ML solution, it does not take into account the possibility of adoption of signatures by the sensors and does not generate new signatures for the sensors and Firewall systems.

In the framework called INTIME [16] and in the iGen proposal presented in the thesis [26], both do not identify IoA because they focus on the identification of IoCs. These works use ML, Deep Learning, and Convolutional Neural Network to detect IoCs, so they do not use sensor signatures and do not generate blocking rules in Firewall systems. However, they propose integration with TISPs.

The case of APIRO [35], which also uses a Convolutional Neural Network, has a different approach. It identifies IoAs but does not propose the generation of rules to block Firewall systems and do not directly use sensor signatures.

CyTIME [14] is another example of a framework that does not identify IoA, but unlike previous works, it does not use ML or AI. Nor does it use its sensors with signatures as a source. In the proposed framework, the collection of records is done in malware bases, TAXII servers, and TISP. Thus, the integration with TISP is for obtaining IoCs. The ultimate goal is to generate rules for organizations' sensors and firewall systems.

In Table 2 we summarized the techniques used by the main current academic works in the study of the Suricata IDS combined or not with the TISP MISP.

Table 3. Main characteristics of the Proposed Methodology

Article	[19]	[22]	[16]	[26]	[14]	[35]	Proposed Methodology
Identifies attacks on Real-time	NO	YES	NO	NO	NO	YES	**YES**
Identifies Compromise	YES	YES	YES	YES	YES	YES	**YES**
Uses IDSsignatures	NO	NO	NO	NO	NO	NO	**YES**
Generates rules for IDS or Firewall	NO	NO	NO	NO	YES	NO	**YES**
Integrates with TISPs	NO	YES	YES	YES	YES	YES	**YES**

Contributions. A feature that stands out in the proposed methodology is that we can identify attacks in real-time, as we consider any record source, whether IDS or even a Honeynet, similar to [22,35]. However, we observed the complexity of implementing both proposals simultaneously to validate the methodology. Thus, we have not yet implemented a Honeynet. Furthermore, we consider Honeynet an excellent source of records with the possibility of identifying Indicators of Attacks (IoA) with reasonably simple deployment, operation, and maintenance, as the daily activities of security incident response teams already include TTP studies, malware analysis, and computer forensics.

Another approach that simplifies the adoption of the proposed methodology is to be based on records in the sensors' JSON format. Thus, it is not necessary to make any API-type interpreter or construction of a specific framework, as proposed by [19,35].

Finally, in almost all frameworks and architecture proposals we observe the use of ML, AI, or Convolutional Neural Network to identify or correlate IoCs. The identification of IoAs when possible is done through the adoption of complex structures and with the use of several different record sources. As we do not use ML or AI, we consider that the active sensors are using updated signatures from several different sources, which makes operation, adjustments, and maintenance very agile. Another need that is present is the integration with a TISP, as security incident response teams need to confront and share collected records.

Implications for Practice. According to the flow of Fig. 2, in the first step, the use of sensor data (IDS/IPS collector) with updated signatures accelerated the collection of quality information, through rules with signatures that provide records of Threat Indicators minimally categorized and organized. Because it involves real-time collection and analysis, it was possible to identify attacks in progress. It is noteworthy that when filtering the records produced by the IDS, the possible IoCs and IoAs were immediately identified. Thus, there was an efficiency gain in the process. The implementation of Honeypots would make it possible to identify the Indicators.

The structure set up as shown in Fig. 3 did not change the functioning of the network and enabled the transparent collection of records. The equipment used in the sensor and the enrichment only depends on the storage capacity of records necessary to process the filtering and enrichment. Filtering before enrichment is necessary as the logs have various sync data and logs enabled for system auditing. Thus, the volume of data for enrichment has decreased considerably. The machine with the greatest storage capacity should be the TISP, as it works as a data persistence point.

In the second stage, processing was carried out to enrich these records with the collection of relevant complementary information, which facilitated the identification of attacks by analysts. At this stage, there was no quality filtering of the information obtained in the enrichment, as it was decided to load all the data obtained in the TISP for further evaluation.

Finally, in the third step, the data were loaded into the MISP. It was evident that the decision not to perform the filtering was correct, as the MISP event base immediately identified the records that already had a history. Thus, indicators and possible false positives were identified. This identification is visual and very simple in MISP. The decision to share the records with other MISP partners is the responsibility of the security analyst, as well as the creation or improvement of rules for application in Firewall systems.

It is observed that the proposed methodology is easily adopted regardless of the tools used as sensors and also adapts well to any network topology. Another interesting feature is the possibility of mounting several sensors in various networks on the Internet to produce Threat Intelligence at a sectoral or even national level, an essential activity carried out by CSIRTs. However, the same methodology could easily be adopted by a single organization, where your security incident response team could mount multiple sensors to identify attacks, audits, and information exfiltration attempts.

5 Conclusion

The proposed architecture developed from a methodological framework for the generation and systematic analysis of records with the identification of anomalous pat-terns and the derivation of detection rules encoded using signatures. As a proof of concept, the implementation of the architecture was carried out, reaching the objectives of the proposal satisfactorily. The proof of concept demonstrated that the proposed architecture makes it possible to accelerate the identification of threats and possible incidents before they are widely reported by the community, contributing to the anticipation of security actions and cyber threat intelligence.

All systems used in the proof of concept are OpenSource, adopting standards and protocol formats standardized by the Threat Intelligence community. These characteristics facilitate its use and make its use more flexible, making clear the applicability of the proposed architecture. Thus, the methodology can be widely adopted by organizations, being adaptable to tools and solutions already installed, enabling the composition without changing the network design and with reduced costs. The adaptability of the sensor (IDS system) to the various protocols and networks already established would be an additional incentive for its adoption. In addition, the importance of a TISP sharing information with several databases is observed.

In future work, issues related to integration with other types of sensors, as well as in other TISPs, can be considered.

References

1. Abdullahi, M., et al.: Detecting cybersecurity attacks in internet of things using artificial intelligence methods: a systematic literature review. Electronics **11**(2), 1–28 (2022). https://doi.org/10.3390/electronics11020198
2. Albasheer, H., et al.: Cyber-attack prediction based on network intrusion detection systems for alert correlation techniques: a survey. Sensors **22**(4), 1494 (2022). https://doi.org/10.3390/S22041494
3. Alcantara, L., Padilha, G., Abreu, R., D'Amorim, M.: Syrius: synthesis of rules for intrusion detectors. IEEE Trans. Reliab. **71**, 1–12 (2021). https://doi.org/10.1109/TR.2021.3061297
4. Bhati, N.S., Khari, M., García-Díaz, V., Verdú, E.: A Review on Intrusion Detection Systems and Techniques (2020). https://doi.org/10.1142/S0218488520400140
5. Burger, E.W., Goodman, M.D., Kampanakis, P., Zhu, K.A.: Taxonomy model for cyber threat intelligence information exchange technologies. In: WISCS 2014: Proceedings of the 2014 ACM Workshop on Information Sharing and Collaborative Security, pp. 51–60 (2014). https://doi.org/10.1145/2663876.2663883
6. Cheswick, W.R., Bellovin, S.M.: Firewalls and Internet Security: Repelling the Wily Hacker. Addison-Wesley (1994). https://archive.org/details/firewallsinterne00ches
7. de Melo e Silva, A., Gondim, J.J.C., de Oliveira Albuquerque, R., Villalba, L.J.G.: A methodology to evaluate standards and platforms within cyber threat intelligence. Future Internet **12**(6), 1–23 (2020). https://doi.org/10.3390/fi12060108
8. DTF: Date Time Format Info. Universal Sortable Date Time Pattern. http://shorturl.at/kWZ25
9. Elmellas, J.: Knowledge is power: the evolution of threat intelligence. Comput. Fraud Secur. **2016**(7), 5–9 (2016)

10. Ferrag, M.A., Babaghayou, M., Yazici, A.: Cyber security for fog-based smart grid SCADA systems: solutions and challenges. J. Inf. Secur. Appl. **52**, 102500 (2020). https://doi.org/10.1016/j.jisa.2020.102500
11. Hoepers, C., Steding-Jessen, K., Montes, A.: Honeynets applied to the CSIRT scenario. In: FIRST, p. 9 (2003). http://www.honeynet.org/alliance/
12. Irfan, A.N., Ariffin, A., ri Mahrin, M.N., Anuar, S.: A malware detection framework based on forensic and unsupervised machine learning methodologies. In: ACM International Conference Proceeding Series, pp. 194–200 (2020). https://doi.org/10.1145/3384544.3384556
13. Kalogeraki, E.M., Papastergiou, S., Panayiotopoulos, T.: An attack simulation and evidence chains generation model for critical information infrastructures. Electronics **11**(3), 404 (2022). https://doi.org/10.3390/electronics11030404
14. Kim, E., Kim, K., Shin, D., Jin, B., Kim, H.: Cytime: cyber threat intelligence management framework for automatically generating security rules. In: ACM International Conference Proceeding Series Part F1377 (2018). https://doi.org/10.1145/3226052.3226056
15. Klock, A.C.T., Gasparini, I., Pimenta, M.S.: 5W2H framework. In: Proceedings of the 15th Brazilian Symposium on Human Factors in Computing Systems, pp. 1–10. ACM, New York (2016). https://doi.org/10.1145/3033701.3033715
16. Koloveas, P., Chantzios, T., Alevizopoulou, S., Skiadopoulos, S., Tryfonopoulos, C.: inTIME: a machine learning-based framework for gathering and leveraging web data to cyber-threat intelligence. Electronics **10**(7), 818 (2021). https://doi.org/10.3390/electronics10070818
17. Kristiansen, L.M., Agarwal, V., Franke, K., Shah, R.S.: CTI-Twitter: gathering cyber threat intelligence from twitter using integrated supervised and unsupervised learning. In: Proceedings - 2020 IEEE International Conference on Big Data, Big Data 2020, pp. 2299–2308 (2020). https://doi.org/10.1109/BigData50022.2020.9378393
18. Marchio, J.: Analytic tradecraft and the intelligence community: enduring value, intermittent emphasis. Intell. Natl. Secur. **29**(2), 159–183 (2014). https://doi.org/10.1080/02684527.2012.746415
19. Masip-Bruin, X., et al.: Cybersecurity in ICT supply chains: key challenges and a relevant architecture. Sensors **21**(18) (2021). https://doi.org/10.3390/S21186057
20. Mavroeidis, V., Jøsang, A.: Data-driven threat hunting using sysmon. In: ACM International Conference Proceeding Series, pp. 82–88 (2018). https://doi.org/10.1145/3199478.3199490
21. McAuliffe, N., Wolcott, D., Schaefer, L., Kelem, N., Hubbard, B., Haley, T.: Is your computer being misused? A survey of current intrusion detection system technology. In: Proceedings - Annual Computer Security Applications Conference, ACSAC, pp. 260–272 (1990). https://doi.org/10.1109/CSAC.1990.143785
22. Mironeanu, C., Archip, A., Amarandei, C.M., Craus, M.: Experimental cyber attack detection framework. Electronics **10**(14) (2021). https://doi.org/10.3390/ELECTRONICS10141682
23. Nam, K., Kim, K.: A study on SDN security enhancement using open source IDS/IPS Suricata. In: 9th International Conference on Information and Communication Technology Convergence: ICT Convergence Powered by Smart Intelligence, ICTC 2018, pp. 1124–1126 (2018). https://doi.org/10.1109/ICTC.2018.8539455
24. Nash, A.: Demystifying cyber threat intelligence sharing platforms: an evaluation of data quality issues and their effects on cyber attribution. Master degree in science, Faculty of Utica College (2021). http://shorturl.at/bdgRX

25. OISF: Suricata — Open Source IDS/IPS/NSM engine (2020). https://suricata-ids.org/. https://github.com/OISF/suricata/
26. Panwar, A., Ahn, G.J., Doupé, A., Zhao, Z.: iGen: toward automatic generation and analysis of indicators of compromise (IOCs) using convolutional neural network. Master of science, Arizona State University (2017). https://hdl.handle.net/2286/R.I.44216
27. Riesco, R., Villagrá, V.A.: Leveraging cyber threat intelligence for a dynamic risk framework. Int. J. Inf. Secur. **18**(6), 715–739 (2019). https://doi.org/10.1007/s10207-019-00433-2
28. Roopak, M., Tian, G.Y., Chambers, J.: An intrusion detection system against DDoS attacks in IoT networks. In: 2020 10th Annual Computing and Communication Workshop and Conference, CCWC 2020, pp. 562–567 (2020). https://doi.org/10.1109/CCWC47524.2020.9031206
29. Sander, T., Hailpern, J.: UX aspects of threat information sharing platforms. In: Proceedings of the 2nd ACM Workshop on Information Sharing and Collaborative Security, pp. 51–59. ACM, New York (2015). https://doi.org/10.1145/2808128.2808136
30. Schlette, D., Böhm, F., Caselli, M., Pernul, G.: Measuring and visualizing cyber threat intelligence quality. Int. J. Inf. Secur. **20**(1), 21–38 (2021). https://doi.org/10.1007/s10207-020-00490-y
31. Schreiber, J., Meehan, M., Langston, R.: 2021 Open Source IDS Tools: Suricata vs Snort vs Bro (Zeek) — AT&T Cybersecurity (2020). http://shorturl.at/oPS37
32. Shafiq, M., Yu, X., Bashir, A.K., Chaudhry, H.N., Wang, D.: A machine learning approach for feature selection traffic classification using security analysis. J. Supercomput. **74**(10), 4867–4892 (2018). https://doi.org/10.1007/s11227-018-2263-3
33. Siebert, E.: Indicadores de ataque versus indicadores de comprometimento. Technical report, CrowdStrike Holdings, Inc, Austin, Texas (2020). http://shorturl.at/bru49
34. de Sousa, C.E., Gondim, J.J.C., Albuquerque, R.d.O.: ENRICHER: ferramenta de enriquecimento de dados integrada à plataforma MISP. Dissertation completion graduation, Universidade de Brasília (2021)
35. Sworna, Z.T., Islam, C., Babar, M.A.: APIRO: a framework for automated security tools API recommendation. ACM Trans. Softw. Eng. Methodol. 41 (2022). https://doi.org/10.1145/3512768
36. Wendt, D.W.: Exploring The Strategies Cybersecurity Specialists Need To Improve Adaptive Cyber Defenses Within The Financial Sector: An Exploratory Study. D.c.s, Colorado Technical University (2019). https://shorturl.at/ouV46
37. Zhou, Y., Tang, Y., Yi, M., Xi, C., Lu, H.: CTI view: APT threat intelligence analysis system. Secur. Commun. Netw. **2022** (2022). https://doi.org/10.1155/2022/9875199

Multi-agent Architecture for Passive Rootkit Detection with Data Enrichment

Maickel Trinks[1(✉)], João Gondim[2], and Robson Albuquerque[1]

[1] Professional Post-Graduate Program in Electrical Engineering - PPEE - Department of Electrical Engineering, University of Brasília, Brasília, DF, Brazil
maickeljosue@yahoo.com.br, robson@redes.unb.br

[2] Department of Computer Science, Universidade de Brasília, Brasília, DF, Brazil
gondim@unb.br

Abstract. The added value of the information transmitted in a cybernetic environment has resulted in a sophisticated malicious actions scenario aimed at data exfiltration. In situations with advanced actors, like APTs, such actions use obfuscation techniques of harmful activities as persistence assurance on strategic targets. The MADEX and NERD architectures proposed flow analysis solutions to detect rootkits that hide network traffic; however, it presents some operational cost, either in traffic volume or due to lack of aggregated information. In that regard, this work changes and improves user flow analysis techniques to eliminate impacts on network traffic, with data enrichment on local and remote bases, detection of domains consulted by rootkits and aggregation of information to generate threat intelligence, while maintaining high performance. The results show the possibility of aggregating information to data flows used by rootkits in order to have effective cyber defense actions against cybernetic threats without major impacts on the existing network infrastructure.

Keywords: Rootkit detection · data enrichment · threat intelligence · cybersecurity

1 Introduction

The last few years have brought significant changes in the cyber environment. In addition to the constant increase in users connected to the Internet [6] and migration of services to online platforms [3], the transformation caused by the COVID-19 pandemic has forced the large-scale adoption of information technology to overcome the consequences caused by the disease [5] and reduce its proliferation. As examples, there are various initiatives to adopt remote work regimes, distance learning, teleconferences, and migration of services to digital platforms, among others.

In this scenario, in addition to the growing trend of threat actors using advanced technological resources [5], rapid adaptation imposed by pandemic

M. V. Garcia and C. Gordón-Gallegos (Eds.): CSEI 2022, LNNS 678, pp. 29–41, 2023.
https://doi.org/10.1007/978-3-031-30592-4_3

made it challenging to adopt cybersecurity controls in appropriate places, making information systems even more vulnerable to cyberattacks. Morgan [10] has estimated a global cost of cybercrime of $10.5 trillion annually in 2025, up from $6 trillion in 2021.

The increase of sensitive information in digital media has also rised the sophistication of malicious techniques aimed at data exfiltration [13]) and digital extortion, especially with the advent of ransomware attacks. Targeted phishing (spearphishing) techniques for installing malware on targets of interest, such as Advanced Persistent Threats (APTs), partly sponsored by nation states, have been used for cyber espionage on large corporations [1].

This work aims to propose and describe the Passive Rootkit Detector With Enriched Data (PARDED) system, an architecture for detecting rootkits based on hiding network packets passively, which allows feeding other defense systems through centralization and enrichment of information (footprints). The relevance lies in the fact that rootkits, malware distinguished by their high ability to hide their presence on infected systems, can be used by various cyber actors in malicious actions. Their characteristics make them of great value to APT groups and ransomware operators. When adequately executed, attacks resulting from rootkits are highly efficient and difficult to detect. Thus, it is necessary to develop defense techniques that allow its detection and subsequent removal.

PARDED was created based on the extension of the MADEX architecture, proposed by Marques et al. [9] and NERD, presented by Terra and Gondim [14]. As a differential from previous architectures, PARDED has as its main characteristic the improvement of blocking techniques in communication network infrastructures without degrading the data transmission rate in network equipment, in addition to providing an interface for integration with other defense systems previously existing systems such as firewalls and intrusion detection systems. In addition, another important innovation is the possibility of enriching information about malicious destinations used by rootkits, notably for command and control (C2) systems, through local or external information bases. The enrichment process helps, for example, detecting the domain names used by C2 systems, which allows subsequent intelligence analysis with greater detail, maintaining the context of the evaluation object, avoiding unnecessary enrichments that pollute the analysis environments.

The organization of this work contains, in addition to this introduction, the Concepts section, which defines the critical expressions for understanding the proposal presented; the Related Work section, witch briefly explains the architectures used as a basis for the project; the Proposed Architecture section, which describes the PARDED system; and the Results section, which presents data on the performance and enrichment of flows obtained in the laboratory. Finally, the conclusion exposes the work's synthesis and possible future paths.

2 Concepts

There is great relevance in some concepts for a better understanding of the PARDED system and the malicious actions that PARDED proposes to detect:

2.1 Advanced Persistent Threat (APT)

A high-sophisticated attack to an organization that continues its activities and yet remains undetected for an extended period, for control and long-term data collection, through a slow and stealthy approach to avoid detection [1].

2.2 Data Enrichment

Refers to the process of appending or otherwise enhancing collected data with relevant context obtained from additional sources [8]. Typically, data enrichment is achieved using external databases.

2.3 Multi-agent Systems

A specific type of distributed system, where components of the system are autonomous and selfish, seeking to satisfy their objectives. In addition, these systems also stand out for being open systems without a centralized design [7,15].

2.4 Onion Network (TOR)

Onion routing is an infrastructure for private communication over a public network. It provides anonymous connections strongly resistant to eavesdropping and traffic analysis. Applications connect through a sequence of machines called onion routers to reach a responding device. Anonymous connections hide who is connected to whom and for what purpose [12].

3 Related Works

The PARDED architecture derives from a multi-agent base composed of elements that work at different points of a communication network infrastructure and was based on previous work described below:

3.1 MADEX

Multi-Agent Data Exfiltration Detection Architecture (MADEX) [9] is a multi-agent architecture whose functionality is to detect rootkits that perform traffic obfuscation by altering the connection table of the infected terminal. The system comprises a Collector Agent, responsible for data collection, and an Auditor Agent, which detects all traffic from the infected terminal. The Auditor Agent then interacts with the Collector to verify if the Collector has noticed the incoming traffic. The perception of legitimate communication depends on the connection being present in the operating system's connection table. If not, it may indicate the presence of malware that hide traffic from the infected terminal, where the Collector is installed. Figure 1 represents MADEX architecture.

The system's functioning depends on the Collector having the connection table information in a timely manner to respond the Auditor's request. In case of a high number of packets, considering that the Collector cannot update the known connections before the polling time expires, all traffic would be marked as malicious.

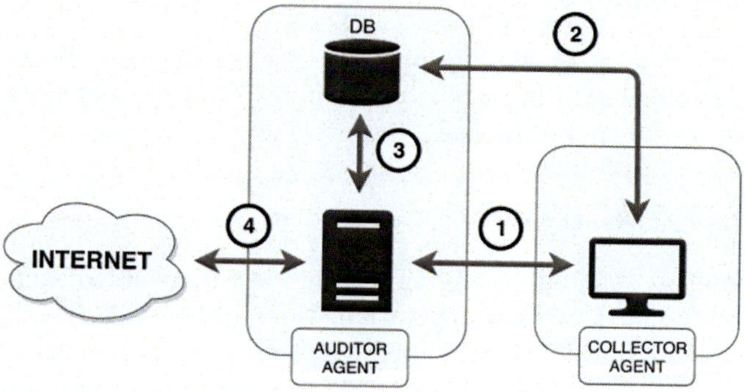

Fig. 1. MADEX architecture [9].

3.2 NERD

Network Exfiltration Rootkit Detector (NERD) [14], like MADEX, is a multi-agent architecture for rootkit detection purposes. However, instead of using the terminal connections table to check for possible obfuscation, it uses traffic capture through the *libpcap* library, avoiding constant calls to the operating system. After receiving the traffic transmitted by the terminal, the Auditor Agent interacts with the Collector to verify if the Collector perceived it. As with the MADEX architecture, traffic received at the Auditor and not perceived by the Collector may indicate the presence of malware that obfuscates detection on the infected terminal. Figure 2 represents NERD architecture.

The NERD architecture reduced false positives and detected all malicious traffic generated in lab tests. However, there was a degradation of approximately 50% in the packet transmission rate in low-performance networks (which have download rates of 50 Mbps) and about 90% in high-performance networks (with download rates close to 300 Mbps) [14], although it has shown capacity gains concerning the MADEX architecture [9].

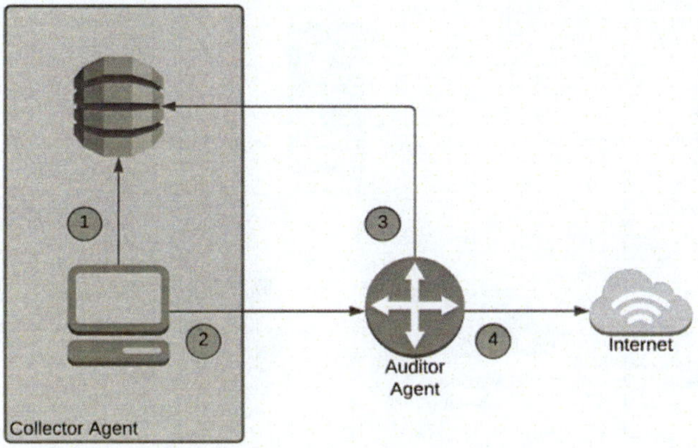

Fig. 2. NERD architecture [14].

4 Proposed Architecture

The proposed architecture of Passive Rootkit Detector With Enriched Data (PARDED) implements conceptual changes foreseen in the MADEX and NERD architectures, mainly in how the Auditor Element acts to detect and block malicious communication passively. This approach avoids degradation in the network's transmission rate and provides integration with other defense systems against malware. For such behavior, the auditor was divided into three components, with the first being composed of an inline element, which has the function of mirroring the traffic as an input to the analysis of the second component, out-of-band, which receives a copy of the traffic and compares it with the data of the Collector Agent for decision making, similarly to the NERD architecture. The third element, inline, receives information from the second element about connections considered malicious and which of them should be blocked.

This modification aims to eliminate the performance bottleneck caused by the characteristics of the active Auditor Agent existing in the other architectures, allowing the treatment of packages, the storage, and the enrichment of information without traffic delays. In this way, the proposed changes result in a structure that enables intelligence analysis of rootkit behavior on the network, either through the integration of data from multiple terminals, information history, and detection of used malicious domains, or through the aggregation of new verification ways with external databases support. Creating a database with suspicious behavior characteristics also allows feeding other network defense systems, such as systems that use blocking rules by IP, domain, or YARA rules. [11].

Figure 3 illustrates the proposed architecture and the following topics explain in detail each of the elements that compose it.

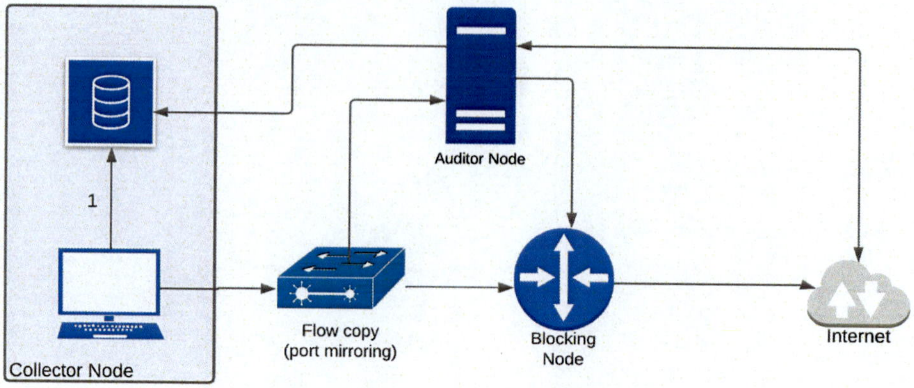

Fig. 3. PARDED - Proposed architecture.

4.1 Detection Strategy

The detection of suspicious traffic is performed by comparing the traffic observed at the terminal, obtained by the Collector Element (ColEl), and the traffic received by the Auditor Element (AudEl). Packets copied to AudEl that ColEl did not observe may indicate evasion of traffic visualization on the terminal by a malicious action of a rootkit. This detection, together with information from other endpoints and external databases, defines the action for each potentially malicious flow.

Collector Element (ColEl). It works passively through the analysis of traffic replica from the terminal. ColEl uses the approach foreseen in the NERD architecture due to the results presented, without changes, where the Collector performs the copy of the traffic through sockets linked to the active network device with the *libpcap* library. The difference presented by PARDED is the use of port mirroring to feed the Auditor Element, just like in an Intrusion Detection System (IDS) [1]. This strategy allows the auditor to passively process the information without interfering with the performance of the network.

Duplicator Element (DupEl). Its only function is to mirror the network traffic from the analyzed terminal to the Auditor Element (AudEl). This strategy allows AudEl works without interfering on network performance and to process information passively, as in an IDS. It can be any kind of hardware that duplicates information, like a network hub or a switch with port mirroring.

Auditor Element (AudEl). Unlike the MADEX and NERD architectures, AudEl does not perform the blocking function internally, allowing it to work out-of-band. This approach can perform more complex analyses before blocking traffic without impacting terminal communication performance.

The information enrichment is initially performed with the capture and store of DNS queries received by the Auditor. In this way, if a DNS request precedes malicious communication, the domain consulted by malware is detected and is included in the information to be enriched. Then, AudEl performs queries to ColEl (and other local or external systems), enabling anomalous or malicious activities detection.

For AudEl to consider the flow malicious, it must reach certain detection thresholds. The definition of initial parameters for each threshold is based on the hypotheses below:

- Six flows that are not detected by the same ColEl (same terminal) are considered malicious action. We used these premises because the NERD architecture obtained excellent false-positive results with this configuration [14].
- Two terminals whose collectors do not detect two flows to the same destination are enough for packets to be considered malicious. This premise is also based on the results obtained by NERD, added to the improbability that false positives, already rare in this architecture, occur in different collectors for the same destination IP.
- Two flows that are not detected and shares same destination, in the following situations:
  - The destination belongs to a TOR network;
  - The queried domain is considered malicious by at least two security companies;
  - The IP is considered malicious by at list four security companies.

  These parameters depend on knowledge of the network infrastructure used to deploy the system. For example, if the network infrastructure allows or predicts user access to TOR network nodes, low threshold values can increase the number of false positives.

The functionalities were split modularly to allow easy evolution of the AudEl and prevent system interruption in case of failure in any step, as shown in Fig. 4. In this way, the system becomes generalizable and adaptable.

After receiving a data packet, the Initial Analysis System checks if it is from an already known flow (it's in Temporary Base). This strategy prevents the AudEl from performing unnecessary queries to the ColEl, making the analysis more efficient by reducing the traffic generated between auditor and collector and reducing processing in the Auditor and the Collector Elements. When AudEl does not know the flow, it consults ColEl. If ColEl has perceived the packet, it is considered legitimate traffic and stores the flow on the Temporary Base; otherwise, it is considered suspicious and the Auditor stores (in the Connection Database) the relevant flow data. This data, along with enrichment information and data from the other ColEls, is used to decide if the flow will be considered malicious.

The Enrichment System verifies, through the Connection Database, which data has not yet been enriched and queries the configured bases. The enrichment system is implemented in such a way that permits external base addition without layout changes.

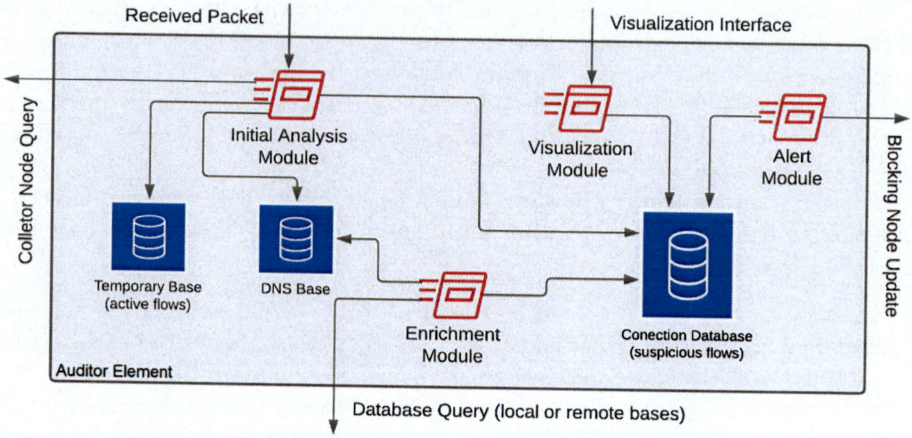

Fig. 4. PARDED - Auditor Element.

The Visualization System allows a security analyst to check the flows marked as suspicious and their characteristics, in addition to an overview of the tool's operation, facilitating its use in data monitoring environments and integration with other security systems.

Through the Connection Database, the Alert Module verifies which destinations have reached the configured thresholds for a flow to be considered malicious and informs the Blocking Element. The Alert Module is scalable; it can send warnings to different blocking devices.

Blocking Element (BlkEl). It is an inline device that checks whether it should block the received network packet based on the warning received by the AudEl. This element can be any device that allows blocking rules update or YANA rules update remotely, such as routers and firewalls.

The elapsed time between detection of the first suspicious packet, enrichment of the information, sending block notification by the Alert Module, and the effective blocking action by the BlkEl depends on several factors, such ColEl query time, remote base query time for enrichment, updating the connection base and reaching the thresholds defined by the auditor.

5 Results

The tests were carried out to verify the architecture's viability and the execution time of each stage of the package analysis process by the Auditor Element. AudEl and ColEl were directly connected without using any intermediary network elements. Softwares run on Linux Debian 11 Operating System. The configuration of the Collector Element is as described in NERD architecture [14]. AudEl uses a relational database (PostgreSQL) for the Connection Database. The prototype used in the tests reflects the architecture proposed in this work and is fully functional.

The development of the Initial Analysis System, belonging to AudEl, was in C language, as it is the most critical module. Since they don't have the same performance characteristics, the Enrichment, Alert, and Visualization Modules were made in Python3 language, facilitating the compatibility with external databases and blocking elements and graphics libraries for visualization. The tests were performed through traffic simulation, with packets sent by the terminal containing a Collector Element copied to the Auditor Element by port mirroring technique.

Two test scenarios were designed. The first, without the active presence of a rootkit, where all data streams are legitimate (not obfuscated), was carried out to verify that all transmitted packets would be correctly processed and also the performance of each system element. The test, containing between 600 and 1200 data streams with 10 packets each, was designed to obtain the average packet processing time after 5 complete executions.

The second, with the active presence of the rootkit, where legitimate (unobfuscated) and malicious (obfuscated) flows were generated, was carried out to verify that all transmitted packets would be correctly processed, the detection rate of malicious flows, and the performance of each system element, for both malicious and legitimate flows. The test, containing between 500 and 1000 legitimate data streams and 100 malicious streams with 10 packets each, was designed to obtain the average packet processing time after 5 executions.

Both scenarios were repeated with additional data to verify performance with high load. The transmission rate chosen for the execution of the tests was 50 Mbps, the same used in the NERD architecture during tests. The load (packet flooding) was made with the Iperf3 software.

The results took into account only TCP flows and "DNS Response" packets. There was no performance change in the network since the AudEl does not interfere with the inline elements; there was only a slight increase in the transmission rate (less than 2%) generated by the queries made to ColEl and external databases.

5.1 Legitimate FLows

In this scenario, the rootkit was not active (without traffic obfuscation). In simulated data transmission, all packets received in AudEl were perceived by ColEl; thus, there were no false positives. On average, packets from unknown flows by AudEl (which requires ColEl queries) were processed in 0.5 s. The Initial Analysis System processed other packets in the flow (in the Temporary Base and representing 95% of the total throughput) in less than 0.0012 ms each. The "DNS response" packets, locally treated to detect possible domains used by rootkits, were processed in approximately 0.004 ms.

Table 1 shows the obtained results with legitimate flows only.

Table 1. System Response Time (without rootkit active operation).

Process Type	Flow	Average Packet Count	Process Time (ms)	Process Time with 50 Mbps load (ms)
Without ColEl verification	In Temporary Base	55,393	0.00110	0.00100
	DNS Response	2030	0.00407	0.00734
With ColEl verification	Legitimate Flows	983	516.78	551.05

5.2 Suspicious Flows

In this scenario, the rootkit was active and obfuscated traffic was transmitted, corresponding to approximately 2% of no-load traffic (100 flows of 10 packets each). In the case of packages that required queries to ColEl (and were considered suspicious), the response time was, on average, 1.6 s. The increase in processing time is due to the treatment and updating of the base of suspicious connections (Connection Database). There was no significant change in the processing time of DNS Response packets or packets present in the Temporary Base.

The results were considered satisfactory for out-of-band solutions that depend on other factors, such as access to external bases and network protocols, such as TCP. The performance of the ColEl is similar to that presented by NERD [14]. That was expected because this project maintains the exact characteristics of the Collector. Table 2 shows the obtained results with suspicious flows.

Table 2. System Response Time (with rootkit active operation).

Process Type	Flow	Average Packet Count	Process Time (ms)	Process Time with 50 Mbps load (ms)
Without ColEl verification	In Temporary Base	54,058	0.00114	0.00100
	DNS Response	1,329	0.00370	0.00730
With ColEl verification	Legitimate Flows	684	536.15	551.05
	Suspicious Flows	100	1,642.64	1,709.62

5.3 Enrichment System

The local enrichment tests were performed with a local database of nodes from the TOR network. It uses a database obtained from the website dan.me.uk [4], locally stored. The Enrichment System processes these queries (local database) in approximately 42 ms each. The remote enrichment tests were performed in real-time using the IP and Domain detection base provided by the Virustotal platform [2]. As this test depends on factors not controlled by the AudEl, such as the internet connection used, in addition to the processing time and response of the researched system, process time results showed more significant variance. On average, it took about 1.2 s to query the Enrichment System on this base, as shown in Table 3.

Table 3. Enrichment System actions.

Process Type	Database Description	Response Time (ms)
Local Base Query	Onion (TOR) node	42.18
Remote Base Query	VirusTotal Platform	1,241.14

5.4 Visualization

The Visualization System was structured to illustrate the data of the flows considered suspicious, the enrichment performed by the system, the status of the Collector Element, and the queried databases, in addition to related data to each suspicious flow. The interface presents two modules: general data and suspicious flow data. In general data, it graphically presents the query results in the external and local databases, besides the solution's working status and a summary of all flows detected, enriched, and marked as malicious, as shown in Fig. 5. In suspicious flow data, each flow is detailed with information from the source terminal, the destination IP, blocking information, and enrichment data, besides the domains used to obtain the destination IP by the source terminal, as shown in Fig. 6. By clicking on a flow, it is possible to see detailed information about each enrichment, like the ones shown in Fig. 7 and Fig. 8.

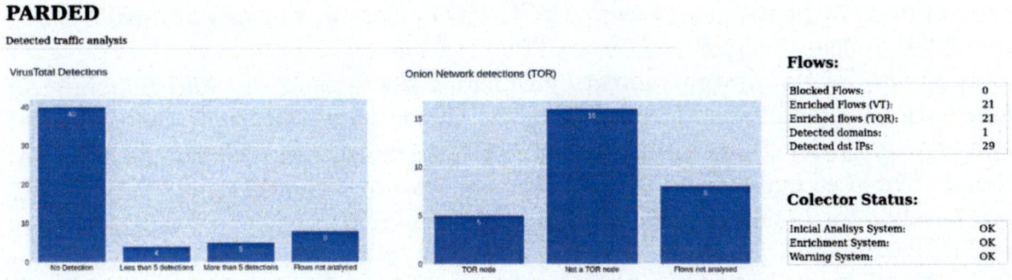

Fig. 5. PARDED - Auditor Element (general data).

5.5 Conclusions

In this work, the developed architecture could detect rootkits that use communication obfuscation techniques in the infected terminal, in a passive, scalable, generalizable, and adaptable way, without impacting the performance of the

Fluxos suspeitos:

id	Source IP	Destination IP	VT Enriched	TOR Enriched	Blocked	Domain list
32	192.168.190.11	200.84.244.50	true	true	false	
39	192.168.190.10	23.108.123.73	true	true	false	ruyicallback.771658.com
40	192.168.190.10	199.232.114.132	true	true	false	debian.map.fastlydns.net
41	192.168.190.10	40.117.131.251	true	true	false	packages.microsoft.com
49	192.168.190.10	201.16.50.19	true	true	false	r3.o.lencr.org
50	192.168.190.10	2.16.15.9	true	true	false	r3.o.lencr.org
51	192.168.190.10	104.18.32.68	true	true	false	ocsp.sectigo.com
52	192.168.190.10	108.158.121.188	true	true	false	ocsp.scalb.amazontrust.com
53	192.168.190.10	104.18.21.226	true	true	false	ocsp.globalsign.com
54	192.168.190.10	217.23.15.46	false	false	false	

Fig. 6. PARDED - Auditor Element (suspicious flow data).

Malicious Detection	Domain	IP	Date
0		23.108.123.73	2022-07-17T00:07:06.564395
0	ruyicallback.771658.com		2022-07-17T00:07:07.706740

Fig. 7. PARDED - Enrichment data from an external base.

network infrastructure. Furthermore, it adds data enrichment and incorporates information derived from multiple endpoints, allowing integration with other existing defense systems, such as IDS and firewalls. This way, it adds a new technique for real-time monitoring of malicious actions to organizations.

The detection of domain names used by suspicious flows, with enrichment data from local and external sources, in addition to the historically and visually presentation, adds threat intelligence and allows further analysis in greater detail, maintaining the context of the object of the evaluation, avoiding unnecessary enrichments that pollute the analysis environments.

About 95% of packets were processed in less than 0.0012 ms, even on heavily loaded systems, without any legitimate packet loss. Since they are executed in parallel, the enrichment data do not interfere with the legitimate data flow. These results prove the possibility of using PARDED in existing network infrastructures without significant impacts.

However, as the system detects suspicious flows passively and depends on information collected and processed after the first packet from the analyzed data stream, at least the initial packets of suspicious flows will not be blocked, even if treated as malicious. Future works can optimize the threshold values used as a baseline for blocking packets, define ciphertext packets (such as DNS-over-HTTPS) strategies, integrate auxiliary detection techniques and databases, and verify the payloads transmitted by suspicious flows in malicious hash bases.

Name	Flags	Version	Date
Unnamed	FGHRSDVX	Tor 0.4.5.8	2022-07-17T00:36:24.120958

Fig. 8. PARDED - Enrichment data from a local base.

Acknowledgements. M.T. gratefully acknowledge Mateus Berardo de Souza Terra for providing the source code of the Collector Element used in the Network Exfiltration Rootkit Detector (NERD). R.A. gratefully acknowledges the technical and computational support of the Laboratory of Technologies for Decision Making (LATITUDE) of the University of Brasília, the General Attorney's Office (Grant AGU 697.935/2019), the General Attorney's Office for the National Treasure - PGFN (Grant 23106.148934/2019-67), and the National Institute of Science and Technology in Cyber Security - Nucleus 6 (grant CNPq 465741/2014-2); The authors thankfully acknowledge the support of Brazilian Intelligence Agency - ABIN grant 08/2019.

References

1. Basin, D.: The cyber security body of knowledge. University of Bristol, ch. Formal Methods for Security, version (2021). https://www.cybok.org
2. Chronicle Security: Virustotal. http://www.virustotal.com. Accessed 15 July 2022
3. Costa, H., Nicoletti, G., Pisu, M., Von Rueden, C.: Are digital platforms killing the offline star? Platform diffusion and the productivity of traditional firms. OECD Economics Department Working Papers, vol. 1, no. 1682, pp. 1–27 (2021). https://doi.org/10.1787/18151973
4. Dan: Tor Node List. https://www.dan.me.uk. Accessed 11 July 2022
5. ENISA: The year in review - ENISA Threat Landscape (2020). https://www.enisa.europa.eu/publications/year-in-review/@@download/fullReport
6. ITU: Individuals using the Internet (2021). https://www.itu.int/en/ITU-D/Statistics/Pages/stat/default.aspx. Accessed 04 July 2022
7. Julian, V., Botti, V.: Multi-agent systems. Appl. Sci. **9**(7) (2019). https://doi.org/10.3390/app9071402
8. Knapp, E.D., Langill, J.T.: Exception, anomaly, and threat detection. Ind. Netw. Secur. 323–350 (2015). https://doi.org/10.1016/B978-0-12-420114-9.00011-3
9. Marques, R.S., et al.: A flow-based multi-agent data exfiltration detection architecture for ultra-low latency networks. ACM Trans. Internet Technol. **21**(4), 1–30 (2021). https://doi.org/10.1145/3419103
10. Morgan, S.: Cybercrime To Cost The World $10.5 Trillion Annually By 2025. Cybersecurity Ventures, p. 1 (2020). https://cybersecurityventures.com/cybercrime-damages-6-trillion-by-2021/. Accessed 20 July 2022
11. Naik, N., et al.: Embedded YARA rules: strengthening YARA rules utilising fuzzy hashing and fuzzy rules for malware analysis. Complex Intell. Syst. **7**(2), 687–702 (2021). https://doi.org/10.1007/s40747-020-00233-5
12. Reed, M., Syverson, P., Goldschlag, D.: Anonymous connections and onion routing. IEEE J. Sel. Areas Commun. **16**(4), 482–494 (1998). https://doi.org/10.1109/49.668972
13. Sowinski, D.: The Growing Danger of Data Exfiltration by Third-Party Web Scripts. https://securityintelligence.com/posts/growing-danger-data-exfiltration-third-party-web-scripts/. Accessed 04 July 2022
14. Terra, M.B., Gondim, J.J.: NERD: a network exfiltration rootkit detector based on a multi-agent artificial immune system. In: 2021 Workshop on Communication Networks and Power Systems, WCNPS 2021 (2021). https://doi.org/10.1109/WCNPS53648.2021.9626241
15. Wooldridge, M.: An Introduction to Multiagent Systems. Wiley, England (2009)

OSINT Methods in the Intelligence Cycle

Roberto Tanabe[1(✉)], Robson de-Oliveira-Albuquerque[1], Demétrio da-Silva-Filho[1,2], Daniel Alves-da-Silva[1], and João-Jose Costa-Gondim[1,3]

[1] Professional Post-Graduate Program in Electrical Engineering, Department of Electrical Engineering, University of Brasília, Brasília 70910-900, Brazil
rtanabe@gmail.com, {robson,daniel.alves}@redes.unb.br, {dasf,gondim}@unb.br
[2] Institute of Physics, University of Brasilia, Brasilia, DF 70910-900, Brazil
[3] Department of Computer Science, University of Brasilia (UnB), Brasília, DF 70910-900, Brazil

Abstract. The process for producing intelligence is traditionally represented by a series of steps forming a cycle. Collection is one of these stages and is characterized by the application of a set of disciplines to obtain information that will be analyzed for the production of an intelligence product. These disciplines are characterized according to the type of source, its methods and techniques. Open-Source Intelligence (OSINT) is the collection discipline focused on publicly available information. Open-source collection methods are usually represented by working diagrams that draw a flow according to the type of information. This paper makes a comparative study of the intelligence cycles of some relevant actors in the international OSINT scene, locates open-source collection in the intelligence cycle, and presents a workflow that combines this cycle with the techniques that form a method for OSINT.

Keywords: OSINT · Intelligence · Methodology

1 Introduction

Intelligence obtained from open-sources is critical to the production of knowledge to support decision-making. Open-source intelligence (OSINT) is a discipline of gathering intelligence on publicly available information. The web and social media platforms have increased the amount of available data and facilitated access to information by promoting online searches and the rapid development of tools to facilitate information retrieval.

OSINT is usually related to a set of tools or a list of online services. Tools can be discontinued or cease to function and tend toward an automation that abstracts from understanding the techniques they enable. A technique is a specific way of performing a task, composed of processes that transform an input into an output. The information paths formed by techniques is usually called

M. V. Garcia and C. Gordón-Gallegos (Eds.): CSEI 2022, LNNS 678, pp. 42–54, 2023.
https://doi.org/10.1007/978-3-031-30592-4_4

a method. To be durable over time, OSINT methodological practice must be based on its techniques and not only on tools

Considering the above, the main contributions of this paper are:

a) Situates open-source intelligence in the intelligence cycle;
b) A comparative study of the OSINT intelligence cycles, methods and techniques of some relevant players in the global OSINT landscape; and
c) Presents a workflow that combines an intelligence cycle with the techniques that form a method for OSINT.

This paper is organized as follows: Sect. 2 defines some necessary background concepts and major related work in OSINT; Sect. 3 presents techniques and methods indicated for working with OSINT; Sect. 4 proposes a method for OSINT that combines a workflow of techniques with the intelligence cycle; Finally, Sect. 5 concludes the main observations of this paper.

2 Related Work

Intelligence refers primarily to the activities of state intelligence agencies or services that collect, analyze, and disseminate information to meet the requirements of a decision-maker.

The intelligence cycle is the process for producing intelligence. In general, it is formed by steps that: a) define the information requirements of a user; b) plan the fulfillment of these requirements; c) collect the necessary information to develop the final product; d) transform the collected information into a usable format; e) analyze the information to obtain meaning; f) create the intelligence product; g) transmit the intelligence product to the user who demanded it; and h) evaluate all these actions constantly. These general steps are usually presented as a continuous cycle (Fig. 1).

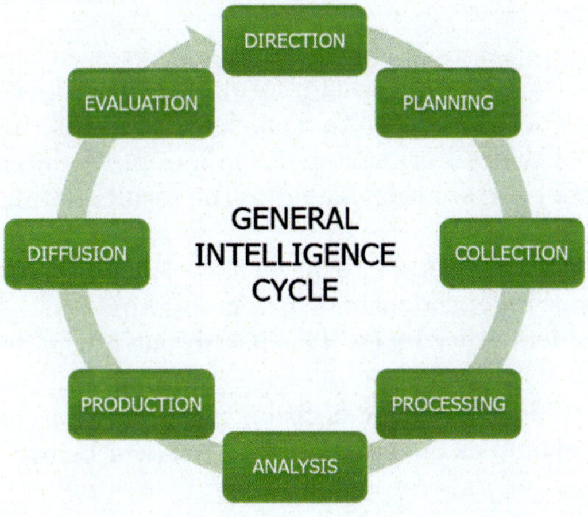

Fig. 1. A general intelligence cycle and its steps.

Collection is a step in the intelligence cycle, and is performed to collect information related to the intelligence gathering disciplines or intelligence sources in order to provide analysis of all available sources (all-source intelligence) [9].

A Discipline of Intelligence Collection (INT) is characterized by the specific methods, techniques, and type of sources for collecting information. Five disciplines are listed as classic INTs [13]. In addition to OSINT we have those of Table 1:

Table 1. Disciplines of Intelligence Collection descriptions

Disciplines of Intelligence Collection	Description
Human Intelligence (HUMINT)	The collection of information provided by a human source, where the collector interacts directly with the source, controls the discussion topics, and directs the source's actions [7]
Image Intelligence (IMINT)	Connected to the collection and analysis of images and geospatial information to describe, evaluate and represent georeferenced activities [14]
Signal Intelligence (SIGINT)	The intelligence form of collecting and processing various forms of electronically transmitted information; these forms are the communication of human language material and data derived from electronic transmission devices [15]
Measurement and Signature Intelligence (MASINT)	Produced by quantitatively and qualitatively analyzing the physical attributes of targets to characterize and identify them [12]. It uses various types of sensors, such as radiation meters

There are many ways for collecting information from open-sources. Thus, an ordering of actions is necessary to obtain efficient results [6]. In [5,10,14], OSINT is described as one of the collection step disciplines of the intelligence cycle. The results of the collected one are then added to the results obtained from the other INTs (see Fig. 2).

Some authors have developed their own cycles for OSINT that have the advantage of working independently of being contained in a larger intelligence cycle that also considers other INTs. This is also one of the drawbacks since the absence of the other INTs makes an all-source product impossible. These cycles detail the path from identifying the decision maker's requirements to delivering the product. Some examples of these cycles are shown below:

2.1 Williams and Blum OSINT Cycle

Williams and Blum's cycle [18], has 4 stages. The first is the Collection stage, which includes identifying and obtaining potentially useful information and pre-

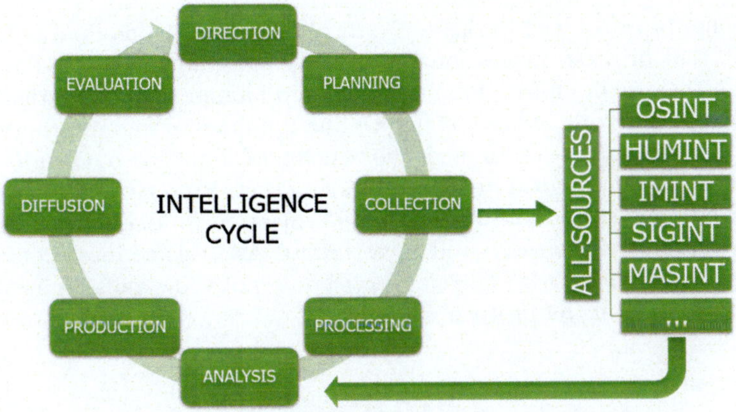

Fig. 2. OSINT as part of the Intelligence Cycle.

serving what has been collected. The second, Processing, involves the translation, conversion and aggregation of the information into a usable format for the other steps of the cycle. In Exploration occurs the analysis that verifies the reliability of the information and relates it to the interest of the decision-maker. In Production, there is the evaluation of whether the information produced should be classified and the delivery to the user. Figure 3 refers to their cycle.

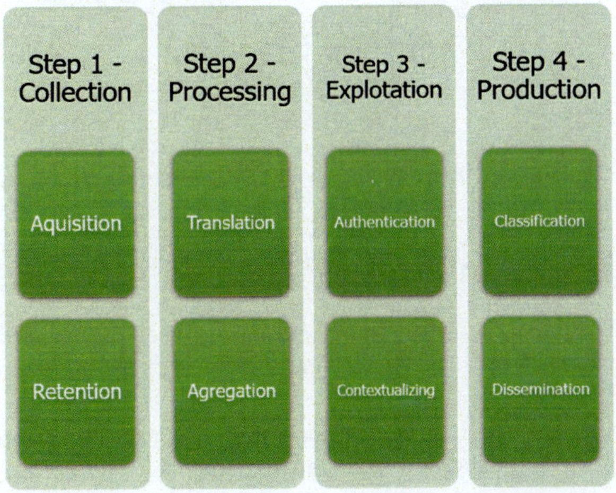

Fig. 3. Williams and Blum OSINT cycle [18].

2.2 Berkeley Protocol OSINT Cycle

The Center for Human Rights at the University of California, Berkeley, School of Law develops, in partnership with the United Nations Human Rights Office, an international protocol, the Berkeley Protocol on Digital Open Source Investigations [17], that provides professional standards and guidelines aimed at improving its effective use in international criminal and human rights investigations [3].

The Berkeley Protocol has 6 steps: Online Inquiries to find information; Preliminary Assessment to determine if the benefits of collection outweigh the risks; Collection to capture digital items from the internet; Preservation to ensure that the information is stored and retrievable; Verification to assess the reliability of the sources and their content; and Investigative Analysis to interpret the data. The Berkeley Protocol OSINT cycle (Fig. 4) does not show a dissemination step, but it is described in the protocol.

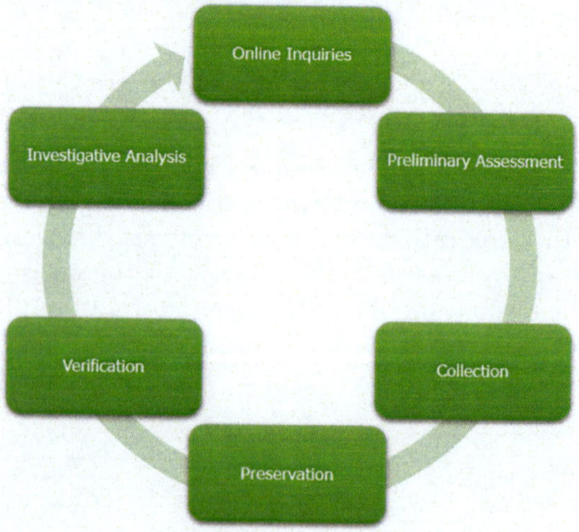

Fig. 4. Berkeley Protocol OSINT cycle [17].

2.3 The Bellingcat OSINT Cycle

Bellingcat is an independent international group of researchers, investigators and journalists using open-sources to report on a variety of conflicts around the world [1].

The Bellingcat cycle begins in Identification with locating sources of information and determining their scope and depth. Next, Collection and Preservation occur to obtain the information and protect it from tampering and destruction. Next, in Verification, geospatial analysis tools, reverse image search, and visual data analysis are employed. In Analysis, the main questions of the case are answered. Next, in Review and Confirmation, content cross-referencing and incident classification are performed. Finally, the information is presented to the user and stored in a database. Figure 5 shows the Bellingcat OSINT cycle. This cycle shows that it tends to start with a large amount of not very relevant information, which in its 7 stages, would deliver to the intelligence user a small amount of very relevant information [2].

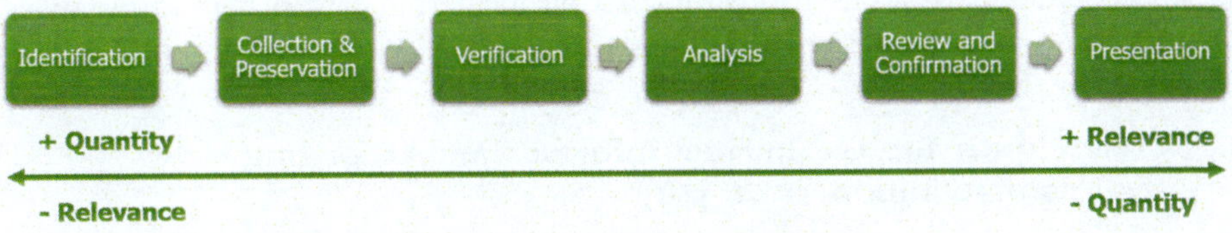

Fig. 5. Bellingcat OSINT cycle [2].

2.4 Pastor-Galindo *et al.* OSINT Approaches

J. Pastor-Galindo *et al.* [16] propose a workflow that expands the amount of data about a target from social networks, email address, username, real name, location, IP address, and domain name. The execution of techniques specific to each piece of data generates an output that can be leveraged as input for another process (data transfer) that exploits another technique to generate more data. The Fig. 6 illustrates their approach.

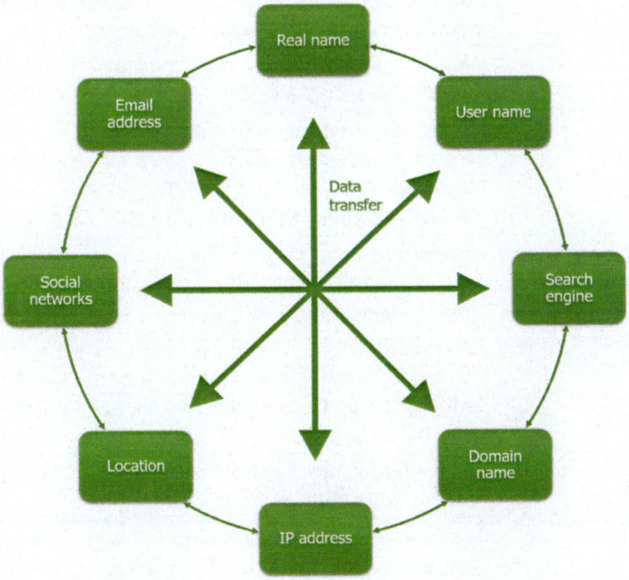

Fig. 6. From any starting point, the dataset about the target increases [16].

3 Techniques and Methods

The authors of the cycles presented in Sect. 2 also describe their techniques. Some techniques are focused on information gathering and others on analysis. Other authors represent the techniques through workflows or diagrams.

3.1 Techniques for Gathering Information According to Pastor-Galindo *et al.* [16]

J. Pastor-Galindo *et al.* present a list of techniques with a focus on collection according to the type of information one has, as shown in Table 2.

Table 2. Description of techniques for gathering information by Pastor-Galindo *et al.* [16].

Techniques	Description
Search engines	Services within the web receive a query trying to provide information that matches the input, returning information to the collector such as Google, Bing and DuckDuckGo;
Social networks	A lot of information about people or organizations can be found on social media platforms like YouTube, LinkedIn, and TikTok;
Email	An email address is unique and acts as the input for numerous web services;
Username	They are used for online services and are also a good way to collect information because the same username can be used in different web services and in each of them reveal information;
Real name	Searching for a target's real name can also reveal social media, home addresses, phone numbers, email, usernames, and more;
Location	Searching the locations related to a target can give us indications about a person's behavior. Photos, addresses, and GPS coordinates are all data that can be obtained;
Internet Protocol (IP)	IP addresses are important for digital forensics to collect information from an event; and
Domain name	They are related to the name of web services. They can reveal information about a target, such as the person who registered the site

3.2 Techniques for Analysis Described in the Berkeley Protocol [17]

The Berkeley Protocol [17] segments the techniques by types of analysis, as shown in Table 3.

Table 3. Description of techniques for analysis described in the Berkeley Protocol [17]

Types of analysis techniques		Description
Technical Analysis	Metadata	The data about other data, such as the latitude and longitude of where an image was taken
	Source Code	The programming language behind a web page (HyperText Markup Language - HTML) or software (Python, Java, PHP, etc.)
Content Analysis	Geolocation	The identification of the whereabouts of an object or an activity
	Chronolocation	The identification of dates and times of an event. This can be done by analyzing the shadow projection of buildings to identify the time of day in a photo, for example.
Investigative Analysis	Image/Video comparison	The process of comparing characteristics of objects, people and/or places when at least one of them is an image
	Image/Video interpretation	Analysis of visual clues of objects and places;
	Spatial analysis	It involves examining different landscape objects and checking them against satellite or other imagery, geodata, and maps;
	Actor mapping	It is to identify the key players and their relationships;
	Social network analysis	It is the mapping and measuring of the relationships between the nodes of a network in the context of social media platforms;
	Incident mapping	It is used to establish the temporal and geographical relationships between different events.

3.3 Techniques Used in Social Media Content Analysis Based on Williams and Blum Approach [18]

Williams and Blum [18], on the other hand, bring in the social network techniques applied to social media platforms (Table 4).

Table 4. Description of Social Network Analysis techniques by Williams and Blum [18]

Social Network Techinques	Description
Lexical analysis	It can cluster vast amounts of text and show the most searched terms in a search engine or which words appeared most frequently. At another level, it can infer information about the people involved including demographic characteristics. Common ways of doing this include Sentiment Analysis, Natural Language Processing and Machine Learning;
Social Network Analysis (SNA)	It attempts to explain relationships between individuals as a series of exchanges that can be mapped to understand the network of connected actors. SNA in the internet age has created an exponential supply of new data points in the study of network interactions. Among its techniques are: Degree, Density, Betweenness and Betweenness Centrality;
Geospatial	Social media platforms allow you to automatically link a post to a specific location. It includes Geotagging, Geolocating, Geo-inference and Georeferencing

3.4 Methods Described as Workflows of Techniques

Some authors prefer to describe methods as workflows of techniques through diagrams that show the path to be taken to work with each type of information.

Bazzell [8]'s diagrams map actions and tools that can show which username or social network is related to an email. Bazzell [8] also presents diagrams for real names, locations, usernames, phone numbers, and domain names.

These workflows by information type decrease the abstraction of what should be done in the collection phase, while detailing a way of doing things that takes into account the environment of each search, regionalisms, and the technical capacity of the OSINT collector.

Sinwindie [4] also uses the model presented by Bazzell to create his own diagrams. For example, Sinwindie [4] developed the diagrams for usernames, email, IP, image, person and website.

Figure 7 shows examples of Bazzell's and Sinwindie's diagrams.

The continuous concatenation of the output obtained by one technique into an input for the next technique from several diagrams would make it possible, from one type of information, to arrive at another type. This can be illustrated in Fig. 8 by the use of Hoffman's [11] graph analyzer.

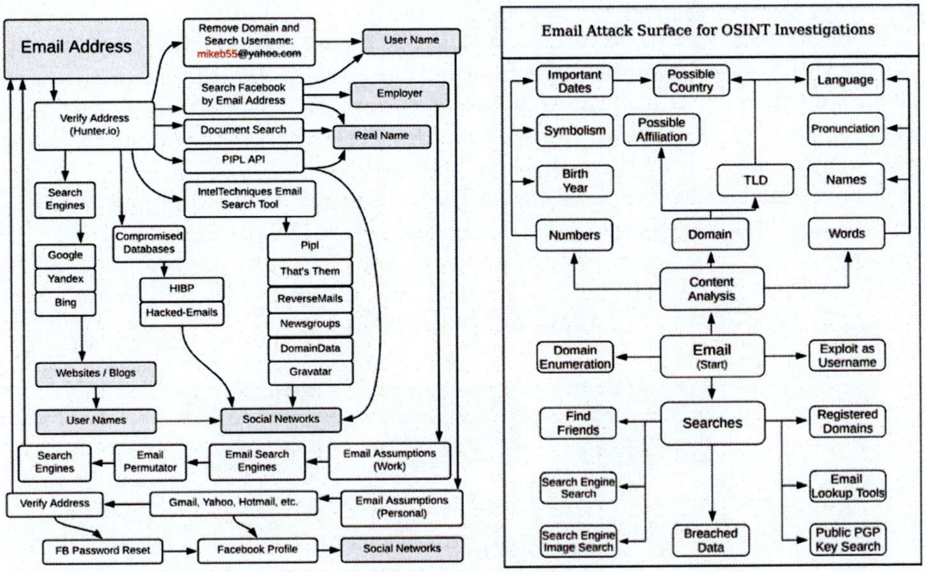

Fig. 7. Bazzell's (left) [8] and Sinwindie's (right) [4] workflows for email.

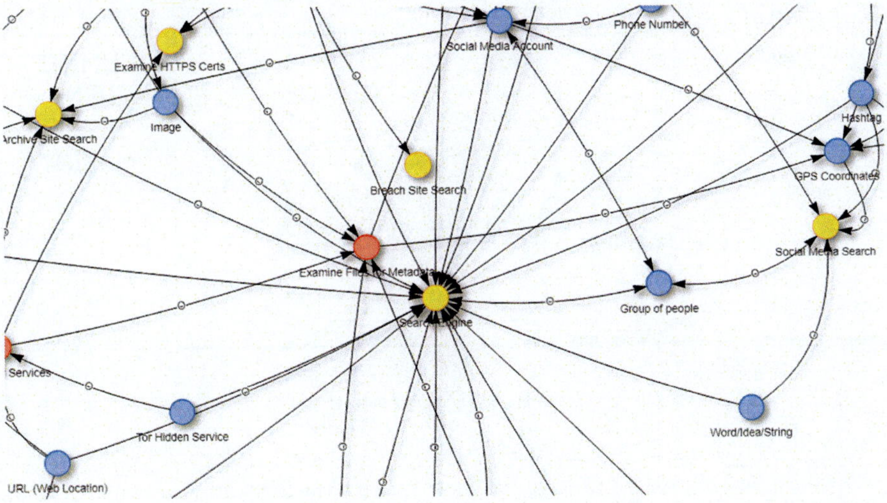

Fig. 8. Hoffman's OSINT Graphical Analyzer [11].

4 Proposed Method

Each type of information corresponds to several types of techniques. The combination of all the information paths within the OSINT discipline and the logical sequence of each technique constitutes a method for OSINT.

Open sources need to be verified, and this is an analysis function. Sometimes, for this verification to happen, the information needs to be converted, either by translating it or by transforming it into a format that can be used. Although traditionally INTs are known as collection disciplines, in practice the Processing and Analysis steps are indispensable in their execution.

Figure 9 illustrates the detailing of the Collection, Processing and Analysis steps of the intelligence cycle with the respective OSINT techniques.

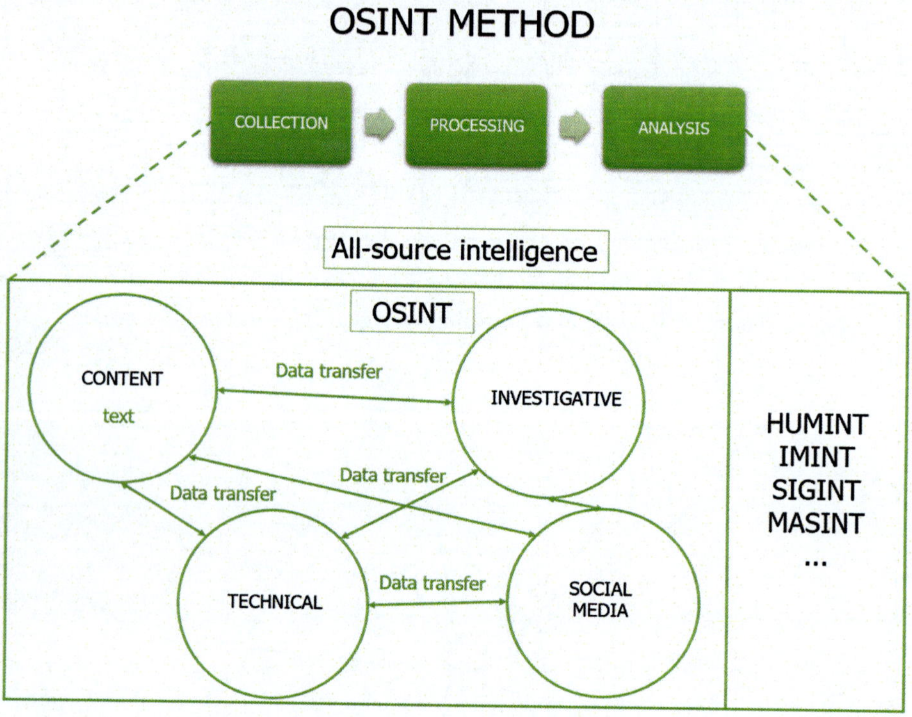

Fig. 9. OSINT Method

As we begin this major step of Collection, Processing and Analysis, we make use of all available INTs (all-source intelligence), starting with OSINT, since publicly available information is most easily obtained. The techniques presented in Sect. 3 as analysis techniques are actually also collection and processing techniques at the same time.

In this paper, we consider the text type to be the most frequent and the most used as a starting point in OSINT collections. In this way, the text type is transformed into other types of information that then go on to technical, investigative and social media analysis.

Compared to the cycles unique to OSINT, the method proposed here excludes the Steering, Planning, Production, Dissemination and Evaluation phases, focusing only on the steps that obtain, transform and analyze information. It also places OSINT execution ahead of other INTs because of the free access characteristic of open sources. Mainly, this method contemplates the other INTs, allowing for a more complete Intelligence product considering other available sources of information.

5 Conclusion

Versions of the classic intelligence cycle are often presented as methods for OSINT which, as a collection discipline, are in the Collection phase. However, its analysis techniques and the eventual need to transform information into a useful format make OSINT, in practice, also part of Processing and Analysis, as shown in Fig. 9.

OSINT techniques depend on the type of information you have, and when chained together to get more information about a target they draw paths that describe a method. Content, technical, investigative, and social media analysis techniques exchange information with each other until the publicly available information is addressed and then follow with the other INTs.

The cycles presented by J. Pastor-Galindo et al., Berkeley Protocol, Bellingcat and Williams and Blum work independently of the other INTs. The method presented in Sect. 4 unifies the workflow of the OSINT techniques with the general intelligence cycle and links it with the other collection disciplines.

Acknowledgments. The authors thankfully acknowledge the support of Brazilian Intelligence Agency - ABIN grant 08/2019; R.d.O.A. and D.A.d.S gratefully acknowledge the technical and computational support of the Laboratory of Technologies for Decision Making (LATITUDE) of the University of Brasília; the General Attorney's Office (Grant AGU 697.935/2019); the General Attorney's Office for the National Treasure - PGFN (Grant 23106.148934/2019-67); the National Institute of Science and Technology in Cyber Security - Nucleus 6 (grant CNPq 465741/2014-2); D.d.S.F gratefully acknowledges the financial support from the Edital DPI-UnB No. 02/2021, from CNPq (grants 305975/2019-6, and 420836/2018-7) and FAP-DF (grants 193.001.596/2017 and 193-00001220/2021-37); D.A.d.S gratefully acknowledges the National Department of Audit of the SUS (Grant DENASUS 23106.118410 /2020-85); the Deans of Research and Innovation and Graduate Studies at the University of Brasília (Grant 7129 FUB/AMENDA/DPI/COPEI/AMORIS).

References

1. Bellingcat - the home of online investigations: about page. https://www.bellingcat.com/about/. Accessed 23 Apr 2022
2. Bellingcat: workflow page. https://yemen.bellingcat.com/methodology/workflow. Accessed 20 May 2022
3. Human rights center about us page. https://humanrights.berkeley.edu/about/about-us. Accessed 22 June 2022

4. Sinwindie: Github - sinwindie/osint: Collections of tools and methods created to aid in osint collection. https://github.com/sinwindie/OSINT. Accessed 16 Sept 2017
5. National doctrine on intelligence activity - doctrinal foundations (2016). https://www.gov.br/abin/pt-br/centrais-de-conteudo/publicacoes/Legislao3V5.pdf. "Portaria n.o 244-ABIN/GSI/PR", Coletânea de Legislação
6. Editorial standards & practices (2020). https://www.bellingcat.com/app/uploads/2020/09/Editorial-Standards-Practices.pdf. Accessed 22 June 2022
7. Althoff, M.: The Five Disciplines of Intelligence Collection, Chap. Sage, Human Intelligence (2015). ISBN: 9781452217635
8. Bazzell, M.: Open source intelligence techniques: resources for searching and analyzing online information, 9th edn. (2022). https://Inteltechniques.com. ISBN-13: 979-8794816983
9. Fingar, T.: A guide to all-source analysis. Intelligencer. J. US Intell. Stud. Assoc. Former Intell. Officers **19**, 1–4 (2012). https://www.afio.com/publications/Fingar_All_Source_Analysis_in_AFIO_INTEL_WinterSprg2012.pdf
10. Gibson, H.: Acquisition and preparation of data for OSINT investigations. In: Akhgar, B., Bayerl, P.S., Sampson, F. (eds.) Open Source Intelligence Investigation. ASTSA, pp. 69–93. Springer, Cham (2016). https://doi.org/10.1007/978-3-319-47671-1_6
11. Hoffman, M.: Introducing OSINT yoga. https://webbreacher.com/2018/06/24/introducing-osint-yoga/. Accessed 05 June 2022
12. Hughes, P.M.: MASINT. Am. Intell. J. **36**(2), 7–10 (2019). Accessed 15 June 2022
13. Kamiński, M.A.: Intelligence sources in the process of collection of information by the us intelligence community. Secur. Dimensions: Int. Nat. Stud. **32**, 02–105 (2019). https://doi.org/10.5604/01.3001.0014.0988
14. Office of the Director of National Intelligence, O.: Us national intelligence: An overview (2013). www.dni.gov/files/documents/USNI%202013%20Overview_web.pdf. Accessed 10 June 2022
15. Nolte, W.N.: The Five Disciplines of Intelligence Collection, Chap. Sage, Human Intelligence (2015). ISBN: 9781452217635
16. Pastor-Galindo, J., Nespoli, P., Mármol, F.G., Pérez, G.M.: The not yet exploited goldmine of OSINT: opportunities, open challenges and future trends. IEEE Access **8**, 10282–10304 (2020). https://doi.org/10.1109/ACCESS.2020.296525
17. United nations office of the high commissioner for human rights: berkeley protocol on digital open source investigations. United Nations, New York, NY (2022). www.ohchr.org/Documents/Publications/OHCHR_BerkeleyProtocol.pdf. Accessed 17 June 2022
18. Williams, H., Blum, I.: Defining second generation open source intelligence (OSINT) for the defense enterprise. Technical report (2018)

Mobile Marketing as a Communication Strategy in Politics 2.0

César-A. Guerrero-Velástegui[1], Cristina Páez-Quinde[1,2](✉), Carlos Mejía-Vayas[1], and Josué Arévalo-Peralta[3]

[1] Facultad de Ciencias Administrativas, Grupo de investigación Marketing C.S., Universidad Técnica de Ambato, Ambato, Ecuador
{ca.guerrero,mc.paez,carlosvmejia}@uta.edu.ec

[2] Instituto Superior Tecnológico España, Madrid, Spain
maria.paez@iste.edu.ec

[3] Facultad de Ciencias Humanas y de la Educación, Universidad Técnica de Ambato, Ambato, Ecuador
jarevalo3676@uta.edu.ec

Abstract. This research titled Mobile Marketing as a communication strategy in poli-tics 2.0 aims to determine the impact of M-Marketing as a communication strategy in the transformation to a context of politics 2.0 in the province of Tungurahua, considering the great importance of marketing in political management and dissemination, which is why it is of great contribution to political parties who have identified the need to establish links with voters. An exploratory and experimental cross-sectional investigation has been developed, based on a sample of 385 students and teachers from the Faculty of Administrative Sciences of the Technical University of Ambato. As techniques for the collection of information, two surveys were applied, the first as an analysis for the development of an informative App within the 2.0 policy and the TAM Model that measures the acceptability of the App, for the validation of the instruments the statistic of Cronbach's alpha with the reliability of 0.828. The verification of the hypothesis is carried out by Kolmogorov Smirnov, where it is confirmed that Mobile Marketing contributes as a communication strategy in the 2.0 policy. As a result of this research, there is good acceptability of the App by the general public, and the information, management, and navigation are adequate and user-friendly. In conclusion, it allows political parties to achieve a better position, since it gives way to active participation strategies and access to current information in real-time, creating proximity through bidirectional communication and interaction between politicians and citizens.

Keywords: Web 2.0 · Mobile Applications · Mobile Marketing · Politics 2.0 · Political communication strategies

Supported by organization x.

M. V. Garcia and C. Gordón-Gallegos (Eds.): CSEI 2022, LNNS 678, pp. 55–69, 2023.
https://doi.org/10.1007/978-3-031-30592-4_5

1 Introduction

Otero. [21] explain that due to the growing presence of smartphones, political communication seeks to gain a representative space on these devices as part of their communication strategies. The authors also determine that currently applications focused on politics manifest themselves as an important innovation of party structures and political management to mobile digital environments, presenting actions that range from direct access to web pages of candidates or parties to experiences informative and differentiating that give way to positioning in the mind of the voter.

This study concludes that, for political actors, effectively reaching the personal devices of their target market (voting population) offers them the opportunity to establish a more individualized, close, and constant communication through notifications, political parties can also access relevant information on voters actively (subscriptions) and passively.

On the other hand, [26] specify that the increase in smartphones has established a new mobile ecosystem that defines a mobile society and culture that establishes a key territory for contemporary political communication. Within this culture, mobile applications have become a common meeting space for organizations and citizens interested in participating in political affairs through the direct experience that these platforms offer. However, despite the increase in technology in this field of study, it is difficult to find a complete and reliable taxonomy of applications in the academic or professional literature that analyzes how these relationships impact political communication.

The authors introduce an analysis of the systematic taxonomy of political communication, collecting in detail the variables necessary to understand the nature of this type of application available for any smartphone. A methodological classification is made based on categories such as promoting agent, target app, level of interaction, level of autonomy, and predominant tone. The author [11] explains how fundamental it is that the policy adapts to the growing digital transformation since it allows better organization and communication between both.

For this, it is necessary to understand that the mobile society represents a cultural challenge to which democratic politics must respond to be useful and practical for citizens by being present in their daily activities. As part of his study to establish his work, the author determines that the importance of the development of mobile internet lies in the possibility offered to users to transform themselves from spectators to actors in the situation, facilitating the autonomy of online political activism.

For their part, the authors [4,9] explain the use of WhatsApp by local administrations that political managers have detected the need to be present in mobile applications, such as the WhatsApp platform, thus responding to new habits of citizens, who currently carry out most of their daily activities from their smartphones. The authors determine that taking advantage of the characteristics of this type of App promotes direct and immediate communication as well as creating better relationships with citizens.

After this study was carried out with a qualitative approach based on interviews, the authors conclude that the WhatsApp application does not allow taking advantage of all the potentialities of the digital environment, mainly due to the lack of strategic planning in the immersion of the political activities of the municipalities studied. Regarding the use of applications on mobile devices, the information collected is extremely positive since it leaves a precedent for the communication networks that can be established between local and national governments and the population if the benefits of technology are taken advantage of, especially the immediacy of the Apps.

Finally, the digital newspaper [15] in its article They launch an app to connect people with politicians talks about the mobile app 'Politicapp', a participatory political social network developed in Argentina that allows political actors to share content with citizens, who can actively participate by sharing opinions and contacting candidates directly. In the article, the creator of the app explains that it offers politicians the option of improving their relationship with the population through interaction and thus collecting useful information for decision-making.

In the context of the evolution of politics 2.0 and its immersion with technological tools, it is understood that this approach has gone through several factors that have modified the communication between citizens and political actors, who have identified the need to establish links with voters that allow them to better understand their needs and expectations, establishing strategies based on it. Therefore, the importance of developing an instrument that establishes fast, agile, effective, and reliable communication has been raised. Indeed, the developers of political campaigns take as their main focus the application of communication strategies that allow them to reach the population of voters with a differentiating message.

In the modern political campaign, the candidates and their communication team have used telephones, faxes, newspapers, and electronic mail to spread their messages massively to potential voters to win the electoral process. In the 90s, advertising campaigns were incorporated on the internet, focusing on the creation of a web page by the politician Bod Dole in 1996, he was the first to carry out political advertising through this medium [16].

Barack Obama's first electoral campaign is considered one of the most important examples of political marketing through the web, since, after the announcement of his candidacy, his campaign was mostly conducted online, generating greater acceptance in the population by taking advantage of tools such as e-mail, social networks, web pages, and electronic media. Former Dean technology employees were recruited who directly influenced the direction of the online strategy, implemented what worked, and learned from their previous mistakes.

Furthermore, by the time the 2008 election period began, more people were using Facebook, MySpace, and YouTube, during the first presidential contest, as these sites had a large voter presence. Without social media, likely, Obama would not have been able to be a viable candidate in the primaries against Senators Clinton and Edwards because he would not have been able to find and connect with his audience, fund his candidacy or motivate them to support him. His campaign [20].

Barack Obama was considered the great pioneer of social networks in presidential campaigns during the year 2008 with his famous slogan "Yes We Can". This innovative phenomenon marked a true milestone in political communication. It developed its campaign strategically in the 50 states, targeting the disaffected center, that is, targeting the undecided and disillusioned. His campaign was based on small donations of money, the collaboration of small amounts, integration of the decisions of the digital campaign, and transforming the offline campaign into an online one by segmenting the messages broadcast [9].

In Ecuador, the use of applications has been increasing, three out of four people have a smartphone, 90.08% access 3G + 4G technology, 94.68% use social networks on their phones, 91.98% internet, and 76.22% GPS. At the international level, Ecuador is ranked 97 out of 176 in the development of ICTs, while at the regional level it occupies the eighth position, with a difference of 2.32 points from Uruguay, first in the ranking. Given the high level of market penetration of these devices, mobile applications have a high development margin. This situation was observed when carrying out exploratory research on Google Play on the available tourist applications [18,24].

The application development market in Ecuador is expanding, some companies are dedicated to the development of mobile apps based on the ideas of entrepreneurs. The Andean region stands out with the highest number of registered Apps (65), with the province of Pichincha being the one with the most Apps in the region (36). At the local level, Quito has the largest number of Apps, possibly because it is one of the most visited cities at the national level.

Digital tools are the most used to get their messages to society in a closer way and through which users can participate in the democratic process. However, Ecuador and other countries of the world, do not understand the real importance of having a set of well-structured digital strategies based on the real demand of the voter. Currently, in several Latin American countries, there is certain digital illiteracy, which occurs because there is no structured planning of the strategic use of social networks, during the development of a political campaign, which if raised by countries like the USA [1,8,22].

2 State of the Art

2.1 Web 2.0

According to [10] web 2.0 is defined as the second technological generation of the web that is created from networks and user communities and has instruments such as social networks, forums, chats, blogs, etc., these promote the rapid exchange of information dynamically and interactively that enhances collective intelligence by expanding collaborative work. The author also determines that together with the arrival of web 2.0 there was a social change that is evident in the relationship of users with information and communication by making them an active part of it.

For its part, [5,14] determines that web 2.0 refers to a way of taking advantage of the benefits of the internet with the predominance of web applications, which facilitate the exchange of information, with a user-centered design, giving move to online collaboration. In the same way, the authors [7] explain that web 2.0 arises as a result of brain-storming and is defined as the second generation of innovative applications. Web 2.0 is based on 7 basic principles.

2.2 Apps

The author [15] defines "Apps" as a software application for use on mobile devices, the term is an abbreviation of the English word "application" and is used to mention an informative connection. Apps appeared in the 90s, but their development in the market was in 2008 with the launch of the Apple company's App Store and then the Google Play platform. From these events the term App began to be used especially for installation on mobile devices, currently, the Apps market is experiencing a boom due to its great reception.

According to [19] a mobile application is designed to be functional and run from a smartphone or tablet mobile device, seeking to provide users with quality experiences and services without using integrated software systems. Likewise, the author [17] explains that mobile applications are software tools that are written in various programming languages for smartphones or tablets. Their goal is to be useful, dynamic, and easy to install and operate. He also comments that it is possible to develop all kinds of applications, games, RRSS, real-time chats, etc. They seek to create satisfying experiences for users in real-time.

2.3 Mobile-Marketing

Mobile Marketing is generated based on carrying out advertising and marketing actions through mobile devices. It generates a personalized experience for customers, since communication is individualized, the user has the option of a private space and bidirectional contact. It suits people who are interested in the content provided by the company having access anytime, anywhere.

Mobile marketing is the set of actions that allows companies to communicate interactively with their audience through mobile devices or networks, this channel between the advertiser and users aims to promote products or services through a ubiquitous network in which users are constantly connected from their personal mobile devices. The use of technologies has made it possible to generate marketing and sales solutions: SMS, MMS, Mobile Advertising, content sales and App [13].

[6] considers that Mobile Marketing is one of the most intrusive advertising channels that can be well received by users but can also generate intimacy and cause effects contrary to what is desired. It is important to consider that the key to successful marketing on mobile devices is to know the tastes and preferences of customers and trends created at the time, the customer must feel that the marketing message is a service considering the form of communication [10].

2.4 Politics 2.0

According to [12], politics 2.0 refers to the evolution of political communication to a bidirectional, participatory, and voluntary model where the citizen has control of the processes, so in this scenario, it is easier to have contact. Direct, give visibility, encourage discussions, or notify about future activities.

Politics 2.0 emerged from primary scenes, moments that drew attention and produced a change in the traditional atmosphere of political communication. Several of these occurred in the United States election campaign in 2004, where there was already a significant number of users connected to high-speed Internet with few applications for interaction with the public. Specifying the interaction between users via text message as part of the 2.0 policy [25].

The authors [2] refer to the concept of politics 2.0 as the effort that citizens make to seek participation in the formulation, development, and evaluation of collective intelligence policies that are embodied in social networks and digital media addressed for that purpose. Through politics 2.0, the electorate has the option of resorting to the Internet to participate and organize massively in the political campaigns of the candidates of their choice to win the elections as a result.

2.5 Political Communication Strategies

Political communication is defined as that discipline from which spaces for dialogue are generated in the different stages of a political process. It is essential that traditional strategies and those directed at digital media be combined to reach a larger audience [23].

For [3] communication and politics are two actions that must go hand in hand as a basis for the establishment of democracy, since the main objective is to convene and persuade a sufficient majority of voters, thus allowing assume the can. To achieve this, it is essential to know the main effective strategies and current trends to reach new audiences. It is essential that the 3 C's are taken into account: consistency, credibility, and coherence.

2.6 Political Marketing

Political marketing is executed through a partisan vehicle, therefore, for the concept of a political party, we will take the minimum definition, the one that is essential for the idea of a political party as an institution to subsist. We will say that it is "any political group identified by a social label that presents itself to the elections can elect candidates for public office" [2].

According to [25] Political marketing consists of the use of the tools of traditional marketing in the context of political activities in any field that allows identifying their needs and proposing a set of strategic offers to satisfy them. It is about the adaptation of administrative tools in the world of politics.

2.7 Apps in Politics

Technology and mobile applications are a great ally for different social and political actions. With the help of various platforms and apps, the public administration of each political party is effective, the most outstanding applications to date facilitate transparency, access to candidate information and public information and, most importantly, citizen interaction and participation [13]. Apps are tools that help us facilitate processes with audiences through mobile devices, unlike websites, apps serve and have very specific functions.

The apps in politics are a mobile tool that allow the geolocation of the activities in real time of the candidates, knowing which sectors it visits, which people it is aimed at and the acceptance of the public, within the app in politics you can create spaces For the dissemination of important data of each of the candidates, their biography and data that are generally unknown, it is considered that an app in politics is like carrying the campaign on cell phones or tablets for communication and interaction with users [3].

3 Methodology

For this research, it was decided to use qualitative and quantitative approaches. The qualitative approach studied reality in its natural context through the survey technique and as an instrument the structured questionnaire that allowed measuring the perceptions and preferences of the study population; The TAM model (Technological Acceptance Model) was also applied, which is made up of qualitative values that, when tabulated, presented the acceptance trend of the proposed technological tool through the analysis of tables and graphs. The quantitative approach of the research used the statistical data collection technique, the results obtained through the survey have been tabulated allowing to determine the acceptance trend that the application will have.

Experimental research: a mobile application was developed, which was implemented on a sample of the active voting population to identify the acceptance of technology during the management of electoral campaigns in politics 2.0 through the TAM model (Technological Acceptance Model). An experiment was proposed, and it was verified if it is accepted or not.

Exploratory research: the research was based on the development of a mobile app focused on politics 2.0 within the province of Tungurahua, it was a new study because the analysis environment did not have previous mobile applications focused on this field. In other words, both the independent and dependent variables had not been studied together, which is why the development of this research was proposed to determine the level of impact that Mobile marketing has as a communication strategy in politics 2.0.

In this project, a survey was applied consisting of a structured questionnaire with 15 questions, 1 dichotomous question, 9 closed questions, 4 multiple-choice questions, and 5 single-choice answers, finally, it has 5 questions on a Likert scale. Of which 6 correspond to the independent variable, 8 correspond to the dependent variable and the rest collect information from the study population. The instrument was validated by 3 experts on the subject of study and its reliability was verified by Cronbach's Alpha, where a result of 0.828 was obtained, that is, the questions that made up the survey were reliable within the investigative process.

After the descriptive analysis of the 6 most representative questions of the survey, it is concluded that the null hypothesis is rejected, as can be seen in Fig. 1. To verify it statistically, the Kolmogorov Smirnov test was applied, since it is the statistical analysis that best fits the investigation and the distribution of the data according to the calculation of the sample, allowing to identify how the data were distributed within the investigation and at the same time if the study variables according to the most representative questions influenced or not in the development of the proposal.

		What features do you look for when downloading a mobile app?	Which of the following mobile marketing benefits would you like to get from downloading an app?	How reliable do you think is the information that circulates on the internet to learn about political issues?	How often have you used your social networks or internet pages to obtain information about events, political issues, or politics?	How likely is it that you download a mobile application focused on the political management of the province?	From the following benefits that downloading a mobile application specialized in local politics will provide, select those that most attract your attention.
N		385	385	385	385	385	385
normal parameters	Media	2,19	4,16	3,01	3,24	3,10	2,87
	Deviation	,950	1,742	,888	1,078	1,310	2,013
More extreme differences	absolute	,433	,188	,246	,181	,215	,247
	Positive	,433	,188	,246	,181	,165	,247
	Negative	-,299	-,188	-,216	-,167	-,215	-,176
Z of Kolgomorov Smirnov		8,500	3,691	4,829	3,548	4,210	4,851
Sig. Asintot. (bilateral)		,000	,000	,000	,000	,000	,000

Fig. 1. Kolmogorov Smirnov test

According to the questions that have been selected as the most representative (6, 7, 9, 11, 14 and 15) of the data collection tool, by allowing to define the main aspects on which the App was developed, the reason why which were considered, the application of the selected statistical test and taking into account the statistical constant of P-value: 0.05, it is determined that the value obtained from the 6 questions previously detailed is less than 0.05, the rejection of H0 is established: Mobile marketing does not contribute as a communication strategy in politics 2.0 "accepting H1: Mobile marketing contributes as a communication strategy in politics" In other words, proposing and developing an informative mobile application for local politics will contribute to the transformation of a 2.0 political context in the province of Tungurahua.

4 Results

Based on the identified need for a digital tool focused on local politics in the province of Tungurahua, the development of an informative mobile application was proposed that offers two-way communication between users and the candidate or elected authority of their choice, allowing the report of political activities and provide access to information in real-time in an interactive way. The app comes from the influence of information and communication about the ideologies of the political parties of Ecuador. With its development, it is intended to reach both young people and adults, generating confidence in users by perceiving reliable information in real-time.

It is a tool that controls, reports, follows up, and organizes all the activities of the province's political campaigns, keeping citizens updated on issues such as the progress of the political projects proposed through the actions carried out, access to the profile of the candidate or authority of your choice through a structure that allows quick, easy, and simple interaction.

It can be used for the structured analysis of the political activity of the candidates or authorities of the different legislatures of the province of Tungurahua. The name "PoliticsEC" has been selected as the name of the mobile application, this is made up of the term politics in English <Politics> and the open extension EC that belongs to Ecuador and allows to identify of individuals and national companies. The combination of both terms refers to the fact that users who download the app will find information on Ecuadorian politics.

Fig. 2. App Logo

The App logo Fig. 2 was inserted on the first screen, the upper right part has additional elements that can be added to the screen to extend content if necessary.

Fig. 3. Registers

Next, the second screen Fig. 3 was generated, where the user registers and logs in; in the test version this data is stored in the Firebase console (Fig. 4).

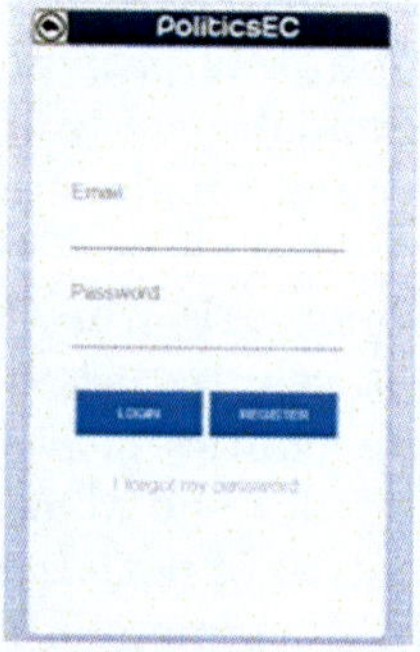

Fig. 4. Log in

Subsequently, the third menu screen was created, this redirects users to the destination screens that the application contains. On the right side, the action that each button will execute, and which screen the user will go to is displayed.

Finally, the information content screens were developed where users will find the description of the province of Tungurahua, and each of the authorities of the province and the cantons that comprise it, this will be expanded with information on the candidates during the election campaign season.

4.1 Application TAM Model

Finally, 60 students, belonging to the study sample, were asked to download the App from Google PlayStore, later the TAM Model was applied to measure their perceptions about PoliticsEC, and the Google Forms link to access the TAM model (Figs. 5 and 6).

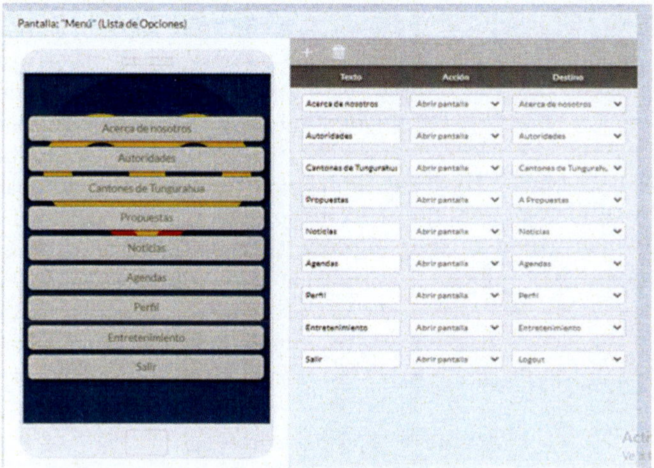

Fig. 5. Main Menu

Fig. 6. Information

Perceived ease of use Question 3. It was easy to get the App to show the information in a general way (Table 1).

Table 1. perceived ease of use

Option	Frequency	Accumulate Percentage
Totally agree	14	23,3%
Agree	17	28,3%
Indecisive	15	25%
Disagree	11	18,3%
Totally disagree	3	3%

According to the results of question 3 of the total of 60 respondents, it can be seen that 41 people have selected the first 3 options; In other words, there is an evident tendency for the sample to be between totally in agreement and in agreement that the App delivers the general information in a simple way to the user; this means that PoliticsEC meets the expectations of the sample regarding the information it offers. However, it is important to take into account that there is the possibility of going deeper and providing more detailed and specific information, which would further improve the user experience in terms of the quality and quantity of information that could be found when downloading the App.

Question 16. The App was recommended by a worker (teacher/ administrative) of the institution (Table 2).

Table 2. App recommended

Option	Frequency	Accumulate Percentage
Totally agree	20	33,3%
Agree	16	26,7%
Indecisive	7	11,7%
Disagree	12	20,0%
Totally disagree	5	8,3%

Regarding this section, it is evident that the majority of those surveyed agree that the mobile application was virilized by teachers and administrators of the Technical University of Ambato and that their classmates also downloaded it, which shows that the faculty recognizes the importance of having this tool, a result that was also obtained from the application of the survey; that is, a proposal was developed based on the needs and that meets the expectations as evidenced in the other sections.

Prior Knowledge

Question 18. I will understand better how to use the App if it has a help guide (Table 3).

Table 3. App recommended

Option	Frequency	Accumulate Percentage
Totally agree	10	16,7%
Agree	22	36,7%
Indecisive	9	15%
Disagree	9	15%
Totally disagree	10	16,7%

It can be seen that the surveyed sample divides their opinions on the importance of a user guide for the App; where an evident majority of people consider a help guide necessary since it is a tool with features that have not been presented before, this need must be taken into account, which should be implemented as an improvement in another version of the App; to provide an even more satisfying user experience.

Question 22 The visual components of the App are interactive (Table 4).

Table 4. App recommended

Option	Frequency	Accumulate Percentage
Totally agree	12	20,0%
Agree	19	31,7%
Indecisive	10	16,7%
Disagree	12	20,0%
Totally disagree	7	11,7%

We can show that in terms of the visual elements that make up the mobile application, the vast majority of the sample has selected the first 3 options, which means that they agree and agree that it shows organized, clear, and interactive elements; Therefore, the idea of providing an intuitive and attractive application that, at the same time as informing about local political management, is also interactive, is fulfilled; building user confidence. Thanks to this, active participation in local politics will be possible, generating the desired links between political actors and citizens.

5 Conclusion

It is concluded that the development of politics responds to the need to have a digital tool for local politics and once users have downloaded and tested the proposed mobile App, trends are evident that demonstrate satisfaction with the quality of the information. And the clear and understandable interface experience, yet the demand for even more detailed and in-depth information is recognized. In addition, the ease of use of the tool was perceived and its usefulness was recognized by members of the community.

However, it is important to take into account that the implementation of a help guide is considered necessary to improve the experience since it is a tool with innovative features that can be complex to apply, and it would be useful for users to have prior knowledge for an experience satisfactory.

On the other hand, it is stated that the visual elements are orderly, attractive, and interactive, so PoliticsEC fulfills its objective of being attractive and interactive; its interface was developed based on the results obtained from the survey and this is evidenced in the design and usability section, as these meet the expectations of users.

Finally, there is the statement of the majority of those surveyed that they intend to use the App frequently, which shows that, despite requiring certain minimal improvements, it generates confidence in the user, providing a positive user experience.

Acknowledgment. Thanks to the Technical University of Ambato, to the Directorate of Research and Development (DIDE acronym in Spanish) for supporting the research group: Marketing C.S. and the project Aplicación del marketing digital como herramienta de transformación en la Política 2.0 dentro de la provincia de Tungurahua: predicción y toma de decisiones mediante web semántica.

References

1. Alvarez, K., Reyes, J.: Reconfigurable manufacturing system based on the holonic paradigm for the die-cutting process in a sports shoes company. In: García, M.V., Fernández-Peña, F., Gordón-Gallegos, C. (eds.) Advances and Applications in Computer Science, Electronics and Industrial Engineering. AISC, vol. 1307, pp. 19–36. Springer, Singapore (2021). https://doi.org/10.1007/978-981-33-4565-2_2
2. Barandiarán Irastorza, X., Unceta SaBaranrústegui, A., Peña Fernández, S.: Comunicación política en tiempos de nueva cultura política (2020)
3. Brossi, L., Dodds, T., Passeron, E.: Inteligencia artificial y bienestar de las juventudes en América Latina. LOM Ediciones (2019)
4. Cabrera-Abad, K., Pinos-Úrgiles, P., Jara-Diaz, O., Duque-Córdova, L., Escobar-Segovia, K.: Ergonomic working conditions in workers under the modality of "homeoffice" due to a Covid-19 pandemic, in a bottling company in Ecuador. In: Garcia, M.V., Fernández-Peña, F., Gordón-Gallegos, C. (eds.) CSEI 2021. LNNS, vol. 433, pp. 41–56. Springer, Cham (2022). https://doi.org/10.1007/978-3-030-97719-1_2
5. Caiza, G., Salazar-Moya, A., Garcia, C.A., Garcia, M.V.: Lean manufacturing tools for industrial process: a literature review. In: Yang, X.-S., Sherratt, S., Dey, N., Joshi, A. (eds.) Proceedings of Sixth International Congress on Information and Communication Technology. LNNS, vol. 236, pp. 27–35. Springer, Singapore (2022). https://doi.org/10.1007/978-981-16-2380-6_3
6. Calvo, L.: ¿qué es una app, para qué se utiliza y qué tipos existen?, December 2022. https://es.godaddy.com/blog/que-es-una-app-y-para-que-se-utiliza/
7. Corral, R.: El espíritu de los tiempos: la web 2.0 en la construcción de la política. Polémika **3**(8) (2012)
8. Garces-Salazar, A., Manzano, S., Nuñez, C., Pallo, J.P., Jurado, M., Garcia, M.V.: Low-cost IoT platform for telemedicine applications. In: Zhang, Y.-D., Senjyu, T., So-In, C., Joshi, A. (eds.) Smart Trends in Computing and Communications. LNNS, vol. 286, pp. 269–277. Springer, Singapore (2022). https://doi.org/10.1007/978-981-16-4016-2_26
9. García, S.M., Fabregat, H.D., Ripollés, A.C.: La plataformización de la comunicación política institucional. el uso de whatsapp por parte de las administraciones locales. Revista Latina de Comunicación Soc. (79), 100–126 (2021)
10. García Peña, M.: Tendencias actuales en estrategia y acción de marketing: Marketing móvil, February 2018

11. Gutiérrez-Rubí, A.: La transformación digital y móvil de la comunicación política. Fundación Telefónica Madrid, España (2015)
12. Hauncher, Á.: Mobile Marketing. Editorial Elearning, S.L (2019). https://books.google.com.ec/books?id=F3flDwAAQBAJ
13. Herazo, E.: ¿qué es una aplicación móvil?: Anincubator - blog, October 2022. https://anincubator.com/que-es-una-aplicacion-movil/
14. Latorre, M.: Historia de la web, 1.0, 2.0, 3.0 y 4.0. https://marinolatorre.umch.edu.pe/historia-de-la-web-1-0-2-0-3-0-y-4-0/
15. Lpo: Lanzan una app para conectar a la gente con los políticos, September 2021. https://www.lapoliticaonline.com/nota/136269-lanzan-una-app-para-conectar-a-la-gente-con-los-politicos/
16. Merlo, C.: La biblia del marketing político. Anderson Publishing (2021)
17. Molina, G.: Tecnolandia. Wanceulen Editorial, March 2021
18. Muñoz, H., Ortiz, D., Naranjo, I., Pazmiño, A.: Optimization of routes for the collection of solid waste. In: Garcia, M.V., Fernández-Peña, F., Gordón-Gallegos, C. (eds.) CSEI 2021. LNNS, vol. 433, pp. 57–70. Springer, Cham (2022). https://doi.org/10.1007/978-3-030-97719-1_3
19. Nabor, O.A., Villegas, M.P.G., Covarrubias, A.C.R., Solis, A.I., Arciniega, L.A.L., Luna, A.L.A.: Uso de aplicaciones de la web 2.0 para la evaluación del aprendizaje significativo (use of web 2.0 applications for the evaluation of significant learning). Pistas Educativas **40**(130) (2018)
20. Ordoñez Cevallos, J.I., Zúñiga Rodríguez, J.D.: Enfoque del marketing político en las diferentes redes web de la Provincia de Tungurahua. B.S. thesis, Universidad Técnica de Ambato, Facultad de Ciencias Administrativas, Carrera (2019)
21. Piñeiro-Otero, T., Rolán, X.M.: Understanding digital politics-basics and actions. Vivat Academia **23**(152), 19–48 (2020)
22. Revelo Benalcázar, K.V., et al.: Uso político de las redes sociales: estrategias y contenidos utilizados por los candidatos César Montúfar y Jorge Yunda en las elecciones a la Alcaldía de Quito, período del 5 de febrero al 20 de marzo de 2019. Master's thesis, Universidad Andina Simón Bolívar, Sede Ecuador, Quito, EC (2021)
23. Texeira, R.G.: Política 2.0, las redes sociales (Facebook y Twitter) como instrumento de comunicación política: estudio: caso Uruguay. Ph.D. thesis, Universidad Complutense de Madrid (2017)
24. Urvina Alejandro, M.A., Lastra-Bravo, X.B., Jaramillo-Moreno, C., et al.: Turismo y aplicaciones móviles. preferencias de turistas y prestadores de servicios en el cantón tena, napo, ecuador (2022)
25. Vallvé, A.M.: El mobile marketing y las apps: Cómo crear apps e idear estrategias de mobile marketing (epub), vol. 493. Editorial UOC (2017)
26. Zamora-Medina, R., Losada-Díaz, J.C., Vázquez-Sande, P.: A taxonomy design for mobile applications in the spanish political communication context. El Profesional de la información **29**(3) (2020)

Spatial Concentration of Deaths from Chronic-Degenerative Diseases in the Province of Tungurahua (2016–2020), Ecuador

Kleber-H. Villa-Tello[1](✉) and Juan-F. Torres-Villa[2]

[1] Unidad de Registros Administrativos del Instituto Nacional de Estadística y Censos Centro, Pasaje Velastegui SN y Av Manuelita Saenz, Ambato, Ecuador
khvillat@gmail.com

[2] Ministerio de Salud Pública, Av José Peralta y Pompillo Llona, Ambato, Ecuador
jtorres0873@outlook.es

Abstract. In this study, the Spatial concentration of Chronic-Degenerative Diseases under study is quantified: Diabetes Mellitus, Ischemic Heart Diseases and Malignant Tumors, in the province of Tungurahua. For this, a methodology is proposed that can be applied in the rest of the provinces of Ecuador. For the development of the research, the databases of General Deaths from the year 2016 to 2020 of the National Institute of Statistics and Censuses (INEC) referring to Tungurahua are used, taking as reference the Administrative Political Division at the parish level as a geographic unit. As the cause of death is the object of study, the variable habitual residence of the deceased and the basic cause of death are used. Quantitative geographic procedures that analyze spatial concentration are used, such as the Global Spatial Concentration Index (ICEG) and the Areal Spatial Concentration Index (ICEA), for graphic representation the Lorenz Curve and Geographic Information Systems are used through thematic cartography. The study focuses on the province of Tungurahua, performing an analysis at both rural and urban parish levels. The results present spatial concentration of these diseases in certain parishes of Tungurahua, which would be a point of reflection and starting point for decision-making at the health level to face structural causes, in said jurisdictions, demonstrating in turn the validity of the analysis techniques spatial.

Keywords: Spatial Concentration · General Deaths · Geographic Analysis · Health Geography

1 Introduction

Health inequity leads to inequalities at both social and economic levels [30]. The countries generate efforts to extend health care, to curb inequity and its consequences, although this inequity does not affect a single health problem, this

M. V. Garcia and C. Gordón-Gallegos (Eds.): CSEI 2022, LNNS 678, pp. 70–85, 2023.
https://doi.org/10.1007/978-3-031-30592-4_6

study intends to pay attention to chronic degenerative diseases in the province of Tungurahua. With the objective that these are measured and monitored in places where they have a concentration, using quantitative analysis methods of the Geographic Information Systems included in the Geography of Health, through the study of patterns of spatial distribution of diseases [18].

In Ecuador, it is necessary to carry out a study of the main causes of death, especially those that have been established as the main ones at a general level over the years [13]. This study aims to collaborate by establishing through Quantitative Geography methods, the areas of concentration of chronic degenerative diseases in order to develop strategies to prevent both morbidity and mortality through public policy, understanding the inequalities that occur in the geographic space [32] and the spatial distributions between population and disease [7], in such a way as to support effective decision-making in the field of public health, ranging from disease prevention to access to health services in the health-disease-care process [28].

Within chronic-degenerative diseases, the study focuses on diabetes mellitus, ischemic heart disease and malignant tumors, since these are consolidated among the main causes of death worldwide, and similarly in the country [21]. Even in countries with high national income, high quality of governments, high care and health spending [17,27]. In addition to being chronic - degenerative they cause suffering, generating a great economic expense for the state and especially for those who suffer from them [16,32]. In Ecuador, the development of these diseases has a genetic and environmental basis, families with a history of chronic diseases, additionally, unhealthy lifestyles, diets with high concentrations of carbohydrates, sugars, salt and the chronic use of additives in food, excessive use of preservatives, condiments and harmful substances such as alcohol, tobacco, finally the poor attention that people pay to their health, the lack of health services, and the inequity to access health services generates an increase in the complications of these pathologies [26].

These diseases as a whole represent the main health problems, already established some years ago, which should make it necessary to study their geographical distribution in order to design and implement public and private policies to improve health systems in the areas in which they concentrate. Chronic diseases are largely preventable, either in the medium or long term, by combating their common determinants and risk factors, hence the importance of locating in which geographical areas they are concentrated [21,29].

To combat these risk factors, it is necessary to establish public policies that allow the in-depth analysis of medical history in the family, the study and analysis of the environmental factors present in the daily development of the population. Carry out a timely and close follow-up of compliance with treatments and home care that these patients must maintain [9]. Additionally, in those deceased patients, a pathological study of the organs affected by chronic-degenerative diseases is necessary, in order to identify common injuries, severity of injuries and their evolution, allowing a better application of treatments and drugs to prevent death in other patients [14].

For the study, Indexes belonging to Quantitative Geography are used to define areas of spatial concentration, such as the Global Spatial Concentration Index (ICEG) and the Areal Concentration Index (ICEA). To represent them graphically, the Lorenz Curve and thematic maps at the parish level are used, which is the unit of analysis. This combination of methods has been used in several investigations such as in the investigation of socio-spatial determinants of health [31], in the investigation on urban socio-spatial segregation [22], in the analysis of segregation and concentration of services [3], in the study of chronic degenerative diseases [32], in the spatial distribution and segregation of foreigners [4], in the study on the spatial concentration of social determinants of health [6], in the study of population concentrations [4]. In all these investigations, this combination has proven its validity and consistency, making it possible to conclude the existence of concentration when using it, demonstrating an aptitude for urban-regional analysis at different scales.

2 Study Area

Tungurahua is the smallest and most densely populated province in Ecuador [8,23], it was created on May 21, 1861 and became independent on November 12, 1820. Its population is 504,583 inhabitants, of whom 48.51% are men (244,783) and 51.49% are women (259,800). It is distributed over an area of 3369.4 km^2. It limits to the north with the province of Cotopaxi, to the south with the province of Chimborazo, to the east with the provinces of Pastaza and Napo, and to the west with the provinces of Cotopaxi and Bolívar. It is divided into nine cantons that are Ambato (provincial capital), Baños, Cevallos, Mocha, Patate, Pelileo, Píllaro, Quero and Tisaleo, with 53 parishes, 44 rural parishes and 9 urban ones [10].

3 Methodology

In health systems there are persistent inequalities, so it is necessary to measure and monitor them, a person must have their full health potential within their reach and no one should be excluded or disadvantaged from reaching this full potential, either by social position or another determiner [21]. These inequalities are determined by different natural, demographic, cultural, economic and social factors [11,32].

The concentration or segregation that produces inequality can be measured based on five dimensions: Equality (differences in distribution), Exposure (potential contact), Concentration (relative sum of physical space occupied), Centralization (degree of location of a group in the center of an area) and Clustering (groups that are disproportionately in contiguous areas) [12]. Spatial distributions do not occur randomly, but can be linked to forms of behavior in general that lead to the formulation of scientific laws that can explain the organization of the territory [15], in the present study we will base ourselves on the concentration to analyze the mode of spatial distribution of a vital event (death) in a territory (the province of Tungurahua) [32] (Fig. 1).

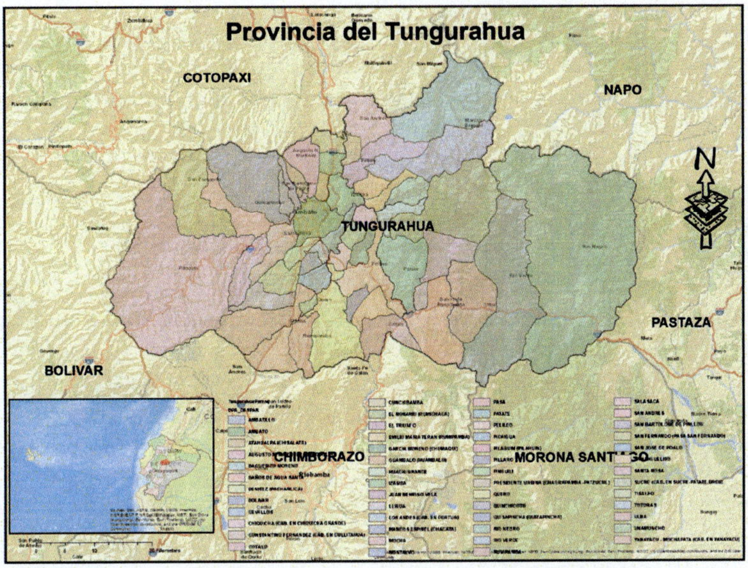

Fig. 1. Location map of the study area with its parishes.

The Geography of Health has been divided into medical geography and geography of health services, and based on this division its different methodologies have been classified, in short, it is applied geography that encompasses the application of geographical knowledge and skills for problem solving social, economic and environmental [15]. In the present study for the spatial analysis of health, it is carried out using Medical Geography methods, which addresses the study of the spatial distribution of diseases through the analysis of spatial distribution by thematic cartography, exploratory analysis of spatial data and autocorrelation analysis [7], within this field the use of the Global Spatial Concentration Index (ICEG) and the Areal Concentration Index (ICEA), have been used.

In the present study, three processes were carried out:

The first is the compilation and classification of the General Deaths databases from 2016 to 2020, which are freely downloadable from the website of the National Institute of Statistics and Censuses (INEC) (https://www.ecuadorencifras.gob.ec/defunciones-generales/), filtering the deaths whose province of habitual residence of the deceased was the province of Tungurahua. The variable for analysis of the cause of death is the Basic Cause of Death, grouped under the terms of the Short List of 103 diseases adopted by the World Health Assembly in 1990 for data tabulation [20], grouping on this list the causes referring to malignant tumors, with this a new variable was generated with the categories under study and all the others as the rest. The geographical unit used is the parish, which appears as a field within this database as the parish of habitual residence.

The second process was the generation of summaries at the level of the parish of habitual residence for each chronic-degenerative disease under study, with which the calculation of the Indicators is made, both of the Global Spatial Concentration Index (ICEG) and the Concentration Index Area (ICEA). These indices are used to calculate the spatial concentration of all population characteristics, combining the use of calculation templates with their graphic representation, and Geographic Information Systems for carrying out thematic cartography [5].

The third process was carried out with the alphanumeric base, which is integrated with the INEC 2012 cartography, at the parish level. The INEC 2012 cartography is in the coordinate system WGS 1984 UTM Zone 17 South at a scale of 1:50,000 and can be downloaded publicly on the INEC website (https://www.ecuadorencifras.gob.ec/clasificador-geografico-estadistico-dpa/). The databases were processed and integrated using the Statistical Package for the Social Sciences (SPSS) version 23, with the publication format of the databases being SPSS. Once the information was extracted at the geographic level of the parish of habitual residence, it was joined to the cartography at the same level using the ArcMap 10.5 tool.

The classification methods used in thematic cartography group the spatial units by natural cuts, equal intervals or quantiles, in these cases there will always be a class that will have greater data [4]. With the incorporation of quantitative methods of geography, in this study with the indices measures are found on the behavior of a variable within a geographic area of study [5], and it is extremely useful to start an investigation from having a first approach through the study of spatial differentiations in a given area [2].

Thus, the ICEG is calculated with the following formula:

$$ICEG_{sup,b} = 0.50 \sum_{i=1}^{n} |sup_i - b_i| \tag{1}$$

where:
$ICEG_{sup,b}$ = Global Spatial Concentration Index for population category b with respect to the total area.
sup_i = Percentage of surface that contains each spatial unit
b_i = Percentage of population group under study
0.50 = Constant that allows only positive or negative values to be used.

When the surface value is not used and instead the total population is considered, the C for concentration is modified by S for segregation, obtaining the Global Spatial Segregation Index [4]. When there is no spatial concentration, the ICEG would be zero, and the maximum 100% will denote a maximum concentration [32].

To calculate the ICEA, the following formula is used:

$$ICEA_i = \frac{b_i(\%)}{sup_i(\%)} \tag{2}$$

where:
$ICEA_i$ = Areal Spatial Concentration Index of the spatial unit under study.
$b_i(\%)$ = Percentage of population group under study
$sup_i(\%)$ = Percentage of surface that contains each spatial unit.

The ICEA will be less than 1 when the proportion of surface is greater than that of the population group under study, if the ICEA is equal to 1 it means that the proportions are distributed in a similar way and ICEA will be greater than 1 where there is population concentration in the object variable study.

For the presentation of the information and the analysis process, the Lorenz Curve was used for each disease under study, since it is the most common way of representing inequality [18]. When using the Lorenz Curve, it should be analyzed that the closer the curve is to the diagonal of the square, the more evenly distributed the population under study will be, and the opposite will indicate a higher level of concentration [1] being 100% the maximum concentration possible [25].

4 Results

The study of spatial distribution allows us to observe how the facts are distributed on the earth's surface and the relationships they maintain with neighboring facts [32], and the use of Geographic Information Systems incorporated the spatial dimension into their studies in order to study the relationship between a human component and a physical-natural component [2].

Based on the study carried out, it is concluded that, for the ICEG, high values of concentration of these diseases are presented (Table 1.). Diabetes Mellitus being the highest, however, the high rate found for both ischemic heart disease and malignant tumors is mentioned. In other words, at least a population redistribution of 64.78% must be carried out in the case of ischemic heart disease.

Table 1. Global Spatial Concentration Index (ICEG).

	Mellitus diabetes	Ischemic heart	Malignant tumors
ICEG	75,20	64,78	68,19

With the results obtained, the graphs of the Lorenz Curves were generated, which represent the concentration of each group of diseases under study.

The Lorenz curves confirm the high level of concentration, in Fig. 2 it can be seen that in Diabetes Mellitus 20% of the surface concentrates more than 90% of the population, in Fig. 3 in Ischemic heart disease 30% of the surface concentrates more than 90% of the population, while Fig. 4 referring to Tumors shows that more than 90% of the population is concentrated in 25% of the surface.

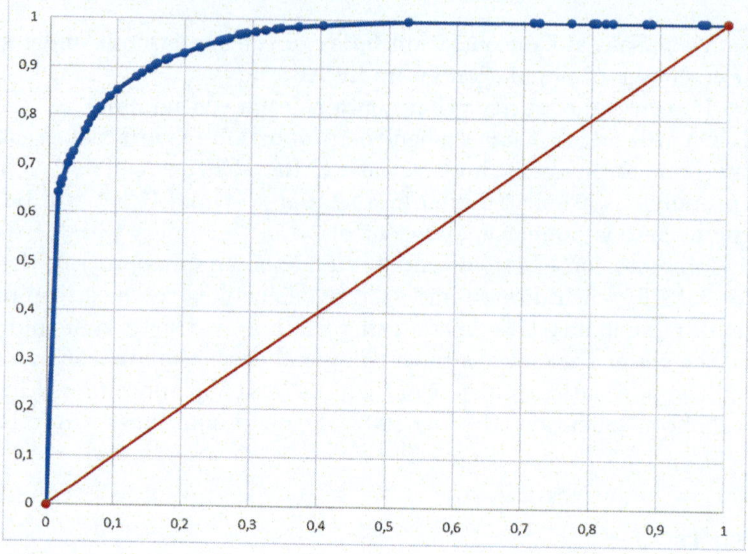

Fig. 2. Lorenz curve: Diabetes mellitus. Tungurahua 2016–2020.

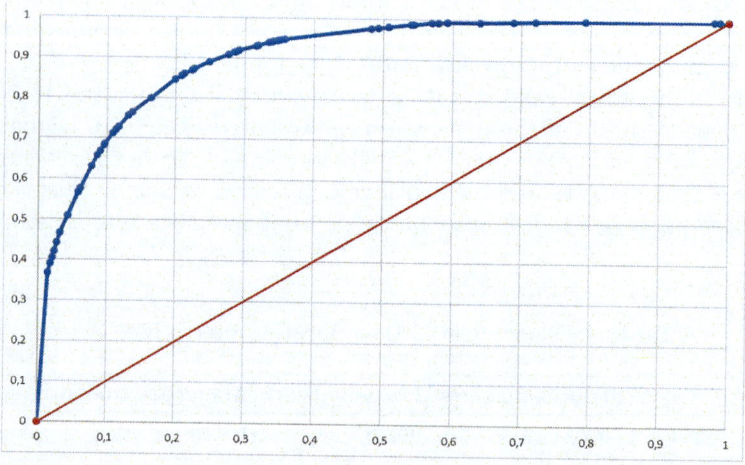

Fig. 3. Lorenz curve: Ischemic heart disease. Tungurahua 2016–2020.

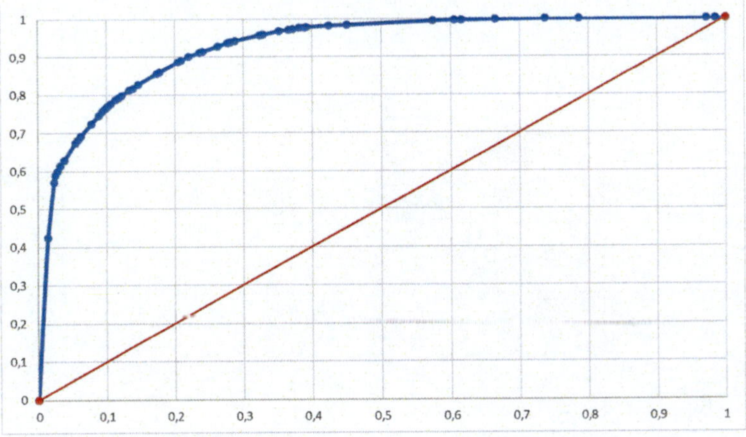

Fig. 4. Lorenz curve: Malignant tumors. Tungurahua 2016–2020.

As for the ICEA, it allows us to identify those urban or rural parishes that have higher concentrations of these chronic degenerative diseases. We present Fig. 5 thematic cartography, with the concentration of the population general in the province of Tungurahua, for analysis based on the concentration of the population in general.

In the thematic cartography developed with the ICEA, ranges from 0 to 0.99 are used, which indicates the lowest population relative to the proportion of surface, from 1 to 1.99 when the proportion of population is equal to or slightly higher than the range and 2 to 2.99 when the proportion is double and almost triple the surface, and the upper range greater than or equal to 3 when the concentration is triple or more.

The results referring to the ICEA are presented in summary form in Table 2 and in detail in Annex 1 (Figs. 6, 7 and 8).

Table 2. Parishes of Tungurahua with the highest concentration of ICEA

Chronic - Degenerative Disease	Parishes of Tungurahua with the highest incidence
Mellitus diabetes	Ambato, San Bartolomé de Pinllog, Atahualpa (Chisalata), Izamba, Totoras
Ischemic of the heart	Ambato, Santa Rosa, San Bartolomé de Pinllog, Atahualpa (Chisalata), Izamba, Píllaro, Picaigua, Totoras, Montalvo, Salasaca, Pelileo
Malignant tumors	Ambato, Atahualpa (Chisalata), Izamba, Picaigua, Totoras

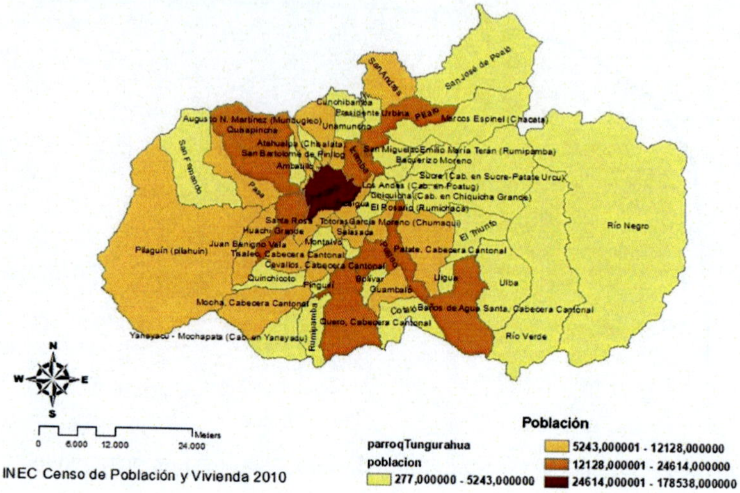

Fig. 5. Population concentration in the Province of Tungurahua

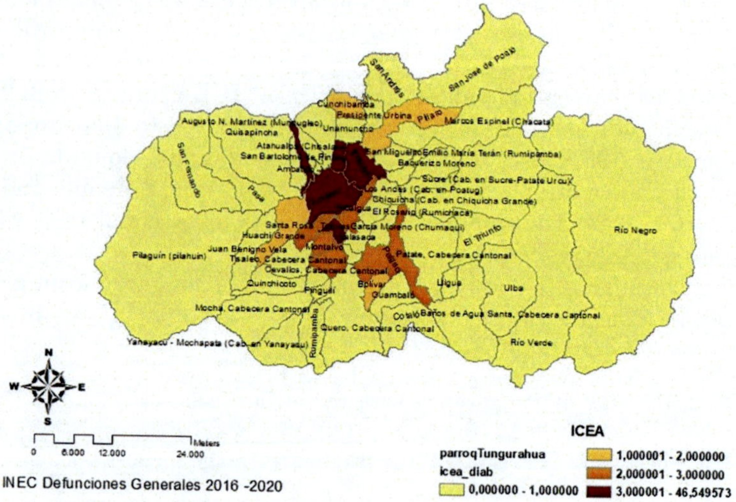

Fig. 6. Areal Spatial Concentration Index: Diabetes Mellitus

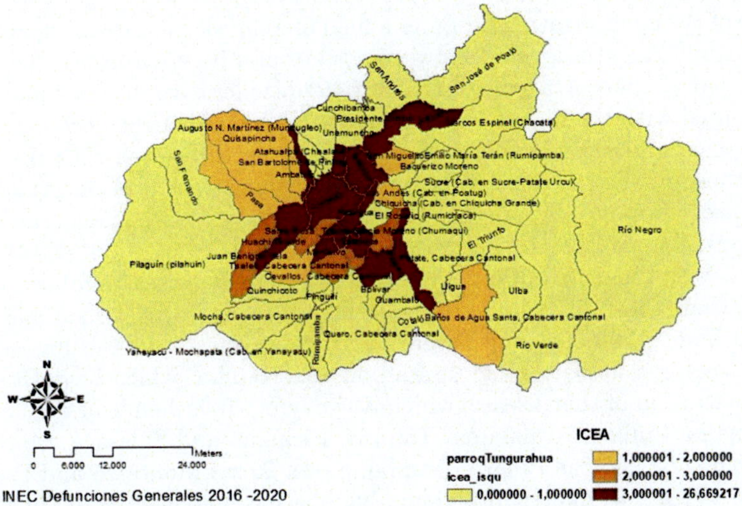

Fig. 7. Areal Spatial Concentration Index: Ischemic heart disease

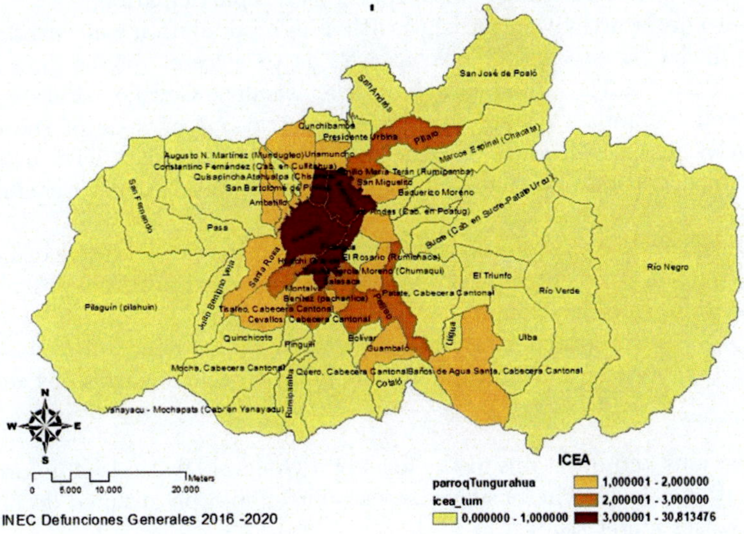

Fig. 8. Spatial Areal Concentration Index: Malignant Tumors

5 Discussion

Generally, the distributions that occur in the geographical space are not random, but are linked to the general behavior that explains the organization of the territory [15], as well as its social differences and its temporal evolution, so that the measure of the concentration becomes a field of analysis [3], providing alternative ways to study the relationship between society and its environment [2].

The spatial concentrations of the chronic degenerative diseases under study are presented with the highest concentration in the parish with the highest population concentration, which is Ambato, this being the parish with the smallest area of those in the province. Observing the concentration of diseases in other more densely populated parishes, they do not have this behavior in terms of concentration.

This is why the distribution of chronic-degenerative diseases follows different patterns than those of population concentration (Fig. 5), for example in Diabetes we find Atahualpa and Totoras as the parishes with the highest ICEA without leaving San Bartolomé de Pinllog and Izamba, which have three times the concentration of this disease, which is worrying. In ischemic of the heart, San Bartolomé de Pinllog, Atahualpa, Totoras, Picaigua and Salasaca are the highest, and the rest such as Pillaro, Izamba, Santa Rosa, Montalvo and Pelileo are equally worrying. In malignant tumors Izamba has a concentration of 17 times, that is, the percentage participation of the population with malignant tumors is 17 times greater than the percentage participation of the surface, Atahualpa, Picaigua and Totoras with six times the expected participation.

It is also noteworthy that parishes with a high population concentration such as Quisapincha, Quero and Baños, as well as to a lesser degree, San Andrés, Pilaguin, Pasa, Mocha, Juan Benigno Vela, Tisaleo, Cevallos, Patate, Garcia Moreno and Guambalo do not present concentrations of none of the chronic-degenerative diseases studied. On the other hand, Montalvo, with a low population concentration, has a high concentration of deaths from ischemic heart disease.

It is also observed that the concentration of chronic-degenerative diseases has a general tendency to centralize around Ambato, in a general way different from the population concentration.

To apply the proposed methodology to another province of Ecuador, and even a certain jurisdiction in a different country, the following steps should be followed:

- The sections and questions in the form of a General Death, is standard at the international level. The basic cause variable should be grouped according to the diseases under study.
- Public cartography at the administrative political division (DPA) level will always be available, as well as the death database must contain the DPA variables.
- The tools used such as ArcMap, SPSS and Excel, used for standard processes. Others could be used without difficulty such as QGis and Python for example.

The proposed methodology allows to easily identify the spatial concentrations of chronic-degenerative diseases, as well as their spatial distribution.

The indices used allow us to first observe at a general level the existence of concentration at a general level (ICEG) and how they are concentrated in certain parishes (ICEA). It allows graphically visualizing through the use of Lorenz Curves as well as thematic maps, the spatial distribution of the concentrations of chronic degenerative diseases.

There are other indices to be applied in the study of Spatial Concentration, such as that of Cumulative Participations, but this does not take into account the entire distribution and considers all zones equally (although these are very different), The Herfindahl Index, which similarly considers all regions as homogeneous, The Hirschmann-Herfindahl Index that was born as an alternative to the Gini Index [24] although it has more use in economics for market concentration, the Morán I statistic that measures the degree of association between an attribute and the weighted average considering nearby locations, but it is used more to measure autocorrelation rather than concentration [19], so the Gini index and the Lorenz Curve are the most appropriate and used techniques to graph spatial concentration.

The methodology used has potential to find and highlight the spatial concentrations found, it can be increased by looking for auto-correlation relationships between contiguous surface units in search of conglomerates, and it could be increased with the inclusion of Moran's Index I in it.

6 Conclusions

The Indices, both the Global Spatial Concentration Index and the Areal Spatial Concentration Index, show based on the surface, the concentration of chronic-degenerative diseases at the level of parishes in Tungurahua. The graphic representation using Lorenz curves and thematic cartography are the correct ones to visualize and highlight the concentration of diseases, presenting several elements that allow a correct diagnosis of the spatial concentration of deaths due to said diseases.

The methodology visualizes the parishes in which deaths due to chronic-degenerative diseases occur more intensely, and allows presenting the magnitude of the intensity of this concentration.

The cartography used provides reliability for representation as well as for integration with the General Death database, since geographic fields such as the administrative political division (DPA) are used in the database.

The proposed methodology demonstrates its applicability at the level of larger areas, since the geographical units where concentration occurs depend on the study area.

Spatial concentration is a dynamic process, it generally has a cause and effect related to the socio-spatial structure of the province of Tungurahua. The concentrations of diseases found could be due to socioeconomic aspects of the parishes or to the life-style of the people in them.

The proposed methodology is useful in the search for spatial solutions to social problems.

Visualizing spatial concentrations will help decision-making on public health and social prevention policies, which would aim to reduce this type of disease in those parishes with the highest incidence.

This paper is the starting point to highlight areas where a certain type of disease is concentrated, intending that these areas be the subject of health policies by the institutions in charge of them. The next step is to find an entity that finances research to continuously monitor the spatial concentration of chronic-degenerative diseases and extend to the search for causes. As well as evaluating the result of the establishment of health policies in the geographical areas of concentration.

Appendix A Annexes

(See Table 3).

Table 3. Annex 1

Name	Diabetes	Ischemic	Tumors
Quinchicoto	0,00	0,09	0,24
Tisaleo, Cabecera Cantonal	0,94	1,08	1,17
San Miguelito	0,62	1,52	1,94
San Jose de Poalo	0,00	0,03	0,03
San Andres	0,42	0,47	0,71
Presidente Urbina	0,42	0,56	0,95
Marcos Espinel (Chacata)	0,00	0,02	0,05
Emilio María Terán (Rumipamba)	0,33	0,44	0,41
Baquerizo Moreno	0,22	0,00	0,21
Pillaro	1,42	3,81	2,14
Salasaca	0,44	5,59	2,27
Guambalo	0,87	0,68	1,35
Garcia Moreno (Chumaqui)	0,36	2,07	1,34
El Rosario (Rumichaca)	0,00	0,37	0,62
Chiquicha	0,79	1,05	1,17
Cotalo	0,24	0,05	0,33
Bolivar	1,42	0,63	0,59
Benítez (pachanlica)	0,00	2,44	1,64
Pelileo	2,68	3,20	2,61
Yanayacu - Mochapata	0,00	0,00	0,00
Rumipamba	0,00	0,21	0,04
Quero, Cabecera Cantonal	0,48	0,69	0,95
Sucre	0,00	0,03	0,02
Los Andes	0,00	0,23	0,32

(*continued*)

Table 3. (*continued*)

Name	Diabetes	Ischemic	Tumors
El Triunfo	0,22	0,19	0,00
Patate Cabecera Cantonal	0,52	0,81	0,62
Pinguili	0,00	0,41	0,46
Mocha, Cabecera Cantonal	0,49	0,50	0,37
Cevallos, Cabecera Cantonal	2,41	2,53	2,69
Ulba	0,13	0,14	0,09
Río Verde	0,00	0,02	0,02
Río Negro	0,01	0,01	0,00
Lligua	0,00	0,49	0,14
Baños de Agua Santa, Cabecera Cantonal	0,98	1,33	1,08
Unamuncho	0,00	0,16	1,28
Cunchibamba	1,77	0,39	0,44
Totoras	4,83	5,80	4,62
Santa Rosa	1,35	3,87	1,98
San Fernando	0,00	0,18	0,10
San Bartolome de Pinllog	3,22	6,31	1,83
Quisapincha	0,09	1,29	0,41
Pilaguín (pilahuin)	0,07	0,22	0,09
Picaigua	2,47	5,30	3,06
Pasa	0,23	1,90	0,60
Montalvo	2,28	3,29	1,84
Juan Benigno Vela	0,71	2,19	0,88
Izamba	3,85	3,15	16,91
Huachi Grande	2,70	2,56	2,39
Constantino Fernandez	0,47	0,84	0,94
Augusto N. Martínez (Mundugleo)	0,88	0,84	1,20
Atahualpa (Chisalata)	5,32	5,76	6,46
Ambatillo	0,45	1,98	1,44
Ambato	46,55	26,67	30,81

References

1. Alcañiz, M., Pérez, A., Marín, J.: Concentración: Curva de Lorenz e indice de Gini. In: Estadística Descriptiva (2018)
2. Buzai, G.: Geografía global: la dimensión espacial en la ciencia y la sociedad. Anales de la Sociedad Científica Argentina (2018)
3. Gustavo, B.: Mapas Sociales Urbanos. Lugar Editorial, Argentina (2014)

4. Buzai, G., Baxendale, C.: Distribución y segregación espacial de extranjeros en la ciudad de Luján. Un análisis desde la Geografía Cuantitativa. In: Signos Universitarios (2003)
5. Buzai, G., Santana, M.: Métodos Cuantitativos en Geografía Humana. Instituto De Investigaciones Geográficas (INIGEO) (2019)
6. Buzai, G., Villerías, I.: Concentración espacial de los determinantes sociales en la cuenca del río Luján, Provincia de Buenos Aires, Argentina. In: Huellas (2018)
7. Fuenzalida, M., Buzai, G., Jiménez, A.M.: Geografía, Geotecnología y Análisis Espacial: Tendencias, Métodos y Aplicaciones. Editorial Triángulo, (2015)
8. Galleguillos-Pozo, R., Jordan, E.P., Tigre-Ortega, F., Garcia, M.V.: Integration and application of balanced scorecard with diffuse ANP in an SME; [integración y aplicación de balanced scorecard con ANP difuso en un PyMe]. RISTI - Revista Iberica de Sistemas e Tecnologias de Informacao **2021**(E42), 500–509 (2021)
9. Goldman, L., Schafer, A: Tratado de Medicina Interna. Elsevier Health Sciences, Spain (2017)
10. Honorable Gobierno Provincial de Tungurahua HGPT. Agenda Tungurahua desde la Visión Territorial. Gobierno Provincial de Tungurahua (2017)
11. Hojas-Mazo, W., Simón-Cuevas, A., de la Iglesia Campos, M., Ruíz-Carrera, J.C.: Semantic processing method to improve a query-based approach for mining concept maps. In: Nummenmaa, J., Pérez-González, F., Domenech-Lega, B., Vaunat, J., Oscar Fernández-Peña, F. (eds.) CSEI 2019. AISC, vol. 1078, pp. 22–35. Springer, Cham (2020). https://doi.org/10.1007/978-3-030-33614-1_2
12. Iceland, J., Weinberg, D.H., Steinmetz, E.: Racial and Ethnic Residential Segregation in the United States: 1980–2000. U. S, Goverment Printing Office (2002)
13. Instituto Nacional de Estadística y Censos INEC. Boletín técnico. registro estadístico de defunciones generales. Boletín Técnico. Registro Estadístico de Defunciones Generales. (2021). https://www.ecuadorencifras.gob.ec/documentos/web-inec/Poblacion_y_Demografia/Defunciones_Generales_2020/boletin_tecnico_edg_2020_v1.pdf
14. Kumar, V., Abbas, A.K., Aster, J.C.: Robbins Patología Humana. Elsevier, España (2018)
15. Linares, S.: Soluciones espaciales a problemas sociales urbanos: Aplicaciones de Tecnologías de la Información Geográfica a la planificación y gestión municipal. Universidad Nacional del Centro de la Provincia de Buenos Aires, Facultad de Ciencias Humanas (2016)
16. López, S., Lema, F., Rosero, C., Sánchez, C., López, J., Tigre, F.: Management by integrated processes with biosafety parameters. Case study SMEs manufacturing rest footwear in the province of tungurahua. In: Garcia, M.V., Fernández-Peña, F., Gordón-Gallegos, C. (eds.) Advances and Applications in Computer Science, Electronics, and Industrial Engineering. CSEI 2021. Lecture Notes in Networks and Systems, vol. 433, pp. 107–123. Springer, Cham (2022). https://doi.org/10.1007/978-3-030-97719-1_6
17. Mackenbach, J., et al.: Determinants of the magnitude of socioeconomic inequalities in mortality: a study of 17 European countries. Health Place **47**, 44–53 (2017)
18. Medina, F.: Consideraciones sobre el Índice de Gini para medir la concentración del ingreso. CEPAL (2001)
19. Moreno, M.: La distribución espacial de las comunidades peruanas en los estados unidos. Debates en Sociología (2011)
20. Organización Panamericana de la Salud OPS. Clasificacion Estadística Internacional de Enfermedades y Problemas Relacionados con la Salud. Organización Panamericana de la Salud (2015)

21. Organización Panamericana de la Salud OPS. Salud en las Américas. Organización Panamericana de la Salud (2017)
22. Orellana, D., Osorio, P.: Segregación socio-espacial urbana en cuenca, ecuador. In: Analítika (2014)
23. Ospina, P., et al.: Tungurahua rural: el territorio de senderos que se bifurcan. Documento de Trabajo N 70. RIMISP (2011)
24. Paez, P., Sanchez, G., Saenz, J.: Concentración de la industria manufacturera en colombia, 2001–2010: una aproximación a partir del índice de herfindahl-hirschman. In: Diálogos de Saberes (2014)
25. Ramirez, L., Falcon, V.: Analisis de segregación y concentración espacial de los servicios sanitarios en el area metropolitana del gran resistencia. ResearchGate (2014)
26. Rozman, C., Cardellach, F.: Farreras Rozman Medicina Interna. GEA Consultoria Editorial S. I. (2020)
27. Saltos, L.F., Zavala-Calahorrano, A., Ortiz-Villalba, P., Mayorga-Valle, F., García, M.V.: Comparative study of the level of depression in older adults in urban and rural areas; [estudio comparativo del nivel de depresión de adultos mayores en zonas urbanas y rurales]. RISTI - Revista Iberica de Sistemas e Tecnologias de Informacao **2021**(E42), 522–533 (2021)
28. Santana, G.: Variabilidad espacial de la mortalidad general y características económicas en el estado de méxico. Semestre Económico (2020)
29. Villa Tello, K.: Selection of optimal lodging site in the city of Baños, Ecuador. In: García, M.V., Fernández-Peña, F., Gordón-Gallegos, C. (eds.) Advances and Applications in Computer Science, Electronics and Industrial Engineering. AISC, vol. 1307, pp. 53–65. Springer, Singapore (2021). https://doi.org/10.1007/978-981-33-4565-2_4
30. Fondo de las Naciones Unidas para la Infancia UNICEF. Informe sobre Equidad en Salud 2016: Análisis de las inequidades en salud reproductiva, materna, neonatal, de la niñez y de la adolescencia en América Latina y el Caribe para guiar la formulación de políticas. Resumen, Fondo de las Naciones Unidas para la Infancia UNICEF (2016)
31. Villerías, I.: Asociación espacial de la mortalidad por enfermedades isquémicas del corazón en Guerrero, México. Rev. Geogr. Venezolana (2020)
32. Villerías, I., Buzai, G.: Concentración espacial de defunciones por enfermedades crónico-degenerativas (diabetes mellitus, isquémicas del corazón y tumores malignos) en el Estado de Guerrero, México. In: Geografía y Sistemas de Información Geográfica (GEOSIG) (2018)

Methodology to Improve the Quality of Cyber Threat Intelligence Production Through Open Source Platforms

Rogerio Machado da Silva(✉), João José Costa Gondim, and Robson de Oliveira Albuquerque

Professional Post-Graduate Program in Eletrical Engineering - PPEE, Department of Eletrical Engineering, University of Brasília, Distrito Federal, Brasília 70910-900, Brazil
rogerio.machado@aluno.unb.br, gondim@unb.br, robson@redes.unb.br

Abstract. In cyberspace, boundaries are constantly being crossed in the name of progress and convenience, and invariably result in new vulnerabilities and potential attacks. Traditional security approaches are not able to contain the dynamic nature of new techniques and threats, which are increasingly resilient and complex. In this scenario, the sharing of threat intelligence is growing. However, the vast majority of data is shared in the form of unstructured textual reports, or extracted from blogs and social media. These data sources have been imposing great limitation on security analysts due to the high volume and low quality of Cyber Threat Intelligence (CTI). Among the various aspects that impose limitations on the use of CTI, we focus on data quality. Inaccurate, incomplete or outdated information makes actions reactive, in no way different from traditional approaches. However, quality threat intelligence has a positive impact on incident response time. In this work we propose an Indicator of Compromise enrichment process to improve the quality of CTI, based on the intelligence production cycle, we conduct research to define metrics capable of evaluating the CTI produced through open source licensed threat intelligence platforms.

Keywords: Quality of Cyber Threat Intelligence · Intelligence production cycle · Open Source

1 Introduction

The advancement of connectivity introduces significant advantages to society [32]. As new technologies are created, a wide range of new vulnerabilities are also incorporated [32]. As a consequence cyber threats grow in volume and sophistication [1,11,16,18,21,25,39], and constitute complicating factors for cyber defense professionals [17,21,28,38]. In addition, attackers have collaborated with each other, sharing tools and services to increase the effectiveness of their attacks [30].

Supported by ABIN TED 08/2019.

M. V. Garcia and C. Gordón-Gallegos (Eds.): CSEI 2022, LNNS 678, pp. 86–98, 2023.
https://doi.org/10.1007/978-3-031-30592-4_7

As a mitigation measure institutions have adopted proactive defense mechanisms against cyber attacks [2,27]. In this context Cyber Threat Intelligence (CTI) is one of the measurements used, it refers to the set of information collected and organized about cyber threats that can be used to predict, prevent or defend cyber attacks [2,26].

We identified, through literature review, the factors that influence the quality of CTI, as well as the selection of the Threat Intelligence Platform (TIP) to be used. We defined the hypothesis that this problem can be addressed by applying the intelligence cycle to improve quality using the proposed methodology. First we apply procedures to improve the planning and data collection phase by determining the gaps that need to be addressed, and then we improve the processing and analysis of the data.

The main contribution of this research is to propose an Indicator of Compromise (IoC) enrichment process to improve the quality of CTI, based on the intelligence production cycle. In this way the security analyst will be able to be more assertive and proactive as the CTI is based on an intelligence production process, which throughout its flow, describes the event, identifies the author, positions the event in a timeline and geographically, and describes the mechanisms employed.

The rest of this paper is structured as follows: Sect. 2, related work; Sect. 3, definitions; Sect. 4, methodology; Sect. 5, procedures; Sect. 6, discussions; and Sect. 7, conclusion.

2 Related Works

There are many works around the CTI theme that approach various aspects, of this universe, few approach the production and sharing platforms or the quality of the sources, data and intelligence produced.

Despite the increase in technology and the growing usage of threat intelligence platforms, there are still limitations. There is a large volume of threat information produced and shared in many different formats [2,42]. This contributes to TIPs providing information with little or no processing. Thus, security analysts have difficulty finding relevant and quality intelligence [2]. In this context, existing approaches remain mostly reactive [4,27,39,41].

Research points to the lack of a defined process, besides not considering the intelligence cycle [28,39], however, there are studies that point to the need for the implementation of the intelligence cycle in the process of obtaining CTI. [34] Nonetheless, no studies have been observed that contemplate the intelligence cycle in its scope [28].

Another challenge related to CTI is the assessment of the quality of shared information [1,2,12,20,22,30,33,39,43,44]. There is no consensus on the characteristics that indicate the quality of the information. However, four characteristics are among the most cited [2,3,43]:

- Opportunity: it is related to the origin of an event and the reaction time or use of certain information;

- Relevance: indicates the relationship of the information with the organization's service and network assets;
- Accuracy: measures how much the information allows to improve the response to an incident and;
- Completeness: indicates the information's ability to describe an incident.

There is conclusive evidence that inaccurate, incomplete, or outdated threat information is a important challenge [2,6,39]. To ensure the quality of CTI throughout the collaboration process is crucial to its success. The exchange and use of meaningful threat information depends on measuring and ensuring its quality. This need is reinforced when it is stated that the quality of shared information has an impact on the time required to respond to an incident [31].

3 Definitions

3.1 Intelligence Cycle

A generalized definition of intelligence is described as the conversion of a subject from a completely unknown stage to a state of complete understanding, that is, based on a defined framework, random and general data is filtered to achieve a more relevant data set, when it is processed and converted into information [23].

Most TIPs focus on the data collection phase, thus, the other phases of the intelligence cycle take a back seat [1,29], generating little or no CTI, as a consequence many TIPs are just replicators and repositories of IoC.

Although, the efforts to employ the intelligence production cycle, few phases are supported by TIP [23], especially the planning phase [28]. The few studies that address the intelligence cycle in CTI, indicate the planning phase as the stage of data source selection [3,9]. However, the contributions of the planning phase go much further. In this phase the scope, objectives and deadlines are established, as well as the parameters and techniques that will be used [7], and the resources that may be used. Based on the known aspects, the questions that need answers are verified [7]. This phase is crucial to the quality of the CTI [15].

3.2 Threat Intelligence Platform

The collect, treatment and processing of data can be very time consuming [41], in face of the large volume of data [37], especially when performed by humans, besides being financially burdensome. To overcome this challenge, organizations have adopted tools that manage the flow of information, convert it into knowledge [39], and facilitate sharing [36].

The selection of the platform used in this research was based on recent studies that point to Malware Information Sharing Platform (MISP) as the most complete and flexible Threat Intelligence Platform available in open source [9,12,20,23,34,39]. These studies took into account aspects such as integration capacity, support for consolidated standards, availability of documentation and community responsiveness. Although the MISP is the most complete TIP, it does not cover all phases of the intelligence cycle.

3.3 Enrichment

The use of appropriate processes, combined with the automation power of TIP, increases the capacity in CTI production, and also contributes to unburdening security analysts [5]. The challenge is to handle the large daily volume of new Indicator of Compromise, which require evaluation to verify possible relationships [22].

In this context, data enrichment is related to obtaining context information derived from a set of raw data, apparently unrelated [24], increasing the value of the information to later transform it into knowledge [9].

The most common method to gain knowledge from a specific data is by cross-checking it with IoC from different external sources [2,6,13,21], in order to take advantage of the enrichment capacity of the different communities that made this data available [2].

Otherwise the correlation of data from various sources can be augmented with a set of data collected within the organization itself [12,14]. Comparison with internal data allows for identifying relevance and priority of the resulting data in the form of IoC, as well as producing situational awareness through additional context [12,14,40].

Through data enrichment, it is possible to contribute to three aspects that influence the quality of CTI:

- Relevance: as the relationship between external and internal events increases, the greater the relevance of information that makes sense to the organization;
- Accuracy: when security analysts gain situational awareness through understanding contexts, supported by data enrichment, the response to an incident if can be improved;
- Completeness: as data enrichment produces more comprehensive information, the ability to describe an incident increases.

On the other hand, many approaches expect to find context from raw data [2,43], however, from an intelligence cycle perspective, the data needs to be organized in a way that supports answers that complement a predefined context [12,14]. Without context there are no elements needed to underpin decision making. Without action CTI has no impact and proves useless, further burdening security analysts and adding no actionable intelligence.

4 Methodology

Based on the definitions presented, we verified the importance of knowing the context in which the organization is inserted. Thus, we employed the knowledge construction matrix, allied to the 5W3H method, during the intelligence cycle to create situational awareness and clarify the objectives at each stage of the process. The construction of knowledge goes through four stages [8] that lead from ignorance to actionable intelligence as represented in Fig. 1.

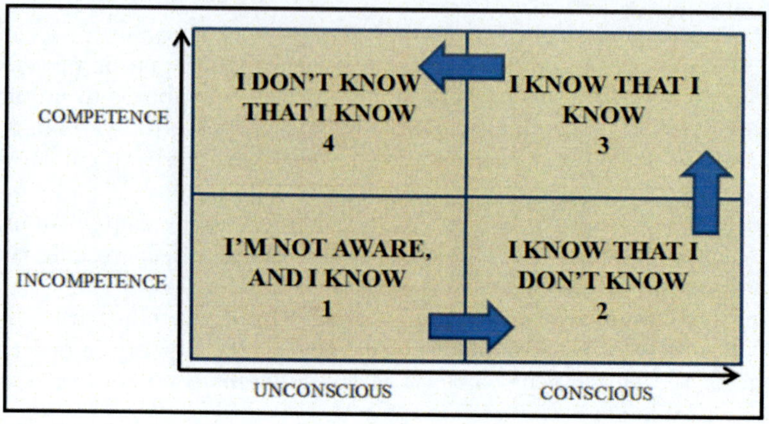

Fig. 1. Matrix of Knowledge Construction

What I don't know that I don't know represents the state of ignorance, the total absence of knowledge about a certain object. In this phase we do not know our vulnerabilities, capabilities, nor the evident threats. We can correlate it with the Planning and Direction Phase of the intelligence cycle.

What I know that I don't know refers to the awareness that there is something to be discovered, however, we do not have this knowledge. It is the knowledge of a certain vulnerability, but without knowing who, what, or how it can be exploited. It expresses the state of consciousness about the goal to be reached, following the premise that this goal must be useful, that is, it needs to be translated into action. It is the most labor-intensive step of the CTI process due to the large amount of data. We can correlate it with the result of processing the collected data.

What I know that I know corresponds to awareness and mastery over a certain subject, the initial stage for standardizing and expanding knowledge. In this stage the objective is to massify the knowledge. We can associate this phase with the result of the analysis phase of the intelligence cycle, as well as the planning of future actions.

What I don't know that I know is the apex of knowledge, at this point knowledge is strongly rooted, so that certain processes are automated to the point of going unnoticed. This way, when we verify a certain threat or incident, we automatically trigger defense mechanisms without the need for effort or the search for new knowledge in order to mitigate the effects of the threat. It is about proactivity, so measures to eliminate vulnerabilities are taken before they can be exploited by a threat. We associate it with the outcome of the deployment and dissemination phase of the intelligence cycle. Here are also the consequences, inferences and deductions from what is known that have not yet been made explicit.

We adopted the 5W3H method as a guide. It originates from Aristotle's seven circumstances [35]. This method is related to a set of eight questions: what, who, why, when, how, how much, and how long. It is widely applied in several areas in order to obtain contextualization of a theme in its completeness [34]. The 5W3h questions in conjunction with the phases of knowledge construction, contribute to the awareness of the maturity level in relation to what is known and the goals for generating actionable intelligence.

The 5W3H method will initially be employed in the planning phase. The advantage of this approach is that it is easy to see which questions are unanswered. In the following phases we seek answers to the questions that were not answered in the planning phase, so in each cycle of collection, processing and analysis we check the completeness of the context, seeking answers to all questions of the 5W3H method. The more questions in the 5W3H that are answered, the greater the completeness of the CTI and consequently higher the quality. If at the end of this process there are answers to most of the questions, we probably have actionable intelligence [34].

The "what" defines the object to be studied, which in the context of threat intelligence, refers to threats or incidents. The "where" refers to the geographical location where the event originated, and may also be the path taken to the destination. The "When" determines the date and time when the event occurred. The "How" defines the tactics, techniques, and procedures employed. The "How much" refers to the ability to cause damage, it may also be related to funding. "How long" indicates the duration of the event, incident, or threat. "Who" associates the threat or incident organization or individual responsible. "Why" explains the motivations of the person responsible for the event. In Fig. 2 you can see the relationship of the elements of the 5W3H method to entities involved in an incident or threat.

About the selected TIP, besides the advantages already mentioned, the MISP platform has integration with several enrichment tools, however, many of them are not Open Source, or their free versions impose limitations that make their large scale use unfeasible. Since in this paper we propose to work with open source software, we use only open source plugins and integrations. On the other hand, the methodology employed should be independent of TIP, despite the limitations described above.

We apply the cycle of intelligence production using open source data as shown in the Fig. 3 and following paragraphs.

In the direction and planning phase, the first step is to become aware of the current scenario of the organization, this phase reflects on all subsequent processes and can be taken up again when necessary. Through a survey of the technologies used, main applications and applications, as well as the topology and subsequent risk analysis, the requirements and priorities of the organization are listed. In this stage we leave the stage of ignorance, "I don't know what I don't know", to the stage of partial awareness, "I know that I don't know".

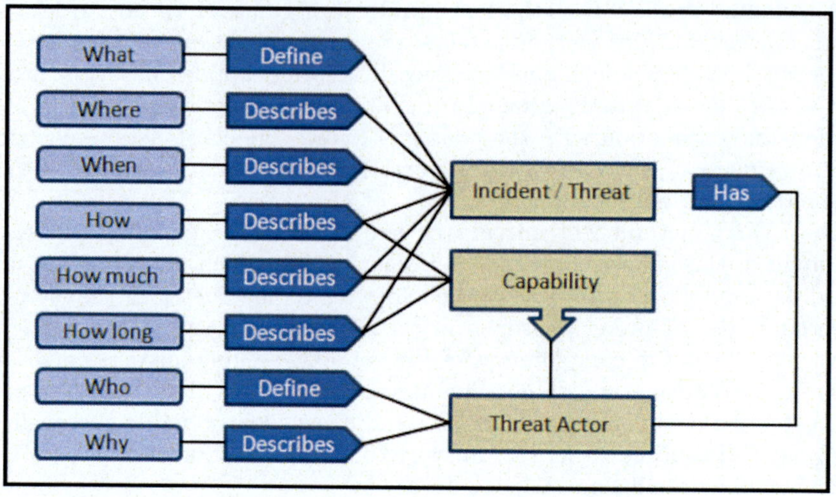

Fig. 2. Relationship Diagram [34]

At this point it is already possible to understand the internal scenario and based on the definition of priorities, assess the existing vulnerabilities for each asset, whether software or hardware. It is crucial to understand which external factors have an influence on the organization and the elements that may be of interest to threat actors. It is also fundamental to collect internal data for comparison with data from external sources, to prove the existence or not of adverse actions.

The 5W3H method helps to delineate the relevant aspects that need to be known in order to build the context. This brings us to the stage where we are aware of the unanswered questions. Then we start planning to define the relevant data sources to complement the context by means of the unanswered questions of the 5W3H method.

After defining the sources and collecting data, it is necessary to verify credibility and validity, as well as evaluate the possibility of enrichment according to the type of data and plugins available. This phase also needs prior planning, in order to define what kind of enrichment can contribute to the completeness of the context.

Supported by the 5W3H model and based on the information created through the enrichment, correlation and patterns of the data from external sources and also the presence or absence of relationships with the organization's own data, we become more aware of the context. At this point new planning is established with the purpose of proposing pertinent actions, that include safeguarding guidelines, security patching, and incident mitigation, among others.

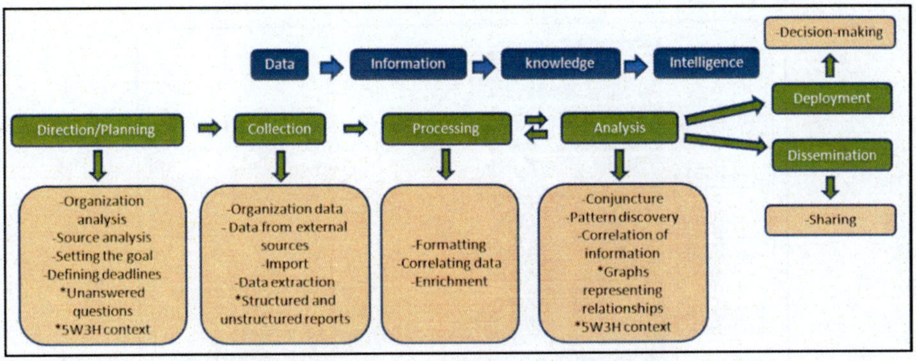

Fig. 3. Cycle for intelligence production

5 Procedures

Collecting information from various sources is fundamental to generate solid knowledge, however, the credibility of the data and the suitability of the source, under the aspects of authenticity, trust, and competence, must be observed.

In this way, our approach is based on structured and unstructured reports of the Pegasus case, which for the purposes of our experiment we admit are from reliable sources in view of having already been widely discussed by companies with expertise in the field of cyber security, in addition to the fact that they have differentiated resources compared to other institutions, having, for example, the advantage of receiving numerous feedbacks from applications installed in the infrastructure of their customers.

Pegasus is spyware developed by Israel's NSO Group. It has been used as a surveillance tool against high-ranking government officials, human rights activists, journalists, and heads of State [10].

6 Discussions

Based on the proposed methodology we verify the results achieved sustained by the 5W3H model (see Fig. 4)

The information presented was extracted only from eight selected reports. Due to the characteristic of the case analyzed, other reports could result in differences mainly in "where" and "who".

Considering that the reports and data used in the use case are not recent, most domains no longer exist or have a registration date later than the report date. The average number of expired domains (available for re-registration) is 68%, ranging from 58% to 80% in reports from August 2016, July 2021, September 2021, December 2021, February 2022, April 2022, and June 2022. This is of concern, given that throughout this research we have identified tools for detecting Pegasus spyware based on IoC sets from these reports. Consequently, occurrences of false positives may be observed.

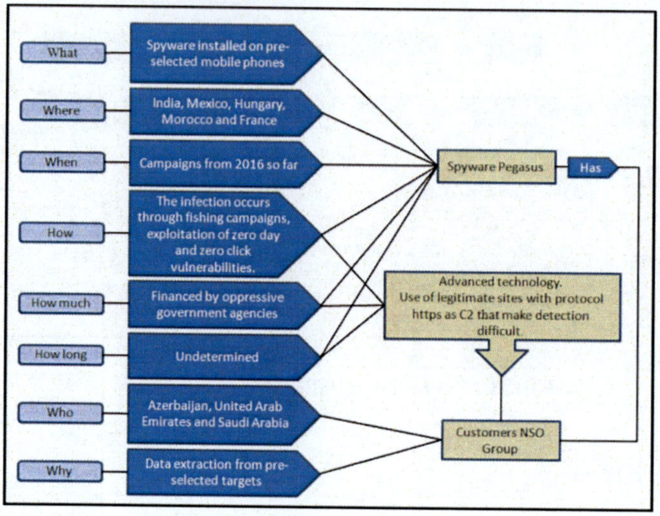

Fig. 4. Result Analysis

We observed that identical reports, when analyzed by different organizations, produce different results. For example, the report "The Million Dollar Dissident - Citizen lab report" [19], available in pdf format, was analyzed by more than one security expert organization, however, the amount of IoC generated by each organization was different. We imported and enriched this report in MISP and observed that the analysis is limited due to the inherent computational complexity of natural language processing (NLP). We also verified that relevant information contained in reports available in natural language was not extracted by the tools employed for collection.

We conclude that the type of enrichment that makes the most sense for "expired" data is to look for relationships with data spotted in the same period. In addition, when querying domain data, we noted the scarcity of Open Source tools that provide history of the registration data, which could contribute to the temporal definitions of the event. However, we were able to determine some of the tactics, techniques, and procedures employed, as well as identify the vulnerabilities exploited. Thus, we verified the ability to generate intelligence and propose actions, based on the use of the proposed method.

7 Conclusions and Future Work

Threat intelligence alone cannot protect an organization, but rather complements security components related to detection, response, and prevention in order to reduce the potential damage done by increasing the effectiveness of security components, decreasing response time, reducing damage recovery time, and reducing the time the adversary remains in the organization's environment.

As security components become more robust through information originating from CTI, threat prediction becomes closer to reality, since threat mitigation is more inherent to the security state of the organization than to the study of data transiting the network.

Analyzing the possible relationships of known data, it is necessary to verify what type of enrichment can generate usable results. Disordered enrichment often generates a lot of information that is not used and does not add to knowledge. Just as indiscriminately importing reports from multiple sources does not guarantee that you will get actionable CTI.

Although studies show that MISP is the most complete platform, it has limitations, but it was possible to generate Actionable Intelligence by applying the proposed methodology.

The use of the knowledge construction matrix, coupled with the 5W3H method, throughout the intelligence production cycle proved to be effective in creating situational awareness of the organization and clarifying the objectives of each stage of the process. In this way unnecessary data collection and enrichment is avoided.

The main advantage of the proposed methodology is that it contemplates the entire cycle of intelligence production.

Future work can be addressed to integrate tools or platforms that complement each other in order to solve limitations such as employment of the full intelligence cycle and relationship discovery and visualization.

Acknowledgement. R.d.O.A. gratefully acknowledges the technical and computational support of the Laboratory of Technologies for Decision Making (LATITUDE) of the University of Brasília; the General Attorney's Office (Grant AGU 697.935/2019); the General Attorney's Office for the National Treasure - PGFN (Grant 23106.148934/2019-67); the National Institute of Science and Technology in Cyber Security - Nucleus 6 (grant CNPq 465741/2014-2).

The authors thankfully acknowledge the support of Brazilian Intelligence Agency - ABIN grant 08/2019.

References

1. Abu, M.S., Selamat, S.R., Ariffin, A., Yusof, R.: Cyber threat intelligence - issue and challenges. Indonesian J. Electr. Eng. Comput. Sci. **10**(1), 371–379 (2018). https://doi.org/10.11591/ijeecs.v10.i1.pp371-379 10.11591/ijeecs.v10.i1.pp371-379 10.11591/ijeecs.v10.i1.pp371-379
2. Azevedo, R., Medeiros, I., Bessani, A.: Automated solution for enrichment and quality IoC creation from OSINT. In: Simpósio de Informática (INForum 2018), p. 12 (2018). http://disiem-project.eu/wp-content/uploads/2018/11/INForum2018_enr-IoC.pdf
3. Basheer, R., Alkhatib, B.: Threats from the dark: a review over dark web investigation research for cyber threat intelligence. J. Comput. Netw. Commun. **2021**, 1–21 (2021). https://doi.org/10.1155/2021/1302999
4. Berndt, A., Ophoff, J.: Exploring the value of a cyber threat intelligence function in an organization. In: Drevin, L., Von Solms, S., Theocharidou, M. (eds.) WISE

2020. IAICT, vol. 579, pp. 96–109. Springer, Cham (2020). https://doi.org/10.1007/978-3-030-59291-2_7
5. Bromander, S.: Understanding Cyber Threat Intelligence - Towards Automation. Ph.D. thesis, University of Oslo (2021). http://urn.nb.no/URN:NBN:no-87408
6. Bromander, S.: Investigating sharing of cyber threat intelligence and proposing a new data model for enabling automation in knowledge representation and exchange. Digit. Threats Res. Pract. **3**(1), 1–22 (2022). https://doi.org/10.1145/3458027
7. Bubach, R., Herkenhoff, H.G., Herkenoff, L.S.B.: O cilclo da inteligência e os requisitos para a produção do conhecimento. Ph.D. thesis, Universidade Vila Velha (2019). https://repositorio.uvv.br//handle/123456789/570
8. Businessballs: Conscious competence learning model. https://www.businessballs.com/self-awareness/conscious-competence-learning-model/#theories_models_change_learning
9. Chantzios, T., Koloveas, P., Skiadopoulos, S., Kolokotronis, N., Tryfonopoulos, C., Bilali, V.G., Kavallieros, D.: The quest for the appropriate cyber-threat intelligence sharing platform. In: Proceedings of the 8th International Conference on Data Science, Technology and Applications, pp. 369–376. SCITEPRESS - Science and Technology Publications (2019). https://doi.org/10.5220/0007978103690376
10. Chawla, A.: Pegasus spyware - "a privacy killer". SSRN Electron. J. (2021). https://doi.org/10.2139/ssrn.3890657. https://www.ssrn.com/abstract=3890657
11. Check, P.: Cyber security report 2021. Technical report, Check Point, San Carlos, CA (2021). https://pages.checkpoint.com/cyber-security-report-2021.html?utm_source=cp-home&utm_medium=cp-website&utm_campaign=pm_wr_21q1_ww_security_report
12. Faiella, M., Gonzalez-Granadillo, G., Medeiros, I., Azevedo, R., Gonzalez-Zarzosa, S.: Enriching threat intelligence platforms capabilities. In: ICETE 2019 - Proceedings of the 16th International Joint Conference on e-Business and Telecommunications, vol. 2, pp. 37–48 (2019). https://doi.org/10.5220/0007830400370048
13. Gao, Y., Li, X., Peng, H., Fang, B., Yu, P.S.: HinCTI: a cyber threat intelligence modeling and identification system based on heterogeneous information network. IEEE Trans. Knowl. Data Eng. **34**(2), 708–722 (2022). https://doi.org/10.1109/TKDE.2020.2987019. https://ieeexplore.ieee.org/document/9072563/
14. González-Granadillo, G., Faiella, M., Medeiros, I., Azevedo, R., González-Zarzosa, S.: ETIP: an enriched threat intelligence platform for improving OSINT correlation, analysis, visualization and sharing capabilities. J. Inf. Secur. Appl. **58**, 102715 (2021). https://doi.org/10.1016/j.jisa.2020.102715
15. Hettema, H.: Rationality constraints in cyber defense: incident handling, attribution and cyber threat intelligence. Comput. Secur. **109**, 102396 (2021). https://doi.org/10.1016/j.cose.2021.102396
16. Husari, G., Al-Shaer, E., Ahmed, M., Chu, B., Niu, X.: TTPDrill: automatic and accurate extraction of threat actions from unstructured text of CTI sources. In: ACM International Conference Proceeding Series. Part F1325, pp. 103–115. Association for Computing Machinery, New York, Orlando (2017). https://doi.org/10.1145/3134600.3134646
17. Koloveas, P., Chantzios, T., Alevizopoulou, S., Skiadopoulos, S., Tryfonopoulos, C.: InTIME: a machine learning-based framework for gathering and leveraging web data to cyber-threat intelligence. Electronics (Switzerland) **10**(7), 818 (2021). https://doi.org/10.3390/electronics10070818

18. Korte, K.: Measuring the quality of open source cyber threat intelligence feeds. Master's thesis, JAMK University of Applied Sciences - Finland (2021). https://www.theseus.fi/handle/10024/500534. http://urn.fi/URN:NBN:fi:amk-202105178967
19. Marczak, B.B., Scott-Railton, J.: The million dollar dissident: NSO group's iPhone zero-days used against a UAE human rights defender. Technical report, Citizen Lab - University of Toronto, Toronto, Ontario - Canada (2016). https://citizenlab.ca/2016/08/million-dollar-dissident-iphone-zero-day-nso-group-uae/
20. Martins, C., Medeiros, I.: Generating quality threat intelligence leveraging OSINT and a cyber threat unified taxonomy. ACM Trans. Priv. Secur. **25**(3), 1–39 (2022). https://doi.org/10.1145/3530977
21. Nikolaienko, B., Vasylenko, S.: Application of the threat intelligence platform to increase the security of government information resources. Informatyka, Automatyka, Pomiary w Gospodarce i Ochronie Środowiska **11**(4), 9–13 (2021). https://doi.org/10.35784/iapgos.2822
22. Oosthoek, K., Doerr, C.: Cyber threat intelligence: a product without a process? Int. J. Intell. Counter Intell. **34**(2), 1–16 (2020). https://doi.org/10.1080/08850607.2020.1780062
23. Papaioannou, F.: Threat intelligence platforms evaluation. Ph.D. thesis, University of Piraeus (2021). https://dione.lib.unipi.gr/xmlui/handle/unipi/13346
24. Park, Y., Choi, J., Choi, J.: An extensible data enrichment scheme for providing intelligent services in internet of things environments. Mobile Inf. Syst. **2021**, 1–18 (2021). https://doi.org/10.1155/2021/5535231. https://www.hindawi.com/journals/misy/2021/5535231/
25. Preuveneers, D., Joosen, W.: Sharing machine learning models as indicators of compromise for cyber threat intelligence. J. Cybersecur. Priv. **1**(1), 140–163 (2021). https://doi.org/10.3390/jcp1010008. https://www.mdpi.com/2624-800X/1/1/8/htm
26. Rahman, M.R., Mahdavi-Hezaveh, R., Williams, L.: A literature review on mining cyberthreat intelligence from unstructured texts. In: 2020 International Conference on Data Mining Workshops (ICDMW). IEEE (2020). https://doi.org/10.1109/ICDMW51313.2020.00075
27. Samtani, S.: Developing proactive cyber threat intelligence from the online hacker community: a computational design science approach. Ph.D. thesis, The University of Arizona (2018). http://hdl.handle.net/10150/628454
28. Sauerwein, C., Fischer, D., Rubsamen, M., Rosenberger, G., Stelzer, D., Breu, R.: From threat data to actionable intelligence: an exploratory analysis of the intelligence cycle implementation in cyber threat intelligence sharing platforms. In: ACM International Conference Proceeding Series (2021). https://doi.org/10.1145/3465481.3470048
29. Sauerwein, C., Sillaber, C., Mussmann, A., Breu, R.: Threat intelligence sharing platforms: an exploratory study of software vendors and research perspectives. In: The 13th International Conference on Wirtschaftsinformatik, pp. 837–851 (2017). https://wi2017.ch/images/wi2017-0188.pdf
30. Schaberreiter, T., et al.: A quantitative evaluation of trust in the quality of cyber threat intelligence sources. In: ACM International Conference Proceeding Series (2019). https://doi.org/10.1145/3339252.3342112
31. Schlette, D., Böhm, F., Caselli, M., Pernul, G.: Measuring and visualizing cyber threat intelligence quality. Int. J. Inf. Secur. **20**(1), 21–38 (2020). https://doi.org/10.1007/s10207-020-00490-y

32. Accenture Security: Cyber Threatscape Report. Technical report, Accenture Security (2020). https://www.accenture.com/_acnmedia/pdf-107/accenture-security-cyber.pdf
33. Sillaber, C., Sauerwein, C., Mussmann, A., Breu, R.: Data quality challenges and future research directions in threat intelligence sharing practice. In: WISCS 2016 - Proceedings of the 2016 ACM Workshop on Information Sharing and Collaborative Security, co-located with CCS 2016, pp. 65–70 (2016). https://doi.org/10.1145/2994539.2994546
34. de Melo e Silva, A., Gondim, J.J.C., de Oliveira Albuquerque, R., Villalba, L.J.G.: A methodology to evaluate standards and platforms within cyber threat intelligence. Future Internet **12**(6), 1–23 (2020). https://doi.org/10.3390/fi12060108
35. Sloan, M.: Aristotle's as the original locus for the septem circumstantiae. Class. Philol. **105**, 236–251 (2010). https://doi.org/10.1086/656196
36. Stojkovski, B., Lenzini, G., Koenig, V., Rivas, S.: What s in a cyber threat intelligence sharing platform? In: ACM International Conference Proceeding Series, pp. 385–398 (2021). https://doi.org/10.1145/3485832.3488030
37. Sun, T., Yang, P., Li, M., Liao, S.: An automatic generation approach of the cyber threat intelligence records based on multi-source information fusion. Future Internet **13**(2), 1–19 (2021). https://doi.org/10.3390/fi13020040
38. Tekin, U., Yilmaz, E.N.: Obtaining cyber threat intelligence data from twitter with deep learning methods. In: ISMSIT 2021 - 5th International Symposium on Multidisciplinary Studies and Innovative Technologies, Proceedings, pp. 82–86 (2021). https://doi.org/10.1109/ISMSIT52890.2021.9604715
39. Tounsi, W., Rais, H.: A survey on technical threat intelligence in the age of sophisticated cyber attacks. Comput. Secur. **72**, 212–233 (2018). https://doi.org/10.1016/j.cose.2017.09.001
40. Vielberth, M.: Human-as-a-security-sensor for harvesting threat intelligence. Cybersecurity **2**(1), 1–15 (2019). https://doi.org/10.1186/s42400-019-0040-0
41. Wagner, T.D., Mahbub, K., Palomar, E., Abdallah, A.E.: Cyber threat intelligence sharing: survey and research directions. Comput. Secur. **87**, 101589 (2019). https://doi.org/10.1016/j.cose.2019.101589
42. Zhao, J., Yan, Q., Liu, X., Li, B., Zuo, G.: Cyber threat intelligence modeling based on heterogeneous graph convolutional network. In: 23rd International Symposium on Research in Attacks, Intrusions and Defenses (RAID 2020), pp. 241–256 (2020). https://www.usenix.org/conference/raid2020/presentation/zhao
43. Zibak, A., Sauerwein, C., Simpson, A.C.: Threat intelligence quality dimensions for research and practice. Digit. Threats **3**, 1–22 (2021). https://doi.org/10.1145/3484202
44. Zibak, A., Simpson, A.: Cyber threat information sharing: perceived benefits and barriers. In: Proceedings of the 14th International Conference on Availability, Reliability and Security, pp. 1–9. ACM (2019). https://doi.org/10.1145/3339252.3340528

Evaluation of Web Accessibility in the State Citizen Service Portals Most Used by Disabled People

Enrique Garcés-Freire(✉), Verónica Pailiacho-Mena, and Joselyn Chucuri-Yachimba

Pontificia Universidad Católica del Ecuador-Ambato, Ambato, Ecuador
{egarces,vpailiachojoselyn.a.chucuri.y}@pucesa.edu.ec

Abstract. A website is accessible means that all types of users have a satisfactory experience in navigating it, regardless of age, if they have a disability, or the type of device and operating system they use. With the exponential increase in the use of the web it is necessary to know if the websites have this characteristic so this article presents the evaluation of web accessibility to the state sites of citizen services most used by people with visual and hearing disabilities of Tungurahua-Ecuador, and their compliance with the current standard NTE INEN-ISO/IEC 40500, using the TAW tool, our findings in-dicate that the most used state sites are the Ministry of Labor, the Ecuadorian Institute of Social Security, the Internal Revenue Service, and the Autonomous Government Decentralized Municipal of Ambato, the investigation shows as a result that none of the 4 websites complies with current regulations, however, the SRI website presents least problems and the GADMA website presents the most accessibility problems, In a cross-sectional analysis, it was determined that the Operable principle is the one that presents the most problems in all websites, in its criterion 2.4.4 Purpose of the links this measurement was made based on the AA criterion, which is the one that should be fulfilled for Ecuador until August 8, 2020 according to the approved regulations.

Keywords: Web accessibility Evaluation · state citizen websites · disabled people

1 Introduction

In recent years, the use of the web for business, administrative, and information services has grown exponentially; so that, the term web accessibility has gained strength and can be defined as the extent to which a website can be used by everyone [1], including children, youth, adults, seniors, and people with disabilities. According to [2], it is necessary that the contents, the authoring tools, and browsers are accessible, only in this way, this communication way will allow the

M. V. Garcia and C. Gordón-Gallegos (Eds.): CSEI 2022, LNNS 678, pp. 99–110, 2023.
https://doi.org/10.1007/978-3-031-30592-4_8

integration of anybody into the digital society, Therefore, an evaluation of web accessibility is required.

In a global context, there are the Web Content Accessibility Guidelines, it is the Web Content Accessibility Guidelines (WCAG) 2.0. In [3–6], it is determined that this guide contains the guidelines for accessible web design, it must consider more people with disabilities, who may have blindness or low vision, deafness, and hearing loss, learning disabilities, cognitive limitations, reduced mobility, language deficiencies, photosensitivity or a combination of all these. WCAG 2.0 is organized into 4 principles, 12 guidelines, 61 compliance criteria, 3 compliance levels A, AA, and AAA, as well as techniques and recommendations, to develop and evaluate web content. To carry out the accessibility evaluation it is necessary to understand the 4 principles, the first is Perceivable, which refers to the fact that all the elements of the interface must be perfectly identifiable, in any situation, for this special attention must be paid to color, images, buttons, audiovisual content; the second is Operable, it means that the website must provide different means to execute actions or search for content, for example, to navigate you can use a mouse or keyboard or adequate time to read and understand the information; the third is Understandable, it describes the information must be legible, clear, evident, special attention in the fonts used and the function of the website must be predictable, that is, that users do not have to guess what makes on the website; Finally Robust, it refers to the site must be compatible with any browser, device, operating system, and support technology applications. For this, the developers must respect the HTML standards and style sheets.

In Ecuador, the current Organic Law on Disabilities [7], ensures the prevention, timely detection, empowerment, and rehabilitation of disability and guarantees the full validity, dissemination, and apply the rights of people with disabilities, in Art. 65 Priority attention in web portals affirms that public and private institutions that provide public services will include in their web portals, an access link for people with disabilities so that they can access information and specialized and priority attention. In addition, there are the Ecuadorian Standard NTE INEN-ISO/IEC 40500 Information Technology - W3C Web Content Accessibility Guidelines (WCAG) 2.0 which is an identical translation of ISO/IEC 40500:2012 Information Technology - W3C Web Content Accessibility Guidelines (WCAG) 2.0 [8].

As mentioned, there are web accessibility standards both globally and nationally, and there are several web accessibility evaluation studies, with interesting results. In [9], it is mentioned that the web is used as a marketing and business platform, a comparison of the results is made between the web accessibility evaluations of the Portuguese websites that use version 1 and version 2 of the W3C Web Content Accessibility Guidelines this means that in Portugal the standards are considered in the design of websites. Based on [10] the discussion is established that the web educational content must be accessible to all students, regardless of their abilities or disabilities, they present a review that identifies the authoring tools most used by teachers in exact sciences, with an emphasis on mathematics, physics, and geometry to then be evaluated in accessibility, and recommendations regarding web accessibility are determined. [11] evaluates Web Accessibility in selected software products, and the results show that in general accessibility guidelines are not considered in the design and development of technological products. According to [12], e-government portals in Latin America do not comply with accessibility standards. Based on [2,3] evaluation of the web portals of the universities in Ecuador, as well as the accredited institutes of education, [13] does not comply with current accessibility regulations, this shows that there is still no total inclusion in the digital society and the digital divide in people with disabilities is much more evident Well so that they can navigate through the websites in an easy and simple way, many choose to ask for help from family, friends or even unknown people to guide them or give them the required process, this is the reason why they are affected in their social and labor integration due to the lack of access [5,6], and high-security risk.

Understanding that web accessibility is important for the digital inclusion of all people, the objective of this article is to report the results of the evaluation of web accessibility of the state websites of citizen services most used by people with visual and hearing disabilities in Tungurahua-Ecuador, and compliance with the Ecuadorian Standard NTE INEN-ISO/ IEC 40500 and seeks to answer the following questions: Do the citizen service websites most used by people with disabilities comply with the Ecuadorian web accessibility standard NTE INEN-ISO/IEC 40500 with an AA conformity level?

The document is organized as follows: Sect. 2 presents the method used in the research, Sect. 3 details the results obtained, and finally, Sect. 4 presents the conclusions.

2 Method

This research has as its counterpart the Labor Integration Service - Tungurahua (SIL), this organization belongs to the National Council for Equality of Disabilities (CONADIS), the SIL is responsible for providing care and assistance to people with disabilities through different support mechanisms, to that they can enroll in working life or create enterprises that allow them to support themselves.

Within the established method, as a first activity, the citizen service websites that people with disabilities use the most for their activities in Tungurahua-Ecuador, through interviews and focus groups, identify the sites they use most frequently for their work or productive activities

The next step was to carry out a documentary search on the Regulations and Standards that must be complied with regarding web accessibility in Ecuador, it was found that in Ecuador the Ecuadorian Standard NTE INEN-ISO / IEC 40500, Then, based on the search in the literature regarding accessibility, a tool that allows evaluating accessibility was identified. the selected tool was TAW (Test Accessibility Web), which has been used in several investigations [3,8, 14–17], and evaluation research on tools that are coupled to the WCAG 2.1 standard was found as a highly valued tool for its flexibility and adherence to the standard, [18–21], TAW is a free tool, its analysis format is a web and can be configured to make the evaluation based on the reliability levels established by the WCAG standard, for this analysis, the AA level was defined as an evaluation parameter, which is the level suggested by the standard established for Ecuador.

The following part describes the results obtained for this accessibility research in the selected portals. This allowed us to answer the question: Do the citizen service websites most used by people with disabilities comply with the Ecuadorian web accessibility standard NTE INEN-ISO/IEC 40500 with an AA conformity level? the evaluation was applied to all sites on July 21, 2022, to the index pages of each website, this is because it is the first page that is accessed in each portal and from which it is derived to different applications, this measurement to the main page allows to have a general idea of how each website handles accessibility.

3 Results

As a first task, we work with the members of SIL Tungurahua to identify the websites they use for their commercial or productive activities in Tungurahua-Ecuador, this activity had the support of 22 members, with whom information was obtained through interviews and focus groups; among all the options that they mention, it is determined that four are the sites with the highest frequency of use, this is due to the nature of their activities both as employees and employers, being mentioned between 17 and 20 times among the 22 participants:

- The Ministry of Labor, this portal is used to carry out all hiring and legalization activities of employees in organizations at the national level, labor inspections, and approval of regulations, among others [22].
- Ecuadorian Social Security Institute (IESS), the portal offers affiliates, retirees, employers, and the public several online services, such as queries and requests for the services and benefits they require [23].
- Internal Revenue Service (SRI) in this portal, Ecuadorians who have registered an economic activity must comply with their tax obligations as taxpayers, including filing tax returns and tax compliance certifications [24].
- Autonomous Government Decentralized Municipal of Ambato (GADMA) in this portal belongs to the municipality of the city of Ambato, capital of the province of Tungurahua, which is an online service system where digital service transactions are carried out, such as domain transfers, property taxes, municipal taxes, vehicle registration, among others [25].

On the other hand, in the internet search of regulations that must be complied in Ecuador, it was found that in Ecuador the Ecuadorian Standard NTE INEN-ISO / IEC 40500 "Information Technology - W3C Web Content Accessibility Guidelines (WCAG) 2.0 (ISO/IEC 40500:2012, IDT)" [26]. And the Ecuadorian Technical Regulation RTE INEN 288 "Accessibility for web content" regulates compliance with the standard [27], this standard should apply until August 8, 2020.

In the search for an accessibility evaluation tool, the TAW tool is selected, because it is widely regarded as a WCAG 2.0 compliant application, the standard on which the NTE INEN-ISO/IEC 40500 is based, and must be fulfilled in Ecuador, TAW is a tool of free access, which evaluates the HTML code, CSS, and in its new update to the JavaScript code, has been working for some time in the area of accessibility and is supported by the Technological Center of Information and Communication of Asturias (CTIC) [28,29].

The NTE INEN-ISO/IEC 40500 standard, like WCAG, establishes a qualification on the fulfillment of accessibility criteria measured in A, AA, AAA, TAW does the review of the site in two parts: Automatic and Manual, it can configure the level of analysis as follows:

- WCAG 2.1 A. Meets all priority checkpoints 1
- WCAG 2.1 AA. Meets all priority checkpoints 1 y 2
- WCAG 2.1 AAA. Meets all priority checkpoints 1, 2 y 3

The TAW tool for this analysis was configured with an AA conformance level, The parameters measured by the tool at this level are those that are aligned with the NTE INEN-ISO/IEC 40500 Standard as acceptable for Ecuador and are the following:

For a better organization of the results, the acronyms of each entity have been designated in Table 2 as codes to identify each website used in this research, in addition, the web address corresponding to the evaluated page is added, which for all is the index or the main page of the website, this is because it is the first impression of a visitor to the portal, for this study, it is considered that the main page has accessibility problems, the rest of the pages will have them in the same way, the order in which the information is displayed has no hierarchy:

Once the most used websites have been determined, the research question can be answered, and information will be disaggregated as appropriate.

Do the citizen service websites most used by people with disabilities comply with the Ecuadorian web accessibility standard NTE INEN-ISO/IEC 40500 with an AA conformity level?

An automatic review was applied to each of the web portals through the TAW web tool, an AA conformity level, the summary of the observations detected by the tool is shown in Table 3, based on the 4 principles of the WCAG 2.1 Standard on which TAW is based, suitable for verifying compliance with the NTE INEN-ISO/IEC 40500 standard. For a better presentation of the information, each of the principles is coded as follows: Perceivable (P), Operable (O), Understandable

Table 1. Parameters used in TAW organized by principles.

PERCEIVABLE	OPERABLE	UNDERSTANDABLE	ROBUST
1.1. Text alternatives 1.1.1. Non-text content	2.1. Keyboard accessible 2.1.1. Keyboard 2.1.2. No keyboard trap	3.1. Readable 3.1.1. Language of page 3.1.2. Language of parts	4.1. Compatible 4.1.1. Parsing 4.1.2. Name, role, value
1.2. Time-based media 1.2.1. Audio-only and video-only 1.2.2. Captions 1.2.3. Audio description or media alternative 1.2.4. Captions 1.2.5. Audio description	2.2. Enough time 2.2.1. Timing adjustable 2.2.2. Pause, stop, hide	3.2. Predictable 3.2.1. On focus 3.2.2. On input 3.2.3. Consistent navigation 3.2.4. Consistent identification	
1.3. Adaptable 1.3.1. Info and relationships 1.3.2. Meaningful sequence 1.3.3. Sensory characteristics	2.3. Seizures 2.3.1. Three Flashes or below threshold	3.3. Input assistance 3.3.1. Error identification 3.3.2. Labels or instructions 3.3.3. Error suggestion 3.3.4. Error prevention (legal, financial, data)	
1.4. Distinguishable 1.4.1. Use of color 1.4.2. Audio control 1.4.3. Contrast (minimum) 1.4.4. Resize text 1.4.5. Images of text	2.4. Navigable 2.4.1. Bypass blocks 2.4.2. Page titled 2.4.3. Focus order 2.4.4. Link purpose (in context) 2.4.5. Multiple ways 2.4.6. Headings and labels 2.4.7. Focus visible		

Table 2. Selected portals.

Identifier	Institution	URL
MT	Ministry of Labor	https://www.trabajo.gob.ec/
IESS	Ecuadorian Institute of Social Security	https://www.iess.gob.ec/
SRI	Internal Revenue Service	https://srienlinea.sri.gob.ec/sri-en-linea/inicio/NAT
GADMA	Autonomous Government Decentralized Municipal of Ambato	https://ambato.gob.ec/

(C), Robust (R), plus a grouping of results concerning problems that are serious web accessibility errors, detected warnings that represent problems that are not very serious, but must be corrected and unverified criteria, as they require a manual review.

Table 3. TAW Analysis Summary.

Identifier	Problems					Warnings					Not reviewed				
	P	O	C	R	Total	P	O	C	R	Total	P	O	C	R	Total
MT	7	3	1	0	11	84	57	12	0	153	4	8	5	1	18
IESS	12	26	5	2	45	76	101	12	0	189	3	5	4	0	12
SRI	0	0	0	0	0	47	14	18	0	79	4	7	2	1	14
GADMA	29	50	5	6	90	91	47	24	0	162	4	8	2	1	15

After this review of the four state service websites, none one meets the levels of compliance required by the WAI-W3C WCAG 2.0 standard, which is referred to in the Ecuadorian technical regulation RTE INEN 288, to the respective compliance there should be no problems or warnings in the web accessibility principles of the standard.

It can be determined that the SRI is the website with the best web accessibility features, although it does not present serious problems, it must work on the warnings, on the other hand, the GADMA website is the one with the most problems and warnings that require attention. In general, the 4 websites must work on some corrections, to offer web accessibility to their users with disabilities.

From the information obtained, it is possible to analyze the principle of web accessibility that presents the most problems in the state sites of citizen services analyzed

Therefore, it can be concluded that to improve the web accessibility of the sites under analysis, aspects such as color contrast, images, buttons, audiovisual content, allowing browsing the site with the use of a mouse, keyboard, or information technology must be improved reading.

Each principle of Web accessibility has criteria that must be met, for each criterion evaluated, TAW shows 4 possible evaluations as a result:

- No problems found
- There are problems
- Requires manual review
- Impossible to perform an automatic check

With this explanation, we proceed to show the results obtained by the four web-sites with the most information, mentioning the criteria in which web accessibility problems are found, and a summary of the relevant findings of each web portal analyzed.

Ministry of Labor (MT)

- PERCEIVABLE:
  Criterion 1.1.1 - Non-text Content, 6 Problems
  Criterion 1.3.1 - Info and Relationships, 1 Problem

Table 4. Summary of analysis with TAW of problems and warnings.

Identifier	Problems				Warnings			
	P	O	C	R	P	O	C	R
MT	7	3	1	0	84	57	12	0
IESS	12	26	5	2	76	101	12	0
SRI	0	0	0	0	47	14	18	0
GADMA	29	50	5	6	91	47	24	0
Total	48	79	11	8	298	219	66	0

- OPERABLE
  Criterion 2.4.4 - Link Purpose (In Context), 3 Problems

- UNDERSTANDABLE
  Criterion 3.3.2 - Labels or Instructions, 1 Problem

- ROBUST
  Of the 2 criteria evaluated in this Principle, the website has no problems

  On the other hand, criterion 3.1.1 - Language of Page, it has an A level rating.

  Ecuadorian Institute of Social Security Social (IESS)
- PERCEIVABLE:
  Criterion 1.1.1 - Non-text Content, 6 Problems
  Criterion 1.3.1 - Info and Relationships, 6 Problems

- OPERABLE
  Criterion 2.4.4 - Link Purpose (In Context), 26 Problems

- UNDERSTANDABLE
  Criterion 3.1.1 - Language of Page, 1 Problem.
  Criterion 3.2.2 - On Input, 2 Problems.
  Criterion 3.3.2 - Labels or Instructions, 2 Problems.

- ROBUST
  Criterion 4.1.2 - Name, Role, Value, 2 Problems.

Internal Revenue Service (SRI)

The SRI site does not present problems, but there are 79 Warnings in total, which are beyond the analysis of this question. On the other hand, in the criteria 1.3.1 - Info and Relationships, 3.1.1 - Language of Page, y 3.3.2 - Labels or Instructions No problems are found being criteria qualified with level A.

Autonomous Government Decentralized Municipal of Ambato (GADMA)

- PERCEIVABLE:
  Criterion 1.1.1 - Non-text Content, 7 Problems

Table 5. Summary of problems detected by each principle.

Web sites	PERCEIVABLE		OPERABLE	UNDERSTANDABLE			ROBUST
	1.1.1 Non-text Content	1.3.1 Info and relationships	2.4.4 Link purpose	3.1.1 Language of page	3.2.2 On Input	3.3.2 Labels or instructions	4.1.2 Name, role, value
MT	6	1	3	–	–	1	–
IESS	6	6	26	1	2	2	2
SRI	–	–	–	–	–	–	–
GADMA	7	22	50	–	–	5	6

Criterion 1.3.1 - Info and Relationships, 22 Problems

- OPERABLE
  Criterion 2.4.4 - Link Purpose (In Context), 50 Problems

- UNDERSTANDABLE
  Criterion 3.3.2 - Labels or Instructions, 5 Problems

- ROBUST
  Criterion 4.1.2 - Name, Role, Value, 6 Problems

On the other hand, in criterion 3.1.1 - Language of page, no problems are found, being a criterion qualified with level A.

The following table presents a summary of the criteria with accessibility problems for each Principle

As can be seen, the greatest problems are shown in criterion 2.4.4. Link Purpose, refers to the help that the link gives users to understand the purpose of each link and they can decide whether to activate it or not. The next criterion 1.3.1. Info and Relationships, ensures that the information and relationships of the visual or auditory format are maintained when the format of the presentation changes, for example when using a screen reader, all this confirms what was mentioned above that the website of the Autonomous Government Decentralized Municipal of Ambato (GADMA), It is the site that offers less web accessibility, in which it is necessary to work in a more inclusive way, so this new workspace is open for future research.

4 Conclusions

This article shows that the four citizen service portals that are most used by people with disabilities for their activities are the Ministry of Labor, the Ecuadorian Institute of Social Security, the Internal Revenue Service, and the GAD Municipality of Ambato, but the four websites do not comply with the acceptance levels set forth by the W3C accessibility standard on which the ISO / IEC 40500: 2012 standard applied in Ecuador it must be met on state websites, it has been shown that there are a significant number of problems and warnings that must be taken into account. so that the portals are accessible by people with visual and hearing disabilities, and in this way they can be inclusive portals, it must be remembered that in Ecuador the websites of public institutions had to comply with accessibility standards until August 8, 2020, as stated in the Ecuadorian technical regulation RTE INEN 288, it is evident in the information presented that no managed to meet this requirement, although the SRI website is the one that demonstrates having better management of accessible information by not presenting problems at the time of this investigation, only in 3 criteria does it meet a level A, on the other hand, the GADMA website it is the one that reports the most problems with a value of 90 problems, and 162 warnings.

Once the TAW tool has been applied to the four websites, it has been detected that the greatest number of problems are found in the Operable principle, two websites are the ones that report the highest values, the GADMA website with 50 problems, and the website web of the IESS with 26, this principle refers to the components of the user interface and navigation must be operable using a keyboard, a mouse or a technology that allows the visitor to navigate and find the information within the web-site. Regarding the warnings, the principle that presents the most errors is Perceptible, where the four sites present high values, which implies that work should be done mainly on color contrast, images with descriptions, buttons, and audiovisual content.

After the analysis with the TAW tool, it has been found that the criteria that are repeated with more problems in the four websites are the following: In the Principle, Perceivable are criterion 1.1.1 - Non-text Content, 1.3.1 - Info and Relationships. In the Operable principle is found the criterion 2.4.4 - Link Purpose (In Context). In the Principle, Understandable the criteria that are repeated are 3.3.2 - Labels or Instructions. And finally, in the Robust principle, the criterion that is repeated is 4.1.2 - Name, Role, Value, It is worth mentioning that within the findings it is also found that the criterion 3.1.1 - Language of Page, is satisfactory on three websites, and as well as the criteria 1.3.1 - Info and Relationships and 3.3.2 - Labels or Instructions.

Finally, there is still a lot of work to be done to develop accessible websites. Programmers and designers need to know the standards required to make inclusive web-sites. of web design and programming, of web design and programming, introduce concepts of accessibility within their curricula to train professionals

Proof. Proofs, examples, and remarks have the initial word in italics, while the following text appears in normal font.

For citations of references, we prefer the use of square brackets and consecutive numbers. Citations using labels or the author/year convention are also acceptable. The following bibliography provides a sample reference list with entries for journal articles [1], an LNCS chapter [?], a book [?], proceedings without editors [?], and a homepage [?]. Multiple citations are grouped [1, ?,?], [1, ?,?,?].

References

1. Chanchí, G.E.G., Campo, W.Y.M., Pérez-Medina, J.: Defini-tion of minimum accessibility criteria for the construction of web applications, RISTI - Revista Iberica De Sistemas e Tecnologias De Informacao, pp. 424–436 (2021)
2. Castro J.L.F., Normand, L.M.: Accesibilidad Web. TRANS. Rev. Traductología (11), 135–154 (2007). https://doi.org/10.24310/TRANS.2007.v0i11.3103
3. Campoverde-Molina, M., Luján-Mora, S., Valverde, L.: Análisis de accesibilidad web de las universidades y escuelas politécnicas del Ecuador aplicando la norma NTE INEN ISO/IEC 40500:2012, Web Accessibility Analysis of the Universities and Polytechnic Schools of Ecuador applying the standard NTE INEN ISO/IEC 40500:2012, ago. (2019). Accedido: 30 de julio de 2022. [En línea]. Disponible en: http://rua.ua.es/dspace/handle/10045/99754
4. W3C, Web Content Accessibility Guidelines (WCAG) 2.0. https://www.w3.org/TR/WCAG20/. (accedido 30 de julio de 2022)
5. Francisco, V.: Garcés, Enrique; Pailiacho, Verónica, [No title found], en ACTAS DEL VI CONGRESO INVESTIGACIÓN, DESARROLLO E INNOVACIÓN DE LA UNIVERSIDAD INTERNACIONAL DE CIENCIA Y TECNOLOGíA IDI - UNICyT 2021, Panamá, ene, p. 809 (2022)
6. Agüero, A.L., Guzmán, A.E., Gramajo, S.C., Varas, V.D.: Beneficios e implementación de accesibilidad web en la plataforma EVA UNLaR, Virtu@lmente, vol. 5, no. 1, Art. no. 1, (2017). https://doi.org/10.21158/2357514x.v5.n1.2017.1863
7. Asamblea Nacional del Ecuador, LEY ORGÁNICA DE DISCAPACIDADES (2012). Accedido: 30 de julio de 2022
8. Acosta, T., Luján-Mora, S., Acosta, T., Luján-Mora, S.: Análisis de la accesibilidad de los sitios web de las universidades ecuatorianas de excelencia, Enfoque UTE, vol. 8, pp. 46–61 (2017). https://doi.org/10.29019/enfoqueute.v8n1.133
9. Gonçalves, R., Martins, J., Pereira, J.: Web accessibility: Portuguese web accesibility with WCAG-1.0 and WCAG-2.0. https://n9.cl/i3tfi. (20 de septiembre de 2022)
10. Baldiris, S., Mancera, L., Vargas, D., Velez, G.: Accessibility eval-uation of web content that support the mathematics, geometry and physics's teaching and learning. https://n9.cl/n5pya (20 de septiembre de 2022)
11. Mariño, S.I., Alfonzo, P. L.: Evaluación de la accesibilidad web. Una mirada para asegurar la formación en la temática, Campus Virtuales, vol. 6, no. 2, Art. no. 2 (2017)
12. Rodríguez, Y.S., Pérez, L.B., Calderón, E.Á., Anlas, C.A.S.: Estado de la accesibilidad web de los portales de gobierno electrónico en América Latina, Bibliotecas. Anales de investigación, vol. 16, no. 1, Art. no. 1 (2021)
13. Naranjo-Villota, D., Guaña-Moya, J., Acosta-Vargas, P., Muirragui-Irrazabal, V.: Evaluación de la accesibilidad web en institutos acreditados de educación superior del Ecuador, Revista ESPACIOS, vol. 41, no. 04 (2020). [On line]. http://revistaespacios.com/a20v41n04/20410405.html
14. Balsells, L.A.C., González, J.C.G., Balsells, M.A.C., Chamorro, V.A.P.: La accesibilidad de los portales web de las universidades públicas andaluzas, Revista Española de Documentación Científica, vol. 40, no. 2, Art. no. 2 (2017). https://doi.org/10.3989/redc.2017.2.1372
15. Ortiz Ruiz, Y.T.: Accesibilidad en sitios web del Ministerio de Educación de Chile, Tendencias pedagógicas, (2019). https://doi.org/10.15366/tp2019.33.008

16. Campoverde Molina, M.: La accesibilidad web. Un reto en el entorno educativo ecuatoriano, RCTU-UPSE, vol. 3, no. 3, pp. 90-98 (2016). https://doi.org/10.26423/rctu.v3i3.172
17. Moreira-Cevallos, C.L.: Accesibilidad web en universidades de Manabí para usuarios con discapacidad visual según norma NTE INEN-ISO/IEC 40500:2012, Revista Científica de Informática ENCRIPTAR - ISSN: 2737-6389., vol. 2, no. 3, Art. no. 3 (2019)
18. Ochoa-Urrego, R.: índice de accesibilidad para cibermedios mexicanos, Revista Española de Documentación Científica, vol. 42, no. 3, Art. no. 3 (2019). https://doi.org/10.3989/redc.2019.3.1541
19. Pagnoni, V., Mariño, S.I.: Calidad de contenidos en dominios de educación. Evaluación de la Accesibilidad Web mediada por validadores automáticos, EDMETIC, vol. 8, no. 1, Art. no. 1, (2019). https://doi.org/10.21071/edmetic.v8i1.10221
20. Villagra, M., Falcó, L.: Accesibilidad en los portales web de las universidades paraguayas, REVISTA PARAGUAYA DE EDUCACIÓN A DISTANCIA (REPED), vol. 3, no. 2, Art. no. 2 (2022)
21. Serrano Mascaraque, E.: Accesibilidad vs usabilidad web: evaluación y correlación, IB, vol. 23, no. 48 (2010). https://doi.org/10.22201/iibi.0187358xp.2009.48.16970
22. Ministerio del Trabajo - Ecuador. https://www.trabajo.gob.ec/. (accedido 30 de julio de 2022)
23. IESS - INSTITUTO ECUATORIANO DE SEGURIDAD SOCIAL. https://www.iess.gob.ec/. (accedido 30 de julio de 2022)
24. Portal intersri, Servicio de Rentas Interna. https://www.sri.gob.ec/web/intersri/home. (accedido 30 de julio de 2022)
25. GADMA, GAD Municipalidad de Ambato, GAD Municipalidad de Ambato. https://ambato.gob.ec/. (accedido 30 de julio de 2022)
26. Instituto Ecuatoriano de Normalización, NTE INEN-ISO/IEC 40500 TECNOLOGíA DE LA INFORMACIÓN - DIRECTRICES DE ACCESIBILIDAD PARA EL CONTENIDO WEB DEL W3C (WCAG) 2.0 (2014)
27. Servicio Ecuatoriano de Normalización, REGLAMENTO TÉCNICO ECUATORIANO RTE INEN 288 "ACCESIBILIDAD PARA EL CONTENIDO WEB" (2016). https://www.normalizacion.gob.ec/buzon/reglamentos/RTE-288.pdf
28. Montalvo, W., Ibarra-Torres, F., Garcia, V, M., Barona-Pico, V.: Evaluation of whatsapp to promote collaborative learning in the use of software in university professionals. In: Applied Technologies (ICAT 2019), PT II. Communications in Computer and Information Science, vol. 1194, pp. 3-12 (2020). https://doi.org/10.1007/978-3-030-42520-3_1
29. TAW — Servicios de accesibilidad y movilidad web. https://www.tawdis.net/. (accedido 30 de julio de 2022)

Failure of Tech Startups: A Systematic Literature Review

José Santisteban[1], Vicente Morales[2(✉)], Sussy Bayona[3], and Johana Morales[4]

[1] Universidad Privada Norbert Wiener, Av. Arequipa 440 con Jr. Larrabure y Unanue 110. Urb. Santa Beatriz, Lima 15056, Peru
jose.santisteban@uwiener.edu.pe

[2] Universidad Técnica de Ambato, Los Chasquis y Río Payamino, Ambato 180103, Ecuador
jvmorales99@uta.edu.ec

[3] Universidad Autónoma del Perú, Autopista Panamericana Sur Km 16.3, Lima 15842, Peru

[4] Universidad de las Fuerzas Armadas - ESPE, Av. Gral. Rumiñahui S/N, Sangolquí 171103, Ecuador
mmorales0011@uta.edu.ec

Abstract. Tech Startups are exposed to multiple challenges and are known for being inserted into uncertain and risky scenarios. Initially, new businesses face great uncertainty and have high failure rates, but a minority of them go on to become successful and influential. The purpose of this study is to determine the main causes of failure of Tech Startups in their early stage, for which a systematic review of empirical studies regarding why Tech Startups fail was carried out. The search strategy in the databases: ScienceDirect, IEEE Xplore, SpringerLink, Emerald and EBSCO identified 1996 studies of which 36 were identified as empirical studies and after applying the inclusion and exclusion criteria, 23 primary studies were selected that classify to the factors in 3 categories: organizational, technological, and environmental. Among the main factors for failure are the characteristics of the owner, poor location of the business, products/services that do not meet the needs, high costs of ICT, lack of skills in the entrepreneurial team, external and competitive pressure, few benefits perceived ICT use, low resources, low government support.

Keywords: Tech Startups · Early-Stage Startups · Factors Failure

1 Introduction

Tech startups are exposed to multiple challenges. The discussion about entrepreneurship associated with the phenomenon of new companies seems to have taken on a character of novelty as something new. In fact, as pointed out by Vivas [26], the concepts of entrepreneurship and the entrepreneur himself have been studied for a long time.

M. V. Garcia and C. Gordón-Gallegos (Eds.): CSEI 2022, LNNS 678, pp. 111–126, 2023.
https://doi.org/10.1007/978-3-031-30592-4_9

Although Tech Startups are important for the development of the economies of the countries [6,23,25], these types of companies have a high "mortality rate" [10,22,27]. According to Van Gelderen [4,9], Tech Startups are those temporary organizations that focus on the creation of high technology and innovative products, with little or no history of operations, with the aim of aggressively growing their businesses in highly scalable markets. Tech Startups around the world enable the rise of new markets, accessible technologies, and venture capital [1,21].

Despite many success stories, many Tech Startups fail before they have fulfilled their commercial potential [8,20,21]. However, the failures of Tech Startups receive little attention, despite the rapid proliferation of Startup communities, which have been able to learn how to build a Startup [2,21][?]. On the other hand, more than 90% of Tech Startups fail, mostly due to self-destruction rather than competition [5]. According to Ng [16], the failure of Tech Startups has been determined from the point of view of their CEOs, who have broad perspectives on their Startup company. The results show that Tech Startups did not follow coherent strategies to understand the problem they were trying to solve [20], thus diluting their focus on running in the wrong direction [21].

With this approach, this study will identify the main causes of the failure of Tech Startups, so that in the future it can help mitigate these risks to maximize the opportunities behind each business action of Tech Startups. This systematic review aims to assess, synthesize, and present the empirical results on the main causes of failure of Tech Startups to date, and provide an overview about Tech Startups, its findings, conclusions and implications for research and practice. We believe this overview will be important for practitioners who want to stay up to date with the state of research. This opinion will also help the scientific community and the State that works with the growth and development of Startups. The results of this research will be relevant for the development of technological entrepreneurship in developing countries.

This article is organized into five sections. Section two describes the methodology used in the present study. Section three presents the results of the review. Section four shows the discussion of the study. Finally, in section five, the conclusions are shown.

2 Methodology

In this section, the development of the review protocol, the identification of inclusion and exclusion criteria, a search for relevant studies, the evaluation of quality criteria, data extraction and synthesis were carried out.

2.1 Protocol Planning

An outline of the review can be seen in Table 1 illustrating the planning, conducting, and reporting processes on a timeline and the results produced as part of each process.

The planning activity deals with developing the review protocol as well as deciding how researchers should interact and work to carry out the systematic review. In addition, the improvements in the review process are shown. The overview activity reflects the completion of the steps taken in the process of conducting the systematic review. The reporting activity shows how the pilot report and the final report evolved. Finally, the results are described in terms of protocols, forms, and how the number of relevant papers changed as the systematic review process progressed

Table 1. Systematic Review Planning [Author]

Planning	Realization	Reports	Results
Protocol development			Review protocol
	Data recovery		Repository with articles
	Selection of title studies		
	Selection of studies on abstracts		
	Consensus review		Primary studies reviewed
	Data extraction pilot, 3 papers (all)		3 articles reviewed
Process Improvement			Draft data extraction form
	Examine reviewed papers		
	Data extraction pilot, 10 articles		13 papers reviewed
			Refining of the data extraction form
	Examine reviewed papers		
	Data extraction pilot, 5 articles		23 articles reviewed
	Data synthesis pilot		
		Report pilot	
Data synthesis			

First, the systematic review protocol was developed, including research questions, search strategy, evaluation, inclusion/exclusion criteria, data extraction form, and synthesis methods. The protocol was revised and refined in iterations after piloting each of the related revision steps. The second activity of conducting the review breaks down into four main steps: data retrieval, study selection, data extraction, and data synthesis.

2.2 Protocol Development

A protocol for the systematic review of the literature has been developed, as can be seen in Fig. 1. With this protocol, the search for primary articles that are directly related to the research problem can be carried out.

2.3 Data Recovery

During data retrieval, the limits of the systematic review were established. First, the keywords for the search were selected. These aimed at finding empirical research on Tech Startups and the reasons why they fail.

On the other hand, to search in full text he made sure to cover all the synonyms that exist about Tech Startups. The papers were chosen if they reported on any type of empirical evidence related to the reasons for the failure of Tech Startups. The final search strings are based on the experience of the experimental searches and consisted of a Boolean expression: (A1 or A2 or A3 or A4) and (B1 or B2 or B3 B4 OR), where:

A1 - Tech Startups B1 - Fail
A2 - New Technology-Based Ventures B2 - Early-Stage
A3 - New Venture B3 - Software
A4 - Venture Firms B4 - Case study

The results of the search in the Databases are shown in Table 2. Works published before 1990 were not included in the search, since studies carried out after 1990 are considered more relevant.

Table 2. Systematic Review Planning [Author]

Data banks	Results
Springer Link	4
Emerald	4
IEEE Xplore	3
ScienceDirect	6
EBSCO	6
Total Link	23

The general search resulted in a high proportion of papers. Therefore, it was not enough to use the search strings as the only criteria to decide whether to include or exclude a specific paper. Therefore, the researchers jointly decide the limits based on agreed criteria. This is also one of the reasons why it is crucial that several researchers are involved in a systematic review. All search results were carefully documented. According to Fig. 2, in recent years the interest of researchers has been growing in knowing the factors that lead to the failure of Tech Startups.

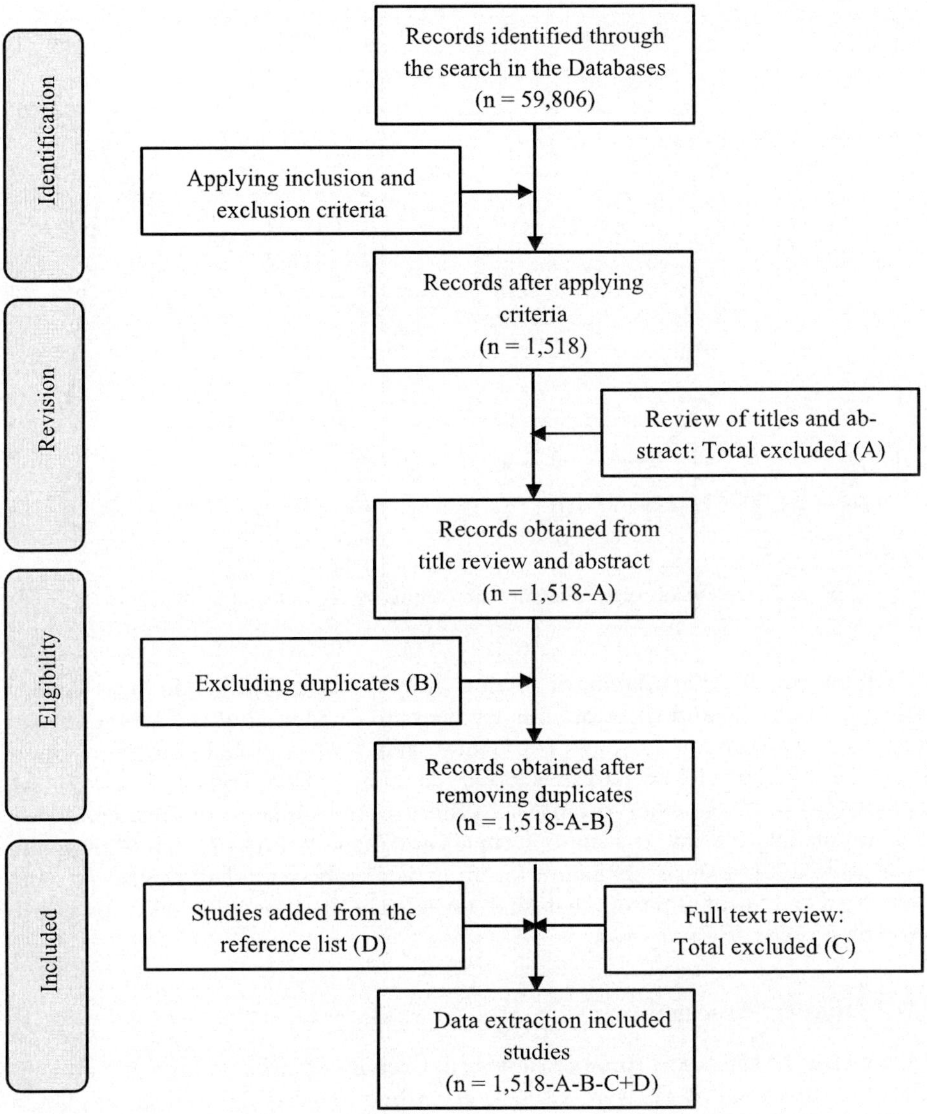

Fig. 1. Flow of the search process [Author].

2.4 Study Selection

The objective of the study selection process was to identify the relevant articles for the objectives of the systematic review according to the scope of the research. The search strings, analyzed above, were quite broad and therefore it was to be expected that not all the studies identified would make it to the final stage of the selection process.

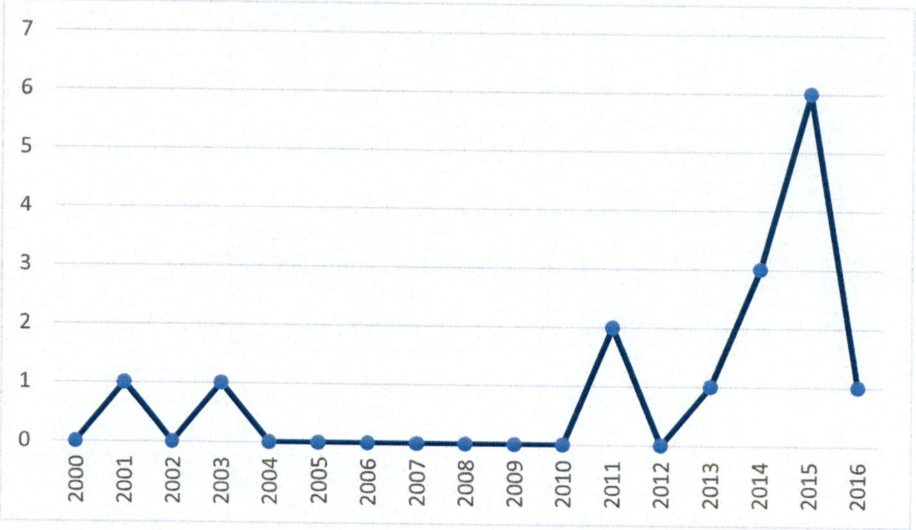

Fig. 2. Evolution of studies on the failure of tech startups [Author].

All papers in the systematic review have been published in peer review indexed databases and therefore quality is assured. Several of the articles found refer to Tech Startups business companies that have a global context of operations in both developed and developing countries. These types of articles are included in the study, since, despite the different environments of these countries, it is important to know the reasons that lead to the failure of Tech Startups in both contexts. On the other hand, for an article to be included, the study must be carried out in an environment that reveals the reasons why Startups fail in their first years of life.

2.5 Quality Assessment

Each of the 23 selected primary studies that remained after applying the inclusion and exclusion criteria were assessed according to 6 quality criteria. To choose these criteria we rely on Kitchenham's study [12] for the principles of good practice for conducting empirical research.

These criteria are:

- Is the article based on research (or is it a learned injury study or report based on expert opinion)?
- Are the objectives of the research clear?
- Is there an adequate description of the context in which the research was carried out?
- Was the research design appropriate to reach the research objectives?
- Was the data analysis rigorous enough?
- Is there a clear statement of the research results?

These criteria include 3 important issues, related to quality, which were considered for this review:

- Rigor.
- Credibility.
- Relevance.

For the evaluation of each of the 6 criteria, it was classified on a dichotomous scale ("yes" or "no"), receiving a score of "1" for "yes" and "0" for "no".

2.6 Data Extraction

For data extraction, a template was used to facilitate the synthesis process of the collected data. The template evolved from analysis of a selected segment of the data being carried out as a pilot prior to the systematic review conducted in this article. The template includes data capture regarding the empirical relevance and focus of the submitted study. In addition, a qualitative evaluation of the articles was made to compare the point of view of the principal investigator and the supporting investigator to carry out the review.

All articles obtained were evaluated by the principal investigator and the supporting investigator. The extracted data was documented using the template and then the MS Excel tool was used, which was additionally used in the data synthesis phase. Table 3 shows the search string used in the Data Banks to obtain the 23 primary articles (Search performed on 04/30/2021).

2.7 Data Synthesis

The data extracted from the reviewed articles were analyzed to answer the research questions. The classification form used in the data extraction helped for the categorization of data with respect to the study population, empirical background, methodology used, year, type of study company and the results of the studies.

To illustrate the data, we use bar charts supplemented with references to the included papers. Finally, it should be noted that although the 36 articles were judged relevant to the objectives of the systematic review, since the articles are related to the research problem. Some of these 36 articles were classified as irrelevant to the specific investigation. For example, articles reporting studies of Tech Startups in developed and developing countries were considered relevant when discussing the reasons why they fail.

3 Results

The results of the systematic review based on the 23 papers finally selected are presented below.

Table 3. Systematic Review Planning [Author]

Databases	Search string	Found papers	Papers with inclusion and exclusion criteria	Selected papers
Springer Link	'Factors AND Fail AND Startups AND Small Business AND New Technology-Based Firm AND Early-Stage'	795	31	4
IEEE Xplore Digital Library	(((("Document Title": Startups) AND "Document Title": Factors) AND "Document Title": Fail) OR "Document Title": New Technology-Based Firms)	1633	15	3
ScienceDirect	TITLE-ABSTR-KEY (Factors Fail Startups) or TITLE-ABSTR-KEY (Fail Early-Stage Startups)	156	39	6
ScienceDirect	Tech Startups AND Factors AND Fail OR New Technology-Based Firms OR Fail New Firm OR Fail Small Business	230	64	6
			Selected Papers	23

3.1 Overview of Studies

Regarding the research methods of the Studies, according to the results shown in Fig. 3, the most used method is the qualitative method, followed by the quantitative method and others that used multiple case studies.

Regarding the databases of publications with more studies on the failure of Tech Startups, it is observed that EBSCO and Springer Link are the ones that host more studies regarding this problem, as observed in Fig. 4.

The Fig. 5 shows that among the countries where there are more studies dealing with the reasons for failure of Tech Startups are the US, Italy, Australia, and Brazil.

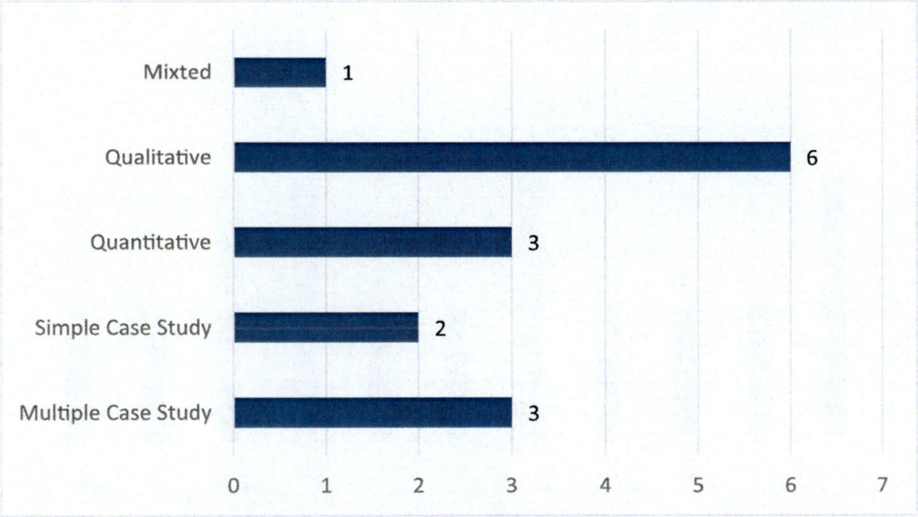

Fig. 3. Studies conducted by research methods [Author].

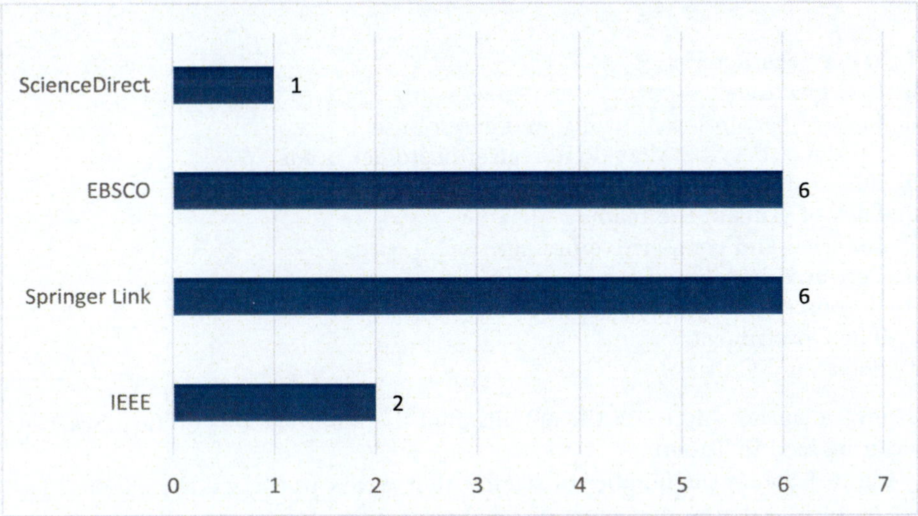

Fig. 4. Database with studies on the failure of tech startups [Author].

3.2 Factors of Tech Startup Failure

Eleven main factors that influence the failure of Tech Startups were identified. These factors have been studied in different studies and their influence has been empirically demonstrated, which are shown below and in Table 4.

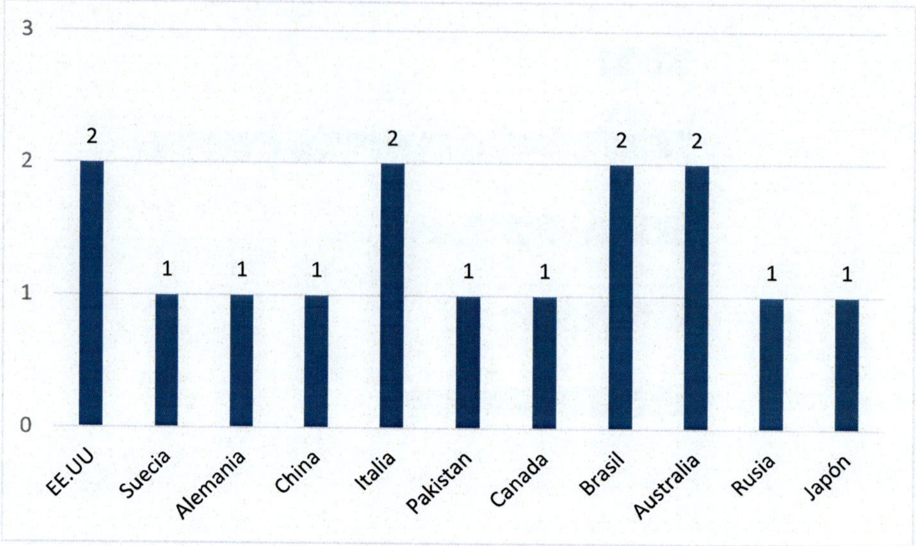

Fig. 5. Studies on the failure of tech startups by country [Author].

1. Owner characteristics
2. Bad location
3. Lack of perception of utility and ease of use
4. Products/Services that do not meet customer needs
5. ICT costs
6. Lack of skills in the team
7. External and competitive pressure
8. Perceived benefits of the use of ICT
9. Resources
10. The Government
11. Social and Cultural

Several authors agree on the definition of the failure factors of Tech Startups as can be seen in Table 5.

Figure 6 shows the number of studies that have categorized their failure factors of Tech Startups, and most of them correspond to the organizational category.

It is important to point out that, of the 11 factors identified, there are 2 factors for which all the studies consulted include in their investigations. These factors are those related to government support and the factors that have to do with products/services that do not meet the needs of the market. Figure 7 shows the number of times the factors have been mentioned in the studies.

3.3 Limitations of This Review

The keywords and search terms that allowed us to identify the existing literature have been identified. However, it is important to recognize that Tech Startup

Table 4. Startup failure factors [Author]

References														
Factor No	[21,23]	[7]	[27]	[10]	[9]	[18]	[1]	[19]	[5]	[16]	[19]	[13]	[13,20,22]	[14]
1	✓	✓	✓		✓		✓	✓		✓	✓		✓	
2			✓		✓							✓		
3	✓					✓	✓					✓		
4	✓	✓	✓	✓	✓	✓	✓	✓	✓	✓	✓	✓	✓	✓
5	✓	✓	✓	✓			✓		✓		✓		✓	✓
6	✓			✓		✓	✓		✓					
7			✓		✓		✓		✓			✓		✓
8			✓	✓	✓		✓			✓	✓	✓	✓	✓
9		✓			✓			✓	✓				✓	
10	✓	✓	✓	✓			✓		✓		✓		✓	✓
11		✓				✓	✓	✓		✓		✓	✓	✓

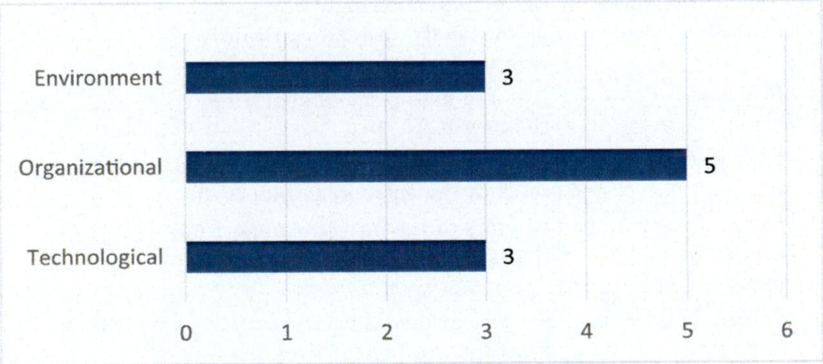

Fig. 6. Categorization of the failure factors of tech startups [Author].

keywords are not standardized, therefore due to our choice of keywords and search strings per database, there is a risk that relevant studies have been missed. Also, as this study focuses on empirical research, lessons learned articles and articles based solely on expert opinion were excluded.

Many articles lack sufficient information for us to be able to document them satisfactorily in the data extraction form. More specifically, we frequently found that methods were not adequately described, that bias and validity issues were not always addressed, that data collection and analysis methods were often not well explained. Therefore, there is a possibility that the extraction process may be the result of some inaccuracy in the data.

Table 5. Definitions of Startup Failure Factors [Author]

No.	Owner Characteristics	Definition	Reference
1	Owner Characteristics	Aspects that stand out about the owners of Startups, such as: gender, age, education, knowledge, abilities and skills, etc.k	[1,7,21,27]
2	Bad location	Lack of attention to providing effective communication resources would have led companies to a lack of planning	[13,21]
3	Lack of perception of usefulness and ease of use	Absence of the perception of usefulness and ease of use that the owners of Tech Startups have	[13,21]
4	Products/Services that do not meet Customer needs	Ignore the wants and/or needs of customers, knowingly or unknowingly, instead constantly seeking customer feedback	[1,7,20,27]
5	ICT costs	It is the high price of acquiring the new ICT	[1,7,27]
6	Lack of team skills	Lack of business and technological skills of the entrepreneurial team	[1,6,21]
7	External and competitive pressure	It refers to the intense competition that startups experience every day	[13,21,27]
8	Perceived benefits of ICT use	Set of expected advantages that Startups can benefit from having adopted ICT	[1,13,27]
9	Resources	They are the financial, technical, management and information resources of the Startup	[7,21]
10	The Government	Financing support provided by government institutions in favor of the Startup	[1,7,13,21,27]
11	Social and Cultural	Are all those trends that can affect the future of the company	[15,21,27]

4 Discussion

This systematic review shows that there are many more empirical studies on startup failure factors in general than have been previously recognized. This review used an explicit search strategy combined with inclusion and exclusion criteria. This study identified 22 studies that deal with the failure factors in Tech Startups, in which we found that several authors have discrepancies when

Fig. 7. Failure factors of tech startups included in the research [Author].

deciding whether the factor is included in the failure of Tech Startups. We now turn to our research question, beginning by discussing what we found regarding factors that many authors disagree with. Regarding the characteristics of the owner in the study of [5,21,27], they indicate that this factor is a facilitator for Startups to be successful in their operations, however [7,23]conclude that the characteristics of the owner do influence the moment of the failure of the Startup, since they argue that the owners do not have solid knowledge of how to run a company, much less of IT management. Regarding the factor of poor location in the investigations of [3,10,13,18], they affirm that they influence the failure of Tech Startups.

The lack of perception of usefulness and ease of use has been widely studied and authors such as [1,23] indicate that the owner, not knowing that ICT can help the survival of his company, simply does not use it, and therefore it fails. Regarding the products and/or services offered by Startups, there is broad agreement in the studies pointing out that, if the Startup does not offer a quality product, it will gradually lose market niche and consequently fail [11,15,22].

The cost of ICT is the factor by which studies such as [17,19] show that the high price of acquiring ICT that supports the operation of Tech Startups, makes them lose the battle against large companies and therefore their prices fall sales until closing its operations due to low employment in the market. In the studies of [1,10,16], the lack of skills in the team is considered in several investigations, which indicate that, if the team ignores the use of ICT, the competition that uses ICT will be taking advantage. External and competitive pressure makes Tech Startups keep up with Technologies, but if they are not aligned with the rest, it is a fact that they will tend to fail [5,27]. On the other hand, not perceiving the benefits of the use of ICT means that the Startup does not adopt ICT in its business and, consequently, it could not operate optimally until it fails [1,10,27].

It is well known that Startups, being small companies, do not have enough financial resources to survive at least the first 5 years. This is what the studies of [1,7,9] do. Government support through regulations or laws helps the Startup to operate, survive and compete with medium or large companies [15,21,24]. Finally, the social factor and culture, according to [1,7,18] state that the culture of a country can make businesses prosper, due to the type of entrepreneurship found, or they can fail in the attempt.

5 Conclusions

1990 primary studies dealing with the failure factors of Tech Startups were identified from the literature search, of which 36 turned out to be research studies of acceptable rigor, credibility, and relevance. 30 of the 36 identified studies were primary studies, while 6 were secondary studies. The Failure Factors of Tech Startups identified in the investigations are 11, of which 4 studies make a classification of factors: technological, organizational, and environmental. All 23 studies agreed that the factors of product/service not meeting customer needs and government support are the most influential for Tech Startups to fail. Through the literature search, in South America the country where more research is being carried out regarding Tech Startups is Brazil and in North America it is the US. Both countries are very interested in supporting this type of company, since it allows to improve its GDP, in addition to generating more employment and development. There is a need to increase both the number and the quality of studies in development of Tech Startups. In this context, there is a clear need to establish a research center to support the development and survival of the Startup.

In this study, it has been shown that Tech Startups fail in their first years of existence due to the lack of knowledge of mechanisms that allow the existence, growth and sustainability of this type of company. In the case of the Peruvian government, for more than 8 years it has been developing initiatives to provide non-reimbursable financing to Tech Startups and training that allow their leaders to improve their business management. Many Tech Startups fail because the members of their founding team do not have the technological and entrepreneurial skills to sustain the growth of the company. Part of these shortcomings is the fact that they do not adapt to the constant change of the market where they operate, a very important characteristic that has led to the success of companies that knew how to adapt to the changing environment of their target market.

Future Works. As future work based on the factors identified in this study, an IT tool should be developed to predict the failure of startups, as well as to propose strategies to mitigate the factors of failure.

Acknowledgments. The gratitude for the support provided in the research to the Technical University of Ambato, Research and Development Directorate (DIDE), and to the Norbert Wiener Private University.

References

1. Al Sahaf, M., Al Tahoo, L.: Examining the key success factors for startups in the kingdom of Bahrain. Int. J. Bus. Ethics Gov. **4**(2), 9–49 (2021)
2. Bajwa, S.S., Wang, X., Nguyen Duc, A., Abrahamsson, P.: "failures" to be celebrated: an analysis of major pivots of software startups. Empir. Softw. Eng. **22**(5), 2373–2408 (2017)
3. Cantamessa, M., Gatteschi, V., Perboli, G., Rosano, M.: Startups' roads to failure. Sustainability **10**(7), 2346 (2018)
4. Cepeda Vaca, F., et al.: Controlled high pressure grinding roll by model predictive control. In: 2017 IEEE 3RD Colombian Conference on Automatic Control (CCAC) (2017)
5. Chong, Z., Luyue, Z.: The financing challenges of startups in China (2014)
6. Escolano, V.J.C.: Successes and failures of startups in the Philippines: an exploratory study. In: 2022 7th International Conference on Business and Industrial Research (ICBIR), pp. 610–615. IEEE (2022)
7. Finkelstein, S.: Internet startups: so why can't they win? J. Bus. Strateg. **22**(4), 16–16 (2001)
8. Galleguillos, R., Altamirano, S., Garcia, M.V., Perez, F., Marcos, M.: Low cost CPPs for industrial control under FAHP algorithm. In: 2017 22ND IEEE International Conference on Emerging Technologies and Factory Automation (ETFA). IEEE International Conference on Emerging Technologies and Factory Automation-ETFA (2017)
9. van Gelderen, M., Thurik, R., Bosma, N.: Success and risk factors in the pre-startup phase. Small Bus. Econ. **24**(4), 365–380 (2005)
10. Hyder, S., Lussier, R.N.: Why businesses succeed or fail: a study on small businesses in Pakistan. J. Entrep. Emerg. Econ. (2016)
11. Kalyanasundaram, G.: Why do startups fail? A case study based empirical analysis in Bangalore. Asian J. Innov. Policy **7**(1), 79–102 (2018)
12. Kitchenham, B.A., et al.: Preliminary guidelines for empirical research in software engineering. IEEE Trans. Software Eng. **28**(8), 721–734 (2002)
13. Ko, C.R., An, J.I.: Success factors of student startups in Korea: from employment measures to market success. Asian J. Innov. Policy **8**(1), 97–121 (2019)
14. LeBrasseur, R., Zanibbi, L., Zinger, T.J.: Growth momentum in the early stages of small business start-ups. Int. Small Bus. J. **21**(3), 315–330 (2003)
15. Nam, G.J., Lee, D.M., Chen, L.: An empirical study on the failure factors of startups using non-financial information. Asia-Pacific J. Bus. Ventur. Entrep. **14**(1), 139–149 (2019)
16. Ng, A.W., Macbeth, D., Southern, G.: Entrepreneurial performance of early-stage ventures: dynamic resource management for development and growth. Int. Entrep. Manag. J. **10**(3), 503–521 (2014). https://doi.org/10.1007/s11365-014-0303-x
17. Okrah, J., Nepp, A., Agbozo, E.: Exploring the factors of startup success and growth. Bus. Manag. Rev. **9**(3), 229–237 (2018)
18. Onetti, A., Pepponi, F., Pisoni, A.: How the founding team impacts the growth process of early stage innovative startups. Sinergie Italian J. Manag. **33**(May-Aug), 37–53 (2015)

19. Prohorovs, A., Bistrova, J., Ten, D.: Startup success factors in the capital attraction stage: founders' perspective. J. East-West Bus. **25**(1), 26–51 (2019)
20. Santisteban, J., Inche, J., Mauricio, D.: Critical success factors throughout the life cycle of information technology start-ups. Entrep. Sustain. Issues **8**(4), 446 (2021)
21. Santisteban, J., Mauricio, D.: Systematic literature review of critical success factors of information technology startups. Acad. Entrep. J. **23**(2), 1–23 (2017)
22. Santisteban, J., Mauricio, D., Cachay, O., et al.: Critical success factors for technology-based startups. Int. J. Entrep. Small Bus. **42**(4), 397–421 (2021)
23. Skaleckii, E.V., Nagapetyan, A.R., Khamdamov, J.K., Glupak, A.S.: Problems of surviving and growth of new ventures: literature review. Mediterr. J. Soc. Sci. **7**(2), 212 (2016)
24. Triebel, C., Schikora, C., Graske, R., Sopper, S.: Failure in startup companies: why failure is a part of founding. In: Kunert, S. (ed.) Strategies in Failure Management. MP, pp. 121–140. Springer, Cham (2018). https://doi.org/10.1007/978-3-319-72757-8_9
25. Vimos, V.H., Benalcázar, M., Oña, A.F., Cruz, P.J.: A novel technique for improving the robustness to sensor rotation in hand gesture recognition using sEMG. In: Nummenmaa, J., Pérez-González, F., Domenech-Lega, B., Vaunat, J., Oscar Fernández-Peña, F. (eds.) CSEI 2019. AISC, vol. 1078, pp. 226–243. Springer, Cham (2020). https://doi.org/10.1007/978-3-030-33614-1_16
26. Vivas, G.V., Pazos, J.S., Tito, V.B., Ordoñez, M.C.: Revisión de modelos para identificar los factores de adopción de tic en pymes. Revista Ibérica de Sistemas e Tecnologias de Informação **E40**, 496–511 (2021)
27. Yin, Y., Wang, Y., Evans, J.A., Wang, D.: Quantifying the dynamics of failure across science, startups and security. Nature **575**(7781), 190–194 (2019)

Augmented Reality System as a 5.0 Marketing Strategy in Restaurants: A Case Study in Ambato Ecuador

Pablo-R. Paredes(✉) and Leonardo-Gabriel Ballesteros-Lopez

Universidad Tecnica de Ambato, UTA, Ambato 180103, Ecuador
{pparedes2177,lg.ballesteros}@uta.edu.ec

Abstract. The Covid-19 pandemic affected several productive sectors. However, the tourism sector was and continues to be one of the most damaged. Companies were forced to evolve technologically in order to cope with this condition. More than 90% of businesses lost their investment, as well as their human resources. After more than two and a half years after this crisis's beginning, establishments are still looking for a solution to their financial debacle. This article presents the development and implementation of an augmented reality system as a 5.0 marketing strategy focused on a restaurant in the city of Ambato, Ecuador. The results showed encouraging data. The evaluation of user experience through the System Usability Scale showed a value above 70. The comparison between before and after the implementation of this system showed a percentage increase in sales of 47.45%. Finally, the T-student test for independent samples was used to verify the existence of a statistically significant difference.

Keywords: Marketing 5.0 · Augmented Reality · Covid-19 · Food industry

1 Introduction

Marketing 5.0 refers to applying techniques that mimic and even enhance human abilities to market a product or service [10]. DWithin this context, what is known as "next tech" appears as a group of technological tools that, together with the skills of a human being, generate an efficient and updated approach to marketing. Among the technologies present in marketing 5.0, we can mention artificial intelligence (AI), the internet of things (IoT), augmented reality (AR), and virtual reality (VR) [13].

Of the technologies mentioned above, AI has been the one that has taken up most of the marketing field. Several world-renowned companies such as AB InBev, Chase, and Lexus have taken this generational leap and reduced human participation in developing their advertising [5,23]. AR is not far behind. Although its understanding and development may seem challenging for marketers, its accessibility and benefits have involved them in an unprecedented

M. V. Garcia and C. Gordón-Gallegos (Eds.): CSEI 2022, LNNS 678, pp. 127–137, 2023.
https://doi.org/10.1007/978-3-031-30592-4_10

digital world. This is the case of IKEA and Sephora, which provide a unique experience to customers by allowing them to try products before they buy them [7,24].

Marketing 5.0 provides an unparalleled customer experience. Its impact has been such that, according to a survey conducted by Gartner, 89% of companies put customer experience as the cornerstone of business success, while 86% of buyers are willing to live an unprecedented practice in exchange for an above-average payment [6].

The impact of Covid-19 forced various sectors worldwide to reinvent themselves and look for current and low-cost ways of doing business. In this era of confinement, social networks came into a boom, not only as a means of entertainment and information but also as a tool to satisfy basic human needs through online shopping. At this point, small and medium-sized enterprises (SMEs) realized that they had to adapt to this reality, promoting their businesses, offering discounts, or simply creating a page to build customer loyalty [15].

In consequence, [12], separate the current pandemic into two phases, the first one inherent to the confinement and the second one corresponding to the recovery phase. Here, several strategies are mentioned. For example, in April 2020, a YouTube project known as Greece from Home was launched with the slogan: "It's time to stay home and take a break".

Professionals from all over the country, dedicated to art, gastronomy, fashion, music, and sports, published several short videos to encourage potential consumers to book their trips for when things stabilize. On the other hand, [11] poses an experiment where they study the digital marketing employed by certain SMEs after being hit by the coronavirus. The social networks analyzed include Instagram, Facebook, and WhatsApp. This research was conducted in Jakarta, Indonesia, where the sales reports of three domestic companies that specialized in selling traditional snacks were analyzed. The data was segmented according to the digital marketing tool used. The results showed that the social network with the highest consumer presence was Instagram.

Digital marketing is the best tool to reach potential customers. This was established by [2,4]. The objective of their research focused on analyzing how this tool has evolved and changed the pace of several establishments in the city of Barceló, Portugal. A qualitative and quantitative approach was employed where 106 respondents and three interviewees, who were marketing and tourism professionals, participated. It was concluded that digitizing advertising in global health crises, where human beings cannot have an everyday life and interaction, allows for better management, attracting and retaining customers, and increasing sales volume.

In Latin America, specifically in Ecuador, the coronavirus forced the modality of exchange of goods and services to change; in this country, people were used to going to any establishment in person to carry out a transaction. However, by learning the facilities and advantages that today's technological world offers, new consumer culture was adapted, and the use of marketing 5.0 became relevant. In addition, both the industrial sector and SMEs sought, according to their possibilities, new marketing strategies to help them recover from the devastating economic blow caused by this pandemic [1,8].

In this way, [16], establishes a mixed approach experiment, where the importance of digital marketing in Cuenca, Ecuador, was examined. Data obtained from the National Institute of Statistics and Census (INEC) and the Central Bank of Ecuador were used. The results show that, of the group surveyed, there was a percentage increase of 110% in online purchases, and this was done through text messages. The social network with the most significant impact in this study was WhatsApp.

Finally, [21], established the advantages of applying digital marketing in times of COVID-19 in Ecuador. Among the points to highlight is the ease of advertising a product through social networks without needing a physical establishment or offices. In addition, it is concluded that the 5.0 marketing tools are and will be a crucial point for SMEs' transformation and digital evolution.

For this reason, the problem that arises is the lack of knowledge of new technological tools by restaurants in the city of Ambato, as well as the use of conventional marketing strategies that, in the digitalized world in which we find ourselves, it is not enough to have a profitable business. The expected result of this research work is the optimization of service and increased sales during the COVID-19 pandemic, adapting to the organizational culture of the establishment under study. This will be achieved by applying a digital marketing strategy to social networks, which have become the primary means of information and communication during the health crisis the whole world is going through.

The importance of this research work is based on the use of new technologies to eradicate, to a certain extent, the negative effects of the COVID-19 pandemic in the tourism sector of the city of Ambato, specifically in the field of gastronomy. Therefore, the objective of this article is to develop an augmented reality system as a marketing strategy 5.0 in the service sector of the city of Ambato-Ecuador, to improve the presence in social networks of the establishment, acquire a portfolio of new customers, and, at the same time, to improve profits.

This article is divided into five sections, including the introduction. Section 2 presents the case study, i.e., the history and baseline of the facility under investigation. Section 3 details the AR system developed, while Sect. 4 shows the results. Finally, Sect. 5 presents the conclusions and future work.

2 Case Study

According to INEC's Business Directory, in 2018, the date of the latest registration, a total of 49918 companies engaged in the restaurant and mobile food service activities were registered. On the other hand, analyzing the geographical location of each of these companies, it was noted that 21.1% are located in Quito, 12.9% are located in Guayaquil, 5.7% are located in Cuenca, 3.3% are located in Ambato, and 2.6% are located in Santo Domingo [9].

As can be seen, Quito and Guayaquil are where most of the food establishments are concentrated; however, due to the scarce availability of economic and human resources, a convenience sampling was used, limiting the study to the province of Tungurahua, specifically the city of Ambato. Finally, since the application of marketing 5.0 technologies involves an arduous and extensive process of planning and implementation, this study focuses on a single restaurant to set a precedent for applying these methodologies in the service and tourism sector of the city.

The establishment under study specializes in all types of roasts and cuts of beef. Its owner, Uruguayan by birth but Ecuadorian at heart, is a former professional soccer player who found his passion in the kitchen once he retired from playing soccer. It has more than five years of trajectory, and, in addition, it has become a tourist attraction since different former glories of Ecuadorian and international sports meet in this restaurant to taste the exquisite dishes that are prepared.

On the other hand, for the development of this research, the following steps were used i) the marketing need was established: the establishment's owner only has a Facebook page, in which he makes sporadic publications, without a defined plan. ii) The appropriate product was chosen to promote it, in this case, a cut of meat: rib eye. iii) The Unity 3D graphics engine generated the environment, initially created to develop video games. iv) The AR system was evaluated through a survey known as the usability scale (SUS). Finally, v) the impact of the proposed marketing strategy was analyzed.

Figure 1, shows the digital marketing used before the pandemic by the restaurant under study. At this stage, there was simply a Facebook page created, with sporadic publications on topics not related to the business. The information about the establishment was not adequate. Location, contact number, and menu were not available. For this reason, the interaction with its customers was shallow. It is also worth mentioning that the profile and cover image did not represent the main idea of the business.

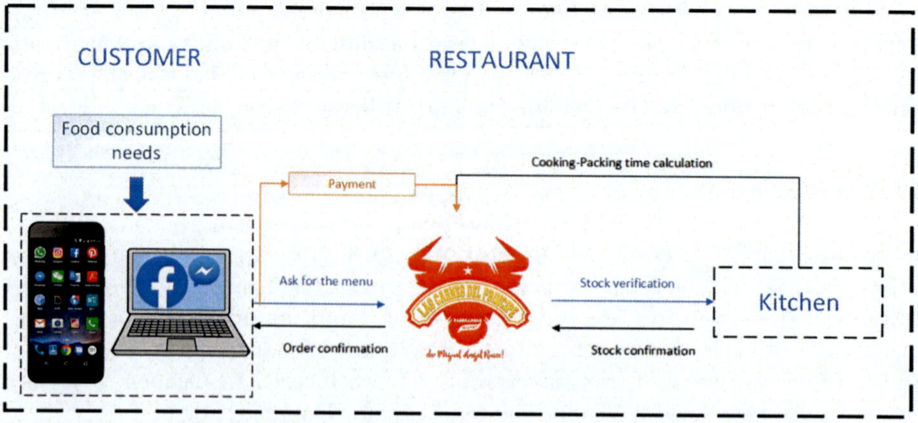

Fig. 1. Digital marketing before the healthcare crisis.

On the other hand, Fig. 2, shows the proposed digital marketing strategy. It is necessary to mention that this goes together with the conventional advertising methodology. The application of innovative technologies opens a wide range of possibilities since different traditional businesses, which have always been in their comfort zone, can have a presence in the digital world. In addition, the possibility of targeting content to the ideal audience avoids bias or loss of resources such as time and money. The owner of this establishment had to make an important decision: reinvent or die. Despite having a Facebook page created, there was no adequate interaction between the business and its customers, concluding that it is not the same to belong to a social network as to be present on it.

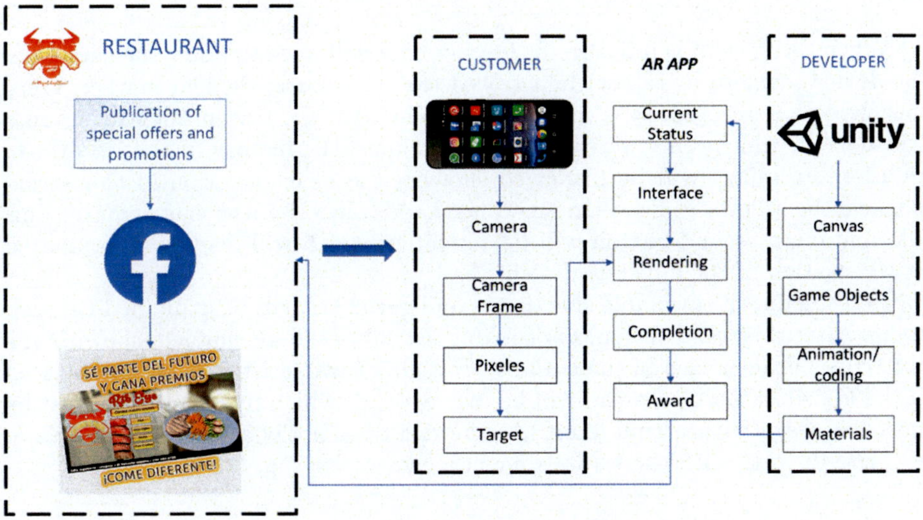

Fig. 2. AR Digital marketing.

3 AR System

The graphic environment delimits the space where the developer creates and interacts with digital elements [3]. AR can be defined as a technology inherent to Industry 4.0 and Marketing 5.0 that allows for generating interactive experiences through the combination of digital elements in the physical or real world [17,20]. In Fig. 3, it can be understood that the AR system consists of three modules: i) History, where, through voice prompts, the user will be able to learn the origin of the rib eye; ii) Dish composition, which details the components of the dish mentioned above, its sides and the cooking time in which it can be ordered; and iii) Reward, an option that allows the user to answer several questions about points I and II. The customer will have access to several prizes according to the score obtained.

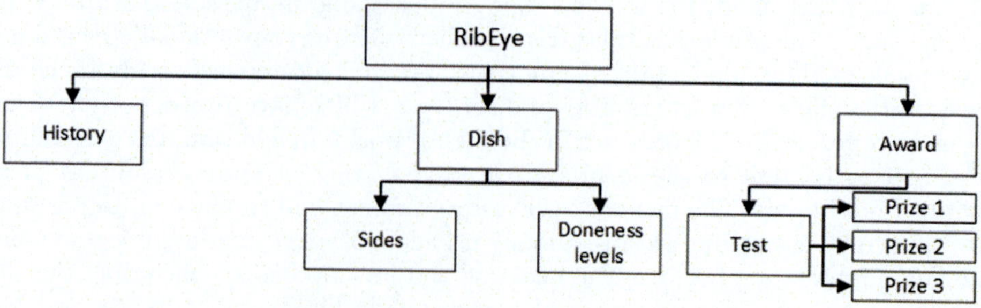

Fig. 3. System block diagram.

When the system is initialized, the customer will have to point the camera of the device (camera or tablet) being used to a previously defined marker. Once this element is recognized, a menu will appear with the restaurant's logo and a first button, which, once activated, will announce the history of the rib eye. In addition, an image will be displayed showing the exact location of this steak. This can be seen in Fig. 4a. On the other hand, once the user has acquired this information, a second button will be activated, which will display an animated representation of the advertised dish.

Here, again, the available side dishes and garnishes will be explained through audio. After this, the five types of cooking available will be enabled, showing the difference between each of them and their main characteristics. This can be seen in the Fig. 4b. Finally, a menu will be displayed, in which five questions must be answered, each with a time limit. At the end of this trivia, the user can claim his prize directly with the establishment's owner. See Fig. 4c.

4 Results

4.1 System Usability Scale

As in other research projects [18, 22], the acceptance evaluation was carried out through a survey based on a measurement methodology similar to the Likert scale. The System Usability Scale (SUS) measures the usability of an application, system, or device. It consists of 10 questions, which can be calculated from 1 to 5, where one means completely disagree, and five refers to agree. Despite being a straightforward tool, several studies consider it a reliable and accurate technique.

There are several interpretations of the results because several authors have established different ranges for rating the system or device. A range of 0 to 50 means that the system developed is unacceptable and can be significantly improved. An interval of 50 to 68 is a marginal result, meaning that the system is acceptable; however, there are still some shortcomings. Finally, a range of 68 to 100 is a proper system [14].

To determine if the AR application developed for this research is adapted to the service and gastronomy environment, a pilot test was conducted with 25 diners. The following items were included in the interview:

(a) Module I: History.

(b) Module II: Dish.

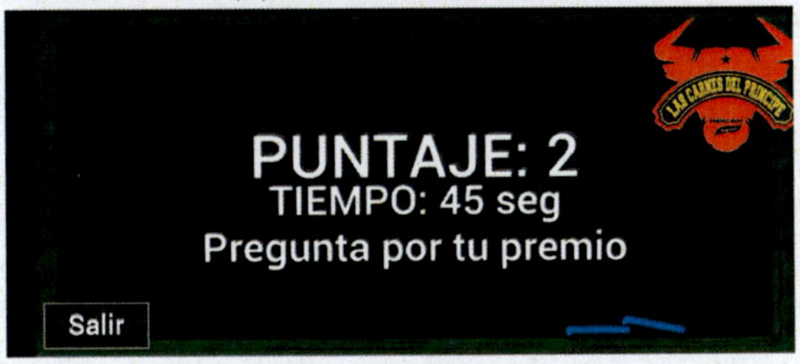

(c) Module III: Award.

Fig. 4. AR Interfaces

1. I think I would use the AR system frequently.
2. I find the AR system excessively complex.
3. The AR system was easy to manipulate.
4. The help of a person specialized in the subject is necessary for the use of the AR system.
5. The AR system functions are well integrated.
6. I consider the AR system to be inconsistent.
7. I think most customers would learn to use this AR system quickly.
8. I find the AR system difficult to use.
9. I feel safe and confident using the AR system.
10. I needed to acquire a lot of information to use this AR system.

It is important to note that, to obtain the result of this instrument, the averaged results obtained from the questionnaires will be added together. For the odd questions, one is subtracted from the value assigned by the user, while the result obtained for the even questions is subtracted from 5. Finally, the value obtained is multiplied by 2.5. The average obtained for the developed AR system was 71.5, which is acceptable according to [19].

4.2 Marketing Strategy

Once the external situation of the restaurant under study was analyzed, the augmented reality system was developed and validated. After this, the social network Facebook was defined as the ideal medium to advertise the establishment and its products using the owner's social page. Here, it was decided to use paid advertising.

The advertising image was disseminated for a period of two weeks, and the public was segmented under the following characteristics: i) population located in Tungurahua, Chimborazo, and Cotopaxi, ii) interests in cooking, barbecues, and fast food, iii) interests focused on popular sports such as soccer and basketball; and, iv) age range between 18 and 60 years.

The objectives were focused on stabilizing the drop in sales due to the endemic crisis of Covid-19 and improving sales during the two weeks of the publication. The advertising reached 1200 people, while there were 55 interactions. Compared to the results of the other publications on this page, which have an average reach of 950 people, there is an increase of 26.31%. However, the percentage difference in interactions corresponded to 3.77%.

It is necessary to consider that the restaurant provides its services from Monday to Saturday, with uninterrupted hours from 12:00 to 22:00. As seen in Fig. 5, there is a percentage difference of 47.45% between the average sales before and during the application of the marketing strategy. It is worth mentioning that the days that most influence sales are: Thursday, Friday, and Saturday, the latter being the critical point of the week.

However, although there was a difference between these two groups numerically, it is essential to verify whether it is statistically significant, so the T-student statistical tool for independent samples was used. For this calculation, the IBM SPSS Statistics 25 statistical program was used. Two columns of data were entered. In the first one, the criteria corresponding to before and after the application of the marketing strategy were entered, while in the second one the total daily sales were entered.

After this, the descriptive statistics were computed, verifying that there were no missing cases. 100% of the data was helpful for the analysis. The mean, median, variance, skewness, and kurtosis were shown for both groups, proving that the percentage difference was correct. At this point, several graphs were also obtained (normal Q-Q without sales trend and box-and-whisker plot), which better represent the difference between the two groups.

Two hypotheses were established: Ho) There is no statistically significant difference between the average sales prior to the application of the marketing strategy and the average sales during the application of the marketing strategy, and H1) There is a statistically significant difference between the average sales prior to the application of the marketing strategy and the average sales during

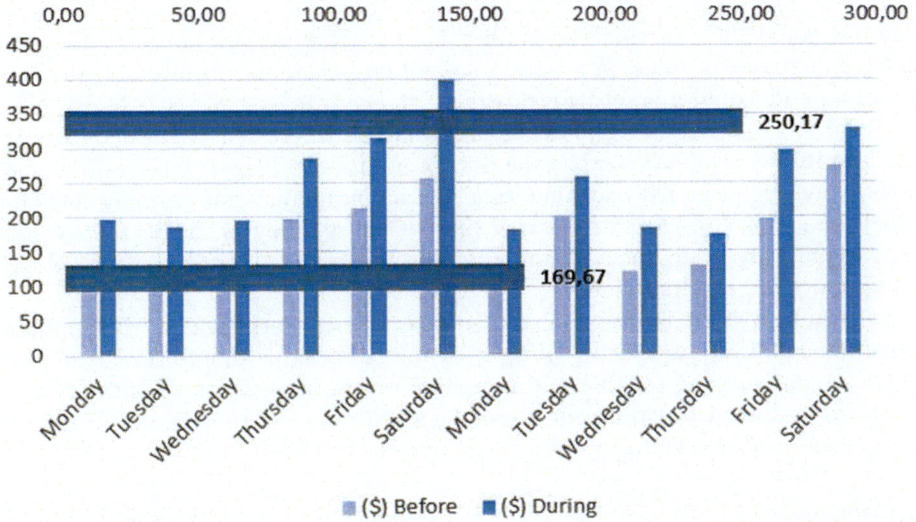

Fig. 5. Difference in sales before and during the application of the AR system.

the application of the marketing strategy. A significance level of 5% was selected. After this, the normality of each of the groups was determined. However, due to the robustness of this test, it is not a prerequisite. Following the number of elements that make up each group, the Shapiro-Wilk test was used since it is adequate when working with less than 30 data. The significance for the control group was 0.079, and for the experimental group, it was 0.070, both values being higher than the alpha level (0.05), which proved normality.

Next, Levene's test was applied. This inferential test evaluates two hypotheses: Ho-1) The population variances are equal, H1-1) The population variances are not equal. A value of 0.416 was obtained, which, being more significant than 0.05, determines that it is very likely that the differences calculated in the sample have been generated in a random population with equal variances. Therefore, Ho-1 is accepted, concluding that there is no difference between the variances of the selected groups.

Finally, a significance of 0.008 was obtained. Since this value is less than 0.05, the null hypothesis was rejected, and the alternative hypothesis was accepted, i.e., There is a statistically significant difference between the average sales prior to the application of the marketing strategy and the average sales during the application of the marketing strategy.

5 Conclusions and Future Work

This article has presented the implementation of a 5.0 marketing strategy in a restaurant in the city of Ambato, Ecuador. An analysis of the developed system has been carried out through the evaluation of its usability, obtaining encouraging results but with room for improvement. For the period selected in this case study, it could be seen that the application of AR generates considerable percentage increases in sales, in addition to building customer loyalty and turning it into an additional marketing strategy.

To propose a marketing plan requires that all interested parties contribute with significant elements so that the flow of information is not interrupted, and in this way, it can be fed back, entering into a continuous improvement. On the other hand, it is considered necessary that this case study be replicated so that the variables involved, such as time, population, sample, service, and product, can support the results found.

In future work, it is proposed to use several social networks to disseminate promotions and essential news inherent to the restaurant. In addition, it is proposed to compare the organic and inorganic reach, analyze its advantages and disadvantages, and use different technologies such as virtual reality, artificial intelligence, and Big Data.

References

1. Acevedo-Duque, Á., Gonzalez-Diaz, R., Vega-Muñoz, A., Fernández Mantilla, M.M., Ovalles-Toledo, L.V., Cachicatari-Vargas, E.: The role of B companies in tourism towards recovery from the crisis COVID-19 inculcating social values and responsible entrepreneurship in Latin America. Sustainability **13**(14), 7763 (2021). https://doi.org/10.3390/su13147763
2. Arantes, L., Sousa, B.: Digital marketing and tourism : case study applied to barcelos. In: 2021 16th Iberian Conference on Information Systems and Technologies (CISTI), pp. 1–6. IEEE (2021). https://doi.org/10.23919/CISTI52073.2021.9476556
3. Caiza, G., Riofrio-Morales, M., Robalino-Lopez, A., Toscano, O.R., Garcia, M.V., Naranjo, J.E.: An immersive training approach for induction motor fault detection and troubleshooting. In: De Paolis, L.T., Arpaia, P., Bourdot, P. (eds.) AVR 2021. LNCS, vol. 12980, pp. 499–510. Springer, Cham (2021). https://doi.org/10.1007/978-3-030-87595-4_36
4. Caiza, G., Riofrio-Morales, M., Veronica Gallo, C., Santiago Alvarez, T., Lopez, W.O., Garcia, M.V.: Virtual reality system for training in the detection and solution of failures in induction motors. In: 33rd European Modeling and Simulation Symposium, EMSS 2021, pp. 199–207 (2021). https://doi.org/10.46354/i3m.2021.emss.027
5. Draganov, M., Panicharova, M., Madzhirova, N.: Marketing 5.0. transactions of artificial intelligence systems in the digital environment. In: International Conference on High Technology for Sustainable Development, HiTech 2018 - Proceedings, pp. 1–3 (2018). https://doi.org/10.1109/HiTech.2018.8566547
6. Fink, M., Koller, M., Gartner, J., Floh, A., Harms, R.: Effective entrepreneurial marketing on Facebook - a longitudinal study. J. Bus. Res. **113**, 149–157 (2020). https://doi.org/10.1016/j.jbusres.2018.10.005
7. González-Ferriz, F.: El marketing 5.0 y su efecto en la estrategia empresarial del sector industrial en España. Redmarka. Revista de Marketing Aplicado **25**(1), 1–20 (2021). https://doi.org/10.17979/redma.2021.25.1.7848
8. Hoyos-Estrada, S., Sastoque-Gómez, J.D.: Marketing Digital como oportunidad de digitalización de las PYMES en Colombia en tiempo del Covid - 19. Revista científica anfibios **3**(1), 39–46 (2020). https://doi.org/10.37979/afb.2020v3n1.60

9. INEC: Restaurantes y servicio móvil de comida (2018)
10. Janani, S., Christopher, R.M., Nikolov, A.N., Wiles, M.A.: Marketing experience of CEOs and corporate social performance. J. Acad. Mark. Sci. **50**(3), 460–481 (2021). https://doi.org/10.1007/s11747-021-00824-9
11. Karjo, C.H., Hermawan, F., Napitupulu, B.: The impact of digital marketing media on the household business sales during COVID-19 Pandemic. In: 2021 3rd International Conference on Cybernetics and Intelligent System (ICORIS), pp. 1–4. IEEE (2021). https://doi.org/10.1109/ICORIS52787.2021.9649491
12. Ketter, E., Avraham, E.: #StayHome today so we can #TravelTomorrow?: tourism destinations' digital marketing strategies during the Covid-19 pandemic. J. Travel Tourism Mark. **38**(8), 819–832 (2021). https://doi.org/10.1080/10548408.2021.1921670
13. Kotler, P., Kartajaya, H., Setiawan, I.: Marketing 5.0: Tecnología para la humanidad. Wiley, Hoboken (2020)
14. Lewis, J.R.: The system usability scale: past, present, and future. Int. J. Hum. Comput. Interact. **7**, 577–590 (2021). https://doi.org/10.1080/10447318.2018.1455307
15. Martínez, L.M.: Riesgos psicosociales y estrés laboral en tiempos de COVID-19: instrumentos para su evaluación. Revista de Comunicación y Salud **10**(2), 301–321 (2020). https://doi.org/10.35669/RCYS.2020.10(2).301-321
16. Mogrovejo Lazo, A., Cabrera Espinoza, C.: Marketing digital en el Ecuador tras la crisis sanitaria de la Covid-19. Sociedad & Tecnología 5(2), 226–240 (2022). https://doi.org/10.51247/st.v5i2.209
17. Naranjo, J.E., Ayala, P.X., Altamirano, S., Brito, G., Garcia, M.V.: Intelligent oil field approach using virtual reality and mobile anthropomorphic robots. In: De Paolis, L.T., Bourdot, P. (eds.) AVR 2018. LNCS, vol. 10851, pp. 467–478. Springer, Cham (2018). https://doi.org/10.1007/978-3-319-95282-6_34
18. Naranjo, J.E., Urrutia Urrutia, F., Garcia, M.V., Gallardo-Cardenas, F., Franklin, T.O., Lozada-Martinez, E.: User experience evaluation of an interactive virtual reality-based system for upper limb rehabilitation. In: 2019 Sixth International Conference on eDemocracy & eGovernment (ICEDEG), pp. 328–333. IEEE (2019). https://doi.org/10.1109/ICEDEG.2019.8734389
19. Pal, D., Vanijja, V.: Perceived usability evaluation of Microsoft Teams as an online learning platform during COVID-19 using system usability scale and technology acceptance model in India. Child. Youth Serv. Rev. **119**, 105535 (2020). https://doi.org/10.1016/j.childyouth.2020.105535
20. Regenbrecht, H., Baratoff, G., Wilke, W.: Augmented reality projects in the automotive and aerospace industries. IEEE Comput. Graph. Appl. **25**(6), 48–56 (2005). https://doi.org/10.1109/MCG.2005.124
21. Rengel, M.D., Suconota, D.G., Moscoso, A.E.: Ventajas del Marketing Digital en el sector comercial de Ecuador, en tiempos de COVID-19. Espacios, **42**(03), 43–52 (2). https://doi.org/10.48082/espacios-a22v43n03p05
22. Riofrio-Morales, M., Garcia, M.V.: Training virtual reality-based system for detection and simulation of motors failures. J. Phy. Conf. Ser. **1983**(1), 012099 (2021). https://doi.org/10.1088/1742-6596/1983/1/012099
23. Salgues, B.: Society 5.0 and the Management of the Future. Society 5.0, pp. 91–119 (2018). https://doi.org/10.1002/9781119507314.ch6
24. Yagnik, A., Thomas, S., Suggala, S.: Creativity centred brand management model for the post-covid marketing 5.0 world. J. Content Community Commun. **12**, 227–236 (2020). https://doi.org/10.31620/JCCC.12.20/21

Kinect-Enabled Electronic Game for Developing Cognitive and Gross Motor Skills in 4-5-Year-Old Children

Carlos Núñez(✉), Eddy López, Jenrry-Patricio Nuñez, and David-Sebastian González

Universidad Técnica de Ambato, Av. Los Chásquis, Ambato, Ecuador
{ci.nunez,elopez6422,jnunez5291,dgonzalez6696}@uta.edu.ec
http://uta.edu.ec

Abstract. This paper presents the results of a Kinect enabled electronic game implemented for stimulating the development of cognitive and gross motor skills in 4–5-year-old children. The game consists of catching as many moving objects as possible within a virtual environment using all different parts of the body (torso, head, legs, and arms). A study was carried out with the participation of children of an Ambato school, in the Tungurahua province. After four attempts of using the game, the children participating in the experiment showed satisfactory results. Concerning the actions related to gross motor skills, an increase of 70% on the success level was attained while the cognitive skills showed a 35% increase on the level of success. These positive results corroborate that the proposed study is an efficient tool for improving the cognitive and gross motor skills in 4–5-year-old children.

Keywords: Technology · Kinect · Motor · Cognitive · Education · Electronic game

1 Introduction

Kinect is a legacy product originally released for Xbox 360, and then for the Xbox One series with the Kinect Adapter. Kinect has many built-in cameras that allow to track the user movements and gestures for playing interactive games, taking pictures, and more.

The use of technology in education has exponentially increased and has become crucial in modern society. More specifically, in the context of individuals with special needs, technology has proven to be very beneficial for rehabilitation and learning processes when implemented in a didactic manner [6].

The progress of patients using Kinect-enabled environments were documented in previous studies such as the Design and development of a playful environment using Kinect as support tool for children with physical disabilities in the Pediatric Rehabilitation area of José Carrasco Arteaga Hospital. In this

M. V. Garcia and C. Gordón-Gallegos (Eds.): CSEI 2022, LNNS 678, pp. 138–149, 2023.
https://doi.org/10.1007/978-3-031-30592-4_11

paper 95% of 5–8-year-old children improved their cognitive and motor skills in a 15 day period [14].

Kinect based games have been used in many educational institutions worldwide, showing significant progress in children. For this reason, Kinect is considered a useful tool in teaching [1,4]. However, in Ecuador the use of this alternative is still incipient. That is why this paper proposes to evaluate the use of Kinect technology for improving the cognitive and gross motor skills in 4–5 year old kids.

On the one hand, gross motor skills are abilities that let us do tasks that involve large muscles in our torso, legs, and arms. They involve whole-body movements and the coordination of the muscles and the neurological system. [3,5] Gross motor skills are related to other abilities. These include:

- Balance
- Coordination
- Body awareness
- Physical strength
- Reaction time

On the other hand, cognitive development means how children think, explore and figure things out. It is the development of knowledge, skills, problem solving and dispositions, which help children to think about and understand the world around them. Cognitive development is very important during the growing stage of children, that is, in early and middle childhood [9,10,16].

Children aged 3–7 years old are in the stage called early childhood. Early childhood is a time of tremendous physical, cognitive, socioemotional, memory, and language development. The aim of this research is to measure the development level of cognitive and motor skills in 3–7 year old who used the proposed game for a period of one month with four attempts each child.

The Kinect technology was chosen due to its low cost in contrast to the rest of technologies available in the market. Kinect has the specific features required for collecting data of moving body parts (head, torso, upper and lower extremities).

1.1 Related Work

In year 2020 the study [12] carried out an analysis on the use of Kinect technology for improving the motor skills of 4–5 year olds. Detailed information was provided on the muscles, body segments, balance, displacements, and attitude involved with certain movements.

Kinect technology has been used in previous studies For instance, assessment Tasks and Virtual Exergames for Remote Monitoring of Parkinson's Disease An Integrated Approach of the Motor impairments are among the most relevant, evident, and disabling symptoms of Parkinson's disease that adversely affect quality of life, [2].

In [18] an educational, technology based proposal was created based on a historic, evolutive, theoretical research on Body Language within a preschool educational context. The study depicts the reality of real issues of children, the benefits and the way the proposal is carried out in the school. It suggests developing an interactive facility as a complementary educational tool for effectively meeting the educational goals for this age. Interactivity introduces an innovative, fun, and useful approach to the Body Language classroom [8]. To verify the pertinence of the educational modules in interactive classrooms, the curriculum of the Ministry of Education for ages 3 to 5 was considered during the design and implementation phase [15].

In the study [17] year 2018, the Kinect device and a Kinect compatible Scratch pro-gram were used for creating a learning interactive system. By means of motion capture, the system allowed to develop and improve children body language and motor skills. The results were positive as 86% of the children were able to acquire the abilities they were working on.

In 2016 the study [13] concluded that with the Kinect technology it was possible to recognize and interpret some gestures of the Colombian sign language. This will be a great contribution for kids with hearing disabilities who lack of didactic tools for learning the sing language.

In [7] the adequate use of the Kinect sensor for implementing educational interactive systems for children is explained. It includes numbers, animals, and letters for a more fun and constructive experience. In addition, the combined use of hardware and software was proven to be helpful for educational purposes without any significant limitations. The basic robotic platform is an example of an industrial tool turned into an educational platform for didactic and not experimental purposes.

2 Materials and Methods

The present work started with the definition of the software architecture for the de-velopment and installation of the electronic game. The software development plat-forms used were: Visual Studio .NET with C Sharp as programming language and Kinect technology for movement recognition along with a management program. The Ki-nect sensor represents an attractive feature for the electronic game, making it more dynamic and effective for attaining the learning goals.

A comparison between the Kinect 2.0 and other available technologies for developing the project was made.

3.0 USB Connection

- Recognition of specific body parts in humans and 6 users at the same time. 1920 × 1080-pixel resolution
- Price around 200
- Movement recognition in dark places.
- Background noise suppression.
- Asus Xtion PRO LIVE

2.0 USB Connection

- Whole body detection of a human user
- 640 × 480-pixel resolution
- Price around 100
- System predefined movements
- Two built-in microphones.

[11]

The Kinect 2.0 technology was selected among the other options because it makes it possible to recognize specific body parts in humans allowing to collect more information on the movement of upper and lower extremities. Extremities intervene in the games related to cognitive development as well as in those related to gross motor development. Children must use them for catching moving objects or for selecting a correct answer in a game. Additionally, Kinect 2.0 has better resolution and better connectivity through the 3.0 USB ports.

3 Electronic Game

3.1 Architecture

Figure 1 depicts the architecture of the application. As it is shown Visual Studio 2019 was used for developing the game. Information is stored on a SQL Server 2012 data base. There is a monitor showing the player immersed in the virtual environment. Kinect V2.0 sensor recognizes the user movements from a 3-m distance and connects to the computer which maps and processes the body parts (torso, head, arms, legs) and gestures by means of the Kinect SDK.

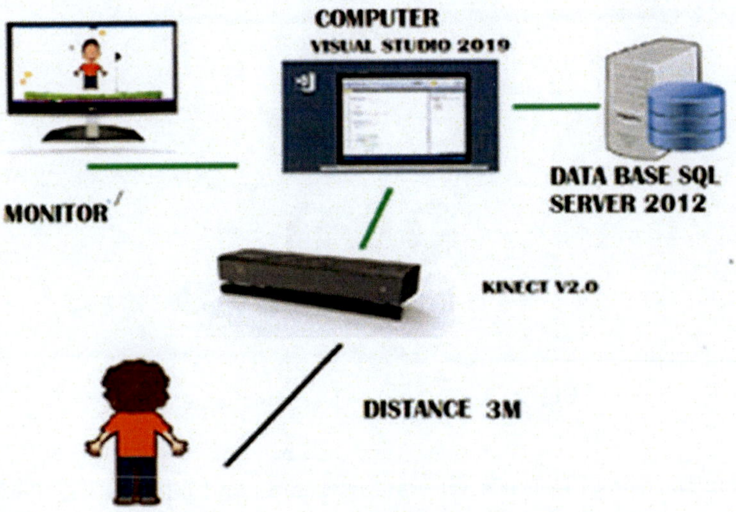

Fig. 1. Architecture of the application.

3.2 System Functioning

The electronic game is designed to develop cognitive and gross motor skills in 4–5-yearolds. The aim of the game is using any of both hands for selecting the correct answer to a random sum of two one-digit numbers shown in the screen. The player will have three attempts before the next sum appears. If the result is correct the user will progress in the game and will then have to catch as many moving objects as possible with different parts of the body. The children will be immersed in the game's virtual environment by means of the Kinect sensor. The gross motor is assessed by counting the rights and wrongs when catching the objects. The cognitive skills are evaluated according to the number of correct sums.

For starting the game, the child 's username and password must be entered. This in-formation is retrieved from the child's profile, registered before the game. If the user is not registered on the database, the Register New User option is available for filling out all the information: name, last name, age, level, sex, username, and password. In order to check the progress, the child must be registered on the game for seeing his/her playing history on a given game and date

The children's personal data are protected by the Organic Law of Personal Data Protection (in Spanish LOPD) for ensuring privacy, private life and personal honor. This legislation regulates the conflicts of interest between the peoples' right to privacy and honor and the right to the freedom of information.

It must be clarified that before installing and using the application the authorities of each educational institution participating in the study must give their approval. Furthermore, no personal data will ever be published; just the results obtained and the aggregate results.

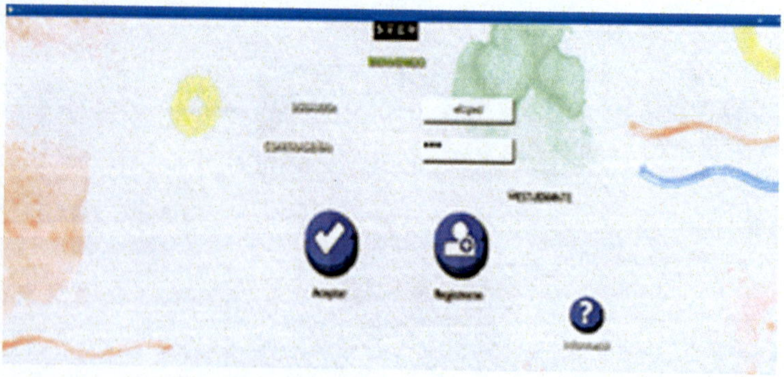

Fig. 2. Accesing the electronic game.

For starting the game, the user enters a username and password as depicted in Fig. 2. Next, a random sum with images representing the amount indicated by the summands is shown. The child is given some time to think and solve the calculation, having cognitive stimuli in this stage of the game. See Fig. 3.

Fig. 3. Visualization of the random sum.

Once the time is over, a group of possible results for the previous sum is shown. The child will be given three attempts to select the correct answer. This is carried out through the Kinect device which allows the virtualization of the child's hands accord-ing to his /her relative position with respect to the space delimited by the screen. The child moves a hand and selects an answer. Next, the program in the computer deter-mines if it was correct or incorrect and registers on the database the rights and wrongs for each sum as well as the number of attempts. Figure 3 shows the virtualization of the child's hand when playing (Fig. 4).

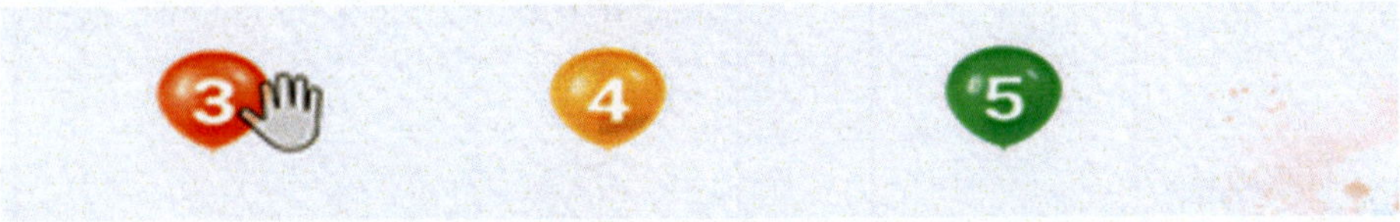

Fig. 4. Electronic Game.

After having selected the correct answer, another screen will be shown for collecting data on the child's gross motor. In this window the child will see the result of the previous sum in the form of a balloon along with two other incorrect numbers. With the help of Kinect, the child is virtualized and rendered inside the game. Then, 60 s will be given to the child for playing and "touching" the answers on a few screens. The player will have to jump, move to the left, to the right, or to the front inside a 3 m × 3 m area for selecting the balloon with the correct answer. This will be repeated during the period of 60 s and the child accumulates points. In case of touching the wrong balloon, a point will be subtracted from the score for every incorrect answer (Fig. 5).

Fig. 5. Virtualization of the player in the Electronic Game.

On the one hand, a goal of the game is to measure the gross motor skills through the action of catching as many objects as possible showing the correct sum result. On the other hand, another objective is to measure the cognitive skills by using sums involving two one-digit summands. In both cases, the total of successful or unsuccessful attempts of touching the correct object was registered. In addition, the body part used by the child to answer was also observed and documented in the database.

As part of the research work, an experimental evaluation of the results was carried out. The game was installed on a school in the heart of downtown Ambato, in Tungurahua province. A total of 10 children, ages 4–5 participated in the study. The game was used by the participants for a one-month period during the Covid pandemic, having one session per week.

4 Results

The experiment results corroborate that children take more interest on learning when using electronic games as the one proposed. A group of 10 children with 4 attempts per child obtained the following results after having used the game:1) 70% increment on the rights between the first and last attempt related to the gross motor skills actions; 2) 17.5% increment on the improvement on actions involving cognitive skills. This last result was obtained using as a reference the Tepsi test results (Test de Desarrollo Psicomotor de 2-5 años (TEPSI)). The Tepsi test is performed during a meeting with the child where some actions are proposed for determining his/her psychomotor level of development. The results obtained during the Tepsi test were compared with the results of the game, using the same criteria and actions to evaluate the child's performance in both, real-life and virtual environments.

Concerning cognitive skills improvement, there is a permanent trend of 65% of children who answered correctly on the first attempt for solving the sum. Then, after many attempts it was verified that on the fourth attempt 100% of players got the sum right, thus a 13% increase on progress was accomplished.

The teachers were asked on which strategies they used for making progress on motor skills. They explained that the key was playing games that included physical activities such as jumping and extremity movements and constantly changing the activities. This strategy works well for children who get easily bored or distracted.

The results of rights and wrongs during the study were progressively improving from the first attempts until the last ones. This progress made by the children demonstrates the effectiveness of technology as a motivation for improving their motor skills.

Representative information of the experiment a sample of 10 children of the school were analyzed. The results were:

For the Analysis Section of the Cognitive Skills the Results were: As depicted in Fig. 6, the number of successful attempts executed between the first and fourth try proves that the proposed tool really helps on improving cognitive learning in children.

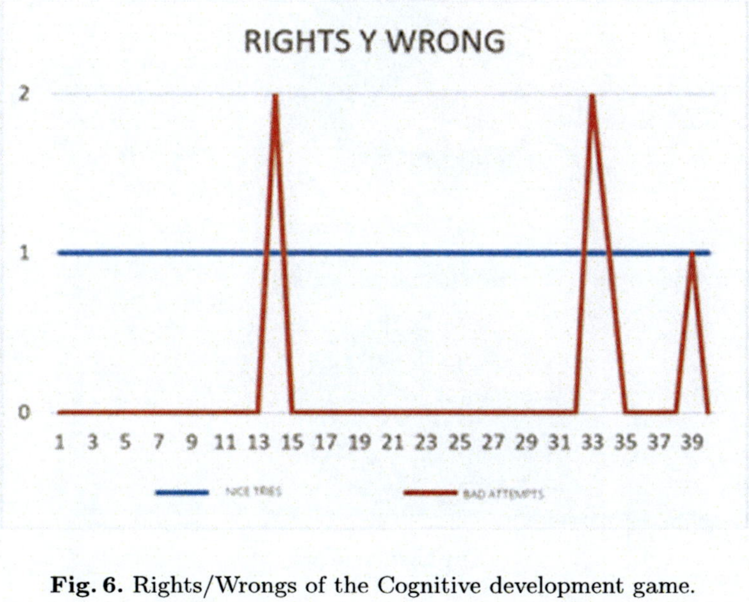

Fig. 6. Rights/Wrongs of the Cognitive development game.

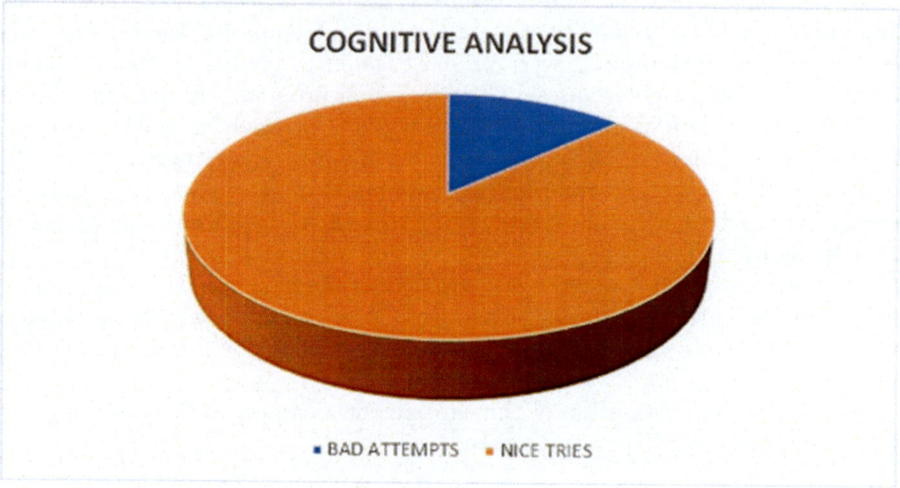

Fig. 7. Rights/Wrongs of the Cognitive development game First Attempt.

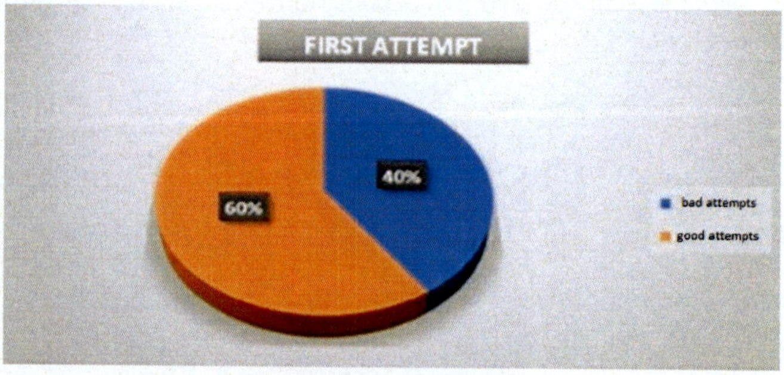

(a) First Attempt

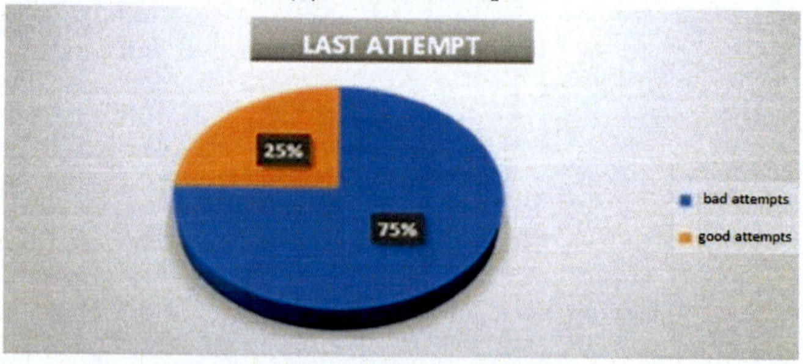

(b) Last Attempt

Fig. 8. Rights/Wrongs of the Gross Motor Skills game

For measuring usability in educational games and electronic devices for children, the observation method is frequently used. The usability test is conducted in the children's own familiar environment which is the classroom. Data are

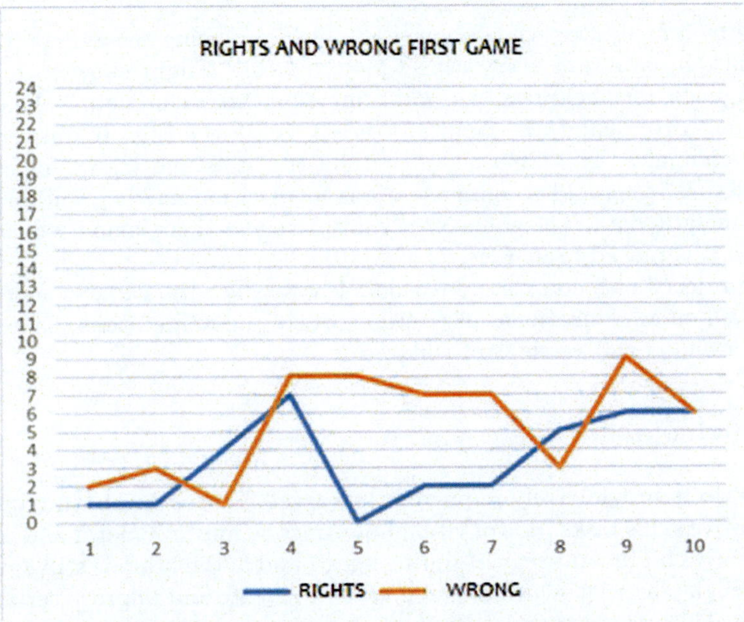

Fig. 9. Rights and wrongs of the first game.

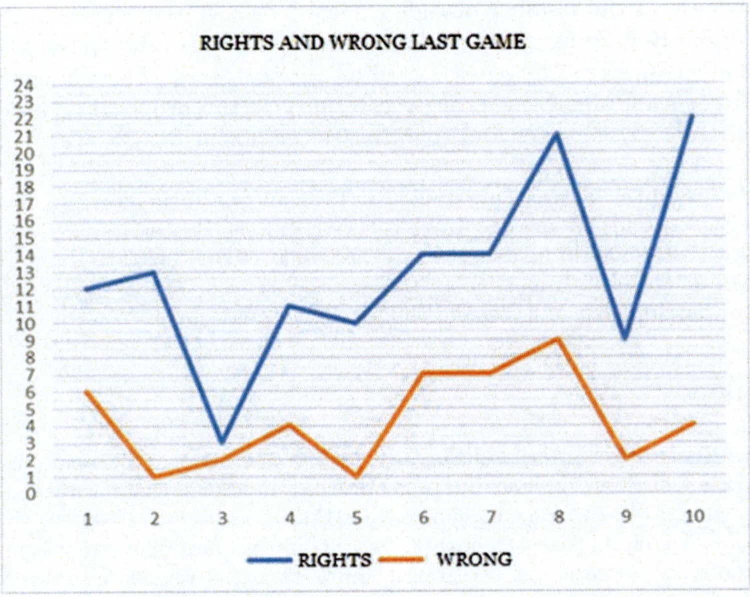

Fig. 10. Rights and wrongs of the last game.

collected by a facilitator while observing the child playing the games. The facilitator must be aware of every aspect that the child might show such as their body language, facial expressions, and comments. After completing the usability test session with a child, the facilitator needs to ask the child to answer a post-task questionnaire. The observation method is simple and allows for exploring unexpected aspects, add recommendations for decision-making, and presenting facts as analytic data. The game is very intuitive and there is no need to ask the child how it is played (See. Figs. 7, 8, 9, 10).

The project has a completely reasonable cost since the game is conceived to be installed on all educational institutions where the main interest is improving the motor and cognitive skills of children.

5 Conclusions

Study subjects progressively improved their level of success in both gross motor and cognitive skills tasks. It is of vital importance to improve motor and cognitive skills in early childhood because, during ages 4 and 5, the brain is still developing having a high plasticity which makes it easier to create and improve certain brain functions. This stage is crucial for the future years of the child.

The use of technology represents a helpful tool for improving the quality of teaching and learning in a more fun and didactic manner. With the present study, a 35% increase in the number of rights related to cognitive development tasks was obtained. Regarding gross motor skills, an increase of 75% was attained in the number of right answers concerning this category. These increments in the children's level of success prove the validity of the proposal as an effective support tool for gross motor and cognitive development.

Acknowledgments. The authors would like to thank the Technological University of Ambato (Universidad Técnica de Ambato UTA) and the Research and Development Department (Dirección de Investigación y Desarrollo DIDE) for their support on the execution of this project. In particular, through the project "Kinect Technology in the gross motor development of 4 year-old children", sffisei05.

References

1. Acosta Blanco, V.E., Arrúa Ginés, J.L., Ayala Díaz, K.A.: Aplicación lúdica utilizando la tecnología de kinect para los procesos de enseñanza y aprendizaje de la lengua guaraní. Portal de conocimiento (2015)
2. Amprimo, G., et al.: Assessment tasks and virtual exergames for remote monitoring of parkinson's disease: An integrated approach based on azure kinect. Sensors **22**(21), 8173 (2022). https://www.scopus.com/
3. Garcia, C.A., Garcia, M.V., Irisarri, E., Perez, F., Marcos, M., Estevez, E.: Flexible container platform architecture for industrial robot control. In: 2018 IEEE 23RD International Conference on Emerging Technologies and Factory Automation (ETFA), pp. 1056–1059. IEEE International Conference on Emerging Technologies and Factory Automation-ETFA (2018)

4. Garcia, C.A., Lanas, D., Alvarez M, E., Altamirano, S., Garcia, V, M.: An approach of cyber-physical production systems architecture for robot control. In: IECON 2018–44TH Annual Conference of the IEEE Industrial Electronics Society, pp. 2847–2852. IEEE Industrial Electronics Society (2018)
5. Gizatdinova, Y., et al.: PigScape: an embodied video game for cognitive peer-training of impulse and behavior control in children with ADHD. In: ASSETS 2022 - Proceedings of the 24th International ACM SIGACCESS Conference on Computers and Accessibility (2022). https://www.scopus.com/
6. Hernandez, R.M.: Impacto de las tic en la educación: Retos y perspectivas. Propósitos y representaciones **5**(1), 325–347 (2017)
7. Ilvay Taday, R.B.: Sistema de educación para niños de 3 a 5 años, mediante un robot controlado por el sensor kinect. DSpace ESPOCH (2014)
8. León, B.C.: Desarrollo psicomotor. Revista mexicana de medicina física y rehabilitación **14**(2–4), 58–60 (2002)
9. Mariño, C., Vargas, J.: Ergonomic postural evaluation system through non-invasive sensors. In: Nummenmaa, J., Pérez-González, F., Domenech-Lega, B., Vaunat, J., Oscar Fernández-Peña, F. (eds.) CSEI 2019. AISC, vol. 1078, pp. 274–286. Springer, Cham (2020). https://doi.org/10.1007/978-3-030-33614-1_19
10. Martínez, L.G.: Desarrollo cognitivo y educación formal: análisis a partir de la propuesta de ls vygotsky. Universitas philosophica **34**(69), 53–75 (2017)
11. Núñez, C., Chicaiza, D., Larrea, A., Acosta, S.: Comparison of kinect v1, v2 and azure technology focused on children's motor skills. RISTI - Revista Iberica de Sistemas e Tecnologias de Informacao 2021(Special Issue E43), 338–350 (2021). https://www.scopus.com/
12. Núñez, C., Nuñez-Nuñez, J., Fernández-Peña, F.: Longitudinal evaluation of the gross motor improvement in 4 to 5 year old children by using kinect technology. RISTI Revista Iberica de Sistemas e Tecnologias de Informacao **2020**(E37), 194–204 (2020). https://www.scopus.com/
13. Pacheco, L.P.P., Martínez, S.P.R., Narváez, E.L., Leal, N., Miranda, C.H.: Aplicación integrada a la tecnología kinect para el reconocimiento e interpretación de la lengua de señas colombianas. Escenarios **14**(2), 7–19 (2016)
14. Pérez Muñoz, Á.A.: Diseño y desarrollo de una aplicación lúdica basada en el dispositivo kinect como herramienta de apoyo para niños con discapacidad física en el área de rehabilitación de pediatría del hospital josé carrasco arteaga. Repositorio Institucional de la Universidad Politécnica Salesiana (2018)
15. Ruiz, A., Ruiz, I.: Madurez psicomotriz en el desenvolvimiento de la motricidad fina. COMPAS, Guayaquil (2017)
16. Sagastibeltza, N., et al.: Preliminary study on the detection of autonomic dysreflexia using machine learning techniques. In: Garcia, M.V., Fernaández-Peña, F., Gordón-Gallegos, C (eds.) Advances and Applications in Computer Science, Electronics, and Industrial Engineering. CSEI 2021. Lecture Notes in Networks and Systems, vol. 433, pp. 341–351. Springer, Cham (2022). https://doi.org/10.1007/978-3-030-97719-1_20
17. Villarreal Terán, M.T.: Sistema de aprendizaje interactivo enfocado al desarrollo de la expresión corporal y motricidad de niños de entre 3 a 4 años de edad del centro infantil "la primavera". Repositorio Digital Universidad Técnica del Norte (2018)
18. Vinueza Jácome, X.E., et al.: emotion: instalación interactiva para educar en expresión corporal a niños de 3 a 5 años. Repositorio Digital USFQ (2018)

Optimizing User Information Value in a Web Search Through the Whittle Index

German Mendoza-Villacorta[1](✉) and Yony-Raúl Santaria-Leuyacc[2]

[1] Universidad Privada del Norte, Lima, Peru
german.mendoza@upn.edu.pe
[2] Universidad Nacional Mayor de San Marcos, Lima, Peru
ysantarial@unmsm.edu.pe

Abstract. Attention economy is a rich information management approach in order to get only significant information. In this work, we analyze the problem of optimizing the value of information presented in an electronic device to users who seek information on the web and whose attention is a priori limited and considered as a scarce and valuable resource. The optimization problem is posed as a dynamic and stochastic prioritization problem and is modeled as a dual-speed multi-armed restless bandit problem (RMABP) in a finite state-space and discrete-time setting. In addition, Adaptive-Greedy algorithm (AG) is used to approximate their solution, this algorithm assigns the value of Whittle's index to each piece of information, which determines whether or not it is favorable to be presented to the user at a given time. Computational experiments based on Monte Carlo modeling are presented, which show that this methodology substantially improves Greedy index policy and asymptotically approximates the optimization solution to Whittle benchmark.

Keywords: Attention Economy · RMABP · Whittle Index

1 Introduction

Regarding the concept of *attention economy*, Simon [20] proposes that an information rich world leads to scarcity of what that information consumes, in this case, people attention. In order to manage the gap between available information and the scarcity of attention, researchers and professionals work on ways to avoid information overload [5,8,19,20] and improve information allocation processes through applications aimed at improving the control or personalization of information presented to a user, see for instance [1,9,18]. In line with this, we highlight the theoretical model of competition for attention developed by Falkinger [6], which allows to claim that globalization and the development of information technologies tend to decrease global diversity and people's attention levels. We also mention the work of Brynjolfsson and Oh [2] who developed a new framework for measuring the value of information presented to users of digital services, although they do not pay with cash, must pay attention or time.

M. V. Garcia and C. Gordón-Gallegos (Eds.): CSEI 2022, LNNS 678, pp. 150–165, 2023.
https://doi.org/10.1007/978-3-031-30592-4_12

In a web environment, search engines such as Google and Yahoo, among others, it is considered the available display space on screen of a electronic device, prioritize and show search results on consecutive pages with an assumed decreasing information value for users. Even Google has implemented a feature called "I'm Feeling Lucky", which when given a set of search terms automatically chooses the "best" result for users. This approach deals with two problems described by Huberman and Wu [3,9]. The first lies with the content provider, who needs to decide what to prioritize in order to gain the attention of users. This decision may be made based on objective criteria (site popularity, number of recommendations for the software, news prominence, etc.) or a heuristic rule developed by the content provider. In any case, it is not evident that these procedures maximize user value. For instance, while an algorithm such as PageRank places the links most closely related to the search terms on the first page of a search result, meanwhile, other links with incipient information are not shown to the users in the first instance.

The second problem arises from the number of links that users can handle in a given time interval. There is empirical evidence that users rarely visit pages beyond the first page in a search result, making them less likely to visit the last pages of a search engine (see [4]). This behavior tends to reinforce the leadership position of "top" links while also increasing their popularity, which in turn makes it more difficult to be access new content that is not yet well known. Therefore, it is easier for one link to stay on the "top" list and harder for another from the bottom to rise up, although the latter could be more valuable. According to Pandey [16], it is important to break this distortion reinforcement by encouraging users to explore more items, which will increase on average the information value they can obtain.

This article presents an alternative solution to the difficulties described in the previous paragraphs by configuring a mechanism that maximizes the benefit for users searching for information on the Web. In this work, we extend the results found in [9], to obtain this goal it is formulated the problem of optimizing the information obtained from browsing web pages or any other digital content as a *dynamic and stochastic prioritization problem* for a series of projects (links in this case), which compete to be displayed, this formulation will follow the RMABP framework. It was observed in [17] that RMABP is in general computationally irresolvable. The approach introduced by Whittle [22] will be applied to approximate the optimization goal, which uses the *Whittle index* assigned to each link competing to be displayed, followed by using *Whittle index policy*, "to choose the links with the highest current Whittle index."

The Whittle index is implicitly defined. Therefore, a computationally efficient algorithm called the AG algorithm introduced by Niño-Mora [11,12] will be used to calculate the *marginal productivity index* (MPI), which coincides with Whittle index under certain assumptions that are satisfied in the model presented herein. This work will address cases of *link indexing* and *multi-link indexing*, performing computational experiments to analyze the degree of *sub-optimality* of Whittle index policy, comparing its performance with that of the Greedy index policy and *Whittle benchmark.*

The authors had limitations to obtain real data, for that reason in the present work Monte Carlo modeling was used with data created by simulations which resembles the real behavior of web users. The remainder of this paper is structured as follows. In the second section, the problem addressed herein will be formulated as a *dual-speed* RMABP and the indexing approach will be reviewed. In the third section, adjustments are made to link parameters to use the AG algorithm. The fourth section outlines the results of the computational simulation experiments, in which the Whittle index policy is compared with the *Greedy index policy* and Whittle benchmark. Finally, in sections five and six, future work and conclusions are presented.

2 Formulation of RMABP, and the Indexation Approach

2.1 One-Armed Restless Bandit Problem and RMABP

In this subsection we follow the concepts introduced and explained in [15, 22]. Let X be a *project* (the terms project and link will be used interchangeably) whose evolution over time periods $t = 0, 1, 2, \ldots$ is controlled by a content provider who decides at the beginning of each time period whether the link should be active (shown) or passive (not shown). If at the beginning of period t, the link occupies the state $X(t) = i \in \mathbb{X}$ (in which $\mathbb{X}$ indicates the finite state-space representation of the link) is displayed, that is, the *active action* $a(t) = 1$, it yields an immediate reward $R(i, 1)$, which moves to the state $X(t+1) = j \in \mathbb{X}$ in a Markovian manner with transition probability $P(i, j|1)$. If the link is not shown the *passive action* is taken at $a(t) = 0$, this produces an immediate reward $R(i, 0)$, and its state moves to $X(t + 1) = k \in \mathbb{X}$ in a Markovian manner, with probability $P(i, k|0)$.

The *one-armed restless bandit problem* is finding a policy π^* that indicates when to display the link, which maximize the expected value total reward, which is discounted through the time by a factor of $0 < \beta < 1$. In the infinite horizon version starting at i_0 is formulated as:

$$\max_{\pi \in \Pi} E_{i^0}^{\pi} \left[\sum_{t=0}^{\infty} R\left(X(t), a(t)\right) \beta^t \right],$$

in the formulation, Π indicates the class of admissible policies, among which, an optimal policy is required. This class consist of policies that make nonanticipa decisions, namely, base each actions $a(t)$ on the state of the project and action history $X(0), \ldots, X(t), a(0), \ldots, a(t-1)$. Moreover, $E_{i_0}^{\pi}[\cdot]$ indicates the expected value under the policy conditions to which the initial project state is equal to $X(0) = i^0$.

On the other hand, RMABP arises when a system (e.g., Google search engine or any web page) in periods of time $t = 0, 1, 2, \ldots$ needs to show $N \geq 2$ links to a user. Nevertheless, the screen size of the electronic device used can only show M, with $(N < M)$.

The index of the link $n = 1, \ldots, N$ will be incorporated in the following equation to indicate a simple link, and we denote X_n, $R_n(i, a)$, $p_n(i, j|a)$, $X_n(t)$, and $a_n(t)$ with the usual notation. In this case, In such a setting, the project portfolio manager observes at the start of each period t the joint state $\mathbf{X}(t) = (X_n(t))_{n=1}^N$ and takes a joint action $\mathbf{a}(t) = (a_n(t))_{n=1}^N$, which need to be based on the history of joint states and actions, and satisfy the following:

$$\sum_{n=1}^{N} a_n(t) = M.$$

The choice of action is based on adopting a *scheduling policy* π, which is to be taken from a resulting class of *admissible scheduling policies* $\Pi(M)$, which comprises non-anticipatory policies. The transition of portfolio states laws are the result of the transition of individual projects.

An important assumption is that the state transition of a link is stochastically independent of other state transitions of the projects. As for the joint reward, it is assumed it is assumed to be additive across projects.

The *infinite horizon* problem consist on to find an admissible scheduling policy π^* prescribing which projects to engage at each time, if any, which maximizes the expected total discounted net reward earned (where future rewards are geometrically discounted by a factor of $0 < \beta < 1$), maximizing the expected total discounted rewards earned. The problem is formulated according to the discounted criteria as:

$$\max_{\pi \in \Pi(M)} E_{i_0}^{\pi} \left[\sum_{t=0}^{\infty} \sum_{n=1}^{N} R_n \left(X_n(t), a_n(t) \right) \beta^t \right]$$

where $E_{i_0}^{\pi}[\cdot]$ indicates the expected value under the scheduling policy π, conditional on initial portfolio state being equal to: $\mathbf{X}(0) = i^0 = (i_n^0)_{n=1}^N$.

2.2 Index Policy to Approximate the RMABP Solution

A central issue in the literature on RMABP is identifying efficiently calculable indexes that lead to a priority-index rule that possesses an efficient benchmarking. In general, one can only expect asymptotic optimality of an RMABP index policy, as shown by Weber and Weiss [21]. Since, finding the optimum solution for RMABP is computationally irresovable. We follow the approach introduced by [22], who was the first to propose an approximate solution to this problem by using a *Lagrangian relaxation* approach. Whittle relaxation is utilize to replace the constraint of operating a fixed number of links at each instant of time (an infinite number of constraints) with a single constraint (activating the average number of projects required). By using a Lagrange multiplier, this constraint can be dualized and included in the objective. This simplifies the RMABP solution to solve simple one-armed restless bandit problem.

The value of Whittle relaxation optimization target is known as Whittle benchmark, which approximates the optimization target of the original RMABP. This will use to compare if the benchmarking of other policies are approximate to the optimal value. [22] also proposed an index, known thereafter as the Whittle index, which is implicitly defined for each link as a function of its state and is used by Whittle index policy "to assign the scarce resource to the required projects with the highest current Whittle index." This policy is typically sub-optimal, and is utilize in the present attention economy model.

Whittle also noted that the existence of the indexes is guaranteed only for the RMABP that satisfy the so-called *indexability property*, affirming that:

> ... one would very much like to have simple sufficient conditions for indexability; at the moment, none are known.

Above situations observed by Whittle led to Niño-Mora addressing the same and other issues regarding the indexability of restless bandit projects. Niño-Mora [11–13] introduced methods to determine *a priori* if a restless bandit model is *indexable* and created the AG algorithm based on Klimov [10] to calculate the so-called marginal productivity index (MPI). The MPI is based on objectives established by the *marginal productivity theory* in economics, developed by several authors in the late 19th century. In addition to being quite manageable and applicable, a growing amount of computational data indicates that this index policy often exhibits near-optimal performance while outperforming alternative approach policies.

2.3 AG Algorithm and PLC-Indexability Conditions

Researchers are generally interested in analytically establishing whether a particular model that emerges from an application is indexable under a set of appropriate parameters. It will be useful to have sufficient and general indexability conditions that are widely applicable. In this subsection, the concepts introduced and explained by Niño-Mora [11,13,14] are used to present a methodology for indexing RMABP. These conditions are introduced, developed, and deployed for the first time, together with the AG algorithm, which produces the MPIs. These ideas will be reviewed below.

Given a link, a policy is evaluated by two measurements. The first is the *reward measure*, defined as

$$F_{i0}^{\pi} := E_{i0}^{\pi}\left[\sum_{t=0}^{\infty} R\left(X(t), a(t)\right)\beta^{t}\right];$$

which gives the expected total discounted reward earned on an infinite horizon, starting at state i_0. The second measurement refers to the expenses of the related resource. Thus, if $Q(j, a)$ work units are spent by taking action a in state j, we use the *work measure*, defined by

$$G_{i0}^{\pi} := E_{i0}^{\pi}\left[\sum_{t=0}^{\infty} Q\left(X(t), a(t)\right)\beta^{t}\right];$$

resulting in the corresponding expected total discounted work spent on an infinite horizon, starting at state i_0.

Moreover, for a link, an action at $a \in \{0, 1\}$ and an active set $S \in \mathbb{X}$, which represents the set of states where the link can be displayed (for the active action $a = 1$), we denote by $\langle a, S\rangle$ if the policy that takes the action a in the initial period and next adopts the active policy S. Additionally, the *marginal work measure* is defined as follows:

$$w_i^S := G_i^{\langle 1,S\rangle} - G_i^{\langle 0,S\rangle} = 1 + \beta \sum_{j \in \mathbb{X}} \left(p(i,j|1) - p(i,j|0)\right) G_j^S,$$

and marginal reward measure as

$$r_i^S := F_i^{\langle 1,S\rangle} - F_i^{\langle 0,S\rangle} = R(i,1) - R(i,0) + \beta \sum_{j \in \mathbb{X}} \left(p(i,j|1) - p(i,j|0)\right) F_j^S.$$

Note that w_i^S (respectively r_i^S) measures the marginal increase of the work spent (respectively the value of the reward earned) that results from working instead of resting in the initial period starting at state i, on condition that the active policy S is adopted afterwards. Furthermore, if $w_i^S \neq 0$, the *marginal productivity measure* is defined as:

$$\nu_i^S := \frac{r_i^S}{w_i^S}.$$

In addition, *marginal productivity index* (MPI) is defined as the result of the AG algorithm shown in Table 1. This algorithm generates a family of nested sets:

$$\{S_0, S_1, \ldots, S_n\} \quad \text{where} \quad S_0 := \emptyset,\ S_n := \mathbb{X} \text{ and } S_k := \{i_1, \ldots, i_k\}, \quad \text{for } 1 \leq k \leq n,$$

where each set S_k is added to the state $i_k \in \mathbb{X} \backslash S_{k-1}$ if it has the highest marginal productivity measure.

The output consist of an ordered string $i_1, \ldots, i_n$ of all the project of states with its respective values $v_{i_k}^*$, which form the MPI of the link. The algorithm is well-defined since the computed marginal productivity rates have nonzero denominators.

The AG algorithm will be used to define a certain class of projects. Note that the acronym PCL refers to the *partial conservation laws* introduced by Niño-Mora [7].

Definition 1 (PCL-indexability). *A project is PCL-indexable if it satisfies the following conditions:*

(i) Positive marginal work: $w_i^S > 0$ *for* $i \in \mathbb{X}$, $S \in 2^{\mathbb{X}}$.
(ii) Monotone nonincreasing index computation: The index values produce by algorithm AG satisfy:

$$v_{i_1}^* \geq v_{i_2}^* \geq, \ldots, \geq v_{i_n}^*.$$

Table 1. AG Algorithm.

AG ALGORITHM

Output: $\{i_k, v^*_{i_k}\}_{k=1}^{n}$
$S_0 := \emptyset$
for $k := 1$ **to** n **do**
  pick $i_k \in \arg\max \left\{\nu_i^{S_{k-1}} : i \in \mathbb{X} \setminus S_{k-1}\right\}$
  $v^*_{i_k} := \nu_{i_k}^{S_{k-1}}$; $S_k := S_{k-1} \cup \{i_k\}$
end {for}

Observe that part (i) of Definition 1 ensures that the algorithm is well-defined. The importance of the class of projects with PCL-indexability is based on the following result obtained by Niño-Mora [11–13] in increasingly general settings.

Theorem 1. *If a project is PCL-indexable, then it is indexable and its Whittle index is the MPI produced by the AG algorithm.*

A project is dual speed if the transition state probabilities under the active and passive actions satisfy the following condition

$$P(i,j|0) = \begin{cases} \epsilon_i P(i,j|1), & \text{if } i \neq j, \\ (1-\epsilon_i) + \epsilon_i P(i,i|1), & \text{if } i = j. \end{cases}$$

where $\epsilon_i \in (0,1)$.

Remark 1. Glazebrook et al. [7] showed that a dual-speed project is PCL-indexable. Therefore, according to Theorem 1, it is indexable and its Whittle index is the MPI produced by the AG algorithm.

3 Adjusting Link Parameters to Use the AG Algorithm

Considering a system (e.g., Google search engine or a website) that pretend to show "N" links to a user but can only show M with ($M < N$) in a time period $t = 0, 1, 2, \ldots$ the resulting M links will be called the "top list", and each link is assigned a star rating between 1 and 5. Five stars indicate the highest rating and 1 the lowest. This rating is voluntarily given by users. Each link also has an access level between 1 and 5, which indicates the number of clicks over time, where 5 denotes the highest number of clicks received. The access level can be set quite arbitrarily. In our case, level 1 indicates between 1 and 1000 clicks, level 2 between 1001 and 2000 clicks, level 3 between 2001 and 3000 clicks, etc.

Therefore, there are two-dimensional states that can be represented as a vector $(s, c) \in \{1, 2, 3, 4, 5\}^2$ in which s is the star rating and c is the access level.

There is one additional state to these 25 states, the "d" state, which it will be called the "unknown" state. Each link initially begins in this state as it has never been accessed or evaluated. The assumption is that occasionally a link "dies", and if this occurs, it is immediately replaced by a new link. This is equivalent to assume that there is a small probability of transition from each of the 25 states to the unknown state, which means starting over. The "d" state serves as both a source and a sink.

The reward for user $R(i, a)$ will be taken as follows: $R(i, 1) = V(i)$ and $R(i, 0) = 0.3V(i)$, where $V(s, c)$ is a value function, defined as:

$$V(s,c) = \begin{cases} 0 & \text{in the unknown state;} \\ s \cdot c & \text{in another case.} \end{cases}$$

Moreover, the work performed by the link manager in taking active and passive action in state i will be taken as $Q(i, 1) = 1$ and $Q(i, 0) = 0$, respectively.

The state transition probabilities will be taken when the active action is applied (placing the link in the top list), in the following manner:

$$\begin{array}{lll}
P((s,c),(s+1,c)|1) = 0.1 & \text{if } 1 \le s \le 4, & 1 \le c \le 5, \\
P((s,c),(s-1,c)|1) = 0.1 & \text{if } 2 \le s \le 5, & 1 \le c \le 5, \\
P((s,c),(s,c+1)|1) = 0.2 & \text{if } 1 \le s \le 5, & 1 \le c \le 4, \\
P((s,c),(s,c-1)|1) = 0.1 & \text{if } 1 \le s \le 5, & 2 \le c \le 5, \\
P((s,c),(s+1,c-1)|1) = 0.1 & \text{if } 1 \le s \le 4, & 2 \le c \le 5, \\
P((s,c),(s-1,c+1)|1) = 0.1 & \text{if } 2 \le s \le 5, & 1 \le c \le 4, \\
P((s,c),(s+1,c+1)|1) = 0.2 & \text{if } 1 \le s \le 4, & 1 \le c \le 4, \\
P((s,c),(s-1,c-1)|1) = 0.01 & \text{if } 2 \le s \le 5, & 2 \le c \le 5, \\
P((s,c),d|1) = 0.01 & \text{if } 1 \le s \le 5, & 1 \le c \le 5, \\
P(d,d|1) = 0.01 & \text{if } 1 \le s \le 5, & 1 \le c \le 5, \\
P(d,(s,1)|1) = 0.2 & \text{if } s = 1,3,4,5, & \\
P(d,(s,1)|1) = 0.19 & \text{if } s = 2. &
\end{array}$$

The probability $P((s, c), d|1)$ will be small in relation to the probabilities $P((s, c), j|1)$ with $j \neq d$, because a state dying is considered to have a low probability.

The transition probabilities that are not shown are zero except $P(i, i|1)$ which is equal to one minus the sum of the other values of the i-th row of the stochastic matrix that is generated.

Figure 1 shows the 26 states and the transition probabilities $P(i, j|1)$ among them. The horizontal axis indicates the access level, and the vertical axis denotes the star evaluation. The figure does not show is that each state transits to the unknown state with a small probability of 0.01, nor does it illustrate the probability of remaining in the same state.

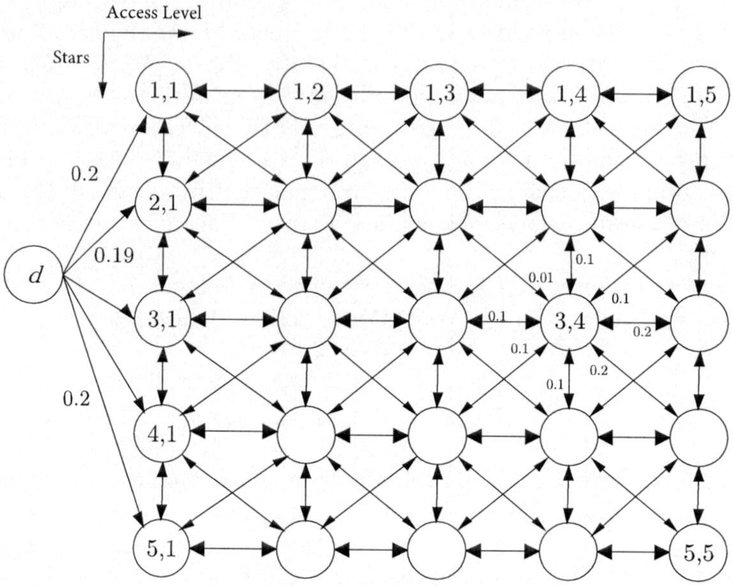

Fig. 1. State Dynamics.

Note that we consider that a state only has the following movement alternatives: moving toward contiguous states or moving toward the unknown state. The other movement options are considered to have zero probability because of the consideration that a user natural behavior is to evaluate the characteristics of the state up or down only one level or to stay in the same level. Moreover, there is another assumption that the access level tends only to grow, and the state of the characteristics tend to grow together. This is due to the natural behavior of the manager who shows increasingly better rated links.

Taking into consideration that a link in a top list encourages more users to access it, the transition probability toward a different state is greater under active action than passive action ($P(i, j|1) > P(i, j|0)$ with $i \neq j$), and the probability of staying in its state under active action is lower than under passive action ($P(i, i|1) < P(i, i|0)$), This is a dual-speed assumption that can be written as follows:

$$P(i, j|0) = \begin{cases} \epsilon_i P(i, j|1), & \text{if } i \neq j, \\ (1 - \epsilon_i) + \epsilon_i P(i, i|1), & \text{if } i = j. \end{cases}$$

where $\epsilon_i \in (0, 1)$. To find the corresponding probabilities when the link is not placed in the top list (passive action), the dual-speed function with $\epsilon_i = 0.1$ will be used for all of states of i.

The objective is finding a policy[1] in the stationary policy space that maximizes the expected total discounted reward earned[2] by the user. In an infinite horizon, this means

$$\max_{\pi \in \Pi} E_{i^0}^{\pi} \left[\sum_{t=0}^{\infty} \sum_{n=1}^{N} R_n(X_n(t), a_n(t)) \beta^t \right],$$

where $0 < \beta < 1$ is the future discount factor, $X_n(t)$, $a_n(t)$ and $R_n(X_n(t), a_n(t))$ are the state, action, and processed reward by n-th link in time period t, which is also conditioned that the initial state of the links portfolio is $i^0 = (X_n(0))_{n=1}^{N}$ and that the active action to M links in each time period is applied, which is as follows:

$$\sum_{n=1}^{N} a_n(t) = M, \quad \text{for} \quad t = 0, 1, 2, \ldots$$

By Theorem 1, using the AG algorithm, the states of a link can be indexed with the Whittle index, and the degree of sub-optimality of Whittle index policy can be analyzed, comparing its performance with that of the Greedy index policy and Whittle benchmark.

4 Simulations and Discussion

This section provides a report of some of the results of computational experiments implemented in MATLAB, developed by the authors.

4.1 Experiment 1

The parameters presented in the prior section and the AG algorithm were used in this experiment to index the states of a link with the Whittle index. Furthermore, a *sensitivity analysis* was realized to order the states by varying the discounted factor β.

Figure 2 shows that for β between 0.05 and 0.9, the highest priority is given to the state $(5, 5)$, that has the highest rewards (active and passive), and the lowest priority is given to state d, which has the lowest rewards. Additionally, the situation is reversed for β values greater than or equal to 0.98. Therefore, the conclusion is that in the short-term, exploitation is valued more than exploration, and in the long-term, exploration is valued more than exploitation, but the latter is not trivial.

On other hand, Figs. 3 and 4 illustrate the state priorities and show that if one state parameter is fixed and the other is increased, priority increases for values equal to 0.05, 0.1, 0.2, while for values equal to 0.999, the priority decreases, which is related to the above conclusion.

[1] This policy gives i the set of links shown in the top list for each time period.

[2] Total indicates the sum over all time periods.

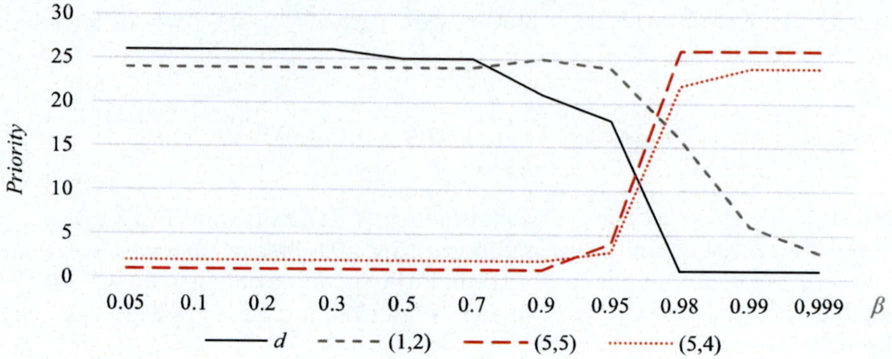

Fig. 2. Sensitivity Analysis.

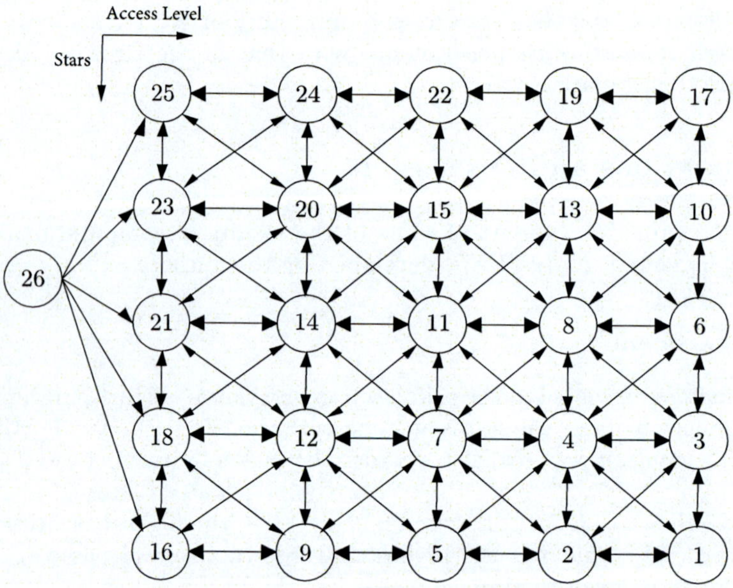

Fig. 3. State priority for $\beta = 0.05, 0.1, 0.2$ y 0.3.

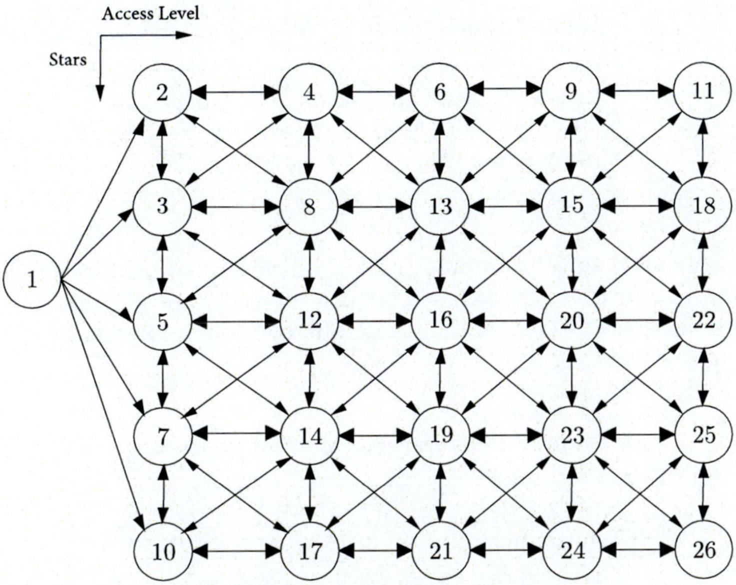

Fig. 4. State priority for $\beta = 0.999$.

In the following experiments, a Monte Carlo simulation was used to average 100 results from the total discounted reward, each with a planning horizon of 74,000 time periods. Furthermore, link transition states were simulated as a discrete random variable for each time period.

4.2 Experiment 2

In this experiment, a homogeneous multi-project is implemented, that is, N links have the same parameters (defined in Sect. 3). It is selected $M = 1$ link from a total of $N = 12$ with $\beta = 0.99$ since we are interested the long-term planning case. The initial state of the link portfolio is arbitrary, in this case, $i^0 = \{i_n^0\}_{n=1}^{12} = \{20, 21, 4, 5, 7, 13, 8, 9, 26, 2, 12, 6\}$.

Table 2 shows that Whittle index policy outperformed the Greedy index policy by 17.6%, and that the 98% confidence intervals are disjunctive. Furthermore, the confidence interval of the mean difference does not contain zero, which ensures improvement.

4.3 Experiment 3

In this experiment, a heterogeneous multi-project is implemented, that is, where the states have different state dynamics, but the same movement alternatives as the homogeneous case. We consider N, M, β and i^0 as in Experiment 2.

Table 2. Homogeneous Multi-project Case.

	$\bar{x}$	C.I
Whittle Index Policy	6285.9	(6191.1, 6380.6)
Greedy Index Policy	5174	(5078.1, 5269.3)

Table 3 shows that Whittle index policy outperformed the Greedy index policy by 30.2%, and that the 98% confidence intervals are disjunctive. Furthermore, the confidence interval of the mean difference does not contain zero, which ensures improvement.

Table 3. Heterogeneous Multi-project Case.

	$\bar{x}$	C.I
Whittle Index Policy	3807.3	(3737.6,3876.9)
Greedy Index Policy	2656.4	(2601.8,2711)

4.4 Experiment 4

In this experiment, Whittle benchmark $W(i^0)$ is calculated. Table 4 shows the percentage variation between the performance of Whittle's index policy and Whittle's benchmark in the homogeneous and heterogeneous case. This indicates that in this model, Whittle index policy provides values close to the real optimum value.

Table 4. Whittle Benchmark

	Whittle index policy	Whittle benchmark	Variation (%)
Homogeneous	6285.5	6359.5	1.16
Heterogeneous	3952.7	4074.3	3.08

4.5 Experiment 5

Whittle conjecture that was proposed by Whittle [22] is studied in this experiment. Table 5 shows that in the homogeneous case, Whittle index policy outperforms the Greedy index policy by 18.3% and is less than 2.5% of Whittle benchmark and Table 6 shows that in the heterogeneous case, Whittle index policy outperforms the Greedy index policy by 31.2% and is less than 3.5% of Whittle benchmark. Furthermore, increasing the total number of links and their respective active links maintain this trend.

Table 5. Whittle Conjecture of the Homogeneous Case.

Active Links	Index Policy		Whittle Benchmark	Percent Variance	
	Greedy	Whittle		Greedy\ Whittle	Whittle\ Whittle Benchmark
1 de 12	5174	6285.9	6359.5	17.60	1.1
2 de 24	10243.0	12489.8	12719.0	17.90	1.8
3 de 36	15355.0	18666.7	19078.5	17.74	2.1
4 de 48	20355.5	24780.9	25438.0	17.85	2.5
5 de 60	25597.2	31218.0	31797.5	18.00	1.8
6 de 72	30680.1	37566.0	38157.0	18.30	1.5

Table 6. Whittle Conjecture of the Heterogeneous Case.

Active Links	Index Policy		Whittle Benchmark	Percent Variance	
	Greedy	Whittle		Greedy\ Whittle	Whittle\ Whittle Benchmark
1 de 12	2883.8	3952.7	4074.3	27.0	2.98
2 de 24	5192.0	7548.2	7825.5	31.2	3.50
3 de 36	8358.7	11700	11716	28.5	0.14
4 de 48	11061	15354	15579	27.9	1.40

5 Future Works

As future work, it is proposed to calculate the link parameters (given in Sect. 3) with real data, using these values as input data in the AG algorithm and applying the methodology of the present work to validate if the results follow the trend of the present work. In addition, verify the results using other $R(i, a)$ reward functions for the user. On the other hand, for further study it is suggested to use computers with higher processing speed to verify Whittle's conjecture using more links and active links.

6 Conclusions

The popularity of the Internet and digital media has resulted in an impressive abundance of information. Search engines such as Google and Yahoo, billions of websites, targeted advertising, and easy access to digital content provide people with many ways to satisfy their most complex information needs.

The overload of information is easily explained in terms of the attention economy. However, the solution to this problem faced by the user remains insufficient. This article provides a mechanism to approximate the objective of optimizing the value of the information that users can obtain. To this end, the problem was formulated as a dual-speed RMABP, and we use the AG algorithm combined with Whittle index policy to approximate its solution.

The computational experiments realized in this paper show that Whittle index policy yields sub-optimal results with a small degree of sub-optimality, and also substantially improves upon the results of the Greedy index policy. In future works, the reward function used could be generalized or the size of the states could be increased. Another interesting research challenge is trying to obtain results without the dual-speed condition.

References

1. Brown, P.J., et al.: The perversion of certainty: choice architecture, digital paternalism and virtual validation in the attention economy. In: Proceedings of the International Annual Conference of the American Society for Engineering Management, pp. 1–7. American Society for Engineering Management (ASEM) (2018)
2. Brynjolfsson, E., Oh, J.: The attention economy: measuring the value of free digital services on the internet. In: ICIS (2012). https://aisel.aisnet.org/icis2012/proceedings/EconomicsValue/9
3. Caiza, G., Bologna, J.K., Garcia, C.A., Garcia, M.V.: Industrial training platform using augmented reality for instrumentation commissioning. In: De Paolis, L.T., Bourdot, P. (eds.) AVR 2020. LNCS, vol. 12243, pp. 268–283. Springer, Cham (2020). https://doi.org/10.1007/978-3-030-58468-9_20
4. Cho, J., Roy, S.: Impact of search engines on page popularity. In: Proceedings of the 13th international conference on World Wide Web, pp. 20–29. ACM (2004). https://doi.org/10.1145/988672.988676
5. Davenport, T., Beck, J.: The Attention Economy: Understanding the New Currency of Business. Harvard Business School Press, Boston (2013)
6. Falkinger, J.: Attention economies. J. Econ. Theory **133**(1), 266–294 (2007). https://doi.org/10.1016/j.jet.2005.12.001
7. Glazebrook, K.D., Niño-Mora, J., Ansell, P.S.: Index policies for a class of discounted restless bandits. Adv. Appl. Probab. **34**(4), 754–774 (2002). https://doi.org/10.1239/aap/1037990952
8. Hinz, O., Hill, S., Kim, J.Y.: TV's dirty little secret: the negative effect of popular TV on online auction sales. MIS Q. **40**(3), 623–644 (2016). https://www.jstor.org/stable/26629030
9. Huberman, B.A., Wu, F.: The economics of attention: maximizing user value in information-rich environments. Adv. Complex Syst. **11**(04), 487–496 (2008). https://doi.org/10.1142/s0219525908001830
10. Klimov, G.: Time-sharing service systems. i. Theory Probab. Appl. **19**(3), 532–551 (1975). https://doi.org/10.1137/1119060
11. Niño-Mora, J.: Restless bandits, partial conservation laws and indexability. Adv. Appl. Probab. **33**(1), 76–98 (2001). https://doi.org/10.1017/s0001867800010648
12. Niño-Mora, J.: Dynamic allocation indices for restless projects and queueing admission control: a polyhedral approach. Math. Program. **93**(3), 361–413 (2002). https://doi.org/10.1007/s10107-002-0362-6
13. Niño-Mora, J.: Restless bandit marginal productivity indices, diminishing returns, and optimal control of make-to-order/make-to-stock m/g/1 queues. Math. Oper. Res. **31**(1), 50–84 (2006). https://doi.org/10.1287/moor.1050.0165

14. Niño-Mora, J.: Dynamic priority allocation via restless bandit marginal productivity indices. TOP **15**(2), 161–198 (2007). https://doi.org/10.1007/s11750-007-0025-0
15. Niño-Mora, J.: Multi-armed restless bandits, index policies, and dynamic priority allocation. Boletín de Estadística e Investigación Operativa **26**(2), 124–133 (2010)
16. Pandey, S., Roy, S., Olston, C., Cho, J., Chakrabarti, S.: Shuffling a stacked deck: the case for partially randomized ranking of search engine results. In: Proceedings of the 31st International Conference on Very Large Data Bases, pp. 781–792. VLDB Endowment (2005)
17. Papadimitriou, C.H., Tsitsiklis, J.N.: The complexity of optimal queueing network control. In: Proceedings of IEEE 9th Annual Conference on Structure in Complexity Theory, pp. 318–322. IEEE (1994). https://doi.org/10.1109/sct.1994.315792
18. Servia-Rodríguez, S., Huberman, B., Asur, S.: Deciding what to display: maximizing the information value of social media. In: Proceedings of the International AAAI Conference on Web and Social Media, vol. 9, pp. 13–21 (2015)
19. Shapiro, C., Varian, H.: Information Rules: A Strategic Guide to the Network Economy. Harvard Business Press, Boston (2013)
20. Simon, H., Laird, J.: The Sciences of the Artificial, Reissue of the Third Edition with a New Introduction by John Laird. MIT Press, Cambridge (2019)
21. Weber, R.R., Weiss, G.: On an index policy for restless bandits. J. Appl. Probab. **27**(3), 637–648 (1990). https://doi.org/10.2307/3214547
22. Whittle, P.: Restless bandits: activity allocation in a changing world. J. Appl. Probab. **25**(A), 287–298 (1988). https://doi.org/10.2307/3214163

Artificial Intelligence

PCAnEn - Hindcasting with Analogue Ensembles of Principal Components

Carlos Balsa[1,2], Murilo M. Breve[1,2], Baptiste André[3], Carlos V. Rodrigues[4], and José Rufino[1,2(✉)]

[1] Research Centre in Digitalization and Intelligent Robotics (CeDRI), Instituto Politécnico de Bragança, Campus de Santa Apolónia, 5300-253 Bragança, Portugal
balsa@ipb.pt, murilo.breve@ipb.pt, rufino@ipb.pt

[2] Laboratório para a Sustentabilidade e Tecnologia em Regiões de Montanha (SusTEC), Instituto Politécnico de Bragança, Campus de Santa Apolónia, 5300-253 Bragança, Portugal

[3] Université de Toulouse - INP - ENSEEIHT, 31071 Toulouse Cedex 7, France
baptiste.andre@etu.inp-n7.fr

[4] Vestas Wind Systems A/S, Design Centre Porto, Centro Empresarial Lionesa, R. Lionesa Edifício B, 4465-671 Leça do Balio, Portugal
calvr@vestas.com

Abstract. The focus of this study is the reconstruction of missing meteorological data at a station based on data from neighboring stations. To that end, the Principal Components Analysis (PCA) method was applied to the Analogue Ensemble (AnEn) method to reduce the data dimensionality. The proposed technique is greatly influenced by the choice of stations according to proximity and correlation to the predicted one. PCA associated with AnEn decreased the errors in the prediction of some meteorological variables by 30% and, at the same time, decreased the prediction time by 48%. It was also verified that our implementation of this methodology in MATLAB is around two times faster than in R.

Keywords: Hindcasting · Analogue ensembles · Principal component analysis · Time series · R · MATLAB

1 Introduction

The meteorological field is used daily for many purposes in our society and has a great impact on many decision-making processes. For instance, renewable energy management often requires information about weather conditions in places without available historical data or weather forecasts.

Weather conditions can be recreated by applying a forecast model to a past starting point, a process known as *hindcasting*. Its main function is to validate the forecast model when comparable past observations are available. It can also be used for *reconstruction* purposes, whereby it derives missing past data (past observations not recorded) from the forecast model.

M. V. Garcia and C. Gordón-Gallegos (Eds.): CSEI 2022, LNNS 678, pp. 169–183, 2023.
https://doi.org/10.1007/978-3-031-30592-4_13

Hindcasting is also a research field aiming to improve methods in other domains of meteorology such as *downscaling* or *forecasting.* Meteorological data reconstruction techniques are essentially based on the Analogue Ensembles (AnEn) method [10,11]. Hindcasting with the AnEn method allows to reconstruct data of a meteorological variable i) based on data of other variable(s) at the same location, or ii) based on data of the same or other variable(s), from one or several nearby locations.

As the AnEn method benefits from large training datasets, there is a great interest in improving its computational efficiency [7,15]. In a previous work, a faster variant of the AnEn method was proposed, based on K-means clustering [3]. In this variant, the determination of the clusters is done through the previous clustering of all possible analogues.

The Principal Components Analysis (PCA) method is widely used in multivariate statistics to reduce the dataset size, retaining only the most relevant information. In the context of forecasting, PCA has already been applied together with post processing techniques like Neural Networks (NN) and the AnEn method, to forecast wind power and solar radiation [5].

The goal of this work is to combine the PCA and the AnEn methods into a new method (PCAnEn). First, PCA is applied on the input data in order to reduce the datasets dimension. Then, AnEn is used to reconstruct the data from a station by means of the neighbouring stations data. This allows the use of more stations/information, but without compromising processing time.

The remaining of this paper is organized as follows. Section 2 presents the dataset used and the correlations between the meteorological variables and stations. Section 3 introduces the PCA technique in the context of the reduction of the dataset dimensionality, and combines it with the AnEn method (PCAnEn). Section 4 is dedicated to the application of the new PCAnEn method to the principal components, for the reconstruction of meteorological variables of a single station; the results of accuracy and computational efficiency tests are presented. Final considerations and future work directions are presented in Sect. 5.

2 Meteorological Dataset

This section characterizes the meteorological dataset used and presents the correlations between the meteorological variables and stations selected.

2.1 Data Characterization

The National Data Buoy Center (NDBC), located in the southern Mississippi, in the United States, operates and maintains a network of data collection buoys and coastal stations, with collected data being publicly available [12]. The buoy network is spread worldwide, with the largest numbers located in North America.

Figure 1 shows weather stations maintained by NDBC in the region near Hampton and Newsport News. In this work, the WDSV2 station (in red) is the predicted station. The predictor stations (in black) are within a radius of

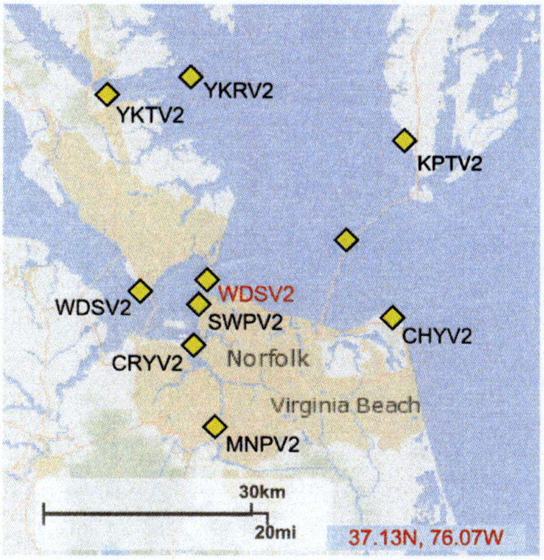

Fig. 1. Geolocation of the meteorological stations [12].

approximately 30 km from the WDSV2 station. For the experiments, the stations were ordered based on their proximity to the predictor station. The closest are SWPV2, CRYV2 and MNPV2, and so were used first in the test setups.

The meteorological variables available in the NDBC dataset, measured at each station, are: air pressure ($PRES$) [bar]; air temperature ($ATMP$) [°C]; wind speed ($WSPD$) [m/s] averaged over a 6 min period; peak gust speed (GST) [m/s] during the same 6 min period.

The characterization of these variables is shown in Table 1. Variables with more than 85% data availability were selected for the analysis. However, variables have different availability at different stations. Thus, for each variable, we choose different stations combinations, to maximize data availability.

2.2 Data Correlation

A correlation study between variables and stations was performed to define the test setups. Variables and stations that are sufficiently correlated with each other can be used together in the PCA technique, as more correlation allows more information to be kept in fewer dimensions.

The original multivariable historical dataset can be represented by the matrix $H_0 \in \mathbb{R}^{m \times n}$, where m is the number of records of n meteorological variables:

$$H_0 = \left[h_0^1 \, h_0^2 \, \cdots \, h_0^n\right] \tag{1}$$

Each column of H_0 includes the historical dataset of one of the n variables.

The matrix of the means of the observations for each variable is given by

$$\bar{H}_0 = \left[\bar{h}_0^1 \, \bar{h}_0^2 \, \cdots \, \bar{h}_0^n\right] \tag{2}$$

Table 1. Meteorological dataset characterization.

WSPD					*GST*			
Station	Min	Mean	Max	Availability (%)	Min	Mean	Max	Availability (%)
WDSV2	0.0	5.7	26.7	**97.5**	0.0	6.6	32.2	**97.5**
YKRV2	0.0	5.9	27.6	**98.0**	0.0	6.9	39.6	**98.0**
YKTV2	0.0	4.3	23.8	**97.7**	0.0	5.4	32.8	**97.7**
MNPV2	0.0	2.6	18.6	**96.4**	0.0	4.1	30.7	**96.5**
CHYV2	0.0	5.4	29.7	**95.5**	0.0	6.9	34.9	**95.5**
DOMV2	0.0	3.9	24.3	**97.5**	0.0	5.3	32.1	**97.5**
KPTV2	0.0	4.7	29.6	**97.4**	0.0	6.0	35.6	**97.5**
SWPV2	*NA*	*NA*	*NA*	0	*NA*	*NA*	*NA*	0
CRYV2	0.0	4.1	22.2	82.5	0.0	15.6	30.5	80.5
PRES					*ATMP*			
Station	Min	Mean	Max	Availability (%)	Min	Mean	Max	Availability (%)
WDSV2	970.1	1017.4	1044.9	**93.6**	−12.7	16.5	44.4	**87.9**
YKRV2	972.6	1017.4	1043.9	**98.6**	−12.8	15.9	36.3	**98.5**
YKTV2	974.7	1017.3	1044.3	**98.4**	−13.5	16.0	37.8	**98.2**
MNPV2	968.5	1017.5	1044.1	**97.9**	−13.8	16.8	37.3	**97.7**
CHYV2	985.2	1017.0	1042.7	31.1	−12.2	16.1	36.5	**97.0**
DOMV2	972.8	1017.8	1044.5	**98.3**	−12.6	16.1	37.2	**98.2**
KPTV2	*NA*	*NA*	*NA*	0	*NA*	*NA*	*NA*	0
SWPV2	972.0	1017.7	1044.1	**96.1**	*NA*	*NA*	*NA*	0
CRYV2	970.3	1017.6	1044.3	82.8	−10.5	16.5	36.3	34.3

where $\bar{h}_0^i \in \mathbb{R}^{m\times 1}$, for $i = 1, \ldots, n$, is a constant vector with value $\bar{h}_0^i$. Including the standard deviation s_i of each variable $i = 1, \ldots, n$, in a diagonal matrix $S \in \mathbb{R}^{n\times n}$ and subtracting the means and dividing by the standard deviation of each observation leads to the matrix of the scaled meteorological variables:

$$H = S^{-1}\left(H_0 - \bar{H}_0\right) = \left[\frac{h_0^1 - \bar{h}_0^1}{s_1}\ \frac{h_0^2 - \bar{h}_0^2}{s_2}\ \cdots\ \frac{h_0^n - \bar{h}_0^n}{s_n}\right] = [h_1\, h_2 \cdots h_n] \quad (3)$$

The matrix H is then used to obtain the correlation matrix, given by

$$C = \frac{1}{m} H^T H \quad (4)$$

where each (i, j)-entry of the matrix C is the correlation between the meteorological variables h_0^i and h_0^j.

Figures 2 and 3 present the correlations between different meteorological variables within the same station. In stations KPTV2 and CHYV2 (Fig. 2) there are only records of two ($WSPD$ and GST) and three ($WSPD$, GST and $ATMP$) variables, respectively. In the other stations (Fig. 3) there are records of four variables: $WSPD$, GST, $ATMP$ and $PRES$.

In all stations there is a high correlation between $WSPD$ and GST. A mild inverse correlation is observable between $ATMP$ and $PRES$. The other variable interactions showed low and inconsistent correlation among the stations.

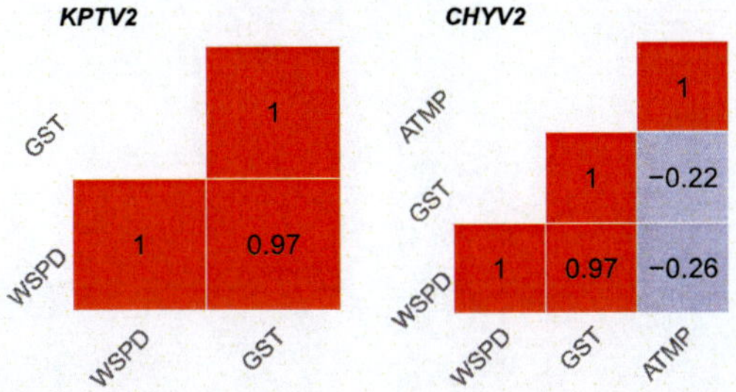

Fig. 2. Correlation between variables at stations with less data available.

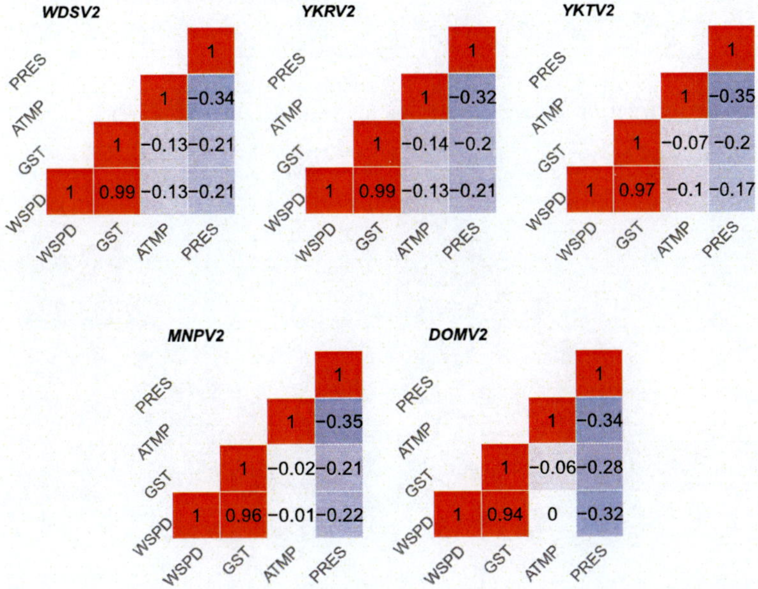

Fig. 3. Correlation between variables at stations with more data available.

Complementarily, Figs. 4 and 5 show the correlations for the same variable between different stations. It can be observed that, with minor exceptions, correlations are high, with variables $PRES$ and $ATMP$ showing the highest correlations between different stations.

3 Methods

In this section the PCA technique is introduced and applied in the context of the reduction of the dimensionality of the dataset used in this study. A brief revision of the AnEn method is also provided, along with a clarification on how it is combined with the PCA technique.

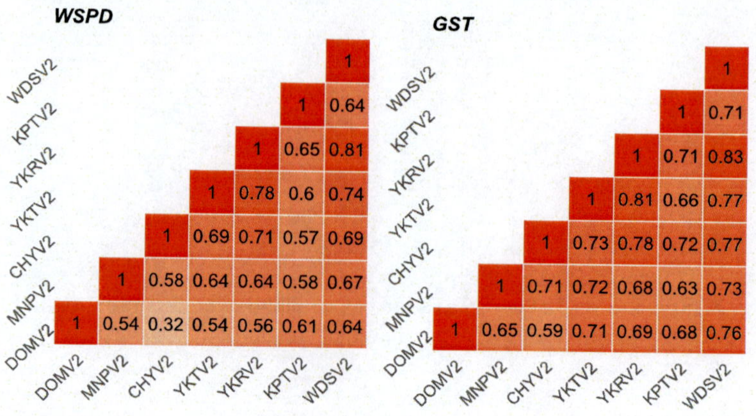

Fig. 4. Correlation between stations for the variables $WSPD$ and GST.

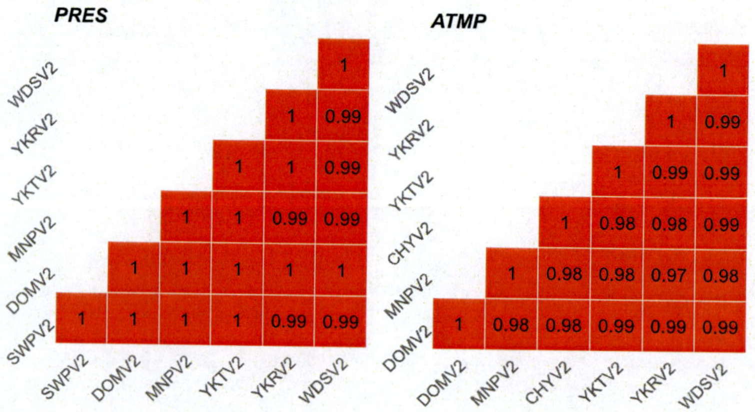

Fig. 5. Correlation between stations for the variables $PRES$ and $ATMP$.

3.1 Principal Components Analysis

The PCA technique identifies the dimensions along which the data are most dispersed. In this way, we can identify the dimensions that best differentiate the dataset under analysis, that is, its principal components. This can be achieved by the thin singular value decomposition of the scaled data matrix $H \in \mathbb{R}^{m\times n}$, previously introduced in Eq. (4), that is given by

$$H = U\Sigma V^T \tag{5}$$

where $U \in \mathbb{R}^{m\times n}$, $\Sigma \in \mathbb{R}^{n\times n}$ and $V \in \mathbb{R}^{m\times n}$ (see [6]). The diagonal matrix Σ contains the singular values σ_i of H, for $i = 1, \ldots, n$, where $\sigma_1 > \sigma_2 > \ldots > \sigma_n$. The right singular vectors v_i are the *principal components directions* of H.

The vector

$$z_1 = Hv_1 \tag{6}$$

has the largest sample variance, given by σ_1^2/m, amongst all normalized linear combinations of the columns of H. Vector z_1 represents the first new variable and it is called the first principal component (PC_1). The second principal component (PC_2) is $z_2 = Hv_2$ because v_2 corresponds to the second largest variance (σ_2^2/m), and the remaining principal components are defined similarly. The new variables are linear combinations of the columns of H, i.e., they are linear combinations of the normalized variables $h_1, h_2, \ldots, h_n$,

$$z_i = v_{1i}h_1 + v_{2i}h_2 + \ldots + v_{ni}h_n \quad \text{for} \quad i = 1, 2, \ldots, n \tag{7}$$

where the coefficients v_{ji}, $j = 1, 2, \ldots, n$ (called *loadings*) are the elements of the vector v_i. The magnitude of a coefficient is related to the relative importance of the corresponding variable in the principal component.

The substitution criterion of the original variables by a few of the new variables must take into account the influence of the new variables on the variance of the original data. This influence is directly proportional to the magnitude of the corresponding singular values. It is expected that the first few principal components, corresponding to the largest singular values, account for a large proportion of the total variance, so they are all that is needed for future analyses [14].

3.2 Dimension Reduction of the Dataset

A decomposition into principal components (PCs) of the original meteorological variables, coming from different stations, was performed. Tables 2 and 3 show the standard deviations of each PC for different amount of input stations. In Table 2, PCs were calculated from the variables $WSPD$ and GST coming from a number of neighboring stations between 2 and 6. In Table 3, PCs were calculated from the variables $PRES$ and $ATMP$ from a number of neighboring stations ranging from 2 to 5. The variables from the WDSV2 station were not included in the original variables because WDSV2 was used only as the predicted station.

In Table 2 the PCA is performed from the data matrix that includes two meteorological variables, GST and $WSPD$, coming from different stations. This is because GST and $WSPD$ are highly correlated (recall Sect. 2.2) and so it is possible to use them together in the PCA. In this table, standard deviations above 1 are highlighted. When this occurs, the corresponding PC has a higher variance than the original scaled variables and, consequently, more information.

It can also be seen in Table 2 that, for most cases, the standard deviation is greater than 1 for PC_1 and PC_2, meaning that these two new variables concentrate the information contained in all the original variables (included in the data matrix H). As in [5], we choose the PCs with standard deviations higher than 1 to represent the original dataset. It can be seen that for $WSPD$ and GST, PC_1 and PC_2 showed values higher than 1, except for the 2-station configuration. As expected, by increasing the amount of input stations, more components are needed to represent the original dataset.

Table 3 shows the standard deviations of the PCs computed from the $ATMP$ and $PRES$ variables. It is important to note that, unlike Table 2, these variables

Table 2. Standard deviation of the PCs generated from the variables $WSPD$ and GST coming from different stations together.

Standard Deviation						
Stations	WSPD and GST					
	PC_1	PC_2	PC_3	PC_4	PC_5	PC_6
2	**1.771**	0.881	0.234	0.169	–	–
3	**2.059**	**1.027**	0.782	0.207	0.172	0.148
4	**2.386**	**1.030**	0.807	0.688	0.207	0.178
5	**2.689**	**1.038**	0.848	0.691	0.609	0.207
6	**2.913**	**1.047**	0.885	0.834	0.652	0.607

Table 3. Standard deviation of the PCs generated from the variables $PRES$ and $ATMP$ coming from different stations together.

Standard Deviation								
Stations	PRES				ATMP			
	PC_1	PC_2	PC_3	PC_4	PC_1	PC_2	PC_3	PC_4
2	**1.414**	0.039	–	–	**1.408**	0.137	–	–
3	**1.731**	0.040	0.039	–	**1.721**	0.154	0.126	–
4	**1.998**	0.063	0.039	0.028	**1.987**	0.154	0.129	0.102
5	**2.233**	0.086	0.050	0.037	**2.220**	0.185	0.139	0.102

were analyzed separately, because they do not correlate sufficiently with each other. The same pattern of values was observed for both variables, that is, PC_1 was sufficient to represent the data in all configurations.

3.3 Combining PCA with AnEn (PCAnEn)

The AnEn method can be used to reconstruct missing or incomplete data of a weather station. The reconstruction is done with data from nearby predictor stations. This method allows using more than one predictor station (or variable). In this case, the data from the predictor stations (or variables) can be used either dependently or independently (i.e., with the analogues selected in different predictor series having to overlap in time, or not – see [1] and [2]).

Figure 6 is a representation of the AnEn classical method combined with the PCA technique, in the experimental scenario considered in this work. The combined PCAnEn method uses the time series with the chosen principal components (PCs) to reconstruct a period (prediction period) of another times series (station WDSV2). Both time series are complete in the training period. In the prediction period, only the PCs generated by neighbouring stations have data (actually, data from the predicted station along the prediction period are also available and so are used to measure the accuracy of the reconstruction method).

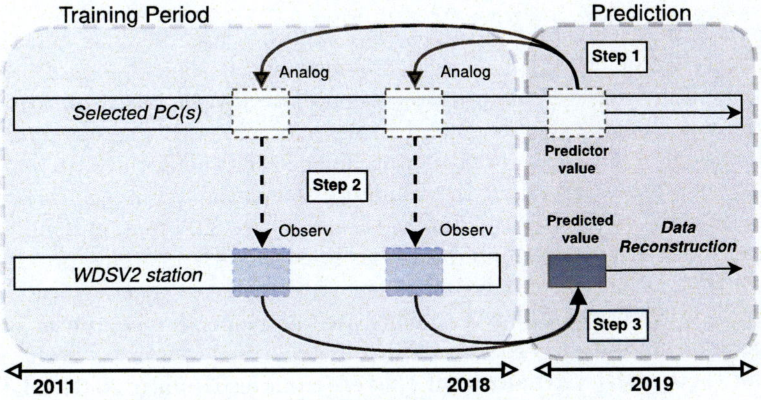

Fig. 6. Hindcasting with the PCAnEn method.

The methodology starts by selecting analogue values according to similarity with the predictor value (step 1). Both the predictor and the analogues are vectors of $2k+1$ elements, where each element is the value of a weather variable at successive $2k+1$ instants of the same time window, and $k > 0$ is an integer representing the width of each half-window (into the past, and into the future) around the central instant.

In step 2, each analogue value has a corresponding observed value, at the same time instant as the central analogue value. The comparison of vectors, rather than single values, accounts for the evolutionary trend of the weather variable around the central instant of the time window. It thus allows the selection of analogues to take into account weather patterns, rather than single isolated values. For the experiments, a value of $k = 1$ was used, that is, each vector represents 12 min (data are distributed every 6 min).

Finally, in step 3, the observations selected in the training period are used to predict (hindcast) the missing value, through its average, weighted or not. When this value is actually available as real observational data (as it happens in this work), it then becomes possible to assess the prediction/reconstruction error.

In this study, the available historical data ranges from 2011 to the last hour of 2018, and the reconstruction period is 2019. Because of the high resolution (6 min) and large amount of data, we opted to make predictions only between 10 am and 4 pm, every 6 m. For the classical AnEn experiments, the original data is used instead of the PCs.

4 Experiments with the PCAnEn Method

In this section, the PCAnEn method is applied to a hindcasting problem with the dataset presented in Sect. 2. Several experiments were conducted in order to evaluate the effects of using principal components instead of the original historical data. The accuracy of the reconstructed values is assessed by comparison to the exact values recorded at the WDSV2 station during the prediction period.

The error metrics used to evaluate the accuracy are the Bias and the Root Mean Square Error (RMSE), as recommended by [4]. The Bias is a basic indicator of the *systematic error* in a prediction. The RMSE is an indicator of the corresponding global error, i.e., the RMSE error includes the *systematic error* and the *random error*. In the artificial intelligence (AI) field, the *systematic error* is the *approximation error* and the *random error* is the *estimation error* [9].

All tests were performed in duplicate, using two different implementations of the methods, one in R [13] and another in MATLAB [8]. This provided confidence on the numerical results obtained (which were expected, and verified, to be identical) and also allowed to compare the respective computational performance. The computer system used in the experiments was a virtual machine hosted on the CeDRI virtualization cluster, running Ubuntu 20.04.4 LTS. The resources associated with the virtual machine were 16 virtual cores of an Intel Xeon W-2195 CPU, 64 GB of RAM and 256 GB of SSD-based secondary storage.

The tests were divided between different amounts of predictor stations. In addition, the results obtained with the classical AnEn method applied to the original variables are also presented, allowing a comparison with the results obtained with the PCAnEn methodology. Note that for AnEn, the same stations as the PCAnEn (2-station configuration) were chosen for the prediction; the variable is predicted from the same variable located in the two closest stations, in order to ensure the most favourable configuration of the AnEn method.

The Subsect. 4.1 presents and discusses the accuracies obtained from the experiments. The Subsect. 4.2 shows a comparison of performance between the AnEn and PCAnEn methods, and between R and MATLAB implementations.

4.1 Comparing Accuracy

Figure 7 allows to compare the accuracy of the AnEn and PCAnEn methods, with different amounts of stations, for the four meteorological variables considered in this study. For each combination, the number of PCs used is 1 or 2, as indicated in Tables 2 and 3. The chart is based on the accuracies provided by the R implementation; however, they are very similar to those of the MATLAB implementation, as may be seen in Table 4; this table provides the full accuracy results, including also the Bias in addition to the RMSE.

The smallest errors were obtained by the PCAnEn method in the configurations with 4 or 5 stations. For the variables $ATMP$, GST and $WSPD$, 5 stations showed better RMSE. In contrast, for $PRES$, the 4-station configuration provided the most accurate prediction.

For instance, as shown in Table 4, the predictions of $WSPD$ and GST with the PCAnEn method generated 30% and 21.8% lower RMSE errors compared to the classical AnEn method, respectively. To a lesser effect, the reconstructions of the $PRES$ and $ATMP$ variables showed a reduction of 13.6% and 16.7% with the PCAnEn method. Moreover, the lowest Bias measurements were obtained with 4 stations in all variables.

The non-$PRES$ variables showed better predictions with all configurations of PCAnEn compared to AnEn, except for the one with 2 stations. This is expected

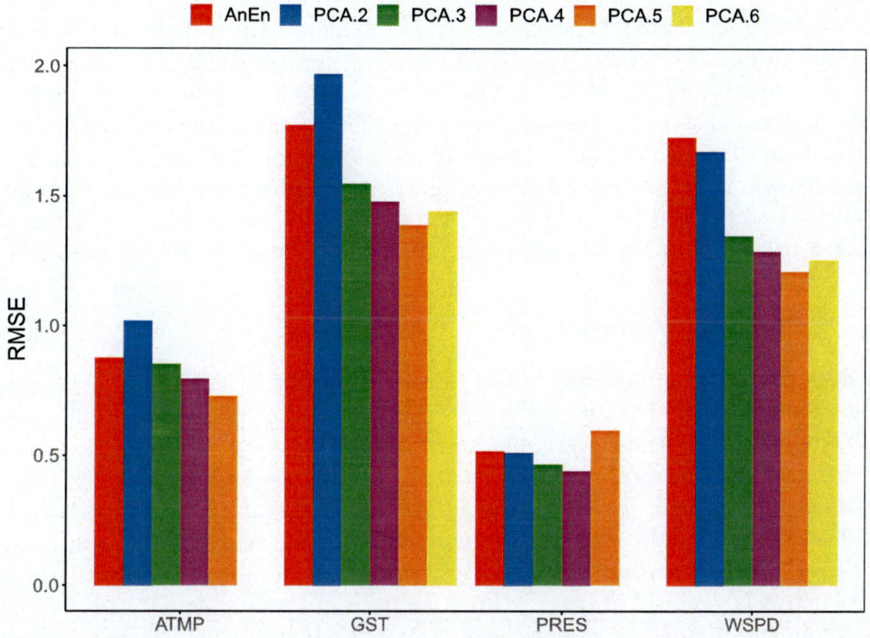

Fig. 7. Comparison of the RMSE for different variables and number of stations.

Table 4. Comparison between the PCAnEn and AnEn methods.

Method	#St	Errors	R				MATLAB			
			WSPD	*GST*	*ATMP*	*PRES*	*WSPD*	*GST*	*ATMP*	*PRES*
PCAnEn	2	Bias	−0.086	−0.349	0.204	0.444	−0.085	−0.347	0.204	0.444
		RMSE	1.674	1.972	1.020	0.512	1.675	1.972	1.019	0.512
	3	Bias	−0.065	−0.327	0.179	0.367	−0.065	0.327	0.179	0.367
		RMSE	1.347	1.548	0.851	0.467	1.347	1.549	0.851	0.467
	4	Bias	**−0.038**	**−0.295**	**0.139**	**0.357**	**−0.038**	**−0.295**	**0.139**	**0.357**
		RMSE	1.287	1.478	0.795	**0.442**	1.288	1.479	0.795	**0.442**
	5	Bias	−0.048	−0.297	0.119	0.508	−0.049	−0.297	0.119	0.508
		RMSE	**1.210**	**1.388**	**0.729**	0.597	**1.210**	**1.389**	**0.729**	0.597
	6	Bias	−0.066	−0.313	–	–	−0.066	−0.313	–	–
		RMSE	1.255	1.442	–	–	1.256	1.443	–	–
AnEn	2	Bias	−0.147	−0.353	0.164	0.410	−0.147	−0.353	0.164	0.410
		RMSE	1.728	1.774	0.875	0.516	1.728	1.774	0.875	0.516

since these stations are the same as the ones used in the tests with AnEn. The *PC*s can represent the data in fewer components, but there is always some information that is lost.

Regarding the issue of *dependency*, Fig. 8 shows the values of the RMSE obtained in the prediction of $WSPD$ and GST by the PCAnEn method used in a *dependent* and *independent* way, with 3 or more stations (for details on these two variants of the method see [3]). The results show clearly that an *independent* PCAnEn did not improve the results in any configuration or station, in comparison with the *dependent* version. It is also observed that increasing the number of stations up to 5 leads to a reduction in the RMSE error, but the increase to 5 stations no longer brings advantages, since the RMSE increases.

4.2 Comparing Performance

Figure 9 shows the processing times obtained by our MATLAB and R implementations of the PCAnEn and AnEN methods, with different station quantities. The processing times were measured when using 14 CPU-cores (above that number, the decrease in overall execution time was negligible – see Fig. 10). As previously mentioned, for the AnEn method only 2 stations were used, and so Fig. 9 only provides two execution times (one for each AnEn implementation).

The PCAnEn method significantly reduces the total processing time compared with the classical AnEn method, for both of the implementations. This is evident in the 2 stations scenario: using MATLAB, the PCAnEn method consumes 38% (30.4/79.3) of the time spent by the AnEn method (a speedup of 2.6x); in turn, using the R implementations, the PCAnEn method runs in 28% (51.3/179.6) of the time needed by the AnEn method (a speedup of 3.5×).

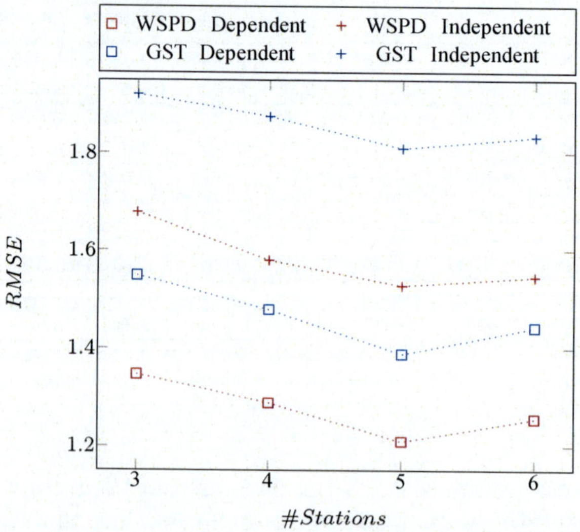

Fig. 8. RMSE of PCAnEn used in a dependent and independent way.

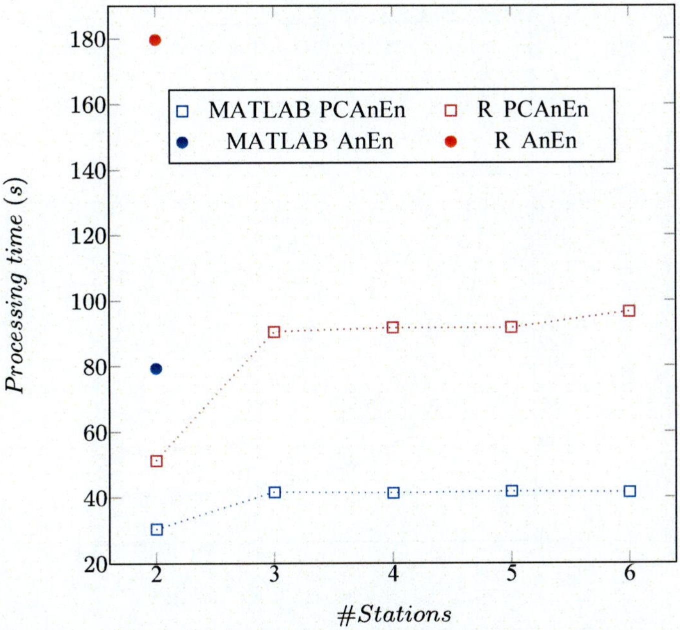

Fig. 9. Processing time for different number of stations and different methods.

Focusing only on the PCAnEn method, the processing times varies little with the number of stations, in both implementations (the exception is the 2 stations scenario, where the processing time is visibly smaller than with more stations).

For any number of stations used, the MATLAB implementation was always found to be faster than the R implementation. For instance, with 6 stations, PCAnEn in MATLAB was 2.3× (96.6/41.5) faster than in R, though with 2 stations the speedup was only 1.7× (51.3/30.4). In turn, also with 2 stations, AnEn in MATLAB was 2.3× (179.6/79.3) faster than its implementation in R.

Finally, the processing times in function of the number of CPU cores used was also evaluated. Figure 10 shows the processing times for the PCAnEn method with 6 stations (using *PCs* generated from the variables *WSPD* and *GST*), when varying the number of CPU-cores from 1 up to 14. It can be observed that both implementations scale reasonably well, though with diminishing returns past 8 CPU cores. Again, the MATLAB implementation offers superior performance and slightly better scalability. It should be noted, however, that MATLAB is known to be particularly optimized to take advantage of Intel CPUs (as the one where this evaluation was performed), once it relies on the Intel MKL library.

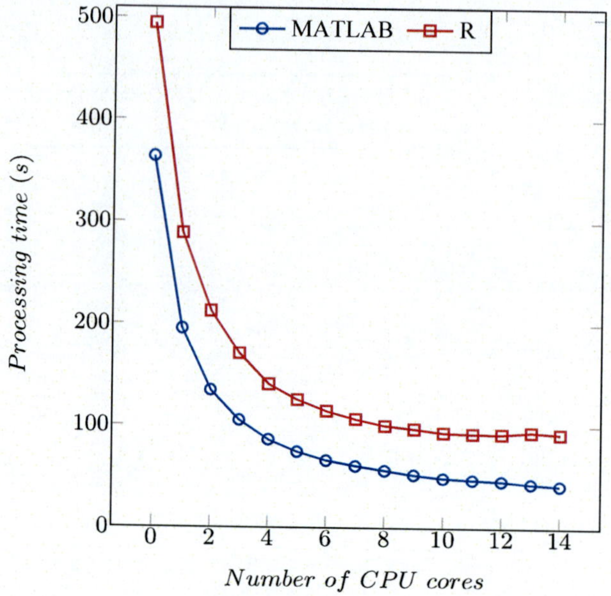

Fig. 10. Processing time for different number of CPU cores used by PCAnEn.

5 Conclusion

This study presents a methodology where PCA is applied to the classical AnEn method. Hence, information from several stations is combined in a reduced number of time series, corresponding to the principal components, which are then submitted to the AnEn method. In general, the location of the stations is of great importance, because it promotes more correlation between variables and stations, and thereby greater effectiveness of the PCA technique.

The combination of the PCA technique with the AnEn method results in a new methodology (PCAnEn) that proved, in our experiments, to offer better hindcasting accuracy than the classical AnEn method. In the present study, the data reconstruction of the WDSV2 station by means of the 5 nearest station seems to be optimal. However, the choice of predictor stations must take into account the proximity and correlation between them (which needs to be assessed prior to the determination of the PCs).

In terms of computational performance, the PCAnEn method allows to reduce the processing time considerably, compared to the classical AnEn method. It was also verified that the implementation in MATLAB is faster (and by which magnitude) than the implementation in R. This information may then be considered in the choice between a proprietary non-free platform and an open-source free one, to solve the same kind of hindcasting problems.

In the future, we plan to apply the same methodology in different regions and with a different number of neighboring stations available. This will allow us to better evaluate the virtues of this new approach, especially regarding the balance between closeness, correlation and quantity of stations.

Acknowledgements. The authors are grateful to the Foundation for Science and Technology (FCT, Portugal) for financial support through national funds FCT/MCTES (PIDDAC) to CeDRI (UIDB/05757/2020 and UIDP/05757/2020) and SusTEC (LA/P/0007/2021).

References

1. Balsa, C., Rodrigues, C.V., Lopes, I., Rufino, J.: Using analog ensembles with alternative metrics for hindcasting with multistations. ParadigmPlus **1**(2), 1–17 (2020). https://journals.itiud.org/index.php/paradigmplus/article/view/11
2. Balsa, C., Rodrigues, C.V., Araújo, L., Rufino, J.: Hindcasting with cluster-based analogues. In: Guarda, T., Portela, F., Santos, M.F. (eds.) ARTIIS 2021. CCIS, vol. 1485, pp. 346–360. Springer, Cham (2021). https://doi.org/10.1007/978-3-030-90241-4_27
3. Balsa, C., Rodrigues, C.V., Araújo, L., Rufino, J.: Cluster-based analogue ensembles for hindcasting with multistations. Computation **10**(6), 91 (2022). https://doi.org/10.3390/computation10060091
4. Chai, T., Draxler, R.R.: Root mean square error (RMSE) or mean absolute error (MAE)? – arguments against avoiding RMSE in the literature. Geosci. Model Dev. **7**(3), 1247–1250 (2014). https://doi.org/10.5194/gmd-7-1247-2014
5. Davò, F., Alessandrini, S., Sperati, S., Monache, L.D., Airoldi, D., Vespucci, M.T.: Post-processing techniques and principal component analysis for regional wind power and solar irradiance forecasting. Solar Energy **134**, 327–338 (2016). https://doi.org/10.1016/j.solener.2016.04.049
6. Eldén, L.: Matrix Methods in Data Mining and Pattern Recognition. SIAM, Philadelphia (2007)
7. Hu, W., Vento, D., Su, S.: Parallel analog ensemble - the power of weather analogs. In: Proceedings of the 2020 Improving Scientific Software Conference, pp. 1–14. NCAR (2020). https://doi.org/10.5065/P2JJ-9878
8. MATLAB: version 7.10.0 (R2010a). The MathWorks Inc., Natick, Massachusetts (2010)
9. de Mello, R.F., Ponti, M.A.: Machine Learning. Springer, Heidelberg (2018). https://doi.org/10.1007/978-3-319-94989-5
10. Monache, L.D., Eckel, F.A., Rife, D.L., Nagarajan, B., Searight, K.: Probabilistic weather prediction with an analog ensemble. Mon. Weather Rev. **141**(10), 3498–3516 (2013). https://doi.org/10.1175/mwr-d-12-00281.1
11. Monache, L.D., Nipen, T., Liu, Y., Roux, G., Stull, R.: Kalman filter and analog schemes to postprocess numerical weather predictions. Mon. Weather Rev. **139**(11), 3554–3570 (2011). https://doi.org/10.1175/2011mwr3653.1
12. National Weather Service: National Data Buoy Center. https://www.ndbc.noaa.gov
13. R Core Team: R: A Language and Environment for Statistical Computing. R Foundation for Statistical Computing, Vienna, Austria (2022). https://www.R-project.org/
14. Spence, L., Insel, A., Friedberg, S.: Elementary Linear Algebra: A matrix Approach. Pearson Education Limited (2013)
15. Vannitsem, S., et al.: Statistical postprocessing for weather forecasts: review, challenges, and avenues in a big data world. Bull. Am. Meteorol. Soc. **102**(3), E681–E699 (2021). https://doi.org/10.1175/bams-d-19-0308.1

Prediction Models for Car Theft Detection Using CCTV Cameras and Machine Learning: A Systematic Review of the Literature

Joseph Ramses Méndez Cam[1], Félix Melchor Santos López[2(✉)], Víctor Genaro Rosales Urbano[3], and Eulogio Guillermo Santos de la Cruz[3]

[1] School of Science and Engineering, Pontifical Catholic University of Peru, Lima, Peru
a20170407@pucp.edu.pe

[2] Department of Engineering, Pontifical Catholic University of Peru, Lima, Peru
fsantos@pucp.edu.pe

[3] Faculty of Industrial Engineering, National University of San Marcos, Lima, Peru
{vrosalesu,esantosd}@unmsm.edu.pe

Abstract. Car theft is a constant problem in parking lots and places where cars are left unattended. Car theft detection is a time-consuming task due to the human resources that are required. Therefore, the task of checking closed circuit television (CCTV) cameras can be automated using machine learning techniques. The implementation of such a system would mean an optimization of the current technology. Even if a CCTV camera is installed, it requires human labor to supervise the area, which is a repetitive and time-consuming task. A machine learning algorithm could simplify the task and attend to many cameras without decreasing the attention on each one. In this context, a systematic review of the literature on machine learning was conducted based on four research questions using the PRISMA methodology. The research method may help to find the current methods used in similar applications and possible ways to implement the proposed automatic solution. This scientific study retrieved 384 articles from Web of Science, Scopus, and IEEE databases. The number of studies used to answer the research questions was 58. Finally, analyzing the most frequent models and metrics, Convolutional Neural Networks and Accuracy were the most referenced, with 30 and 42 mentions, respectively.

Keywords: Machine learning · car theft · model · recognition · prediction · video analysis

1 Introduction

Car theft is one of the most common problems in parking lots. Moreover, various problems can occur in places where cars are left alone. In the worst-case scenario, the whole car can be stolen. It is easy for a thief to steal valuable objects or car

M. V. Garcia and C. Gordón-Gallegos (Eds.): CSEI 2022, LNNS 678, pp. 184–200, 2023.
https://doi.org/10.1007/978-3-031-30592-4_14

parts without the owner noticing. According to the International Association of Auto Theft Investigators, in 2020 there were 810,400 vehicles stolen, which summed to 7.4 billion dollars in vehicle thefts. That is the reason why many parking lots use closed circuit television (CCTV) cameras for surveillance [20].

However, the task of constantly checking the security cameras demands resources and it is possible for the person in charge to omit some incidents if many cameras need to be checked at the same time. Moreover, some people do not have a good impression of traditional video surveillance because it invades their privacy. Hence, the task of improving and automating the surveillance arises. The labor of analyzing and judging whether a car is being stolen can be automated using machine learning algorithms. That is the reason why it is important to find different algorithms that could potentially help automate the CCTV camera surveillance to avoid car theft. With the development of machine learning models and artificial intelligence applications, there have been improvements in solving the current problem. Even though machine learning models involving car theft are not common, related models such as recognition of general theft, violence, or suspicious activity have been developed. Accordingly, the objective of this paper is to analyze the most updated literature connected to the topic using a systematic review.

The following article will be structured as follows: first, materials and methods are presented in the beginning, including the search methodology, criteria for inclusion and exclusion, research questions, and the studies involved. The research questions were R1 (What models exist that can detect theft using a CCTV camera?), R2 (What metric do they use to evaluate the detection models?), R3 (What kind of activity can be detected using the models?), R4(How is the data pre-processed for the machine learning model to work properly?). Then, the next section presents the results and discussion, including the answer to each research question and data sources. After, an analysis of the studies is presented, including additional information for each research question. Finally, the article presents the conclusion and references.

2 Materials and Methods

This study applies the methodology of Preferred Reporting Items for Systematic Reviews and Meta-Analyses (PRISMA) as a guideline for the literature review. Using this methodology allows the reader to assess the article in a simple manner. Also, the standard structure and format enables other people to replicate the analysis.

2.1 Search Methodology

The search using keywords was conducted in three different databases: Scopus, IEEE, and Web of Science. The search included the following keywords: "theft AND recognition AND artificial intelligence". The second search included the following keywords: "security AND video AND surveillance AND artificial intelligence". Using filters, only studies that were public access and published in the last 5 years were included. In the first search, there were 151 scientific articles, of which 71 were from Scopus, 4 were from IEEE, and 76 were from Web of Science. In the second search, there were 233 scientific articles, of which 151 were from Scopus, 50 were from IEEE, and 32 were from Web of Science. Afterward, the important data was extracted from specific characteristics.

2.2 Criteria for Inclusion and Exclusion

It was necessary to review the obtained articles using certain criteria to ensure that they were relevant to the topic of theft detection in cars. The chosen articles were included if they accomplished the following eligibility criteria: (1) Detection/Prediction/Analysis or similar terms in their title and abstract (2) The article included at least one artificial intelligence algorithm (3) The article included information on suspicious or violent activities (4) The document works with any kind of surveillance using video (5) The document uses at least one metric for the Detection/Prediction/Analysis. Criteria of exclusion: (1) Duplicated documents (2) Articles that were general and not completely relevant, such as study articles (3) Articles that did not show helpful information to answer the research questions.

2.3 Research Questions

The research questions presented in this literature review will help to retrieve important characteristics from the articles that can highlight the crucial information that is required to apply artificial intelligence methods to the current problem of detecting car theft using CCTV cameras. The research questions are expressed as follows:

- R1: What models exist that can detect theft using a CCTV camera?
- R2: What metric do they use to evaluate the detection models?
- R3: What kind of activity can be detected using the models?
- R4: How is the data pre-processed for the machine learning model to work properly?

2.4 Studies Selection

Following the steps that the PRISMA methodology indicates in the search with keywords, the important documents were obtained. First, duplicate investigations were removed, so 170 articles were excluded. After, the 214 remaining articles underwent a second evaluation. Considering the title, abstract, and content, 106 articles were removed because they did not meet the inclusion and exclusion criteria. Most of the included articles included topics related to image recognition models. The excluded articles mostly involved related topics that were not relevant, such as license plate recognition. After applying the eligibility criteria, 108 articles remained in the third stage. Because of the previous analysis, some of the articles did not meet the criteria for inclusion. Finally, 58 articles were used to answer the research questions. The process of selection according to PRISMA methodology is represented in Fig. 1.

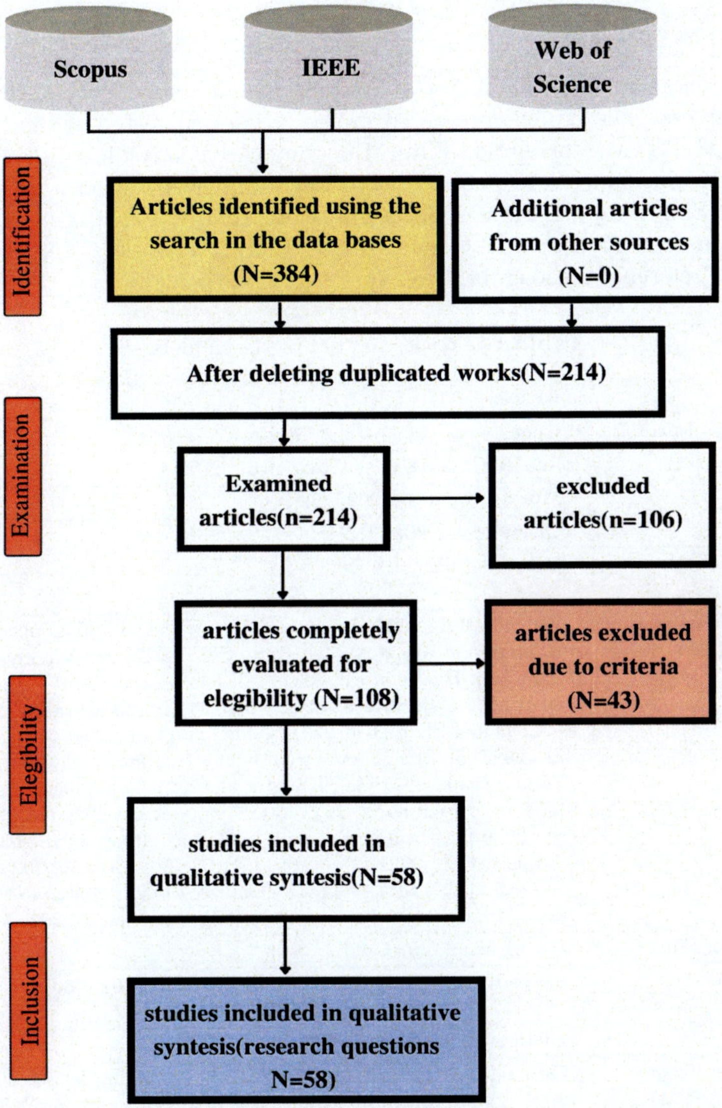

Fig. 1. Diagram using PRISMA methodology for the studies included in the systematic review.

3 Results and Discussion

The systematic review includes 58 articles related to prediction models for car theft detection using CCTV cameras and machine learning. The articles were from the three following sources: 24 articles from Scopus (41% of the articles),12 articles from IEEE (21% of the articles), and 22 articles from Web of Science (38% of the articles).

3.1 Answer to RQ1

The first question asks whether the existent models can detect theft using a CCTV camera. The models presented in the articles consisted of the following: CNN, SVM, LSTM, Correlation Filter, Boosting, Adversarial learning, SSD, DS-GRU, Convolutional Autoencoder, Auto-Encoder, Deep Generative Network, Temporal Shift Module, Radial Basis Function, BFFN, RBM, and NNC. The most common method was the Convolutional Neural Network (CNN), which was used in 30 references, shown in Table 1.

Table 1. Articles related to the models.

Model	Articles
Adversarial learning	Cheng et al., 2020 [13]; Chen et al., 2018 [12]
Auto-Encoder	Sun et al., 2018 [46]; Wang et al., 2022 [53]
BFFN	Murugesan et al., 2022 [32];
Boosting	Cheng et al., 2020 [13]
Convolutional autoencoder	Duman et al., 2019 [17]
Convolutional Neural Network (CNN)	Ahmed et al., 2021 [1]; Ata-ur-Rehman et al., 2021 [5]; Ronquillo et al., 2020 [38]; Bibi et al., 2020 [10]; Srividya et al., 2021 [43]; Haque et al., 2021 [18]; Islam et al., 2021 [21]; Apon et al., 2021 [4]; Shreyas et al., 2020 [41]; Sudhakaran et al., 2017 [44]; Peipei Zhou et al., 2017 [60]; Lou et al., 2021 [29]; Atto et al., 2020 [6]; Shoaib et al., 2021 [40]; Rendon et al., 2021 [37]; Sanam et al., 2021 [33]; Kang et al., 2021 [23]; Vieira et al., 2022 [50]; Abid Mehmood, 2021 [30]; Cheng et al., 2020 [13]; Baba et al., 2019 [8]; Brahmaiah et al.,2021 [11]; Atzori et al.,2021 [7]; Ahmed et al., 2021 [2]; Ye et al., 2021 [56]; Li et al.,2020 [26]; Alsaedi et al., 2021 [3]; Fath et al., 2019 [49]; Hussain et al., 2022 [19]; Abid Mehmood, 2021 [31]; Singh et al., 2020 [42]
Correlation Filter	Cheng et al., 2020 [13];
Deep Generative Network	Saypadith et al., 2021 [39]; Reinolds et al., 2022 [36]
DS-GRU	Ullah et al., 2021 [48]
Long Short Term Memory (LSTM)	Duman et al., 2019 [17]; Wang et al., 2022 [53]; Islam et al., 2021 [21]; Sudhakaran et al., 2017 [44]; Lou et al., 2021 [29]; Lejmi et al., 2020 [25]; Shoaib et al., 2021 [40]; Rendon et al., 2021 [37]; Kang et al., 2021 [23]
Nearest neighbor classifier (NNC)	Chou et al., 2018 [14]
Radial Basis Function	Liang et al., 2020 [57]
RBM	Vu, 2017 [51]
Single Shot Detection (SSD)	Brahmaiah et al., 2021 [11]; Ke et al., 2021 [24]; Cob-Parro et al., 2021 [15]
Support Vector Machine (SVM)	Bibi et al., 2020 [10]; Zhang et al., 2020 [58]
Temporal Sift Module	Liang et al., 2021 [28]

3.2 Answer to RQ2

The second question asks about the metrics used in the detection models. Most of the metrics are standard and can be used in every model. Table 2 shows the metrics used for the different studies. The metrics consist of Accuracy, AUC, ROC, Precision, Recall, F1 score, Confusion matrix, Sensitivity, Mean Average Precision, F-measure, MSE, ERR, and Loss. The metric that was used in most of the studies was Accuracy, with 42 mentions in Table 2.

Table 2. Articles related to every metric.

Metric	Articles Related
Accuracy	Tahir et al., 2021 [9]; Waheed et al., 2022 [52]; Ahmed et al., 2021 [1]; Yan et al., 2021 [55]; Brahmaiah et al., 2021 [11]; Jaen-Vargas et al., 2021 [22]; Ke et al., 2021 [24];]; Ullah et al., 2021 [48];]; Atzori et al., 2021 [7]; Xie et al., 2021 [54]; Liang et al., 2021 [28]; Dong et al., 2020 [16]; Nikouei et al., 2018 [35]; Ahmed et al., 2021 [2]; Ata-ur-Rehman et al., 2021 [5]; Chen et al., 2018 [12]; Bibi et al., 2020 [10]; Sun et al., 2018 [46]; Wang et al., 2022 [53]; Srividya et al., 2021 [43]; Alsaedi et al., 2021 [3]; Haque et al., 2021 [18]; Islam et al., 2021 [21]; Shreyas et al., 2020 [41]; Sudhakaran et al., 2017 [44]; Peipei Zhou et al. 2017 [60]; Ye et al., 2021 [56]; Lou et al., 2021 [29]; Lejmi et al., 2020 [25]; Fath et al., 2019 [49]; Baba et al., 2019 [8]; Liang et al., 2020 [57]; Atto et al., 2020 [6]; Hussain et al., 2022 [19]; Abid Mehmood, 2021 [30]; Rendon et al., 2021 [37]; Murugesan et al., 2022 [32]; Sanam et al., 2021 [33]; Kang et al., 2021 [23]; Vieira et al., 2022 [50]; Abid Mehmood, 2021 [31]; Chou et al., 2018 [14]
Area under the curve (AUC)	Ul Amin et al., 2022 [47]; Cheng et al., 2020 [13]; Ullah et al., 2021 [48]; Vu, 2017 [51]; Ahmed et al., 2021 [2]; Ata-ur-Rehman et al., 2021 [5]; Duman et al., 2019 [17]; Sun et al., 2018 [46]; Li et al., 2020 [26]; Saypadith et al., 2021 [39]; Shreyas et al., 2020 [41]; Shoaib et al., 2021 [40]; Kang et al., 2021 [23]; Abid Mehmood, 2021 [30]
Confusion matrix	Ullah et al., 2021 [48]; Bibi et al., 2020 [10]; Duman et al., 2019 [17]; Li et al., 2020 [26]; Haque et al., 2021 [18]; Peipei Zhou et al., 2017 [60]; Lejmi et al., 2020 [25]; Liang et al., 2020 [57]; Atto et al., 2020 [6]; Shoaib et al., 2021 [40]; Murugesan et al., 2022 [32]
Equal Error Rate (EER)	Saypadith et al., 2021 [39]; Wang et al., 2022 [53]
F1 score	Tahir et al., 2021 [9]; Waheed at al., 2021 [52]; Yan et al., 2022 [55]; Jaen-Vargas et al., 2021 [22]; Ye et al., 2021 [56]; Alsaedi et al., 2021 [3]; Apon et al., 2021 [4]; Baba et al., 2019 [8]; Liang et al., 2020 [57]; Hussain et al., 2022 [19]; Abid Mehmood., 2021 [30]; Murugesan et al., 2022 [32]
F-measure	Minh et al., 2021 [34]; Sun et al., 2018 [46]
Loss	Shreyas et al., 2020 [41]
Mean Average Precision	Yan et al., 2022 [55]
MSE	Nikouei et al., 2018 [35]; Shreyas et al., 2020 [41]; Vieira et al., 2022 [50]
Precision	Tahir et al., 2021 [9]; Cheng et al., 2020 [13]; Waheed at al., 2021 [52]; Yan et al., 2022 [55]; Cob-Parro et al., 2021 [15]; Ye et al., 2021 [56]; Ahmed et al., 2021 [2]; Duman et al., 2019 [17]; Alsaedi et al., 2021 [3]; Apon et al., 2021 [4]; Baba et al., 2019 [8]; Liang et al., 2020 [57]; Abid Mehmood, 2021 [30]
Recall	Tahir et al., 2021 [9]; Yan et al., 2022 [55]; Cob-Parro et al., 2021 [15]; Ye et al., 2021 [56]; Ahmed et al., 2021 [2]; Alsaedi et al., 2021 [3]; Apon et al., 2021 [4]; Zhang et al., 2020 [58]; Baba et al., 2019 [8]; Liang et al., 2020 [57]; Abid Mehmood, 2021 [30]
Receiver operating characteristic (ROC)	Ul Amin et al., 2022 [47]; Ullah et al., 2021 [48]; Duman et al., 2019 [17]; Sun et al., 2018 [46]; Li et al., 2020 [26]; Saypadith et al., 2021 [39]; Wang et al., 2022 [53]; Shreyas et al., 2020 [41]; Shoaib et al., 2021 [40]; Murugesan et al., 2022 [32]
Sensitivity	Waheed at al., 2021 [52]

3.3 Answer to RQ3

The third question asks about what kind of activity can be detected using the models in the articles. Most of the tests performed in the literature were related to violent actions, anomalies, abnormal behavior, and general theft. Even though most of the studies were not about car theft, the suspicious behavior and surveillance articles support the development of automatic surveillance using CCTV cameras. The most important activities displayed in Table 3 were Violent actions (16 articles) and Anomaly detection (14 articles).

Table 3. Articles related to the activities detected.

Activity	Articles Related
Abnormal behavior	Minh et al., 2021 [34]; Xie et al., 2021 [54]; Yan et al., 2022 [55]; Abid Mehmood, 2021 [30]
Anomaly detection	Ul Amin et al., 2022 [47]; Brahmaiah et al., 2021 [11]; Atzori et al., 2021 [7]; Vu, 2017 [51]; Ata-ur-Rehman et al., 2021 [5]; Duman et al., 2019 [17]; Sun et al., 2018 [46]; Li et al., 2020 [26]; Saypadith et al., 2021 [39]; Wang et al., 2022 [53]; Shreyas et al., 2020 [41]; Hussain et al., 2022 [19]; Murugesan et al., 2022 [32]; Singh et al., 2020 [42]
Behaviour recognition	Dong et al., 2020 [16]
Cooperative surveillance	Lian et al., 2020 [27]
Human activity	Jaen-Vargas et al., 2021 [22]; Ullah et al., 2021 [48]; Srividya et al., 2021 [43]; Alsaedi et al., 2021 [3]; Apon et al., 2021 [4]; Atto et al. 2020 [6]; Waheed at al., 2021 [52]; Bibi et al., 2020 [10]; Chou et al., 2018 [14]; Reinolds et al., 2022 [36]
Motion detection	Cheng et al., 2020 [13]
Parking occupancy	Ke et al., 2021 [24]
People tracking	Cob-Parro et al., 2021 [15]; Nikouei et al., 2018 [35]; Chen et al., 2018 [12]; Abid Mehmood, 2021 [30]
Theft detection	Zhang et al., 2020 [59]; Haque et al., 2021 [18]
Violent action	Ahmed et al., 2021 [2]; Liang et al., 2021 [28]; Islam et al., 2021 [21]; Zhang et al., 2020 [58]; Sudhakaran et al., 2017 [44]; Peipei Zhou et al., 2017 [60]; Ye et al., 2021 [56]; Lou et al., 2021 [29]; Lejmi et al., 2020 [25]; Fath et al., 2019 [49]; Baba et al., 2019 [8]; Liang et al., 2020 [57]; Shoaib et al., 2021 [40]; Rendon et al., 2021 [37]; Kang et al., 2021 [23]; Vieira et al., 2022 [50]
Weapon detection	Tahir et al., 2021 [9]; Brahmaiah et al., 2021 [11]; Ahmed et al., 2021 [2]; Sanam et al., 2021 [33]

3.4 Answer to RQ4

The fourth question asks about how the data is pre-processed in order for the model to work properly. For instance, there is a difference between image and video analysis. For video, it was necessary to split it into different frames. Once separated it could be treated like images. However, some models took recurrence into consideration, so more than one image was analyzed at the same time for only one prediction. The common pre-processing techniques can be seen in Table 4, which include Frame extraction, Region of interest extraction, Data augmentation, Image filtering, and others. From the Table, 8 articles used a Frame extraction method and 8 used the Region of interest extraction method.

Table 4. Articles related to every pre-processing technique.

Pre-processing	Articles Related
Background identification	Shreyas et al., 2020 [41]; Hussain et al., 2022 [19]
Data augmentation	Tahir et al., 2021 [9]; Yan et al., 2022 [55]; Ahmed et al., 2021 [2]; Alsaedi et al., 2021 [3]; Vieira et al., 2022 [50]
Dimensionality reduction	Zhang et al., 2020 [59]; Shreyas et al., 2020 [41]; Fath et al., 2019 [49]
Feature extraction	Ye et al., 2021 [56]; Zhang et al., 2020 [58]; Liang et al., 2021 [28]; Chou et al., 2018 [14]
Frame compression	Sultana et al., 2019 [45]
Frame extraction	Waheed at al., 2021 [52]; Ahmed et al., 2021 [2]; Alsaedi et al., 2021 [3]; Apon et al., 2021 [4]; Shreyas et al., 2020 [41]; Ye et al., 2021 [56]; Hussain et al., 2022 [19]; Abid Mehmood, 2021 [30]
Grayscale	Alsaedi et al., 2021 [3]
Image filtering	Tahir et al., 2021 [9]; Cheng et al., 2020 [13]; Ahmed et al., 2021 [2]; Cob-Parro et al., 2021 [15]; Haque et al., 2021 [18]; Vieira et al., 2022 [50]
Noise filtering	Waheed at al., 2021 [52]; Cob-Parro et al., 2021 [15]; Haque et al., 2021 [18]
Normalization	Waheed at al., 2021 [52]; Cob-Parro et al., 2021 [15]; Apon et al., 2021 [4]
Optical flow maps	Duman et al., 2019 [17]; Rendon et al., 2021 [37]
Region of interest extraction	Waheed at al., 2021 [52]; Cob-Parro et al., 2021 [15]; Islam et al., 2021 [21]; Shreyas et al., 2020 [41]; Fath et al., 2019 [49]; Hussain et al., 2021 [19]; Shoaib et al., 2021 [40]; Chou et al., 2018 [14]
Rescaling	Duman et al., 2019 [17]; Shreyas et al., 2020 [41]; Abid Mehmood, 2021 [30]; Shoaib et al., 2021 [40]; Vieira et al., 2022 [50]
Selective smoothing	Shreyas et al., 2020 [41]
Video compression	Sultana et al., 2019 [45]

3.5 Data Source

Table 5 indicates the source types (Journal and Proceeding) of the articles from the different databases. Most of the proceedings came from the Web of Science (22 studies). Additionally, the table shows that most of the studies came from Journals (47 in total, 81% of the articles).

Table 5. Different source type per database.

Type	Web of Science	Scopus	IEEE	Total	Percentage
Journal	22	13	12	47	81%
Proceedings	0	11	0	11	19%
Total	22	24	12	**58**	**100%**

4 Analysis of the Studies

In addition to answering the questions, further analysis is made for every research question. Figures that indicate the count of references per research question and some important remarks are presented in the following subsections.

4.1 Models (Q1)

The most used model were Convolutional Neural Networks (CNNs) due to their effectiveness and simplicity. CNNs use only the most important aspects of the information to create a neural network. CNN models are simpler than other algorithms. CNNs can be the basis to create more complex models such as 3D CNNs. Also, another important aspect to mention is that many studies used recurrent networks. LSTM is a network that takes memory into consideration. The inputs in an LSTM network are groups of information from more than one frame. Also, CNNs, such as the 3D CNN, use the recurrence aspect when the third dimension is related to a sequence of frames. In Fig. 2, a bar graph represents the number of mentions per model according to Table 1.

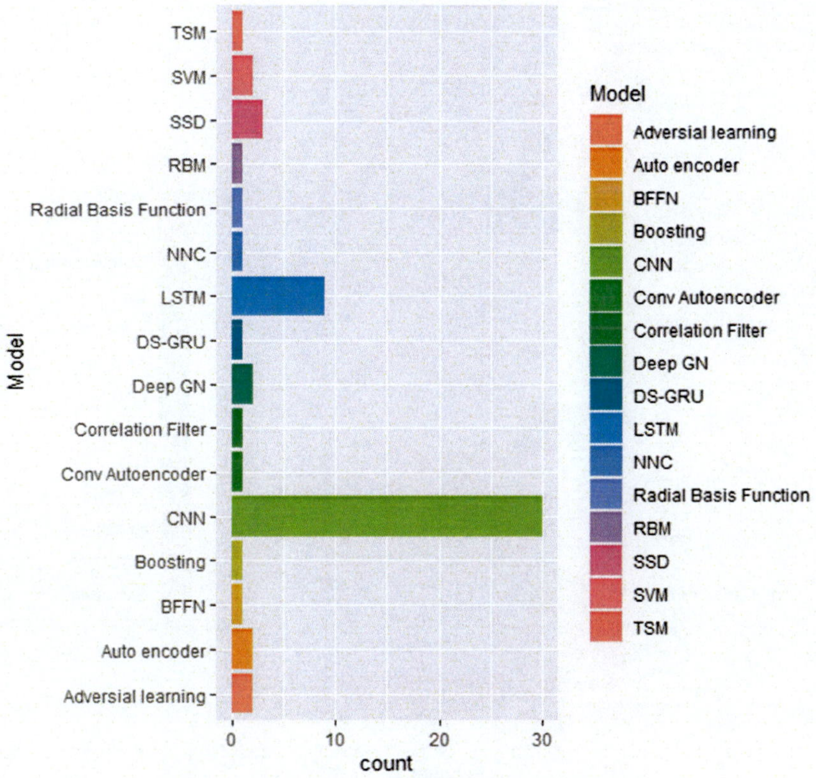

Fig. 2. Count of Models in the systematic review.

4.2 Metrics (Q2)

Most of the metrics could be applied to every model used in the scientific studies of the review. It was simply the choice of the authors to consider and include certain metrics. However, the most essential metric included in most of the articles was Accuracy, with 42 mentions. Even though the metrics are quantifiable, not all of them are completely relevant and easy to compare. For example, high accuracy could be due to a bias in the dataset, and problems such as overfitting can be bothersome. That is the reason why using different metrics at the same time is useful. In Fig. 3, a bar graph represents the number of mentions per metric according to Table 2.

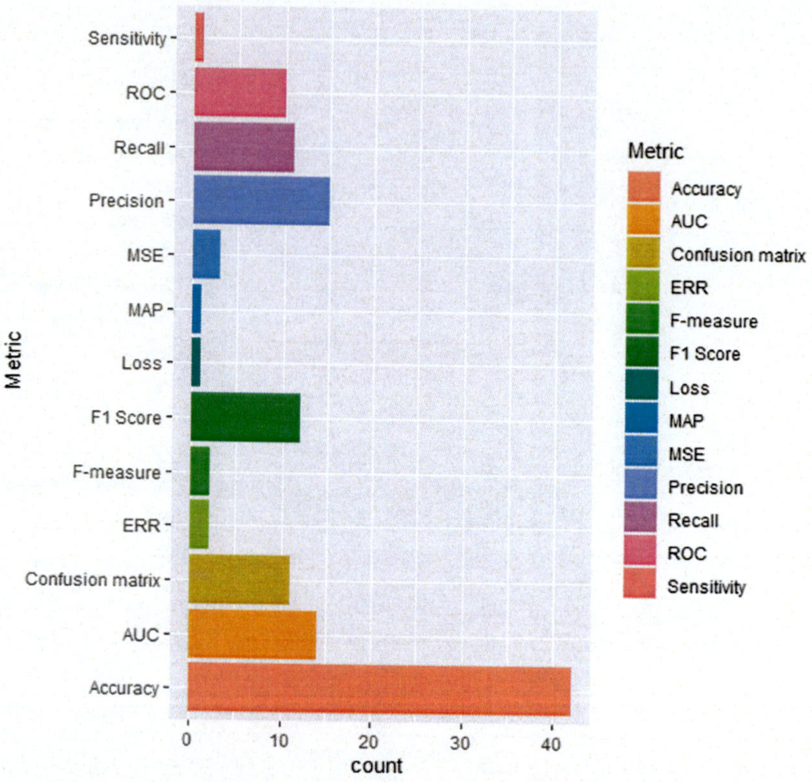

Fig. 3. Count of metrics in the systematic review.

4.3 Activities Detected (Q3)

The activities being detected consisted of violent action, anomalies, Abnormal behavior, Theft, Human activity, People tracking, Motion detection, Weapon detection, Behavior recognition, Parking occupancy, and Cooperative surveillance. The most important activities according to Table 3 were Violent actions (16 articles) and Anomaly detection (14 articles). Notably, the term Anomaly detection was interpreted differently in some cases. For example, some studies generally consider something an anomaly if it is different, like if someone was riding a bike while a crowd was walking. However, other studies specifically consider anomalies only in potentially dangerous situations. Most of the videos used to detect the activities were from different countries. For example the Internet Movies Firearms Database [9] is from Spain, the MuHAVi Dataset [14] is from the United Kingdom and The UCSD Anomaly Detection Dataset [17] is from the US. Also, many datasets included YouTube videos from various countries such as the Surveillance Camera Fight [23]. In Fig. 4, a bar graph represents the number of mentions per activity according to Table 3.

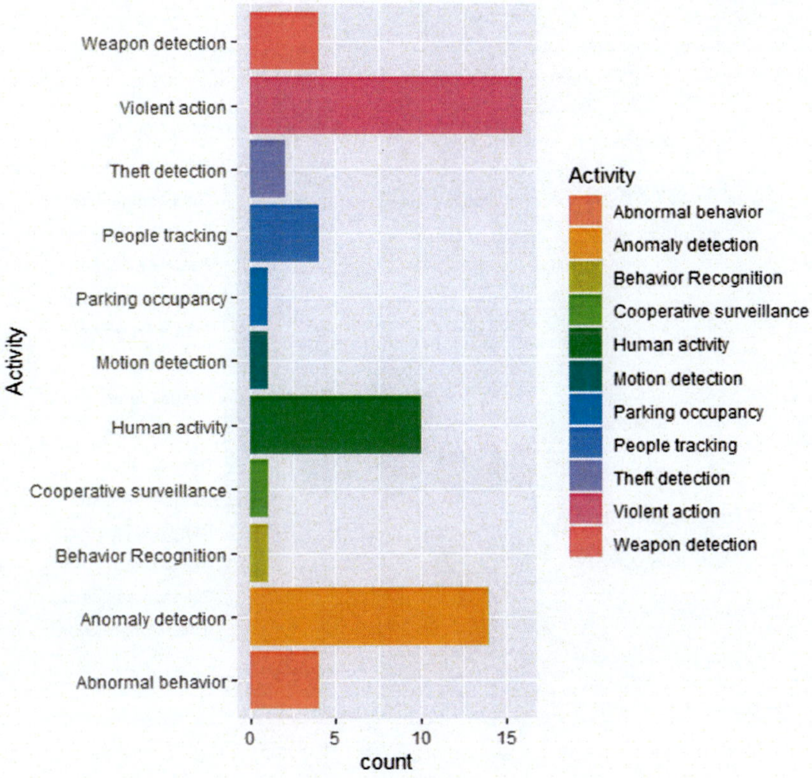

Fig. 4. Count of type of actions in the systematic review.

4.4 Pre-processing Techniques (Q4)

Pre-processing techniques in the literature review consisted of Frame extraction, Region of interest extraction, Data augmentation, Image filtering, Rescaling, Frame compression, Video compression, Feature extraction, Dimensionality reduction, Normalization, Noise filtering, Optical flow maps, Selective smoothing, Background identification, and Grayscale. The purpose of using a CCTV camera is to obtain information in video format. That is the reason why most of the studies applied Frame extraction, because it was necessary to section the video into multiple frames for the network to work properly. Some studies did not mention Frame extraction as a pre-processing step because it was done manually or it was omitted for other reasons. In Fig. 5, a bar graph represents the number of mentions per pre-processing technique according to Table 4.

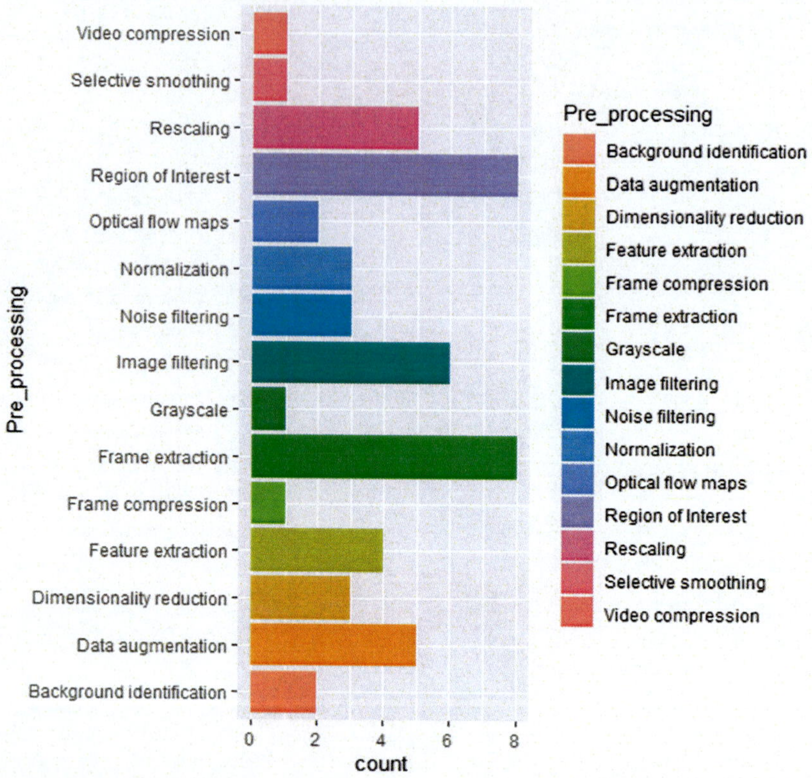

Fig. 5. Count of Pre-processing techniques included in the systematic review.

5 Conclusion

The current study aimed to review the literature regarding machine learning and artificial intelligence models that may aid in the creation of a car theft detection algorithm. The reviewed articles were obtained from Scopus, IEEE, and Web of Science. The articles were chosen using several criteria, including their potential to help answer the 4 research questions posed by this review. In the first search, there were 151 scientific articles, of which 71 were from Scopus, 4 were from IEEE, and 76 were from Web of Science. In the second search, there were 233 scientific articles, of which 151 were from Scopus, 50 were from IEEE, and 32 were from Web of Science. Then, after deleting the duplicate articles, only 214 remained. In the end, only 58 studies passed the eligibility criteria that were initially set.

Most of the studies were not directly related to the application of car theft. However, most of the articles were concerned with recognizing suspicious activities, and the way the activities are recognized can also be applied to cases of car theft. The literature showed that there are many options when it comes to artificial intelligence. From the 58 articles, the most popular models were CNN (used in 30 articles) and LSTM (used in 9 articles). For the metrics used to assess the models, Accuracy (42 articles), AUC (14 articles), F1 score (12 articles), Recall (11 articles), and ROC (10 articles) were the most useful. Most of the activities detected by the models were related to Violent actions (15 articles), Anomalies (11 articles), and Abnormal Behavior (4 articles). Finally, the most common pre-processing techniques of the input data were Frame extraction (8 articles), ROI extraction (8 articles), and Data augmentation (5 articles).

References

1. Ahmed, A.A., Echi, M.: Hawk-eye: an AI-powered threat detector for intelligent surveillance cameras. IEEE Access **9**, 63283–63293 (2021). https://doi.org/10.1109/ACCESS.2021.3074319
2. Ahmed, M., et al.: Real-time violent action recognition using key frames extraction and deep learning. CMC-Comput. Mater. Continua **69**(2), 2217–2230 (2021). https://doi.org/10.32604/cmc.2021.018103
3. Alsaedi, M.A., Mohialdeen, A.S., Albaker, B.M.: Development of 3D convolutional neural network to recognize human activities using moderate computation machine. Bull. Electr. Eng. Inform. **10**(6), 3137–3146 (2021). https://doi.org/10.11591/eei.v10i6.2802
4. Apon, T.S., Chowdhury, M.I., Reza, M.Z., Datta, A., Hasan, S.T., Alam, M.G.R.: Real time action recognition from video footage (2021). https://doi.org/10.1109/STI53101.2021.9732601
5. Ata-Ur-Rehman, Tariq, S., Farooq, H., Jaleel, A., Wasif, S.M.: Anomaly detection with particle filtering for online video surveillance. IEEE Access **9**, 19457–19468 (2021). https://doi.org/10.1109/ACCESS.2021.3054040
6. Atto, A.M., Benoit, A., Lambert, P.: Timed-image based deep learning for action recognition in video sequences. Pattern Recognit. **104** (2020). https://doi.org/10.1016/j.patcog.2020.107353
7. Atzori, A., Barra, S., Carta, S., Fenu, G., Podda, A.S.: HEIMDALL: an AI-based infrastructure for traffic monitoring and anomalies detection, pp. 154–159 (2021). https://doi.org/10.1109/PerComWorkshops51409.2021.9431052
8. Baba, M., Gui, V., Cernazanu, C., Pescaru, D.: A sensor network approach for violence detection in smart cities using deep learning. Sensors **19**(7) (2019). https://doi.org/10.3390/s19071676
9. Bhatti, M.T., Khan, M.G., Aslam, M., Fiaz, M.J.: Weapon detection in real-time CCTV videos using deep learning. IEEE Access **9**, 34366–34382 (2021). https://doi.org/10.1109/ACCESS.2021.3059170
10. Bibi, S., Anjum, N., Amjad, T., McRobbie, G., Ramzan, N.: Human interaction anticipation by combining deep features and transformed optical flow components. IEEE Access **8**, 137646–137657 (2020). https://doi.org/10.1109/ACCESS.2020.3012557
11. Brahmaiah, M., Madala, S.R., Chowdary, C.M.: Artificial intelligence and deep learning for weapon identification in security systems. In: Journal of Physics: Conference Series, vol. 2089 (2021). www.scopus.com
12. Chen, X., Qing, L., He, X., Su, J., Peng, Y.: From eyes to face synthesis: a new approach for human-centered smart surveillance. IEEE Access **6**, 14567–14575 (2018). https://doi.org/10.1109/ACCESS.2018.2803787

13. Cheng, X., Song, C., Gu, Y., Chen, B.: Learning attention for object tracking with adversarial learning network. EURASIP J. Image Video Process. **2020**(1), 1–21 (2020). https://doi.org/10.1186/s13640-020-00535-1
14. Chou, K.P., et al.: Robust feature-based automated multi-view human action recognition system. IEEE Access **6**, 15283–15296 (2018). https://doi.org/10.1109/ACCESS.2018.2809552
15. Cob-Parro, A.C., Losada-Gutiérrez, C., Marrón-Romera, M., Gardel-Vicente, A., Bravo-Muñoz, I.: Smart video surveillance system based on edge computing. Sensors **21**(9), 2958 (2021). https://doi.org/10.3390/s21092958
16. Dong, W.: Research on character behavior recognition based on local spatio-temporal relationship in surveillance video, vol. 1982 (2021). https://doi.org/10.1088/1742-6596/1982/1/012009
17. Duman, E., Erdem, O.A.: Anomaly detection in videos using optical flow and convolutional autoencoder. IEEE Access **7**, 183914–183923 (2019). https://doi.org/10.1109/ACCESS.2019.2960654
18. Haque, M.R., et al.: Crime detection and criminal recognition to intervene in interpersonal violence using deep convolutional neural network with transfer learning. Int. J. Ambient Comput. Intell. **12**(4), 154–167 (2021). https://doi.org/10.4018/IJACI.20211001.oa1
19. Hussain, A., et al.: Anomaly based camera prioritization in large scale surveillance networks. CMC-Comput. Mater. Continua **70**(2), 2171–2190 (2022). https://doi.org/10.32604/cmc.2022.018181
20. International Association of Auto Theft Investigators: Car theft statistics 2022. https://www.iaati.org/news/entry/car-theft-statistics-2022. Accessed 28 June 2022
21. Islam, Z., Rukonuzzaman, M., Ahmed, R., Kabir, M.H., Farazi, M.: Efficient two-stream network for violence detection using separable convolutional LSTM, vol. 2021-July (2021). https://doi.org/10.1109/IJCNN52387.2021.9534280
22. Jaén-Vargas, M., et al.: A deep learning approach to recognize human activity using inertial sensors and motion capture systems. Front. Artif. Intell. Appl. **340**, 250–256 (2021). https://doi.org/10.3233/FAIA210196
23. Kang, M.S., Park, R.H., Park, H.M.: Efficient spatio-temporal modeling methods for real-time violence recognition. IEEE Access **9**, 76270–76285 (2021). https://doi.org/10.1109/ACCESS.2021.3083273
24. Ke, R., Zhuang, Y., Pu, Z., Wang, Y.: A smart, efficient, and reliable parking surveillance system with edge artificial intelligence on IoT devices. IEEE Trans. Intell. Transp. Syst. **22**(8), 4962–4974 (2021). https://doi.org/10.1109/TITS.2020.2984197
25. Lejmi, W., Ben Khalifa, A., Mahjoub, M.A.: A novel spatio-temporal violence classification framework based on material derivative and LSTM neural network. Traitement Signal **37**(5), 687–701 (2020). https://doi.org/10.18280/ts.370501
26. Li, Z., Li, Y., Gao, Z.: Spatiotemporal representation learning for video anomaly detection. IEEE Access **8**, 25531–25542 (2020). https://doi.org/10.1109/ACCESS.2020.2970497
27. Lian, D., Xu, A., Chen, S., Xu, X., Jiang, Y., Hong, W.: Cooperative training video surveillance technology under the edge computing, vol. 1575 (2020). https://doi.org/10.1088/1742-6596/1575/1/012132
28. Liang, Q., Li, Y., Chen, B., Yang, K.: Violence behavior recognition of two-cascade temporal shift module with attention mechanism. J. Electron. Imaging **30**(4) (2021). https://doi.org/10.1117/1.JEI.30.4.043009

29. Lou, J., Zuo, D., Zhang, Z., Liu, H.: Violence recognition based on auditory-visual fusion of autoencoder mapping. Electronics **10**(21) (2021). https://doi.org/10.3390/electronics10212654
30. Mehmood, A.: Abnormal behavior detection in uncrowded videos with two-stream 3D convolutional neural networks. Appl. Sci.-Basel **11**(8) (2021). https://doi.org/10.3390/app11083523
31. Mehmood, A.: Efficient anomaly detection in crowd videos using pre-trained 2D convolutional neural networks. IEEE Access **9**, 138283–138295 (2021). https://doi.org/10.1109/ACCESS.2021.3118009
32. Murugesan, M., Thilagamani, S.: Bayesian feed forward neural network-based efficient anomaly detection from surveillance videos. Intell. Autom. Soft Comput. **34**(1), 389–405 (2022). https://doi.org/10.32604/iasc.2022.024641
33. Narejo, S., Pandey, B., Esenarro Vargas, D., Rodriguez, C., Anjum, M.R.: Weapon detection using YOLO V3 for smart surveillance system. Math. Probl. Eng. **2021** (2021). https://doi.org/10.1155/2021/9975700
34. Nguyen, M.T., Truong, L.H., Le, T.T.H.: Video surveillance processing algorithms utilizing artificial intelligent (AI) for unmanned autonomous vehicles (UAVs). MethodsX **8** (2021). https://doi.org/10.1016/j.mex.2021.101472
35. Nikouei, S.Y., Chen, Y., Song, S., Xu, R., Choi, B.Y., Faughnan, T.: Smart surveillance as an edge network service: from harr-cascade, SVM to a lightweight CNN, pp. 256–265 (2018). https://doi.org/10.1109/CIC.2018.00042
36. Reinolds, F., Neto, C., Machado, J.: Deep learning for activity recognition using audio and video. Electron. (Switz.) **11**(5) (2022). https://doi.org/10.3390/electronics11050782
37. Rendon-Segador, F.J., Alvarez-Garcia, J.A., Enriquez, F., Deniz, O.: ViolenceNet: dense multi-head self-attention with bidirectional convolutional LSTM for detecting violence. Electronics **10**(13) (2021). https://doi.org/10.3390/electronics10131601
38. Ronquillo-Freire, P.V., Garcia, M.V.: Measurement of work as a basis for improving processes and simulation of standards: a scoping literature review. In: Arai, K. (ed.) FICC 2021. AISC, vol. 1363, pp. 77–92. Springer, Cham (2021). https://doi.org/10.1007/978-3-030-73100-7_6
39. Saypadith, S., Onoye, T.: An approach to detect anomaly in video using deep generative network. IEEE Access **9**, 150903–150910 (2021). https://doi.org/10.1109/ACCESS.2021.3126335
40. Shoaib, M., Sayed, N.: A deep learning based system for the detection of human violence in video data. Traitement Signal **38**(6), 1623–1635 (2021). https://doi.org/10.18280/ts.380606
41. Shreyas, D.G., Raksha, S., Prasad, B.G.: Implementation of an anomalous human activity recognition system. SN Comput. Sci. **1**(3), 1–10 (2020). https://doi.org/10.1007/s42979-020-00169-0
42. Singh, V., Singh, S., Gupta, P.: Real-time anomaly recognition through CCTV using neural networks. Procedia Comput. Sci. **173**, 254–263 (2020). https://doi.org/10.1016/j.procs.2020.06.030
43. Srividya, M., Anala, M., Tayal, C.: Deep learning techniques for physical abuse detection. IAES Int. J. Artif. Intell. **10**(4), 971–981 (2021). https://doi.org/10.11591/IJAI.V10.I4.PP971-981
44. Sudhakaran, S., Lanz, O.: Learning to detect violent videos using convolutional long short-term memory (2017). https://doi.org/10.1109/AVSS.2017.8078468

45. Sultana, T., Wahid, K.A.: IoT-guard: event-driven fog-based video surveillance system for real-time security management. IEEE Access **7**, 134881–134894 (2019). https://doi.org/10.1109/ACCESS.2019.2941978
46. Sun, J., Wang, X., Xiong, N., Shao, J.: Learning sparse representation with variational auto-encoder for anomaly detection. IEEE Access **6**, 33353–33361 (2018). https://doi.org/10.1109/ACCESS.2018.2848210
47. Ul Amin, S., et al.: EADN: an efficient deep learning model for anomaly detection in videos. Mathematics **10**(9) (2022). https://doi.org/10.3390/math10091555
48. Ullah, A., Muhammad, K., Ding, W., Palade, V., Haq, I.U., Baik, S.W.: Efficient activity recognition using lightweight CNN and DS-GRU network for surveillance applications. Appl. Soft Comput. **103** (2021). https://doi.org/10.1016/j.asoc.2021.107102
49. Ullah, F.U.M., Ullah, A., Muhammad, K., Ul Haq, I., Baik, S.W.: Violence detection using spatiotemporal features with 3D convolutional neural network. Sensors **19**(11) (2019). https://doi.org/10.3390/s19112472
50. Vieira, J.C., Sartori, A., Stefenon, S.F., Perez, F.L., de Jesus, G.S., Leithardt, V.R.Q.: Low-cost CNN for automatic violence recognition on embedded system. IEEE Access **10**, 25190–25202 (2022). https://doi.org/10.1109/ACCESS.2022.3155123
51. Vu, H.: Deep abnormality detection in video data, pp. 5217–5218 (2017). https://doi.org/10.24963/ijcai.2017/768
52. Waheed, M., et al.: An LSTM-based approach for understanding human interactions using hybrid feature descriptors over depth sensors. IEEE Access **9**, 167434–167446 (2021). https://doi.org/10.1109/ACCESS.2021.3130613
53. Wang, L., Tan, H., Zhou, F., Zuo, W., Sun, P.: Unsupervised anomaly video detection via a double-flow convLSTM variational autoencoder. IEEE Access **10**, 44278–44289 (2022). https://doi.org/10.1109/ACCESS.2022.3165977
54. Xie, Y., Zhang, S., Liu, Y.: Abnormal behavior recognition in classroom pose estimation of college students based on spatiotemporal representation learning. Traitement Signal **38**(1), 89–95 (2021). https://doi.org/10.18280/TS.380109
55. Yan, K., et al.: Deep learning-based substation remote construction management and AI automatic violation detection system. IET Gener. Transm. Distrib. **16**(9), 1714–1726 (2022). https://doi.org/10.1049/gtd2.12387
56. Ye, L., Liu, T., Han, T., Ferdinando, H., Seppänen, T., Alasaarela, E.: Campus violence detection based on artificial intelligent interpretation of surveillance video sequences. Remote Sens. **13**(4), 1–17 (2021). https://doi.org/10.3390/rs13040628
57. Ye, L., Shi, J., Ferdinando, H., Seppänen, T., Alasaarela, E.: A multi-sensor school violence detecting method based on improved relief-F and D-S algorithms. Mob. Netw. Appl. **25**(5), 1655–1662 (2020). https://doi.org/10.1007/s11036-020-01575-7
58. Zhang, L., Ruan, X., Wang, J.: WiVi: a ubiquitous violence detection system with commercial WiFi devices. IEEE Access **8**, 6662–6672 (2020). https://doi.org/10.1109/ACCESS.2019.2962813
59. Zhang, Y., et al.: A new intelligent supermarket security system. Neural Netw. World **30**(2), 113–131 (2020). https://doi.org/10.14311/NNW.2020.30.009
60. Zhou, P., Ding, Q., Luo, H., Hou, X.: Violent interaction detection in video based on deep learning, vol. 844 (2017). https://doi.org/10.1088/1742-6596/844/1/012044

Optimization of Vortex Dynamics on a Sphere

Carlos Balsa[1(✉)], Raphaelle Monville-Letu[2], and Sílvio Gama[3]

[1] Research Centre in Digitalization and Intelligent Robotics (CeDRI), Laboratório para a Sustentabilidade e Tecnologia em Regiões de Montanha (SusTEC), Instituto Politécnico de Bragança, Campus de Santa Apolónia, 5300-253 Bragança, Portugal
balsa@ipb.pt

[2] Université de Toulouse - INP - ENSEEIHT, 31071 Toulouse Cedex 7, France
raphaelle.monvilleletu@etu.toulouse-inp.fr

[3] Mathematics Center of the Porto University (CMUP), Mathematics Department, Faculty of Sciences, University of Porto, R. Campo Alegre s/n, 4169-007 Porto, Portugal
smgama@fc.up.pt

Abstract. Vortex points on a sphere can be considered as simplified models of atmospheric circulation. The use of these models allows the simulation of displacements of passive particles advected by vortex flow. In this study, a strategy is proposed to determine the optimal trajectory between two given points on the sphere, taking into account that the displacement occurs due to a vortex flow. It is an alternative numerical strategy to the methods proposed by the theory of optimal control. The original problem is discretized into a constrained optimization problem. The solution of this problem by two alternative numerical optimization methods shows that the strategy is feasible and leads to optimal or quasi-optimal solutions.

Keywords: Vortex · Passive Particle · Spherical Motion · Control Problem · Nonlinear Optimization Problem · Interior Point · Active Set

1 Introduction

Point vortices are finite-dimensional approximations to the two-dimensional vortex dynamics of incompressible ideal fluids (zero viscosity). This research topic, initiated by Helmholtz [4] and continued a few years later by Kelvin [14] and Kirchhoff [7], continues to lead to a great deal of work using theories of dynamical systems, differential geometry, numerical analysis, optimal control, and so on. Thus, point vortices have been studied in many types of surfaces, such as plane [1], sphere [15], or hyperbolic sphere [5, 10].

The present study is concerned with the motion of point vortices on a sphere. Point vortices on the sphere are relevant because they represent a simplified approximation to the behavior of certain geophysical flows for which the curvature of the Earth is important and which persist over long periods of time [15]. Indeed, many questions related to the fundamental dynamics of atmospheric flows are answered by vortex point models [12]. Conceptual models of point vortices are also used to identify and evaluate physical phenomena affecting the structure and interaction of atmospheric and oceanic

M. V. Garcia and C. Gordón-Gallegos (Eds.): CSEI 2022, LNNS 678, pp. 201–213, 2023.
https://doi.org/10.1007/978-3-031-30592-4_15

vortices [9]. More recently, point vortices are also used to model pesticide dispersion in agricultural problems [6,17].

This work focuses on optimizing the displacement of a passive particle interacting with multiple vortices on the surface of a sphere. More precisely, we are interested in optimally controlling the displacement of the passive particle between two fixed points by minimizing the energy spent on the displacement, taking into account that the time to perform the displacement is fixed. This problem can be viewed as a simplified version of the displacement of a glider moving between two points using atmospheric circulation to consume as little energy as possible.

To solve this problem, the displacement of the passive particle is converted into a control problem, which is solved using a direct numerical approach. A similar approach has been used in previous work to solve a vortex problem in the infinite plane [2,3]. The time T available to perform the displacement is discretized into n subintervals, where the controls are constant. The resulting nonlinear programming (NLP) problem is solved numerically using the solver `fmincon` of the Matlab Optimization Toolbox [8]. This solver allows to solve constrained NLPs by different optimization methods. In previous work, we found that the most suitable methods are the interior point method and the active set method [2,3]. In this study, we present the results obtained by these two optimization methods to make a comparison.

This article is organised as follows. Section 2 is devoted to the introduction of the control problem underlying the determination of the optimal trajectory. The transformation of the control problem into an NLP problem in the case of a single vortex ($N = 1$) is analysed in Sect. 3. In Sect. 4, the problem of a flow induced by multiple vortices ($N = 2$, 3 and 4) is treated. Finally, in Sect. 5 some concluding considerations are given.

2 Statement of the Control Problem

We begin by introducing the equations of the vortex on a nonrotating sphere whose centre is the origin and whose radius is a R. The representation of the vortex and the passive particle can be in different coordinate systems, e.g., spherical coordinates or Cartesian coordinates (see, e.g., [11]). In this study we use only Cartesian coordinates.

The position of a vortex on a sphere is given by the vector $\mathbf{x}_j$ that points from the centre of the sphere to the vortex location $\mathbf{x}_j = (x_j, y_j, z_j)$ on the spherical surface $||\mathbf{x}_j|| = R$. The position of the vortex is given by

$$\dot{\mathbf{x}}_i = \frac{1}{2\pi R} \sum_{\substack{j=1 \\ j \neq i}}^{N} k_j \frac{\mathbf{x}_j \times \mathbf{x}_i}{||\mathbf{x}_i - \mathbf{x}_j||^2}, \qquad i = 1, 2, \ldots, N, \tag{1}$$

with the respective initial conditions, where k_j is the circulation of vortices j, and N is the total number of vortex on the sphere. We can observe that chord distance between vortex i and j is given by

$$||\mathbf{x}_i - \mathbf{x}_j||^2 = 2\left(R^2 - \mathbf{x}_i \cdot \mathbf{x}_j\right). \tag{2}$$

A passive particle is, by definition, a point vortex with circulation $k = 0$. Thus, the dynamics of a system with P passive particles advected by N point vortices is given by Eq. (1) together with the equations for the passive particles

$$\dot{\mathbf{x}}_p = \frac{1}{2\pi R} \sum_{j=1}^{N} k_j \frac{\mathbf{x}_j \times \mathbf{x}_p}{||\mathbf{x}_p - \mathbf{x}_j||^2}, \qquad p = N+1, N+2, \ldots, N+P, \tag{3}$$

with the respective initial conditions.

Considering a single controlled passive particle ($P = 1$) moving in a spherical fluid induced by N, the corresponding equation is

$$\dot{\mathbf{x}} = \frac{1}{2\pi R} \sum_{j=1}^{N} k_j \frac{\mathbf{x}_j \times \mathbf{x}}{||\mathbf{x} - \mathbf{x}_j||^2} + \mathbf{U}(t), \tag{4}$$

with the respective initial conditions. In the r.h.s. of Eq. (4), $\mathbf{U}(t)$ is the control vector function.

Since the motion of the vortices and particles takes place on a sphere with radius R, this means that the control $\mathbf{U}$ must be such that the position vector of the passive particle $\mathbf{x}$ has a norm equal to R, that is, $\mathbf{x} \cdot \mathbf{x} = R^2$. Applying $\mathbf{x}\cdot$ to both sides of the Eq. (4), we obtain:

$$\mathbf{x} \cdot \dot{\mathbf{x}} = \mathbf{x} \cdot \mathbf{U},$$

or, equivalently,

$$\frac{1}{2} \frac{d(\mathbf{x} \cdot \mathbf{x})}{dt} = \mathbf{x} \cdot \mathbf{U} \quad \Longleftrightarrow \quad \frac{1}{2} \frac{d(R^2)}{dt} = \mathbf{x} \cdot \mathbf{U},$$

Thus, $\mathbf{x} \cdot \mathbf{U} = 0$. In other words, for the particle's motion to occur on the sphere, the control exerted in the particle must be orthogonal to the particle's position vector, i.e. $\mathbf{U}(t) \perp \mathbf{x}(t)$, $\forall t \geq 0$. If we let $\mathbf{x}(t) = (x(t), y(t), z(t))$, we then get

$$\mathbf{U}(t) = \alpha(t)\,(y(t), -x(t), 0) + \beta(t)\,(0, z(t), -y(t)),$$

where $\alpha(\cdot)$ and $\beta(\cdot)$ are two scalar control functions. These controls allow the particle to move in any direction on the surface of the sphere. The sum of the square of their magnitude, $\alpha(\cdot)^2 + \beta(\cdot)^2$, by definition corresponds to the energy expended for the displacement. Since the control $\mathbf{U}(\cdot)$ depends on $\alpha(\cdot)$ and $\beta(\cdot)$ from now on we assume $\mathbf{U}(\cdot) = (\alpha(\cdot), \beta(\cdot))$, and Eq. (4) is rewritten as

$$\dot{\mathbf{x}} = \frac{1}{2\pi R} \sum_{j=1}^{N} k_j \frac{\mathbf{x}_j \times \mathbf{x}}{||\mathbf{x} - \mathbf{x}_j||^2} + \alpha\,(y, -x, 0) + \beta\,(0, z, -y), \tag{5}$$

In this work, we consider that a given single passive particle ($P = 1$) is advected by $N = 1, 2, 3$ or 4 point vortices. The control problem is to move this particle between two given points ($\mathbf{x}_0$ and $\mathbf{x}_f$), on the surface of the sphere, in a given fixed time (T), while consuming as little energy as possible.

3 Conversion to an Optimization Problem

The control problem presented in the previous section is solved by a direct approach based on numerical optimization of the corresponding discretized problem. In previous work, the same methodology was used to control the displacement of a passive particle in the plane [3]. The control function $\mathbf{U}(\cdot)$ is replaced by n control vectors $\mathbf{u}_0, \mathbf{u}_1, \cdots, \mathbf{u}_{n-1}$. Numerical calculations were performed in Matlab using the nonlinear programming solver `fmincon`. This solver provides some limited optimization algorithms, such as the interior point or the active set (see [8]).

We start by solving this problem for a single passive particle in a spherical vortex flow, and then in the Sect. 4, we treat the cases with up to four vortices.

The equation that describes the motion of a passive particle induced by a single vortex on the sphere is

$$\dot{\mathbf{x}} = \frac{k}{2\pi R}\frac{\mathbf{x}_1 \times \mathbf{x}}{||\mathbf{x}-\mathbf{x}_1||^2}, \tag{6}$$

with the given initial condition $\mathbf{x}(0) = \mathbf{x}_0$ and the control problem, previously introduced, is then defined as follow:

$$\begin{array}{ll} (\mathscr{P}) & \text{Minimize :} \\ & \quad \int_0^T ||\mathbf{U}(t)||^2 dt \\ \text{subject to :} & \\ & \quad \dot{\mathbf{x}} = \frac{k}{2\pi R}\frac{\mathbf{x}_1 \times \mathbf{x}}{||\mathbf{x}-\mathbf{x}_1||^2} + \mathbf{U}(t) \\ & \quad \mathbf{x}(0) = \mathbf{x}_0 \\ & \quad \mathbf{x}(T) = \mathbf{x}_f \\ & \quad ||\mathbf{U}|| \leq u_{\max} \end{array}$$

with $\mathbf{U}(t) \in \mathbb{R}^3$, and $\mathbf{x}_0, \mathbf{x}_f \in \mathbb{R}^3$, $T > 0$ and $u_{\max} > 0$ given. In the control problem $(\mathscr{P})$, $\mathbf{x}(0)$ is the initial position (when $t = 0$) and $\mathbf{x}(T)$ is the final position of the passive particle (when $t = T$).

For example, in this optimization problem $(\mathscr{P})$, the objective function (cost function) represents the energy expended by the controller $\mathbf{U}(\cdot)$ to drive the passive particle from the starting point $\mathbf{x}_0$ to the end point $\mathbf{x}_f$. The first restriction corresponds to the state equation that determines the position $\mathbf{x}(t)$ of the particle as a function of time. The vectorial control function $\mathbf{U}(t)$ is introduced into this equation to move the particle from $\mathbf{x}_0$ to $\mathbf{x}_f$ in a fixed time value $T > 0$. The points $\mathbf{x}_0$ and $\mathbf{x}_f$ are predefined, as is the time T available to reach the destination $\mathbf{x}_f$. In addition, the fourth restriction requires that the norm of the control vector is not greater than a given value $u_{\max}$.

To solve this problem, we proceed to discretize the control function $\mathbf{U}(\cdot)$ into n (discrete) vector variables defined as ($t_0 = 0,\ t_n = T$):

$$\begin{array}{ll} \mathbf{U}(t) = \mathbf{u}_0 & \text{if } t_0 \leq t < t_1, \\ \mathbf{U}(t) = \mathbf{u}_1 & \text{if } t_1 \leq t < t_2, \\ \mathbf{U}(t) = \mathbf{u}_2 & \text{if } t_2 \leq t < t_3, \\ & \vdots \\ \mathbf{U}(t) = \mathbf{u}_{n-1} & \text{if } t_{n-1} \leq t \leq t_n. \end{array}$$

Thus, each $\mathbf{u}_i$ $(i = 0, 1, 2, \cdots, n-1)$ corresponds to the vector of control exercised in the subinterval $[t_{i-1}\ t_i)$. All these subintervals have amplitudes equal to $\Delta t = (t_n - t_0)/n$. The discretization of the objective function by the trapezoidal rule leads to the approximation

$$\int_0^T \|\mathbf{U}\|^2 dt \approx \frac{\Delta t}{2}\left(\|\mathbf{u}_0\|^2 + \|\mathbf{u}_{n-1}\|^2 + 2\sum_{j=1}^{n-2}\|\mathbf{u}_j\|^2\right) \equiv f_n. \tag{7}$$

The control problem $(\mathscr{P})$ is then replaced by its discretized version:

$(\mathscr{D}\mathscr{P}_n)$ Minimize :

$$f_n = \frac{\Delta t}{2}\left(\|\mathbf{u}_0\|^2 + \|\mathbf{u}_{n-1}\|^2 + 2\sum_{j=1}^{n-2}\|\mathbf{u}_j\|^2\right)$$

subject to :

$$\dot{\mathbf{x}} = \frac{k}{2\pi R}\frac{\mathbf{x}_1\times\mathbf{x}}{\|\mathbf{x}-\mathbf{x}_1\|^2} + \mathbf{u}_0,\ \ \mathbf{x}(0) = \mathbf{x}_0,\ \ \|\mathbf{u}_0\| \le u_{\max},\ \ t_0 \le t < t_1$$
$$\dot{\mathbf{x}} = \frac{k}{2\pi R}\frac{\mathbf{x}_1\times\mathbf{x}}{\|\mathbf{x}-\mathbf{x}_1\|^2} + \mathbf{u}_1,\ \ \mathbf{x}(t_1) = \mathbf{x}_{t_1},\ \ \|\mathbf{u}_2\| \le u_{\max},\ \ t_1 \le t < t_2$$
$$\vdots$$
$$\dot{\mathbf{x}} = \frac{k}{2\pi R}\frac{\mathbf{x}_1\times\mathbf{x}}{\|\mathbf{x}-\mathbf{x}_1\|^2} + \mathbf{u}_{n-1},\ \ \mathbf{x}(t_{n-1}) = \mathbf{x}_{t_{n-1}},\ \ \|\mathbf{u}_{n-1}\| \le u_{\max},\ \ t_{n-1} \le t < t_n$$
$$\mathbf{x}(t_n) = \mathbf{x}_f$$

First, we solve the discretized problem $(\mathscr{D}\mathscr{P}_n)$ where n is the number of control vectors (in our study from 1 to 4), considering a single vortex with circulation $k = 2$ at the north pole of the sphere ($\mathbf{x}_1 = (0,\ 0,\ 1)$). A set of vectors $\mathbf{u}_0, \mathbf{u}_1, \ldots, \mathbf{u}_{n-1} \in \mathbb{R}^2$ is sought that drives the passive particle from $\mathbf{x}_0$ to $\mathbf{x}_f$ in exactly $T = 10$ (natural) time units and minimizes the objective function f_n defined by (7). This optimization problem is solved numerically using the Interior Point [16] and the Active Set [13] optimization algorithms, which are included in the `fmincon` Matlab solver. In the numerical calculations, $R = 1$ is assumed.

The constrained optimization problem $(\mathscr{D}\mathscr{P}_n)$ provides a set of n control vectors $\mathbf{u}_0$, $\mathbf{u}_1$, ..., $\mathbf{u}_{n-1}$, subject to the constraint that the final position of the particle must be equal to x_f. In the implementation of this problem, we assume that the target position is reached when $\left\|\mathbf{x}(T) - \mathbf{x}_f\right\| \le 10^{-4}$. To verify this condition, it is necessary to solve the ordinary differential equations (ODEs) sequentially

$$\dot{\mathbf{x}} = \frac{k}{2\pi R}\frac{\mathbf{x}_1\times\mathbf{x}}{\|\mathbf{x}-\mathbf{x}_1\|^2} + \mathbf{u}_i, \qquad i = 0, 1, \ldots, n-1. \tag{8}$$

The initial condition of each initial value problem (IVP) is given by the final position of the previous IVP. For each of these IVPs, the ODE is solved numerically using Matlab's built-in function `ode45`, the 4th and 5th order Runge-Kutta methods.

In Eq. (8), $\mathbf{u}_i \in \mathbb{R}^2$ is the control vector used in each of the subintervals corresponding to $i = 0, 1, \ldots, n-1$. As already mentioned, these vectors cannot be arbitrary, because the particle can move only on the surface of the sphere. Thus, for any particle position $\mathbf{x} = (x, y, x)$, the control vector $\mathbf{u}_i = (u_{ix}, u_{iy}, u_{iz})$ must be tangent to the sphere (orthogonal to the position vector $\mathbf{x}$), i.e. $\mathbf{x} \cdot \mathbf{u}_i = 0$ for $i = 0, 1, \ldots, n-1$. This condition implies that the control vector has the following structure.

$$\mathbf{u}_i = (u_{ix}, u_{iy}, u_{iz}) = (\alpha_i y, -\alpha_i x + \beta_i z, -\beta_i y), \tag{9}$$

where α_i and β_i are two real constants. For this reason, finding the optimal controls $\mathbf{u}_0$, $\mathbf{u}_1$, ..., $\mathbf{u}_{n-1}$ is equivalent to finding the optimal control parameters α_i and β_i for each of the subintervals $i = 0, 1, \ldots, n-1$.

As a consequence of (9), Eq. (8) can be written row wise

$$\begin{cases} \dot{x} = \frac{k}{2\pi R} \frac{-y}{x^2+y^2+(z-R)^2} + \alpha_i y \\ \dot{y} = \frac{k}{2\pi R} \frac{x}{x^2+y^2+(z-R)^2} - \alpha_i x + \beta_i z \,, \qquad i = 0, 1, \ldots, n-1 \\ \dot{z} = -\beta_i y \end{cases} \tag{10}$$

The results of solving the problem $(\mathscr{D}\mathscr{P}_n)$ obtained with the Interior Point algorithm are shown in Table 1 and the results obtained with the Active Set algorithm are shown in Table 2. In each of these tables, the controls obtained by the optimization algorithms for the displacement of the passive particle from the original position $\mathbf{x}_0 = (0.8860, 0.0, -0.5)$ to the target position $\mathbf{x}_f = (-0.8860, 0.0, 0.5)$ are shown. Additionally, the corresponding values of the objective function f_n and the computation times (CPUt), in seconds, are also included.

The results obtained with the Interior Point method (Table 1) show that there is no regular variation of f_n as a function of the number of controls n. The same is true for the CPUt computation times.

Comparing the results obtained with the Interior Point method and the Active Set method, we can see that the results are similar, except for $n = 4$, where the Active Set is more efficient in finding the controls that minimize the objective function and in reducing the computation time.

Figure 1 shows the trajectories obtained with the controls shown in Table 1, and Fig. 2 shows the trajectories obtained with the controls shown in Table 2.

It can be seen that the trajectories generated by the two methods (Fig. 1a) and Fig. 2b)) are very close for $n = 1, 2$, and 3 and differ a little for $n = 4$.

4 Flow Created by Several Vortices

In this section, we treat the problem of displacement of a passive particle entrained by several vortices (N). As before, we want to find the vectors $\mathbf{u}_0$, $\mathbf{u}_1$, ..., $\mathbf{u}_{n-1}$ which pull the particle from $\mathbf{x}_0 = (0.8860, 0.0, -0.5)$ to $\mathbf{x}_f = (-0.8860, 0.0, 0.5)$. We also consider the time $T = 10$, and the same circulation for all vortices $k_i = 2$, for $i = 1, 2, 3, 4$. As before, to solve the three optimization problems corresponding to $N = 2, 3, 4$, we used the Interior Point and Active Set methods described in the `fmincon` Matlab solver (see [8]).

Table 1. Optimal controls obtained with the Interior Point optimization algorithm.

n	$\mathbf{u}_0$		$\mathbf{u}_1$		$\mathbf{u}_2$		$\mathbf{u}_3$		f_n	CPUt
	α_0	β_0	α_1	β_2	α_3	β_3	α_4	β_4		
$N=1$										
1	−0.096	−0.156							1.83	0.71
2	−0.094	−0.317	0.000	0.000					0.83	3.95
3	−0.126	−0.474	0.000	0.000	0.020	−0.065			0.93	1.07
4	−0.177	−0.568	0.000	0.000	0.000	0.000	−0.013	−0.207	1.00	4.73
$N=2$										
1	−0.213	0.044							2.18	1.47
2	0.000	0.000	−0.386	−0.072					0.98	3.12
3	0.000	0.000	0.000	0.000	−0.540	−0.209			0.97	4.13
4	0.000	0.000	0.000	0.0000	0.000	0.000	−0.685	−0.349	0.96	4.73
$N=3$										
1	−0.130	−0.153							2.00	2.50
2	−0.307	−0.108	0.000	0.000					0.81	5.80
3	−0.424	−0.112	0.000	0.000	−0.017	−0.005			0.76	5.16
4	−0.537	−0.133	0.000	0.000	0.000	0.000	−0.442	−0.011	0.73	7.03
$N=4$										
1	−0.144	−0.102							1.76	1.68
2	−0.230	−0.095	−0.070	−0.008					0.80	6.22
3	−0.310	−0.126	0.000	0.000	−0.114	−0.025			0.75	5.16
4	−0.403	−0.151	0.000	0.000	0.000	0.000	−0.146	−0.039	0.73	9.98

4.1 Flow Created by Two Vortices ($N = 2$).

The position of the two vortices ($\mathbf{x}_1$ and $\mathbf{x}_2$) are governed by

$$\begin{cases} \dot{\mathbf{x}}_1 = \frac{k}{2\pi R} k_2 \frac{\mathbf{x}_2 \times \mathbf{x}_1}{||\mathbf{x}_1 - \mathbf{x}_2||^2} \\ \dot{\mathbf{x}}_2 = \frac{k}{2\pi R} k_1 \frac{\mathbf{x}_1 \times \mathbf{x}_2}{||\mathbf{x}_2 - \mathbf{x}_1||^2} \end{cases} \tag{11}$$

with the respective initial conditions $\mathbf{x}_1(0)$ and $\mathbf{x}_2(0)$. In this case, the initial positions of the vortices are $\mathbf{x}_1(0) = (0.2185, 0.2815, 0.9511)$, and $\mathbf{x}_2(0) = (0.4330, 0.74, -0.5)$, and the circulations are $k_1 = k_2 = 2$. The passive particle, initially at $\mathbf{x}(0) = \mathbf{x}_0$, is governed by the equation

$$\dot{\mathbf{x}} = \frac{k}{2\pi R}\left(k_1 \frac{\mathbf{x}_1 \times \mathbf{x}}{||\mathbf{x} - \mathbf{x}_1||^2} + k_2 \frac{\mathbf{x}_2 \times \mathbf{x}}{||\mathbf{x} - \mathbf{x}_2||^2}\right) + \alpha\,(y, -x, 0) + \beta\,(0, z, -y), \tag{12}$$

where α and β are the controllers.

Table 2. Optimal controls obtained with the Active Set optimization algorithm.

n	$\mathbf{u}_0$		$\mathbf{u}_1$		$\mathbf{u}_2$		$\mathbf{u}_3$		f_n	CPUt
	α_0	β_0	α_1	β_2	α_3	β_3	α_4	β_4		
$N = 1$										
1	−0.096	−0.156							1.83	0.49
2	−0.101	−0.312	0.000	0.000					0.82	0.77
3	−0.117	−0.487	0.000	0.000	0.013	−0.047			0.92	1.09
4	−0.139	−0.652	0.000	0.000	0.016	−0.063	0.000	0.000	0.10	1.62
$N = 2$										
1	−0.213	0.044							2.18	0.63
2	0.000	0.000	−0.384	−0.073					0.98	3.40
3	0.000	0.000	0.000	0.000	−0.537	−0.209			0.96	3.58
4	0.000	0.000	0.000	0.000	0.000	0.000	−0.681	−0.348	0.96	8.68
$N = 3$										
1	−0.130	−0.153							2.00	0.93
2	−0.305	−0.108	0.000	0.000					0.81	2.86
3	−0.423	−0.110	0.000	0.000	−0.014	−0.004			0.75	5.20
4	−0.545	−0.121	0.000	0.000	0.000	0.000	−0.019	−0.007	0.72	9.02
$N = 4$										
1	−0.144	−0.102							1.76	1.50
2	−0.230	−0.095	−0.070	−0.008					0.80	5.32
3	−0.313	−0.126	0.000	0.000	−0.108	−0.022			0.76	7.52
4	−0.404	−0.152	0.000	0.000	0.000	0.000	−0.140	−0.035	0.72	12.52

This control problem is solved in the same way as the problem with a single vortex ($N = 1$) presented in Sect. 3. The problem is discretized in time and transformed into an optimal control problem similar to $(\mathscr{DP}_n)$. Since the control vectors $\mathbf{u}_1, \ldots, \mathbf{u}_{n-1}$ have the same form defined by (9), the problem reduces to determining the control parameters α_i and β_i for each of the subintervals $i = 0, 1, \ldots, n-1$.

The results obtained with the Interior Point method are given in Table 1 and the results obtained with the Active Set method are given in Table 2. The corresponding trajectories are shown in Figs. 1b) and 2b). It can be seen that the results obtained with the two methods are very similar. There is a tendency for the objective function to decrease as the number of subintervals n increases. The computation times also increase with n, but this is more pronounced for the Active Set method. Since the control parameters are very similar, the trajectories obtained with the two methods are practically the same.

4.2 Flow Created by Three Vortices ($N = 3$).

The position of the three vortices ($\mathbf{x}_1$, $\mathbf{x}_2$ and $\mathbf{x}_3$) are governed by

$$\begin{cases} \dot{\mathbf{x}}_1 = \frac{k}{2\pi R}\left(k_2 \frac{\mathbf{x}_2\times\mathbf{x}_1}{||\mathbf{x}_1-\mathbf{x}_2||^2} + k_3 \frac{\mathbf{x}_3\times\mathbf{x}_1}{||\mathbf{x}_1-\mathbf{x}_3||^2}\right) \\ \dot{\mathbf{x}}_2 = \frac{k}{2\pi R}\left(k_1 \frac{\mathbf{x}_1\times\mathbf{x}_2}{||\mathbf{x}_2-\mathbf{x}_1||^2} + k_3 \frac{\mathbf{x}_3\times\mathbf{x}_2}{||\mathbf{x}_2-\mathbf{x}_3||^2}\right) \\ \dot{\mathbf{x}}_3 = \frac{k}{2\pi R}\left(k_1 \frac{\mathbf{x}_1\times\mathbf{x}_3}{||\mathbf{x}_3-\mathbf{x}_1||^2} + k_2 \frac{\mathbf{x}_2\times\mathbf{x}_3}{||\mathbf{x}_3-\mathbf{x}_2||^2}\right) \end{cases} \tag{13}$$

with the respective initial conditions $\mathbf{x}_1(0)$, $\mathbf{x}_2(0)$ and $\mathbf{x}_3(0)$. The initial position of the vortices are $\mathbf{x}_1(0) = (0.2185, 0.2815, 0.9511)$, $\mathbf{x}_2(0) = (0.4330, 0.74-0.5)$, and $\mathbf{x}_3(0) = (-1.0, 0.0, 0.0)$, and the circulations are $k_1 = k_2 = k_3 = 2$. The passive particle position is given by the equation

$$\dot{\mathbf{x}} = \frac{k}{2\pi R}\sum_{i=1}^{3} k_i \frac{\mathbf{x}_i\times\mathbf{x}}{||\mathbf{x}-\mathbf{x}_i||^2} + \alpha\,(y,-x,0) + \beta\,(0,z,-y), \tag{14}$$

with the given initial condition $\mathbf{x}(0) = x_0$. This control problem is solved by a direct approach in a similar way as the previously described problem with one ($N = 1$) and two ($N = 2$) vortices.

The results obtained with the Interior Point method are given in Table 1 and the results obtained with the Active Set method are given in Table 2. The corresponding trajectories are shown in Figs. 1c) and 2c). As in the previous problem ($N = 2$), it can be seen that the results of the two methods are very similar. The objective function also decreases slightly as n increases. The computation times also increase with n and are higher then the computation times of the problems with $N = 1$ and $N = 2$. The trajectories obtained with the Interior Point and Active Set methods, shown in Figs. 1c) and 2c), are practically identical.

4.3 Flow Created by Four Vortices ($N = 4$).

The position of the four vortices ($\mathbf{x}_1$, $\mathbf{x}_2$, $\mathbf{x}_3$ and $\mathbf{x}_4$) are governed by

$$\begin{cases} \dot{\mathbf{x}}_1 = \frac{k}{2\pi R}\left(k_2 \frac{\mathbf{x}_2\times\mathbf{x}_1}{||\mathbf{x}_1-\mathbf{x}_2||^2} + k_3 \frac{\mathbf{x}_3\times\mathbf{x}_1}{||\mathbf{x}_1-\mathbf{x}_3||^2} + k_4 \frac{\mathbf{x}_4\times\mathbf{x}_1}{||\mathbf{x}_1-\mathbf{x}_4||^2}\right) \\ \dot{\mathbf{x}}_2 = \frac{k}{2\pi R}\left(k_1 \frac{\mathbf{x}_1\times\mathbf{x}_2}{||\mathbf{x}_2-\mathbf{x}_1||^2} + k_3 \frac{\mathbf{x}_3\times\mathbf{x}_2}{||\mathbf{x}_2-\mathbf{x}_3||^2} + k_4 \frac{\mathbf{x}_4\times\mathbf{x}_2}{||\mathbf{x}_2-\mathbf{x}_4||^2}\right) \\ \dot{\mathbf{x}}_3 = \frac{k}{2\pi R}\left(k_1 \frac{\mathbf{x}_1\times\mathbf{x}_3}{||\mathbf{x}_3-\mathbf{x}_1||^2} + k_2 \frac{\mathbf{x}_2\times\mathbf{x}_2}{||\mathbf{x}_3-\mathbf{x}_2||^2} + k_4 \frac{\mathbf{x}_4\times\mathbf{x}_3}{||\mathbf{x}_3-\mathbf{x}_4||^2}\right) \\ \dot{\mathbf{x}}_4 = \frac{k}{2\pi R}\left(k_1 \frac{\mathbf{x}_1\times\mathbf{x}_4}{||\mathbf{x}_4-\mathbf{x}_1||^2} + k_2 \frac{\mathbf{x}_2\times\mathbf{x}_4}{||\mathbf{x}_4-\mathbf{x}_2||^2} + k_3 \frac{\mathbf{x}_3\times\mathbf{x}_4}{||\mathbf{x}_4-\mathbf{x}_3||^2}\right) \end{cases} \tag{15}$$

with the respective initial conditions $\mathbf{x}_1(0)$, $\mathbf{x}_2(0)$, $\mathbf{x}_3(0)$ and $\mathbf{x}_4(0)$. In this case, the initial vortex positions are $\mathbf{x}_1(0) = (0.2185, 0.2815, 0.9511)$, $\mathbf{x}_2(0) = (0.4330,$

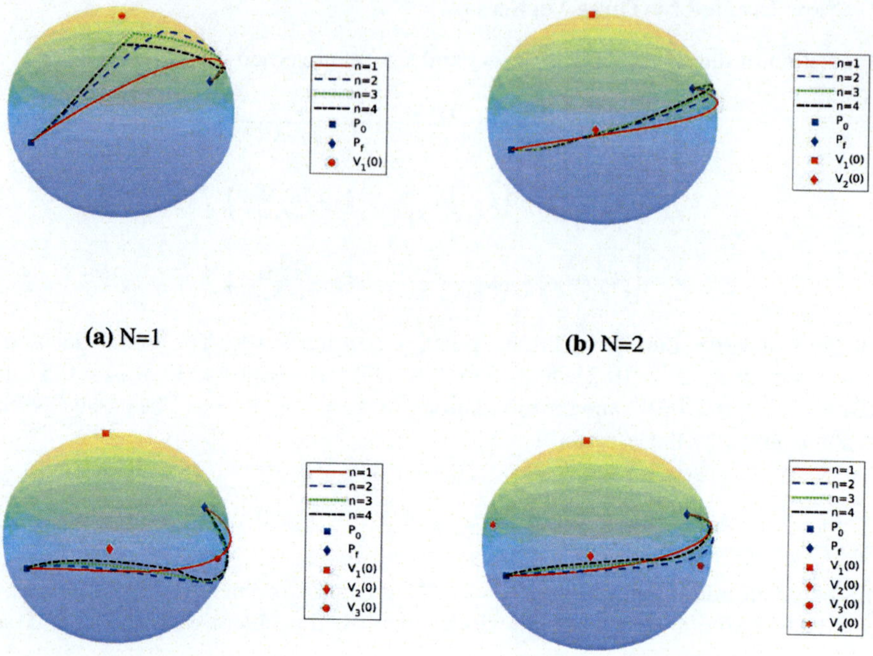

(a) N=1 **(b)** N=2

(c) N=3 **(d)** N=4

Fig. 1. Optimal trajectories determined with the Interior Point optimization method.

$0.74 - 0.5)$, $\mathbf{x}_3(0) = (-1.0, 0.0, 0.0)$, and $\mathbf{x}_4(0) = (1.0, 0.0, 0.0)$, and the circulations are $k_1 = k_2 = k_3 = k_4 = 2$.

The dynamic of the passive particle induced by the four vortices is

$$\dot{\mathbf{x}} = \frac{k}{2\pi R} \sum_{i=1}^{4} k_i \frac{\mathbf{x}_i \times \mathbf{x}}{||\mathbf{x} - \mathbf{x}_i||^2} + \alpha\,(y, -x, 0) + \beta\,(0, z, -y)\,, \tag{16}$$

with a given initial condition $\mathbf{x}(0) = \mathbf{x}_0$. The same direct numerical approach used in the previous problems and described in Sect. 3 is used here to solve this control problem.

The results obtained with four vortices are shown in Tables 1 and 2 and the corresponding trajectories are shown in Figs. 1d) and 2d). As in the previous problems ($N = 2, 3$), the results of the two methods are very similar and the corresponding value of the objective function, which also decreases slightly with increasing n. The computation times also increase with n and are higher then to the computation times obtained for the problems with $N = 1, 2, 3$. The corresponding trajectories obtained by the two methods (see Figs. 1c) and 2c)) are practically identical.

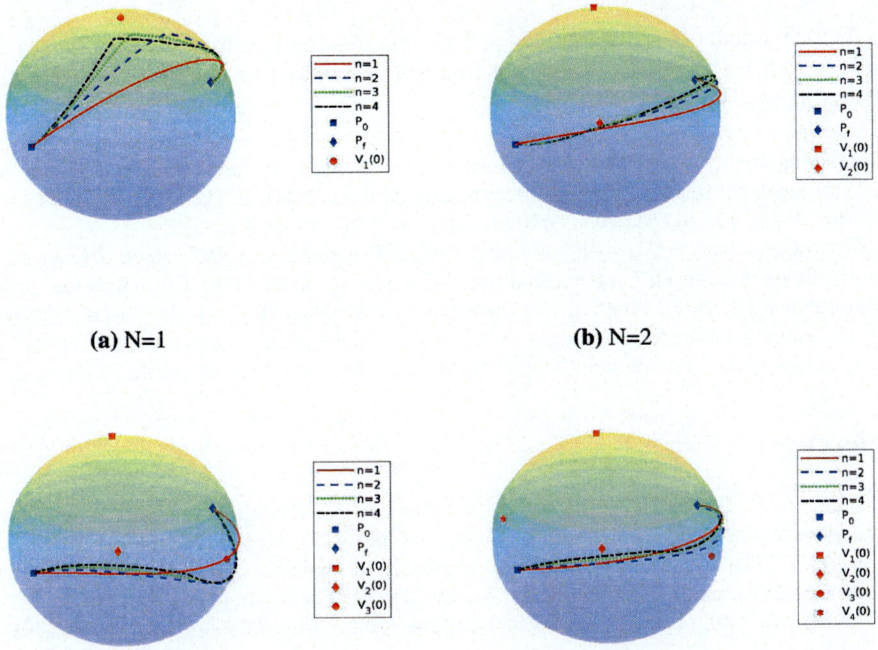

(a) N=1 **(b)** N=2

(c) N=3 **(d)** N=4

Fig. 2. Optimal trajectories determined with the Active Set optimization method.

5 Conclusion

This study is concerned with the determination of the optimal trajectory of a passive particle, moving in a flow induced by one or more vortices, between two points on the surface of a sphere. The numerical strategy presented is an alternative to the conventional methods of optimal control theory. The control problem is transformed into a constrained optimization problem that can be solved by a numerical method.

Two different optimization methods were tested, the Interior Point and Active Set methods. Both were able to find optimal trajectories regardless of the time discretization level of the problem and the number of vortices involved. The computation times required by both methods are of the same order of magnitude. However, the Active Set requires more time when the number of control parameters to be determined is higher.

For each of the four problems analyzed, the value of the objective function tends to decrease with the increase in the number of control parameters. This result shows that increasing the number of control parameters allows greater freedom in the choice of the trajectory, which minimizes the energy used for the displacement. On the other hand, the computation time is higher.

As the number of vortices increases, the computation time also increases. This is due to the fact that a larger number of calculations is required to solve the corresponding ODE systems.

Acknowledgements. Carlos Balsa was supported by Foundation for Science and Technology (FCT, Portugal) for financial support through national funds FCT/MCTES (PIDDAC) to CeDRI (UIDB/05757/2020 and UIDP/05757/2020) and SusTEC (LA/P/0007/2021).

Sílvio Gama was partially supported by (i) CMUP, member of LASI, which is financed by national funds through FCT - Fundação para a Ciência e a Tecnologia, I.P., under the project with reference UIDB/00144/2020, and (ii) project SNAP NORTE- 01-0145- FEDER-000085, co-financed by the European Regional Development Fund (ERDF) through the North Portugal Regional Operational Programme (NORTE2020) under Portugal 2020 Partnership Agreement.

References

1. Babiano, A., Boffetta, G., Provenzale, A., Vulpiani, A.: Chaotic advection in point vortex models and two-dimensional turbulence. Phys. Fluids **6**(7), 2465–2474 (1994)
2. Balsa, C., Gama, S.M.A.: A numerical algorithm for optimal control problems with a viscous point vortex. In: Brito Palma, L., Neves-Silva, R., Gomes, L. (eds.) CONTROLO 2022. LNEE, vol. 930, pp. 726–734. Springer, Cham (2022). https://doi.org/10.1007/978-3-031-10047-5_64
3. Balsa, C., Gama, S.M.: The control of the displacement of a passive particle in a point vortex flow. J. Comput. Methods Sci. Eng. **21**(5), 1215–1229 (2021). https://doi.org/10.3233/jcm-204710
4. Helmholtz, H.: Über integrale der hydrodynamischen gleichungen, welche den wirbelbewegungen entsprechen. Journal für die reine und angewandte Mathematik **55**, 25–55 (1858), http://eudml.org/doc/147720
5. Hwang, S., Kim, S.C.: Point vortices on hyperbolic sphere. J. Geom. Phys. **59**(4), 475–488 (2009). https://doi.org/10.1016/j.geomphys.2009.01.003
6. King, J., Xue, X., Yao, W., Jin, Z.: A fast analysis of pesticide spray dispersion by an agricultural aircraft very near the ground. Agriculture **12**(3), 433 (2022). https://doi.org/10.3390/agriculture12030433, https://www.mdpi.com/2077-0472/12/3/433
7. Kirchhoff, G.R.: Vorlesungenbër mathematische physik. Mechanik (1876)
8. MathWorks: Matlab Optimization Toolbox: User's Guide (R2020a). The MathWorks Inc, Natick, Massachusetts, United State (2020)
9. Mokhov, I.I., Chefranov, S.G., Chefranov, A.G.: Point vortices dynamics on a rotating sphere and modeling of global atmospheric vortices interaction. Phys. Fluids **32**(10), 106605 (2020)
10. Nava-Gaxiola, C., Montaldi, J.: Point vortices on the hyperbolic plane. J. Math. Phys. **55**, 102702 (2014). https://doi.org/10.1063/1.4897210
11. Newton, P.K.: The N-Vortex Problem: Analytical Techniques, vol. 145. Springer, New York (2001). https://doi.org/10.1007/978-1-4684-9290-3
12. Polvani, L.M., Dritschel, D.G.: Wave and vortex dynamics on the surface of a sphere. J. Fluid Mech. **255**(-1), 35 (1993). https://doi.org/10.1017/s0022112093002381
13. Schmid, C., Biegler, L.: Quadratic programming methods for reduced hessian SQP. Comput. Chem. Eng. **18**(9), 817–832 (1994). https://doi.org/10.1016/0098-1354(94)e0001-4
14. Thomson, W.: On vortex motion. Trans. R. Soc. Edinb. **25**, 217–260 (1869)
15. Vankerschaver, J., Leok, M.: A novel formulation of point vortex dynamics on the sphere: geometrical and numerical aspects. J. Nonlinear Sci. **24**(1), 1–37 (2013). https://doi.org/10.1007/s00332-013-9182-5

16. Waltz, R.A., Morales, J.L., Nocedal, J., Orban, D.: An interior algorithm for nonlinear optimization that combines line search and trust region steps. Math. Program. **107**(3), 391–408 (2006)
17. Zhang, B., Tang, Q., ping Chen, L., Xu, M.: Numerical simulation of wake vortices of crop spraying aircraft close to the ground. Biosyst. Eng. **145**, 52–64 (2016). https://doi.org/10.1016/j.biosystemseng.2016.02.014, https://www.sciencedirect.com/science/article/pii/S1537511015302993

Home Automation System for People with Limited Upper Limb Capabilities Using Artificial Intelligence

Ronnie Martínez[1(✉)], Rubén Nogales[1,2], Marco E. Bencázar[2], and Hernán Naranjo[1]

[1] Universidad Técnica de Ambato, Ambato, Ecuador
{rmartinez9951,re.nogales,hf.naranjo}@uta.edu.ec
[2] Escuela Politécnica Nacional, Quito, Ecuador
{ruben.nogales,marco.benalcazar}@epn.edu.ec

Abstract. People with physical disabilities (PD) have problems carrying out daily activities, affecting their independence. In this context, a person with PD can communicate through hand gestures or facial gestures, among others. However, selecting the features and patterns that separate one gesture from another is not a trivial problem. In this sense, we propose a real-time domotic system (DS) that works with three subsystems. The first subsystem recognizes hand gestures using a machine learning model and infrared information. The machine learning model consists of pre-processing, feature extraction, classification and post-processing modules. The second subsystem relays the message between the subsystems. Finally, the third subsystem activates the operation of the actuators by gestures. The Hand Gesture Recognition (HGR) model was trained using 6720 observations and tested offline with 1680 observations, giving an accuracy rate of 92.759%. Additionally, the DS was tested online with 1500 observations from 10 users whose persons were not part of the input dataset or model testing. Online testing gives an accuracy rate of 84.07%. Once the hand gesture is recognized, it is sent wireless to an Esp8266 board through the MQTT protocol. The Esp8266 board activates the operation of several actuators. The mosquitto broker embedded in a Raspberry Pi manages the sending and receiving of messages between the computer where the gesture is recognized and the Esp8266 board. The theoretical response time of the DS is 118.87 ms.

Keywords: Domotic System · Hand Gesture Recognition · MQTT · Leap Motion Controller

1 Introduction

The DS gives people with PD autonomy. In this context, a DS helps a person with PD to do daily living activities without the need for another person to assist them. Freedom to do everyday activities increases self-esteem and improves the

M. V. Garcia and C. Gordón-Gallegos (Eds.): CSEI 2022, LNNS 678, pp. 214–231, 2023.
https://doi.org/10.1007/978-3-031-30592-4_16

quality of life of people with PD [8,10]. The flexibility of DS allows the use of many peripherals, both inputs to receive information and output peripherals (actuators) to execute a specific action, such as opening a door or switching on a light bulb. The DS are not limited to people with particular PD. On the contrary, many people with different kinds of PD can benefit by having a great variety of input peripherals. In the same way, the Human-Computer Interface (HCI) of a DS can be adapted to the specific needs of each PD, allowing control of the environment and communication [19].

In [9] states that until January 2022 in Ecuador, there are 471,205 people with disabilities, of which 45.66% correspond to PD. In people with PD, we find pathologies such as afasia, spasticity and motor deficits caused by Cerebrovascular Diseases (CVD), Spinal Cord Injury (SCI), among others. CVD is classified as ischemic and hemorrhagic [5], among which ischemic CVD is the cause of limited movement in people suffering from that problem. SCI cause sensation loss and motor function in upper, lower, or whole-body extremities. According to the World Health Organization (WHO), people with a spinal cord injury are between 2 and 5 times more likely to die within the first year [18]. Typical movement performance is rarely restored after this kind of pathology and after a long rehabilitation period, limiting the performance of activities of daily living [14,21,26,27]. Therefore, people with PD communicate through body movements, also known as gestures.

The generation of gestures is a way of expressing feelings towards other human beings [11–13]. Unconsciously, people need to communicate with the world around them from the moment they are born. In this sense, people with PD can use gestures to do everyday activities. These gestures provide enriched data that can be used for recognition, allowing information to be captured across different devices. Devices can be classified into sensors covering parts of the body and devices not requiring direct contact.

The DS proposed aims to help people with PD recover independence through hand gestures. Also, the DS captures the gestures executed by the user with the Leap Motion Controller (LMC). The LMC provides a sequence of images and spatial positions representing the movement performed [16,24]. In this case, the spatial positions are the information to recognize the OPEN HAND, CLOSE HAND, WAVE IN, WAVE OUT and PINCH gestures. Once the signal produced for the motion has been captured, it will be preprocessed for subsequent recognition through Machine Learning algorithms. Finally, the label resulting from the recognition will be sent via the Wi-Fi network to the IoT device that will trigger the corresponding actuators.

2 Related Works

In this section, we carry out a systematic review of the literature. We used scientific databases like IEEE Xplore, ACM digital library, Springer and Science Direct for the investigation. The search strings are crated with keywords such as MQTT, LMC, Artificial Intelligence and IoT computers. According to the systematic literature review, we define 6 publications where the authors created DS using MQTT, Machine Learning and IoT computers. However, there is no evidence of DS that integrates the mentioned methods and LMC data into a single cyber-physical model.

In [3] proposes a Brain-Computer Interface (BCI) to control home automation devices through a P300 speller. The proposed system works with EEG signals used for feature extraction and, finally, classification. The captured signal contains the user's desired action and the BCI system decides which of all the objects to control. The BCI 2000 g. Tec system was used to acquire a complete record of P300 evoked potentials. They use a band pass filter with cut-off frequencies of 0.1–20 Hz. For each intensification, the data segments were grouped by channel. This generates a single feature vector equal to each stimulus. The features obtained derive the weights from the stepwise linear discriminant analysis (SWLDA), which was used as a classification method. 2 stages were used for the experiment. The offline phase was conducted in 2 sessions over 2 weeks with 3 users. One session was for training and the other for testing, with 10 runs each. The online phase was carried out with a single user who performed 6 runs. They report an average value of 83% in the offline stage, while it reaches 90% accuracy in the online setting.

In [20], the bracelet Myo-Armband is a simple, wireless way to control a home environment. In addition, electrical and electronic devices in the home are controlled by Zigbee modules. The activities done in the DS are opening and closing blinds, switching lights on and off and locking the whole house. The above actions respond to each signal sent by the bracelet. Signals were acquired from 10 randomly selected individuals and each signal was repeated 5 times. 60 samples are taken from each electrode of the cuff and each coordinate axis of both linear and angular accelerations. The values of the 8 electrode signals are then added together to produce a single signal. The exact process is used for linear and angular acceleration signals. Finally, they use a Feed-Forward neural network to classify with the Levenberg Marquardt algorithm for training. The neural network consists of 3 layers. An input layer with 18 neurons, a hidden layer with 8 neurons and an output layer with three neurons. The linear recognition uses a sliding window and the algorithm was implemented in C++ with the synaptic weights previously trained in MATLAB. The neural network has an accuracy of 83.33% in the recognition and classification of electromyographic signals.

In [6,7], the authors propose in 2 parts a DS controlled by voice commands. The voice commands use an acoustic and language model trained in the first part of this project. The user's voice is captured in a mobile application developed in Android Studio. Audio files are sent by sockets using the FTP protocol. Raspberry Pi was selected as the FTP server and Arduino Uno as the acquisition and signal sending board. Once the audio file is received in .MP4 format, the system transforms this file to WAV. In this case, the recognition process is based on Bayesian Networks, specifically Hidden Markov Models (HMM). The model used for recognition in version 2.1 reports 100% accuracy in the training and 83% in the test phases. The system allows the acquisition and recognition of voice commands, light control, temperature measurement, emulation of refrigerator operation, real-time notification and database querying.

In [4], they implement a voice-controlled DS. The DS uses open-source libraries for keyword detection, voice recognition and smart device integration. Moreover, the MQTT protocol allows communication between devices. The DS uses the Raspberry Pi 3 for the Intelligent Recognition Centre (IRC) and a microphone with a USB audio card to capture voice commands. The IRC performs keyword detection using the Snowboy library. Also, the authors use CMUSphinx to train an acoustic and language model with the Portuguese language. They do not report on the results of the training. The trained model takes care of speech recognition and audio-to-text transformation. The obtained text is indexed to an actuator that will perform a specific action. Finally, the IRC converts the text with the recognized action into audio as a response for the user. Multiple voice commands evaluated the system's ability to interpret. They report 95% accuracy in the preliminary testing phase.

In [17] proposed a system for home automation control through hand gesture recognition focused on immobile and bedridden people. First, the system records a video of the gesture through a camera with a resolution of 1280×720. The video is captured for processing by a CCTV camera installed in the patient's room. Secondly, the gesture is extracted and the same is recognized. They use a convolutional neural network with 3 convolutional layers and 2 Max Pooling layers for recognition. The output of the last layer feeds a Softmax. OpenCV provided the background subtraction function allows segmenting the hand and extracting the mask from the moving foreground. A dataset is created with images of the left and right hand at different depths. The dataset consists of 6 gestures, with a total of 12000 64-pixel images with specific labels. In this case, 80% of the dataset is used for classifier training and the remaining 20% for testing. In the end, the model is adjusted and the number of epochs and steps is defined, with a total of 50 epochs with 6000 steps for training and 3000 steps for testing with an average accuracy of 99.97%.

In [22], they develop a predictive model for optimizing and planning intelligent homes. The model uses data from both outside and inside the home. The simulation of different scenarios of a scaled house provides this information. The outdoor data are temperature and humidity, while the indoor data are the states of the actuators. The authors process the collected data to automatically extract the information the system needs to respond to external events. A central system controls the interrelationship between 2 subsystems with specific functions. The first subsystem allows the control of fundamental smart home properties such as air conditioning and automated devices, among others. In this context, the MQTT protocol performs the sending and receiving of messages between the host system and the control subsystem. The second subsystem help in the management and configuration of experiments through a remote interface, data collection and an agent in charge of obtaining scenarios from the real world. The management subsystem communicates by REST requests because it does not need to stay close to the physical simulation environment. They trained 4 machine learning regressors for the resulting predictive model. Keras and scikit-learn libraries provided LSTM, KNN and SVM regressor models for training. Furthermore, the authors create the DNN model, which consists of 5 layers and 128 neurons in each layer, Huber loss and Relu as an activation function. The first training and testing have a dataset of 300 records. They report 79% accuracy for DNN, 70% for KNN, 64% for LSTM and 42% for SVR. The second training and testing increase the dataset to 4000 records. In the same sense, they report 95% accuracy for KNN, 89% for DNN, 87% for LSTM and 22% for SVR. Training and testing with the two SVR registers always obtained the lowest accuracy percentage.

The main problems we could find with these DS are as follows. Firstly, the DS that works with cloud services stores a large amount of information that can be violated. External companies collect this information by constantly monitoring households. Employees within the companies that provide these services have access to this information, as is the case of AMAZON [2]. Access to this information violates the privacy of each user's data [2]. Secondly, voice-controlled DS has implementation and usability problems. This kind of DS needs a trained speech recognition model to correct functioning. The recognition model works with extensive data for training and testing in a single language. The plenty of data required for the recognition model makes it challenging to implement this type of DS [23]. Also, voice-controlled DS are obsolete for people suffering from aphasia because of CVD. Thirdly, DS that works with wearable sensors such as Myo Armband and EEG sensors, among others, are unsuitable for people with PD. The main reason is the need for direct skin contact with the sensor surface, which causes sweating and discomfort during prolonged use. Finally, DS controlled by CCTV cameras improve the problem with wearable sensors but are limited to control from a specific area due to their static installation as well as the cost of implementation, unlike LMC. In this context, existing DS generates problems for people with PD.

For these reasons, we propose a DS focused on hand gesture recognition with LMC to control the environment for people with PD. The LMC is a low-cost accurate device dedicated to capturing hand movement without affecting the simplicity of movement [1,29]. Also, LMC has an interaction range of 150° wide by 60 cm high with an accuracy of 0.01 mm [9]. Firstly, the DS will use the LMC to capture the information about the user's hand gestures. Secondly, the Machine Learning algorithms will get the gesture's report for recognition. Then, MQTT will enable secure communication between devices and will be implemented on a Raspberry Pi 3. Finally, the Esp8266 board will control the different actuators according to the label it receives.

3 Methodology

We use three stages to implement the DS. In the first stage, the system captures and recognizes hand gestures. In the second stage, the communication protocol is established through MQTT. And the third stage consists of the Esp8266 board programming to control light, blind and door according to the gesture developed by the user. Figure 1 shows the operating scheme of the DS.

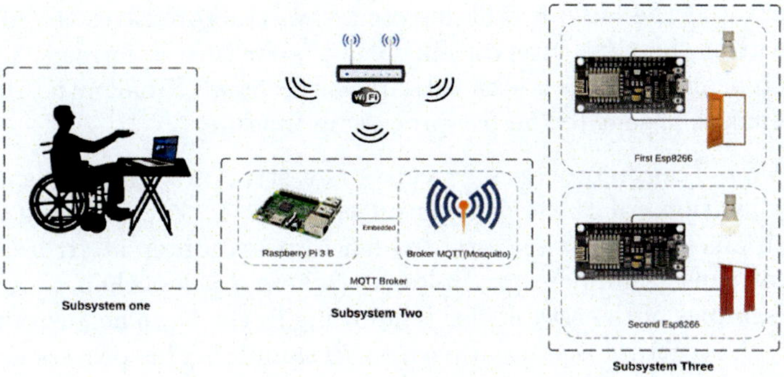

Fig. 1. Domotic system scheme

3.1 HGR Model

First Stage. We use machine learning models for hand gesture recognition. The models work with the dataset held from [25] for training and testing. The dataset contains 8400 samples from 56 users. Each user executes five different gestures 30 times each. In this sense, the models consist of 5 modules, as shown in Fig. 2. These are data acquisition, pre-processing, feature extraction, classification and post-processing.

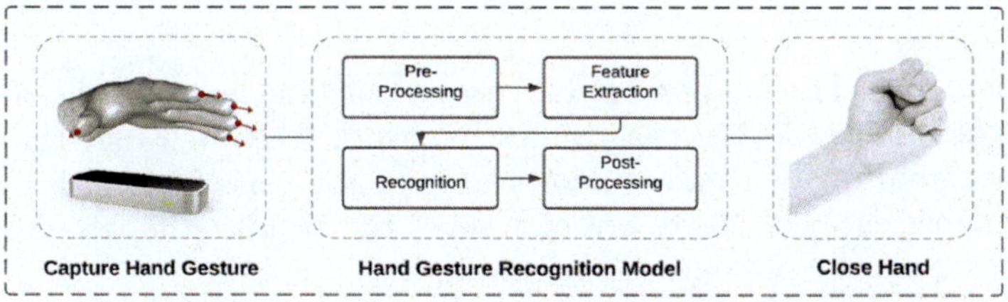

Fig. 2. Hand gesture capture and recognition scheme

Data Acquisition. We use LMC for data acquisition. The LMC constantly tracks objects, but it starts the recording when the hand is within its field of view. LMC has a sampling rate 200 Hz per second. The information captured is the spatial positions and directions of the hand at different time instants. This information is interpreted by MEX code, sent to MATLAB code and stored in a MATLAB structure. The structure formed is limited to a window of 300 samples. The average time for 300 samples is 4.609 s. In this sense, two scenarios are considered to form the structure. First, if the captured samples are less than the set window and the user withdraws the hand, the structure saves the information obtained within this time interval. Second, if the captured samples are equal to the settings window and the hand is still detected, the structure saves the information captured within the time interval and a new structure is formed. The second scenario is repeated until the user withdraws the hand. Finally, the structure in the two scenarios is sent to the pre-processing module.

Pre-processing. The preprocessing receives the structure with the spatial positions and directions of the fingertips on their X, Y, Z coordinate axes. The spatial positions and directions are divided into two different structures. Each structure contains a number of samples (n), where n is less than or equal to the number of samples of the established windowing in the Data acquisition module. However, each structure is standardized to 70 samples. This process is described in the "Zone Values" and "Image Selector" sections of a previous job [28]. The change to the number of samples per structure preserves the characteristics of the incoming signals. Next, we normalize the new resulting signal with the min-max method to reduce their amplitude. The normalization is applied independently for each structure of positions and directions of each user. Then, the signals pass to the feature extraction module.

*Feature Extraction.*The spatial position and directions structures are organized by channels. Each finger's coordinate axes make up the channels. Each channel is divided into windows of 20 with steps of 15. In this sense, we extract the features of each window per channel. This module works with four feature extraction functions. These are the Variance (VAR), Slope Sign Change (SSC), Enhanced Wavelength (EWL) and Standard Deviation (SD) [15]. Using these functions increases the performance and accuracy of the model because it better characterizes the hand gesture signal. The extracted features enter the classification module.

Classification. The classification module uses a machine learning algorithm. The selected algorithm is an Artificial Neural Network (ANN). First, the classifier receives the signal divided into windows. Hence, the classifiers return one label for each window entered. Next, a vector saves the resulting labels. Finally, the label vector enters the post-processing module.

Post-processing. The post-processing module works with an algorithm that receives the label vector. The label vector contains as many labels l_i as the number of windows divided in each channel at different instants of time (t). The algorithm goes through the entire vector and compares the label with its subsequence. Therefore, l_{it} is compared to l_{it+1}. If the two labels are equal, both labels retain their initial values. However, if the two labels are different, l_{it} is compared to l_{it+2}. If the two labels are the same, then l_{it+1} will equal l_{it}. Contrary case, l_{it} and l_{it+1} keep their initial values. This process is repeated until it goes through the entire vector. The label representing the gesture is the mode of the label vector. The resulting label will be sent through the communication protocol.

3.2 Communication Protocol

Second Stage. We establish wireless communication between PC, Raspberry Pi 3 and Esp8266, as shown in Fig. 3. The three devices communicate through a dedicated router. In addition, the DS works with MQTT because it is a protocol specialized in communication between IoT devices. The MQTT architecture works with publishers and subscribers. Also, Raspberry Pi 3 is used as an intermediary device. A broker has been installed on this device. The broker manages the sending and receiving of messages between the publisher and the subscriber.

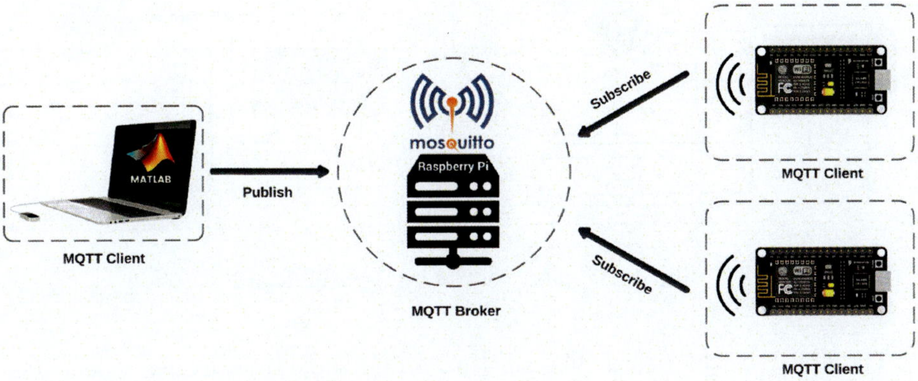

Fig. 3. MQTT Publish/Subscribe Architecture

Broker. The implemented broker is Mosquitto. We selected Mosquitto because the DS does not work in a clustered environment. In this case, Mosquitto efficiently sends messages in non-clustered server environments. Subsequently, the broker contains a configuration of users and passwords. Each user's password is encrypted as a security measure.

Publisher. The PC is the publisher of the message to the broker. The PC uses the "MQTT in MATLAB" complement. This complement allows communication between the PC and the broker through MQTT. The publishing machine sends the user, password and port as parameters. If the connection is successful, the device where the gesture is recognized transmits the label, the subscription topic and the quality of service (QoS) to the broker. The QoS set by the publisher is 0 by the DS approach.

Subscribers. The subscribers to the broker are the boards Esp8266. The subscriber boards use the Arduino library "PubSubClient". This library allows the connection between the Esp8266 board and the broker via MQTT. Subscribers send their username, password and the topic they wish to subscribe to. If the data entered is correct, the broker allows the connection. Once the connection is accepted, the subscriber waits for a message from the broker.

3.3 Control of Bulb, Blind and Door

Third Stage. We programmed two Esp8266 boards to control different actuators. Each board is subscribed to a specific existing topic within the broker. One Esp8266 board controls two different actuators, each actuator work with a particular gesture as described in the experimentation. Figure 4 shows how the different actuators were connected to the Esp8266 boards.

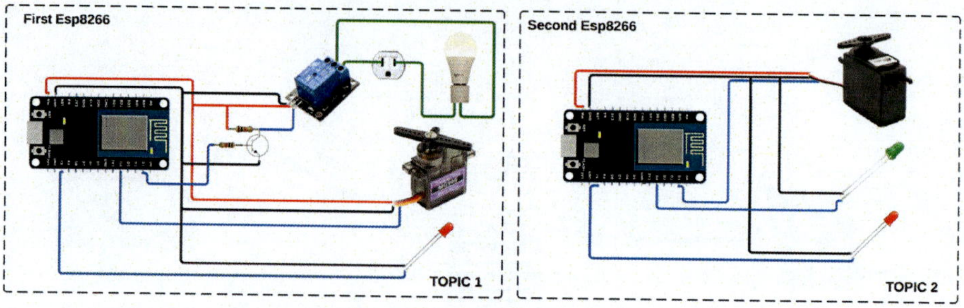

Fig. 4. Actuator control by Esp8266 board

The first Esp8266 board subscribed to a first topic receives two labels. The first label controls the switching on and off of 110 V bulbs. The second label controls an MG90S servomotor. The servomotor mimics the operation of an actual door. Similarly, the second Esp8266 board subscribed to a second topic receives two labels. The first label controls the switching on and off of an LED in a gradual form. The second label activates a servomotor. The servomotor simulates the opening and closing of electric curtains.

4 Experimentation

Three subsystems form the overall DS. Additionally, a dedicated router provides the WI-FI signal for communication between subsystems. The router has 802.11n standard, WPA2-Personal security, 2.4 GHz network bandwidth and 72 Mbps as sending/receiving speed.

First Subsystem. The first subsystem recognizes the gesture and publishes the label representing it. This system is developed in the R2019a version of the MATLAB IDE. A DELL PC with WINDOWS 10 operating system hosts the IDE. This computer has a 7th generation INTEL CORE i7 processor, 16 Gb RAM, four logical processes of 2.70 GHz and a wireless network card Intel(R) Dual Band Wireless-AC 3165. In this sense, a hybrid USB 2/3 cable connects the LMC sensor to the PC. The PC controls the LMC with version 3.2.1 of its SDK. The HGR model uses the information captured by LMC to feed the classifier. The HGR model uses a multi-class ANN as a classifier. ANN works with two hidden layers and one output layer. The first hidden layer has a Relu activation function and 25 neurons. At the same time, the second hidden layer has a Logsig activation function and 15 neurons. The output layer uses a Softmax function. Finally, optimization functions carry out the weight adjustment technique. The optimization functions are cross-entropy and gradient descent. ANN was trained with 2000 interactions and 1.0e1 as a regularization factor. The HRG model receives the information from the LMC. The information goes through the five modules described in the methodology and returns a label. We use the HGR model with ANN because ANN has a test accuracy of 92.759%.

When subsystem one finishes recognition, it publishes the label representing the gesture. The "MQTT in MATLAB" plug-in enables MATLAB to post the label and the topic via the MQTT protocol. MATLAB uses version 1.5 of "MQTT in MATLAB". The PUBLISH function sends the label and topic over a wireless network. Figure 5 shows the operation of the HGR system.

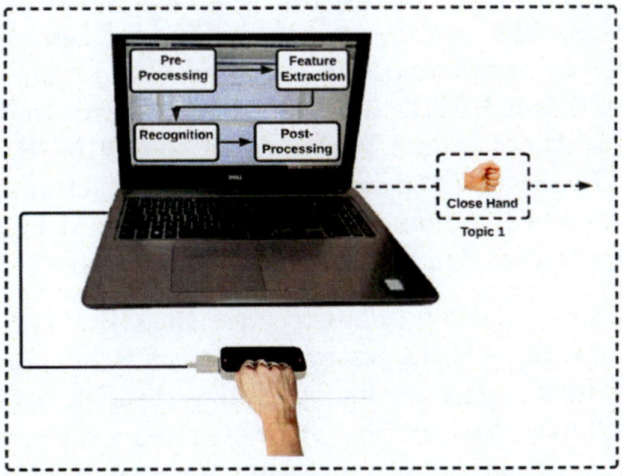

Fig. 5. Scheme of HGR system

Second Subsystem. The second subsystem is on the Raspberry Pi 3 B V1.2. The Raspberry Pi 3 B model has a 64-bit quad-core 1.2 GHz ARMv8 processor, 1GB RAM, Wi-Fi 802.11 b/g/n, 5V/2.5 power supply with micro USB. The Raspberry Pi works with version 3.1.1 of the Mosquitto broker. In this sense, the Raspberry Pi receives the label representing the gesture and subscription theme from subsystem one. Based on the topic, the broker manages the receiving and sending of messages between subsystems one and three. The broker sends the gesture label to all machines subscribed to that topic. Figure 6 shows the operation of the second subsystem.

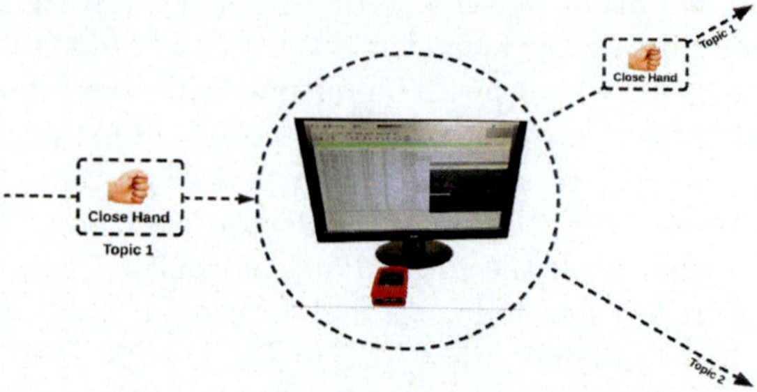

Fig. 6. Scheme of the second subsystem

Third Subsystem. The third subsystem works with two NODEMCU V2 Esp8266 Wi-Fi boards. Each Esp8266 board has 96 KiB of RAM, TCP/IP stack, 802.11 b/g/n Wi-Fi, 15 digital pins for each side, 5V power and 3.3V input/output. The Esp8266 boards are subscribed to different topics. The topic added to the broker by subsystem one is "Room" and "Entry". The "Room" Topic contains the CLOSE HAND and WAVE IN gestures. In the same way, the topic "Entrance" has the gestures WAVE OUT and PINCH. The Open Hand gesture on the two boards defined an emergency stop action. In the same way, both boards have a red LED connected to pin 1. The red LED lights up when the Esp8266 board is successfully connected to the router.

First Esp8266. The first Esp8266 board receives the CLOSE HAND and WAVE IN gestures. The CLOSE HAND gesture controls the switching on and off of a 110 V bulb. The light activates its operation through a 5V relay KY-019. This relay controls 110V through the 5V it receives from the Esp8266 board. Between the board and the relay is a 2N3940 transistor. The transistor switches the voltage sent by the Esp8266 to the 5 V needed by the relay. Similarly, the WAVE IN gesture controls an MG90S servomotor through pin 2. The servomotor mimics the operation of an actual gate.

Second Esp8266. The second Esp8266 board receives the WAVE OUT and PINCH gestures. The WAVE OUT gesture controls a DS04-NFC servomotor. The pin 5 sends instructions to the servomotor to turn clockwise and counter-clockwise. The servomotor has an operating voltage between 4.8 and 6 V. The Esp8266 board powers the servomotor through its VIN pin, which provides 5V. The servomotor simulates the opening and closing of electric curtains. Similarly, the PINCH gesture controls an LED's switching on and off. The pin 2 of the Esp8266 board feeds the LED. The LED lights up gradually. Figure 7 shows the operation of the third subsystem.

Fig. 7. Scheme of the third subsystem

In addition, we calculate the theoretical label delivery time for the entire DS. The propagation time and the transmission time form the sector delivery time. In this case, the propagation time is calculated from the router characteristics. The division between the router range and the propagation speed generates the propagation time. The indoor router range is 50 m according to the 802.11n standard. Meanwhile, the propagation speed in air is equal to 3×10^8. In the same sense, the division between the average packet size and the transmission rate generates the transmission time. The average packet size is calculated from the data captured by the Wireshark software. Next, we set the average transmission speed for the PC, the ROUTER and the Raspberry Pi to 72 Mbps. Finally, we summed the delivery times of the system's four sections to obtain the entire DS's theoretical delivery time. Table 3 shows the theoretical delivery time of the DS.

5 Results

Ten users tested the online DS. A user executed each gesture 30 times. We obtain a data set of 1500 samples. Testing achieves an average of 84.07% accuracy in the operation of the DS. Table 1 shows the test results.

Table 1. Gesture evaluation results: amount of correctly classified instances per gesture and user.

	Open Hand	Close Hand	Wave In	Wave Out	Pinch	Total by user 150	Accuracy rate by user (%)
User 1	30	27	30	27	21	135	90,00
User 2	30	29	30	26	8	123	82,00
User 3	30	29	27	27	16	129	86,00
User 4	30	24	24	23	11	112	74,67
User 5	30	26	29	28	20	133	88,67
User 6	30	27	30	26	17	130	86,67
User 7	30	28	29	27	9	123	82,00
User 8	30	24	27	27	17	125	83,33
User 9	30	28	30	27	18	133	88,67
User 10	30	28	28	23	9	118	78,67
Total recognized gestures of 300	300	270	284	261	146	1261	84,07
Accuracy rate by gesture (%)	100	90,00	94,67	87,00	48,67	84,07	

In the same sense, online testing generates response times by users for each gesture performed. We have an average time of 88.17 milliseconds in the HGR model. Table 2 shows the average times obtained per gesture and per user.

Table 2. Time evaluation of the HGR model.

	Open Hand	Close Hand	Wave In	Wave Out	Pinch	Average recognition time by user (ms)
User 1	78,30	84,33	73,57	72,93	62,33	74,29
User 2	73,70	111,70	95,27	90,80	79,43	90,18
User 3	75,70	73,37	72,17	95,40	113,67	86,06
User 4	85,00	92,77	69,83	112,13	79,83	87,91
User 5	146,10	89,77	93,53	78,53	81,43	97,87
User 6	85,50	83,53	87,30	109,57	118,20	96,82
User 7	79,30	92,43	98,10	95,43	97,90	92,63
User 8	46,30	91,47	94,83	108,67	95,23	87,30
User 9	72,10	99,83	83,17	75,23	98,00	85,67
User 10	47,30	87,10	78,53	82,00	119,70	82,93
Average recognition time by gesture (ms)	78,93	90,63	84,63	92,07	94,57	88,17

We took the size of 160 packets per gesture when we tested the DS with the ten users. The 160 packages represent 53.33% of the packages sent by the gesture. Each package contains the subscription theme and a label that represents a gesture. Table 3 shows the average packet size per gesture and the average packet size for the entire DS. In addition, the theoretical delivery time of the DS is 30.7 ms.

Table 3. Theoretical delivery time of DS.

Gesture	Topic	Average packet size per gesture (bits)
OPEN HAND	Emergency	548
CLOSE HAND	Entry	528
WAVE IN	Entry	528
WAVE OUT	Room	552
PINCH	Room	552
	Average packet size (bits)	541,6
	Total transmission time of DS (ms)	30.07
	Total propagation time of DS (ms)	0.6
	Theoretical delivery time of DS (ms)	30.07

Figure 8 shows the results of the testing per gesture for each user. The X-axis represents the users. At the same time, the Y-axis represents the number of recognized gestures. The CLOSE HAND, WAVE IN and WAVE OUT gestures do not significantly differ in the number of gestures recognized. In contrast, the PINCH gesture has a low number of recognized gestures.

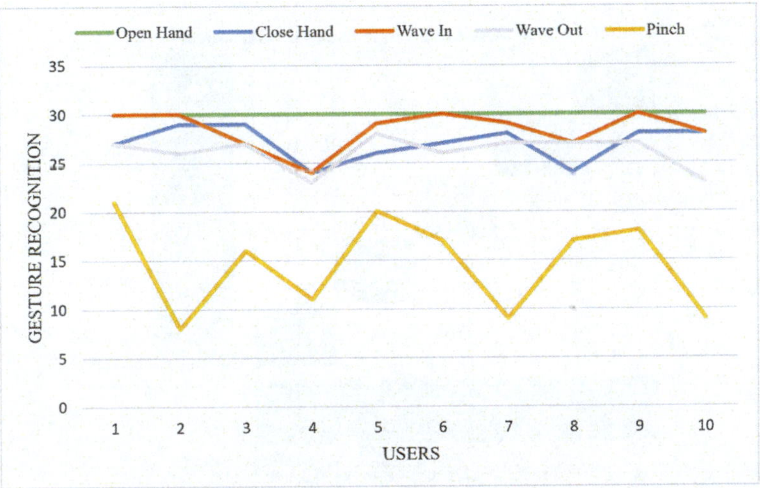

Fig. 8. Gesture recognition by users

Figure 9 shows the percentage accuracy of each gesture recognized by the DS. In this context, OPEN HAND and WAVE IN is the gesture with the highest accuracy percentage at 100% and 94.67%, respectively. On the other hand, PINCH is the gesture with the lowest accuracy percentage at 48.67%. The CLOSE HAND and WAVE OUT gestures have 90% and 87%, respectively.

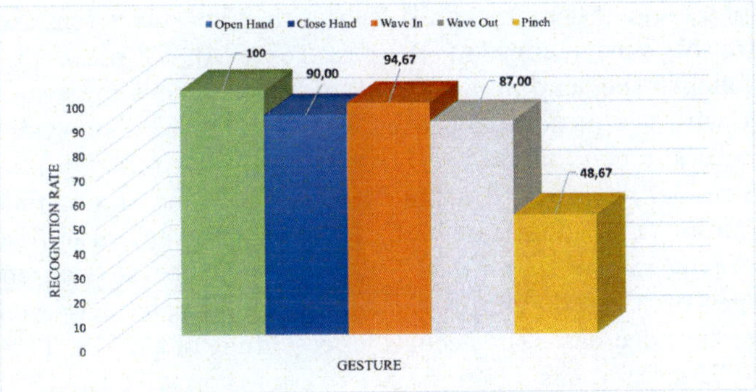

Fig. 9. Gesture recognition rate

Figure 10 shows the confusion matrix of the DS online test. The matrix results from the evaluation applied with the five gestures mentioned in this paper. In addition, the DS uses the ANN as a classifier of the online system.

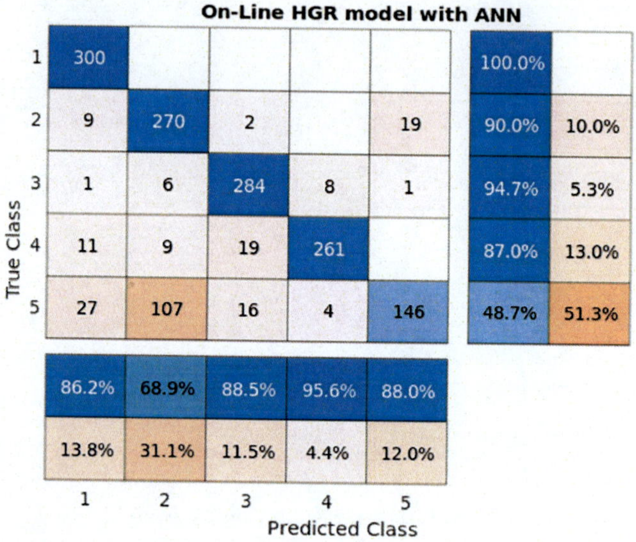

Fig. 10. Confusion matrix of online HGR model with ANN

6 Conclusions

In this work, we developed a DS using hand gesture recognition for people with PD. The DS works with three subsystems wirelessly connected. The first subsystem consists of a hand gesture recognition system. This subsystem uses ANN as a classifier. Also, this subsystem sends the gesture by MQTT protocol. The second subsystem manages the sending and receiving of messages between subsystems one and three with the mosquitto broker. Finally, the third subsystem activates different actuators depending on the gesture the broker sends in real-time.

The HGR model was trained using 6480 observations and tested using 1680 observations offline. The offline test had a 92.759% accuracy rate. Also, we tested the DS online using 1500 observations. This proof returned an accuracy of 84.07% with the five gestures detailed in previous sections. Also, we observed that using the PINCH gesture decreases the accuracy percentage of the DS. This problem is since the PINCH gesture has a high degree of confusion with the CLOSE HAND gesture. Therefore, when the DS works without the PINCH gesture, it increases its accuracy to 92.92%.

In addition, we evaluate the average DS response time. The DS time is obtained from the sum of the theoretical delivery time and the average response time of the HGR model. In this sense, the HGR model generates the label identifying the gesture in 88.17 ms. Furthermore, the theoretical delivery time of the message to the actuator was generated in the most extreme scenario of wireless signal range. The maximum range for wireless router 802.11n standard is 50 m, the transmission rate is 72 Mbps and an average packet weight of 541.6 bits. As a result, the theoretical delivery time of the label between the PC, the Raspberry Pi and the Esp8266 is 30.7 ms. To this value, we add the response time of the HGR model. Consequently, the DS transmits the label in a total time of 118.87 ms.

7 Discussion

The scientific literature reported DS that working with CNN and a CCTV camera to detect hand gestures has a test accuracy of 99.97%. However, the range of the communication signal is limited to that of the Bluetooth HC-05 device. In addition, they do not report an online test or processing time. On the other hand, the scientific literature also reported a DS that works with EEG signals and a DS with voice recognition for the Portuguese language report efficiency scores of 90% and 95% accuracy in online testing, respectively, but only the DS with voice recognition mentions how they obtained the efficiency of their system. However, our DS present the HGR model has a 92.759% offline test accuracy with a response time of 88.17 ms. In the same sense, the DS has an accuracy of 84.07% online with a theoretical response time of 118.87 ms. The accuracy decreases due to the low percentage of PINCH recognition. The HGR has a high degree of confusion between the CLOSE HAND and PINCH gestures, as shown in Fig 10. The confusion is due to the similarity between the direction and position of the fingers in both gestures. In this context, the recognition rate of our DS is very competitive with the mentioned DSs. Since LMC is a device specialized in capturing information from the hand, our DS overcomes the disadvantages of the mentioned DS as an aid for people with PD and gives a real-time response.

References

1. Aditya, K., et al.: Recent trends in HCI: a survey on data glove, LEAP motion and Microsoft Kinect. In: 2018 IEEE International Conference on System, Computation, Automation and Networking (ICSCA), pp. 1–5. IEEE (2018)
2. Balta-Ozkan, N., Boteler, B., Amerighi, O.: European smart home market development: public views on technical and economic aspects across the United Kingdom, Germany and Italy. Energy Res. Soc. Sci. **3**, 65–77 (2014). https://doi.org/10.1016/j.erss.2014.07.007. ISSN 2214-6296
3. Bentabet, N., Berrached, N.-E.: Synchronous P300 based BCI to control home appliances. In: 2016 8th International Conference on Modelling, Identification and Control (ICMIC), pp. 835–838 (2016). https://doi.org/10.1109/ICMIC.2016.7804230
4. Borba, G., Milanés, A., Barbosa, G.: A responsible approach towards user and personal voice assistants interaction. In: Proceedings of the 18th Brazilian Symposium on Human Factors in Computing Systems, pp. 1–4 (2019)

5. Cañizares-Villalba, M.J., Calderón-Salavarría, K., Vásquez-Cedeño, D.: Neurologia Argentina **11**(2), 61–66 (2019). https://doi.org/10.1016/j.neuarg.2019.02.002
6. Núñez, J.C., et al.: Acoustic and language modeling for speech recognition of a Spanish dialect from the Cucuta Colombian region. Ingeniería **22**(3), 362–376 (2017)
7. Celis, J., et al.: Voice processing with Internet of Things for a home automation system. In: 2018 IEEE XXV International Conference on Electronics, Electrical Engineering and Computing (INTERCON), pp. 1–4. IEEE (2018)
8. Collaguazo, H., Córdova, P., Gordón, C.: Communication and daily activities assistant system for patient with amyotrophic lateral sclerosis. In: 2018 International Conference on eDemocracy eGovernment (ICEDEG), pp. 218–222 (2018). https://doi.org/10.1109/ICEDEG.2018.8372373
9. CN Discapacidades. Consejo nacional para la igualdad de discapacidades (2020)
10. Errobidart, J., et al.: Offline domotic system using voice comands. In: 2017 Eight Argentine Symposium and Conference on Embedded Systems (CASE), pp. 1–6 (2017). https://doi.org/10.23919/SASE-CASE.2017.8115370
11. Español, S.: Cómo hacer cosas sin palabras. Antonio Machado (2004)
12. Garcia, C.A., Salinas, G., Perez, V.M., Salazar L., F., Garcia, M.V.: Robotic arm manipulation under IEC 61499 and ROS-based compatible control scheme. In: Botto-Tobar, M., Barba-Maggi, L., González-Huerta, J., Villacrés-Cevallos, P., S. Gómez, O., Uvidia-Fassler, M.I. (eds.) TICEC 2018. AISC, vol. 884, pp. 358–371. Springer, Cham (2019). https://doi.org/10.1007/978-3-030-02828-2_26 ISBN 978-3-030-02828-2
13. Garcia, M.V., et al.: An approach of load balancers for low-cost CPPSs in software-defined networking architecture. In: 2019 15th IEEE International Workshop on Factory Communication Systems (WFCS) (2019). ISBN 978-1-7281-1268-8
14. Gassert, R., Dietz, V.: Rehabilitation robots for the treatment of sensorimotor deficits: a neurophysiological perspective. J. Neuroeng. Rehabil. **15**(1), 1–15 (2018)
15. Gruener, S., Koziolek, H., Rückert, J.: Towards resilient IoT messaging: an experience report analyzing MQTT brokers. In: 2021 IEEE 18th International Conference on Software Architecture (ICSA), pp. 69–79 (2021). https://doi.org/10.1109/ICSA51549.2021.00015
16. Hand Tracking—Gemini Is Here - Ultraleap documentation. Accessed 12 Apr 2022
17. Jayaweera, N., et al.: Gesture driven smart home solution for bedridden people. In: Proceedings of the 35th IEEE/ACM International Conference on Automated Software Engineering Workshops, pp. 152–158 (2020)
18. Lesiones medulares. es. Accessed 12 Apr 2022
19. Lopez, N.M., et al.: From hospital to home care: Creating a domotic environment for elderly and disabled people. IEEE Pulse **7**(3), 38–41 (2016). https://doi.org/10.1109/MPUL.2016.2539105
20. Luna-Romero, S., Delgado-Espinoza, P., Rivera-Calle, F., Serpa-Andrade, L.: A domotics control tool based on MYO devices and neural networks. In: Duffy, V., Lightner, N. (eds.) AHFE 2017. AISC, vol. 590, pp. 540–548. Springer, Cham (2018). https://doi.org/10.1007/978-3-319-60483-1_56
21. Mantilla-Brito, J., Pozo-Espín, D., Solórzano, S., Morales, L.: Embedded system for hand gesture recognition using EMG signals: effect of size in the analysis windows. In: Nummenmaa, J., Pérez-González, F., Domenech-Lega, B., Vaunat, J., Oscar Fernández-Peña, F. (eds.) CSEI 2019. AISC, vol. 1078, pp. 214–225. Springer, Cham (2020). https://doi.org/10.1007/978-3-030-33614-1_15

22. Mendula, M., et al.: Interaction and behaviour evaluation for smart homes: data collection and analytics in the ScaledHome project. In: Proceedings of the 23rd International ACM Conference on Modeling, Analysis and Simulation of Wireless and Mobile Systems, pp. 225–233 (2020)
23. Moore, R.K.: A comparison of the data requirements of automatic speech recognition systems and human listeners. In: Eighth European Conference on Speech Communication and Technology (2003)
24. Moreta, J., Moreno, H., Caicedo, F.: Real-time video transmission and communication system via drones over long distances. In: Garcia, M.V., Fernández-Peña, F., Gordón-Gallegos, C. (eds.) CSEI 2021. LNNS, vol. 433, pp. 323–339. Springer, Cham (2022). https://doi.org/10.1007/978-3-030-97719-1_19
25. Nogales, R., Benalcazar, M.E., Toalumbo, B., Palate, A., Martinez, R., Vargas, J.: Construction of a dataset for static and dynamic hand tracking using a non-invasive environment. In: García, M.V., Fernández-Peña, F., Gordón-Gallegos, C. (eds.) Advances and Applications in Computer Science, Electronics and Industrial Engineering. AISC, vol. 1307, pp. 185–197. Springer, Singapore (2021). https://doi.org/10.1007/978-981-33-4565-2_12
26. Quezada, J., Siguenza-Guzman, L., Llivisaca, J.: Optimization of motorcycle assembly processes based on lean manufacturing tools. In: Nummenmaa, J., Pérez-González, F., Domenech-Lega, B., Vaunat, J., Oscar Fernández-Peña, F. (eds.) CSEI 2019. AISC, vol. 1078, pp. 247–259. Springer, Cham (2020). https://doi.org/10.1007/978-3-030-33614-1_17
27. Segura, E., et al.: Designing an app for home-based enriched music-supported therapy in the rehabilitation of patients with chronic stroke: a pilot feasibility study. Brain Injury **35**(12–13), 1585–1597 (2021)
28. Toalumbo, B., Nogales, R.: Hand gesture recognition using leap motion controller, infrared information, and deep learning framework. In: Narváez, F.R., Proaño, J., Morillo, P., Vallejo, D., González Montoya, D., Díaz, G.M. (eds.) SmartTech-IC 2021. CCIS, vol. 1532, pp. 412–426. Springer, Cham (2022). https://doi.org/10.1007/978-3-030-99170-8_30
29. Weichert, F., et al.: Analysis of the accuracy and robustness of the leap motion controller. Sensors (Switzerland) **13**(5), 6380–6393 (2013). https://doi.org/10.3390/s130506380

Electronics Engineering

Development of a Controller Using the Generalized Minimum Variance Algorithm for a Twin Rotor Mimic System (TRMS)

Luis Sani-Morales and William Montalvo(✉)

Universidad Politécnica Salesiana, UPS, 170146 Quito, Ecuador
lsani@est.ups.edu.ec, wmontalvo@ups.edu.ec

Abstract. The study of unmanned vehicles (UAV) is booming, due to its flexible structure, easy manufacturing and multiple applications, so it is important to start using controllers that ensure minimum power consumption prolonging its working time, additional in this type of system are multiple disturbances such as tracking and random, or the simultaneous existence of disturbances between the actuator and the controlled object. This work describes how the generalized minimum variance control (GMVC) tries to reduce the effect of disturbances at the output or changes in the reference of a system similar to a UAV as is the twin-rotor system (TRMS), trying to have a strengthening effect in the presence of noise improving stability for the desired reference. The GMVC has theoretically demonstrated a good tracking and regulation performance, which is corroborated in the present work where its performance analysis in a real system is carried out. The performance of the GMVC is compared to a conventional PID controller to perform a Benchmark through simulations and field tests. The comparative analysis is statistical using the Wilcoxon test and using performance criteria based on the error as the integral squared error (ISE), which will also determine whether the controller presents a decrease in energy consumption in the control action. The results are interesting and promising from the point of view of control.

Keywords: Generalized Minimum Variance Algorithm · PID Controller · Performance Indices · TRMS

1 Introduction

Aerodynamic systems have taken a certain boom in applications for research in the area of industrial control [5], with the creation of new control systems for unmanned vehicles (UAV) [9], for which, several optimal control techniques have been developed and studied [11,12], to return the stability [10] and response time of these systems [27].

M. V. Garcia and C. Gordón-Gallegos (Eds.): CSEI 2022, LNNS 678, pp. 235–248, 2023.
https://doi.org/10.1007/978-3-031-30592-4_17

The TRMS twin rotor MIMO system is highly complex, this is due to its high nonlinearity [25], significant cross-coupling between its two axes and complex aerodynamics, which makes it an interesting problem for analysis [19], besides being a suitable system to analyze new control strategies for UAVs. The TRMS is a didactic module where several types of controls with certain work rates have been performed [3,23], since it is a highly nonlinear multiple-input multiple-output system.

The MVC has shown in the literature that the control law of minimizing variance, in the case of any variation of the parameters can generate the system to generate an error in steady state or even lead to a divergence to the reference signal [15], however, the GMVC is created with this structure to correct the deficiencies presented by the MVC [22], which ensures that the variance in the output tries to reach zero [6,8]. The main idea of the control is to create an optimization process in the reference tracking, reduce the energy consumption [24] or increase the action performance by reducing the noise in the system, this is achieved by minimizing the cost function.

Controllability studies of GMVC in linear and nonlinear systems have been developed. Wangwang Zhu [29] analyzed the control in systems based on random polynomials and also evaluated the process in a DC-AC converter. Alipouri [1] addressed the noise elimination and stabilization of an electric vehicle in noisy conditions, showing that the controller can track the input reference speed and yaw rate.

In addition, Poshtan [2] works out some shortcomings of the GMVC in nonlinear MIMO systems, such as the need for the exact model of the system, making the controller tend to be very aggressive by designing a GMVC based on a vector autoregressive model (VARX). The GMVC seeks to create a robust system in the face of the slightest perturbation, contributing to the development of this control to cover deficiencies of other controllers in high-precision systems.

This study seeks to make a comparison through the use of performance indices [16], between a PID control and a generalized minimum variance control for a TRMS, where the control is focused on its main engine with its YAW angle and its tail engine with its PITCH angle respectively [4,17].

This project drives future investigations of robust controllers on highly unstable systems such as UAVs, to determine through its main characteristics, the suitability of its use in systems with similar dynamics to cover deficiencies presented by other controllers, with performance index criteria and applying a statistical test such as Wilcoxon's test.

2 Methodology

For data acquisition, a test module from Feedback is used, consisting of a TRMS twin rotor. The diagram shown in Fig. 1 shows the communication of the module for data acquisition of both inputs and outputs of the system, allowing to work with the angle of elevation and the angle of rotation (PITCH and YAW respectively).

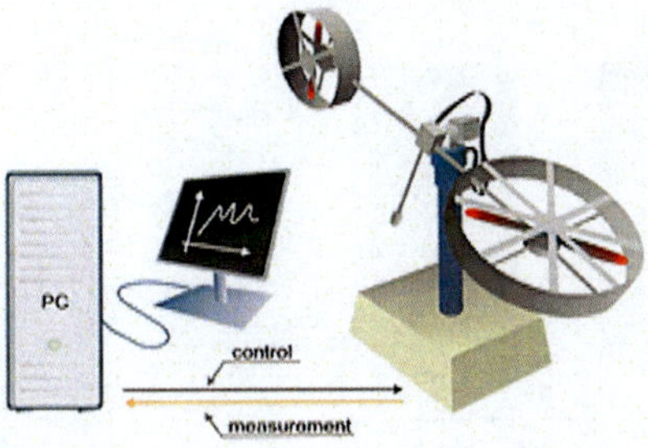

Fig. 1. Twin rotor training module (TRMS).

2.1 System Identification

La Fig. 2 muestra cómo se emplea el motor de cola para determinar el ángulo YAW, introduciendo una señal de excitación a la entrada para así adquirir los valores a la salida, el mismo proceso se realizó con el motor principal como muestra la Fig. 3, el which allows to determine the PITCH angle. It is important to block the respective support and add a zero input to the motor opposite to the data acquisition as it is a multi-input multi-output system.

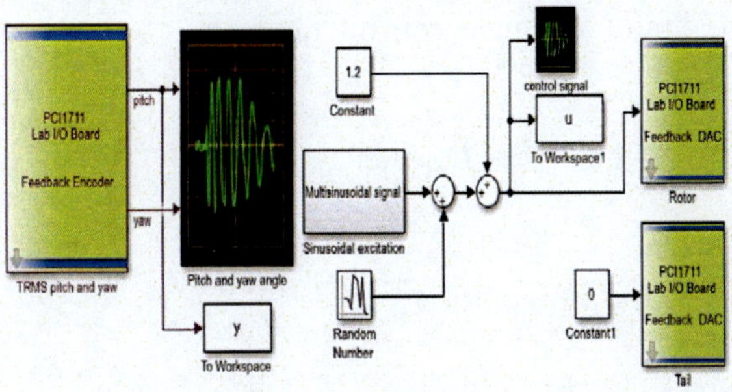

Fig. 2. PITCH data collection connection.

The ARMAX model [18] on the TRMS was used to obtain the mathematical model in its PITCH and YAW angles since it allows to establish the equations needed for the GMVC. In (1) and (2) the transfer functions of the system are shown for PITCH and YAW respectively.

$$Pitch_{tfs} = \frac{0.01009s^2 + 0.04488s + 2.116}{s^3 + 1.194s^2 + 4.508s + 4.9743} \tag{1}$$

$$Yaw_{tfs} = \frac{0.0009585s^2 + 0.05354s + 0.6662}{s^3 + 2.261s^2 + 1.802s + 1.219} \tag{2}$$

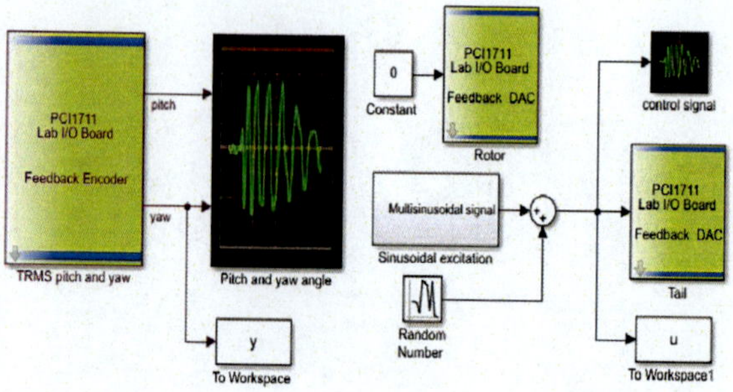

Fig. 3. YAW data collection connection.

2.2 Generalized Minimum Variance Controllers

Clarke [7] proposed the GMVC as a method of real application in the industry, which aims to reduce the effect of disturbances in the output by reducing the J index to predict the future, which could establish actions in advance and adjust them for desired future effects. In some cases, unforeseen conditions are generated, preventing an accurate, but an approximate prediction. This is how the GMVC works: It predicts the future output action and adjusts the current control signal, giving the desired value to the system output [21].

In (3) the minimum variance control algorithm is represented as follows (MVC):

$$J = E(y^2(k + d + \frac{1}{k})) \tag{3}$$

where:

- E is the mathematical expectation.
- k is the instant.
- d is the delay.

Figure 4 shows the block diagram for an MVC, which is the basis on which the GMVC is elaborated.

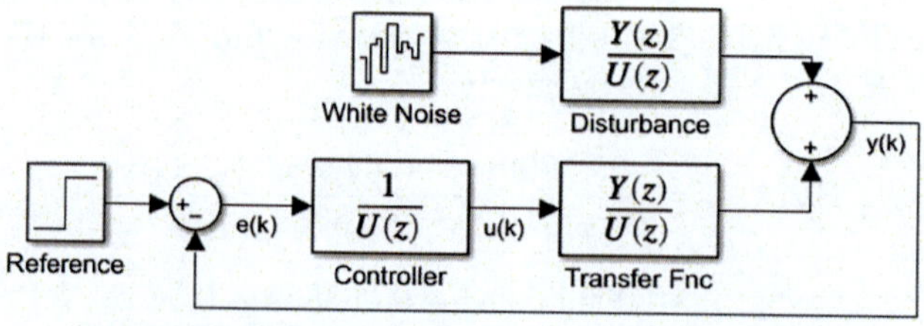

Fig. 4. Conventional MVC block diagram

The MVC is represented by a discrete-time ARMAX model as shown in (4), where the polynomial A represents the denominator, B determines the numerator of the process and C is a polynomial of the system noise.

$$A(z^{-1})y(k) = z^{-d}B(z^{-1})u(k) + C(z^{-1})v(k) \tag{4}$$

Once the system is affected by the input u(k) affects a delay at the output in y(k+d+1), to calculate the prediction of future output. Equation (4) is rewritten as follows:

$$A(k+d+1) = \frac{zB(z^{-1})}{A(z^{-1})}u(k) + \frac{C(z^{-1})}{A(z^{-1})}v(k+d+1) \tag{5}$$

The perturbation can be expressed as a function of the polynomials F which is the combination of the perturbations and G is the effect of the perturbations on the output, G being one degree less than A, this equation is used (6):

$$\frac{C(z^{-1})}{A(z^{-1})} = F(z^{-1}) + z^{d+1}\frac{G(z^{-1})}{A(z^{-1})} \tag{6}$$

The system output signal relating system input, noise and output are shown in Eq. (7) and Eq. (8) represents the control law that minimizes the variance of y(k) of the process.

$$\hat{y}(k+d+1) = \frac{G(z^{-1})}{A(z^{-1})}y(k) + \frac{zB(z^{-1})F(z^{-1})}{C(z^{-1})}u(k) \tag{7}$$

$$u(k) = \frac{G(z^{-1})}{zB(z^{-1})A(z^{-1})}y(k) \tag{8}$$

Now, this minimum variance controller has a steady state error, to correct this characteristic the generalized minimum variance control includes the error in the cost function and a weighting (r) on the control action, resulting in the Eq. (9).

$$J = E\{[y(k+d+1) - w(k++1]^2 + ru^2(k)\} \tag{9}$$

Deriving Eq. (9), and expecting the variance to be zero in the best case (10), the Eq. (11) is obtained, where u(k) is for generalized minimum variance control.

$$\frac{\partial y}{\partial x(k)} = 0 \tag{10}$$

$$u(k) = \frac{1}{zB(z^{-1})A(z^{-1})} + \frac{r}{b_1 B(z^{-1})}[C(z^{-1})w(k+d+1) - G(z^{-1})y(k)] \tag{11}$$

The closed-loop diagram of the GMVC is shown in Fig. 5.

Having determined the transfer functions, the controller set for PITCH and YAW angle is shown in (12) and (13) respectively.

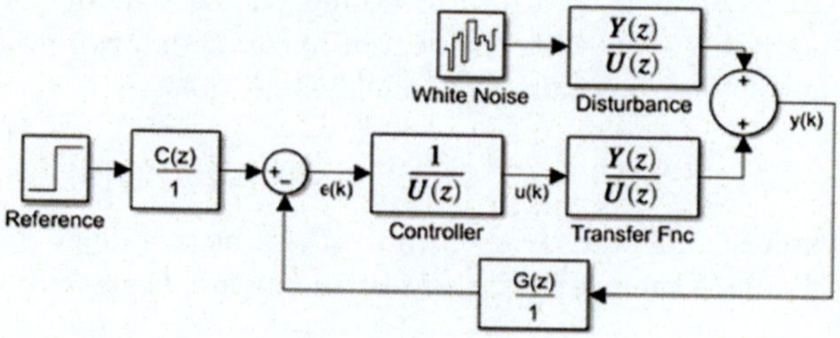

Fig. 5. GMVC block diagram

$$u(k) = \frac{1}{z^{-3} + 0.5002z^{-2} + 0.0001z} \tag{12}$$

$$u(k) = \frac{1}{0.95z^{-3} + 0.4752z^{-2} + 0.0001z} \tag{13}$$

Using GMVC with the pitch and yaw transfer function, with a sampling time of 0.1s, the algorithm is tested by tracking a reference. Figure 6 shows that the control is correcting the signal and acting on the system in pitch. Figure 7 shows the behavior of the output of the YAW rotational motion.

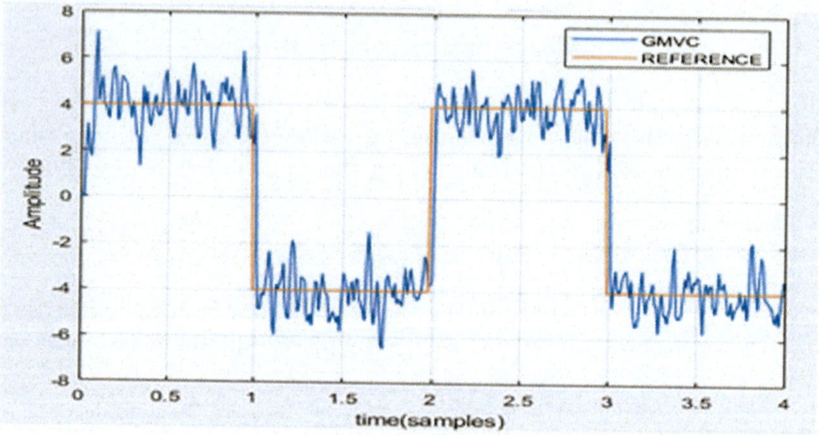

Fig. 6. Reference tracking of the resulting PITCH angle for GMVC

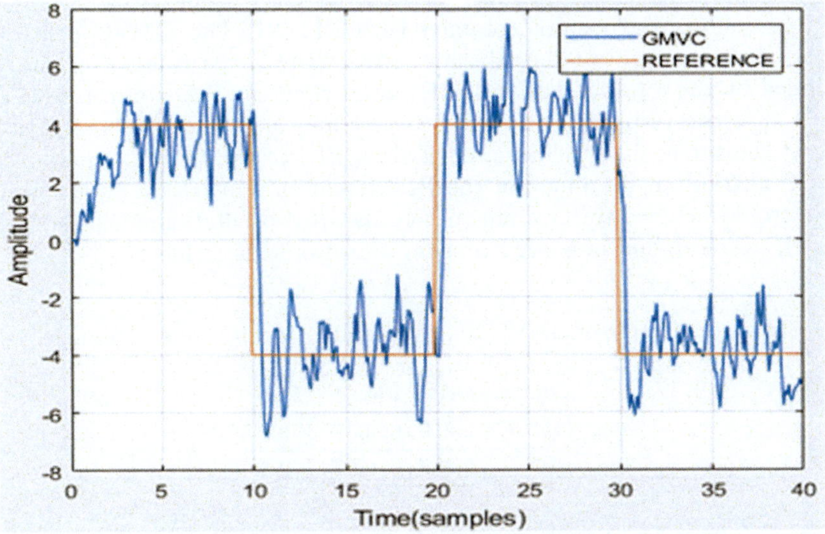

Fig. 7. Reference tracking of the resulting YAW angle for GMVC

3 Analysis and Results

After the tests performed, the data for the evaluation of the controller was identified, as shown in the Simulink block diagram in Fig. 8.

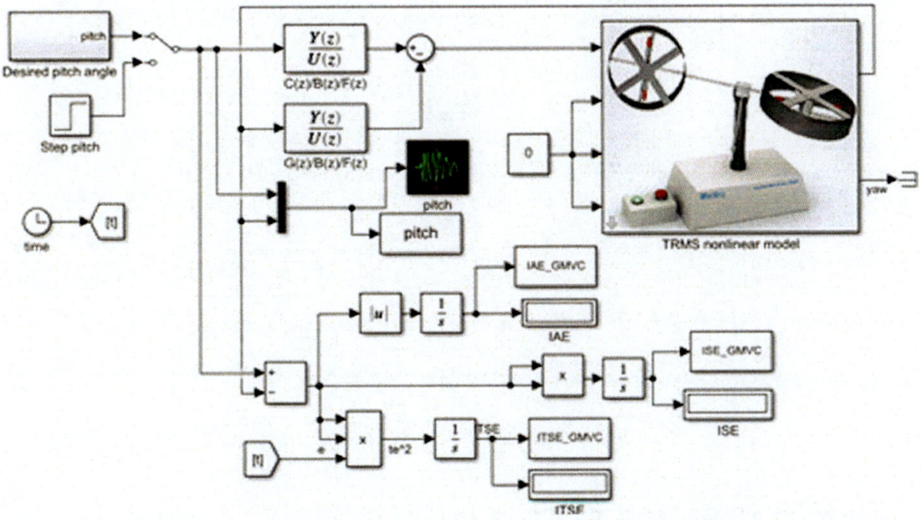

Fig. 8. Block diagram in Simulink for PITCH angle with GMVC.

Table 1 shows the results obtained with PITCH. The GMVC control has indexes of the Integral of the Absolute Error (IAE), the Integral of the Time multiplied by the Squared Error (ITSE) lower than the PID control, so it could be said that the response obtained is better for the two controllers. Also, the index of the Integral of the Squared Error (ISE) of the GMVC shows a lower value (0.4082) than the index of the PID (0.4567). This result shows that the system would achieve to have adequately damped oscillations. Concerning the lower ISE, a reduction in energy consumption would be achieved.

Table 1. Testing of GMVC controller and PID control PITCH angle.

Control	IAE	ISE	ITSE
GMVC	2.210	0.4082	1.299
PID	2.232	0.4567	1.455

It can be observed in Fig. 9 that the settling time (ts) of the MGVC is higher, being 40s while the ts of the PID is 30s, furthermore, an overshoot is observed in the PID while the GMVC has an attenuated response acting satisfactorily in the system. Figure 10 shows the tracking to the reference in the system response.

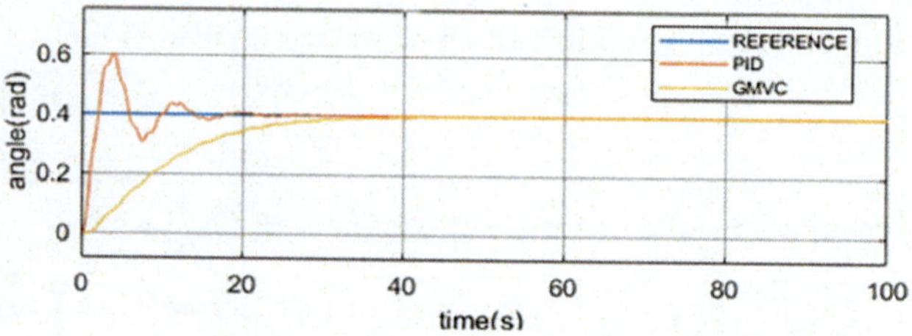

Fig. 9. GMVC vs. PID unit step response in PITCH

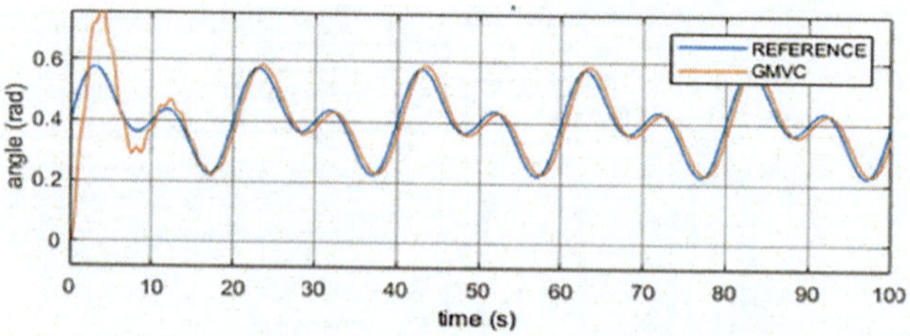

Fig. 10. Resulting graph in reference tracking with GMVC at PITCH angle

3.1 The Wilcoxon Test

The Wilcoxon test is an appropriate tool for nonparametric data analysis of a repeated measures design with two conditions [14]. The data are sorted to produce two total ranks, one for each condition, if there is considerable variation between the two states, then most of the high ranks remain in one condition and most of the low ranks in the other condition [28]. The Wilcoxon test statistic is simply the smallest of the total ranks.

The Wilcoxon test is used with the IBM SPSS Statistics software with a sample of 1000 data, using the ISE performance indexes of Table 1, for which the null hypothesis (H0) is established: "The GMVC is not more effective than the PID control according to its performance indexes" and the alternative hypothesis (H1): "The GMVC is more effective than the PID control according to its performance indexes".

The results of the test are shown in Table 2, it is established by Wilcoxon that when the bilateral asymptotic results in a value less than zero, H0 can be annulled, that is to say, that H1 is true. Therefore, this statistical method proves that the GMVC has a better performance in the TRMS plant.

Table 2. Test de Wilcoxon with PITCH.

	PID	GMVC	DIFERENCE
Media	2.210	0.6247	–0.2713
Standard deviation	0.353411	0.20411	0.11597
Inferior Lim.	0.155033	0.22467	–0.49863
Superior Lim.	0.551789	1.024800	–0.04402
Z	10.63307		
Bilateral asymptotic	2.232		

Figure 11 shows the block diagram of the YAW angle and Table 3 shows the results obtained with the YAW angle. The GMVC control has a lower IAE, ISE and ITSE index than the PID control, so it can be established that the response obtained is the best response of the two controllers. This result indicates that the system could have adequately damped oscillations, and for the ISE, it would have a lower energy consumption.

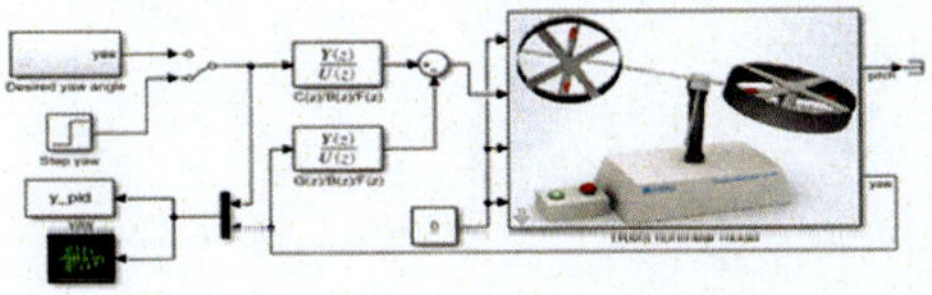

Fig. 11. Block diagram in Simulink for YAW angle with GMVC

Table 3. Testing of GMVC controller and PID control PITCH angle.

Control	IAE	ISE	ITSE
GMVC	1.262	0.4155	0.4242
PID	3.1	0.6457	2.809

It can be observed in Fig. 12 that the settling time (ts) of the GMVC is lower, settling in 10s while the ts of the PID is 23s, in addition, there is a high overshoot in the PID compared to the MGVC. It can be seen in Fig. 13 that the GMVC performs satisfactorily in the system.

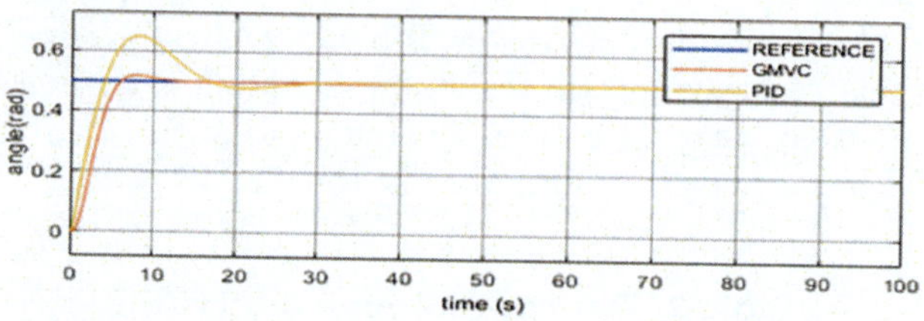

Fig. 12. GMVC unit step response vs. PID in YAW

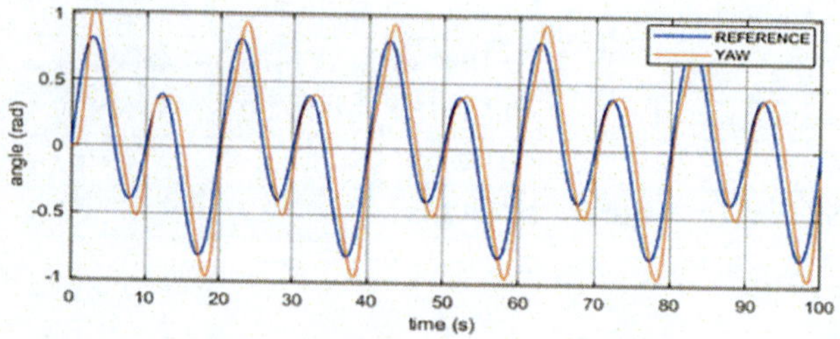

Fig. 13. Resulting plot in reference tracking with GMVC at YAW angle

Performing the Wilcoxon test with the ISE in Table 3, in the same way as for the YAW angle, the data in Table 4 are obtained, which indicate that the bilateral asymptote has a value of 0.001%, that is, by this statistical method, the hypothesis that the MGVCcontrol has a better performance than the PIDcontrol can be considered true.

Table 4. Test de Wilcoxon with PITCH.

	PID	GMVC	DIFERENCE
Media	0.888804	1.9507096247	–1.06190
Standard deviation	0.352148	0.952729	0.60654
Inferior Lim.	1.1986	0.0834	–2.2507
Superior Lim.	1.5790	3.8181	–0.1269
Z	8.63307		
Bilateral asymptotic	0.001		

4 Discussion

Inoue Akira [13] proposes a GMVC of polynomial approach, comparing it with a full order observer, however, for the analysis of the control performance, he uses simulations with random equations for the analysis of the same, which leaves some ambiguity index for the development of this control in real plants or systems, this is something similar to what Minami [20] and Yanou [26] do, where they frame their results with numerical examples. The present work done instead employs real data and assesses the control action of a GMVC in a highly nonlinear MIMO TRMS system, which will serve as an application analysis for future highly complex systems such as those employed in Industry 4.0.

5 Conclusion

The development of a GMVC has shown a better performance than the conventional PID both in the performance indexes ISE, ITSE, IAE as shown in Tables 1 and 3, with an approximate 1% reduction in IAE, an approximate 11.1% in ISE, an approximate 10.1% in ITSE in pitch and with an approximate 55% reduction in IAE, an approximate 35.65% in ISE, an approximate 60% in ITSE in yaw and with an approximate 55% reduction in IAE, an approximate 35.65% in ISE, and an approximate 60% in ITSE in yaw.

These indices indicate that the GMVC has better performance from the point of view of energy consumption. From the control action point of view. It presents better settling time, lower overshoot and adequately damped oscillations, as shown in Fig. 13, being very important factors in unmanned devices, and aircraft that need minimum power consumption. In summary, this controller guarantees stability and solid performance for real systems such as UAVs.

In summary, this controller guarantees stability and solid performance for real systems such as UAVs. For future work, artificial intelligence algorithms can be used to optimize the GMVC, overcoming the limitations of adjustment in the identification of nonlinear systems.

References

1. Alipouri, Y., Alipour, H.: Attenuating noise effect on yaw rate control of independent drive electric vehicle using minimum variance controller. Nonlinear Dyn. **87**(3), 1637–1651 (2016). https://doi.org/10.1007/s11071-016-3139-9
2. Alipouri, Y., Poshtan, J.: A linear approach to generalized minimum variance controller design for MIMO nonlinear systems. Nonlinear Dyn. **77**(3), 935–949 (2014). https://doi.org/10.1007/s11071-014-1352-y
3. Caiza, G., Bonilla-Vasconez, P., Garcia, C.A., Garcia, M.V.: Augmented reality for robot control in low-cost automation context and IoT. In: IEEE International Conference on Emerging Technologies and Factory Automation, ETFA, vol. 2020-September, pp. 1461–1464 (2020). https://doi.org/10.1109/ETFA46521.2020.9212056
4. Cajo, R., Agila, W.: Evaluation of algorithms for linear and nonlinear PID control for twin rotor MIMO system, pp. 214–219. Institute of Electrical and Electronics Engineers Inc., October 2015. https://doi.org/10.1109/APCASE.2015.45
5. Cheng, Y.S., Pai, H.Y., Liu, Y.H.: PID parameter tuning for flyback converter with synchronous rectification using particle swarm optimization. In: 2022 Joint 12th International Conference on Soft Computing and Intelligent Systems and 23rd International Symposium on Advanced Intelligent Systems (SCIS&ISIS), pp. 1–6 (2022). https://doi.org/10.1109/SCISISIS55246.2022.10002077
6. Chiliquinga, S., Manzano, S., Cordova, P., Garcia, M.V.: An approach of low-cost software-defined network (SDN) based internet of things. In: Proceedings - 2020 International Conference of Digital Transformation and Innovation Technology, INCODTRIN 2020, pp. 70–74 (2021). https://doi.org/10.1109/Incodtrin51881.2020.00025
7. Cortes, D., Hoang, V., Tran, D.: Aeta 2019- recent advances in electrical engineering and related sciences: theory and application (2019)
8. Cunha, L.B., Silveira, A.D.S., Barra, W.: Parametric robust generalized minimum variance control. IEEE Access **10**, 75884–75897 (2022). https://doi.org/10.1109/ACCESS.2022.3185020
9. Ghith, E., Tolba, F.: Design and optimization of PID controller using various algorithms for micro-robotics system. J. Robot. Control (JRC) **3**, 244–256 (2022). https://doi.org/10.18196/jrc.v3i3.14827
10. Gualotuna, S.: Estrategia de Control Robusto Descentralizado para una Micro-Red Aislada con Generación Distribuida Acoplada para Mejorar la Estabilidad de Voltaje. Master's thesis (2021)
11. Herrera, V., Ilvis, D., Morales, L., Garcia, M.: Optimization of hoeken mechanism for walking prototypes. In: Garcia, M.V., Fernández-Peña, F., Gordón-Gallegos, C. (eds.) Advances and Applications in Computer Science, Electronics, and Industrial Engineering. CSEI 2021. LNNS, vol. 433, pp. 89–105. Springer, Cham (2022). https://doi.org/10.1007/978-3-030-97719-1_5
12. Inoue, A., Deng, M., Masuda, S., Yoshinaga, S.I.: A predictor based on a modified full-order observer for generalized minimum variance control equivalent to polynomial approach. In: International Conference on Advanced Mechatronic Systems, ICAMechS 2020-Decem, pp. 234–238 (2020). https://doi.org/10.1109/ICAMechS49982.2020.9310083

13. Inoue, A., Deng, M., Sato, T., Yanou, A.: An extended generalized minimum variance control using a full-order observer equivalent to the controller based on polynomials. In: 2021 International Conference on Advanced Mechatronic Systems (ICAMechS), pp. 220–225 (2021). https://doi.org/10.1109/ICAMechS54019.2021.9661498
14. Institute: Proceedings, 2020 International Conference on Industrial Engineering, Applications and Manufacturing (ICIEAM) : Sochi, Russia, 18–22 May 2020
15. Jing, L., Qianchao, L., Hao, L.: UAV penetration mission path planning based on improved holonic particle swarm optimization. J. Syst. Eng. Electron. 1–17 (2023). https://doi.org/10.23919/JSEE.2022.000132
16. Larrea, A., Barambones, O., Ramos-Herran, J.: Design and implementation of a predictive control system for a photovoltaic generator, vol. 2016-Novem (2016). https://doi.org/10.1109/ETFA.2016.7733501
17. Loo, N., Hernandez, C., Mauricio, D.: Decision support system for the location of retail business stores. In: García, M.V., Fernández-Peña, F., Gordón-Gallegos, C. (eds.) Advances and Applications in Computer Science, Electronics and Industrial Engineering. AISC, vol. 1307, pp. 67–78. Springer, Singapore (2021). https://doi.org/10.1007/978-981-33-4565-2_5
18. Mathworks (2022). https://la.mathworks.com/help/ident/ref/armax.html, https://la.mathworks.com/help/ident/ref/armax.html
19. Miranda Orellana, M.E., Solis Delgado, J.E.: Diseño e implementación de algoritmos de control difuso y Pid adaptativo STR para el sistema mimo de doble rotor 33–220 de feedback para el laboratorio de control automático de la universidad politécnica salesiana sede Guayaquil. B.S. thesis (2018)
20. Panepistemio, K.: 2017 22nd IEEE International Conference on Emerging Technologies and Factory Automation : 12–15 September 2017, Limassol, Cyprus (2017)
21. Pinto, E.: Fundamentos de control con MatLab. Pearson Educación, London (2019)
22. Ruderman, M.: Motion control with optimal nonlinear damping: from theory to experiment. Control Eng. Pract. **127** (2022). https://doi.org/10.1016/j.conengprac.2022.105310
23. Sain, D., Praharaj, M., Bosukonda, M.: A simple modelling strategy for integer order and fractional order interval type-2 fuzzy PID controllers with their simulation and real-time implementation. Expert Syst. Appl. **202** (2022). https://doi.org/10.1016/j.eswa.2022.117196
24. Trentini, R., Kutzner, R., Bartsch, A., Baasch, A.: Damping of interarea modes using a GMVC-based WAPSS. vol. 53, pp. 13539–13544. Elsevier B.V. (2020). https://doi.org/10.1016/j.ifacol.2020.12.797
25. Trentini, R., Silveira, A., Bartsch, M.T., Kutzner, R., Hofmann, L.: On the design of stochastic RST controllers based on the generalized minimum variance. In: 2016 UKACC 11th International Conference on Control (CONTROL), pp. 1–6 (2016). https://doi.org/10.1109/CONTROL.2016.7737552
26. Yanou, A.: A Study on Self-Tuning PID Control by Smart Strong Stability System; A Study on Self-Tuning PID Control by Smart Strong Stability System (2019). 10.0/Linux-x86_64
27. Zapata Chancusig, B.R.: Desarrollo e implementación de un control de orientación y elevación mediante control en modo deslizante y control MPC lineal, aplicado a un sistema aerodinámico TRMS (Twin Rotor MIMO System). Master's thesis, Quito, 2018 (2018)

28. Zhang, L., Wang, S., Sun, Q., Li, A.: Remote sensing image segmentation based on Wilcoxon rank sum test and mean absolute deviation, vol. 2016-November, pp. 6340–6343. Institute of Electrical and Electronics Engineers Inc., November 2017. https://doi.org/10.1109/IGARSS.2016.7730657
29. Zhu, W., Zhang, Z., Armaou, A., Hu, G., Zhao, S., Huang, S.: Dynamic data reconciliation to improve the result of controller performance assessment based on GMVC. ISA Trans. **117**, 288–302 (2021). https://doi.org/10.1016/j.isatra.2021.01.047

Metaheuristics of the Artificial Bee Colony Used for Optimizing a PID Dahlin in Arm Platform

Jonathan Cevallos, Juan Gárate, and William Montalvo(✉)

Universidad Politécnica Salesiana, UPS, 170146 Quito, Ecuador
{jcevallosi1,jgarateg}@ups.est.edu.ec, wmontalvo@ups.edu.ec

Abstract. The simplicity, high efficiency and robustness of PID controllers make them a widely accepted alternative in the industry around the world. This paper presents the development of two Dahlin-type Proportional Integral and Derivative (PID) controllers, the first one based on the development of a traditional controller and the second one optimized by employing the metaheuristic ABC algorithm for the speed control of a permanent magnet DC motor in a plant control trainer (CTP). The programming is developed with Matlab software, using the Simulink tool together with the drivers and programmers of the Waijung library, and the ARM STM32F4-Discovery platform of 32-bit ARM-Cortex architecture is in charge of the management of the control process. The performance comparison is carried out using the performance index of the Integral of Absolute Error multiplied in time (ITAE), statistically analyzed by IBM's Statistical Package for the Social Sciences (SPSS), where the performance of each controller is evaluated. The results are interesting from the point of view of automatic control.

Keywords: Artificial Bee Colony Algorithm · Dahlin Control · Integral of the absolute error multiplied in time · STM32F4-Discovery

1 Introduction

The PID (Proportional Integral and Derivative) type controller has become one of the most widely used controllers within applications, and its general characteristics of automation have simplified the use of PID [3]. The great ease of use of the PID controller is largely since it is a simple structural control technique, simple to understand and with general applications [1]. The Dahlin controller extends the features of the Deadbeat controller, which controls processes in dead-time [21]. It features integral parts that are used to adjust and stabilize a system in less time. Compared to the proportional controller (PI), the Dahlin controller has fast dynamics and better steady-state performance [23].

Several metaheuristic methods are implemented for the optimization of a PID controller; however, this article makes use of the Artificial Bee Colony (ABC) Algorithm, which in 2005 was proposed by the Turkish academic Dervis

M. V. Garcia and C. Gordón-Gallegos (Eds.): CSEI 2022, LNNS 678, pp. 249–261, 2023.
https://doi.org/10.1007/978-3-031-30592-4_18

Karaboga. The use of this algorithm has advanced over the years while is implemented in image processing, route planning, combinatorial optimization and other engineering fields [8,12]. ABC replicates the behavior of bees so that they always seek the best food source with abundant nectar for the bees within the swarm. They are divided into three classes: employed bees, onlooker bees and scout bees. For a specific problem, the algorithm states that each food source is a possible solution within the search space [13,22].

The performance indexes play an important role in the development of the research because they quantify the behavior of the control loops based on the error signal, which is the result of the difference between the desired value or Set-Point (SP) and its real value or system response. The ABC algorithm requires a cost function; for that reason, this work is provided by the Integral Performance Index of Absolute Error multiplied in time (ITAE), which is obtained by analyzing the error [2,17]. The development of the ABC algorithm is performed using Matlab software in its 2015 version and its implementation is carried out in the Control Training Plant (CTP). Due to the large amount of data generated, it is necessary to make use of a device capable of processing the information without affecting the control in this research; for this reason, it was decided to make use of an STM32F4-Discovery card.

The STM32F4-Discovery card is a low-cost microcontroller device that has a 32-bit ARM Cortex processor with an FPU core, therefore, it allows to load and store of the variables coming from several logical sequences allowing to reduce the execution time [6,9]. Likewise, the WAIJUNG library is designed to program, communicate and control the STM32F4-Discovery card through Matlab's Simulink tool, which allows obtaining the responses generated by the CTP in real-time.

2 Methodology

2.1 Data Acquisition Using the STM32F4-Discovery Card

The process used to obtain data is performed using three important elements, which play a fundamental role in the development of this research: Matlab where the metaheuristic code is developed and the data obtained is processed, the ctp training module, which has a permanent magnet DC motor for the control process, and the STM32F4-Discovery card required for data collection.

The programming, collect and control the CTP plant using the STM32F4-Discovery card, first install the WAIJUNG library, which was developed to operate in the Simulink environment of Matlab, through the program blocks: Host Serial, PWM Capture, Regular ADC, etc. and additional blocks to condition the received values such as Gain, To Workspace, etc.

Figure 1 shows the elements that make up the process.

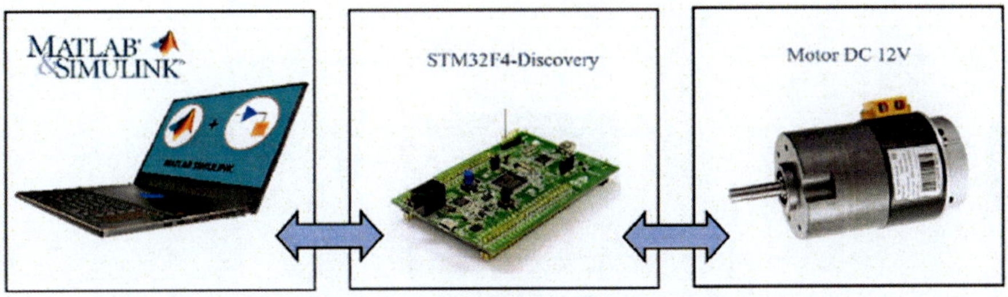

Fig. 1. Equipment used in the control process

Figure 2 shows the blocks used in the data acquisition of the permanent magnet DC motor. As it can be noted, there is a need to condition the signal obtained by the ADC block, responsible for reading the set point with gains. Before sending it to the basic PWM block, which performs the function of controlling the DC motor in the CTP plant.

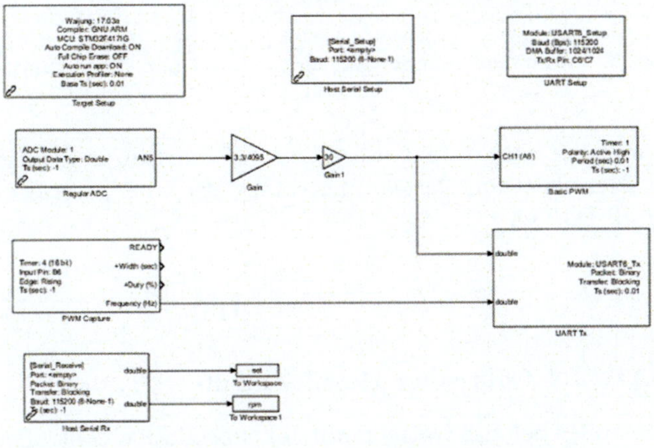

Fig. 2. Block programming for data collection

2.2 Identification of the Transfer Function

Once the data is stored in the workspace, the tool is executed using the command "ident" , and "System Identification" tool is executed to import the data into the time domain. In this case, the data to be analyzed are the Set-Point system input, with a variation between (0–100%). As output data, the RPM of the DC motor of the CTP reaches a maximum value between 2220 to 2500 revolutions per minute.

Figure 3 shows the data obtained.

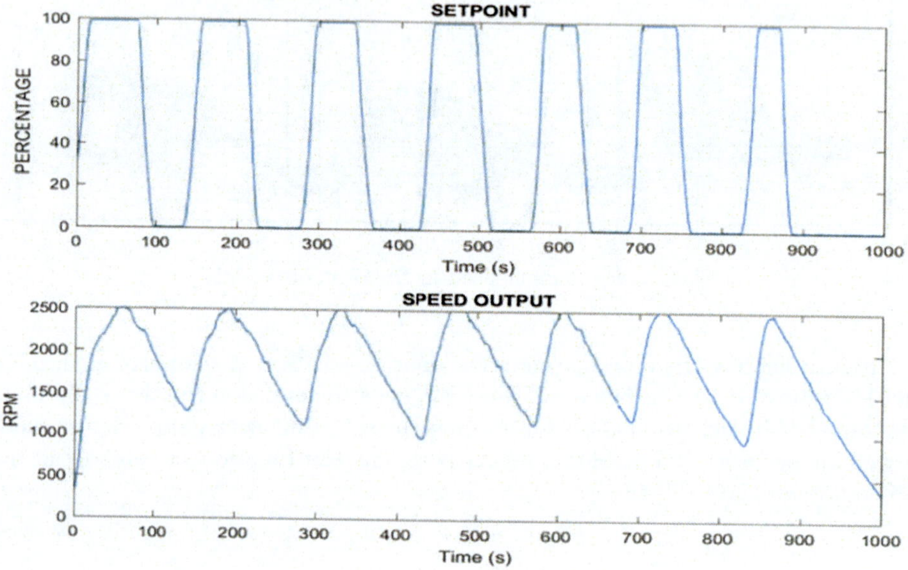

Fig. 3. Permanent magnet DC motor input and output data

For this particular case, a first-order transfer function is required, therefore in the System Identification, the parameters are set with several 0 zeros and 2 poles resulting in Eq. (1).

$$G(s) = \frac{1840}{s^2 + 729s + 1101} \tag{1}$$

2.3 PID-DAHLIN Controller Development

The Dahlin controller belongs to the family of cancellation controllers, its objective is to cancel the dynamics of the system and replace them with a desired dynamic. The behavior in the entire closed loop will be the same as that of a first-order transfer function in dead time [4,15].

In Eq. (1), a delay time is added, resulting in Eq. (2).

$$G(s) = exp(-0.01s) * \frac{1840}{s^2 + 729s + 1101} \tag{2}$$

The transfer function Eq. (2) is discretized using Zero-Order Hold (ZOH) in Matlab resulting in Eq. (3).

$$G(z) = \frac{0.02167z + 0.003412}{z^3 - 0.9857z^2 + 0.0006823z} \tag{3}$$

The development of the Dahlin controller requires the desired transfer function, which is given in Eq. (4).

$$T(s) = \frac{1}{1 + s\lambda} * e^{-} s\tau \tag{4}$$

where τ represents the stabilization time of the plant, while λ represents the approximate constant in the sampling.

For the development of the controller, a Matlab script was implemented that allows the mathematical development of the controller, from which the discretized transfer function of Eq. (5) is obtained, which will be used for the experimental part of the CTP.

$$D(z) = \frac{0.00995z^3 - 0.009808z^2 + 6.789e(-0.6)z}{0.02167z^4 - 0.01805z^3 - 0.003378z^2} \tag{5}$$

The parameters Kp, Ki and Kd are obtained using the scheme shown in Fig. 5, where they are tuned using the "PID Tuner", a tool integrated with Matlab, resulting in the values shown in Table 1.

Table 1. Dahlin PID gain values obtained with "PID Tuner"

Gains	Values
Kp	0.0250
Ki	0.4863
Kd	0

2.4 PID-Dahlin Tuning Using ABC

The artificial bee colony algorithm is inspired by the behavior and ability of bees to find food sources, honeybees possess a collective intelligence that allows them to obtain information such as distance, direction and amount of nectar, which is shared with the rest of the hive through a particular dance [14]. Figure 4 shows how the artificial honey bee hive is composed.

As mentioned and detailed in the article [16] the equations of the algorithm are given as follows. The initial bee population is given by the Eq. (6).

$$x_{ij} = x_{minj} + rand * (0, 1) * (x_{maxj} - x_{minj}) \tag{6}$$

where:

- x_{ij}: Food sources generated.
- x_{maxj}: Maximum food source search limit.
- x_{minj}: Minimum limit search for a food source.

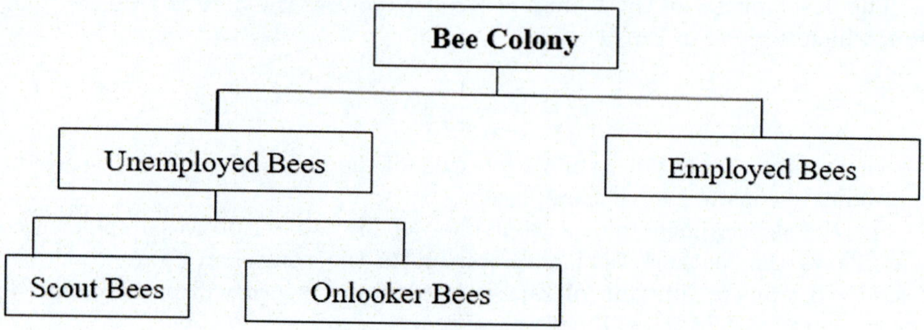

Fig. 4. Member of the artificial bee colony

The employer bee phase is given by Eq. (7).

$$x_{new} = x_{ij} + r * (x_{ij} - x_{kj}) \tag{7}$$

where:

- x_{new}: Search for the solution in the neighborhood for the current position.
- x_{kj}: Randomly selected indicators.

The calculation of possible values is given by Eq. (8).

$$P_i = f_i / \sum_{i=1}^{SN} = f_i \tag{8}$$

The development of the code is done in Matlab, for which it was necessary to couple the response generated by the ITAE performance index, which is useful for tuning the PID-Dahlin, taking this into account, the values presented in Table 2 are obtained.

Table 2. Dahlin PID gain values with ABC tuned by IATE

Gains	Values
Kp	1.2518
Ki	1.5548
Kd	3.2881

3 Results

To obtain the Kp, Ki and Kd values, the first step is to work in continuous time using Eq. (1), which together with Eq. (4) makes up the Dahlin control. Figure 5 shows the program created in Simulink that allows to evaluate and obtain the necessary data such as Set-Point, Time and Response, which will be used by the metaheuristic algorithm created to tune the PID values.

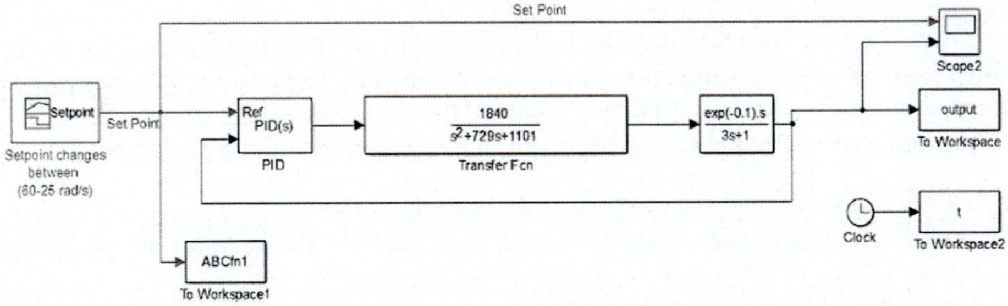

Fig. 5. Dahlin controller schematic in Simulink

3.1 Continuous-Time System Response Optimized with ABC Tuned to the ITAE Performance Index

Prior to the development of the programming in the STM32F4-Discovery card, the respective simulation of the response of the system to a change of the set point of the speed in revolutions per minute (RPM) is carried out. A second program is created in Simulink, which is designed following the scheme of Fig. 5, where the values obtained by the ABC are evaluated.

Once the program has been executed, Fig. 6 shows the result of the behavior in the simulation of the Dahlin controller, with the values Kp, Ki and Kd of Table 1 and 2.

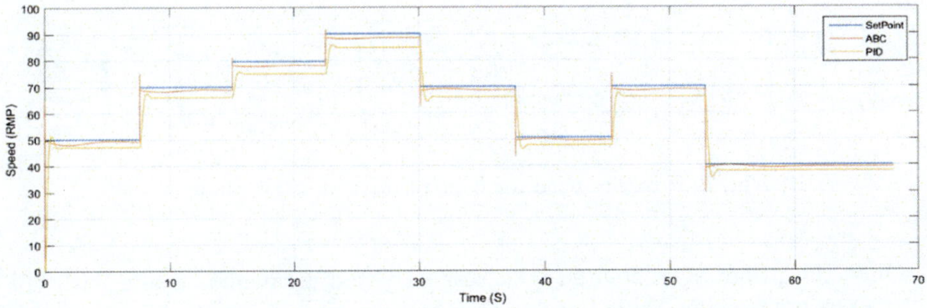

Fig. 6. Dahlin controller simulation response

To check the behavior of the PID-Dahlin controller optimized by the ABC algorithm, it is necessary to develop a program capable of obtaining all the variables used to generate control over the CTP plant, therefore, it is essential to know what data the system requires for this case: Set-Point, RPM of the permanent magnet DC motor.

Once the variables are identified, they are implemented in the Simulink development environment, which must have the WAIJUNG library. This library allows the use of programming blocks for the STM32F4-Discovery platform.

Among the implemented blocks, the use of the following is highlighted:

- Target Setup allows selecting the card model.
- Host Serial Setup identifies the COM port used by the card for its programming.
- Regular ADC reads and selects the port used by the card for Set-Point reading.
- PWM Capture obtains the data generated by the encoder in the plant.
- Basic PWM sends the signal to the plant for DC motor control.
- UART Tx, Host Serial Rx links the closed-loop communication between the computer and the ARM platform.

Figure 7 shows the block program used to perform the tests.

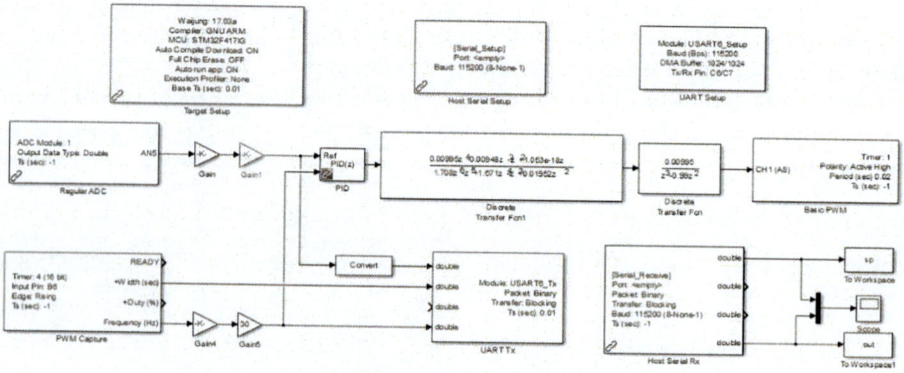

Fig. 7. Program used for control tests on CTP plant

Once the program is created, the respective adjustment is made with the parameters obtained in Table 1 and 2. For the particular case, a Set-Point of 80% RPM is introduced, to observe the responses generated by the control system, it is worth mentioning that within the experimentation slight disturbances or signals out of range are evidenced, therefore, do not generate any problem.

The data generated in Fig. 8 and 9 consists of 8000 samples.

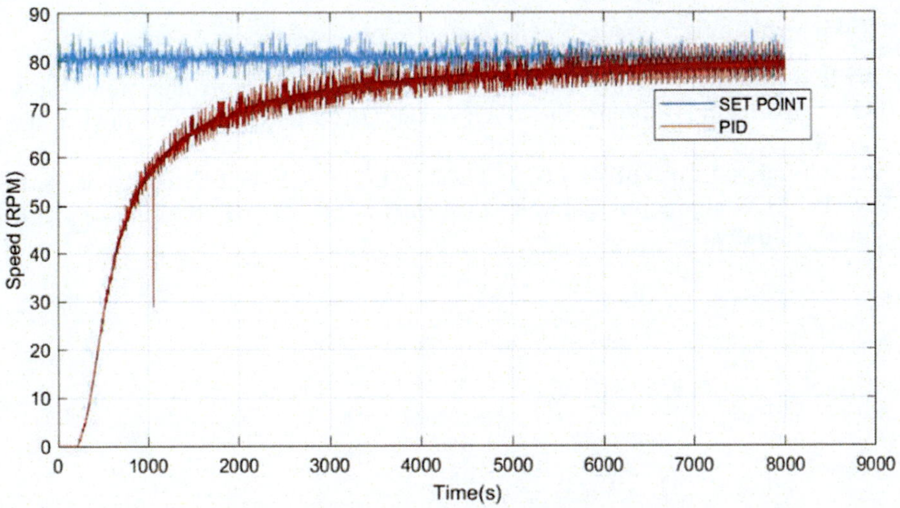

Fig. 8. Plant response tuned by "PID Tuner"

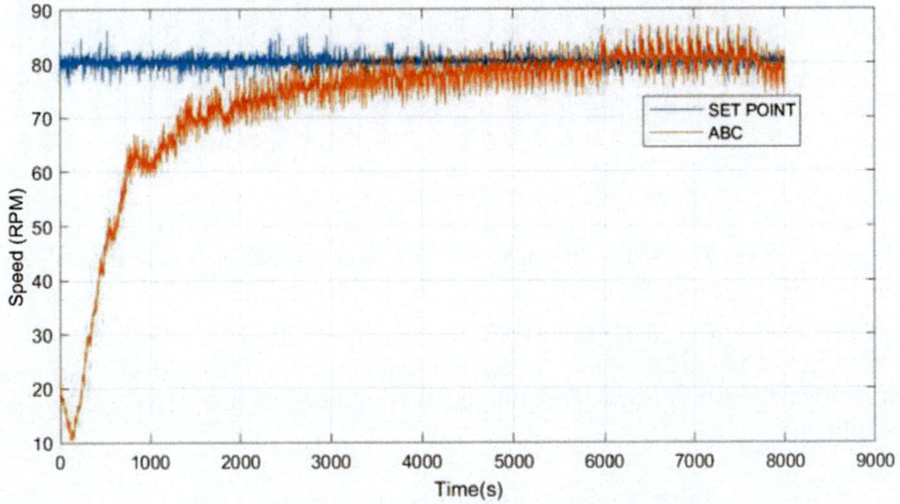

Fig. 9. Plant response tuned by ABC

3.2 Results Obtained Using the Wilcoxon Test

The Wilcoxon test is known as the rank sum test and is generally used for the comparison of two independent sets [20]. The Wilcoxon test is invariant to monotonic transformations, therefore, it is robust to outliers that may occur during sampling [5]. For this case, it was necessary to have the IBM SPSS program.

Taking this into account, the following hypotheses are proposed:

- **Null hypothesis** (H_o): The ITAE of the DAHLIN controller tuned by the ABC algorithm is close to the ITAE of the DAHLIN controller tuned conventionally.
- **Alternative hypothesis** (H_a): The ITAE of the DAHLIN controller tuned by the ABC algorithm is lower than that of the DAHLIN controller tuned conventionally.

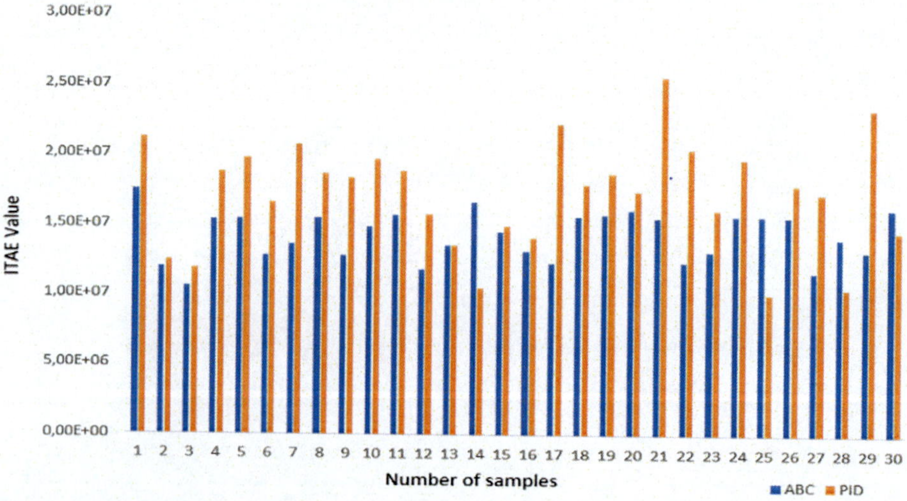

Fig. 10. ITAE Dahlin PID tuned by ABC and traditional Dahlin PID

The values obtained by the SPSS program are reflected in Table 3. As can be seen, the Z is greater than the Z_α, consequently, the alternative hypothesis is accepted. Figure 10 shows the difference graphically in the ITAE of the two controllers.

Table 3. Values obtained by the Wilcoxon test

	N	Average Ranges
Negative Ranges	0.0250	1082.1
Positive Ranges	0.4863	253.9
Tie	0	
Total	1500	
Z_α	1.96	
Z	46.3	

4 Discussion

In the previous investigation [10,18] a tuning work is performed on a traditional PID optimized by ABC and the design of a Dahlin controller respectively, improving the efficiency of the same and demonstrating that the response of the Dahlin controller has a delay in stabilization; however, the combination between the Dahlin PID and ABC improves the behavior of the control system and response times to physical disturbances or changes in the set point. This is because the control of a Dahlin PID optimized by ABC behaves as a positioning control and not as speed control.

With a background in research [19] where the optimization of a Dahlin PID by Ant Colony Optimization (ACO) is performed, the behavior of the algorithm is evidenced, which only performs a search for solutions to find the Kp, Ki and Kd values due to the metaheuristic of the ant colony, while the ABC performs a double search for solutions to find the appropriate tuning of the Kp, Ki and Kd values.

The optimal values of the gain parameters proposed in the research are as follows [11] evidencing a good performance on the controller optimized by the ABC algorithm. The ABC optimization algorithm was developed to solve combinatorial problems in which its response can mutate and rearrange to find a more efficient one [7]. The aforementioned research lacks implementation and experimentation in a physical environment. In this research, the physical implementation evidence that the data obtained in the simulation apply to real control environments.

5 Conclusions

When analyzing the ITAE performance index test (Fig. 10), together with the data obtained by the Wilcoxon test, the alternative hypothesis is accepted since it has a reliability of 95% with tests of up to 25 samples, allowing for extending the life of the elements that make up a control system, especially its actuators (motors, valves, resistors, etc.).

The Dahlin controller, when optimized by the bee colony algorithm, has a response without overshoot (Fig. 9), making it ideal for applications that require extreme precision.

The bee colony algorithm presents a significant advantage since it always provides the best response options due to the double search process it performs, which translates into better performance when designing control systems for temperature, pressure, motor starting, servomotor controllers, etc.

It was evidenced that the STM32F4-Discovery card has a proportionate balance between cost-benefit between flash memory and clock frequency, making it one of the best options for control processes and metaheuristic optimization.

References

1. Ang, K.H., Chong, G., Li, Y.: Pid control system analysis, design, and technology. IEEE Trans. Control Syst. Technol. **13**(4), 559–576 (2005)
2. Orozco, O.A., Ruíz, V.M.A.: Sintonización de controladores pi y pid utilizando los criterios integrales iae e itae (2003)
3. Borase, R.P., Maghade, D., Sondkar, S., Pawar, S.: A review of pid control, tuning methods and applications. Int. J. Dyn. Control **9**(2), 818–827 (2021)
4. Dahlin, E., et al.: Designing and tuning digital controllers. Instr. Control Syst. **41**(6), 77–83 (1968)
5. Fay, M.P., Malinovsky, Y.: Confidence intervals of the mann-whitney parameter that are compatible with the wilcoxon-mann-whitney test. Stat. Med. **37**(27), 3991–4006 (2018)
6. Huapaya, D., Marin, D., Mauricio, D.: TCO app: telemonitoring and control of pediatric overweight and obesity. In: García, M.V., Fernández-Peña, F., Gordón-Gallegos, C. (eds.) Advances and Applications in Computer Science, Electronics and Industrial Engineering. AISC, vol. 1307, pp. 79–97. Springer, Singapore (2021). https://doi.org/10.1007/978-981-33-4565-2_6
7. Joseph, S.B., Dada, E.G., Abidemi, A., Oyewola, D.O., Khammas, B.M.: Metaheuristic algorithms for pid controller parameters tuning: review, approaches and open problems. Heliyon, e09399 (2022)
8. Li, M., Feng, X.: Research on pid parameter tuning based on improved artificial bee colony algorithm. In: Journal of Physics: Conference Series. vol. 1670, p. 012017. IOP Publishing (2020)
9. Margapuri, V., Neilsen, M.: Prototype implementation of temperature control system with can and freertos on stm32f407 discovery boards. EPiC Ser. Comput. **63**, 130–139 (2019)
10. Mohammed, I., Abdulla, A.: Fractional order pid controller design for speed control dc motor based on artificial bee colony optimization. Int. J. Comput. Appl. **179**(24), 43–49 (2018)
11. Mohammed, I.K.: Design of optimized pid controller based on abc algorithm for buck converters with uncertainties. J. Eng. Sci. Technol. **16**(5), 4040–4059 (2021)
12. Montalvo, W., Encalada, P., Miranda, A., Garcia, C., Garcia, M.: Implementation of opc ua into a web-based platform to shop-floor communication integration [implementación de opc ua en una plataforma web para la integración de comunicación en el área de producción]. RISTI - Revista Iberica de Sistemas e Tecnologias de Informacao **2020**(E26), 667–680 (2020)
13. Montalvo-Lopez, W., Catota, P., Garcia, C.A., Garcia, M.V.: Development of a virtual reality environment based on the CoAP protocol for teaching pneumatic systems. In: De Paolis, L.T., Arpaia, P., Bourdot, P. (eds.) AVR 2021. LNCS, vol. 12980, pp. 528–543. Springer, Cham (2021). https://doi.org/10.1007/978-3-030-87595-4_39
14. Öztürk, Ş, Ahmad, R., Akhtar, N.: Variants of artificial bee colony algorithm and its applications in medical image processing. Appl. Soft Comput. **97**, 106799 (2020)
15. Peñafiel, V., Buenaño, H.: Experiments on a mashup web-based platform for increasing e-participation and improving the decision-making process in the university. In: Nummenmaa, J., Pérez-González, F., Domenech-Lega, B., Vaunat, J., Oscar Fernández-Peña, F. (eds.) CSEI 2019. AISC, vol. 1078, pp. 99–110. Springer, Cham (2020). https://doi.org/10.1007/978-3-030-33614-1_7

16. Rahnema, N., Gharehchopogh, F.S.: An improved artificial bee colony algorithm based on whale optimization algorithm for data clustering. Multimedia Tools Appl. **79**(43), 32169–32194 (2020)
17. Ruiz-Navarro, J.A., Santos-López, F.M., Portella-Delgado, J.M., de-la Cruz, E.G.S.: Computer vision technique to improve the color ratio in estimating the concentration of free chlorine. In: Lecture Notes in Networks and Systems, pp. 127–141. Springer, Heidelberg (2022). https://doi.org/10.1007/978-3-030-97719-1_7
18. Tian, X., Peng, H., Luo, X., Nie, S., Zhou, F., Peng, X.: Operating range scheduled robust dahlin algorithm to typical industrial process with input constraint. Int. J. Control Autom. Syst. **18**(4), 897–910 (2020)
19. Torres, S., Melo, M., Montalvo, W.: Pid-dahlin polynomial speed controller optimized by ant colony algorithm on an arm platform. In: International Conference on Innovation and Research, pp. 189–202. Springer, Heidelberg (2022)
20. Turcios, R.S.: Prueba de wilcoxon-mann-whitney: mitos y realidades. Rev. Mex. Endocrinol. Metab. Nutr. **2**, 18–21 (2015)
21. Walz, S., Lazar, R., Buticchi, G., Liserre, M.: Dahlin-based fast and robust current control of a PMSM in case of low carrier ratio. IEEE Access **7**, 102199–102208 (2019)
22. Wang, H., Wang, W., Xiao, S., Cui, Z., Xu, M., Zhou, X.: Improving artificial bee colony algorithm using a new neighborhood selection mechanism. Inf. Sci. **527**, 227–240 (2020)
23. Xu, M., Zhang, Y., Luo, D., Shen, A.: Review of low carrier ratio converter system. Chin. J. Electr. Eng. **7**(1), 79–93 (2021)

Drone Design for Urban Fire Mitigation

Robert Humberto Pinedo Pimentel[1], Felix Melchor Santos Lopez[2](✉), Jose Balbuena[2], and Eulogio Guillermo Santos de la Cruz[3]

[1] School of Science and Engineering, Pontifical Catholic University of Peru, Lima, Peru
a20160684@pucp.edu.pe

[2] Department of Engineering, Pontifical Catholic University of Peru, Lima, Peru
{fsantos,jose.balbuena}@pucp.edu.pe

[3] Faculty of Industrial Engineering, National University of San Marcos, Lima, Peru
esantosd@unmsm.edu.pe

Abstract. Attending urban fires in high buildings is complicated due to challenges in accessing fire areas and the lack of related information. Specifically, in the capital of Peru, Lima, with increases in building floors during recent years and the availability of only one telescopic ladder, such tasks are even more challenging. This article proposes integrating a commercial drone with a fire extinguisher ball mechanism, which stores three balls launched horizontally. This system is connected to a cloud-hosted web interface leveraging an Amazon Web Services (AWS) software architecture, developed with version three of the Attribute Driven Design (ADD) methodology. Drone-related data monitoring and mechanism controls are performed through this web interface. The system also incorporates a camera leveraging the You Only Look Once X (YOLOX) algorithm to detect people's presence within the urban fire's event field.

Keywords: urban fires · fire extinguisher ball · ADD · drone · YOLOX

1 Introduction

Fires cause up to 180,000 deaths globally per year. This is more than triple the annual average number of fatalities due to all-natural hazards [20]. The city of Lima, Peru is widely known as a large and chaotic city, repeatedly exposed to urban fires. According to the Peruvian General Corps of Volunteer Firefighters, the cities of Lima, Callao, and Ica registered 64,574 fires during ten years. Moreover, in recent years, the number of high-rise buildings has increased. From the year 2017 to 2020, the average number of floors in "Center Lima" increased from 10.9 to 13.9, while "Modern Lima" went from a floor average of 11.8 to 12.5 in the same period [17,27]. In 2016, the Lima district of San Isidro bought a modern telescopic ladder to enable firefighters to access high building fires of more than 20 floors within five minutes [16,18]. However, today, not all Lima

M. V. Garcia and C. Gordón-Gallegos (Eds.): CSEI 2022, LNNS 678, pp. 262–277, 2023.
https://doi.org/10.1007/978-3-031-30592-4_19

districts own a modern telescopic ladder, so firefighters continue to experience critical challenges when attending such fires. For example, reaching a high fire area safely is complicated, while obtaining information about the fire relating to scale and potential people trapped is also difficult.

This paper proposes a unique drone system that allows firefighters to mitigate high building fires easier compared to current approaches. A drone is equipped with a mechanism for horizontally launching fire extinguisher balls and integrates a computer vision algorithm to detect the presence of people in the fire area. A cloud-based web application is implemented for technicians to monitor the computer vision results and control the mechanism. The remainder of this paper is organized as follows. First, a literature review synthesizes the findings of various existing firefighter drones. Next, the mechanical and electronic design of the mechanism is outlined, including how it attaches to the selected drone. The developed cloud computing architecture using the ADD 3.0 methodology is described, followed by an explanation of the computer vision algorithm. Finally, the summary from the study is presented.

2 Literature Review

Drones are a popular approach for achieving a broad range of modern technical tasks with multiple approaches for fulfilling a desired function. Common applications include aerial photography, payload carrying, agriculture, live streaming events, search and rescue, meteorology, and firefighting.

For example, a typical application for drones is to perform an evaluation. In [3], the presented drone system analyses fire risk based on observing vegetation structures. The drone in [24] was trained with an object detection algorithm for detecting humans, pets, and inflammable objects, and [25] presented a drone for identifying fire spots.

A variety of drone designs enable the capability of carrying fire extinguisher material for throwing it across fire areas, as described in [7,11,12,14,23–25,35], and [33]. Specifically, one design can carry up to 150 L of firefighting foam [25], and others carry fire extinguisher balls [11,14,23,24,33]. In addition, the drones in [13,14,23,24], and [33] throw these balls from above in a vertical direction, whereas [11] throws balls in a forward direction. To throw fire extinguisher balls from above, [33] applies a payload drop mechanism based on a servo motor, which is controlled through a radio channel of the remote controller.

Other approaches use water from drones for extinguishing fires, such as being equipped with a pump where water is recharged from a ground tank [5] or a drone that remains connected to the ground through a subjection system to supply water and power [32]. IoT technology is integrated in [23] for drones to receive fire signals from a house, directing them to fly to the provided coordinates.

Typically, a drone cannot carry too much weight, such as in [3], which features a carrying capacity of up to a 5 kg payload. However, larger drones with additional motors, as in the hexacopter from [14], can carry up to a 13 kg payload.

3 System Proposal

Unlike previous works, the proposed drone system was only trained to perform people detection. It doesn't use water, but fire extinguisher balls to help mitigate the fire. Besides, the drone is capable of carrying a considerable amount of weight, so it can store the balls and launch them in a forward direction, and not vertically, as most firefighter drones do.

This section first describes the mechanical and electronic designs of the proposed drone mechanism. Second, a commercial drone is selected for the implementation, determined by the weight of the mechanism. Finally, the integration of the mechanism with the selected drone is presented.

3.1 Mechanical Design

Fire Extinguisher Ball Analysis.- A fire extinguisher ball self-activates after about three to five seconds of fire exposure and disperses non-toxic extinguishing chemicals. These balls weight 1.3 kg and have a diameter of 15.2 cm [19]. The purpose of the mechanism is to transport and horizontally launch a fire extinguisher ball. Horizontal throwing makes it possible to the balls to break into the building's windows and reach any floor where the fire is occurring, meanwhile launching the balls vertically limits the range of the balls to the last floor. An important mechanical consideration when moving through the air is the effects of drag forces. This force opposes the forward motion of the projectile ball, so its movement through the air slows [29]. The air drag force is approximated as

$$F_d = KV^2 \tag{1}$$

where

$$K = \frac{1}{2}\rho C_d A \tag{2}$$

$$V = \sqrt{V_x^2 + V_y^2 + V_z^2} \tag{3}$$

A temperature of 50°C was considered, therefore the air density (ρ) has a value of 1.09 kg/m^3. The typical values of the drag coefficient (C_d) are between 0.2 and 1, in this case an intermediate value of 0.5 was chosen. The area (A) is 0.018 m^2 and it was found using the ball's diameter. Finally, the drag force coefficient (K) was calculated using the Eq. (2), with a final outcome of 0.005.

The air drag force (F_d) varies during the complete trajectory of the fire extinguisher ball because it depends on the instant velocities that vary in time. To simplify the calculation, a constant velocity is used for the initial velocity (i.e., a maximum velocity), redefining the air drag force as

$$F_d = Kv_i^2 \tag{4}$$

Two additional considerations for the ball launch include time and distance. A drone should not fly too close to a building during a fire event because the electronic components cannot withstand high temperatures. This limitation suggests that the drone must maintain a minimum distance of ten meters from the fire. Therefore, the launch distance selected for modeling the mechanism is 10 m. In addition, because the ball is launched horizontally, gravity will initiate its fall, so the ball must reach the building as quickly as possible. The fall distance times listed in Table 1 were calculated using parabolic movement equations.

Table 1. Fall distances and corresponding times.

Fall distance (m)	Time (s)
0.1	0.14
0.3	0.24
0.5	0.32
1	0.45
2	0.64

Based on Table 1, a fall distance of 0.3 m is selected because the smallest height standard window size is 0.6 m. The diameter of the ball is 0.152 m, so if appropriately launched, the ball should break into the target building before falling too far due to gravity. In other words, the ball must travel ten meters and impact the building within a maximum of 0.24 s.

Applying these considerations, a kinematics and kinetics analysis of the fire extinguisher ball is developed in the following.

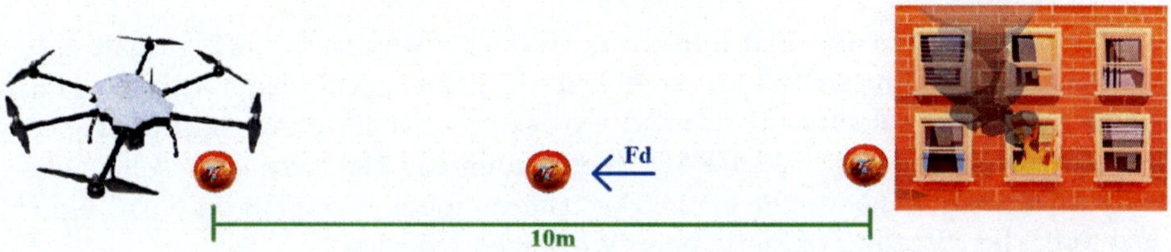

Fig. 1. A trajectory analysis of a fire extinguisher ball.

As illustrated in Fig. 1 and applying Newton's second law, a sum of forces analysis is made when the ball is still in the air affected by air drag force:

$$\sum F = ma \tag{5}$$

$$\frac{-Kv_i^2}{m} = a \tag{6}$$

Moreover, the distance (10 m), time (0.24 s), mass (1.3 kg), and acceleration (Eq. 6) are applied to calculate the initial velocity required by the ball to reach the fire area:

$$d = v_i t + \frac{at^2}{2} \tag{7}$$

$$v_i = 40.66\frac{m}{s} \tag{8}$$

Ball Thrower Wheels Analysis.- The dimensions of the two thrower wheels are selected so sufficient kinetic energy can be stored and transferred to the balls through frictional forces to attain the desired initial velocity. The wheels are composed of aluminum (inner side) and polyurethane (outer side). Aluminum is chosen due to its relatively low density, good resistance, and affordable price, and polyurethane is used for its high friction and affordable price [9]. To join both sides, special glues with an appropriate viscosity are applied.

Table 2. Thrower wheels dimensions.

	Aluminum side	Polyurethane side
Radius (m)	0.05	0.06
Height (m)	0.12	0.12
Volume (m^3)	0.0009	0.0013
Weight (kg)	2.54	0.28

The selected dimensions are listed in Table 2 with the following analysis to verify that these dimensions are sufficient [30]. The combined radius (0.11 m) and desired initial velocity (40.66 m/s) are used to calculate the rotational speed needed by each wheel (3,530 RPM). Approximating the wheels as solid cylinders with a combined weight of 2.83 kg, the moment of inertia is 0.017 kgm^2. Therefore, the rotational kinetic energy for each wheel is

$$E_W = \frac{I\omega^2}{2} = 1169.98\,\mathrm{J} \tag{9}$$

The ball throwing energy is calculated with its mass (1.3 kg) and desired initial velocity (40.66 m/s) by

$$E_B = \frac{mv^2}{2} = 1075.01\,\mathrm{J} \tag{10}$$

Both wheels lose energy transmitted to the ball, so the dimension selection is adequate if this lost energy is smaller than the rotational kinetic energy of each wheel. Assuming an energy transfer efficiency, e, of 0.75,

$$E_L = \frac{E_B}{2e} = 719.67\,\mathrm{J} \tag{11}$$

Therefore, the energy lost per wheel (Eq. 11) is smaller than the rotational kinetic energy per wheel (Eq. 9), so the dimensions selected for the wheels are considered appropriate to deliver the expected functionality.

3.2 Electronic Design

The electronic design consists of connected components to enable recording high-quality videos of an active urban fire, receiving and sending data with the cloud-based web application, rotation of the two wheels, release of the fire extinguisher ball, and launching the ball through the wheels.

The following considerations guided the selection of the electronic components. The motors must attain 3,530 RPM, the rotational speed required by each wheel. The selected battery provides the mechanism sufficient power for at least 30 min of autonomy, determined by an approximation of the operation times per component for energy consumption calculations. Also, the power components, such as relays and regulators, must withstand the high electrical currents used by the motors and solenoid.

Figure 2 presents the interactions between all system components and is divided into five sections of the actuator, power control, main controller, sensor/camera, and communication. First, the components providing or regulating energy, including the batteries, converter, and regulator, are grouped within the power control. Second, the servomotor, solenoid, and DC brushless motors are within the actuators section. The servomotor and the solenoid release the fire extinguisher balls when activated because the servomotor spins a blocking rod, allowing the solenoid to freely push a ball through a ramp. The servomotor is activated by the microcontroller, while the solenoid must be powered by a relay also activated by the microcontroller. On the other side, two DC brushless motors and the corresponding drivers drive the spin of the two launching wheels. The microcontroller, through the RS485 convertor, configures the motor to the desired speed.

Finally, the primary controller section, besides the microcontroller and RS485 convertor, leverages a Raspberry Pi to receive a WiFi-based signal through the WiFi module within the communication section to support interaction with the web application. The data sent include the video captured by an onboard camera compatible with the Raspberry Pi interface, the motor status from the microcontroller via UART communication, and the distance from the drone to the urban fire, as measured by the mini LiDAR. Data received by the Raspberry Pi include instructions to release and launch the fire extinguisher balls, which are forwarded to the microcontroller via UART.

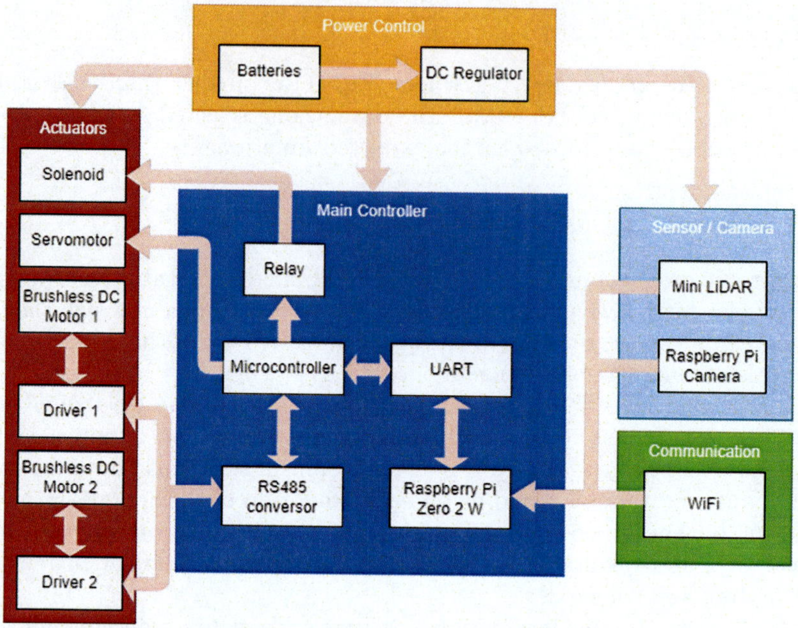

Fig. 2. Electronic block diagram of the proposed drone system.

3.3 Integration with the Mechanism

Most of the electronic components are located in a box in the center of the mechanism, as seen in Fig. 3(a). This box presents holes, so the inside components like the microcontroller or Raspberry Pi are able to connect with the outside components such as the drivers or the camera.

The launch side consists of motors attached to the inferior side of a carbon fiber platform through a base. The wheels utilize a coupling and parallel key for torque transmission with these motors. A thrust ball bearing, as seen in Fig. 3(b), supports the axial forces. A security guard prevents the wheel from making vertical movements.

Further, the store-and-release side design is based on supporting a maximum of three fire extinguisher balls for the mechanism. A servomotor connects to a blocking rod so that when activated, it spins to unblock the path to the ramp. Next, the solenoid, in the position illustrated in Fig. 3(c), is activated to push the ball through the ramp.

The combined weight of the mechanism is 23 kg, so the commercial ASTA AGL-30 drone is selected with its 30 kg payload capacity.

Four aluminum tubes connect the mechanism to this drone. As seen in Fig. 4(b), four landing gears are attached to one extreme of each tube, with the other extremes positioned in the drone, supporting the weight of the drone and the mechanism. Two rectangular aluminum bars provide additional support for holding the weight of the mechanism and are joined with the landing gears. As seen in Fig. 4(a), the mini LiDAR and camera are positioned on the front side of the drone.

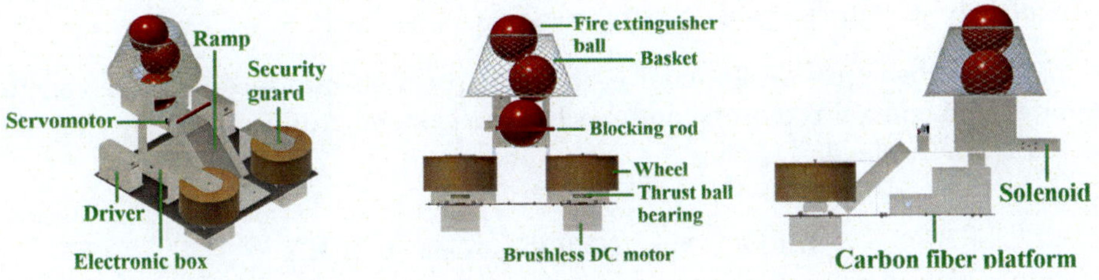

(a) Isometric view. (b) Front view. (c) Side view.

Fig. 3. Multiple views of the 3D design model for the mechanism.

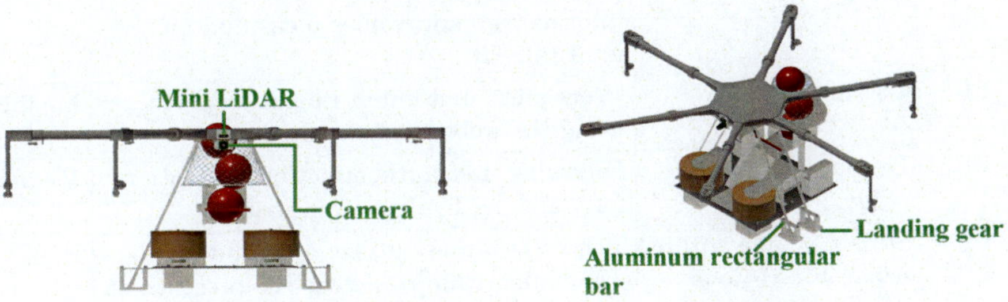

(a) Front view. (b) Isometric view.

Fig. 4. Complete mechatronic system 3D design.

4 Cloud-Based Web Application Design

The system consists of a web interface to enable communication control with the drone through a cloud-based application. The following describes the developed computing architecture, and the computer vision algorithm for analyzing images to search for people.

4.1 ADD Process Design

The ADD Methodology v3 is applied in the architecture software design that highlights a focus on quality attribute concerns. Certain inputs must first be defined to follow an iterative design approach to establish a set of steps for designing the architecture [28]. The methodology's inputs include:

- *Design goal,* the software's primary objective.
- *Functional Requirements*, typically involving the core functionality.

- *Quality attribute drivers,* measurable characteristics of interest to end-users and developers.
- *Constraints,* the known technical limitations and restrictions.
- *Concerns,* known design decisions that should be made, even if not stated explicitly within the goals or requirements.

The architecture's design goal is to develop a web interface for a person to monitor and control the drone and mechanism system. Additional design inputs are listed in Tables 3, 4, 5, and 6.

Table 3. Functional Requirements (FR).

Code	FR	Description	Priority
FR1	Collect data	The software system must receive the images captured by the drone, the image processing output data, motor information, and values measured by the mini LiDAR	High
FR2	Log in	Users must first enter a login page before using the web interface resources	Medium
FR3	Display captured video	Following user authentication, an option for watching live streaming video	Medium
FR4	Detect people within urban fire regions	The system must utilize a computer vision algorithm to detect the presence of a person within the urban fire, as well as display the visual results.	High
FR5	Control mechanism	The web interface must provide options for controlling the mechanism, such as buttons for toggling power and releasing and launching the balls	High

Along with the inputs described in Tables 3, 4, 5 and 6, the following reference architectures are used as inputs: [6,10,21,22,26,31]. [31] shows an example of IoT communication between Amazon Web Services (AWS) serverless components and an ESP32 microcontroller to allow it to publish and subscribe to topics. This means that the device can send any arbitrary information, such as sensor values, into AWS IoT Core while also being able to receive commands. [10] provides two AWS solutions for live streaming video. Both solutions build a highly available architecture that delivers a reliable real-time viewing experience. The architecture in [6] deploys a Machine Learning (ML) application with Amazon SageMaker, while [21] uses an Amazon SageMaker endpoint on a live video stream for activity detection. Moreover, the architecture in [22] hosts a web application in an Amazon S3 bucket, sets up a custom domain and secures the entire connection. Finally, [26] provides an architecture that uses AWS components, such as Amazon Cognito to establish an authentication mechanism.

Table 4. Quality Attributes (QA).

Code	QA	Description	Priority
QA1	Security	Every user must provide a verified username and password before accessing the web interface resources	High
QA2	Security	All user credentials must be encrypted and stored in a database.	Medium
QA3	Performance	Image processing must be performed within five seconds	High
QA4	Performance	Live streaming video must contain at least 20 frames per second with low latency of less than 50 ms.	Medium
QA5	Availability	The system must work continuously with an approximately 10-min recovery time in case of system failure.	Medium
QA6	Scalability	The system must support 30 simultaneous connected users.	Medium
QA7	Testability	During development, the system should facilitate performing tests for all services and features applied in the solution.	Low

Table 5. Architectural Constraints (CON).

Code	Description
CON1	Exclusive use of services provided by AWS.
CON2	Live stream drone video in HLS format.
CON3	Support of up to 50 simultaneous connected devices.
CON4	Web interface access must occur through a commercial web browser (e.g., Google Chrome, Mozilla Firefox, Opera, Microsoft Edge).
CON5	Communication protocols must be IoT oriented

Table 6. Architectural Concerns (CRN).

Code	Description
CRN1	Design the web interface using the React JavaScript library.
CRN2	Establish a general system structure following reference architectures

The iterative steps of the design process were performed and the desired system was obtained with three iterations. The first iteration established the IoT communication and the drone's live streaming video. The second iteration defined the computer vision algorithm integrated within the live streaming video. The third iteration designed the hosting web interface and created a login and display for collected data. Figure 5 presents the final architecture.

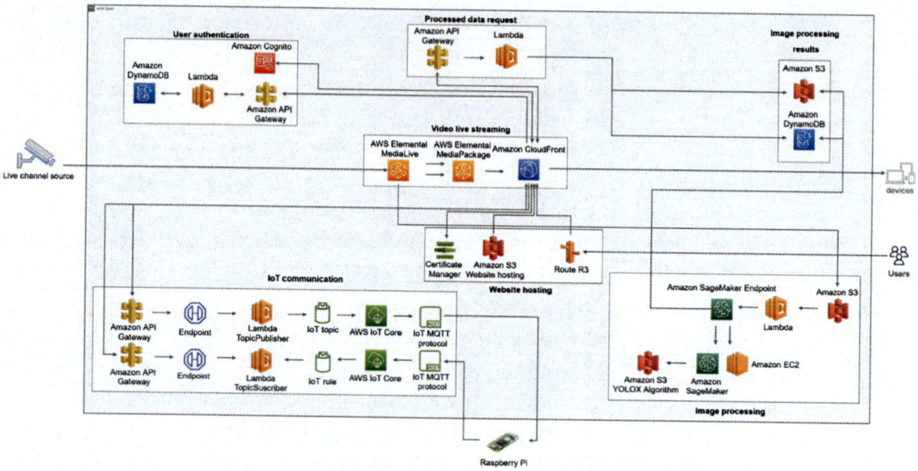

Fig. 5. The designed AWS cloud architecture.

All functions performed by each component service are provided by AWS, as outlined in Fig. 5 and described in Table 7.

4.2 Computer Vision Algorithm

The You Only Look Once X (YOLOX) algorithm was chosen for performing people detection within urban fire events. The process receives an image as an input and generates an output that includes confidence scores of object classes detected within the image and corresponding bounding box coordinates. This algorithm detects these objects in one stage through a simple regression problem [2].

In the first step, images of people, specifically those located within an urban fire event, are collected. Second, each image is manually labeled with an application called "labelme" that labels the images in a JSON format. However, the Common Objects in Context (COCO) format is required for training a YOLOX model, so the Python library "labelme2coco" is applied for the conversion. Third, the training process is initiated by separating the image set into training and validation subsets. At least 300 epochs are recommended for training a YOLOX model.

Table 7. Usage descriptions of AWS services.

AWS service	Description
AWS Elemental MediaLive	Receives video in real-time and transcodes it into HSL (HTTP Live Streaming) and ABR (Adaptive Bitrate Streaming).
AWS Elemental MediaPackage	Packs the HSL video for delivery as an output endpoint.
Amazon Cloudfront	Displays the live streaming video, an interface for controlling the mechanism, and the mini LiDAR and motors data.
Amazon API Gateway	Enables communication between the web interface and other AWS services.
AWS Lambda	–Writes or reads data from an IoT topic –Calls the Amazon SageMaker endpoint when a new image is stored in an Amazon S3 bucket –Requests the image processing results –Executes SQL commands for user authentication.
IoT topic	Identifies the topic to which an IoT message is written.
IoT rule	Invokes a lambda function for extracting data.
AWS IoT Core	Delivers messages from a topic to the Raspberry Pi.
IoT MQTT protocol	Enables Raspberry Pi-MQTT Broker communication.
Amazon S3	Stores video captured by the drone, the image processing results, the computer vision algorithm, and the web interface code.
Amazon SageMaker	Prepares data for training, performs the training and evaluation, and deploys the model.
Amazon EC2	An instance for running the computer vision algorithm.
Amazon DynamoDB	Saves the computer vision algorithm results and registered users.
Route 53	Connects the web browser with the web interface.
Certificate Manager	Protects and secures the web interface with encryption.
Amazon Cognito	Controls the creation and access of authorized users.

For testing purposes, a GeForce GTX 1050 Nvidia GPU, the lightest YOLOX model (YOLOX-s) and 800 images were used for training and validating a model, with an average precision of 40.465 and an average inference time of 61.40 ms.

However, the performance of the model can be strongly improved with a faster GPU, more images and a better YOLOX model.

The images used for testing the model and the average resultant confidence scores are shown in Fig. 6 and Table 8, respectively.

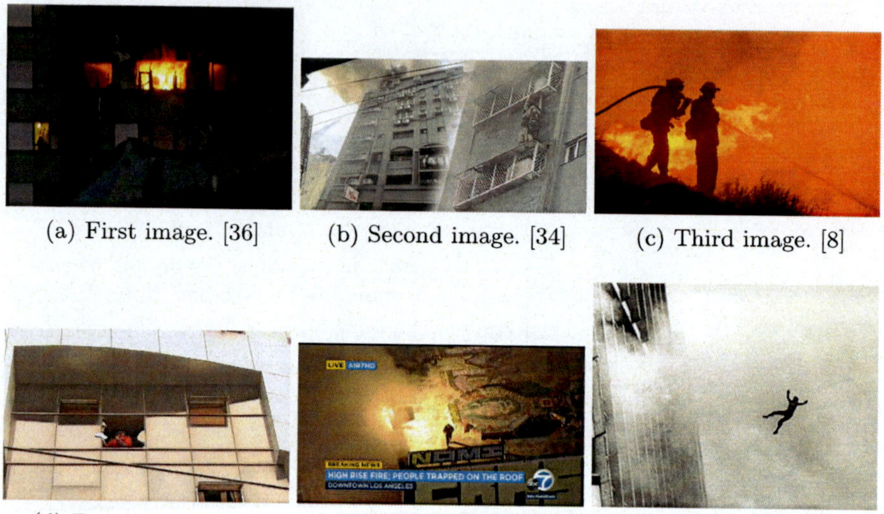

(a) First image. [36] (b) Second image. [34] (c) Third image. [8]

(d) Fourth image. [15] (e) Fifth image. [1] (f) Sixth image. [4]

Fig. 6. Images used for testing.

5 Summary

In this paper, a literature review was performed to learn about existing drone use in fire scenarios. A new drone system consisting of a commercial drone customized with a designed mechanism was proposed for supporting firefighting in urban areas. This mechanism includes several components with motors that rotate with sufficient speed to launch firefighting balls horizontally, considering a minimum vertical deviation. Also, the mechanism weight was calculated at 23 kg, guiding the selection of a drone with a 30 kg payload capacity. This mechanism was attached to the drone with aluminum tubes and rectangular bars.

The seven-step ADD V3 methodology was applied for designing a robust software architecture, completed with three iterations to obtain a final design based on AWS services. The primary goal of the first iteration was to design the IoT communication and the live streaming video process. The second iteration developed the image processing, and the third created the user authentication and web interface.

Finally, a YOLOX-s model was integrated to perform the people detection within images of urban fire events. This algorithm was tested with different

Table 8. Results from the trained model.

Image	Average confidence score for people detection
Fig. 6 (a)	91.2
Fig. 6 (b)	29.45
Fig. 6 (c)	91.4
Fig. 6 (d)	68.6
Fig. 6 (e)	85.3
Fig. 6 (f)	81.4

images and in most cases the resultant average score was superior to 80. Nevertheless, images where the people's body is not clear or overlapped by other body or object present a low score, but it can be improved with more training images and a better YOLOX model.

References

1. abc7: Multiple people left stranded on roof of downtown la building after fire erupts (2021). https://abc7.com/los-angeles-fire-rescue-operation-building-lafd/10446474/. Accessed 18 June 2022
2. Sharma, A.: Introduction to the yolo family. https://pyimagesearch.com/2022/04/04/introduction-to-the-yolo-family/
3. Aerocamaras: Aerohyb, el dron ideal para prevenir incendios.https://aerocamaras.es/aerohyb-dron-prevenir-incendios/
4. Ah Duvido: Os 50 lugares mais assustadores do mundo (2021). https://br.pinterest.com/pin/720716746607824384/. Accessed 30 June 2022
5. Al Jaber, R., Sikder, M.S., Hossain, R.A., Malia, K.F.N., Rahman, M.A.: Unmanned aerial vehicle for cleaning and firefighting purposes. In: 2021 2nd International Conference on Robotics, Electrical and Signal Processing Techniques (ICREST), pp. 673–677. IEEE (2021)
6. Andrewngai: Using aws sagemaker and lambda to build a serverless ml platform. https://towardsdatascience.com/using-aws-sagemaker-and-lambda-function-to-build-a-serverless-ml-platform-f14b3ec5854a
7. Angulo, K., Gil, D., Espitia, H.: Method for edges detection in digital images through the use of cellular automata. In: Nummenmaa, J., Pérez-González, F., Domenech-Lega, B., Vaunat, J., Oscar Fernández-Peña, F. (eds.) CSEI 2019. AISC, vol. 1078, pp. 3–21. Springer, Cham (2020). https://doi.org/10.1007/978-3-030-33614-1_1
8. ARANTEC: Incendios forestales: tres fases en las que internet de las cosas (iot) puede ser útil (2020). https://arantec.com/tecnologia-iot-consecuencias-incendios-forestales/. Accessed 23 June 2022
9. Arslan, C., Arslan, M., Yalçin, G., Kaplan, T., Kahramanli, H.: Ball throwing machine design to develop footballers' technical attributes. Eur. Mech. Sci. **5**(1), 39–43 (2021)

10. AWS: Live streaming on aws. https://aws.amazon.com/solutions/implementations/live-streaming-on-aws/
11. Barua, S., Tanjim, M.S.S., Oishi, A.N., Das, S.C., Basar, M.A., Rafi, S.A.: Design and implementation of fire extinguishing ball thrower quadcopter. In: 2020 IEEE Region 10 Symposium (TENSYMP), pp. 1404–1407. IEEE (2020)
12. Bel Fenellós, C., Flores Hernández, V., Tabares, X., Velastegui, R., García, M.: Executive profile in 5-to-7 year-old children in Ambato (Ecuador), vol. 3129 (2022)
13. Caiza, G., Ibarra-Torres, F., Garcia, M.V., Barona-Pico, V.: Problems with health information systems in ecuador, and the need to educate university students in health informatics in times of pandemic. In: Nagar, A.K., Jat, D.S., Marín-Raventós, G., Mishra, D.K. (eds.) Intelligent Sustainable Systems. LNNS, vol. 334, pp. 119–127. Springer, Singapore (2022). https://doi.org/10.1007/978-981-16-6369-7_11
14. Cervantes, A., et al.: A conceptual design of a firefighter drone. In: 2018 15th International Conference on Electrical Engineering, Computing Science and Automatic Control (CCE), pp. 1–5. IEEE (2018)
15. DAWN: One dead, 12 wounded in lahore building fire (2015). https://www.dawn.com/news/1199807. Accessed 26 June 2022
16. ElComercio: Bomberos tendrán escalera telescópica más moderna de la región
17. Elida Vega Córdova: Crecimiento inmobiliario vertical de lima muestra comportamientos diferenciados. https://elcomercio.pe/economia/negocios/crecimiento-inmobiliario-vertical-de-lima-muestra-comportamientos-diferenciados-mercado-inmobiliario-capeco-tinsa-ncze-noticia/?ref=ecr
18. Felix M.M.C., Renan, A.G., Nancy, P.R., Giovanni, J.H.: An approach to the morphological quality of fruits with applying deep learning, a lustrum of analysis. In: Advances and Applications in Computer Science, Electronics, and Industrial Engineering, pp. 3–40. Springer, Cham (2022). https://doi.org/10.1007/978-3-030-97719-1_1
19. Firetech Global: Fire x: Our fire extinguisher ball. https://www.firetechglobal.com/products/fire-x-fire-extinguisher-ball/
20. Group, W.B.: Urban fire regulatory assessment & mitigation evaluation diagnostic (2020)
21. Poonawala, H., Meharizghi, T.: Activity detection on a live video stream with amazon sagemaker. https://aws.amazon.com/blogs/machine-learning/activity-detection-on-a-live-video-stream-with-amazon-sagemaker/
22. Fibiger, I.: Setup an amazon cloudfront distribution with ssl, custom domain and s3. https://www.stormit.cloud/post/setup-an-amazon-cloudfront-distribution-with-ssl-custom-domain-and-s3
23. Jayapandian, N.: Cloud enabled smart firefighting drone using internet of things. In: 2019 International Conference on Smart Systems and Inventive Technology (ICSSIT), pp. 1079–1083. IEEE (2019)
24. Jewani, K., Katra, M., Motwani, D., Jethwani, G., Bulani, K.: Fire fighter drone (2021)
25. Antunes, J.: Four innovative use cases for drones in fire rescue. https://www.commercialuavnews.com/public-safety/four-innovative-use-cases-for-drones-in-fire-rescue
26. Sauvannet, J.: Secure your serverless app in aws (using cognito, cloudfront, api gateway, and lambda). https://jstw.github.io/serverless-app-with-secured-api/

27. Jurado, F., Donoso, D., Escobar, E., Mayorga, T., Bilous, A.: A prototype electronic toy for the development of mathematical logical reasoning in children from five to seven years old using python. In: García, M.V., Fernández-Peña, F., Gordón-Gallegos, C. (eds.) Advances and Applications in Computer Science, Electronics and Industrial Engineering. AISC, vol. 1307, pp. 3–18. Springer, Singapore (2021). https://doi.org/10.1007/978-981-33-4565-2_1
28. Kazman, R., Cervantes, H.: Add 3.0: rethinking drivers and decisions in the design process. Saturn (2015)
29. Mahapatra, A., Chatterjee, A., Roy, S.S.: Modeling and simulation of a ball throwing machine. In: 14th National Conference on Machines and Mechanisms (NaCoMM09), NIT, Durgapur, India, December, pp. 17–18 (2009)
30. Marquette, A.J.: Design and construction of an omni-directional soccer ball thrower (2013)
31. Zara, M.: Building an aws iot core device using aws serverless and an esp32. https://aws.amazon.com/blogs/compute/building-an-aws-iot-core-device-using-aws-serverless-and-an-esp32/
32. Moore, J.: Uav fire-fighting system, uS Patent App. 13/306,419 (2013)
33. Soliman, A.M.S., Cagan, S.C., Buldum, B.B.: The design of a rotary-wing unmanned aerial vehicles-payload drop mechanism for fire-fighting services using fire-extinguishing balls. SN Appl. Sci. **1**(10), 1–10 (2019)
34. Taiwan English News: Multiple fatalities after residents trapped in apartment building fire: one man survives fall from 6th floor (2022). https://taiwanenglishnews.com/multiple-fatalities-after-residents-trapped-in-apartment-building-fire-one-man-survives-fall-from-6th-floor/. Accessed 22 June 2022
35. Velastegui, R., Flores, F., Velastegui, D., Fenellos, C., Garcia, M.: Evaluation of selective attention in 9-to-11-year-old children, vol. 3129 (2022)
36. VGTV: Fanget på innsiden av blokken som brenner (2017). https://tv.vg.no/video/142735/fanget-paa-innsiden-av-blokken-som-brenner. Accessed 20 June 2022

Drone Collaboration Using OLSR Protocol in a FANET Network for Traffic Monitoring in a Smart City Environment

Franklin Salazar[1](✉), Jesús Guamán-Molina[1], Juan Romero-Mediavilla[2], Cristian Arias-Espinoza[2], Marco Zurita[2], Carchi Jhonny[3], Sofia Martinez-García[4], and Angel Castro[4]

[1] Universidad Técnica de Ambato, Ambato 180103, Ecuador
{fw.salazar,jguaman0585}@uta.edu.ec
[2] CIDFAE, Ambato 180103, Ecuador
{juan_romero,caris,marcozurita}@fae.mil.ec
[3] Universidad Nacional de Chimborazo, Riobamba, Ecuador
jmcarchi.fie@unach.es
[4] Universidad Autónoma de Madrid, Madrid, Spain
{sofia.martinez,angel.decastroe}@unach.es

Abstract. A Flying Ad Hoc Network (FANET) is a new type of network derived from MANET, these networks use as nodes of unmanned aerial vehicles (UAV) that can be equipped with positioning, vision, or other systems. Currently, they present some advantages such as collaborative work between UAVs improving efficiency compared to single UAV systems. FANET's capacity to cover large geographical areas has already positioned it as one of the best alternatives for traffic monitoring and control in smart cities. On the other hand, these networks present some challenges and problems that must be considered at the time of their design. The analysis of these networks using two-dimensional models, in some cases propose the use of a UAV in a stationary way in a specific area. Therefore, in this work we propose the implementation of a 3D model, allowing to generate a realistic environment to the movement of a UAV, using the Gauss Markov mobility model evaluating Ad Hoc routing protocols OLSR, AODV and DSDV. For the analysis of the results, the Network Simulator version 3 (Ns-3) software is used to simulate a FANET network, evaluating the efficiency of the routing protocols. The results obtained from the simulation and implementation of the proposed network showed that the OLSR protocol presents better efficiency maximizing the scope of traffic monitoring. Finally, the FANET network with 4 heterogenous drones, allowed to improve the coverage of a larger geographical area with an adequate performance of the protocol making it suitable for remote traffic monitoring in a smart city environment.

Keywords: UAV's · FANET · drones · OLSR · smart city · traffic · transport

Supported by organization x.

M. V. Garcia and C. Gordón-Gallegos (Eds.): CSEI 2022, LNNS 678, pp. 278–295, 2023.
https://doi.org/10.1007/978-3-031-30592-4_20

1 Introduction

Unmanned aerial vehicles or UAVs are robots that can be remotely controlled from several kilometers away or in confined spaces. New technological advances allow the development and manufacture of these robots with different features, depending on their characteristics they can be used in different environments, especially where human presence is difficult or is compromised by some risk [5]. Technological development has allowed these robots to be used not only in military applications, due to their low cost and performance, today they are applied in many civilian applications. Currently, the use of drone swarms using mobile Ad-Hoc networks (MANETs) has gained more scientific attention [16].

A mobile Ad-Hoc network is the grouping of two or more nodes that communicate through a wireless channel and can be used in different applications, mainly when a conventional network cannot be found due to infrastructure limitations, installation or other reasons. Figure 1 shows the different types of mobile Ad-Hoc networks according to their application, e.g., Vehicular Ad-Hoc Networks (VANETs) and Vehicular Ad-Hoc Networks (FANETs) [19].

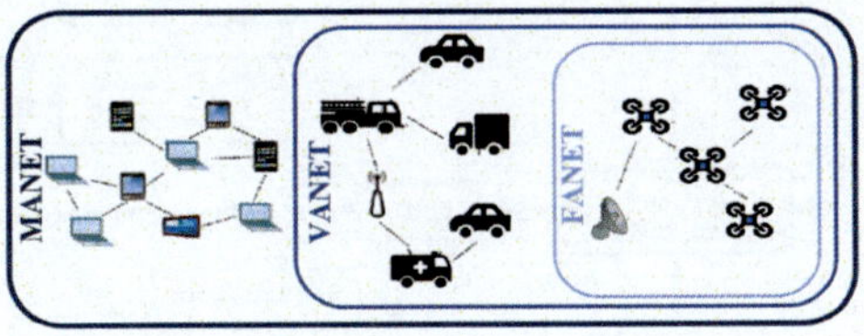

Fig. 1. MANET, VANET and FANET

Airborne Ad-Hoc networks are a new type of mobile Ad-Hoc network where the network nodes are UAVs [9,18,19]. Due to the mobility of UAV's, FANET networks present constant changes in the network topology, causing the established links to be destroyed, generating delays in communications or even the total loss of communication of the network, Fig. 2. Therefore, establishing an efficient routing protocol is one of the main components to be considered when deploying a network of this kind [12].

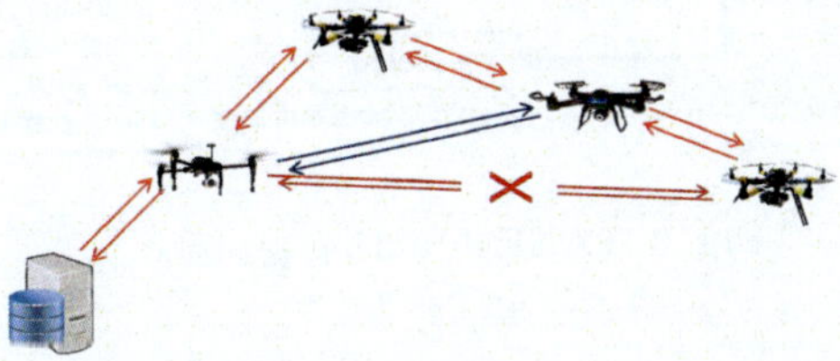

Fig. 2. Flying Ad-Hoc Network

In the different types of networks, whether wired or wireless, the routing protocol is one of the main elements because it is responsible for routing traffic through each node belonging to the network. Compared to infrastructure-based networks, mobile Ad-Hoc networks have nodes that move in a certain area establishing communication in an arbitrary manner. Nodes can act as routers to manage route discovery and maintenance. For this reason, to realize a reliable and effective network with highly dynamic nodes, routing protocols are one of the key problems to be solved in FANET networks. Figure 3 shows the division of FANET protocols [5,8,20].

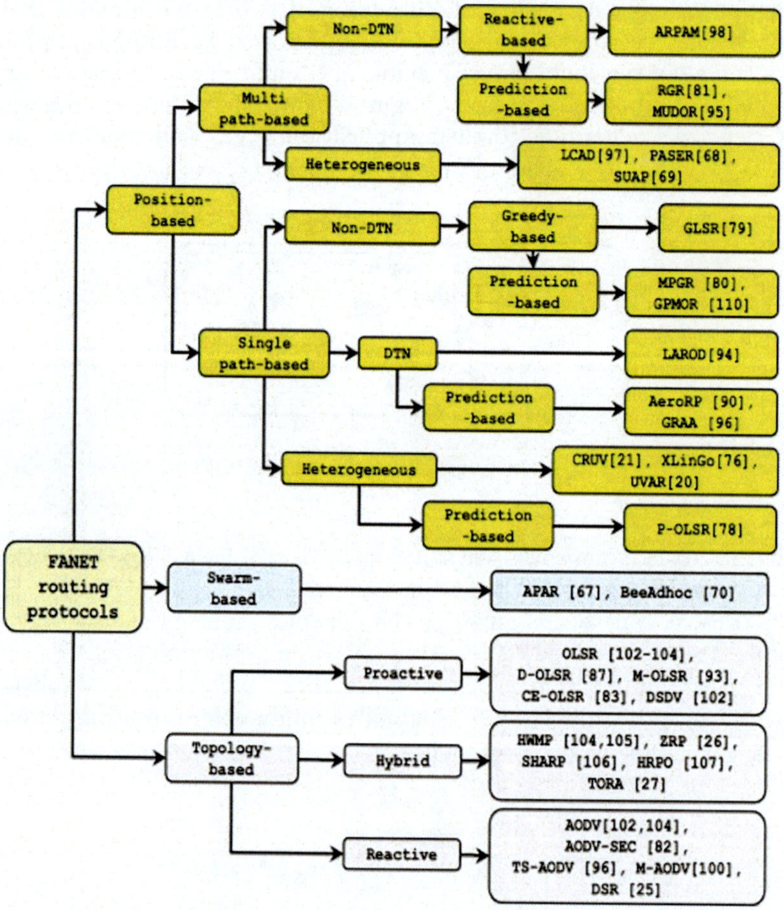

Fig. 3. FANET routing protocol

As shown in Fig. 3, the protocols used in FANET networks ensure communication between the transmitter and receiver and are classified according to their topology as follows:

(a) Reactive routing protocol: it operates under an algorithm that requests route discovery on demand, i.e., when a node needs to co-communicate with another. These protocols maintain the routes if necessary or until they expire, being one of its main advantages since there is no overhead in the network [4,7].
(b) Proactive routing protocol: It is based on routing tables, where the addressing information between nodes is kept up to date. These routing protocols have the advantage of having the routes available at any time they are required [8].
(c) Hybrid routing: This is a combination of proactive and reactive routing protocols, which aims to combine the advantages of the previous protocols [4,10].

Knowing the characteristics, applicability and challenges presented by FANET networks, it is necessary to address the problems recently encountered in this new type of network, such as communication problems due to factors such as the constant change of topology. For this reason, a FANET network requires a routing protocol with adequate performance; moreover, this protocol must be able to handle different scenarios and conditions [2].

In this paper, the analysis of different routing protocols for mobile Ad-Hoc networks that allow communication between UAVs is proposed [18]. Likewise, we propose the modeling of an evaluation scenario that resembles the reality of a physical environment for the deployment of a FANET network for traffic control within a smart city environment, for which we propose a mobility model that is like the movement performed by a drone [11]. Three fundamental metrics are analyzed to determine the efficiency of the routing protocol using the Network Simulator 3 (NS-3) simulator, finally the deployment of a FANET network with three UAVs for traffic control within a smart city environment is performed.

The paper focuses on the analysis of the communication link between UAVs, the main problem in FANET networks. Modeling an evaluation scenario for 3D mobility FANET networks to evaluate different Ad-Hoc routing protocols, simulating a physical environment for traffic control within a smart city through the FANET network, allows to obtain more realistic results of the system. In the research, each Ad-Hoc routing protocol has been analyzed in relation to the number of nodes and area size, and the efficiency of each protocol has been obtained through the evaluation metrics for networks. Finally, the deployment of a FANET network for traffic control in a smart city using one of the analyzed Ad-Hoc routing protocols is presented to contrast the results obtained previously.

This paper is organized as follows: Sect. 1 presents the works related to the subject matter of this study. Section 2 presents the methodology used for the development of this research. Section 3 presents the metrics used in the FANET network analysis process. Section 4 presents the results obtained in the analysis and presents the discussion of the comparative results obtained in the deployment of the FANET network. Finally, Sect. 5 presents the conclusions.

1.1 Related Work

To date, different types of research have been carried out to learn more about applications, design features, present problems, and suggestions that could be considered or applied to improve the performance of FANET networks.

In [18], it is concluded that the main problem in FANET networks is communication, obtained by evaluating the differences between FANET networks and other types of mobile Ad-Hoc networks in terms of energy consumption, node density, mobility and others. Likewise, in [5], some characteristics that should be considered when designing a network of this type are described. Finally, test scenarios are presented, as well as simulators that could be used to study FANET networks. The following lines describe the most relevant works analyzed for the study of FANET networks.

The analysis carried out in [15], studies FANET networks with a static mobility model, considering the use of UAV as relay nodes, for which they established a bidirectional communication between two nodes of the network while the rest worked as data relay nodes. In their work, they simulated the network by applying the Ad-Hoc on-demand routing protocol AODV and the OLSR protocol based on tables that have an IETF RFC certification. These protocols are also standardized and widely studied by different researchers for use in Ad-Hoc networks.

There are new ways that make possible the fast, efficient, and low-cost deployment of future airborne Ad-Hoc networks. In [9], different wireless architectures and technologies that could be used in the communication links between the UAV and the ground station were studied. Through simulator analysis they demonstrated the feasibility of a hybrid communication scheme employing a low power consumption function and high data rate capability.

As in other works found, emphasis is placed on the main problem existing in FANET networks, which is communication. There are methodologies developed to solve this problem since not only the co-communication between UAV's is discussed, but also the communication between UAV's and the ground station (GS). The author Srivastava proposes in [10], the use of FANET as a solution to solve the limitations of the communication range between UAV and GS without infrastructure. The study presents the implementation of a test model that is easy to perform in addition to being economically cost-effective to support studies of FANET networks. That said, it is evident that multi-UAV systems can be very low cost and can be employed for testing where real results can be obtained.

In other research presented in [11,14,17] an analysis of FANET networks is performed through simulators, using the network simulation software NS-2 or NS-3, which allows establishing certain parameters such as the mobility model for the network nodes as considered by the authors, currently there are works that describe each of the mobility models and help to determine the most appropriate according to the study required [18,21]. In [14], the Random Waypoint mobility model (RWP) is applied, and its characteristics also specify the use of AODV and OLSR routing protocols. The analysis performed was based on the density of nodes, demonstrating experimentally that the factor affecting the performance of the airborne Ad-Hoc network is the number of nodes distributed in an area and the communication range. Finally, the authors determined that it is necessary to redesign the simulation model to bring it closer to reality, which leaves several points of study for future work.

2 Methods and Materials

This section proposes an evaluation model for a FANET network to determine an efficient routing protocol for communication between UAVs for traffic control in a smart city. The main objective is the modeling of a scenario that allows the analysis and implementation of a FANET network. The deployment of the UAV's network is performed with a certain number of nodes, these processes are carried out with the aim of making measurements that can be compared with the results obtained by simulation.

For the development of the sketch that allows the evaluation of Ad-Hoc networks, several important aspects must be considered, among them the mobility model that oversees giving a displacement pattern to the nodes through the created evaluation environment. Evaluating different mobility models as well as different routing protocols, in this context the study of the proposed mobility models is proposed looking for the one that favors similar movements of a multi-rotor UAV [3,21]. Therefore, the Gauss-Markov mobility model will be applied in the study because this pattern performs smooth movements, avoiding abrupt changes in speed and direction, in addition, it is dependent on space and time to perform new movements, it is also characterized by creating a three-dimensional environment making this model very similar to the displacement performed by a drone within a city.

Like mobility models, there are several routing protocols for mobile Ad-Hoc networks. FANET networks do not have a network protocol, so in this article we intend to study the routing protocols AODV, DSDV and OLSR, they are also protocols that have an RFC certification and have been studied in FANET networks with different characteristics where excellent results have been obtained, however, it is considered necessary to evaluate and determine the routing protocol that best suits the network scheme proposed above.

2.1 Gauss Markov Mobility Model

The Gauss Markov mobility model was created by Liang B. and Hass Z. to adapt to real node movements (acceleration, deceleration, or progressive turning), the objective of this model is to improve the speed and direction of a mobile terminal as a function of previous time, an example of this mobility model can be visualized in Fig. 4 [1].

This model evaluates the node velocity after a time (t) and is shown as a stochastic Gauss Markov process. As a node approaches the boundary of the created scenario, it changes direction by rotating 180°. With this control presented

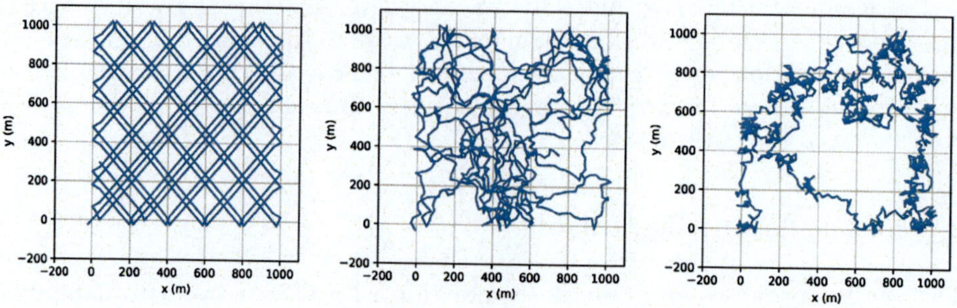

(a) $\alpha = 1.0$ (b) $\alpha = 0.5$ (c) $\alpha = 0.0$

Fig. 4. Example of Gauss Markov mobility model

by the mobility model, the UAVs only approach the boundaries without the possibility of leaving the area. The position of any node in this mobility model is constantly calculated from its past position due to the high speed of movement that the node can exhibit. The update of the velocity and direction is performed periodically by this model, the value of the velocity and direction at a given time can be calculated with the estimation of the velocity and direction [13].

2.2 Simulation Platform

To determine the efficiency of routing protocols in the FANET network, the evaluation process was carried out by implementing an analysis method based on the "third.cc" script found in the Network Simulator 3 NS-3 repository.

The simulation models are implemented in different programming languages, including C++. Network Simulator 3 (NS-3) is a free open-source software under the GNU NS-3 license. It is a simulator controlled by discrete events and its use is focused on the study and analysis of different types of networks, it also allows the development of new algorithms to evaluate, compare and modify various network characteristics, among which are the routing protocols for Ad-Hoc networks. Finally, NS-3 facilitates the export of almost all its API to the Python programming language to import an "NS-3" module in the same way as in C++, this feature makes it possible to use different applicable tools for analysis.

2.3 Simulation Structure and Parameters

The methodology presented in Fig. 5 is used to simulate the system.

As can be seen in Fig. 5, to carry out an adequate simulation of the system, the first step is to establish the main characteristics and parameters of the network, such as the following:

- Communication channel: wireless.
- Node mobility model: Since it is a three-dimensional mobility model, values are established for the X, Y and Z axes.

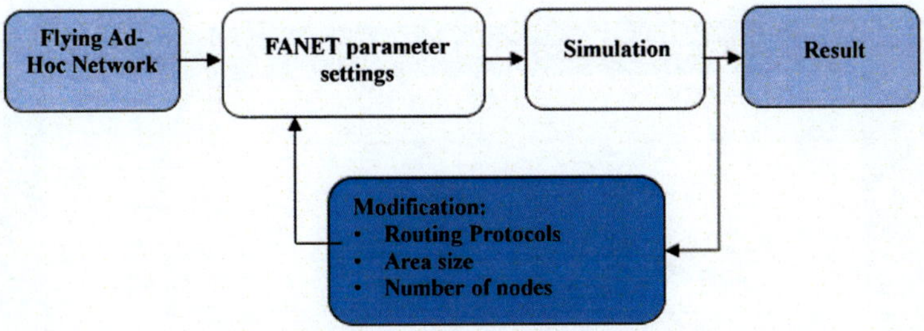

Fig. 5. Example of Gauss Markov mobility model

- Communication system between nodes: TCP.
- Routing protocol: AODV, DSDV and OLSR.

Once the main characteristics and parameters have been configured, the analysis process is carried out to determine the efficiency and effectiveness of the routing protocols, different scenarios are evaluated according to the number of nodes and size of the simulation area. In each evaluation scenario defined by the size of the area, the density of nodes and the routing protocol are varied as shown in Table 1.

Table 1. Configuration of the proposed evaluation scenarios.

Simulation parameters	Scenario 1	Scenario 1
Simulation area	250 m × 250 m Altitude = 120 m	400 m × 400 m Altitude = 120 m
Simulator	Ns-3 (version 3.25)	
Channel type	Wireless	
Protocol	AODV, DSDV, OLSR	
MAC layer protocol	802.11b	
Number of nodes	10, 20, 30, 40	
Node speed	10 m/s	
Mobility model	Gauss-Markov	

In performing the analysis, different quantitative metrics that directly influence network performance were considered, allowing a comparative study of these routing protocols through simulation. Animation is an important tool when performing any type of simulation. Currently NS-3 does not have a predefined animation tool [6], however, the NS-3 package installation contains an integrated animation tool called NetAnim. NetAnim is a stand-alone program that uses XML trace files generated by the animation interface to graphically display the simulation, Fig. 6.

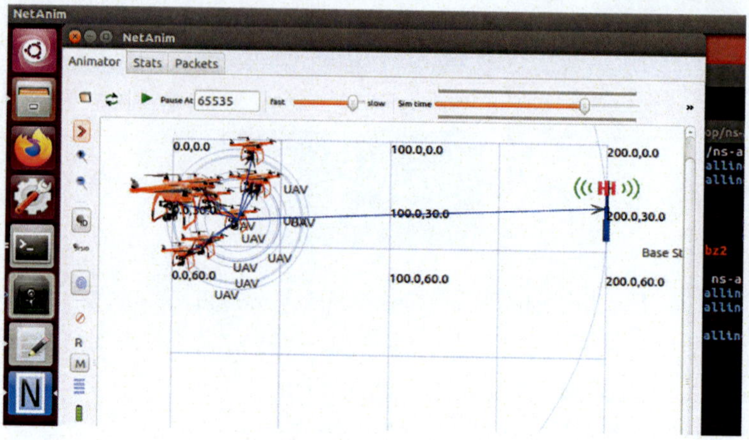

Fig. 6. NetAnim

Another tool available in NS-3 is Flow Monitor, which aims to provide a flexible system for measuring the performance of network protocols. This module uses a type of sensor installed in the network nodes that allows the tracking of the packets exchanged by the nodes, in addition to measuring a series of parameters providing the developer to expose the evaluation results in an effective and efficient way as shown in Fig. 7.

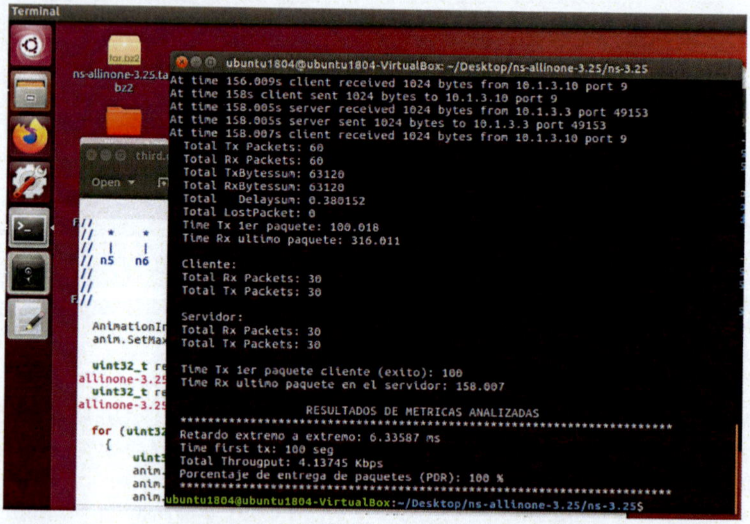

Fig. 7. Terminal output - Flow Monitor

3 Implementation of the Proposal

Finally, the implementation of the routing protocol that has generated the best results in the simulated FANET network is carried out to evaluate the performance of the routing protocol in the deployment of a physical network and to contrast the results obtained through software. The two scenarios evaluated in the simulation are shown in Fig. 8.

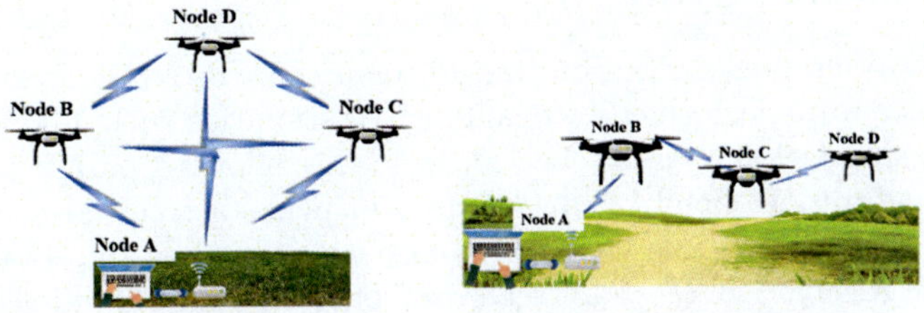

(a)Scenario 1 (b) Scenario 2

Fig. 8. Proposed evaluation scenarios

The proposed scenarios for the deployment of the FANET network consist of two scenarios, each composed of more than two wireless nodes that are interconnected in an Ad-Hoc network, Fig. 8. The first topology is a mesh topology that uses 4 nodes arranged in a range in which each node can be directly connected in a point-to-point communication. In the second scenario, the nodes establish a linear topology in such a way that a multi-hop communication is formed.

This process was performed using Raspberry Pi 3 cards in a system composed of four nodes. An external USB network device using the IEEE 802.11b standard in the 2.4 GHz spectrum was installed on each card, Fig. 9. Node 1 is set as the originating terminal in all scenarios. In both scenarios, measurements of

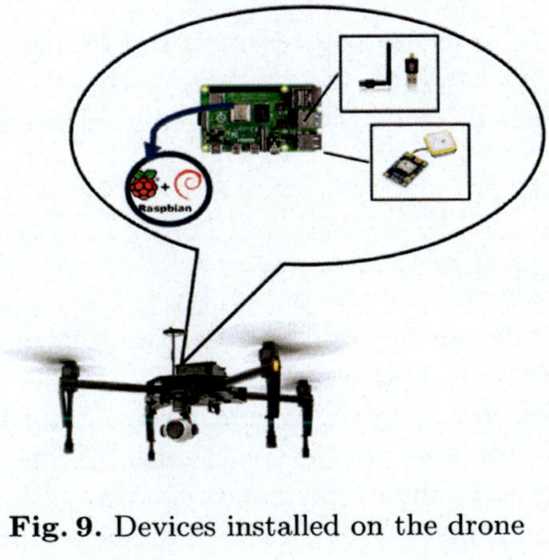

Fig. 9. Devices installed on the drone

the metrics evaluated by the simulator are performed using the communication system that generates traffic in the form of TCP packets. While the communication system is running, ICMP packets are sent from the terminal node to the destination node using the ping command at a certain time. This provides different values that allow direct and indirect measurement of metrics of interest such as End-to-End Delay, Packet Delivery Rate (PDR) and Throughput.

3.1 Performance Metrics

In this work, different Ad-Hoc routing protocols in a FANET network are analyzed, and tests were performed by changing the scenarios with different densities of nodes and simulation area. In each scenario the following parameters are evaluated: Throughput. Defined by the number of packets received at the receiver from the sender in each time [14]. It can also be defined as the total amount of data that the receiver successfully receives from the sender divided by the time it takes the receiver to obtain the last packet, it is measured in bits per second and can be calculated with the help of Eq. (1) [5].

$$Throughput = \frac{bytesreceived \times 8}{Simulationtime} \tag{1}$$

$$\begin{gathered} Throughput = \frac{data}{T_T} \\ 0 < t < T \\ Throughput = \frac{data(t) \times 8}{T_T(t)} \end{gathered}$$

Packet Delivery Ratio (PDR). It is the ratio between the number of data packets sent and the data packets received; normally this measure is represented as a percentage and is calculated using Eq. (2) [14]. The higher the PDR the better the network performance [5].

$$PDR = \frac{Totalpacketsreceived}{Totalpacketstransmitted} \times 100 \tag{2}$$

End-to-end delay. This metric represents the average end-to-end delay and indicates how long a packet took to travel from the sender to the destination [17]; its unit is expressed in seconds and is calculated with the Eq. (3) [5].

$$Retardoextremoaextremo = \frac{TotalDelay}{Totalpacketsreceived} \tag{3}$$

The Ad-Hoc routing protocols were analyzed through a three-dimensional model for FANET, therefore, the Gauss Markov mobility model was considered. In each simulation, different densities of nodes with constant speed were applied to move in the created scenario, in this scenario the simulation area was modified as presented in the previous section; the speed taken as reference is that of a UAV Multirrotor model. Figure 10 shows the results corresponding to the Throughput. As can be evidenced in graphs (a), (b) the OLSR routing protocol maintains the

Throughput value in a more linear way, unlike the AODV and DSDV protocol; the OLSR protocol does not vary drastically by the node density or by the size of the simulation area. In summary, the OLSR routing protocol performs better as the number of nodes and the simulation area increase, i.e. it facilitates the scalability of the network.

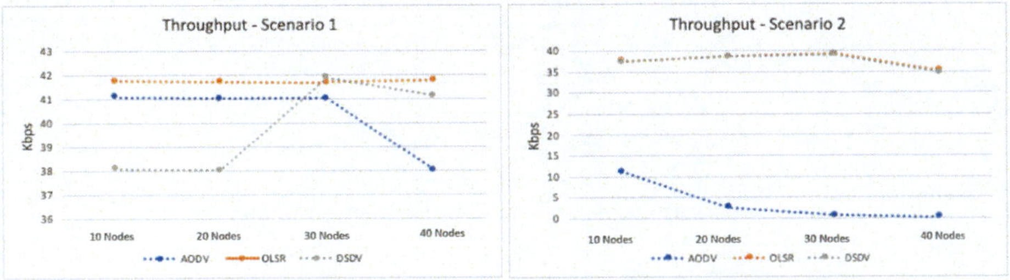

(a) (b)

Fig. 10. Throughput for the two evaluated scenarios

Figure 11(a) shows that the packet delivery rate is higher than 90% for all the protocols studied, while in (b) the AODV protocol decreases radically as the number of nodes in the network increases. Comparing the protocols studied together with the metrics analyzed in the previous paragraph, it can be defined that OLSR is the protocol with the best efficiency in the network.

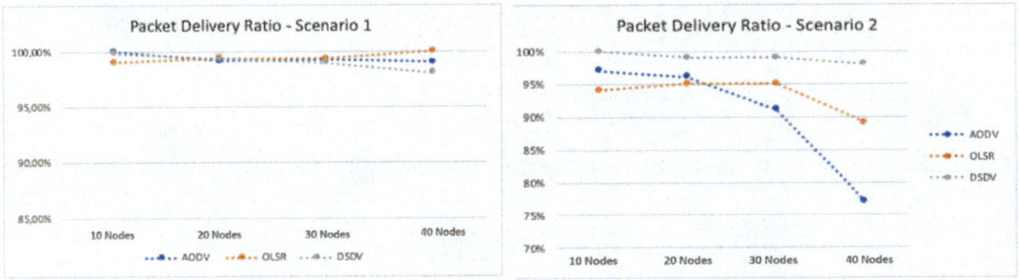

(a) (b)

Fig. 11. Parcel delivery rate for the two evaluated scenarios

Figure 12 shows an increase in the end-to-end delay with respect to the size of the area and number of nodes; this is logical because the nodes are located at a greater separation distance and the packets must pass through a greater number of nodes to reach the destination. However, in graphs (a) and (b) it is observed that the existing delay of the OLSR routing protocol is lower than the other analyzed protocols, establishing the OLSR protocol as the best.

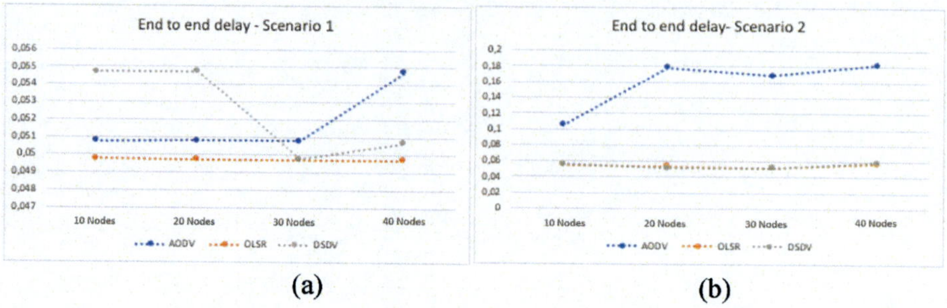

Fig. 12. End to delay for the two evaluated scenarios

4 Implementation Results

In this section we present the real evaluation scenarios implementing an Ad-Hoc wireless network system applying the OLSR routing protocol. For this analysis we implemented a network with 4 drones, defining two evaluation topologies as shown in Fig. 12.

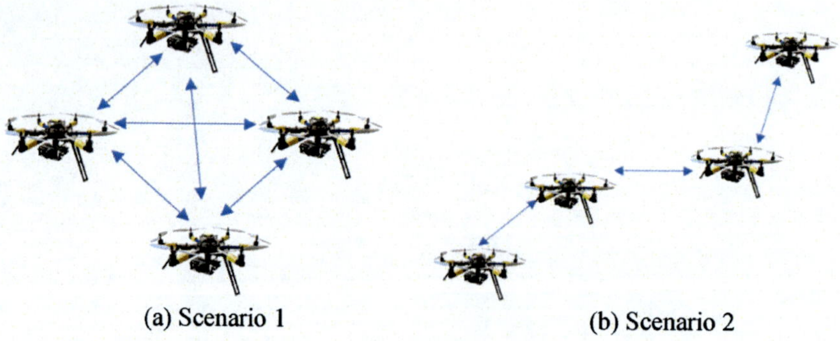

Fig. 13. Evaluation scenarios

The selected protocol was installed, configured, and started through the Raspberry Pi terminal, turning this device into a network node. The execution of OLSR allows visualizing a window like Fig. 14, allowing to observe the links established with the neighboring nodes, in addition to the nodes that maintain an indirect communication (maximum 2 hops) (Fig. 13).

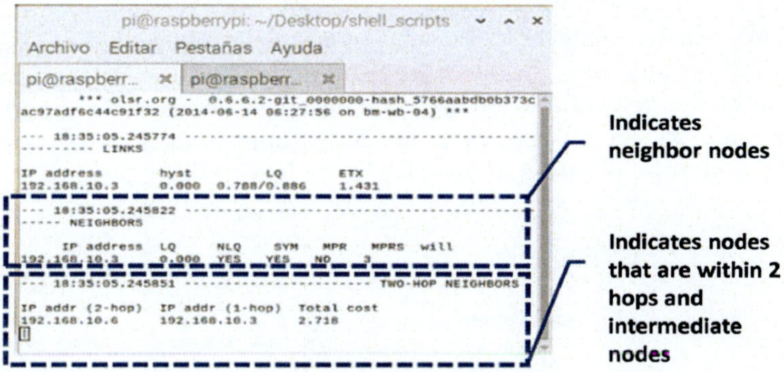

Fig. 14. Operation of the OLSR protocol

The validation of the system was performed with the deployment of the physical network within the premises of the Technical University of Ambato for traffic control at the end of the working day, to determine the traffic generated in the University premises, Fig. 15. In the tests performed, the two scenarios proposed and simulated using the OLSR protocol are evaluated to compare the response of data transmission in a real environment such as traffic analysis, verifying its application in the environment of an intelligent city.

(a) Scenario 1 (b) Scenario 2

Fig. 15. Operation of the OLSR protocol

The first topology analyzed is the linear type, through this topology it was possible to verify the scalability presented by the OLSR routing protocol, since it inserts new network nodes without the need to paralyze the data transmission that allows traffic control in a smart city since the avenues that present greater traffic are extensive in a linear way. In Fig. 16(a) corresponds to the results of the physical network, where the percentage of packet delivery is observed, with an efficiency greater than 90%. The end-to-end delay shows an increase as the number of nodes increases, on the other hand, in Fig. 16(b) corresponding to the simulation of this network, a similarity is observed between the results of this metric. Finally, we can observe that the Throughput in both cases decreases its value as the number of nodes increases.

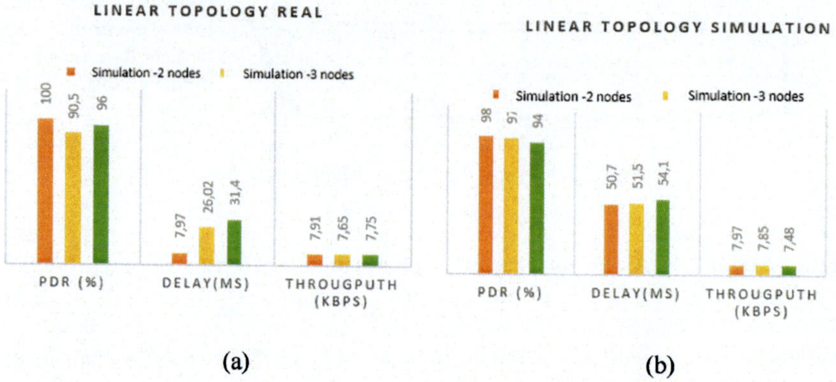

Fig. 16. Comparison of results-Mesh topology

Figure 17 shows the results obtained in the mesh type topology, where a packet delivery percentage very close to 100% is observed. As in the previous scenarios, the communication delay is lower than that obtained in the simulation, and finally the throughput is better compared to the results of the simulation and the deployment of a network with linear topology. In the same way this type of topology can be used in cross streets that are prone to generate a large amount of traffic.

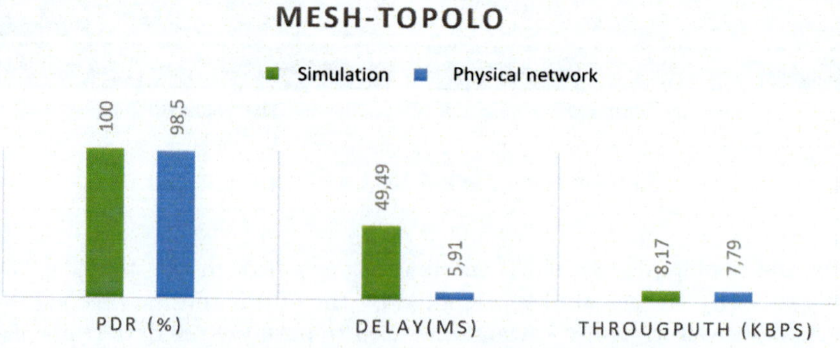

Fig. 17. Operation of the OLSR protocol

5 Conclusions

The analysis of a Flying Ad Hoc Network, through the modeling of an evaluation schedule in NS-3 using the Gauss Markov mobility model, allows to evaluate an environment like the one implemented at the end of our work as an application for traffic monitoring in an intelligent city environment. Some authors in their works perform the analysis of these networks by applying only a 2D model, but in this work a 3D model is considered to corroborate the results obtained by software because most of the work currently performed is carried out in a non-real environment.

The three routing protocols for mobile Ad-Hoc networks compared with each other through different metrics regarding the number of nodes and area size, allowing identifying the protocol with the best response for its implementation. The simulation results demonstrate the efficiency of the OLSR routing protocol in a FANET network. When analyzing the throughput, the OLSR routing protocol stands out, presenting a greater capacity to send information through the network, being little affected by the increase of nodes and simulation area, making it ideal for its application in traffic monitoring in a smart city environment. Finally, one of the parameters that determine OLSR as the best protocol is the end-to-end delay, through the analysis of this metric, the value in OLSR is lower than the other studied protocols, ensuring the delivery of valid packets in a shorter time.

Finally, we have been able to demonstrate that our evaluation model developed is effective, because it evaluates the two scenarios in the simulation and its real implementation, showing a similarity in the behavior of each metric analyzed for the implementation of a traffic monitoring system using the OLSR communication protocol through the collaboration of drones in a FANET network.

Acknowledgments. The authors would like to thank the Investigation and Development Directorate, DIDE, of the Technical University of Ambato for its special help with the development of this proposal, thanks to the financing of the project "Sistema de gestión inteligente para el transporte estudiantil de la Universidad Técnica de Ambat", PFISEI030.

References

1. Alenazi, M.J., Abbas, S.O., Almowuena, S., Alsabaan, M.: RSSGM: recurrent self-similar Gauss-Markov mobility model. Electronics **9**(12), 2089 (2020)
2. Azevedo, M.I.B., Coutinho, C., Toda, E.M., Carvalho, T.C., Jailton, J.: Wireless communications challenges to flying ad hoc networks (FANET). Mob. Comput. 3 (2020)
3. Bujari, A., Palazzi, C.E., Ronzani, D.: FANET application scenarios and mobility models. In: Proceedings of the 3rd Workshop on Micro Aerial Vehicle Networks, Systems, and Applications, pp. 43–46 (2017)
4. Canar, C.P.S., Yépez, J.J.T., López, H.M.R.: Performance of reactive routing protocols DSR and AODV in vehicular ad-hoc networks based on quality of service (QOS) metrics. Int. J. Eng. Adv. Technol. **9**(4), 2033–2039 (2020)

5. Carchi Curimilma, J.M.: Análisis e implementación de protocolos de enrutamiento en una red ad hoc aérea (FANET) para la telegestión de uavs. B.S. thesis, Riobamba, Universidad Nacional de Chimborazo (2022)
6. Font, J.L., Iñigo, P., Domínguez, M., Sevillano, J.L., Amaya, C.: Analysis of source code metrics from NS-2 and NS-3 network simulators. Simul. Model. Pract. Theory **19**(5), 1330–1346 (2011)
7. Garcia, C.A., Naranjo, J.E., Ortiz, A., Garcia, V.M.: An approach of virtual reality environment for technicians training in upstream sector. IFAC Papersonline **52**(9), 285–291 (2019). https://doi.org/10.1016/j.ifacol.2019.08.222
8. Garcia, M.V., Perez, F., Calvo, I., Moran, G.: Building industrial CPS with the IEC 61499 standard on low-cost hardware platforms. In: 2014 IEEE Emerging Technology and Factory Automation (ETFA) (2014)
9. Irisarri, E., Garcia, M.V., Perez, F., Estevez, E., Marcos, M.: A model-based approach for process monitoring in oil production industry. In: 2016 IEEE 21ST International Conference on Emerging Technologies and Factory Automation (ETFA). IEEE International Conference on Emerging Technologies and Factory Automation-ETFA (2016)
10. Khan, M.A., Qureshi, I.M., Khanzada, F.: A hybrid communication scheme for efficient and low-cost deployment of future flying ad-hoc network (FANET). Drones **3**(1), 16 (2019)
11. Khan, M.A., Safi, A., Qureshi, I.M., Khan, I.U.: Flying ad-hoc networks (FANETs): a review of communication architectures, and routing protocols. In: 2017 First International Conference on Latest Trends in Electrical Engineering and Computing Technologies (INTELLECT), pp. 1–9. IEEE (2017)
12. Kim, D.Y., Lee, J.W.: Integrated topology management in flying ad hoc networks: topology construction and adjustment. IEEE Access **6**, 61196–61211 (2018)
13. Korneyev, A., Gorobetz, M., Alps, I., Ribickis, L.: Adaptive traction drive control algorithm for electrical energy consumption minimisation of autonomous unmanned aerial vehicle. Electr. Control. Commun. Eng. **15**(2), 62–70 (2019)
14. Leonov, A.V., Litvinov, G.A.: Simulation-based packet delivery performance evaluation with different parameters in flying ad-hoc network (FANET) using AODV and OLSR. In: Journal of Physics: Conference Series, vol. 1015, p. 032178. IOP Publishing (2018)
15. Litvinov, G., Leonov, A., Korneev, D.: Applying static mobility model in relaying network organization in mini-UAVs based FANET. In: 2018 Systems of Signal Synchronization, Generating and Processing in Telecommunications (SYNCHROINFO), pp. 1–7. IEEE (2018)
16. Mukherjee, A., Keshary, V., Pandya, K., Dey, N., Satapathy, S.C.: Flying ad hoc networks: a comprehensive survey. In: Satapathy, S.C., Tavares, J.M.R.S., Bhateja, V., Mohanty, J.R. (eds.) Information and Decision Sciences. AISC, vol. 701, pp. 569–580. Springer, Singapore (2018). https://doi.org/10.1007/978-981-10-7563-6_59
17. Rahman, M.A., Anwar, F., Naeem, J., Abedin, M.S.M.: A simulation based performance comparison of routing protocol on mobile ad-hoc network (proactive, reactive and hybrid). In: International Conference on Computer and Communication Engineering (ICCCE 2010), pp. 1–5. IEEE (2010)
18. Srivastava, A., Prakash, J.: Future FANET with application and enabling techniques: anatomization and sustainability issues. Comput. Sci. Rev. **39**, 100359 (2021)

19. Sumra, I.A., Sellappan, P., Abdullah, A., Ali, A.: Security issues and challenges in MANET-VANET-FANET: a survey. EAI Endorsed Trans. Energy Web **5**(17), e16–e16 (2018)
20. Yang, H., Liu, Z.: An optimization routing protocol for FANETs. EURASIP J. Wirel. Commun. Netw. **2019**(1), 1–8 (2019)
21. Zafar, W., Khan, B.M.: A reliable, delay bounded and less complex communication protocol for multicluster FANETs. Digit. Commun. Netw. **3**(1), 30–38 (2017)

Solar Panels for Low Power Energy Harvesting

Maritza Nuñez[1], Carlos Gordón[1(✉)], Clara Sánchez[1], and Myriam Cumbajín[2]

[1] GITED Research Group. Facultad de Ingeniería en Sistemas, Electrónica e Industrial, Universidad Técnica de Ambato, UTA, 180207 Ambato, Ecuador
{mnunez4777,cd.gordon,ca.sanchez}@uta.edu.ec

[2] SISAu Research Center, Facultad de Ingeniería y Tecnologías de la Información y la Comunicación, Universidad Tecnológica Indoamérica, UTI, 180103 Ambato, Ecuador
myriamcumbajin@uti.edu.ec

Abstract. Solar panels are widely used nowadays to capture solar radiation and generate voltage, so they are being used for Energy Harvesting applications. The present work carries out the study of low power solar panels for energy storage applications, together with the DC-DC conversion and storage stage. The methodology carried out has been the design, simulation, fabrication and characterization of the elements that form the system. The elements that make up the system are 4 solar panels of 2.4 V and 80 mA, a voltage regulator element and rechargeable batteries. As a result, both in simulation and measurement, the mixed configuration (series-parallel) is the one that provides the best characteristics for its use, with a voltage of 4.57 V and a current of 127.3 mA, obtaining at the converter output a voltage of 19.44 V, concluding that the system meets the design expectations with which it was made, collecting energy, raising it and storing it, providing promising results for future applications.

Keywords: Solar Panel · Low Power · Energy Harvesting

1 Introduction

Currently, energy harvesting elements are a fundamental part for supplying energy to independent devices or systems, besides being an ecological option for the environment, for this reason energy harvesting systems are required in IoT [6,12,17,20]. Nowadays there are several known techniques for energy harvesting [18,19,22], the most known and easy to realize is the derivative of solar panels, arranged in different shapes and distributions according to the needs and coupling of the stages of the system to be realized [11,14].

The architecture of solar energy harvesting systems can have different number of stages as presented in [13,16] and [21] which consist of 3, 4 and 5 stages respectively, where the basic distribution is composed of 3 stages: solar panels,

M. V. Garcia and C. Gordón-Gallegos (Eds.): CSEI 2022, LNNS 678, pp. 296–317, 2023.
https://doi.org/10.1007/978-3-031-30592-4_21

DC-DC converter and storage. The solar panels capture solar energy [7]. The DC-DC converter converts and boosts the input voltage (solar panel) to a stable voltage at its output [5], for storage as the last stage. Other architectures add a fourth MPPT (Maximum Power Point Tracker) regulation stage that tracks the maximum power point, adjusting the switching frequency to obtain maximum power [8,15].

The objective of this work is to design, simulate, fabricate and characterize different configurations of solar panels for an energy harvesting system, identifying the most suitable arrangement to obtain the highest amount of energy with the conversion stage. For this purpose, there are some similar works of low power solar energy harvesting [10], emphasizing that most of the works do not provide all the parameters of operation, design or results obtained in the research [9], therefore, the present work provides, characteristics, calculations, designs and software simulation of the realized system.

This paper presents a solar energy harvesting system composed of 4 mini solar panels in mixed configuration (series-parallel), together with the DC-DC conversion circuit, which raises the measured input voltage of the panel from 4.57 V to 19.44 V in both simulation and measurement, which is stored in rechargeable batteries. The article is distributed as follows: Sect. 2 details the methodology carried out in the design, simulation, fabrication and characterization sections. Section 3 shows the analysis of the results obtained from the simulations and measurements carried out for the panels. Finally, Sect. 4 presents the conclusions of the work.

2 Methodology

The methodology carried out has been the design, simulation, fabrication and characterization of solar panel arrays, together with a DC-DC converter, as shown below:

2.1 Design

The first part of the design is the literature review to establish the stages of a low power solar energy harvesting system, identifying 3 basic stages shown below in Fig. 1.

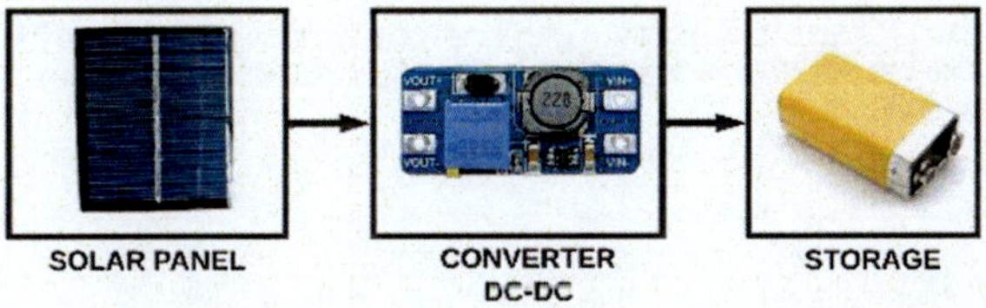

Fig. 1. Stages of a low power energy harvesting system.

As another important part for the design is the selection of the solar panel to be used, for this, after the analysis of different types, the EcoFly Power monocrystalline solar panel has been selected, with 55 mm diameter, high efficiency of 2.4 V, 80 mA and 0.19 W, characteristics described in [1]. The panel is shown in Fig. 2.

Fig. 2. Panel solar EcoFly Power.

In order to select the best panel configuration to be used, 3 types of configurations have been tested: series (3 panels), parallel (3 panels) and mixed (4 panels). Each of the panels has a voltage of 2.4 V and a current of 80 mA.

Series Configuration
The following equations are made for each of the configurations to be simulated:

$$VT = 2.4 + 2.4 + 2.4 = 7.2\,\text{V} \tag{1}$$

$$IT = 80\,\text{mA} \tag{2}$$

$$PT = VT * IT = 7.2\,\text{V} * 80\,\text{mA} = 0.576\,\text{W} \tag{3}$$

$$R = \frac{VT}{IT} = \frac{7.2\,\text{V}}{80\,\text{mA}} = 90\Omega \tag{4}$$

where VT is the total voltage of the panels, IT is the total current, PT is the total power and R is the resistance value used for the simulation of the panels. Figures 3, 4 and 5 show the respective configurations made for each of the calculations.

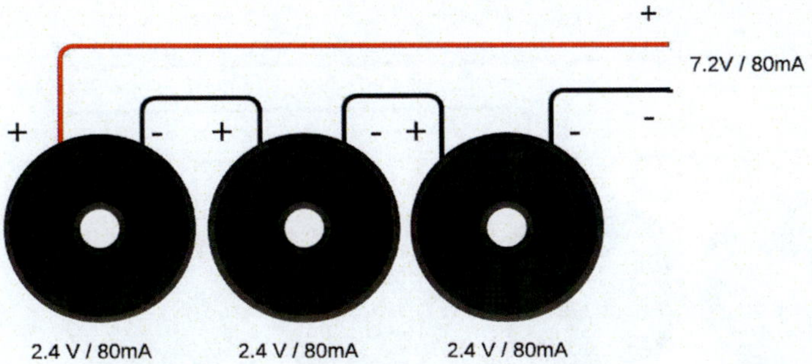

Fig. 3. Series configuration

Parallel Configuration

$$VT = 2.4\,\mathrm{V} \tag{5}$$

$$IT = 80\,\mathrm{mA} + 80\,\mathrm{mA} + 80\,\mathrm{mA} = 0.24\,\mathrm{A} \tag{6}$$

$$PT = VT * IT = 2.4\,\mathrm{V} * 0.24\,\mathrm{A} = 0.576\,\mathrm{W} \tag{7}$$

$$R = \frac{VT}{IT} = \frac{2.4\,\mathrm{V}}{0.24\,\mathrm{A}} = 10\Omega \tag{8}$$

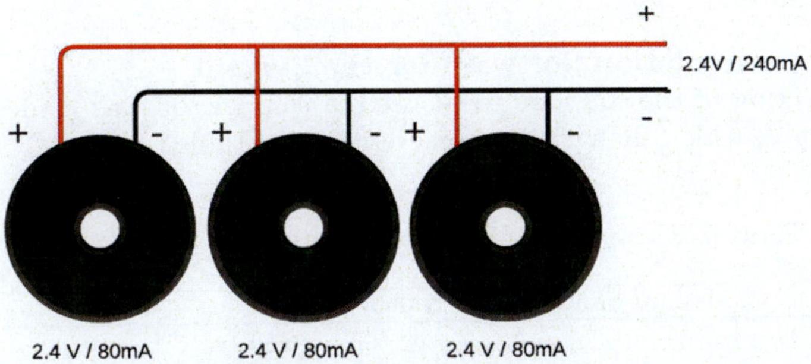

Fig. 4. Parallel configuration

Mixed Configuration

$$VT = 2.4\,\mathrm{V} + 2.4\,\mathrm{V} = 4.8\,\mathrm{V} \tag{9}$$

$$IT = 80\,\mathrm{mA} + 80\,\mathrm{mA} = 160\,\mathrm{mA} \tag{10}$$

$$PT = VT * IT = 4.8\,\mathrm{V} * 160\,\mathrm{mA} = 0.768\,\mathrm{W} \tag{11}$$

$$R = \frac{VT}{IT} = \frac{4.8\,\mathrm{V}}{160\,\mathrm{mA}} = 30\Omega \tag{12}$$

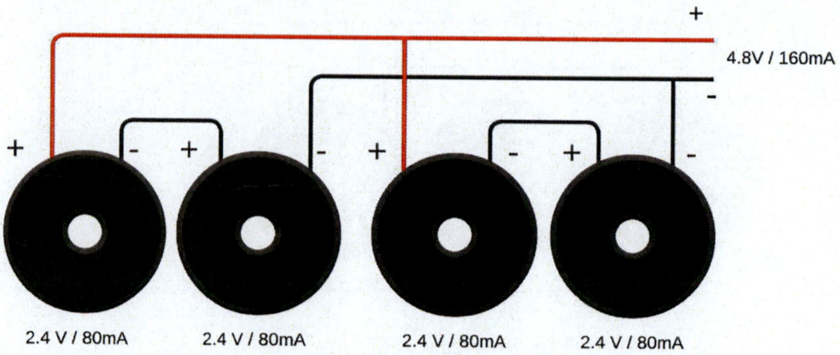

Fig. 5. Mixed configuration

According to the calculations performed, it was determined that the energy collection stage will be composed of 4 solar panels in mixed configuration (series-parallel) as shown in Fig. 5. These panels will provide a voltage of 4.8 V and a current of 160 mA theoretically according to the characteristics of the solar panels.

2.2 Simulation

Selection of Simulation Software for the System

For the selection of the software to be used, a comparison was made between 3 different types, which show their characteristics in Table 1.

Table 1. Comparison of simulation software for solar panels.

	Matlab/ Simulink [3]	Scilab/ Xcos [4]	Homer Pro [2]
Image			
License	Owner	GPL compatible	Owner
Software	Paid out	Free	Paid out
Operating system	Windows/Linux/Mac	Windows/Linux/ Mac	Cloud, SaaS, Web, Windows
Outline Format	Plain text similar to JSON	XML valid	-
Simulation time	10 times greater than Xcos software	10 times smaller than Simulink software	-
Values obtained in simulation	High precision	Medium - high precision	Medium - high precision

The software selected for the simulation of the solar panel is Matlab - Simulink because it is a block diagram environment that is used to design systems with multidomain models, without the need to write code, in addition to being more precise. for the measurement of results in the simulation.

System Simulation

The simulations of the configurations are performed with the calculations obtained in the design.

Series Configuration

Figure 6 shows the simulated voltage, current and power results for the series configuration.

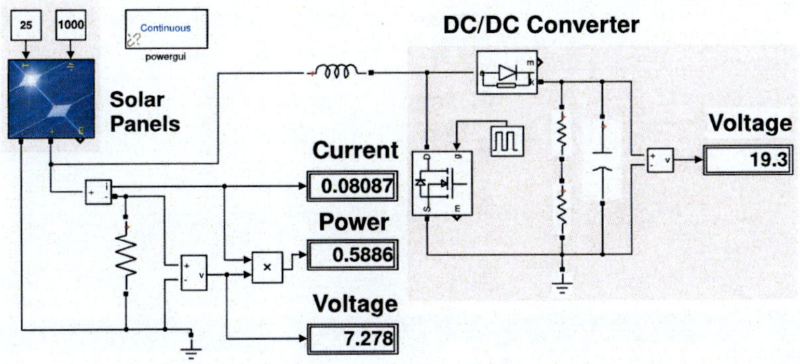

Fig. 6. Simulation of panel configuration in series

Figure 7 shows the configuration within the solar panel parameter block for the series configuration.

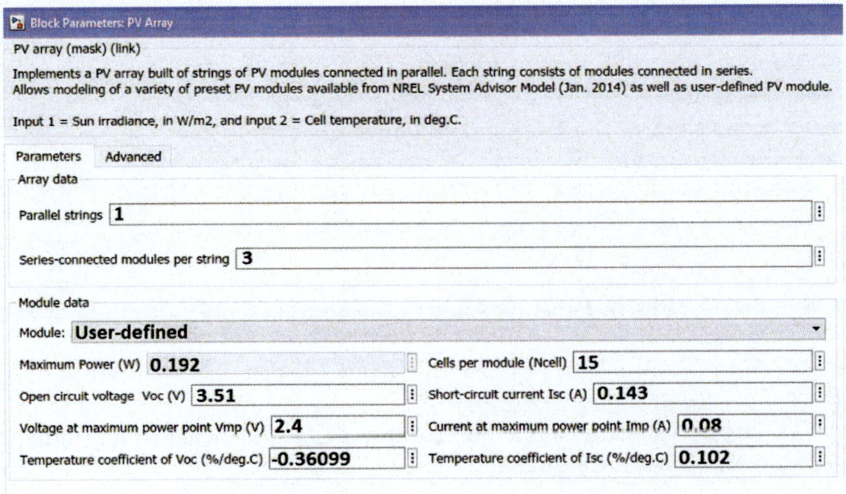

Fig. 7. Series configuration parameter block

Parallel Configuration

The simulation in parallel configuration is shown in Fig. 8 with the respective voltage, current and power results, and the panel configuration parameters in Fig. 9.

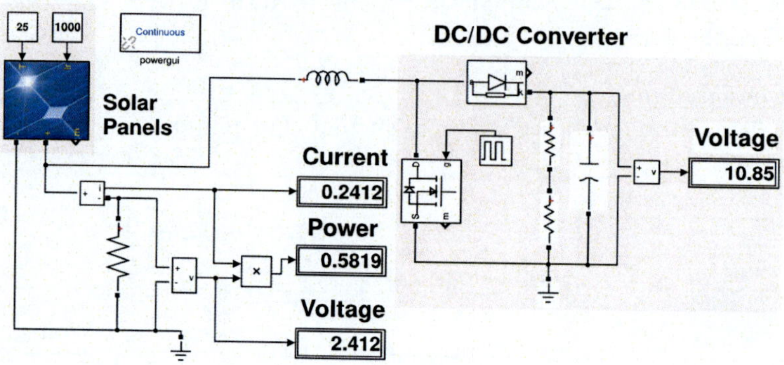

Fig. 8. Simulation of parallel panel configuration

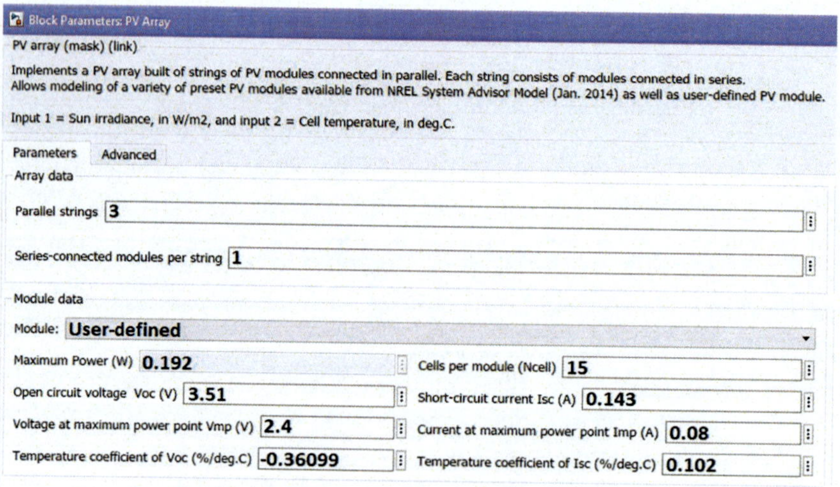

Fig. 9. Panel parameter configuration block

Mixed Configuration

The panel output values, in mixed configuration, are shown in Fig. 10, as well as the parameter block in Fig. 11.

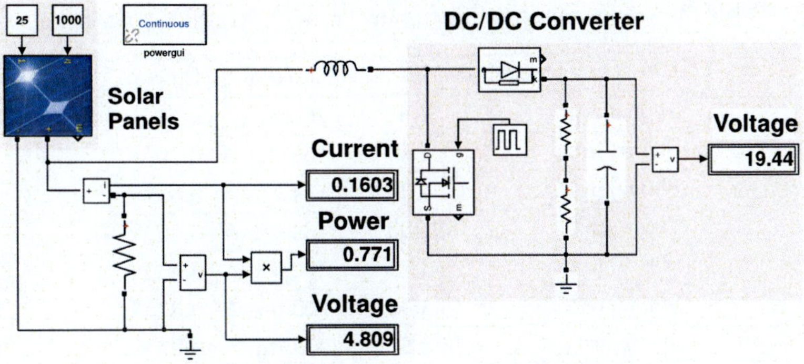

Fig. 10. Simulation of mixed panel configuration

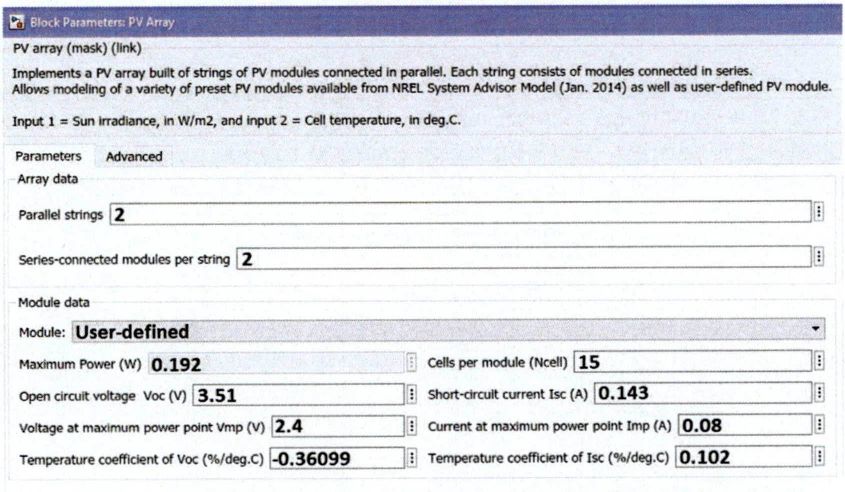

Fig. 11. Parameter block for mixed configuration

In order to visualize the voltage and current variation of the mixed panel configuration, the simulation is performed with different irradiance constants as a representation of the variation of the hours of the day, showing the data obtained in Table 2.

Table 2. Values obtained in simulation with irradiance variation.

Constant Irradiance	Voltage	Current	Power
1000 w/m^2	4.8 V	160 mA	0.77 W
900 w/m^2	4.59 V	153.1 mA	0.70 W
800 w/m^2	4.34 V	144.7 mA	0.63 W
700 w/m^2	4.05 V	134.9 mA	0.55 W
600 w/m^2	3.71 V	123.6 mA	0.46 W
500 w/m^2	3.32 V	110.6 mA	0.37 W
400 w/m^2	2.87 V	95.5 mA	0.27 W
300 w/m^2	2.33 V	77.8 mA	0.18 W
200 w/m^2	1.70 V	56.8 mA	0.09 W
100 w/m^2	0.94 V	31.3 mA	0.03 W

Figure 12 shows the variation of the current with respect to the irradiance, observing that the current varies from 0.01 A to 0.02 A every 1000 W/m^2. Having a current of 160 mA per hour of peak or standard irradiance.

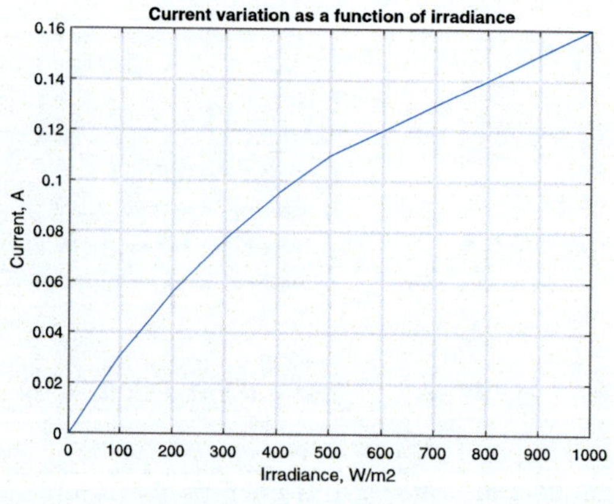

Fig. 12. Variation of current with respect to radiation in simulation

The voltage in simulation as well as the current varies according to the irradiation, as shown in Fig. 13, having as peak voltage 4.8 V at 1000 W/m^2. It should be noted that the 0 W/m^2 irradiation considers the night hours where the panel does not generate any voltage or current.

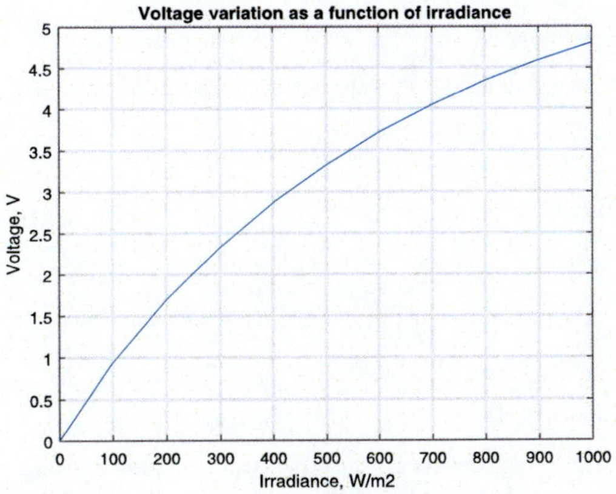

Fig. 13. Variation of voltage with respect to radiation in simulation

2.3 Characterization

The measurements of the practical results are carried out on the physical circuit, using a measuring element, and the values measured for each of the configurations are shown below.

Series Configuration
Table 3 and Fig. 14 show the values obtained from the measurements performed for this configuration.

Table 3. Measurements in series configuration

Voltage	Current
6.87 V	65.9 mA

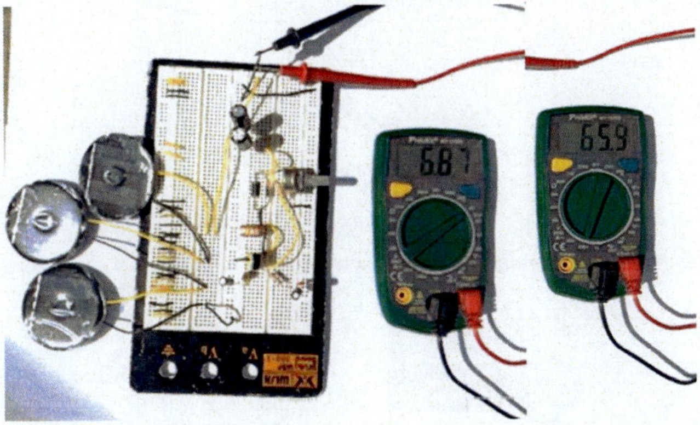

Fig. 14. Measurement of 3 panels in series configuration

Parallel Configuration
The proposed panels in parallels and measurements are shown in Fig. 15 and Table 4 respectively.

Table 4. Measurements in parallel configuration

Voltage	Current
2.34 V	197.7 mA

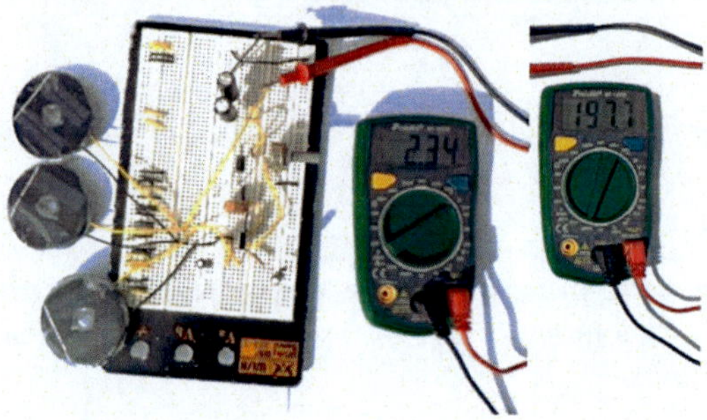

Fig. 15. Measurement of 3 panels in parallel configuration

Mixed Configuration
The mixed circuit adds a panel for its calculations as shown in Fig. 16 and the values obtained are shown in Table 5.

Table 5. Measurements in mixed configuration

Voltage	Current
4.57 V	127.3 mA

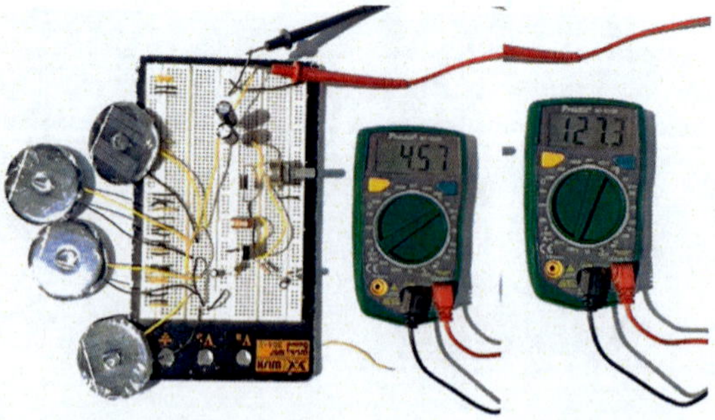

Fig. 16. Measurement of 3 panels in mixed configuration

Additionally, for this configuration, a 24-h characterization was performed for a sunny day and a cloudy day, in order to determine the variation of measurements in both voltage and current. The data obtained for the sunny day are shown in Table 6. It should be noted that measurements have only been obtained from 6:00 am to 18:00 pm because at night the panel does not provide any voltage or current.

Table 6. Measurements obtained for a sunny day

Time	Weather	Voltage	Current	Power
6:00am	12 °C	3.57 V	3.10 mA	0.01 W
7:00am	12 °C	3.70 V	30.97 mA	0.11 W
8:00am	14 °C	4.50 V	98.6 mA	0.44 W
9:00am	16 °C	4.73 V	61.5 mA	0.29 W
10:00am	18 °C	4.54 V	59.6 mA	0.27 W
11:00am	19 °C	4.66 V	136.7 mA	0.64 W
12:00pm	21 °C	4.73 V	138.5 mA	0.66 W
13:00pm	22 °C	4.73 V	139.2 mA	0.66 W
14:00pm	23 °C	4.69 V	113.3 mA	0.53 W
15:00pm	22 °C	4.72 V	107.2 mA	0.51 W
16:00pm	22 °C	4.54 V	36.6 mA	0.17 W
17:00pm	20 °C	4.22 V	11.3 mA	0.05 W
18:00pm	18 °C	4.05 V	5.09 mA	0.02 W

Figure 17 shows the variation of temperature with respect to the hours of sunshine on a sunny day, showing that in the early hours of the day the temperature increases with the hour, reaching a peak temperature in mid-afternoon and then decreasing until the evening hours.

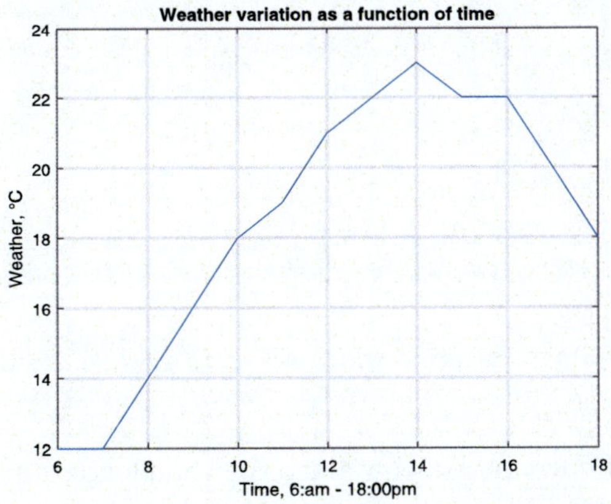

Fig. 17. Temperature variation with respect to the weather on a sunny day

Figure 18 shows the variation of the current with respect to the hours of the day, observing that the highest radiation obtained during the day has been at 11:00 am and 15:00 pm having 0.14 A and 0.114 A respectively, showing that the current varies rapidly.

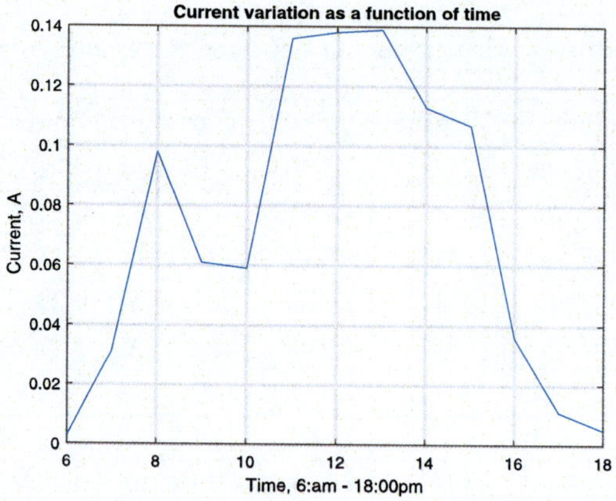

Fig. 18. Variation of the current with respect to the weather on a sunny day

The voltage compared to the current has small variations, remaining without significant changes in the hours of the day as shown in Fig. 19.

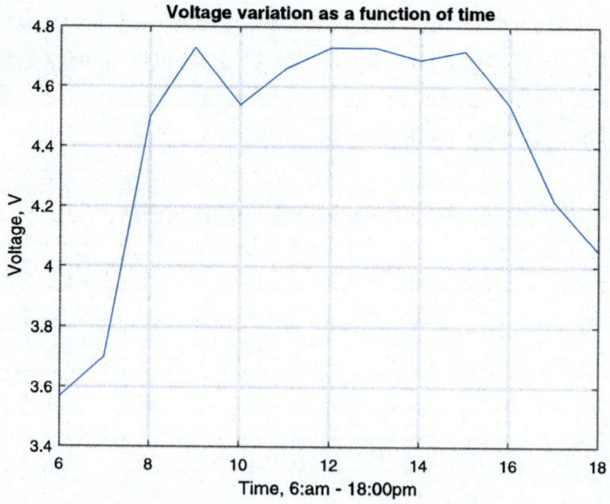

Fig. 19. Voltage variation with respect to time on a sunny day

The voltage and current data obtained for the cloudy day are shown below in Table 7.

Table 7. Measurements obtained for a cloudy day

Time	Weather	Voltage	Current	Power
6:00am	9 °C	1.56 V	0.08 mA	0.12 mW
7:00am	12 °C	1.94 V	0.15 mA	0.29 mW
8:00am	14 °C	4.38 V	37.9 mA	0.17 W
9:00am	14 °C	4.41 V	38.1 mA	0.17 W
10:00am	15 °C	4.33 V	47.6 mA	0.21 W
11:00am	15 °C	4.51 V	57.4 mA	0.26 W
12:00pm	16 °C	4.60 V	125.3 mA	0.58 W
13:00pm	18 °C	4.63 V	90.5 mA	0.41 W
14:00pm	18 °C	4.72 V	65.6 mA	0.31 W
15:00pm	20 °C	4.74 V	114.7 mA	0.54 W
16:00pm	19 °C	4.62 V	89.6 mA	0.41 W
17:00pm	18 °C	4.14 V	25.5 mA	0.11 W
18:00pm	16 °C	3.20 V	0.83 mA	2.66 mW

As the temperature changes for a cloudy day, the diagram is made to visualize how the temperature varies in the hours of the day, observing that the hour of highest temperature of 20 °C is at 15:00 pm and the lowest at 6:00 am with 9 °C, as shown in Fig. 20.

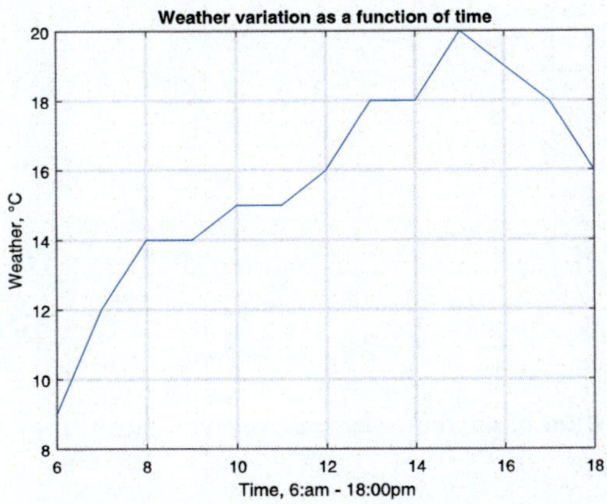

Fig. 20. Temperature variation with respect to the weather on a cloudy day

The voltage measured on a cloudy day as well as on a sunny day does not present large variations in the different hours, having a maximum voltage of 4.74 V at 15:00 pm and the lowest voltage at 6:00 am of 1.56 V. The values are shown in Fig. 21.

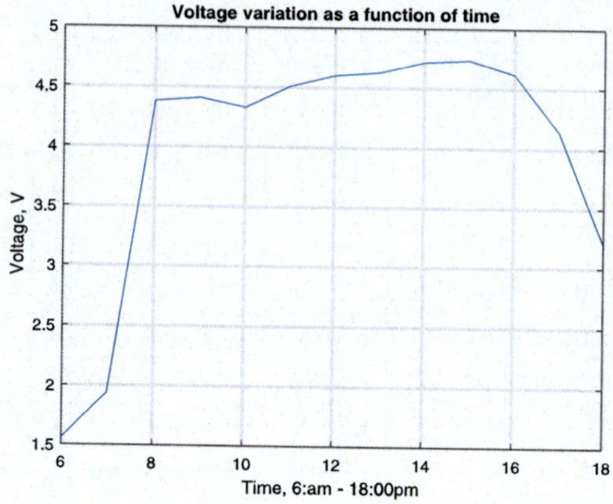

Fig. 21. Voltage variation with respect to weather on a cloudy day

The current compared to the voltage has large variations at all times of the day, having peaks of maximum current at 12:00 pm and 15:00 pm with 16 °C and 20 °C respectively as shown in Fig. 22.

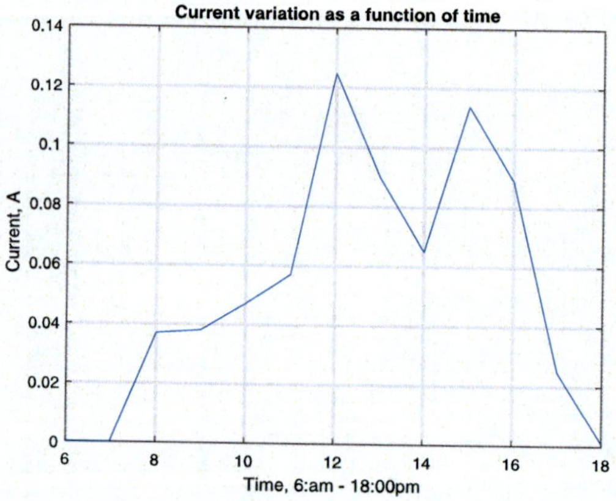

Fig. 22. Variation of current with respect to weather on a cloudy day

2.4 DC-DC Converter Circuit

Design

A DC-DC boost converter is a circuit in charge of transforming and boosting the unstable input voltage into stable output voltage from the voltage generated by the solar panel, it is mainly composed of capacitors, resistors, inductor and integrated circuit. The circuit design has been done using Proteus software, which can be seen in Fig. 23.

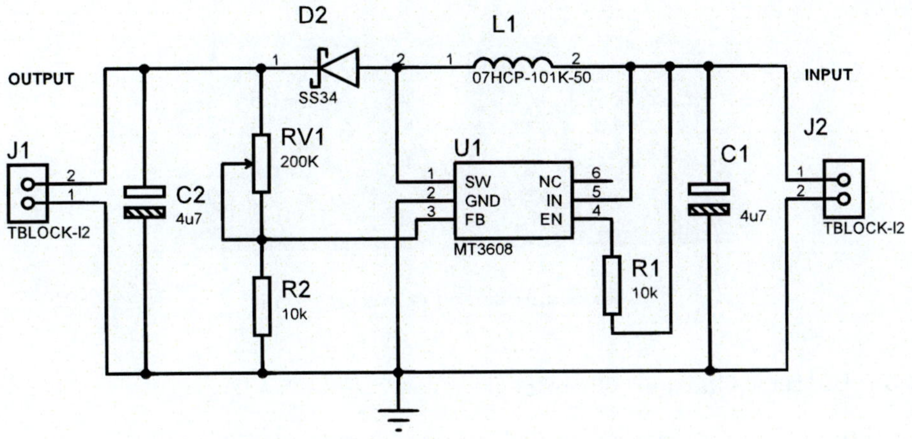

Fig. 23. DC-DC converter circuit design

Manufacture

Once the circuit design is done, we proceed to build it. To do so, we make the PCB circuit in Proteus, selecting the elements, sizes and position of each element of the circuit, as shown in Fig. 24. The circuit is 40 mm wide and 30 mm long.

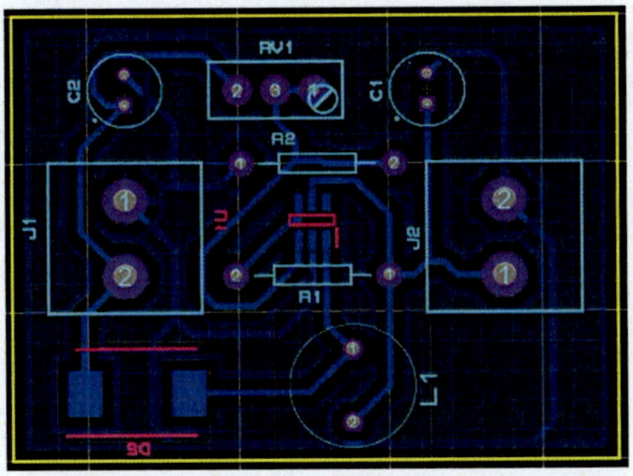

Fig. 24. DC-DC converter PCB circuit

Once the board is fabricated, all components are soldered as appropriate, and the finished and implemented circuit is shown in Fig. 25.

Fig. 25. Implemented converter circuit

2.5 Battery Charger Circuit

Design

For the storage of the energy collected in the rechargeable battery, a higher voltage of 9.5 V is used since the battery to be charged is 9 V. The realized circuit is shown in Fig. 26. It is composed of diodes, resistors, transistor, and relay for its operation.

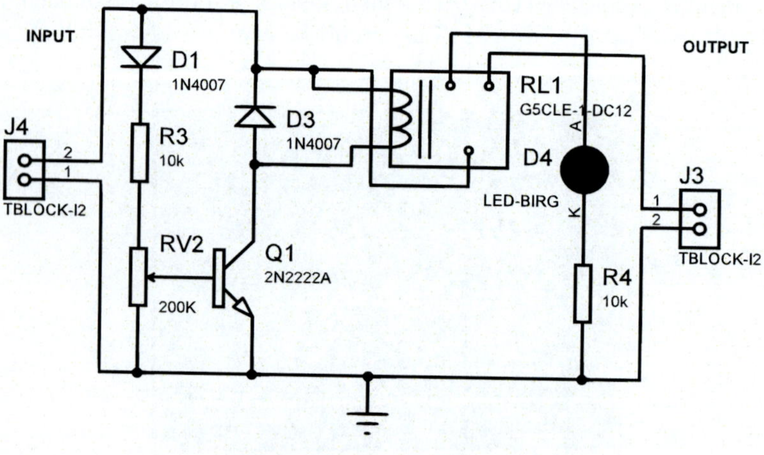

Fig. 26. Battery charger design

Manufacture

With the design made, the PCB of the circuit is generated, aligning the elements as shown in Fig. 27. The circuit is 42.5 mm wide and 35 mm long.

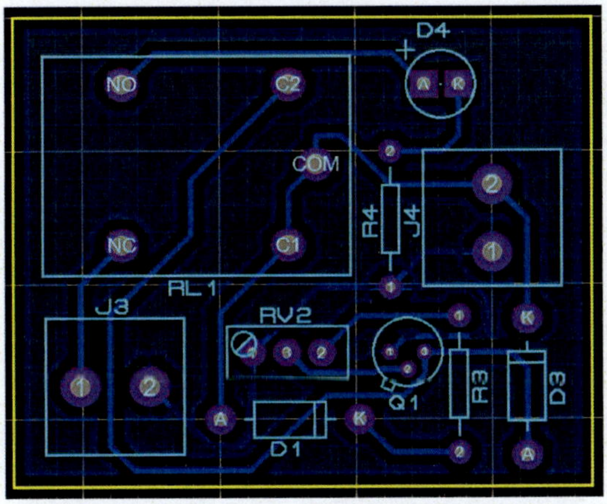

Fig. 27. Battery charger PCB circuit

Finally, the PCB is used to fabricate the board and solder the different elements. Figure 28 shows the implementation of the battery charger circuit.

Fig. 28. Implemented battery charger circuit

3 Analysis of Results

As an analysis of results, the comparison of simulated and measured values of each configuration is performed, in addition to calculating the error percentage for each voltage and current measurement, using Eq. 13.

$$\%error = \frac{|V_{approx} - V_{real}|}{V_{real}} x100\% \tag{13}$$

where: V_{approx} is the value resulting from the simulation and V_{real} is the measured value.

3.1 Series Configuration

The series comparison yielded an error percentage of 0.06% in voltage and 0.21% in current, as shown in Table 8.

Table 8. Characterization of values in series configuration

	Simulation	Measurement	% Error
V_{series}	7.27 V	6.87 V	0.06
I_{series}	0.08 A	0.065 A	0.21

3.2 Parallel Configuration

Table 9 shows the voltage and current errors in parallel configuration of 0.03% and 0.21% respectively.

Table 9. Characterization of values in parallel configuration

	Simulation	Measurement	% Error
$V_{parallel}$	2.41 V	2.34 V	0.03
$I_{parallel}$	0.24 A	0.197 A	0.21

3.3 Mixed Configuration

Table 10 shows the simulation and measurement values of mixed configuration with their respective errors of 0.05% in voltage and 0.26% in current.

Table 10. Characterization of values in mixed configuration

	Simulation	Measurement	% Error
V_{mixed}	4.80 V	4.57 V	0.05
I_{mixed}	0.16 A	0.127 A	0.26

The percentage of error and the measurements performed, allows us to visualize that the current varies with respect to the solar radiation emitted on the solar panels, making the measured values change rapidly compared to the simulation that generates the theoretically calculated values, while the voltage remains almost the same values in simulation and measurement, having a low percentage of error.

4 Conclusions

This paper presents an energy harvesting system designed, simulated, fabricated and characterized, with 4 solar panels in a mixed configuration, together with the DC-DC conversion stage, which raises the voltage of 4.57 V and a current of 127.3 mA to a voltage of 19.44 V for storage in a 9 V rechargeable battery as a third stage. The percentage of error calculated for voltage and current allowed to determine that in the mixed configuration presents an error of 0.05% in voltage and 0.26% in current. The diagrams of practical measurements made showed that the voltage in both a sunny and cloudy day presents small variations during the course of the day, while the current presents greater changes according to the level of radiation received by the solar panels with the change of weather, having peak and low current values, compared to the diagrams obtained in the simulation, where the voltage and current values are directly proportional to the irradiance during the course of the day.

In comparison with the previous simulation works, they do not detail the parameters, characteristics and configuration of the operation of the solar panels used, therefore in this work we provide the calculation parameters and simulations of the array of panels, as a contribution for future researchers.

Acknowledgment. The authors thank the invaluable contribution of the Technological University Indoamerica, for his support in conducting the research project "ESTUDIO DE ALGORITMOS HIBRIDOS DE APRENDIZAJE AUTOMATICO PARA LA PREDICCIÓN DE GENERACIÓN DE ENERGÍAS RENOVABLES", Project Code: 281.230.2022. Also, the authors thank the Technical University of Ambato and the "Dirección de Investigación y Desarrollo" (DIDE) for their support in conducting this research, in the execution of the project "Captación de Energía Limpia de Baja Potencia para Alimentación de Dispositivos de Quinta Generación (5G)", approved by resolution "Nro. UTA-CONIN-2022-0015-R". Project code: SFFISEI 07.

References

1. Ecofly-power store - monocrystalline solar panel (2022). https://es.aliexpress.com/item/1005003206041039.html
2. Homerpro (2022). https://www.homerenergy.com/products/pro/index.html
3. Simulink (2022). https://es.mathworks.com/products/simulink.html
4. Xcos (2022). https://www.scilab.org/software/xcos
5. Baharudin, N., Tunku Mansur, T.M.N., Hamid, F., Ali, R., Irwanto, M.: Topologies of DC-DC converter in solar PV applications. Indones. J. Electr. Eng. Comput. Sci. **8**(2), 368–374 (2017). https://doi.org/10.11591/ijeecs.v8.i2.pp368-374
6. Barona-Oñate, R.V., López-Pérez, S.A., López López, J.P., Mocha-Bonilla, J.A.: Use of E-Learning and audio-lingual method for the development of listening comprehension skills. In: Nummenmaa, J., Pérez-González, F., Domenech-Lega, B., Vaunat, J., Oscar Fernández-Peña, F. (eds.) CSEI 2019. AISC, vol. 1078, pp. 83–98. Springer, Cham (2020). https://doi.org/10.1007/978-3-030-33614-1_6
7. Bhuvaneswari, P., Balakumar, R., Vaidehi, V., Balamuralidhar, P.: Solar energy harvesting for wireless sensor networks. In: 2009 First International Conference

on Computational Intelligence, Communication Systems and Networks, pp. 57–61. IEEE (2009). https://doi.org/10.1109/CICSYN.2009.91

8. Chang, M.C., Liu, S.I.: An indoor photovoltaic energy harvester using time-based MPPT and on-chip photovoltaic cell. IEEE Trans. Circuits Syst. II Express Briefs **67**(11), 2432–2436 (2020). https://doi.org/10.1109/TCSII.2020.2976760
9. Chetto, M.: Optimal scheduling for real-time jobs in energy harvesting computing systems. IEEE Trans. Emerg. Top. Comput. **2**(2), 122–133 (2014). https://doi.org/10.1109/TETC.2013.2296537
10. Dondi, D., Bertacchini, A., Brunelli, D., Larcher, L., Benini, L.: Modeling and optimization of a solar energy harvester system for self-powered wireless sensor networks. IEEE Trans. Ind. Electron. **55**(7), 2759–2766 (2008). https://doi.org/10.1109/TIE.2008.924449
11. Garces-Salazar, A., Manzano, S., Nuñez, C., Pallo, J.P., Jurado, M., Garcia, M.V.: Low-cost iot framework for tele-medicine applications; [plataforma iot de bajo-coste para aplicaciones de telemedicina]. RISTI - Revista Iberica de Sistemas e Tecnologias de Informacao **2021**(E43), 153–166 (2021)
12. Gupta, A.K., Johari, R.: IoT based electrical device surveillance and control system. In: 2019 4th International Conference on Internet of Things: Smart Innovation and Usages (IoT-SIU), pp. 1–5. IEEE (2019). https://doi.org/10.1109/IoT-SIU.2019.8777342
13. Hande, A., Polk, T., Walker, W., Bhatia, D.: Indoor solar energy harvesting for sensor network router nodes. Microprocess. Microsyst. **31**(6), 420–432 (2007). https://doi.org/10.1016/j.micpro.2007.02.006
14. Hassanalieragh, M., Soyata, T., Nadeau, A., Sharma, G.: UR-SolarCap: an open source intelligent auto-wakeup solar energy harvesting system for supercapacitor-based energy buffering. IEEE Access **4**, 542–557 (2016). https://doi.org/10.1109/ACCESS.2016.2519845
15. Jyothy, L.P., Sindhu, M.: An artificial neural network based MPPT algorithm for solar PV system. In: 2018 4th International Conference on Electrical Energy Systems (ICEES), pp. 375–380. IEEE (2018). https://doi.org/10.1109/ICEES.2018.8443277
16. Mabon, M., Gautier, M., Vrigneau, B., Le Gentil, M., Berder, O.: The smaller the better: designing solar energy harvesting sensor nodes for long-range monitoring. Wirel. Commun. Mob. Comput. **2019**, 1–11 (2019). https://doi.org/10.1155/2019/2878545
17. Naranjo, J.E., Robalino-López, A., Alarcon-Ortiz, A., Peralvo, A.E., Romero, R.J., Garcia, M.V.: Augmented reality system for teaching mathematics during covid19's times; [sistema de realidad aumentada para la enseñanza de matemática en tiempos de covid-19]. RISTI - Revista Iberica de Sistemas e Tecnologias de Informacao **2021**(E42), 510–521 (2021)
18. Páez-Quinde, C., Iza-Pazmiño, S., Morocho-Lara, D., Hernández-Domínguez, P.: Gamification resources applied to reading comprehension: projects of connection with society case study. In: Garcia, M.V., Fernández-Peña, F., Gordón-Gallegos, C. (eds.) Advances and Applications in Computer Science, Electronics, and Industrial Engineering. CSEI 2021. LNNS, vol. 433, pp. 205–218. Springer, Cham (2022). https://doi.org/10.1007/978-3-030-97719-1_12
19. Rico-Bautista, D., Medina-Cardenas, Y., Coronel-Rojas, L.A., Cuesta-Quintero, F., Maestre-Gongora, G., Guerrero, C.D.: Smart university: key factors for an artificial intelligence adoption model. In: García, M.V., Fernández-Peña, F., Gordón-Gallegos, C. (eds.) Advances and Applications in Computer Science, Electronics

and Industrial Engineering. AISC, vol. 1307, pp. 153–166. Springer, Singapore (2021). https://doi.org/10.1007/978-981-33-4565-2_10
20. Sanislav, T., Mois, G.D., Zeadally, S., Folea, S.C.: Energy harvesting techniques for internet of things (IoT). IEEE Access **9**, 39530–39549 (2021). https://doi.org/10.1109/ACCESS.2021.3064066
21. Sharma, H., Haque, A., Jaffery, Z.A.: Modeling and optimisation of a solar energy harvesting system for wireless sensor network nodes. J. Sens. Actuator Netw. **7**(3), 40 (2018). https://doi.org/10.3390/jsan7030040
22. Stojčev, M.K., Kosanović, M.R., Golubović, L.R.: Power management and energy harvesting techniques for wireless sensor nodes. In: 2009 9th International Conference on Telecommunication in Modern Satellite, Cable, and Broadcasting Services, pp. 65–72. IEEE (2009). https://doi.org/10.1109/TELSKS.2009.5339410

Effect of Slots in Rectangular Geometry Patch Antennas for Energy Harvesting in 2.4 GHz Band

Danny Merino, Carlos Gordón(✉), Julio Cuji, and Freddy Robalino

GITED Research Group, Facultad de Ingeniería en Sistemas, Electrónica e Industrial, Universidad Técnica de Ambato, UTA, 180207 Ambato, Ecuador
{dmerino2235,cd.gordon,julioecuji,frobalino}@uta.edu.ec

Abstract. This article presents a study about the effects caused by inserting slots in the patch of a rectangular geometry microstrip antenna, by analyzing the graph of the parameter S11 resulting from several simulations conducted with the help of the electromagnetic systems simulation software CST STUDIO SUITE. The microstrip antenna was designed for a resonance frequency $Fr = 2.4$ GHz, with a fiberglass reinforced epoxy resin composite (FR4) substrate of $\epsilon_r = 4.3$ and $h = 1.465$ mm. It is observed in the simulation that the antenna initially works at 2.34 GHz, which represents a difference of 0.06 GHz with respect to the Fr for which it was designed. Simulations were performed adding vertical and horizontal slots of different lengths in the patch, observing that the vertical slots influence the harmonic frequencies of the antenna, displacing them towards the lower side, as the length of the slot increases. With these results, it was possible to design and manufacture a slotted patch antenna for Energy Harvesting of radiofrequency energy in the 2.4 GHz band. words.

Keywords: Antenna · Slotted · Microstrip · Energy Harvesting · Radiofrequency

1 Introduction

In recent years, the use of 5G technology has been gradually increasing due to the great features it has compared to the previous 4G generation [10], especially for Internet of Things IoT applications, features such as: energy efficiency, Reduced latency, faster speeds, among others, make more and more industries lean towards the development of IoT devices based on 5G technology. By 2030, it is estimated that there will be between 40 and 140 billion IoT devices in the world [8], operating in all kinds of applications such as: smart homes, wearables, autonomous vehicles, agriculture, wireless sensor networks, etc. [18].

A great advantage of IoT devices with 5G technology is their low power consumption, even so power is one of the critical elements in IoT systems, because most devices are wireless, for that reason makes the use of batteries necessary,

M. V. Garcia and C. Gordón-Gallegos (Eds.): CSEI 2022, LNNS 678, pp. 318–332, 2023.
https://doi.org/10.1007/978-3-031-30592-4_22

which can represent a problem because the useful lifetime of a battery is limited [4,13,17] and can be reduced depending on several factors such as its time of use, inadequate maintenance, charge and discharge cycles in rechargeable batteries, among others [3,5,9]. in the articles "Energy Efficiency in Wireless Sensor Networks: a top-down survey" [15] and "Energy Efficiency in Smart Buildings: IoT Approaches" [11,14] discusses the implications of this problem and analyzes Energy-efficient mechanisms such as photovoltaic systems, sources of energy harvesting, Sleep/wakeup schemes, among others. Therefore, in recent years, multiple investigations have been conducted focused on ambient energy harvesting, which consists of capturing energy from the environment and converting it into electrical energy. For low-power applications, various investigations have focused on the study of radiofrequency as a source for energy harvesting, because this type of signal prevails all the time in the environment such as TV, radio, and cell phone signals, satellite, WiFi, among others [7].

A radiofrequency energy harvesting system is divided into five stages: Antenna, Coupling, Rectification, Energy Management, and Energy Storage [1,20]. where the most important is the Antenna which captures the electromagnetic waves present in the environment and converts them into electrical energy [12]. Of the various types of existing antennas for energy harvesting, the most used are the Patch or Microstrip type. in the same way within this type of antennas there is a great variety of geometries such as: square, rectangular, circular, elliptical, triangular, among others [2]. The purpose of this article is to determine with a simulation software, how it influences the results of an antenna, if slots of different shapes and sizes are added to its geometry, starting from a simple design of a common patch type antenna and focusing on the characteristics ideal for energy harvesting. This seeks to contribute to future research, with new design options for microstrip antennas that, based on the results obtained during this study, can be applied for different purposes.

2 Methodology

2.1 Antenna Design

For the design of the patch-type antenna, different parameters must be considered, such as: the resonance frequency Fr, the dielectric constant ϵ_r, the thickness of the substrate h and the thickness of the copper layer hc [1], with these data the mathematical calculation can be carried out. To determine the dimensions that the patch antenna is going to have, for the present investigation the resonance frequency is taken as $Fr = 2.4\,\text{GHz}$, the substrate material is FR4 (fiberglass reinforced epoxy resin composite), which has a dielectric constant $\epsilon_r = 4.3$ with a thickness $h = 1.465$ and a thickness of the copper layer $hc = 0.035\,\text{mm}$. For the mathematical calculation were used the Eqs. 1, 2, 3 and 4.

Patch Width

$$W = \frac{c}{2Fr} * \sqrt{\frac{2}{\epsilon_r + 1}} \tag{1}$$

$$W = \frac{30}{2(2.4)} * \sqrt{\frac{2}{4.3 + 1}}$$

$$W = 3.839\,\text{cm}$$

Effective Dielectric Constant

$$\epsilon_{reff} = \frac{\epsilon_{reff} + 1}{2} + \frac{\epsilon_{reff} + 1}{2} * (1 + 12\frac{h}{W})^{-\frac{1}{2}} \tag{2}$$

$$\epsilon_{reff} = \frac{4.3 + 1}{2} + \frac{4.3 + 1}{2} * (1 + 12\frac{0.1465}{3.839})^{-\frac{1}{2}}$$

$$\epsilon_{reff} = 4.128$$

Patch Length

$$\Delta L = (h)0.412 * \frac{(\epsilon_{reff} + 0.3)(\frac{W}{h} + 0.264)}{(\epsilon_{reff} - 0.258)(\frac{W}{h} + 0.8)} \tag{3}$$

$$\Delta L = (h)0.412 * \frac{(4.128 + 0.3)(\frac{3.839}{0.1465} + 0.264)}{(4.128 - 0.258)(\frac{3.389}{0.1465} + 0.8)}$$

$$\Delta L = 07637$$

$$L = \frac{c}{2(Fr)(\sqrt{\epsilon_{reff}})} - 2\Delta L \tag{4}$$

$$L = \frac{30}{2(2.4)(\sqrt{4.128})} - 2(0.07637)$$

$$L = 2.923\,\text{cm}$$

where W = patch width, c = speed of light (3×10^8 m/s), F_r = Resonant frequency, ϵ_r = Dielectric constant of the substrate, ϵ_{reff} = Effective dielectric constant of the patch, h = substrate thickness, and L = Patch length.

Once the calculations of the length and width dimensions of the patch have been made, Fig. 1 shows the final design of the patch antenna with rectangular geometry made in CST ESTUDIO SUITE. The designed antenna has a rectangular copper patch with the calculated dimensions $W * L$, at the rear of the antenna is the ground plane in which the copper layer covers the entire width and length of the FR4 substrate.

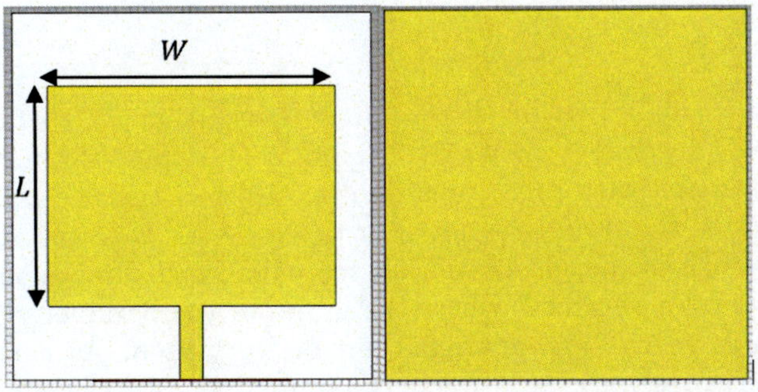

Fig. 1. Design of the patch antenna in CST, front view and rear view.

2.2 Slots

The creation of different types of slots in the plane of the patch of the designed antenna is proposed, starting from a 2 mm square slot in the center as shown in Fig. 2 and modifying their lengths both in the X axis As in the Y axis, the effect of adding more slots to the design is also studied, always keeping the symmetry between them. In this way, it is determined how each type of slot influences the results obtained from the initial patch antenna.

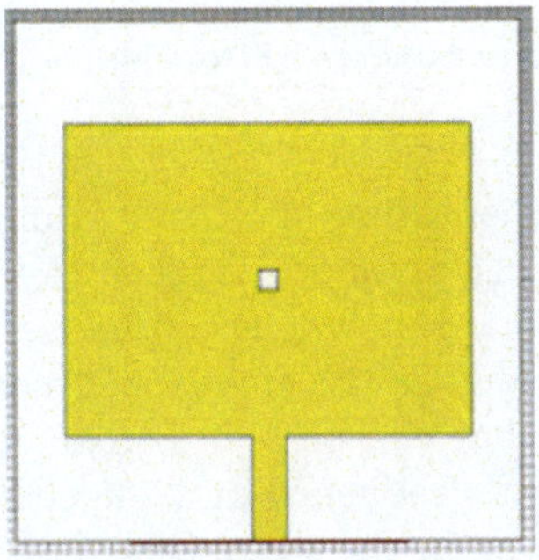

Fig. 2. Patch antenna with square slot in the center.

2.3 Study Parameters

The parameters considered for this study focus on the resonance frequency for which the antenna was designed, which is 2.4 GHz, and the parameter S11, which measures the reflection coefficient of the power entering the antenna [6,19]. It should be considered that the study is carried out focusing on the design of an optimal antenna for a radiofrequency energy harvesting system, so the parameters sought are somewhat different from those of an antenna directed to the radio transmission.

2.4 Simulation

To perform the simulation of the patch antenna, the CST STUDIO SUITE software is used, because it allows the design, analysis and optimization of electromagnetic systems with great precision, it also has the necessary simulation tools to determine the results required for the study. In this way, once the initial parameters were introduced, as well as the calculated dimensions, the results shown in Fig. 3 were obtained, where the graph of the parameter S11, obtained for the common patch type antenna, can be visualized. As can be seen, the antenna initially presents a variation in the simulated resonance frequency, which is 2.34 GHz, compared to the designed 2.4 GHz. In addition, there are two harmonics of the fundamental frequency, resonating at 3.69 GHz and 4.56 GHz.

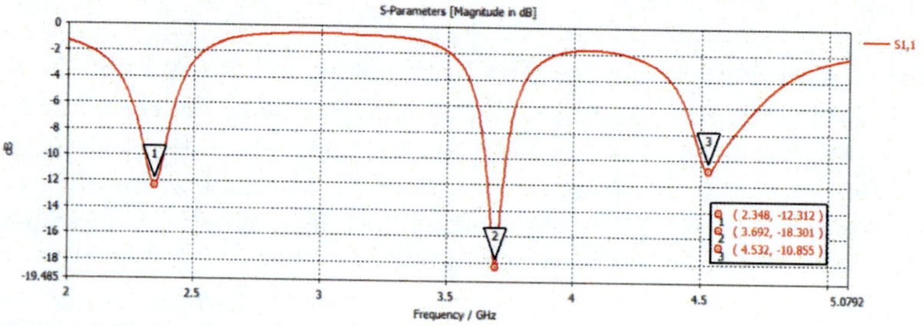

Fig. 3. Graph of the parameter S11 of the common patch antenna.

It starts with the implementation of a 2mm square slot in the center of the patch, this slot is the starting point for further testing. In Fig. 4. The results obtained from the simulation are shown, as can be seen in the graph of parameter S11, with this slot a large change in the resonance frequency of the antenna is not perceived, the only change that can be seen is a small increase in the reflection coefficient for the fundamental frequency of 2.348 GHz and the first harmonic of 3.69 GHz and a slight decrease in the second harmonic of 4.52 GHz.

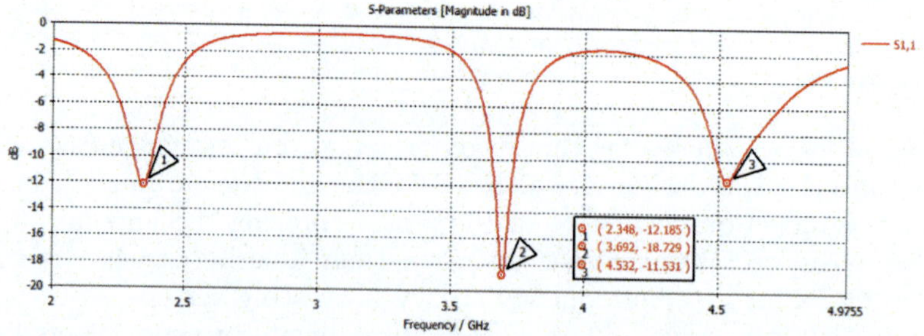

Fig. 4. Results obtained with a 2 mm square slot in the patch center.

For the following simulation, the length Lr of the slot is increased and tests are performed for 3 different values of Lr: 8 mm, 12 mm y 20 mm. Figure 5 shows the results obtained, as can be seen for the resonance frequency that the 2.34 GHz antenna was designed for, no significant changes are recorded, more than small differences in the reflection coefficient. In the first harmonic, which was initially 3.69 GHz, a slight increase can be seen with each iteration, reaching 3.73 GHz with an Lr of 12 mm, likewise, the reflection coefficient decreases in each iteration to −21.82 dB with an Lr of 12 mm. Finally, for the second harmonic that was at 4.53 GHz, a more significant change in the reflection coefficient can be seen with an Lr of 20 mm, the coefficient increased to −9.93 dB,

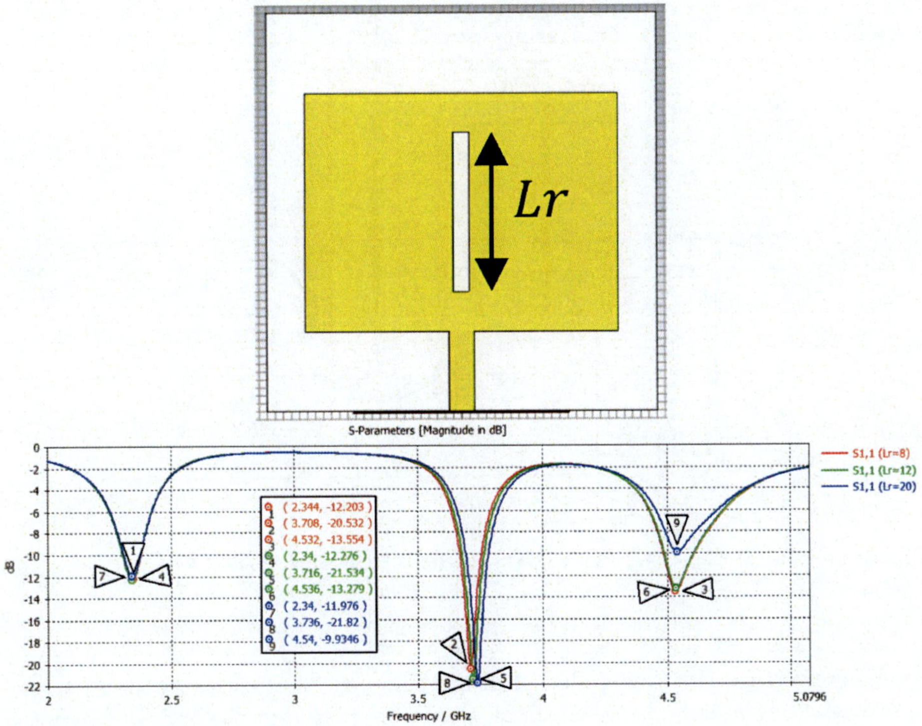

Fig. 5. Results obtained by varying the length of the slot by 8 mm, 12 mm and 20 mm.

Figure 6 shows the results obtained by varying the width of the slot Wr by 8mm, 12 mm and 20 mm. As can be seen in the graph of parameter S11, the results vary for the designed fundamental frequency of 2.34 GHz, decreasing the frequency and increasing the reflection coefficient in each iteration. Likewise, for the third harmonic, the resonance frequency decreases in each iteration from 4.52 GHz in the antenna without a slot to 4.17 GHz with a slot width Wr of

20 mm. Finally, it is observed that for this type of slot, the frequency that does not present significant variations is the one that belongs to the second harmonic of 3.69 GHz.

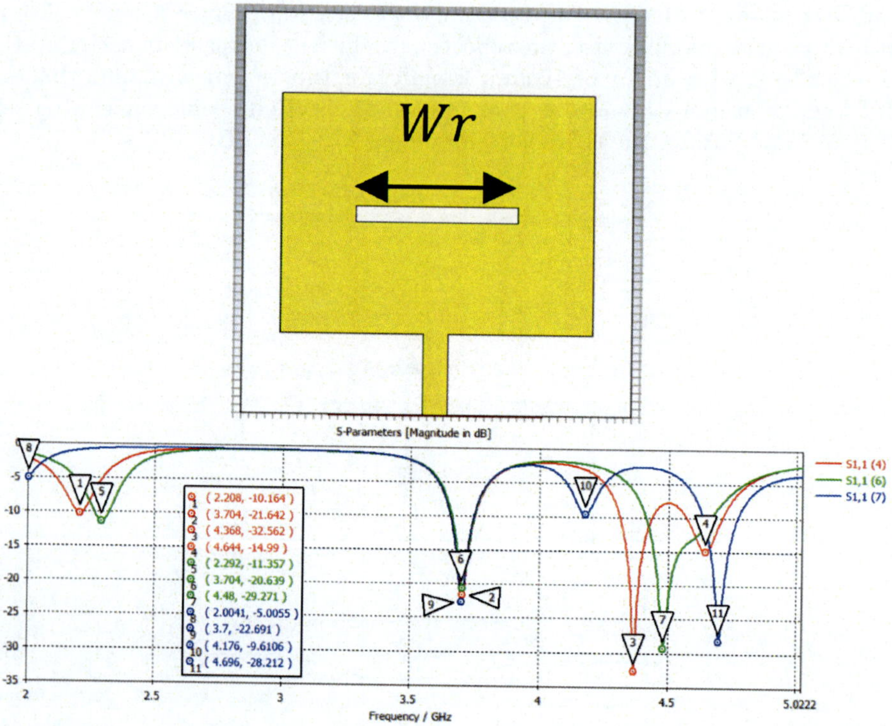

Fig. 6. Results obtained by varying the width of the slot Wr by 8 mm, 12 mm and 20 mm.

A new vertical slot was added with the same characteristics as the first one, to preserve a symmetry between them, they were placed at the ends of the patch on the X axis, leaving 2 mm of distance between each slot and each end. In Fig. 7. The results of the simulation are shown, varying the length Lr of the slots in the same way as in the previous ones. It can be seen in the graph how the fundamental frequency does not present significant changes, while the first and second harmonics decrease their resonance frequency with each iteration in the first going from 3.69 GHz to 2.93 GHz and the second with a smaller variation., from 4.53 GHz to 4.45 GHz. In addition, it is observed how in the second and third iteration a new harmonic appears that follows the pattern of its frequency decrease as the length of the slot increases.

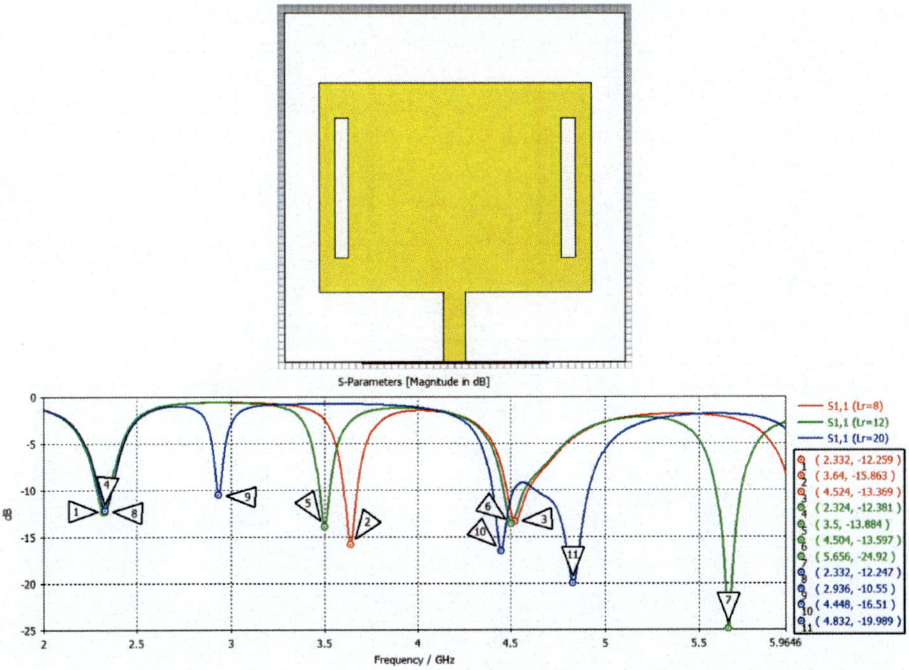

Fig. 7. Results obtained when adding a new slot identical to the first.

Two new slots are added to the antenna patch at 2 mm from the previous ones and with a new length Ly that will be the one that varies by 8 mm, 12 mm and 20 mm, in Fig. 8 the results of this are shown. new design. As can be seen with each iteration the frequencies of the harmonics continue to decrease this time with the second harmonic reaching 2.58 GHz and −14.99 dB.

As what is required is to center the resonance frequency of the antenna at 2.4 GHz for the capture of radiofrequency energy in the WiFi band, so, according to the data obtained through the simulations, what was done was to continue increasing the length Lr of the end slots so that the second harmonic of the fundamental frequency continues to decrease until the desired frequency is obtained. In this way, in Fig. 9. The results obtained with a length Lr of 24 mm are shown, as can be seen in the graph, the frequency of the second harmonic was reduced to 2.43 GHz and its bandwidth was unified with the fundamental frequency. 2.34 GHz, so taking a resonance threshold of −10 dB as a reference, the antenna would work from 2.29 GHz to 2.46 GHz, which makes it perfect to be applied in an energy harvesting system that works in the WiFi band.

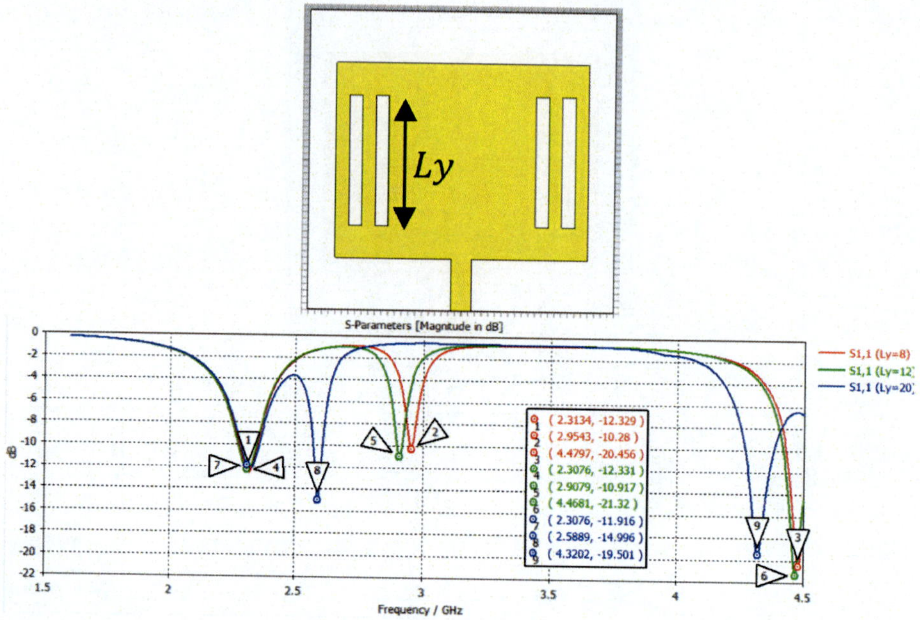

Fig. 8. Results obtained with 4 slots by varying the length Ly of the central slots.

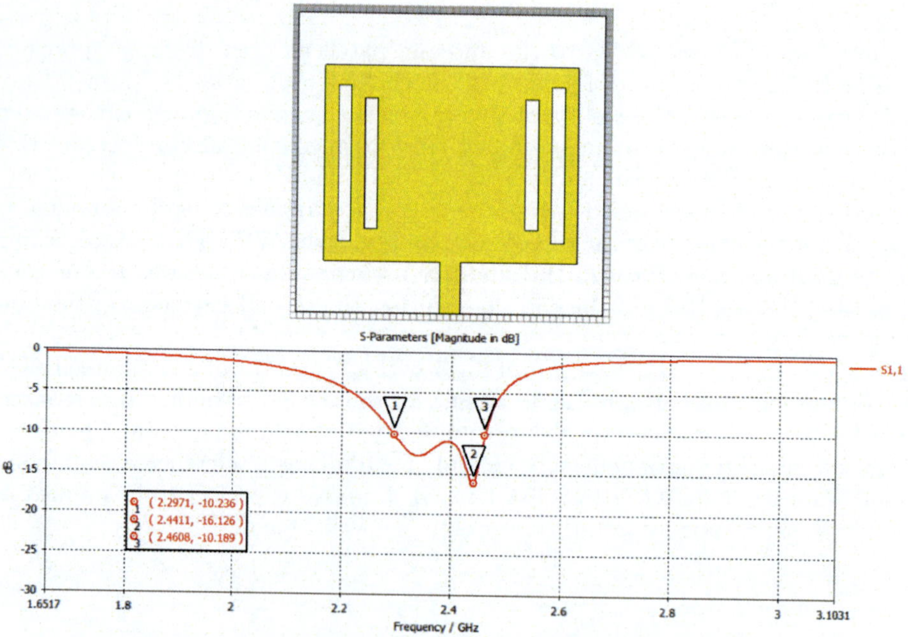

Fig. 9. Results obtained with slot lengths Lr = 24 mm and Ly = 20 mm.

2.5 Fabrication

To manufacture the antenna, the last design was exported to a Drawing Exchange Format (DXF) file, which was transformed to a PNG image and the fill color and size of the image were edited using Photoshop software. Once the image had the design characteristics, it was printed on special thermal transfer paper for PCB manufacturing. Using the ironing technique, the design was transferred to the PCB and introduced in ferric chloride to remove the excess copper. in this way the antenna shown in Fig. 10 was obtained. Additionally, a 50 ohm RP-SMA con-nector was soldered.

Fig. 10. Manufactured slotted patch antenna: Front view and Rear view

2.6 Characterization

Once the antenna is manufactured, we proceed to measure the parameters using the vector network analyzer NanoVNA V2.2, this device allows to measure and graphically visualize the parameters, S11, S21, Impedance, SWR, among others. Figure 11 shows the fabricated antenna connected to channel 0 of the analyzer for its measurements. It also allows to connect it to the computer through its software called Vna_qt, in which the data of the fabricated patch antenna can be analyzed more precisely and it also allows exporting the results obtained in a Touchstone type file that is used to graph it in Matlab and compare it with the simulated results obtained with the CST Studio Suite software.

Fig. 11. Manufactured slotted patch antenna connected to NanoVNA analyzer.

Figure 12 shows the graph of S11 parameter, measured with the NanoVNA V2 analyzer (red line) and simulated using the CST Studio Suite software (blue line). Taking as reference a reflection coefficient threshold of -10 dB, the results show that the antenna works between the frequencies of 2.396 GHz to 2.457 GHz, with a resonance frequency at 2.428 GHz at −15.7 dB.

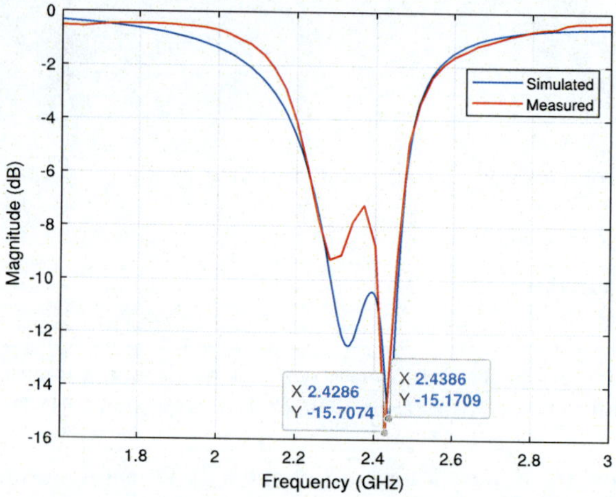

Fig. 12. Graph of S11 parameter measured vs simulated.

3 Analysis of Results

Once the respective simulations are completed, the data obtained is taken and organized in Table 1. Where the number and orientation of the slots are shown, the dimensions of Wr, Lr and Ly with which the tests were conducted and finally the results obtained from parameter S11, according to the variations that were presented both in the fundamental design frequency and in the first and second harmonics.

Table 1. Results obtained from the simulation of the different slots.

No of slots	Orientation	Slot dimensions			Resonance frequency and reflection coefficient (GHz, dB)		
		Wr (mm)	.Lr (mm)	Ly (mm)	Fundamental frequency	First harmonic	Second harmonic
0	–	–	–	–	2.34, −12.31	3.69, −18.30	4.53, −10.85
1	Squared	2	2	–	2.34, −12.18	3.69, −18.72	4.52, −11.53
		8	2	–	2.29, −11.35	3.70, −20.63	4.48, −29.27
1	Horizontal	12	2	–	2.20, −10.16	3.70, −21.64	4.36, −32.56
		20	2	–	<2, −5.21	3.7, −22.69	4.17, −9.61
		2	8	–	2.34, −12.20	3.70, −20.53	4.53, −13.55
1	Vertical	2	12	–	2.34, −12.27	3.71, −21.53	4.53, −13.27
		2	20	–	2.34, −11.976	3.73, −21.82	4.54, −9.93
		2	8	–	2.33, −12.25	3.64, −15.86	4.52, −13.36
2	Vertical	2	12	–	2.32, −12.38	3.5, −13.88	4.5, −13.59
		2	20	–	2.33, −12.24	2.93, −10.50	4.44, −16.51
		2	20	8	2.31, −12.32	2.95, −10.28	4.47, −20.45
4	Vertical	2	20	12	2.30, −12.33	2.90, −10.91	4.46, −21.32
		2	20	20	2.31, −12.25	2.58, −14.99	4.32, −19.501
		2	24	20	2.33, −12.78	2.43, −16.2	4.09, −28.138

When simulating the rectangular patch antenna, a difference of 0.06 GHz is observed between the design frequency of 2.4 GHz and the one obtained with the simulation, which is 2.34 GHz. In addition, the existence of two harmonics, one at 3.69 GHz and another at 4.53 GHz, the three frequencies are below the threshold of −10 dB of reflection coefficient, so the antenna would work for all three fre-quencies without problems [16].

According to the data obtained from the simulations, with a horizontal slot, the result is the displacement towards the lower side of the fundamental frequency and the second harmonic, keeping the frequency of the first harmonic unchanged, these results are not favorable for research, since what is sought is for the anten-na to work in the 2.4 GHz band. With a vertical slot it is observed that the most significant change occurs in the second harmonic where according to as the length of the slot increases, the reflection coefficient also increases. By increasing the number of slots to 2 identical ones at the ends of the patch, the result was a shift towards the lower side of the harmonic frequencies as the length of the slots increased and keeping the frequency unchanged. 2.34 GHz design funda-mental. However, this offset is limited in that the length of the slot cannot exceed the length of the patch. What was solved by adding two new slots of different lengths, this caused the harmonic frequencies to move more allowing in this way that by increasing the length of the end slots by only 4 mm, the first harmonic moves up to 2.43 GHz, at −16.2 dB, which is what is sought for the collection of radiofrequency energy in the 2.4 GHz band.

With the data obtained from the vector network analyzer, the error percentage between the simulation values and the measured values is calculated using Eq. 5. The error percentage of the measured resonance frequency with respect to the simulated resonance frequency is 0.41%, while the error for the reflection coefficient is 3.49%.

$$\%error = \frac{V_{approx} - V_{real}}{V_{real}} * 100 \qquad (5)$$

4 Conclusions

In this article, the effects caused by adding slots of different sizes in the patch of a microstrip antenna were analyzed for both the resonance frequency and the reflection coefficient, with the help of the graph of the parameter S11 resulting from simulations made in the CST STUDIO SUITE software. It was found that inserting 2 identical vertical slots in an antenna patch does not significantly affect to the fundamental resonance frequency for which it was designed, however, it causes the harmonic frequencies to shift toward the bottom side when the length of the slots increases. Starting from this result, it was possible to obtain a slotted antenna design that works according to the characteristics that are sought for the radiofrequency energy harvesting in the 2.4 GHz band. The manufactured slotted patch antenna works at a resonance frequency of 2.428 GHz which represents an error rate of only 0.41% with respect to the simulation, for the reflection coefficient the error rate is 3.49%, this is due to possible errors during manufacturing such as: the dimensions of the substrate or the tin used to solder the RP-SMA connector.

Acknowledgements. The authors thank the Technical University of Ambato and the "Dirección de Investigación y Desarrollo" (DIDE) for their support in carrying out this research, in the execution of the project "Captación de Energía Limpia de Baja Potencia para Alimentación de Dispositivos de Quinta Generación (5G)", approved by resolution "Nro. UTA-CONIN-2022-0015-R". Project code: SFFISEI 07.

References

1. Ali, M., Albasha, L., Qaddoumi, N.: RF energy harvesting for autonomous wireless sensor networks. In: 2013 8th International Conference on Design & Technology of Integrated Systems in Nanoscale Era (DTIS), pp. 78–81 (2013). https://doi.org/10.1109/DTIS.2013.6527782

2. Balanis, C.A.: Antenna Theory: Analysis and Design. Wiley-Interscience (2005)
3. Balarezo, D., Gordón, C., Cuji, J., Salazar, F.: Conditioning system for an electromagnetic energy collection device. In: Garcia, M.V., Fernández-Peña, F., Gordón-Gallegos, C. (eds.) CSEI 2021. LNCS, vol. 433, pp. 219–234. Springer, Cham (2022). https://doi.org/10.1007/978-3-030-97719-1_13
4. Bologna, J.K., Garcia, C.A., Ortiz, A., Ayala, P.X., Garcia, M.V.: An augmented reality platform for training in the industrial context. In: IFAC-PapersOnLine, vol. 53, pp. 197–202 (2020). https://doi.org/10.1016/j.ifacol.2020.11.032
5. Caiza, G., Nuñez, A., Garcia, C.A., Garcia, M.V.: Human machine interfaces based on open source web-platform and OPC UA. Proc. Manuf. **42**, 307–314 (2020). https://doi.org/10.1016/j.promfg.2020.02.089
6. Caspers, F.: RF engineering basic concepts: S-parameters, pp. 69–93 (2012). https://doi.org/10.48550/ARXIV.1201.2346, https://arxiv.org/abs/1201.2346
7. Hassani, S.E., Hassani, H.E., Boutammachte, N.: Overview on 5G radio frequency energy harvesting. Adv. Sci. Technol. Eng. Syst. J. **4**(4), 328–346 (2019). https://doi.org/10.25046/aj040442
8. Khuntia, M., Singh, D., Sahoo, S.: Impact of internet of things (IoT) on 5G. In: Mishra, D., Buyya, R., Mohapatra, P., Patnaik, S. (eds.) Intelligent and Cloud Computing. SIST, vol. 153, pp. 125–136. Springer, Singapore (2021). https://doi.org/10.1007/978-981-15-6202-0_14
9. Kumar, A., Hoque, M.A., Nurmi, P., Pecht, M.G., Tarkoma, S., Song, J.: Battery health estimation for IoT devices using V-edge dynamics. In: Proceedings of the 21st International Workshop on Mobile Computing Systems and Applications, pp. 56–61. Association for Computing Machinery, New York (2020). https://doi.org/10.1145/3376897.3377858
10. Kuppusamy, P.: A comparative study on 4G and 5G technology for wireless applications. IOSR J. Electron. Commun. Eng. (IOSR-JECE) 67–72 (2015). https://doi.org/10.9790/2834-10636772
11. Metallidou, C.K., Psannis, K.E., Egyptiadou, E.A.: Energy efficiency in smart buildings: IoT approaches. IEEE Access **8**, 63679–63699 (2020). https://doi.org/10.1109/ACCESS.2020.2984461
12. Mrnka, M., Vasina, P., Kufa, M., Hebelka, V., Raida, Z.: The RF energy harvesting antennas operating in commercially deployed frequency bands: a comparative study. Int. J. Antennas Propag. **2016**, 1–11 (2016). https://doi.org/10.1155/2016/7379624
13. Nguyen, T., Fraiwan, A., Choi, S.: Paper-based batteries: a review. Biosens. Bioelectron. **54C**, 640–649 (2013). https://doi.org/10.1016/j.bios.2013.11.007
14. Nogales, R., Benalcazar, M.E., Toalumbo, B., Palate, A., Martinez, R., Vargas, J.: Construction of a dataset for static and dynamic hand tracking using a non-invasive environment. In: García, M.V., Fernández-Peña, F., Gordón-Gallegos, C. (eds.) Advances and Applications in Computer Science, Electronics and Industrial Engineering. AISC, vol. 1307, pp. 185–197. Springer, Singapore (2021). https://doi.org/10.1007/978-981-33-4565-2_12
15. Rault, T., Bouabdallah, A., Challal, Y.: Energy efficiency in wireless sensor networks: a top-down survey. Comput. Netw. **67**, 104–122 (2014). https://doi.org/10.1016/j.comnet.2014.03.027
16. Reinoso, F., Gordón, C., Cumbajín, M., Peñafiel, C.: Wide band antennas for energy harvesting. In: Botto-Tobar, M., Montes León, S., Torres-Carrión, P., Zambrano Vizuete, M., Durakovic, B. (eds.) ICAT 2021. CCIS, vol. 1535, pp. 18–33. Springer, Cham (2022). https://doi.org/10.1007/978-3-031-03884-6_2

17. Rodrigues, L.M., Bitencourt, N.L., Rech, L., Montez, C., Moraes, R.: An analytical model to estimate the state of charge and lifetime for batteries with energy harvesting capabilities. Int. J. Energy Res. **44**(7), 5243–5258 (2020). https://doi.org/10.1002/er.5269
18. Sarika, M., Kotak, V., Durafe, A.: A review paper on internet of things (IoT) and its applications. Int. Res. J. Eng. Technol. **6**, 1623–1630 (2019)
19. Singh, R., Mamta, K.: Network analysis of antenna based on scattering parameters, pp. 1–5 (2015)
20. Zea, J.A., Benalcázar, M.E.: Real-time hand gesture recognition: a long short-term memory approach with electromyography. In: Nummenmaa, J., Pérez-González, F., Domenech-Lega, B., Vaunat, J., Oscar Fernández-Peña, F. (eds.) CSEI 2019. AISC, vol. 1078, pp. 155–167. Springer, Cham (2020). https://doi.org/10.1007/978-3-030-33614-1_11

Monitoring and Control System for Energy Harvesting IoT Applications

Cristian Bautista, Santiago Teneda, Patricio Córdova, and Carlos Gordón(✉)

GITED Research Group, Facultad de Ingeniería en Sistemas, Electrónica e Industrial, Universidad Técnica de Ambato, UTA, 180207 Ambato, Ecuador
{cbautista5961,steneda0151,edgarpcordovac,cd.gordon}@uta.edu.ec

Abstract. This work shows the design and implementation of a monitoring system for a hybrid energy harvesting device based on the Internet of Things (IoT), with the purpose of acquiring real-time data to be visualized in a metrics representation platform using Cloud computing. Which is formed by a server or several servers that involve software and hardware whose purpose is to monitor and control energy harvesting systems. For the project we used open-source tools selected on the basis of a technical study, the results of which were verified with the implementation of a dashboard. The monitored voltage signals are processed by a 16-bit ADC (Analog-to-Digital Converter), the values that are sent to the central ESP32 controller, which is responsible for reading the processed data and transmitting them via the MQTT (Message Queuing Telemetry Transport) protocol to the Red Hat Enterprise Linux server located in the Microsoft Azure cloud, and then displayed in Grafana. With the work done we can obtain a control of variables such as antenna voltage, panel and battery which is very promising to provide continuous control of all variables involved in energy storage.

Keywords: Monitoring system · Energy Harvesting · Grafana · Telegraf · InfluxDB

1 Introduction

Nowadays, many of the processes that are carried out in laboratories or in the industry need tools capable of meeting the demands of the world, such as data acquisition systems, as it allows to obtain, store and process this information of some metric that is being measured [12]. These have several advantages such as saving time, money and reducing systematic errors. At the same time, when working on the monitoring of energy harvesting systems, it's foreseen to analyze the feasibility of using this technology to combat climate change. Because, together with global warming, they are a growing problem, due to the excessive use and exploitation of fossil fuels [1].

In general, the progress that exists day by day in information and communication technologies (ICT) in the last decade, has achieved that IoT and Cloud

M. V. Garcia and C. Gordón-Gallegos (Eds.): CSEI 2022, LNNS 678, pp. 333–350, 2023.
https://doi.org/10.1007/978-3-031-30592-4_23

computing are integrated into various solutions in different fields of action such as medicine, environmental monitoring, security and others. Thus, one of the main objectives of the Paris agreement concluded in 2015 is to limit the global average temperature increase to below 2°C through the use of low-carbon energies [1]. Due to this adaptability, it's possible to use these advances in the monitoring and control of Energy Harvesting systems, which will be equipped with sensors adapted to a specific function, and thus have a clean energy harvesting system that would represent a 90% reduction in Co2 emissions worldwide by 2050 [1], obtaining data to be subsequently analyzed in order to better understand the behavior of these systems with respect to the surrounding environment [11].

In such a way, real-time voltage stability monitoring is important in Energy Harvesting Systems to know their efficiency, since nowadays new ecological and autonomous energies are sought that can be used in many applications, such as powering small sensors that do not consume much energy for their operation [18]. The implementation of a data acquisition system allows to know the energy generated by these systems according to their construction and surrounding environment, with constant monitoring and in real time obtaining data needed to be processed and visualized in order to generate reports, configure control instructions based on the data obtained. All this using IoT devices and Cloud computing [5].

In this paper, the implementation of a data acquisition system for monitoring electrical variables from sustainable energy sources for IoT applications EHMS (Energy Harvesting Monitoring System) is presented. This article is constituted as follows: the hardware and software design, the materials and methodology applied in the development of the proposed system are described in Sect. 2. In Sect. 3, the results and measurements are presented, as well as their discussion.

2 Related Work

The capacity of ICTs is fully exploited by IoT smart grids, so that today we have seen an increase in sustainability, quality and performance of energy production thanks to it, with the integration of renewable energy, which have less and less resource consumption.

Thus, the work proposed by [19] is an IoT system that estimates energy losses through the analysis of variables, temperature and irradiance, for the detection of solar cell failures as well as the collation of information in order to generate alerts, In the work proposed by [3, 22] the solution involves the transfer of notifications in a multi-device scenario using MQTT and CoAP (Constrained Application Protocol)technologies, emphasizing on using privacy criteria in relation to the environment and the hierarchy assigned to the user to ensure that the relevant notifications were sent and received according to the defined rules.

In turn, integrating simple networks with the use of sensors and actuators allows the correct monitoring of any environment as presented by [2,6,9,15,21] that through WiFi with MQTT transmits information with a system commanded by an ESP8266 card for analysis with energy monitoring.

It can be noted that, it's important to consider the proper way to store the most amount of energy, with an analysis of the electronic components that leads to this goal, in such a way the work proposed by [4,14] presents a comparison of sustainable energy storage systems in lithium batteries, considering important the evaluation of mass and energy versus parameters such as, power output as a function of time and current.

The work of [6] consists of stages in which communications through the I2C protocol determines the communication between the sensors and controller, to later have cloud services through Wi-Fi, so that the data that are collected are stored in the cloud, with requests based on the client/server structure.

As presented by [8] in his work, it's important to minimize energy consumption and task processing time for IoT scenarios, in this case using an algorithm called TEMS that considers energy, time consumed during processing, cost of data transmission and energy in idle devices. Thus, the results of their research are favorable with savings of up to 51.6% in energy consumption and improvement of task completion time by up to 86.6%.

3 Methodology

In this section we have the development of the monitoring and control system for the energy harvesting systems. The device was designed in EasyEDA software.

3.1 Proposed System Design

In order to achieve the objectives of the solar and electromagnetic energy harvesting monitoring system, a prototype was implemented that provides useful data, such as the voltage of the different components of the Energy Harvesting Systems, which is monitored in real time. Figure 1 shows the general scheme of the monitoring system. It's made up of several stages, which are explained in general terms below.

Stage 1.-Is responsible for the acquisition of the electrical variables of the Energy Harvesting System by means of the voltmeters, with the input of voltage signals from the energy collection and storage, the data is conditioned with an ADC converter to be processed by the microcontroller and its subsequent transmission.

Stage 2.- Describes the Linux RedHat Enterprise 8.2 server installed in Microsoft Azure, where the MQTT broker is hosted, which facilitates the sending of data to Telegraf, InfluxDB and Grafana, which are hosted on the same server.

Stage 3.- consists of a service that is in charge of collecting and reporting metrics, called Telegraf, from the MQTT broker to the InfluxDB database; all the data collected by Telegraf is stored in this database to be later visualized in a metrics visualization and processing platform.

Stage 4.- At this point the data is visualized using the Grafana metrics analysis and visualization tool, which is linked to the InfluxDB database. The data

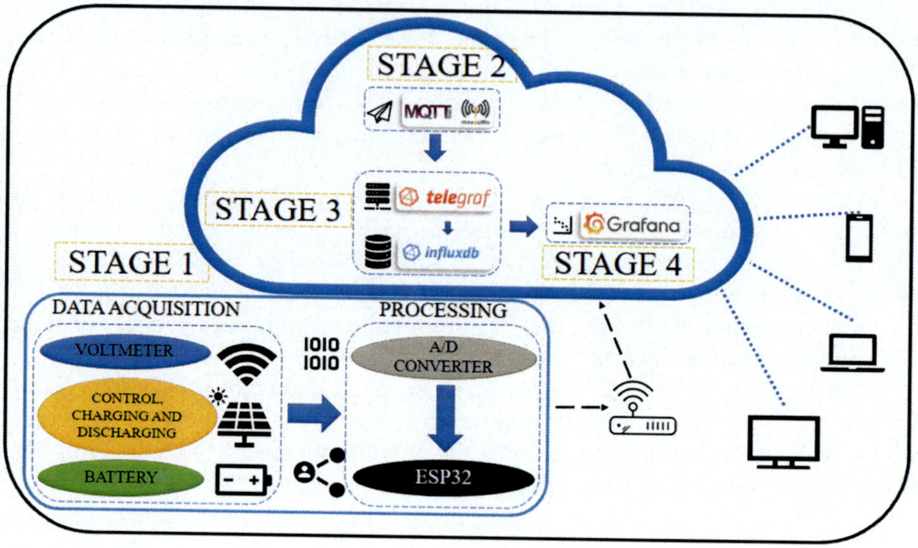

Fig. 1. General EHMS Schematic

will be displayed in an elegant way using the different types of graphs that allow you to observe, query, configure alerts and understand parameters stored with the linked database.

3.2 Hardware Design

ESP32-Based Controller: The ESP32 card features a wide variety of communication protocols. With a power consumption of 80 mA, as well as a higher reception threshold than other cards, higher operating frequency and higher transmission power. Due to these characteristics, it was decided to use the ESP32 card as the main controller in the information processing [23].

Fig. 2. Block diagram for the voltmeter

Voltage Measurement: For signal acquisition, four voltmeters are used to measure different voltages throughout the device. Figure 2 shows the general scheme of each voltmeter, where the processing of the electrical voltage signals, signal conditioning, the A/C conversion stage and subsequent sending of the data are shown by means of blocks.

Signal Processing: ADS1115, it's a 12/16-bit low power analog-to-digital converter with a comparator and a programmable gain amplifier, in addition, 4 multiplexed channels, which means that it can measure two differential inputs or four unipolar inputs and the PGA supports input ranges from ± 256 mV to ± 6,144, thus obtaining measurements of large and small signals with high resolution [24].

Electronic Design of Voltmeters: It's necessary to use a voltage divider for signal conditioning, considering the voltage entering the EMSH, we proceed to attenuate the signal with respect to that magnitude.

The configuration of the divider is shown in Fig. 3, for this case an extra range of values has been added to meet the 30VDC measurement. Table 1 shows the attenuation that resulted in raising the range up to 45VDC. The voltage divider equation used is detailed below.

$$Vout = \frac{Vin * R2}{R1 + R2} \tag{1}$$

Voltmeter for a Maximum Input Value of 45VDC: The maximum voltage supported by the ADC1115 is 4.096VDC, for the resistor values we used R1 = 10000 Ω and R2 = 1000 Ω Ohm, Fig. 3 a) shows the design of the voltage divider.

Table 1. 45 VDC Voltage Divider Calculations.

Vin(V)	Vout(V)
5	0.45
10	0.91
15	1.36
20	1.82
25	2.27
30	2.72
35	3.18
40	3.64
45	4.09

$$Vin = \frac{Vout * (R1 + R2)}{R2} \tag{2}$$

$$Vin = \frac{4,096\,V * (10000\Omega + 1000\Omega)}{1000\Omega} = 45.023\,V \tag{3}$$

Voltmeter for a Maximum Input Value of 5.4 VDC: For the design of the other two voltmeters, the maximum measurement value of 5.337 VDC is established, which will measure the battery voltage and the output of the 5 VDC regulator. Figure 3 b) shows the design of the voltage divider. Thus, R1 = 10000 Ω and R2 = 33000 Ω were used for the resistor values (see Table 2).

Table 2. 5 VDC Voltage Divider Calculations.

Vin (V)	Vout (V)
1	0.77
2	1.535
3	2.30
4	3.0698
5	3.83

$$Vin = \frac{Vout * (R1 + R2)}{R2} \tag{4}$$

$$Vin = \frac{4,096\,V * (10000\Omega + 33000\Omega)}{33000\Omega} = 5.337\,V \tag{5}$$

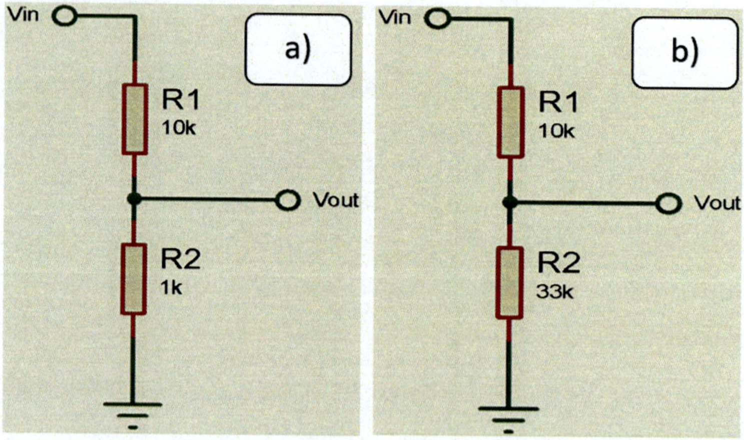

Fig. 3. a) Voltage divider for a maximum voltage of 45 VDC b) Voltage divider for a maximum of 5.3 VDC

Adequacy of the Input Voltage: To power the module a constant voltage of 5VDC is needed, so the LM7805 voltage regulator was used, this is connected to the output of the energy harvesting device. The design of the regulator can be seen in Fig. 4.

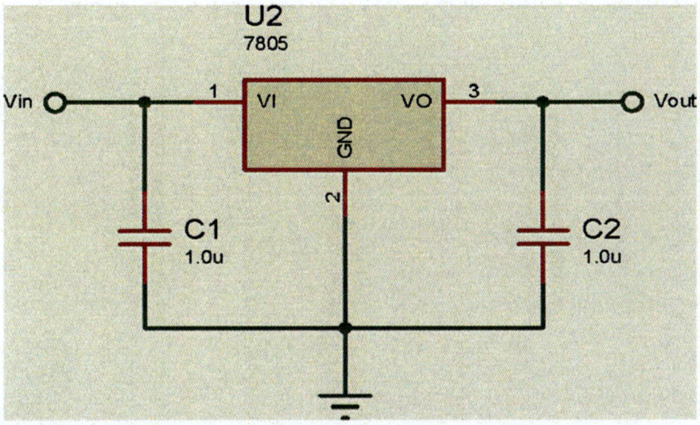

Fig. 4. Schematic of voltage regulator

Battery Charging: TP4056, it's a linear charger, selected for its specific design for 3.7 V to 1 A batteries, either lithium or LiPo, has a charge control algorithm taking as limit value 2.4 V discharge and 4.2 V charge [17]. The wide input voltage range allows handling the variable values provided by the power supplies to be monitored.

General Schematic Design of the Electronic Circuit: Figure 5 shows the schematic diagram developed in the online software EasyEDA, which describes the stages of signal input, signal conditioning by the LM7805 voltage regulator, as well as the reading of the data by the voltmeters for the panel, the output of the regulator (5 V) and the battery. For the reading of the antenna voltage, the value provided is lower than the limits required by the pins of the ESP32 microcontroller, therefore, it's not necessary to make a voltage divider. Finally, the data from the voltmeters is sent to the inputs of the ADS1115, which is a 16-bit ADC converter that sends the information to the ESP32 to process the data.

In addition, the battery charging stage is described by the TP4056 module, with an extra control of charge and discharge of the batteries by a relay module, the circuit described in the scheme has a control stage for the output of the circuit through a source switch (UPS) with this allows to obtain an output voltage most of the time regardless of whether there is power from the main source, thanks to the connection of the backup batteries that has the monitoring system. All data is displayed on a 16×4 LCD, and a battery status indicator is incorporated with an RGB LED module.

PCB Circuit Design: From the schematic, as shown in Fig. 6, the location of the components was determined in the EasyEDA software, taking into account the width of the conductor planes.

Subsequently, Fig. 7 shows the circuit implemented with the soldered electronic components.

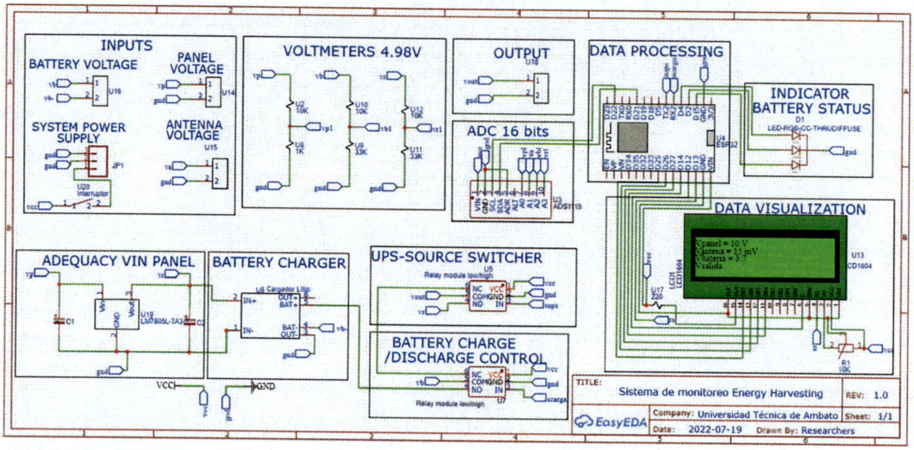

Fig. 5. Schematic diagram of the EHMS developed in EasyEDA

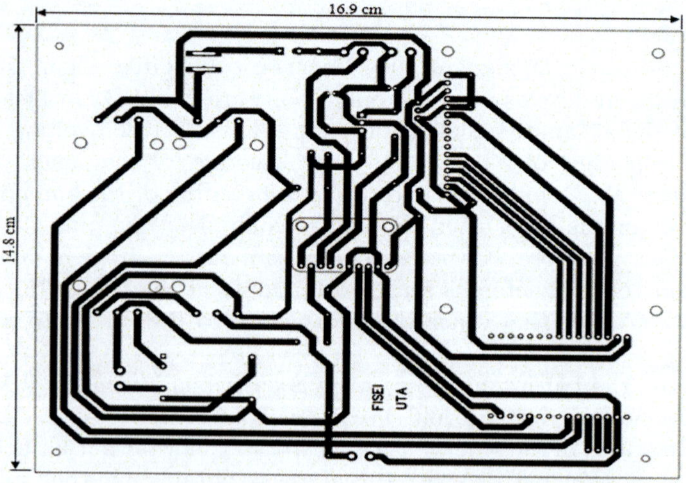

Fig. 6. PCB design in EasyEDA software

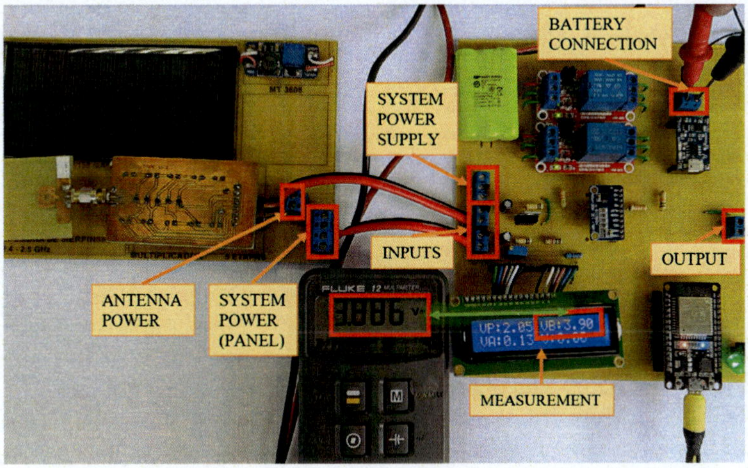

Fig. 7. Implementation of the final circuit of the EHMS

3.3 Software Design

Wireless Technology: Stage 2 considers the need for efficient data transmission with large coverage using WiFi [7], which provides easy internet connectivity for a wide range of equipment and is optimal for IoT applications.

Microsoft Azure: Although it lacks database options, it offers more compatibility oriented to the self-learning framework. The price for its use depends on the use it's given [25]. However, it offers the feature that by registering institutional mail you can have a $100 credit for one year along with the use of free services as part of the incentive for learning, noncommercial research or application design, development, testing and demonstration work, which is important for the realization of this project.

Communication Protocol: MQTT, due to its client/broker architecture and publish/subscribe abstraction, it's a many-to-many communication protocol, with the possibility of messaging between multiple clients through a central exchange being efficient as a communication bus for live data [16]. With this, it has greater flexibility to adapt to IoT scenarios and meets the needs of high volume real-time data transmission.

Data Collection and Storage: InfluxDB, as part of stage 3, InfluxDB is used together with Telegraf, which guarantees the collection of data sent through MQTT and its correct storage, because it's an Open-Source distribution coded in Go programming language.

Data Visualization: Grafana, for stage 4 we have the platform programmed in Go language, variety of data presentation, MQTT metrics, integration with InfluxDB with high compatibility. Optimal to represent data in an easy and pleasant way for users, in addition to having the feature to issue alerts using Telegram.

Process Hierarchy Flowcharts: This section explains the development of all the processes carried out by each of the stages that are part of the monitoring system. ESP32 connection to RedHat Enterprise Linux in Azure: as described in Fig. 8. The process starts with the boot of the microcontroller, in which the libraries are started by programming in Arduino IDE. The Wi-Fi connection of the ESP32 board to the internet network is configured, through this the connection to the MQTT broker hosted in the Azure cloud can be established in a RedHat Enterprise virtual machine, considering the topic designated for each instance. The voltages are the variables that are processed with the ESP32, with the conditioning and sending to the MQTT Broker, where it accepts all the messages published by the clients to the established topic and distributes it to the clients subscribed to it. Finally, the data is presented in Grafana.

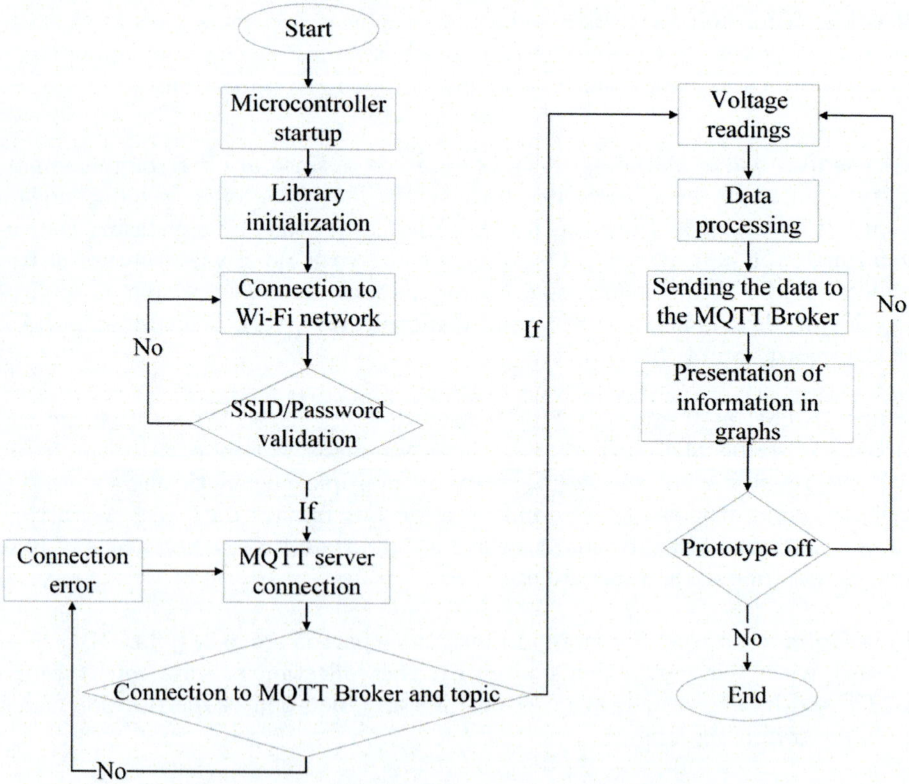

Fig. 8. Flowchart of the connection to the MQTT Broker.

4 Discussion of Results

The design was done in EasyEDA; thus, the circuit is printed in FR4 with copper cladding. Data has been taken at various times of the day, with the measurement of voltage coming from the battery and a hybrid energy harvesting source, which has a 5V solar panel in series with microstrip patch antenna based on the Sierpinski fractal.

4.1 Comparison of Voltage Measurements Between a Fluke Multimeter and the Proposed Energy Harvesting Monitoring System

Voltage measurements were taken during the day (06:00–18:00) and night (18:00–06:00), taking continuous data from a solar panel and a microstrip antenna as power sources, as well as from the battery which is the backup source to power the IoT devices. Figure 9, 10, 11, 12, 13, 14 show the electrical measurements captured in real time displayed on the Grafana dashboards, with the voltage taken with the exact time and date of data collection for 24h.

Voltage Measurement at Daytime. Figure 9 shows the variation of the measured values coming from the panel as a function of the irradiance of the sun, which is an important factor in this type of power supplies.

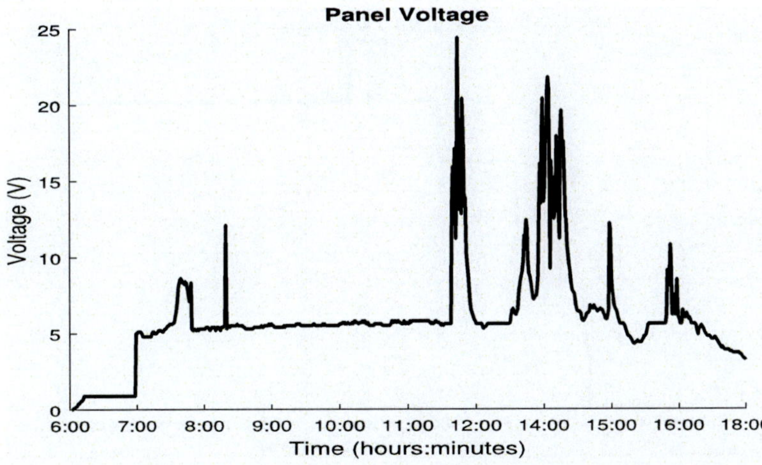

Fig. 9. Panel voltage represented on the Grafana dashboard during the day.

In addition, it must be considered that at the time of the tests, the day was cloudy with small precipitation at certain hours, where the voltage decreases considerably. Nevertheless, the highest peak measured was 22.5 V at 13:00 h.

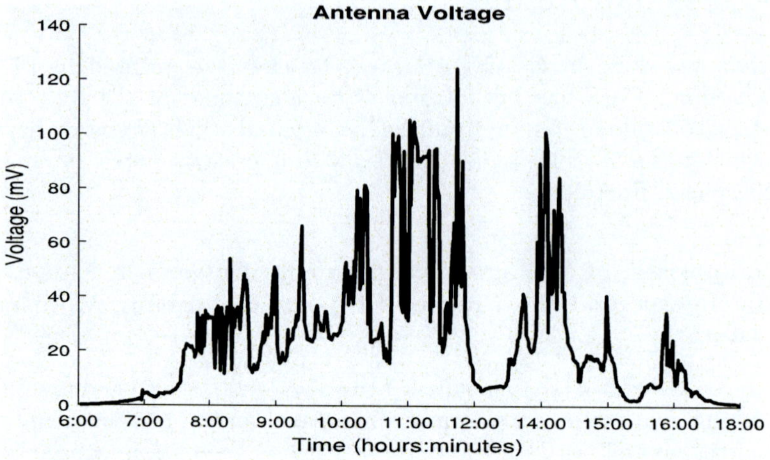

Fig. 10. Microstrip antenna voltage displayed on the Grafana dashboard during the day.

Figure 10 shows the measured voltage values of the antenna that depends on the electromagnetic waves that are at a frequency of 2.4 GHz (Wifi signal), which handles voltage values in the order of millivolts, having the maximum peak voltage measured of 120 mV at 13:00 h.

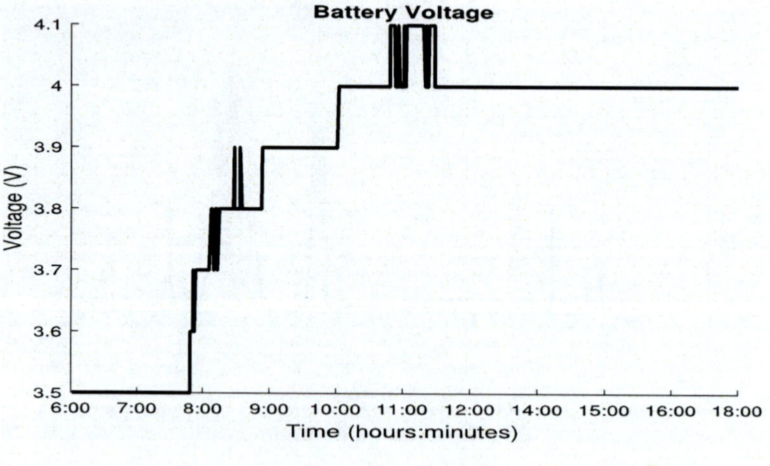

Fig. 11. Battery voltage displayed on the Grafana dashboard during the day.

Figure 11 shows the measured voltage of the battery that at 8:00 has a value of 3.50 V which implies that it needs to be charged, it's so from the panel feed and the antenna has been fully charged at 12:00, from this point a constant voltage value of 4 V is maintained.

Voltage Measurement at Night: Figure 12 shows the measured value of voltage coming from the panel where it can be seen that, since there is no irradiance, the voltage supplied by this source tends to zero.

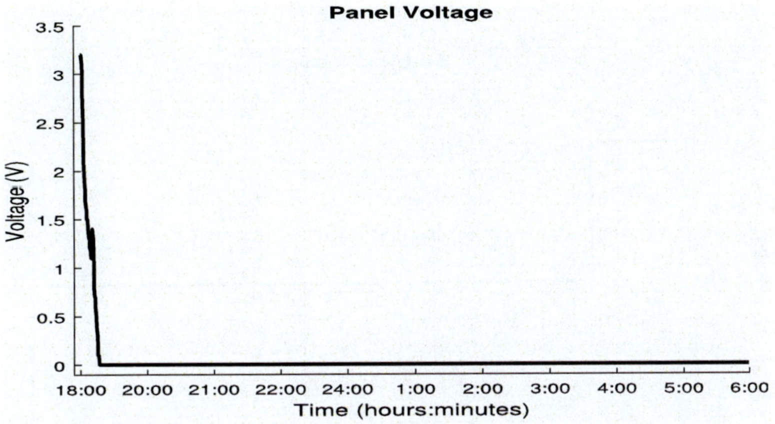

Fig. 12. Panel voltage represented on the Grafana dashboard during the night.

Figure 13 shows the measured antenna voltage values, which mostly oscillate in the range of (100 and 200) mV.

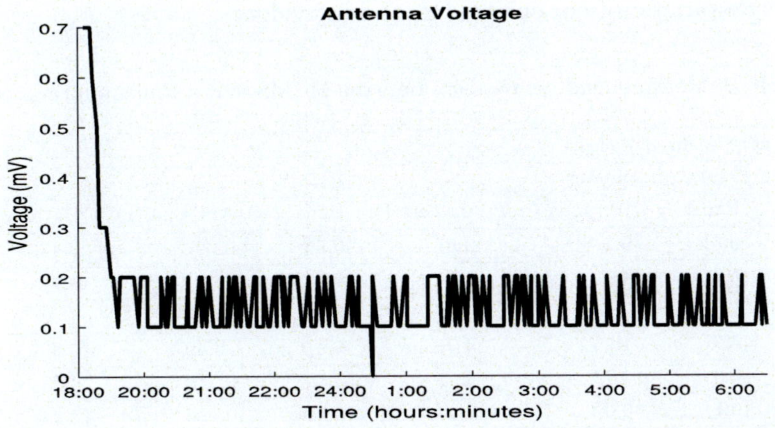

Fig. 13. Microstrip antenna voltage displayed on the Grafana dashboard during the night.

Figure 14 shows the measured voltage of the battery that comes into operation to power the output of the monitoring system, as a backup source when the hybrid power supply does not provide sufficient voltage, where because of power consumption the value it provides drops.

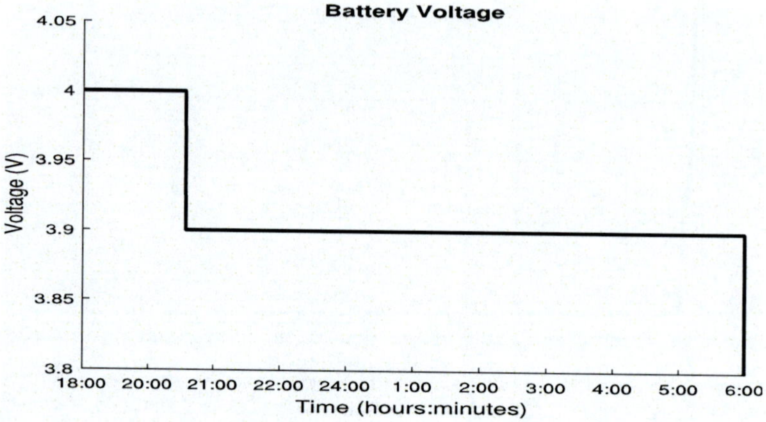

Fig. 14. Battery voltage displayed on the Grafana dashboard during the night.

Table 3 shows the comparison of the voltage measured by a fluke multimeter that has a basic DC accuracy ranging from 0.5% to 0.025% [10] and the EHMS, where little variation from one to the other is evident.

Table 3. Measurement comparison between EHMS and a Fluke multimeter.

Time(h)	Measured Voltage					
	Fluke Multimeter			EHMS		
	Panel (V)	Antenna (mV)	Battery (V)	Panel (V)	Antenna (mV)	Battery (V)
06:00	0.51	0.1	3.5	0.55	0.09	3.47
09:00	5.57	46	3.78	0.51	46.5	3.71
12:00	5.58	33.5	4.08	5.59	33.45	4.02
15:00	6.3	16.5	4.075	6.2	16.49	4.069
18:00	3.3	0.72	4.07	3.4	0.718	4.08
21:00	0.11	0.09	3.98	0.14	0.098	3.94
24:00	0	0.1	3.88	0	0.095	3.81
06:00	0.1	0.1	3.5	0.53	0.097	3.52
09:00	5.57	46	3.78	0.55	46.2	3.79

The following equation is used to calculate the error measured between the multi-meter and the EHMS:

$$\%error = \frac{|V_(aprox) - V_(real)|}{V_(real)} * 100\% \tag{6}$$

Table 4 shows the error obtained from the comparative measurements of the table where it's highlighted that no case exceeds 10% error, which determines the efficiency of the proposed monitoring system.

Table 4. Calculated error of measurements.

Sources and Battery	% Error
Panel	2.49
Antenna	4.13
Battery	1.22

The contributions of this work can be identified by comparing it with the design of the voltmeter in the system [20] is similar to the prototype developed with the difference that supports up to 45 V and with an error rate of less than 5%, so this system can replace it. In addition, it has a web server in the cloud with a powerful graphical platform (Grafana), which can be accessed from anywhere using either a cell phone or a computer, the data are displayed in different graphs as histograms in which you can view the data by date and has alerts which notify of anomalies in the energy storage.

The system [13] developed for monitoring photovoltaic cells uses as a central controller an Arduino Uno along with a Wi-Fi module for transmitting voltage and current monitoring information. EHMS uses as a central controller an ESP32 which is a microcontroller with its own Wi-Fi module which has a lower power consumption, has voltmeters for the panel, antenna and battery to have a total system monitoring, as well as its automatic load control and real-time monitoring on an LCD screen and a visualization platform in the cloud. Therefore, EHMS can replace the previous system.

The monitoring system [6] uses the HTTP communication protocol which is not widely used in IoT applications as it uses more power and capacity compared to MQTT which is more optimized in terms of power consumption and bandwidth. Therefore, this project can replace the previous one and has more inputs to measure the voltage, a control system to switch between the power provided by the panel and the antenna or the battery.

5 Conclusions

The work performed has allowed to have a total control of the variables such as antenna voltage, panel and battery which is very promising to provide a continuous control of these variables involved in energy storage which has allowed us an automatic control of battery charging. In addition, the results obtained in Table 4 demonstrate the reliability of a low cost prototype compared to other measurement instruments. Therefore, the integration of open-source software, processing devices used, a control algorithm and real-time monitoring are suitable for monitoring hybrid Energy Harvesting systems.

Storage in a rechargeable battery by the antenna and the panel for 24 h, resulted in the voltage provided by the solar panel with a maximum peak of 22.5 V during the day and 0 V at night depending on the irradiance of the sun. Similarly, the values obtained by the antenna during the day and night with a voltage that did not exceed 200 mV, the final load was 500 mV in the span of 4 h during the day.

Acknowledgements. The authors thank the Technical University of Ambato and the "Dirección de Investigación y Desarrollo" (DIDE) for their support in carrying out this research, in the execution of the project "Captación de Energía Limpia de Baja Potencia para Alimentación de Dispositivos de Quinta Generación (5G)", approved by resolution "Nro. UTA-CONIN-2022-0015-R". Project code: SFFISEI 07.

References

1. Barón, A., Contreras, I.: Biogás: Importancia y beneficios de energías alternativas en el mundo. Boletín Semillas Ambientales **14**(2), 91–98 (2020)
2. Bengoa, P., et al.: Coordination of two robots for manipulating heavy and large payloads collaboratively: SOFOCLES project case use. In: Lecture Notes in Networks and Systems, pp. 255–271. Springer, Heidelberg (2022). https://doi.org/10.1007/978-3-030-97719-1_15
3. Caiza, G., et al.: Low-cost fuzzy control for poultry heating systems. In: Yang, X.-S., Sherratt, S., Dey, N., Joshi, A. (eds.) Proceedings of Sixth International Congress on Information and Communication Technology. LNNS, vol. 236, pp. 109–118. Springer, Singapore (2022). https://doi.org/10.1007/978-981-16-2380-6_9
4. Caiza, G., Salazar-Moya, A., Garcia, C.A., Garcia, M.V.: Lean manufacturing tools for industrial process: a literature review. In: Yang, X.-S., Sherratt, S., Dey, N., Joshi, A. (eds.) Proceedings of Sixth International Congress on Information and Communication Technology. LNNS, vol. 236, pp. 27–35. Springer, Singapore (2022). https://doi.org/10.1007/978-981-16-2380-6_3
5. Campoverde, A.M., Hernández, D.L., Mazón, B.E.: Cloud computing con herramientas open-source para internet de las cosas. Maskana **6**(Supl.), 173–182 (2015)
6. Cheddadi, Y., Cheddadi, H., Cheddadi, F., Errahimi, F., Es-sbai, N.: Design and implementation of an intelligent low-cost IoT solution for energy monitoring of photovoltaic stations. SN Appl. Sci. **2**(7), 1–11 (2020). https://doi.org/10.1007/s42452-020-2997-4
7. Danbatta, S.J., Varol, A.: Comparison of zigbee, z-wave, wi-fi, and bluetooth wireless technologies used in home automation. In: 2019 7th International Symposium on Digital Forensics and Security (ISDFS), pp. 1–5 (2019). https://doi.org/10.1109/ISDFS.2019.8757472
8. Dos Anjos, J.C., Gross, J.L., Matteussi, K.J., González, G.V., Leithardt, V.R., Geyer, C.F.: An algorithm to minimize energy consumption and elapsed time for IoT workloads in a hybrid architecture. Sensors **21**(9), 2914 (2021). https://doi.org/10.3390/s21092914

9. Escobar, E., Villazón, A.: Sistema de monitoreo energético y control domótico basado en tecnología internet de las cosas. Investigación & Desarrollo **18**(1), 103–116 (2018)
10. Fluke. Fluke 12 multimeter User Manual (2022). https://xdevs.com/doc/Fluke/FLUKE%2012%20User.pdf
11. Gómez, J.E., Castaño, S., Mercado, T., Fernandez, A., Garcia, J.: Sistema de internet de las cosas (iot) para el monitoreo de cultivos protegidos. Ingeniería e Innovación **5**(1) (2018). https://doi.org/10.21897/23460466.1101
12. González, J., Nuñez, B., Viloria, P.: Sistema de monitoreo en tiempo real para la medición de temperatura. Scientia et technica **17**(50), 128–131 (2012)
13. Gopal, M., Prakash, T.C., Ramakrishna, N.V., Yadav, B.P.: IoT based solar power monitoring system. IOP Conf. Ser.: Mater. Sci. Eng. **981**(3), 032037 (2020). https://doi.org/10.1088/1757-899X/981/3/032037
14. Keshan, H., Thornburg, J., Ustun, T.S.: Comparison of lead-acid and lithium ion batteries for stationary storage in off-grid energy systems. In: IET Conference Proceedings, pp. 30(7.)–30 (7.)(1) (2016). https://doi.org/10.1049/cp.2016.1287
15. Morales, F., Mogrovejo, D., González, E., Toasa, R.M.: Monitoring of industrial variables based on LoRa communication protocols. In: García, M.V., Fernández-Peña, F., Gordón-Gallegos, C. (eds.) Advances and Applications in Computer Science, Electronics and Industrial Engineering. AISC, vol. 1307, pp. 201–214. Springer, Singapore (2021). https://doi.org/10.1007/978-981-33-4565-2_13
16. Naik, N.: Choice of effective messaging protocols for IoT systems: Mqtt, coap, amqp and http. In: 2017 IEEE International Systems Engineering Symposium (ISSE), pp. 1–7 (2017). https://doi.org/10.1109/SysEng.2017.8088251
17. NanJing Top Power ASIC Corporation. 1A Standalone Linear Li-lon Battery, Charger with Thermal Regulation in SOP-8 (2022). https://www.mikrocontroller.net/attachment/273612/TP4056.pdf
18. Panatik, K.Z., et al.: Energy harvesting in wireless sensor networks: a survey. In: 2016 IEEE 3rd International Symposium on Telecommunication Technologies (ISTT), pp. 53–58 (2016). https://doi.org/10.1109/ISTT.2016.7918084
19. Ramadhan, J.M., Mardiati, R., Haq, I.N.: IoT monitoring system for solar power plant based on mqtt publisher/subscriber protocol. In: 2021 7th International Conference on Wireless and Telematics (ICWT), pp. 1–6 (2021). https://doi.org/10.1109/ICWT52862.2021.9678503
20. Romadoni, D., Astutik, R.P., Surya, Y.A.: Design of monitoring system and notification hourmeter using web on container crane (cc). INAJEEE (Indon. J. Electr. Electron. Eng.) **4**(1), 1–7 (2021). https://doi.org/10.26740/inajeee.v4n1.p1-7
21. Saeteros, M., Paucar, W., Molina, C., Caiza, G.: Development and analysis of a PID controller and a fuzzy PID. In: Nummenmaa, J., Pérez-González, F., Domenech-Lega, B., Vaunat, J., Oscar Fernández-Peña, F. (eds.) CSEI 2019. AISC, vol. 1078, pp. 143–154. Springer, Cham (2020). https://doi.org/10.1007/978-3-030-33614-1_10
22. Silva, L., et al.: Multi-device notifications: a comparison between MQTT and COAP. In: The Thirteenth International Conference on Mobile Ubiquitous Computing, Systems, Services and Technologies, pp. 9–13. ThinkMind (2016)
23. Škraba, A., Koložvari, A., Kofjač, D., Stojanović, R., Semenkin, E., Stanovov, V.: Prototype of group heart rate monitoring with esp32. In: 2019 8th Mediterranean Conference on Embedded Computing (MECO), pp. 1–4. IEEE (2019). https://doi.org/10.1109/MECO.2019.8760150

24. Texas Instruments. ADS111x Ultra-Small, Low-Power, I2C-Compatible, 860-SPS, 16-Bit ADCs With Internal Reference, Oscillator, and Programmable Comparator (2022). https://www.ti.com/lit/ds/symlink/ads1115.pdf
25. Wankhede, P., Talati, M., Chinchamalatpure, R.: Comparative study of cloud platforms-microsoft azure, google cloud platform and amazon ec2. J. Res. Eng. Appl. Sci **5**(02), 60–64 (2020)

RF Energy Harvesting System Based on Spiral Logarithmic Dipole Rectenna Array

Myriam Cumbajin[1], Patricio Sánchez[1], Darío Pillajo[2], and Carlos Gordón[2(✉)]

[1] SISAu Research Center, Facultad de Ingeniería, Industria y Producción, Universidad Tecnológica Indoamérica, UTI, 180103 Ambato, Ecuador
{myriamcumbajin,patriciosanchez}@uti.edu.ec

[2] GITED Research Group, Facultad de Ingeniería en Sistemas, Electrónica e Industrial, Universidad Técnica de Ambato, UTA, 180207 Ambato, Ecuador
{dpillajo6245,cd.gordon}@uta.edu.ec

Abstract. This paper presents the development, design, simulation and fabrication of a logarithmic spiral antenna array for the electromagnetic energy collection system independent of frequency, compact in microstrip technology, low cost and operating within a range of frequencies. 1 to 4 GHz, are designed using the CST Studio software that allows modeling and simulating essential parameters such as radiation lobes, VSWR, power, gain, parameter S11. From these simulations the printing is done on an RF4 substrate with a thickness of 1.6 mm, a dielectric constant of 4.4 and an input impedance of 50 Ω. The resonator structure antenna with dimensions of 190 × 210 mm. The frequencies obtained with the rectenna matrix, since it offers good performance for electromagnetic energy harvesting applications that operate in standard frequencies such as GSM, WiFi, LTE which are available in the environment and allows us to collect energy. To obtain electrical energy as part of the design of a frequency rectifier which is optimized with the application of the Schottky diode. In addition, its construction of the rectenna matrix was carried out based on the proposed design, then performance tests and validation of the prototype were carried out using the miniVNA Tiny spectrum verifier, and finally energy storage tests were carried out.

Keywords: Energy harvesting · Spiral antenna · Electromagnetic · Microstrip · Multiband

1 Introduction

Significant and obvious advances in wireless communication, power supply and power electricity in recent years has been essential in the field of electronic systems to have a variety of very popular networks to link various devices, the weight or dimensions of any portable equipment. It is for this reason that during

M. V. Garcia and C. Gordón-Gallegos (Eds.): CSEI 2022, LNNS 678, pp. 351–365, 2023.
https://doi.org/10.1007/978-3-031-30592-4_24

the different generations they have begun to propose alternatives in order to improve this aspect, and an applicable model is the energy systems or also called "Energy Harvesting" [9]. Energy harvesting consists of taking advantage of the energy provided by the environment without harming nature to power a low-power electronic system.

The fundamental idea is to capture the energy of radio waves, likewise there are many frequencies in the environment, which must be used. Currently, different investigations have been carried out in the harvesting of RF energy, so this energy harvesting system is a unit that converts the power of RF waves into direct current energy and be usable energy. RF energy is uncountable in the environment, for example: satellite stations, wireless internet, radio stations, wireless digital transmission services, this would provide an opportunity to replace conventional batteries with a limited lifespan, and have the ability to be self-sustaining [1].

They can be grouped into three energy harvesting subsystems. The first collects energy through the use of antennas. Once the energy is collected, the second sub-system employs the use of a rectifier to convert the received energy from alternating voltage to direct voltage. The third subsystem stores the energy obtained through the energy harvester, and this subsystem is critical due to low input power availability [3,4]. The main objective is to enable the creation of a fully independent and self-sustaining platform of low consumption devices for smart cities, homes and IoT applications that could significantly influence our lives. The optimal design of a rectenna remains a challenge [6].

There are projects and works that are related to the geometry and design of the rectenna for energy harvesting systems, which is made up of an antenna and a rectifier, the antenna receives the RF energy, coupled together with a rectifier circuit, whose element main is the diode [5,11]. The construction of the rectenna is made in a patch that consists on one side of the insulating substrate with the respective geometry of the design and its back by a ground plane, the main materials for conductors are copper and gold. When the antenna is made with a thick dielectric substrate and a low dielectric constant, its main advantage is that a performance close to that proposed is obtained, providing good bandwidth, performance and radiation [7,14]. The multiband antenna developed at resonant frequencies from 1 GHz to 4 GHz, with a center frequency of 2.45 GHz, whose measured reflection coefficient values are −19 dB to −17 dB and −13 dB, respectively.

In this article, the development of a rectenna is presented, for a specific purpose of harvesting energy from the environment. The simulation and fabrication of the Archimedean spiral rectenna, its working range is in the most dominant frequency bands that have the highest power density GSM900 and WiFi 2.45. Therefore, it is organized as follows: Its design, materials and methodology applied in the development of the rectenna are described in Sect. 2. The results of the simulation and measurement are presented in Sect. 3. And finally the conclusion of this work.

2 Methodology

This section presents the methodology used for the development of the spiral logarithmic rectenna matrix for capturing electromagnetic energy. The design of an array of antennas that operates from 900 MHz to 4 GHz, specifically at a central frequency of 2.4 GHz, is proposed. The proposed antenna was designed with the CST Studio software. The antenna array design process consists of three key parameters for calculating the dimensions of this array. These parameters are the resonance frequency (fr), the relative permittivity (ϵr) and the height of the substrate (h) (Table 1).

Table 1. Essential parameters for the design of the antenna array.

Parameter	Value for the antenna array
Resonance Frequency	Multiband 1 to 4 GHz
Substrate relative permittivity	4.3
Substrate height	1.6 mm
Input impedance	50 Ω
Loss Tangent	0.02

2.1 Structure for the Logarithmic Spiral Rectenna Matrix

The proposed antenna structure starts from Eq. (1), in which the arms of the spiral antenna are defined using simple polar coordinates and each arm is defined as being an equiangular spiral antenna.

These are centered and set on a separate line.

$$g = 2w + 2r_{in} \tag{1}$$

$$g = [2(1) + 2(0.5)]\,\text{mm} = 3\,\text{mm} \tag{2}$$

Once the frequencies for its operation in the antenna array have been established, the parameters such as inner radius rin and outer radius rout are found. With the low frequency of the spiral antenna the operating point is established and is determined by rout expressed as follows [8,13].

$$f_L = \frac{c}{2\pi r_{out}\sqrt{E_{eff}}} \tag{3}$$

$$r_{out} = \frac{c}{2\pi f_L \sqrt{E_{eff}}} = \frac{3 * 10^8\,\text{m/s}}{2\pi(1\,\text{GHz})\sqrt{4.3}} = 0.0227\,\text{m} = 22.7\,\text{mm} \tag{4}$$

With the high frequency of the spiral antenna, the operating point is established and is determined by rin expressed as follows:

$$f_H = \frac{c}{2\pi r_{in}\sqrt{E_{eff}}} \tag{5}$$

$$r_{in} = \frac{c}{2\pi f_L\sqrt{E_{eff}}} = \frac{3 * 10^8\,\mathrm{m/s}}{2\pi(5\,\mathrm{GHz})\sqrt{4.3}} = 0.0046\,\mathrm{m} = 4.6\,\mathrm{mm} \tag{6}$$

Substituting the values of Eq. (2) and (3) we obtain RM. [2]

$$R_M = \frac{r_{in} + r_{out}}{2} \tag{7}$$

$$R_M = \frac{(4.6 + 22.7)\,\mathrm{mm}}{2} = 13.65\,\mathrm{mm} \tag{8}$$

Considering Eq. (9), the wavelength of the antenna is obtained:

$$\lambda = \frac{c}{f_L} \tag{9}$$

$$\lambda = \frac{(3 * 10^8)\,\mathrm{m/s}}{1\,\mathrm{GHz}} = 0.3\,\mathrm{m} = 30\,\mathrm{mm} \tag{10}$$

where:

c = speed of light
f_L = lower operating frequency

Arm Overall Length:

$$L = 1.25\lambda_L \tag{11}$$

$$L = 1.25\lambda_L = 1.25(300\,\mathrm{mm}) = 375\,\mathrm{mm} \tag{12}$$

Similarly, the number of turns is obtained from the equation

$$N = \frac{L}{2\pi R_M} \tag{13}$$

$$N = \frac{375\,\mathrm{mm}}{2\pi(13.65\,\mathrm{mm})} = 4.37 \tag{14}$$

To calculate the width of the spiral antenna arm it is given by:

$$g = w = \frac{r_{out} - r_{in}}{4N} \tag{15}$$

$$g = \frac{(22.7 - 4.6)\,\mathrm{mm}}{4 * 10} = 0.45\,\mathrm{mm} \tag{16}$$

The antenna is designed in one size and used to build two spiral arms [12]. The parameters included in the design of the spiral and its corresponding spacing are the turns g, the width of the arm w, the inner radius (r_{in}) and the outer radius r_{out}, as shown in the Fig. 1 (Table 2).

Table 2. Results on the calculations made for the parameters of the logarithmic spiral antenna.

Parameter	Abbreviation	Value	Redesigned	Unit
Speed of light	c	$3 * 10^8$	$3 * 10^8$	m/s
High frequency	f_H	3	5	GHz
low frequency	f_L	1	0.5	GHz
Inner radius	r_{in}	4.6	1	mm
Outer radius	r_{out}	22.7	43.5	mm
Number of turns	N	4.37	10	-
Arm width	w	0.45	1	mm
Separation between the arms	g	0.45	3	mm

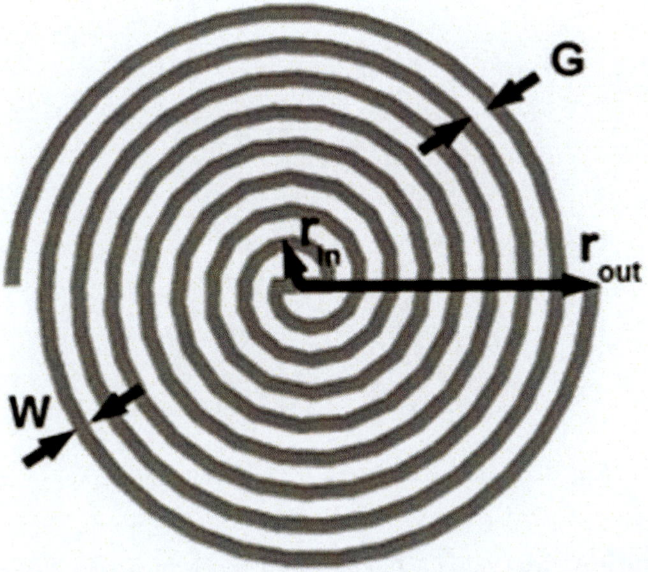

Fig. 1. Geometry of the two-armed Archimedean spiral antenna.

Simulation of the Logarithmic Spiral Antenna Array. With the calculated and redesigned parameters, the design is executed in CST Studio, starting from the creation of a new project, once the program is executed, the application area for the project is chosen (Figs. 2 and 3).

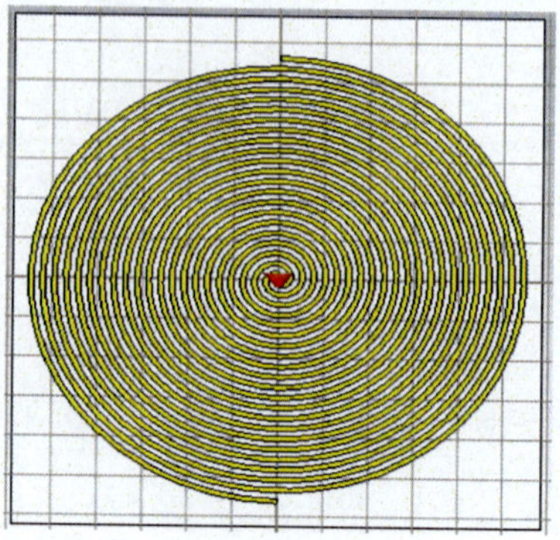

Fig. 2. Modeling of the logarithmic spiral antenna according to the redesigned calculations

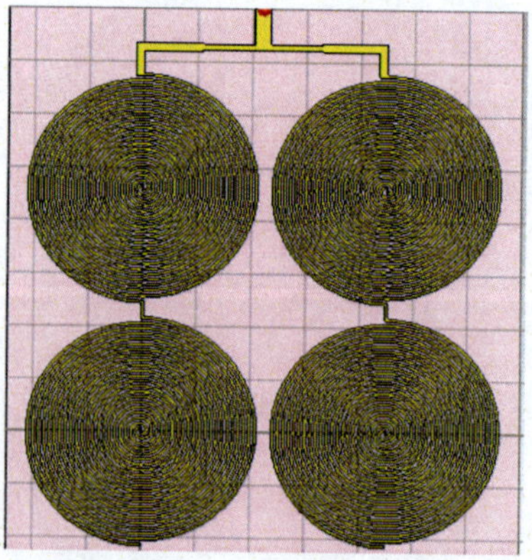

Fig. 3. Modeling of the logarithmic antenna array.

Characterization. Matlab is a fundamental tool which provides comparative graphs, through this software the comparison of the S11 parameters of the simulation in CST Studio and the data obtained from the miniNVA-Tiny analyzer is carried out [10] (Fig. 4).

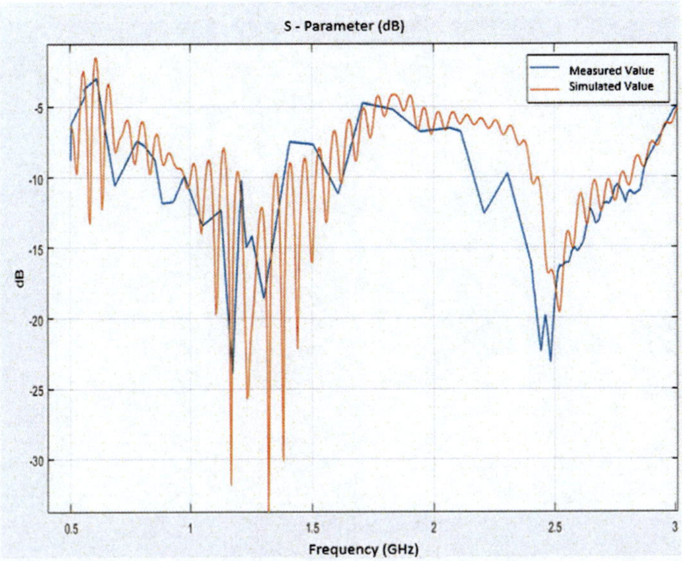

Fig. 4. MPlot of S11 of the logarithmic spiral antenna array.

Manufacture of the Logarithmic Periodic Antenna. The logarithmic spiral multiband antenna array, which the prototype is presented in Fig. 5, mentioned above said manufacturing process.

Structure for Voltage Rectifier. The structure of the rectifier is carried out in order to obtain an appropriate voltage to obtain a level according to the objectives established prior to a study of rectifiers/voltage multipliers. The following figure shows the circuit of the simulated system in Proteus based on the Villard circuit. Its elements are the following:

- SMA male connector
- 470 uF electrolytic capacitors
- 5 Schottky Diodes HSMS-286C
- 2-pin terminal block

Paul Villard's voltage multiplier has a simple configuration, that is, it consists of two capacitors and two diodes, the AC source connected to the input of the multiplier, operation begins when negative high cycles are used to charge the capacitor to the peak AC voltage (Vp), the output is the superposition of the input AC voltage and the capacitor DC voltage. The DC value of the waveform is changed using the circuit effect thereof, since the diode sets the negative peaks of the DC waveform to the value of 0 V and the positive peaks of the DC waveform output have a value of 2 Vp. An advantage of this multiplier circuit is that another stage can be taken and added to the previous one and so on, but

Fig. 5. Fabricated prototype of the logarithmic spiral antenna array.

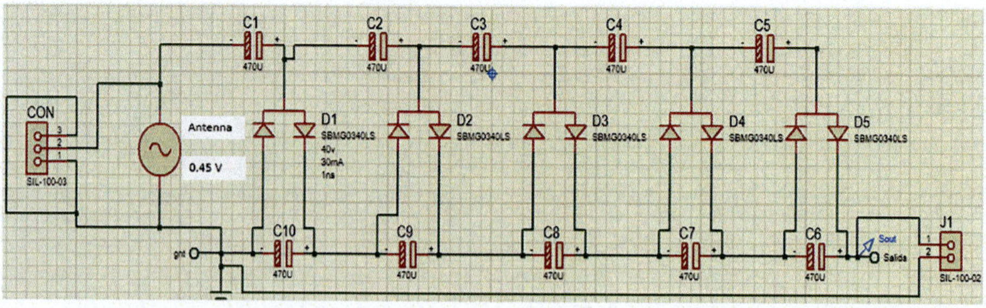

Fig. 6. Scheme of the multiplier circuit designed in Proteus.

it will have a voltage drop given by the elements and the number of stages, in order to reduce the voltage loss, the frequency must be increased, capacitance or both at the same time (Fig. 6).

Each stage in this circuit behaves like a passive voltage changer that generates a DC offset voltage for the next stage [15]. Next, we proceed to make the PCB board with its respective elements of the Schematic capture designed in Proteus, selecting the width for the position of the elements and the size of the board (Fig. 7).

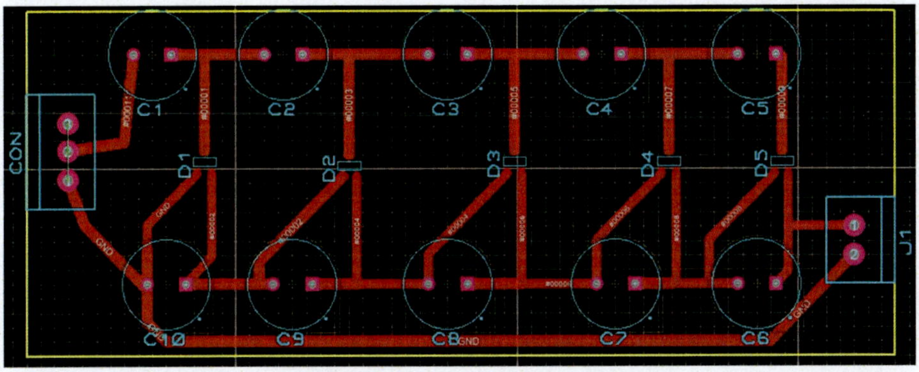

Fig. 7. PCB layout.

Next, we proceed to model the voltage multiplier circuit on a board with the elements mentioned above as shown in the Fig. 8.

Fig. 8. PCB layout.

3 Discussion of Results

The prototype of the proposed antenna array has been designed, simulated and measured. The design and simulation were carried out with the CST Studio software. Printing as a practical process was done on the RF-4 material with copper coating. Once the antenna array was printed, the performance tests were performed using the MiniVNA-Tiny antenna analyzer, it is a device that measures frequencies from 1 MHz to 3 GHz, a computer, USB cable and the miniVNA-TINY device software were used.

3.1 Electromagnetic Energy Harvesting System Performance Tests

(See Table 3).

Table 3. System tests without storage.

Ambient	Rectenna Matrix
Internet modem	104.4 mV
Library internet	72.9 mV
Offset antennas	139.4 mV

3.2 Storage Test

In each test carried out for the storage of voltage, the capacity in the battery was determined and after the energy collection process, for this case a 9 V battery and a 3.6 V battery were used.

The verification of the designed and manufactured rectenna matrix is detailed in the following tables with its timed charging times in the battery and in the cell obtained in practice, thus verifying that the system operates satisfactorily (Table 4).

Table 4. 9 V Battery Storage with Logarithmic Spiral Antenna Array.

Parameter	Initial Voltage (mV)	Time 30 min	Time 60 min	Time 120 min	Stored Voltage
Internet modem	732	773	786	806	74
Library internet	728	741	785	810	72
Offset antennas	6024	6129	6241	6365	341

The tests carried out in the different environments, with the battery, with the antenna array, the rectifier/voltage multiplier, these result in measurements according to the designs made in each simulation. The system coupled to the battery gives slow charge time, it is due to the chemical reaction produced within it, the RF waves have variation due to weather, the completion of the prototype and environmental factors. In general, the rectenna and its designs are promising since they provide a capture of the electromagnetic energy known in Wi-Fi and also in others (Table 5).

Table 5. Storage in the 3.6 V battery with the logarithmic spiral antenna array.

Parameter	Initial Voltage (mV)	Time 30 min	Time 60 min	Time 120 min	Stored Voltage
Internet modem	2654	2690	2697	2702	48
Library internet	3654	2660	3710	3727	73
Offset antennas	2366	2409	2808	3552	1186

3.3 Storage Graphs in the Different Load Environments

Figure 9 shows the battery storage obtained by the spiral logarithmic rectenna matrix in the 3 different environments in which the charge was carried out: Internet modem 48 mV, Library Internet 73 mV and Terrace with offset antennas 1186 mV.

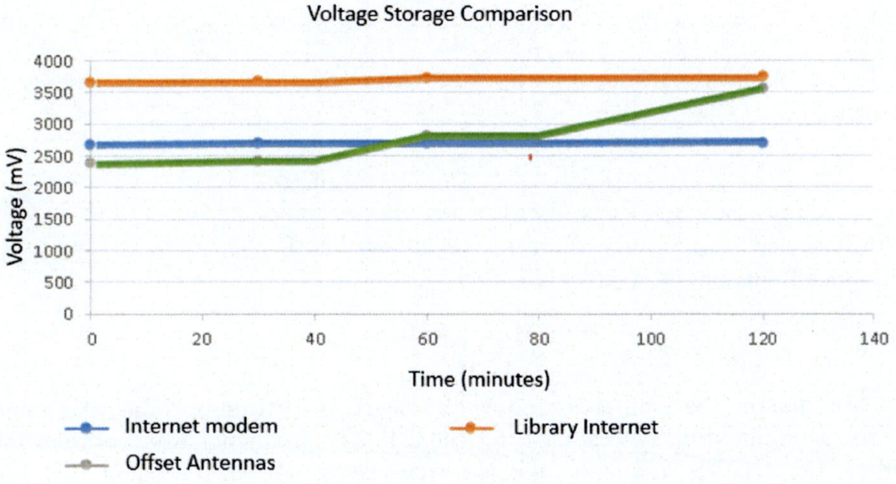

Fig. 9. Comparison of storage in the different environments.

3.4 Comparison of Measured and Simulated Values

Logarithmic Spiral Antenna Array Analysis. In the antenna array structure the reflection coefficient in both the simulated and actual measured value is mainly used for comparison. It is thus that the values obtained from the simulation and the values of the real measurement show that the antenna array operates optimally established in the frequency of 2.4 GHz. It is evident that there will be minimal losses, this is due to the material of the plate, welding and environmental factors.

The value in the CST Studio simulator of up to −10 dB of the antenna array is optimal to implement.

To calculate the error between the simulated value and the actual measured value, the following equation is used.

$$\%error = \frac{|V_(aprox) - V_(real)|}{V_(real)} * 100\% \tag{17}$$

In the simulation, the error percentage with respect to the intended resonance frequency (2.45 GHz) is:

$$\%error = \frac{|2.475\,\text{GHz} - 2.45\,\text{GHz}|}{2.45\,\text{GHz}} * 100\% = 1.02\% \tag{18}$$

In the real mean value, the error percentage with respect to the intended resonance frequency (2.45 GHz) is:

$$\%error = \frac{|2.49\,\text{GHz} - 2.45\,\text{GHz}|}{2.45\,\text{GHz}} * 100\% = 1.63\% \tag{19}$$

In the simulation, the error percentage with respect to the intended resonance frequency (2.4 GHz) is:

$$\%error = \frac{|2.47\,\text{GHz} - 2.4\,\text{GHz}|}{2.4\,\text{GHz}} * 100\% = 2.91\% \tag{20}$$

In the real mean value, the error percentage with respect to the intended resonance frequency (2.4 GHz) is:

$$\%error = \frac{|2.49\,\text{GHz} - 2.4\,\text{GHz}|}{2.4\,\text{GHz}} * 100\% = 3.75\% \tag{21}$$

After making the comparisons, it can be clearly evidenced that the percentage of error is minimal in the real part, for the 2.45 GHz frequency it is less than 2% and for the 2.4 GHz frequency it is less than 4%, taking into account that the percentage error should be less than 5%. Which converts this array of antennas to operate normally at the desired working frequency.

4 Conclusions

This project presents the array of logarithmic spiral antennas designed and manufactured to be used in capturing electromagnetic energy at the 2.4 GHz frequency with considerable bandwidth. This set of antennas is composed of four spiral antennas connected to each other using microstrip technology. Once the calculations have been made: width in each arm of the antennas, space between them, distance between each antenna and coupling parameters, it is concluded that the values in the simulation, the values measured with the miniNVA Tiny instrument and the real measured values present valid results and demonstrating its effectiveness applied in this design. In addition, being a patch type antenna array will make it very useful and easy to incorporate with other devices.

Through the investigations carried out, it was possible to determine each of the parameters and adapt efficiently to the present project, the simulation was carried out with the help of the specialized software CST Studio 2018, being a robust program to design antennas, this tool allowed the construction parameters to be varied of the spiral antenna array, sequentially analyzing the results mainly of the S11 parameter and the value of the VSWR until optimal results are achieved within the centralized frequency of 2.4 GHz, as it is the most explored frequency in the electromagnetic spectrum.

For the design of the rectifier/voltage multiplier, it is carried out correctly with an optimal result, it should be noted that it is directly related to the logarithmic spiral antenna array, the simulation was carried out correctly in the Proteus software, the energy captured by the array of The antennas are usually small, which is why the circuit based on the Villard multiplier is implemented, since it is an excellent topology, therefore it delivers higher output voltages.

Through the design and analysis of the rectenna matrix, the proposed objectives were achieved and it was verified that it works in the 2.4 GHz frequency. The measurements were carried out in different environments, resulting in a load of 48 mV in the environment close to a home internet modem, close to the Internet in the library the load was 73 mV and in offset antennas of the Faculty (FISEI) the charge was 1186 mV after two hours of charging time, in this way it operates in the communication systems in 4G LTE and Wi-Fi technology. This matrix of rectennas are compact and of an acceptable cost.

Acknowledgements. The authors thank the invaluable contribution of the Technological University Indoamerica, for his support in conducting the research project "ESTUDIO DE ALGORITMOS HIBRIDOS DE APRENDIZAJE AUTOMATICO PARA LA PREDICCIÓN DE GENERACIÓN DE ENERGÍAS RENOVABLES", Project Code: 281.230.2022. Also, the authors thank the Technical University of Ambato and the "Dirección de Investigación y Desarrollo" (DIDE) for their support in conducting this research, in the execution of the project "Captación de Energía Limpia de Baja Potencia para Alimentación de Dispositivos de Quinta Generación (5G)", approved by resolution "Nro. UTA-CONIN-2022-0015-R". Project code: SFFISEI 07.

References

1. Barrón, D., Stevens, E., Arce, A., Cárdenas, M., Pineda, U.: Comparacion de circuitos multiplicadores de voltaje para cosecha de energia en rf (comparison of voltage multiplier circuits for rf energy harvest). Pistas Educativas **42**(136) (2020)
2. Barros, O.P.: Antena independiente de la frecuencia, espiral de arquímides impresa. Telemática **16**(3), 41–52 (2017)
3. Caiza, G., Bonilla-Vasconez, P., Garcia, C.A., Garcia, V.M.: Augmented reality for robot control in low-cost automation context and IoT. In: 2020 25th IEEE International Conference on Emerging Technologies and Factory Automation (ETFA), pp. 1457–1460 (2020)
4. Caiza, G., Bustos, S., Garcia, C.A., Garcia, M.V.: Low-cost cyber-physical production systems based on IEC 61499 for analog process. In: Russo, D., Ahram, T., Karwowski, W., Di Bucchianico, G., Taiar, R. (eds.) IHSI 2021. AISC, vol. 1322, pp. 336–342. Springer, Cham (2021). https://doi.org/10.1007/978-3-030-68017-6_50
5. Divakaran, S., Krishna, D., et al.: Dual-band multi-port rectenna for RF energy harvesting. Progress Electromagnet. Res. C **107**, 17–31 (2021). https://doi.org/10.2528/PIERC20100802
6. Ibrahim, H.H., Singh, M.S., Al-Bawri, S.S., Islam, M.T.: Synthesis, characterization and development of energy harvesting techniques incorporated with antennas: a review study. Sensors **20**(10), 2772 (2020)
7. Juan, P., Analuiza, O., Carpio, M., Castillo, W.: Optimization of the mechanical properties responses of SBR 1502 rubber/amorphous silicon dioxide/others by DOE-MSR methodology. In: García, M.V., Fernández-Peña, F., Gordón-Gallegos, C. (eds.) Advances and Applications in Computer Science, Electronics and Industrial Engineering. AISC, vol. 1307, pp. 215–231. Springer, Singapore (2021). https://doi.org/10.1007/978-981-33-4565-2_14
8. Paredes Larroca, F., Saettone Olschewski, E., Quino Favero, J., Rosales Huamaní, J., Castillo Sequera, J.L.: Development of a fuzzy logic-based solar charge controller for charging lead–acid batteries. In: Nummenmaa, J., Pérez-González, F., Domenech-Lega, B., Vaunat, J., Oscar Fernández-Peña, F. (eds.) CSEI 2019. AISC, vol. 1078, pp. 168–183. Springer, Cham (2020). https://doi.org/10.1007/978-3-030-33614-1_12
9. Martínez García, H.: Diseño de una plataforma docente para el estudio de los sistemas de recolección de energía (energy harvesting) en los grados de ingeniería. In: Proceedings of the XIIICHLIE-13th Spanish-Portuguese Conference on Electrical Engineering, pp. 1–4. Universitat Politècnica de València (2013)
10. MiniVNA: Mini-VNA tiny antenna analyzer up to 3 GHz (2019). https://www.wimo.com/en/minivna-tiny
11. Nogales, R., Guilcapi, J., Benalcazar, F., Vargas, J.: A brief literature review of mathematical models of EMG signals through hierarchical analytical processing. In: Garcia, M.V., Fernández-Peña, F., Gordón-Gallegos, C. (eds.) CSEI 2021. LNNS, vol. 433, pp. 273–287. Springer, Cham (2022). https://doi.org/10.1007/978-3-030-97719-1_16
12. Reinoso, F., Gordón, C., Cumbajín, M., Peñafiel, C.: Wide band antennas for energy harvesting. In: Botto-Tobar, M., Montes León, S., Torres-Carrión, P., Zambrano Vizuete, M., Durakovic, B. (eds.) ICAT 2021. CCIS, vol. 1535, pp. 18–33. Springer, Cham (2022). https://doi.org/10.1007/978-3-031-03884-6_2

13. Shafiq, R.R., Elsherbini, M., Omara, A., ElHalawany, B.M.: Green energy scavenging for powering mobile phones. In: 2022 International Telecommunications Conference (ITC-Egypt), pp. 1–6 (2022). https://doi.org/10.1109/ITC-Egypt55520.2022.9855728
14. Suresh Chinnathampy, M., Aruna, T., Muthukumaran, N.: Design and fabrication of micro strip patch antenna for cognitive radio applications. Wirel. Pers. Commun. **121**(3), 1577–1592 (2021). https://doi.org/10.1007/s11277-021-08685-9
15. Zapata, E., López, F., Góez, G.: Dual band rectenna for wireless energy harvesting in the 2.40 GHz and 5.38 GHz. TecnoLógicas **25**(55), e2384–e2384 (2022). https://doi.org/10.22430/22565337.2384

Electronic Bracelet with Artificial Vision for Assisting Blind People

Jessica Tipantuña, Alan Rodriguez, William Oñate, and Gustavo Caiza(✉)

Universidad Politécnica Salesiana, UPS, 170146 Quito, Ecuador
jtipantuna@est.ups.edu.ec, {arodriguez,wonate,gcaiza}@ups.edu.ec

Abstract. In recent years technological progress has enabled the development of new applications in different fields, among which there is social aid as one of the areas in constant growth, where various applications and electronic devices have been developed. Visual disability and blindness are the main types of disabilities that are present in human beings around the world. This document presents the development of a bracelet prototype that uses artificial vision to assist blind people, providing information to the user about the recognition of garments with their corresponding colors, and using the camera for acquiring images in real-time with hearing feedback. The proposed system is constituted by bof2 subprocesses. The first carries out the detection of garments and consists of 5 steps: image processing, training with images, garment detection, comparison, and result. The second subprocess carries out color detection. The method of HAAR feature-based classifiers was used here, which yielded average detections of 88.3% depending on the size of the garment and the distance to it.

Keywords: Artificial Vision · OpenCV · Object detection

1 Introduction

The process of identifying an image is also known as digital image processing and has been carried out in many studies to support works in different fields, such as medicine, industry, home, agriculture, etc. [1]. Since computer vision and bioengineering have become more efficient, researchers in both fields have recently worked together to find a solution for visual disability (VD). There is a considerable number of studies based on electronic mobile devices, to encourage people with visual disabilities to maintain their daily activities without needing them as individuals. Among the approaches based on electronics, systems to avoid obstacles are one of the most studied areas, which are based on the idea of recognizing the object and alerting the user [2].

Visual disability and blindness are the main types of disabilities that are present in human beings around the world and are classified in five degrees of severity. This is an important topic in both the social and scientific society, to find an appropriate solution to this problem. Among the people with special needs, blind people face more difficulties than others in collecting information [3, 4]. Human beings obtain about 70% of their information through the sense of sight. Recent challenges emerged to the scientific

M. V. Garcia and C. Gordón-Gallegos (Eds.): CSEI 2022, LNNS 678, pp. 366–379, 2023.
https://doi.org/10.1007/978-3-031-30592-4_25

surface with the development of prototypes of artificial eyes. New technologies such as artificial vision (AV), process and adapt the dimension, color, and luminosity of the images acquired, which is performed by assessing the fundamental features of the image and obtaining information for different classifications [5].

There are currently various techniques in use for detecting and classifying objects, and the research community is still identifying which method will be the most cost-effective without compromising speed and precision. The different complexities that arise in detection and classification are the variable attributes, such as the figure and magnitude of the object, color, pose, shadows, and background disorder. In addition, the camera angle and the video quality of the frames add layers of complexity. This method uses Haar cascades, a classifier used with the purpose detect which it has been trained [6].

The proposed system was developed to assist blind people in the detection of garments and color, to enable them to be independent when shopping. This document proposes an object detection system that uses the cascade classifier, which is based on automatic learning and is trained using many positive and negative images of the object. The paper is organized as follows: Sect. 2 presents the methodology used for carrying out this work, Sect. 3 indicates the design and development of the proposed prototype, Sect. 4 presents and validates the results and Sect. 5 presents the conclusions.

2 Methodology

2.1 Visual Disability

Visual disability encompasses all the deficiencies and difficulties that a person can not carry out in his/her daily life, such as self-care, and also the different problems that he/she may face in everyday places such as school, work, or the interaction with his/her physical or social environment. When a person with a visual disability does not have access to eye care services to technical aids such as special glasses, guide canes, and mobilization bracelets would drastically increase the burden of the VD and the disability degree in his/her daily environment. Electronic devices for assisting persons with visual disabilities enable them to be independent in activities for which they would need the assistance of third parties to carry them out [7].

2.2 Raspberry Pi

The Raspberry Pi is a low-cost computer that may perform simple tasks, such as making a led flicker, as well as complex tasks, such as image processing [8]. A Raspberry Pi 3B+ is used in this project (Table 1).

2.3 OpenCV (Open-Source Computer Vision Library)

It is a library more frequently used for artificial vision. It was mainly constructed to increase the use of computer vision. This library has available a large number of optimized algorithms, that include a basic algorithm for automatic learning to deep learning

Table 1. Parameters of the raspberry PI 3

Elements	Features
Processor	64 bits broadcom BCM283
Ram	1 GB
Wireless Network	BMC 43438
Bluetooth	Low Energy (BLE)
GPIO	40 pins
Audio	Output type JAR 3.5mm
Output ports	Camera CSI, touch screen DSI
Connections	USB-C 3.0

and various applications of computer vision such as face detection and object detection. The multiplatform feature is the main advantage of the OpenCV library, i.e., that it may be implemented in different programming platforms. Various applications of OpenCV are face recognition, gesture recognition, object recognition, movement tracking and many more [9].

2.4 Cascade Classifier

The cascade classifier is a chain of weak classifiers for efficient classification of the regions of the image. Its objective is to improve the performance of object detection and reduce calculation time. When a filter does not pass the regions of the image, that specific sub-window of the image is eliminated for its further processing. Then it is considered as not being the object to be detected, which implies that the regions of the processed image do not contain the object to be detected. This is very important for the performance of the classifier since all or almost all the sub-windows of negative images will be eliminated in the first stage. On the other hand, when the regions of the image successfully pass the filter, they pass to the subsequent stage, which contains a more complex filter. It is considered that only the regions that successfully pass all filters contain a coincidence of the object. This implies that the regions of the image contain the object subject to detection [10].

2.5 Haar-Cascade Classifier Method

The Haar-cascade method, also known as the Viola-Jones method, is the most frequently used method for object detection. The object detection processes using this method consist of four main parts, i.e., features similar to Haar, integral image, adaptive reinforcement or AdaBoost, and a combination of cascade classifiers [11].

2.6 Image Processing

Carrying out various image processing operations with the OpenCV tool has significantly reduced the difficulties for the development of modules based on image processing or computer vision. It is a programming platform that has a large number of methods that solve a great part of the image processing needs [12, 13].

GaussianBlur. It is an OpenCV method that enables to apply an image smoothing using a Gaussian filter, through a lowpass filter that eliminates high-frequency noise, considered as the pixels that change in a fast manner, it carries out a summation and obtains a solution matrix [14]. The code that implements the aforementioned function is cv2.GaussianBlur(src, dst, ksize): scr is the input, dst is the output and ksize is the dimension of the Gaussian kernel.

Threshold. Segmentation according to threshold has been widely used in image segmentation due to its simplicity, speed, and efficiency, and it is very important for further image processing and recognition; the command that implements such function is [15]. cv2.threshold(src, dst, thresh, maxval): where scr is the input, dst is the output, thresh is the threshold value of type double, maxval is a magnitude used with the threshold types THRESH_BINARY, THRESH_BINARY_INV or cv2.THRESH_OTSU.

CvtColor. This function transforms an input image of a color dimension into another. In the RGB color dimension, the directive of the channels has to be concretely established RGB or BGR). It should be taken into account that the predetermined format is called RGB, but in OpenCV, the bytes are reversed (BGR). Then, for a 24-bit standard color image, the first byte should be an 8-bit blue component, the second is green and the third be red. Then, the fourth, fifth, and sixth bytes would be the second pixel (blue, then green, then red), and so on. The code is cv2.cvtColor(), which uses two parameters. One is obtained using the OpenCV imread() method that represents the reference of the source image and the second argument is used to convert the color format [17].

cv2.COLOR_BGR2GRAY: Convert between RGB and grayscale with a range from 0 to 255. The methods widely used include averaging the values in each BGR color channel, color space desaturation, weighted sum method, and another color-to-gray approach [18].

cv2.COLOR_BGR2HSV: The BGR to HSV conversion (tone saturation value) is a common practice for CV projects. The use of HSV enables separate color and intensity, which helps to reduce the effects of light problems [19].

FindContours. The OpenCV "findContours" method is used to find closed shapes in the image. This method finds all the contours of the digital image using the algorithm known as "Suzuki85". Contours are important to analyze shapes. After finding all contours, it is selected the biggest ten closed shapes among all contours [20]. Code findContours (image, contours, hierarchy): image in the 8-bit input channel, contours are a method for the contours detected, hierarchy is the output vector which stores information of image topology.

2.7 ContourArea

This method determines a contour area. Similar to momentums, the area is determined using Green's formula. Therefore, the area returned and the number of pixels is different than zero, if the contour is drawn using drawContours or fillPoly, they may be different. In addition, the function will surely give incorrect results for contours will self-intersections [20–25, 26].

3 Design

The purpose of the present work is to develop a bracelet to assist blind people; the bracelet is divided into two subprocesses, namely image detection, and color detection. The user will have audio feedback when the device detects the object. Figure 1 shows the general diagram of the final prototype.

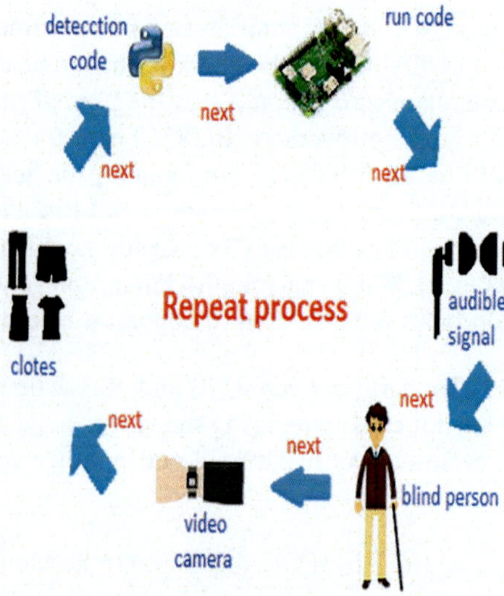

Fig. 1. Operating Diagram

3.1 Generation of Samples

The samples are divided into two parts: positive samples and negative samples. The positive samples are the images that contain the object and the negative samples are the background images. Figure 2 illustrates the process for obtaining the samples. The input image of dimension 300 × 400 is processed (Fig. 2a). First, the image was changed to grayscale (Fig. 2b), which varies in a range from 0 to 255; afterward, the filter GaussianBlur was applied, which uses grayscale and enables blurring the image and eliminating noise scanning each of the pixels in the image (Fig. 2c). Finally, a sum of the Ostu and binary filters are applied, resulting in a binary image without noise (Fig. 2d).

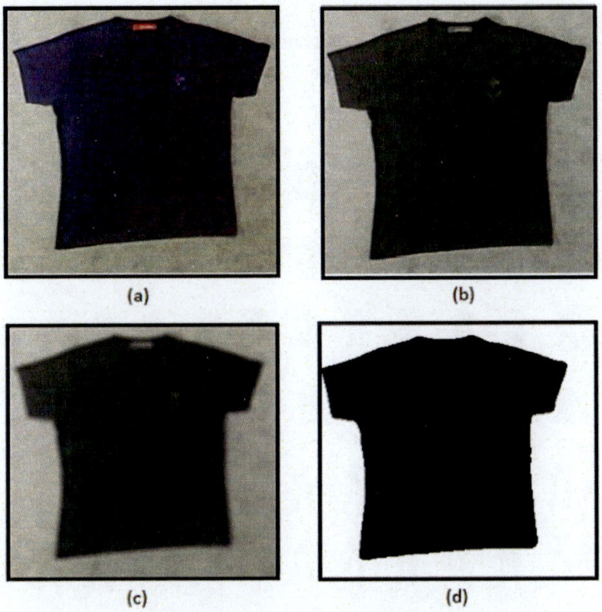

Fig. 2. Image Preprocessing

The processing was carried out to obtain a sharp image, without interferences and noise. Figure 3a shows the histogram of the real-time video which includes the plot of the 3 RGB channels; in Fig. 3b and Fig. 3c it is reduced from three channels to one channel and the image is distorted, and in Fig. 3d the otsu-binary method automatically calculates the threshold value between the two values observed in the plot.

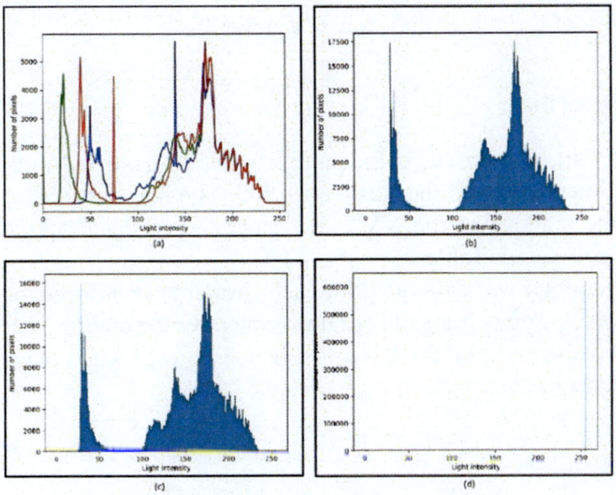

Fig. 3. Image Processing Histograms

Figure 4 shows the storage of 200 positive images of each garment and 250 negative images where the garment was not present.

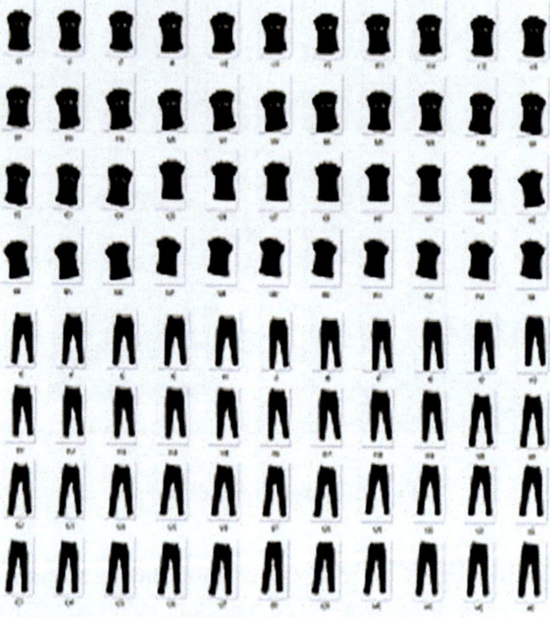

Fig. 4. Examples of Positive Samples

The last step included training each cascade with the positive and negative images of each of the garments to be detected in the cascade trainer guide program, where it is entered the address of the folder that stores the images.

3.2 Garment Detection

For garment detection, it is carried out the same processing in the real-time video captured by the camera is integrated into the Raspberry Pi 3. The features of the video input are compared with the features of the XML file generated for each garment through the detectmultiscale command (Table 2).

In addition, a counter was generated for each garment to reinforce recognition of the object, since for every 2 detections the program generates the audible output for garment recognition, which also reduced the false positives generated when placing the garment in a controlled environment.

3.3 Detection of Colors

The first step for color detection is to select the minimum and maximum ranges to be evaluated, in RGB components, for which such range is stored in a matrix of the

Table 2. Parameters of detect multiscale.

Clothes	ScaleFactor	MinNeighbor	MinSize	MaxSize
T-shirt	1.1	5	240x240	700x700
Pants	1.2	5	240x240	700x700
Short	1.1	5	240x240	700x700
Hoodies	1.2	5	240x240	700x700
Skirt	1.2	5	240x240	700x700
Shirt	1.1	5	240x240	700x700

numpy library with the function np. array. Within an infinite loop, the real-time video in RGB is converted to HSV components using the COLOR_BGR2HSV function, and subsequently, the ranges to be detected are searched using the cv2.inRange function for every color, taking as variables the real-time video matrix, and the minimum and maximum margins of the color.

A function is created that takes as input the matrix with the color ranges in the video and the color to be searched for. An additional step was added to configure the system, which consists of visualizing and grouping the colors recognized using the cv2.findContours function, which requires as variables a matrix with the color ranges in the video, cv2.RETR_EXTERNAL and cv2.CHAIN_APPROX_SIMPLE to thus find the contours of the colors. At this point, the color selection should be filtered according to the area using the cv2.contourArea function and specifying that the detection is only with areas larger than 5000. Once the color is detected the contours are drawn with the cv2.drawContours function, and the cv2.putText text function specifies the color of the real-time video. It is important to clarify that the contours and the text will be erased from the final result. The last step is to use a function for reproducing the sound, which uses the pygame library to upload the audio path in wav format. Afterward, the length

Table 3. Color detection ranges.

Heading level	Example	Font size and style
Blue	[100,100,47]	[112,255,255]
Red	[173,175,20]	[179,255,255]
Yellow	[25,154,163]	[30,255,255]
Purple	[130,75,40]	[149,255,255]
Black	[0,0,0]	[179,255,35]
Green	[37,50,96]	[79,246,229]
Brown	[10,80,0]	[25,130,255]
Wine	[0,90,0]	[10,155,101]

of the sound is calculated to obtain the delay time, and finally, the audio is reproduced, and then comes a delay (Table 3).

3.4 A Design of the Prototype

The base of the prototype was designed and modeled in the Solidworks software with dimensions 12 mm (height) × 35 mm (width) × 33 mm (depth); this base was printed in PLA (polylactic acid) filament, which is a thermoplastic manufactured using renewable resources, the bracelet straps were made with elastic material to enable the prototype to adapt to the wrist (Fig. 5).

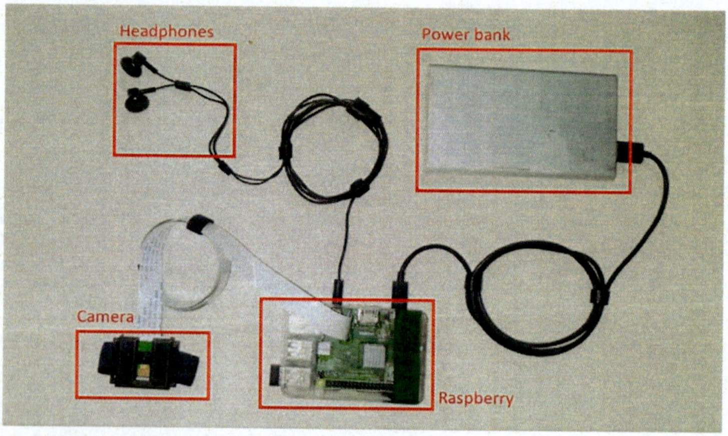

Fig. 5. Prototype

4 Results

The performance tests were conducted in a controlled space, in which the detection height was 1.40 m; this measure was taken as a reference since it corresponds to the height from the floor to the arm completely extended of an average person. The samples of each garment were captured at that height, with a white background. Figure 6 shows the detection of 2 garments and colors.

The accuracy of the prototype was measured with 25 items of each garment, obtaining the results shown in Table 4.

The results obtained through the analysis of the detection of images were significant since 88.3% of the objects were correctly recognized. Thus, the accuracy values shown in Table 4 were achieved; the difference in these results is because the user should present garments similar to the garments with which every cascade was trained, and, in addition, the user should place his/her hand at an approximate distance of 1.40 m to reduce errors due to the size of the garments.

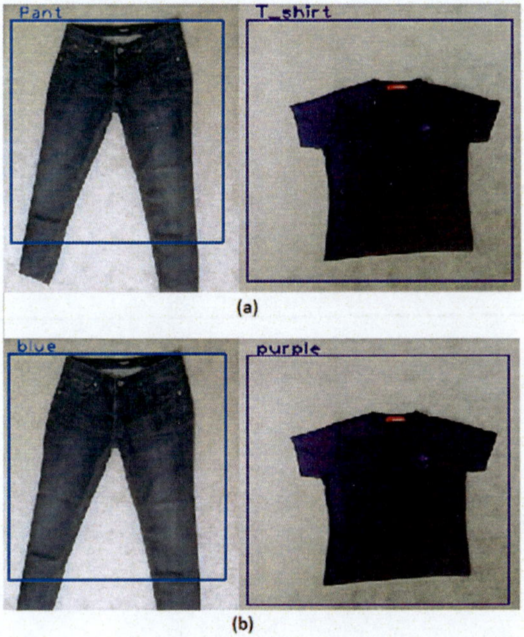

Fig. 6. Detection Example

Table 4. Percentage of accuracy

Clothes	Examples	Accuracy
T-shirt	25	96%
Pants	25	92%
Short	25	90%
Hoodies	25	80%
Skirt	25	84%
Shirt	25	88%

Figures 7 and 8 show the detection tests varying the height between 0.75 m and 1.5 m; this test range was chosen since for smaller values the bigger garments will not fit in the input video square, and for heights greater than 1.5 m the garments of smaller size will be interpreted as noise. It is observed in Fig. 7a that 2 false positives are generated for a scale factor of 1.01, which does not happen for a larger scale factor. In addition, when the scale factor is larger than 1.2, Fig. 8a–b shows that the number of garments detected is reduced.

The color detection results are represented in each of the masks that clean the input to detect the color, as shown in Fig. 8, which act on the input in real-time and determine the color found in the image is within the range established. At this point, environment

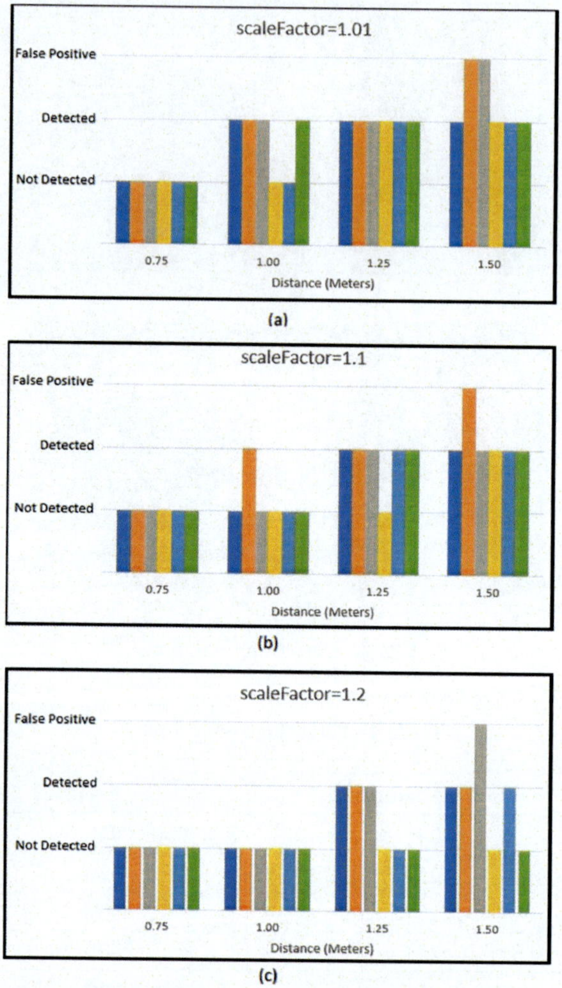

Fig. 7. Accuracy Test with Scale Factors From 1.3 to 1.4

Fig. 8. Color Detection Result

variations should be minimum, otherwise, there will be no detection. For the mean detection height, the optimum range is between 1.50 m and 1.25 m resulting in an average of 1.40 m, since in this range there are more garments detected. A factor to be taken into account is that false positives are generated for values above 1.50.

5 Conclusions

The design of the low-cost bracelet with artificial vision as a prototype for assisting blind people in real-time describes a practical and affordable manner for recognizing garments independently; the total cost of the prototype is 60 USD. For validating the prototype, tests were conducted using different garments in various sizes and colors, and it was observed that the best results in object detection were obtained through Haar feature-based classifiers, which use automatic learning; they achieved an average accuracy of 88.33%.

Height is one of the fundamental parameters for detection, as shown in Figs. 7 and 8; as the scale factor increases accuracy is reduced, whereas if it decreases there is a higher probability that false positives are generated and the processing time will be larger since there will be more samples to be processed by the image pyramid system used in the algorithm, that for every 0.1 tenths in the scale factor the image will reduce 10%, which helps to detect objects at different heights.

References

1. Rahmat, R.F., Azzakirot, Y., Lini, T.Z.: Tree identification to calculate the amount of palm trees using haar-cascade classifier algorithm. In: 2019 3rd International Conference on Electrical, Telecommunication and Computer Engineering, ELTICOM 2019, pp. 36–39 (2019). https://doi.org/10.1109/ELTICOM47379.2019.8943897
2. Pehlivan, S., Unay, M., Akan, A.: Designing an obstacle detection and alerting system for visually impaired people on sidewalks. In: TIPTEKNO 2019 - Tip Teknolojileri Kongresi, pp. 1–4 (2019). https://doi.org/10.1109/TIPTEKNO.2019.8895181
3. Iqbal, A., Farooq, U., Mahmood, H., Asad, M.U.: A low cost artificial vision system for visually impaired people. In: 2009 International Conference on Computer and Electrical Engineering, ICCEE 2009, vol. 2, pp. 474–479 (2009). https://doi.org/10.1109/ICCEE.2009.187
4. de Tristán, G., Arcia, R., Pérez Montes, H.: Aplicación móvil para el monitoreo de personas con discapacidad visual. Univ. Tecnológica Panamá (2016)
5. Pezoa, W.G., Domínguez, M.J.: Combined approach using artificial vision and neural networks for facial recognition, pp. 2–6 (2017)
6. Ashwini, B., Yuvaraju, B.N., Pai, A.Y., Aditya Baliga, B.: Real time detection and classification of vehicles and pedestrians using haar cascade classifier with background subtraction. In: 2nd International Conference on Computational Systems and Information Technology for Sustainable Solution, CSITSS 2017, pp. 187–192 (2018). https://doi.org/10.1109/CSITSS.2017.8447818
7. World Health Organisation: Informe mundial sobre la visión, vol. 214, no. 14 (2019)
8. RaspberryPi: Raspberry Pi 3 Model B+ Datasheet. Datasheet, p. 5 (2016). https://static.raspberrypi.org/files/product-briefs/Raspberry-Pi-Model-Bplus-Product-Brief.pdf

9. Jain, A., Garg, G.: Gun detection with model and type recognition using haar cascade classifier. In: Proceedings of 3rd International Conference on Smart Systems and Inventive Technology, ICSSIT 2020, pp. 419–423 (2020). https://doi.org/10.1109/ICSSIT48917.2020.9214211
10. Guennouni, S., Ahaitouf, A., Mansouri, A.: Face detection: comparing Haar-like combined with cascade classifiers and Edge Orientation Matching. In: 2017 International Conference on Wireless Technologies, Embedded and Intelligent Systems, WITS 2017 (2017). https://doi.org/10.1109/WITS.2017.7934604
11. Setjo, C.H., Achmad, B.: Thermal image human detection using Haar-cascade classifier. In: Proceedings - 2017 7th International Annual Engineering Seminar (2017). https://doi.org/10.1109/INAES.2017.8068554
12. Osimani, C.: Análisis y procesamiento de imágenes para la detección del contorno labial en pacientes de odontología, pp. 1–6 (2014)
13. Kumar, S., Srivastava, A.K., Jha, A., Adarsh, A., Gupta, R., Kumar, B.: Implementation of linear structuring element in OpenCV for blood vessel segmentation from color fundus images. In: 2019 10th International Conference on Computing, Communication and Networking Technologies, ICCCNT 2019, pp. 6–10 (2019). https://doi.org/10.1109/ICCCNT45670.2019.8944610
14. Condo Simbaña, J.A., Chávez Molina, C.E.: Desarrollo de un bastón electrónico para mejorar la movilidad de personas con discapacidad visual apoyado mediante visión artificial. Universidad Politécnica Salesiana (2019)
15. Xiao, L., Ouyang, H., Fan, C.: An improved Otsu method for threshold segmentation based on set mapping and trapezoid region intercept histogram. Optik **196**, 163106 (2019). https://doi.org/10.1016/j.ijleo.2019.163106
16. Wang, J.: Study on the performance identification of OpenCV in cashew nut shell-based activated carbon. IOP Conf. Ser. Earth Environ. Sci. **769**(3), 032030 (2021). https://doi.org/10.1088/1755-1315/769/3/032030
17. Nanduri, A.K., Neelapu, R., Maddumala, V.R., Domathoti, B.R., Gowrisankar, K.: Image recolorization with convolutional neural network using Opencv. J. Nat. Remedies **21**(4), 81–87 (2020)
18. Raveendran, S., Edavoor, P.J., Kumar, N.Y.B., Vasantha, M.H.: Design and implementation of reversible logic based RGB to gray scale color space converter. In: IEEE Region 10 Annual International Conference on Proceedings/TENCON, pp. 1813–1817 (2019). https://doi.org/10.1109/TENCON.2018.8650243
19. Ganesan, P., Rajini, V.: Assessment of satellite image segmentation in RGB and HSV color space using image quality measures. In: 2014 International Conference on Advances in Electrical Engineering, ICAEE 2014 (2014). https://doi.org/10.1109/ICAEE.2014.6838441
20. Garg, U., Goyal, V.: License plate recognition system using OpenCV & PyTesseract. Indian J. Sci. Technol. **3**(12), 31–35 (2020). https://doi.org/10.17485/ijst/2016/v9i12/86631
21. Caiza, G., Garcia, C.A., Naranjo, J.E., Garcia, M.V.: Assessment of engineering techniques for failures simulation in induction motors using numerical tool. In: Iano, Y., Arthur, R., Saotome, O., Kemper, G., Padilha França, R. (eds.) BTSym 2019. SIST, vol. 201, pp. 307–319. Springer, Cham (2021). https://doi.org/10.1007/978-3-030-57548-9_28
22. Caiza, G., Ibarra-Torres, F., Garcia, M.V., Barona-Pico, V.: Use of bots to support management software development and streamline client/producer communication in the 5.0 industry. In: Rocha, Á., Adeli, H., Dzemyda, G., Moreira, F., Ramalho Correia, A.M. (eds.) WorldCIST 2021. AISC, vol. 1367, pp. 401–410. Springer, Cham (2021). https://doi.org/10.1007/978-3-030-72660-7_39
23. Espinosa, R.V., Soto, M., Garcia, M.V., Naranjo, J.E.: Challenges of implementing cleaner production strategies in the food and beverage industry: literature review. In: García, M.V., Fernández-Peña, F., Gordón-Gallegos, C. (eds.) Advances and Applications in Computer

Science, Electronics and Industrial Engineering. AISC, vol. 1307, pp. 121–133. Springer, Singapore (2021). https://doi.org/10.1007/978-981-33-4565-2_8

24. Osorio-Carlozama, J., Llerena-Izquierdo, J.: Utility of computer hardware recycling technique for university learning: a systematic review. In: Garcia, M.V., Fernández-Peña, F., Gordón-Gallegos, C. (eds.) Advances and Applications in Computer Science, Electronics, and Industrial Engineering, pp. 175–189. Springer, Cham (2021). https://doi.org/10.1007/978-3-030-97719-1_10
25. Varela-Aldás, J., Miranda-Quintana, O., Guevara, C., Castillo, F., Palacios-Navarro, G.: Educational robot using lego mindstorms and mobile device. In: Nummenmaa, J., Pérez-González, F., Domenech-Lega, B., Vaunat, J., Oscar Fernández-Peña, F. (eds.) CSEI 2019. AISC, vol. 1078, pp. 71–82. Springer, Cham (2020). https://doi.org/10.1007/978-3-030-33614-1_5
26. Yang, H., Yan, Y., Yu, Z., Ning, Z.: Micro pin header defect detection system based on OpenCV. J. Phys. Conf. Ser. **2137**(1), 1–6 (2021). https://doi.org/10.1088/1742-6596/2137/1/012037

IoT Flowmeter to Obtain the Real Provision of Drinking Water of the Administrative Building at Universidad de Las Fuerzas Armadas ESPE

David-Vinicio Carrera-Villacrés[1,2(✉)], Alejandra-Carolina Cabrera-Torres[1], Grace Chiriboga[1,2], and Holger Chuquin[1]

[1] Universidad de las Fuerzas Armadas - ESPE, Av. General Rumiñahui S/N y, Ambato, Ecuador
dvcarrera@espe.edu.ec

[2] Facultad de Ingeniería en Geología, Minas Petróleos y Ambiental, Carrera de Ingeniería Ambiental, Universidad Central del Ecuador, Av. Universitaria, 170129 Quito, Ecuador
http://www.espe.edu.ec, http://www.uce.edu.ec

Abstract. Nowadays, the actual collection of drinking water helps to determine more precisely the amount of water that enters a facility. Thus, this research work will allow us to know and relate to the NEC 11 the flow that is used in the administrative building of the Universidad de Fuerzas Armadas "ESPE"; For this reason, the objective of this research was to analyze the fluid supply that enters the building in the year 2022 in the period from February 23 to June 1, since in previous years the administrative staff was not working at 100 percent, therefore, data from the established period were used. For this analysis case, the administrative building was considered, which has an IoT flowmeter that records the amount of water used in the building, located on the first floor, consisting of an intelligent system that collects information and sends it to a database where it is stored. This IoT device generates large volumes of data transmission, however, to take advantage of these values it is advisable to identify the results and interpret them correctly. Thus, after analyzing the results obtained by the IoT flowmeter and its population, an average endowment of 25.16 (L/person per day) was obtained and the data according to the NEC-11 is 50–90 (L/person per day). Therefore, it was determined that the calculated endowment corresponds to half of the minimum value suggested by Ecuador design codes.

Keywords: IoT flowmeter · drinking water · smart university

1 Introduction

The use of potable water may vary according to the characteristics of the populations and are considered in the establishment of minimum endowments within

Supported by organization x.

M. V. Garcia and C. Gordón-Gallegos (Eds.): CSEI 2022, LNNS 678, pp. 380–389, 2023.
https://doi.org/10.1007/978-3-031-30592-4_26

the design of the different supply systems in any type of project [3,17]. That is why the correct endowment must take a hand inside the reference of the different uses due to the laws of the NEC-11, Chap. 16 [14]. Taking in mind the norms for projects of development of water usage within the laws of CPE INEN 005-9-1 [10,13], elaborating each year since 1992, which considers a range of populations within the range of where they are established to live. The importance of this is layed between different studies and in ranged climate changes within those lands although is date valuable for our ongoing understanding [9].

On the other hand, to improve data collection, the use of smart objects or, as they are called nowadays, Internet of Things (IoT) is important [11,16]. Which refers to the under-standing and knowledge of different data that can be applied to devices over common uses [8]. Nowadays, the technological advances of mankind have become protagonists in the integration of centralized, distributed and user systems of water resources [1].

This new digital ecosystem allows through a network of sensors, meters and devices with IoT (Internet of Things) technology to obtain a large amount of real-time infor-mation on the status of water resources, in this way the focus on the holistic aspect of the service given can allow to control in the most effective way, which allows a more efficient way of distributing water usage within the population as a primal living nec-cesity of all mankind [7,12].

The flowmeter is able to measure the volume of water that passes through a point of possible drinking water in the base there is a control to measure the liquid and a plat-form that distributes information from the LOT, as well as the use of the a digital platform called THINGER.IO to receive the norms stablished by a professional team which employs data for the different users inside the perception of graphics and dy-namic information which they can see and research about any time they wish [18].

Drinking water projects are designed with population growth in the worst scenario, however, for 40 years it has been shown that the population also decreases, therefore, it is necessary to carry out a detailed study of each population to provide water without a unnecessary financial expense [4,15].

The research had the objective of calculating the average usage of water in the building of Administrative Armed forces "ESPE", contemplating the gathered data of IoT and a population proportioned by the department of human resources. Condensing the gathering of data and information since the 23rd of February until the 1st of June of 2022. We proceeded to deduce values from the information given and portrayed in an EXCEL program we were able to distinguish between the benefits of and the average person with reference to the NEC-11 that was used to process the IoT flowmeter.

2 Materials and Methods

2.1 Zone of Study

The area within the study refers to the administrative building at University ESPE is located if the section of Sangolqui, province of Pichincha, Ecuador, south America in the area coordinates of 0°18′53″S y 78°26′36″ (See Fig. 1).

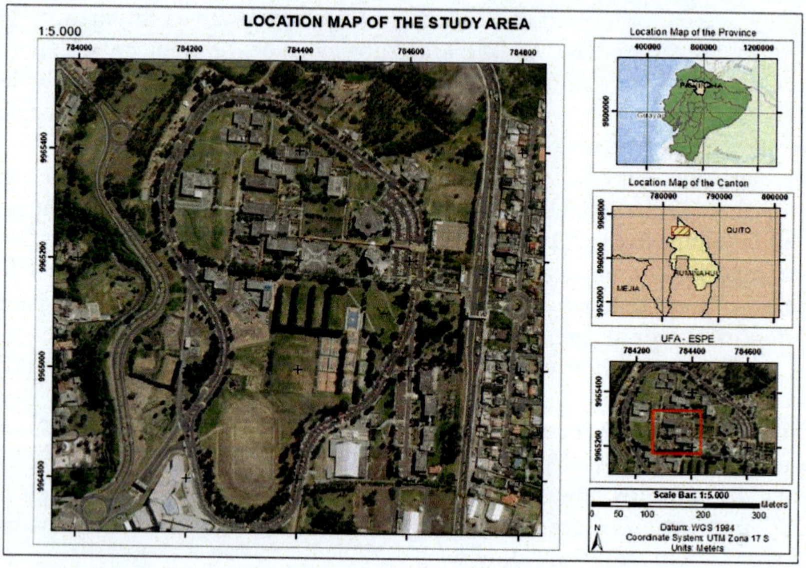

Fig. 1. Area of Analysis, ESPE matriz, Sangolquí.

2.2 Experimental Data

The administrative building consists of 21 units, distributed in 8 floors which welcome closely to 175 persons, being the most habited zone in the human resource center and administrative area. The aforementioned units consist of 30–33 persons, which refers to the high water consumption within the building [2].

Table 1. Information based of the numbers of personal in the administrative building.

Unit	Number
Honorable University Council	**2**
Rectorate	**8**
General Secretary	**12**
Legal Advisory Unit	**13**
Social Communication Unit	**8**
Internal Audit Unit	**6**
Assessment and Academic Quality Assurance Unit	**3**
Strategic Studies Unit	**2**
Research Management Unit	**7**
Logistics Unit	**16**
Graduate Follow-up Unit	**2**
Physical Security Unit	**1**
Human Resources Unit	*33*
Social Liaison Unit	**6**
Financial Unit	**32**
Planning and Institutional Development Unit	**7**
Inter-institutional Cooperation Relations Unit	**4**
Integrated Security Unit	**5**
General Academic Vice-Rector's Office	**3**
Administrative Vice Rector's Office	**2**
Research, Innovation and Technology Transfer Vice-Rector's Office	**3**
TOTAL GENERAL	**175**

2.3 Process

The experimental investigation is categorized by providing for the use of accessible population where a sample is considered to determine the actual water consumption endowment based on data given in an EXCEL format in 2019 and in that way gain graphs and statistics based of real live studies. The next variables can be considered as a part of the actual estimate of the endowment to be calculated (Tablels 1, 5 and 6).

Number of persons: 175 people, obtained by official administrative building register.

Volume accumulated: 96165,26 (L), obtained by the sum of worked days provided by the information maintained by IOT in the THINGER.IO app.

Consumption: 294982,22 units (L), obtained by the ratio of accumulated volume and average hours of discharge, where the result is multiplied by the number of the day, which provides the consumption results for each month in working days.

Total: 4402,720 units (Per/day), obtained by data stored in the thinger.io application, where the consumption is registered in relation to working days.

Table of consumption: obtained by downloading data from the IoT flowmeter and previous calculations.
Mathematic employed:
Calculation of the average flow through arithmetic: Displayed equations are centered and set on a separate line.

$$Q_m = \frac{(\sum_{i=1}^{n}) * Q_i)}{n} \tag{1}$$

Location: Displayed equations are centered and set on a separate line.

$$Qi = Q1 + Q2 + Qn; valuescalculatedwithflowmeterdata[L/day] \tag{2}$$

n = total amount of

$$Q_m = \frac{Q_m}{(N.ofdays)} \tag{3}$$

Location:
Divide Qm by the number of days per month.
Staffing calculation [L/person/day]

$$Doting = \frac{Q_m[l/day])}{N.persons} = Doting[l/person/day] \tag{4}$$

Location Divide Qm by the number of daily occupants of the building.

The base of the information of the flowmeter IoT was downloaded by a digital platform in object connected with WIFI internet where the information compilated since Wednesday 23/02/2022 until Wednesday 01/06/2022.

3 Results and Discussion

In order to obtain the results given in this report, we had to discard the non-working days with reference to the Table 2, Table 3 and Table 4 recognizing the fact that Saturdays and Sundays which are marked with orange color and holidays referred to with green color, according to the schedule of our respected university [19].

In the Table 2 and Fig. 2, we can observe the water consumption within the last 6 d of February, where the consumption was varied, reaching the highest point on Friday and the lowest on Monday, which we can disregard due to the holidays where there were no personnel in the area. On working days, a consumption of 3114.78L was recorded.

Table 2. Data based of real usage of water., 23–28 february 2022.

Day	Flow [L/min]	Volume	Flow [L/da]
23	440.90	440.90	2875.43
24	1119.70	1560.60	3095.94
25	1276.10	2836.70	3372.95
26	30.40	2867.10	1520.00
27	14.50	2881.60	1288.89
28	12.40	2894.00	1293.91

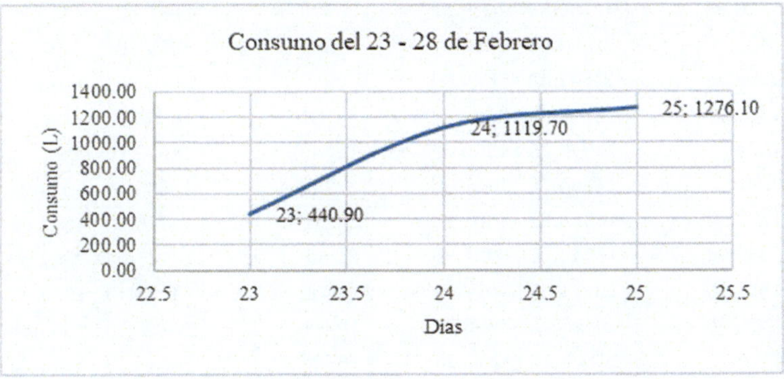

Fig. 2. Graphic usage of water 23–28 February 2022

In this table, we can find different information within the usage o water within the months of March, April and May based of the IoT flowmeter since they 1st until the 31st of each month. Which can relate to holidays or the low usage of water during this festivity times. Within the 22 working days of the month of March the water usage measured about 4043,986 L; in 20 d of workable pace it was measured about 4283,309 L and in the 21st days workable in the month of April the water usage gained a total of 5084,166 L. There is a high variability of water consumption depending on the seasonality of the year, as demonstrated by [5].

In this information, we can base our studies on only one day of usage because the research given by the IoT flowmeter was only made a few hours before this report. [6] describes in more detail the consumption of drinking water at University ESPE.

Average water consumption:

$$Q_m = \frac{294982.22\,L}{(67\,d(noweekends/noholidays)} \tag{5}$$

$$Q_(md) = 4402.720\,L/days \tag{6}$$

Table 3. Usage of real water March y April 2022.

Day	Flow [L/min]	Volume	Flow [L/día]	Flow [L/min]	Volume	Flow [L/día]
1	16.80	16.80	1221.82	1239.50	1239.50	3520.47
2	1557.90	1574.70	4020.39	31.10	1270.60	1735.81
3	1622.50	3197.20	4006.17	19.00	1289.60	1824.00
4	1836.60	5033.80	3942.61	1354.80	2644.40	3750.31
5	294.40	5328.20	2588.13	1275.30	3919.70	3370.84
6	461.50	5789.70	2251.22	1430.80	5350.50	3765.26
7	2062.50	7852.20	5238.10	1524.60	6875.10	4020.92
8	1326.25	9178.45	3629.42	2012.50	8887.60	4830.00
9	2454.60	11633.05	6155.74	21.50	8909.10	1200.00
10	2009.40	13642.45	5824.35	7.60	8916.70	1403.08
11	1320.80	14963.25	3698.86	1478.80	10395.50	4190.22
12	15.00	14978.25	1200.00	1274.60	11670.10	4351.41
13	6.50	14984.75	1560.00	1725.30	13395.40	5006.92
14	1378.80	16363.55	4001.35	1321.00	14716.40	4343.01
15	1380.00	17743.55	3905.66	24.60	14741.00	2361.60
16	1381.10	19124.65	4077.05	14.70	14755.70	1306.67
17	1398.80	20523.45	3996.57	10.00	14765.70	1333.33
18	1175.70	21699.15	3371.18	1872.70	16638.40	5154.22
19	18.70	21717.85	1068.57	1878.50	18516.90	5026.09
20	22.60	21740.45	2358.26	1506.60	20023.50	4141.86
21	1020.50	22760.95	3256.91	1307.10	21330.60	4696.17
22	1598.90	24359.85	4451.69	541.90	21872.50	1946.95
23	1038.90	25398.75	3289.39	19.90	21892.40	1110.70
24	1133.00	26531.75	3846.11	13.90	21906.30	1191.43
25	1370.20	27901.95	3707.42	1349.03	23255.33	4846.80
26	25.60	27927.55	1755.43	1413.50	24668.83	5078.44
27	10.70	27938.25	1712.00	1280.10	25948.93	4599.16
28	1263.20	29201.45	3575.09	1272.20	27221.13	4570.78
29	1444.60	30646.05	3926.43	1240.35	28461.48	4456.35
30	1333.40	31979.45	3575.60	40.20	28501.68	3445.71
31	1219.40	33198.85	3471.60	-	-	-

Average hourly water consumption:

$$Q_(mh) = 183.45\, L/h \tag{7}$$

Real average endowment:

$$Q_m = \frac{4402.720\, L/day}{175 persons} \tag{8}$$

Endowment = 25.16 L/person/day

Table 4. Data based of real usage of water., 23–28 February 2022.

Day	Flow [L/min]	May Volume acum [L]	Flow [L/da]
1	21.00	21.00	1527.27
2	24.20	45.20	1760.00
3	1366.62	1411.82	4910.00
4	1200.75	2612.57	4314.07
5	1222.33	3834.90	4391.61
6	1589.60	5424.50	5711.14
7	159.30	5583.80	3888.00
8	44.10	5627.90	4601.74
9	1600.20	7228.10	5749.22
10	1336.30	8564.40	4801.08
11	1207.60	9772.00	4338.68
12	1345.10	11117.10	4832.69
13	1470.90	12588.00	5284.67
14	12.90	12600.90	737.14
15	20.10	12621.00	2097.39
16	2335.90	14956.90	8392.46
17	1204.90	16161.80	4328.98
18	1337.17	17498.96	4804.19
19	1001.70	18500.66	3598.92
20	1484.90	19985.56	5334.97
21	5.90	19991.46	337.14
22	2.50	19993.96	260.87
23	11.11	20005.08	808.22
24	1666.20	21671.28	5986.35
25	1548.40	23219.68	5563.11
26	1612.86	24832.53	5794.69
27	1791.30	26623.83	6435.81
28	15.37	26639.20	878.11
29	10.80	26650.00	1126.96
30	1766.43	28416.43	6346.46
31	1627.80	30044.23	5848.38

Table 5. Real Usage of water . June 2022.

Day	Flow [L/min]	Volume acum [L]	Flow [L/día]
1	1624.00	1624.00	4236.52

Table 6. Real average endowment.

Real Endowment	(L/person/day)
Minimum	11.13
Average	25.16
Maximum	47.96

4 Conclusion

To distinguish the usage of water that was taken within a population of 175 persons that labored in the administrative building from February 23rd until June 1st excluding holidays and weekends. The usage of water becomes 25,16 liters per person, which due to wasted water would sum up to 25,16 liters of water per person, per day which represents almost 50% of the NEC-11 from Ecuador regulations.

References

1. Al-Kadem, M., Gomaa, M., Al Yateem, K., Al Maghlouth, A.: Multiphase flowmeter health monitoring strategy: maximizing the value of real-time sensors and automation for industrial revolution 4.0. SPE Production & Operations **37**(03), 533–542 (2022). https://doi.org/10.2118/206281-PA
2. Alarcón, A.G., Carrera-Villacrés, D.: Diagnóstico y propuesta de mejora de agua potable y alcantarillado en la Universidad de las Fuerzas Armadas ESPE Matriz implementando caudalímetros inteligentes IoT (2022). http://repositorio.espe.edu.ec/handle/21000/28988
3. Arellano, A., Bayas, A., Antonio, M., Castillo, T.: Los consumos y las dotaciones de agua potable en poblaciones ecuatorianas con menos de 150 000 habitantes. NOVASINERGIA **1**(1), 23–32 (2018). https://doi.org/10.37135/unach.ns.001.01.03
4. Arias, E., Encalada, P., Tigre, F., Granizo, C., Gordon, C., Garcia, V, M.: Convolutional neural network applied to the gesticulation control of an interactive social robot with humanoid aspect. In: Intelligent Systems and Applications, vol. 2, Advances in Intelligent Systems and Computing, vol. 1038, pp. 1039–1053 (2020). https://doi.org/10.1007/978-3-030-29513-4_76
5. Carrera Villacres, D.V., Cajas Quiñonez, J.D., Gálvez Ortiz, K.A., Pinto Villalva, J.L., Santana Flecher, P.A., Zapata Peña, C.J.: Water consumption in portoviejo city from pacific of ecuador, during the period 2017–2020 in the context of the health crisis of sars - cov-2. In: Botto-Tobar, M., Cruz, H., Diaz Cadena, A. (eds.) Recent Advances in Electrical Engineering, Electronics and Energy, pp. 263–274. Springer International Publishing, Cham (2022)
6. Carrera-Villacrés, D., et al.: Dynamic model of sars-cov-2 spread and drinking water consumption impact at universidad de las fuerzas armadas espe, ecuador. Int. J. Adv. Sci. Eng. Inf. Technol. **12**(3), 994–1000 (2022). https://doi.org/10.18517/ijaseit.12.3.14682

7. Carrera-Villacrés, D., Carrera-Villacrés, J.L., Braun, T., Zhao, Z., Gómez, J., Quinteros-Carabalí, J.: Fog harvesting and IoT based environment monitoring system at the ilalo volcano in ecuador. Int. J. Adv. Sci. Eng. Inf. Technol. **10**(1), 407–412 (2020). https://doi.org/10.18517/ijaseit.10.1.10775
8. Carrera-Villacrés, D., et al.: Dynamic model for drinking water consumption in times of sars-cov-2 in corazón city, pangua, cotopaxi, ecuador, south america. Int. J. Adv. Sci. Eng. Inf. Technol. **12**(2), 767–773 (2022). https://doi.org/10.18517/ijaseit.12.2.14681
9. Carrera-Villacrés, D.V., et al.: Relationship between the morphometric and physicochemical parameters of the Urkuhuayku microbasin where the Urku Yaku 2.0 fog collector system tower is located. In: Proceedings of the LACCEI International Multi-Conference for Engineering, Education and Technology 27–31 July (2020). https://doi.org/10.18687/LACCEI2020.1.1.161
10. INEN: Normas para estudio y diseño de sistemas de agua potable y disposición de aguas residuales para poblaciones mayores a 1000 habitantes, p. 291 (1992). https://www.habitatyvivienda.gob.ec/documentos-normativos-nec-norma-ecuatoriana-de-la-construccion/
11. Leon, C.S., Ilvis, D.I., Remache, E.G., Villalba, W.R., Garcia, C.A., Garcia, M.V.: Evaluation of internet of things protocols for shopfloor communication integration. In: Nummenmaa, J., Pérez-González, F., Domenech-Lega, B., Vaunat, J., Oscar Fernández-Peña, F. (eds.) CSEI 2019. AISC, vol. 1078, pp. 199–213. Springer, Cham (2020). https://doi.org/10.1007/978-3-030-33614-1_14
12. Montalvo, W., Ibarra-Torres, F., Garcia, V, M., Barona-Pico, V.: Evaluation of Whatsapp to promote collaborative learning in the use of software in university professionals. In: Applied Technologies (ICAT 2019), PT II. Communications in Computer and Information Science, vol. 1194, pp. 3–12 (2020). https://doi.org/10.1007/978-3-030-42520-3_1
13. Naranjo, A.E., Alarcon-O, A., Amancha-P, G., Ortiz-Espinosa, J., Naranjo, J.E.: Low-cost assistive system for deaf people based on artificial vision. In: García, M.V., Fernández-Peña, F., Gordón-Gallegos, C. (eds.) Advances and Applications in Computer Science, Electronics and Industrial Engineering. AISC, vol. 1307, pp. 249–264. Springer, Singapore (2021). https://doi.org/10.1007/978-981-33-4565-2_16
14. NEC: Norma ecuatoriana de la construcción - capitulo 16: Norma hidrosanitaria nhe, p. 38 (2011). https://www.habitatyvivienda.gob.ec/documentos-normativos-nec-norma-ecuatoriana-de-la-construccion/
15. Paez Rubio, T., Alberti, J., Rezzano Tizze, N.: Tendencia del consumo de agua potable y eficiencia en la inversión en infraestructura de agua y saneamiento: Estudio de caso en América Latina (2020)
16. Parra-Valencia, J.A., Guerrero, C.D., Rico-Bautista, D.: IOT: una aproximación desde ciudad inteligente a universidad inteligente. Rev. Ingenio **13**(1), 9–20 (2017). https://doi.org/10.22463/2011642x.2128
17. Romero, C.V.F., Calle, M.A., Cobos-Torres, J.C.: Automation framework for analog water meters in the city of cañar. In: Lecture Notes in Networks and Systems, pp. 289–305. Springer International Publishing, Cham (2022). https://doi.org/10.1007/978-3-030-97719-1_17
18. The Things Industries: The things network (2022). https://www.thethingsnetwork.org/docs/lorawan/what-is-lorawan/
19. Thinger Io: Thinger Io (2022). https://thinger.io/

Application of the MQTT Protocol for the Control of a Scorbot Robot by Means of EGG Electroencephalographic Signals

Franklin Salazar[1](✉), Jesús Guamán-Molina[2], Cristian Saltos[2], Walter Cunalata[2], and Angel Fernández-S[2]

[1] Universidad Autónoma de Madrid, Madrid, Spain
fw.salazar@uta.edu.ec

[2] Universidad Técnica de Ambato, Ambato 180103, Ecuador
{jguaman0585,cj.saltos,wcunalata6509,afernandez8448}@uta.edu.ec

Abstract. At present, technology is one of the main tools available, its main objective is to contribute to improve the quality of life of people, such as in the medical area. The manipulation of industrial robots through proprietary software promotes the investigation to integrate new communication technologies to the development of a society. This investigation is based on the development of a monitoring system to control the Scorbot Robot based on electroencephalographic (EEG) signals. The EEG signals are monitored to facilitate the operation of the joints by making use of the electrical signals provided by the brain activity of the human being. The Mindwave helmet from the company Neurosky is very useful, as it receives energy impulses, converted into signals that are subsequently processed by the mathematical software Matlab. The technology of the raspberry Pi3 B+ device together with the MQTT protocol enables point-to-point machine-to-machine (M2M) iteration. TCP/IP messaging supports communication between low resource devices such as the raspberry Pi3 B+ and the Scorbot Robot. On the other hand, it is developed a 3D prototype using Unity that allows a monitoring to better visualize the movements in relation to the possible actions in the robot. The data analysis is presented in relation to the recognition times of the commands by the system, using the mindwave helmet. Also, the results of response when executing a given action in the Scorbot Robot and the prototype designed in Unity are presented, showing the degree of speed of each one of them. The results obtained show a maximum response time of 7.01 s and a shortest time of less than 4.20 s for the developed system.

Keywords: Neurosky · Raspberry Pi · MQTT · Electroencephalographic · Scorbot · Unity

Supported by organization x.

M. V. Garcia and C. Gordón-Gallegos (Eds.): CSEI 2022, LNNS 678, pp. 390–411, 2023.
https://doi.org/10.1007/978-3-031-30592-4_27

1 Introduction

The topic of the analysis of the behavior of the electroencephalography (EEG) signal as the main characteristic of human brain activity is a very broad field. There are a lot of variations on the brain activity exposed in different conditions that allow us to consider and understand the complexity and meaning of the results. In this investigation [21,22], a literature review shows many investigations where they have focused on capturing brain signals to transform them into actions [10]. So, it is contributed to topics in the medical field for people with disabilities. In which from the capture of brain signals currently has the necessary tools to achieve control a wheelchair and manipulate a robotic arm allowing to replace an amputated arm or the deterioration of the physical mobility of a patient to improve the quality of life of people. In the same way and using the same principles, it is possible to generate the movement of a robot for the industrial sector, so that a person with a disability can be associated to a work field.

In [26], a prototype is presented where a control with the use of EEG signals to move a robot is shown. This investigation generated a low-cost prototype for the control of the Scorbot robot using the Neurosky helmet. This device is focused on helping in some way integrating people with disabilities to perform activities using a robotic hand. The studies conducted show a training of these brain signals where they allow to have a better management of this type of information converting it into actions. The present investigation is focused on the capture of EEG signals using the Emotiv Xavier Software where it allows to add an action and a condition to move the Scorbot Robot [9,23].

In [19], it uses the MQTT protocol where it is appreciated its simplicity and lightness one of the main pillars of the IoT, this protocol is used in this investigation. MQTT works with a low bandwidth, which is important in wireless networks or connections with internet quality problems.

According to the literature review, taking electroencephalographic signals allows programming actions that can be incorporated into robotic arms for the development of specific activities. Likewise, the development of new communication protocols developed in IoT with hierarchical communications and quality of service mechanisms make it possible to integrate them into these systems, giving them greater robustness. The purpose of this research is to evaluate the integration of the MQTT protocol, for sending response signals to the arm efficiently. For this, the system receives data from the Neurosky helmet, which are analyzed and processed through MatLab to determine the movements that the arm must perform. The information is sent through the MQTT protocol so that the arm performs the programmed actions. This article is structured as follows:

In section two, an analysis of the materials necessary for the implementation of the proposal is presented, as well as the details of the methodology used for the integration of the system. The third section describes the implementation of the methodology based on three parts: sensorization, signal processing and electronic control to integrate the data acquisition system in the Matlab environment for its control. Also, in this section the evaluation of the MQTT protocol for the control of the SCORBT robot is presented. Finally, section four presents the evaluation of the results in terms of the times obtained in the processing and response stage of the implemented system.

2 Methods and Materials

2.1 Materials

The elements that will be selected for the composition of the system have a technical analysis. To be clear which elements make up the system has been divided into modules that are monitoring, control and processing, communication, electronic control interface. The following is the description of these modules and their relevant characteristics:

Sensorization

Mindwave Helmet. This type of helmet works with ThinkGear technology allowing to measure the frequency of brain waves. This headband consists of a sensor in the forehead area where the EEG is captured, reference points placed in the ear clip, and finally, the integrated chip where the data is processed. The brain waves are calculated in the ThinkGear chip where this signal is amplified, eliminated noise and muscle movement and then transferred by Bluetooth [17].

Control and Processing

Single Board Computer (SBC). The information obtained by the Mindwave helmet where the EEG signal is displayed, once processed, must be transmitted by Wi-Fi technology to the MQTT protocol. To transmit this data, the Rasberry Pi 3B+ is used, which acts as a controller of the robotic arm through the GPIO input and output ports [4].

Electronic Board. The importance of having an electronic board is for the actuation of the activities that will be transmitted to the robotic arm. To perform this communication by wireless means, it works with the Esp32 [14]. As devices that have sufficient resources for the control system, in addition to having an affordable cost to develop investigation.

Processing Software
The EEG signal must be subjected to an analysis where the data obtained is evaluated, for this the Matlab software is used, according to the requirements of the system highlighting the signal processing, images, neural networks, and others. It is an accessible software to integrate hardware from any supplier,

design of customized interfaces, development of algorithms for data analysis. It has its own visual programming environment of the software in which it allows to carry out and execute programs with a great demand of continuous data input [24].

Emotiv Xavier Software. The company Emotiv provides its software on data interpretation to display useful and user-friendly information [12].

Comunication

Wireless Technology. Wifi technology in the 2.4 GHz operating band allows the transmission of data wirelessly, reducing the use of wiring, and avoiding errors due to disconnection in a total way, complying with the transmission, range, speed, energy consumption, etc. [7].

Router TP-LINK. The device allows to connect a Wi-Fi wireless network that connects the MQTT protocol with data processing and control. It provides a stable and secure connection to the network. It works under IEEE 802.11 b/g/n protocol guaranteeing its effectiveness [25].

Communication Protocol. In this investigation, the MQTT protocol is used, it is a message queue type, which means that the router generates a unique messaging queue for each client. The router forms a local wireless communication network, which coconnects to a small board computer (Rasberry Pi 3B+) where it acts as a hub of the MQTT protocol, for the case study we have the ESP32 board, Unity [15].

Scorbot Robotic Arm Power Supply

The 5 axes and the gripper of the robot are manipulated by servo motors using DC cocurrent. The polarity of the voltage determines the direction of the motor, the positive DC voltage rotates in one direction, while the negative voltage rotates in the opposite direction. As the voltage of the motors, we have 12V DC which consumes 900 mA.

Total Current = Servomotor Current + Bridge Current H

Total Current = 900 mA + 300 mA

Total Current = 1.2 A

Motor Driver

For this investigation, the L298N H-bridge module was used, which allows the control of direct current servo motors. The driver allows the mobility of the joints of the Scorbot robotic arm [5].

SCORBOT-ER 4u

The SCORBOTER 4 robot is a versatile robotic arm for educational use suitable for various standalone operations for integration into automated work cell applications such as robotic welding, CNC machine positioning and other operations [8].

2.2 Method

This investigation emphasizes the creation of a BCI system "Brain Computer Interface" [1]. It starts with the measurement and capture of electroencephalographic signals EGG capturing the data with the help of the Mindwave helmet of the Neorosky Company. Once these signals are captured, they will be processed in the Matlab tool [13]. Once the EGG signal has been captured, the communication between Matlab and the Scorbot-ER 4U Robot is carried out using a TCP/IP socket. It is necessary to use a machine-to-machine (M2M) protocol, in this case MQTT (Message Queue Telemetry Transport) of the Internet of Things type [15]. For the movements of the Scorbot Robot, the Python programming language will be used. Through the input and output pins of the Raspberry Pi the data are sent to their respective amplifier modules L298N that later reach the D50 connector. In order to provide the user with an intuitive and dynamic system, a 3D prototype has been created with the help of Unity software where the movements of the Scorbot Robot can be observed, allowing the system to be monitored. Brain Computer Interface, proposes a methodology whose purpose is to capture brain signals using sensors, process the signal in order to obtain the data of interest, and finally interact with the environment according to brain activity [12]. Figure 1 shows a schematic representation of the methodology implemented in the development of this investigation.

2.3 General System Design

The development of investigation is carried out with a study of brain waves. When these signals are captured, a communication medium is generated which connects to the Matlab tool where the sensor signal from the front of the brain enters a process of analysis and training of nerve impulses. The data will be displayed on a control panel where you will have the movement activity of the robot [16]. The block diagram in Fig. 2 shows how the system is structured in three stages. The first part consists of the acquisition of electroencephalographic signals from the Neurosky helmet generated by brain activity. Thanks to wireless technology, it allows linking to the second stage where these signals are processed, filtered, and visualized through algorithms. To transmit the data to the third stage, it uses the MQTT protocol, where the architecture for the actuation of the joints of the Scorbot Robot is located.

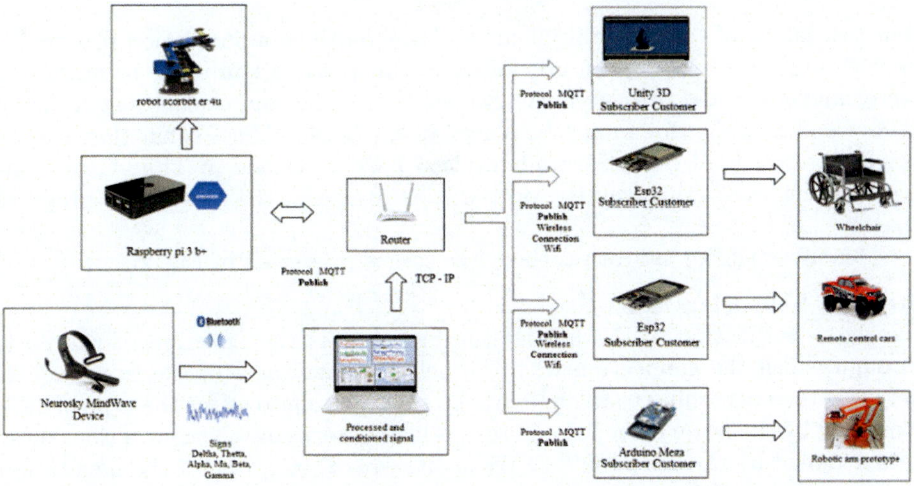

Fig. 1. General scheme of the implemented system (Architecture)

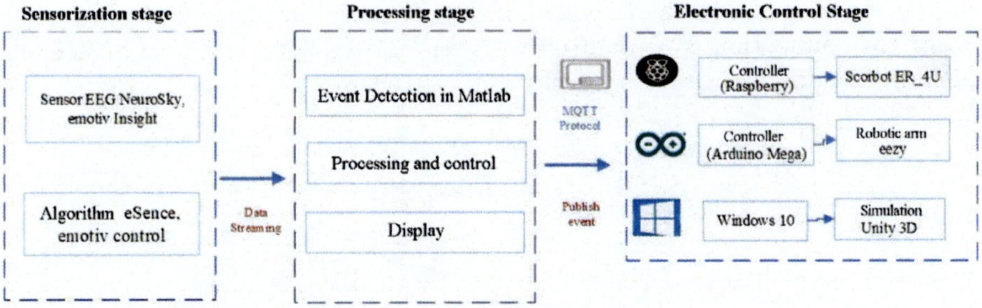

Fig. 2. General scheme of the implemented system.

2.4 Implementation of the Methodology

To implement the proposed methodology, it is necessary to review the connectivity of the communication ports. ThinkGear64 is a library that is very necessary for applications to connect to the NeuroSky helmet. TGC is installed as necessary because certain operating systems do not have direct permission to the computer's COM ports enabled. This software specifically connects to the COM port of the NeuroSky helmet allowing applications to connect to the NeuroSky helmet [3]. This investigation was carried out using the Matlab tool, which provides its M programming language for an integrated development environment. This software allows matrix processing, data and function graphing, implementation of algorithms, design of an excellent user interface (GUI) available for Windows, Mac, and Linux systems. To install the helmet, NeuroSky's own wind-Wave Mobile must be downloaded to obtain the ThinkGear Connector library, which goes together with the files that will be executed in Matlab [18]. For the communication of the NeuroSky helmet a Bluetooth communication protocol is used through the COM port previously enabled and we proceed to download the elements (.h and .dll) necessary to establish a communication between the helmet and Matlab, these must be placed in the same folder so that they can be executed. As a final step, the configuration port is verified in ThinkGear Connector Preferences, this way the port will be placed in the Matlab .m program [2].

This investigation addresses the three stages presented in Fig. 2:

Stage 1: Sensorization
For the sensorization of the helmet the first step is the selection of the port, through which the connection with the helmet is established, the thinkger .dll library is used according to the baud rate, as a next step to select the type of data requested by the library, then the library is loaded verifying the connection, these data are read by the TG_DATA_RAW signal where the graph of the meditation and attention signals can be observed.

First, the communication is established through the bluetooth protocol and the method is sent with a login, for which a library released by Neuroskyn is declared. Figure 3 shows the connection of the electrode to the PC port. Additionally, the connection is configured with respect to speed and packet size, it is important to add the library with the "dll" extension.

```
loadlibrary('thinkgear64.dll','thinkgear64.h');
fprintf ('thinkgear64.dll loaded\n');
dllVersion = calllib ('thinkgear64', 'TG_GetDriverVersion');
fprintf ('ThinkGear DLL version: %d\n', dllVersion);
connectionId1 = calllib ('thinkgear64', 'TG_GetNewConnectionId');
if (connectionId1 < 0) error (sprintf ('ERROR: TG_GetNewConnectionId
() returned %d.\n', connectionId1));
end;
errCode = calllib ('thinkgear64', 'TG_Connect', connectionId1,
comPortName1, TG_BAUD_57600, TG_STREAM_PACKETS);
if (errCode < 0)
end
fprintf ('Connected.  Reading Packets...\n');
if (calllib ('thinkgear64','TG_EnableBlinkDetection', connectionId1,1)
==0)
disp('Blinkdetection');
end
```

Fig. 3. Connection link between Mindwave and Matlab.

For the reception of EEG signals, the Mindwave NeuroSky headset has been selected, which provides brainwave detection with a set of accessible and open tools that will be useful when designing the prototype. To acquire the raw wave packet data, the code shown in Fig. 4 must be considered. First the condition asking if the packet has already arrived, in the second condition it must be checked if the packet that arrived is the correct one [20].

```
i=0;
j=0;
disp ('Reading Brainwaves');
t = timer ('TimerFcn', 'stat=false; disp(''FIN'')','StartDelay',40);
errCode = calllib ('thinkgear64', 'TG_EnableAutoRead', connectionId1,-
1);
start(t)
stat=true;
while (stat==true)
if (calllib ('thinkgear64','TG_GetValueStatus', connectionId1,
TG_DATA_RAW) ~= 0)
data_att(j,1) = calllib ('thinkgear64','TG_GetValue', connectionId1,
TG_DATA_ATTENTION);
data_att(j,2) = calllib ('thinkgear64','TG_GetValue', connectionId1,
TG_DATA_MEDITATION);
data_att(j,3) = calllib ('thinkgear64','TG_GetValue', connectionId1,
TG_DATA_RAW);
disp(data_att(j,1));
Plot Graph
plot(data_att(:1),'-r','Linewidth',1);
title('Attention');
disp(data_att(j,2));
Plot Graph
```

Fig. 4. Data acquisition with Matlab through the Mindwave sensor.

With the help of an algorithm called esence and the Neurosky Mindwave helmet, which helps to obtain the attention signal, meditation, and raw signal as shown in Fig. 5. The signals are captured and analyzed and then stored according to the type of action to be performed. All the individual brain waves are represented, this test is performed in 60 s which allows to know the behavior of the signal. It should be considered that there are several ways to be distracted such as wandering thoughts, anxiety, excitement and feeling stimulation can decrease the level of meditation.

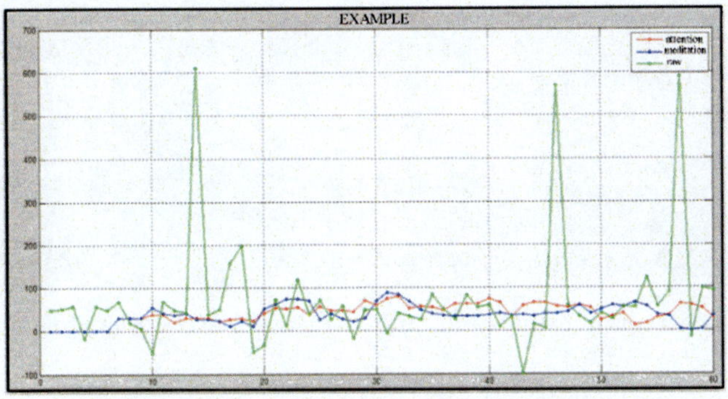

Fig. 5. Brainwaves attention, meditation and raw.

The Neurosky Mindwave helmet can capture and process the raw brain signal, which contains the data as acquired from the sensor. This signal is broken down into: alpha, tetha, gamma, beta, as can be seen in Fig. 6, in which the generation of these brainwaves, evaluated in 60 s, can be visualized.

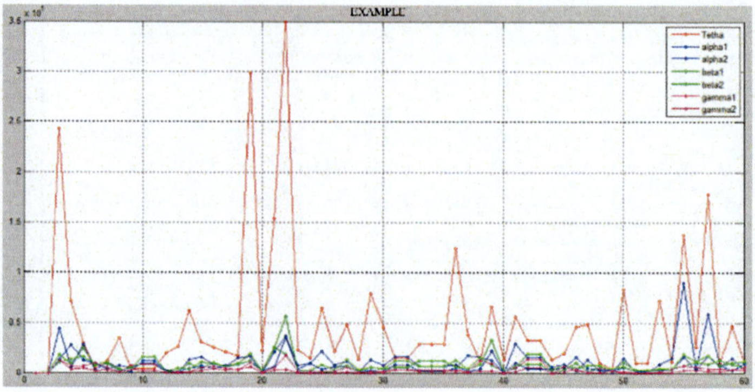

Fig. 6. Brainwaves tetha, alpha, beta, gamma.

The Emotiv Xavier Control panel software allows the interpretation of the data in order to display useful and accessible information to the user. In Fig. 7.a and 7.b you can see how the data obtained allow training brain activity signals by identifying reactions to visual or physical stimuli to subsequently perform an action on a real or virtual object [9]. The tool interface has 5 types of rotational movements and 6 types of translational movements, and an extra action related to the visibility of the virtual object as shown in Fig. 7.b.

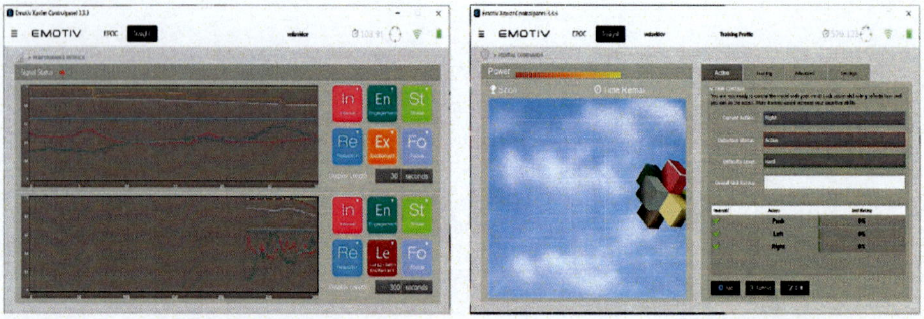

(a) Brain waves attention,meditation and raw. (b) Brain waves tetha,alpha, beta, gamma

Fig. 7. Brain waves obtained using the Neurosky Mindwave

Once the data has been obtained using the Neurosky Mindwave helmet, data processing is performed.

Stage 2: Processing

After the acquisition of the EEG signals to move on to the processing stage, the signal is cleaned to extract a unique pattern of each of the movements of the robotic arm joints, by means of the attention and meditation signals. In this way, each time the signals exceed the thresholds, these movements will be identified. The function sends a message to the control panel with the result according to the threshold value matched. Table 1 shows the values to which each movement belongs.

Table 1. Configuration of the proposed evaluation scenarios.

Condition	Threshold Value Conditions	Signal identification
MOVE LEFT	The maximum peaks of this attention signal must be exceeded by a threshold value greater than 80 uV.	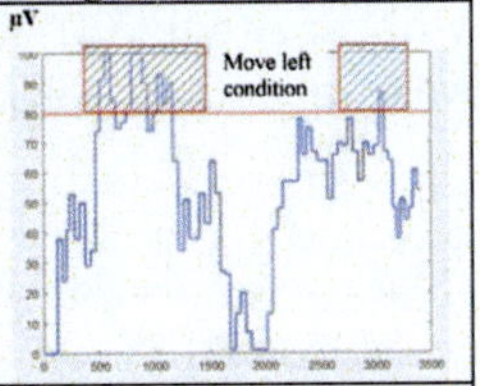
MOVE RIGHT	The maximum peaks of this attention signal should be in the range of 60 uV and 80 uV.	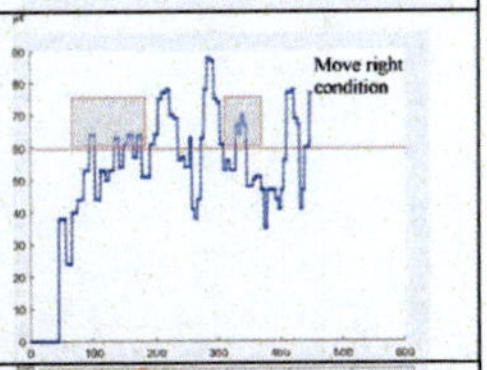
GO UP	The control signal used for this action is the meditation signal, because the relaxation of the person generates changes in this signal. The rise condition of this signal produces a peak greater than 80 uV.	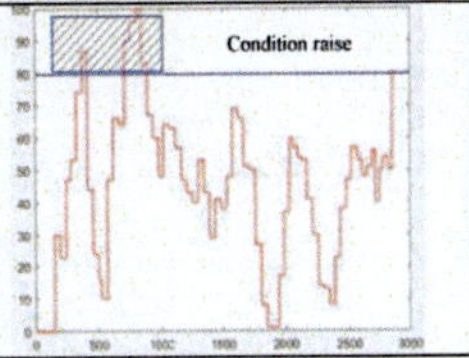
GO DOWN	The maximum peaks of this meditation signal should be in the range of 60 uV to 80 uV.	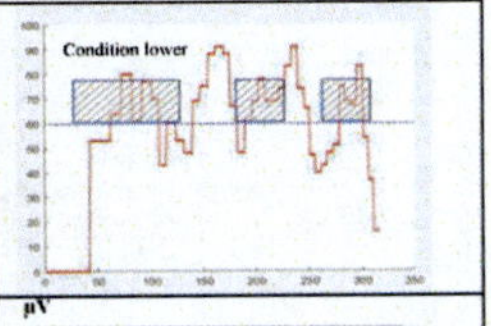
OPEN GRIPPER	For the opening of the clamp, the maximum of the peaks of this attention signal should be in the range of 40 uV and 60 uV.	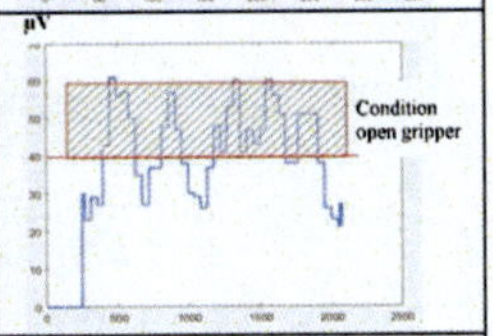
CLOSE GRIPPER	The maximum peaks of this meditation signal should be in the range of 40 uV to 60 uV.	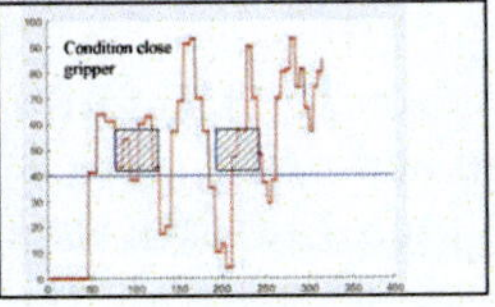

Once the EGG signals have been identified together with their condition and values at the corresponding threshold, these signals are assigned in the Emotiv Xavier Control panel software, presented above, which allows training to assign an action to a specific event as shown in Fig. 8. The software is composed of two parts, the "Add Rule" rules and "Add Condition" conditions. The rules define which action will be sent to the input queue of the operating system. These rules can contain different conditions, which will define what must be fulfilled for the rule to be automatically triggered by a Trigger and perform the programmed action [27]. The conditions will be recognized by the emotiv software control panel through Emokey as shown in Fig. 8, directed to the Matlab control interface. In the configuration of the character sending, use is made of the emotiv library (edk.dll), since it contains the APIs allowing communication between two applications, proceed by double clicking on Target Application [27].

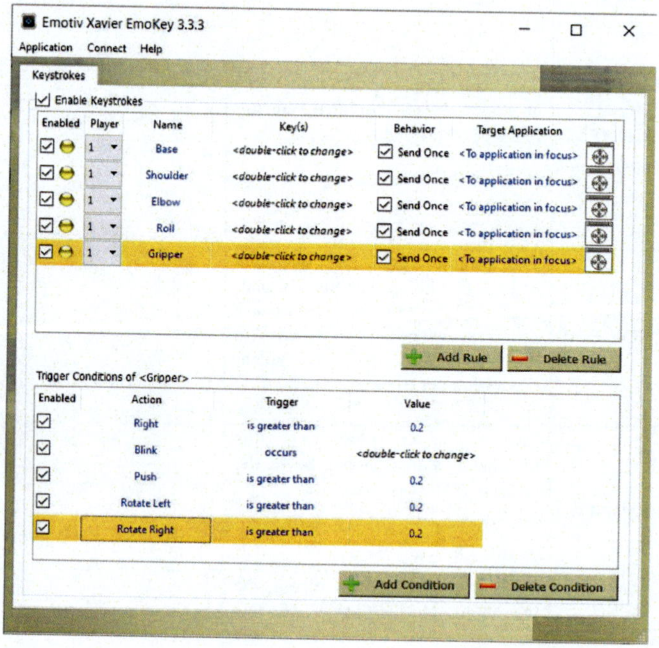

Fig. 8. Emotiv Emokey panel.

Training

Now the person thinks of an action according to the movements to the left or right, or up and down, open or close the robot's gripper. A test was performed with 20 captures of each signal obtaining as a result 85.83% accuracy. This analysis can be seen in Table 2, in the results it can be affirmed that the command with the highest precision is to go up and ascend while the one with the lowest precision are movements of a higher degree of complexity such as opening and closing the robot's gripper.

Table 2. Accuracy of all commands.

Commands	Number of unsuccessful attempts of 20 data captures.	Percentage
left	15	85%
right	15	85%
go up	16	95%
go down	14	80%
shut down	13	90%
open up	13	80%
Overall accuracy		85.83%

Visualization

The visualization of the data captured from the Neurosky headset is shown in this monitoring interface that was made in GUIDE of Matlab, the panel receives information about: communication data, motion activity signals, flicker signals. The motion activity signals received through the headset belong to the front part of the brain and can be seen in Fig. 9 [6].

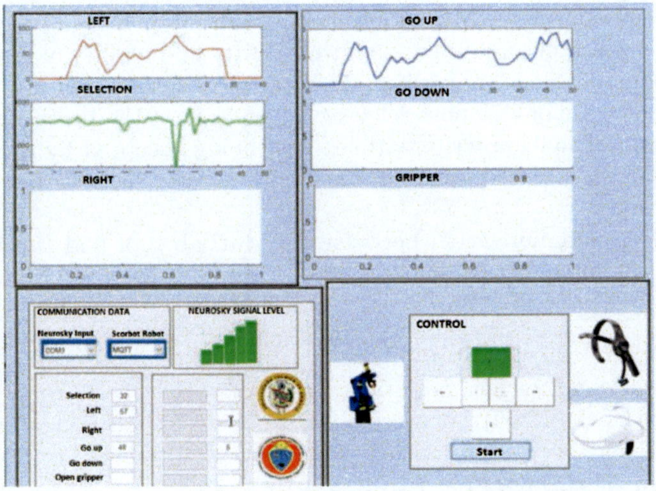

Fig. 9. Main Panel.

In Fig. 10, it is possible to observe the EGG signals generated by the brain when focusing the thought on the left, right, up, down, open or close movements of the robot gripper.

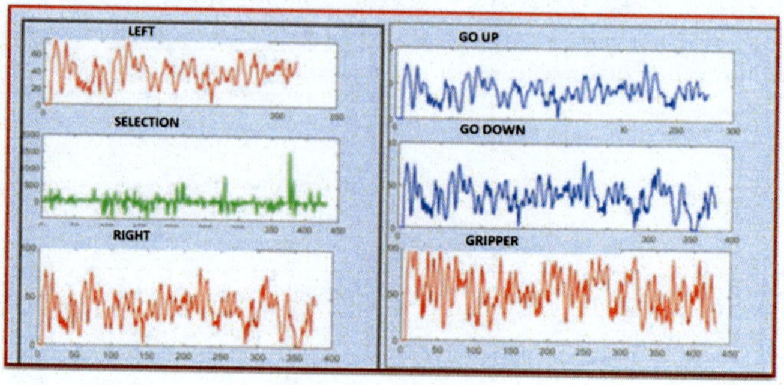

Fig. 10. Main Panel.

Communications Panel. The communication between the Matlab software and the EEG helmet signal is generated by communication through the COM3 port. To proceed with the communication the Raspberry is used together with the MQTT protocol, according to the signals of attention, meditation and Raw, the taking of the ranges corresponds to 600mv that provides the Blink signal with reference to the flicker, to proceed with the signals of attention and meditation. MQTT guarantees the transmission of the information in a safe way. The principal advantages of the MQTT protocol are its simplicity and lightness, low power consumption, minimum bandwidth, security and quality of service (QoS), providing robustness and reliability. This protocol is one of the most widely used at present. One of the most relevant advantages of the protocol is that the client node cannot make a write without first being accepted by the subscriber node, consequently providing the protocol with greater security in sending and receiving data [2].

To establish communication between the MatLab tool and the SCORBOT robotic arm, the following is done:

- Initialization of the MQTT Publisher.
- Communications Interface Selection.
- Installation of the MQTT protocol on Raspberry PI 3 B+.

Initialization of the MQTT Publisher

The method used in the MQTT configuration to initialize the connection channel is done with the help of the javaaddpath ('jars/iMqttClient.jar') library, the topic (PUB_TOPIC_1 = 'home'), the broket ip (mqttinterface = MqttInterface ('matlab_mqtt_node', '172.88.1.104', 1883, 10) and the message(pub$_m sg_1$).

Communications Interface Selection

Consisting of the protocol and the transmission medium resulting from the transfer of data between the computer and the controller card. The communication system has been realized with Wifi technology allowing a higher data transmission speed.

Installation of the MQTT Protocol on Raspberry PI 3 B+
The information is received by Raspberry PI through the address '172.88.1.104' mentioned above acting as a broker. It is important that as a first step the Raspberry Pi system is updated in Table 3, it is shown in detail the code corresponding to the package update and installation of the Mosquito Broker [11].

Table 3 shows the classes used for their respective Pi control, these classes contain properties and methods, it is important to note that there is a composition relationship between the "control" class and the rest of the classes. When executing the following lines of code, a link is created to start sending and receiving data from the 'house' class, which contains the conditionals for identifying the signals and executing the actions programmed in the SCORBOT.

Table 3. Lines of code to establish a link.

Code	Description
sudo apt-get update	Updating repositories and packages.
sudo apt-get install -y mosquito mosquito-clients	Install Mosquitto Broker on Raspbian (Raspbe-rry Pi 3 B+)
sudo systemctl enable mosquitto.service	Runs Mosquito Broker

Stage 3: Electronic Control
To process the data sent from the Neurosky headset and by the application developed in Matlab software, an electronic board is required for this process of sending their respective movements and actions. The action generated by Raspberry Pi 3 B+ basically focuses on the activation of the GPIO pins used to send and receive digital signals in which will have control of each GPIO by coding classes in the Python language. The MQTT communication is represented in the Raspberry Pi 3 B+ which has mqtt attributes, also of String type. These attributes define the IP address of the Broker where the method used is to create a connection (on connect), receive the message (on_message) and execute the client (mqtt.client).

The robotic arm is connected to a controller through wiring with the D50 terminal. It is a 50-pin connector where all the elements such as power wiring, motors, and axis limit switches will be linked. The wiring that will use the servomotors must be flexible for greater mobility of the robotic arm.

In Fig. 11, it can be seen how the electronic connection scheme of the H-bridges is elaborated, each L298N integrated is powered with 12V DC. Each one of the integrated units manages the rotation of a motor, where it should be considered that the angular velocity of these motors is controlled by pulse width odulation (PWM) and the control signals are connected directly to the Raspberry Pi 3 B+.

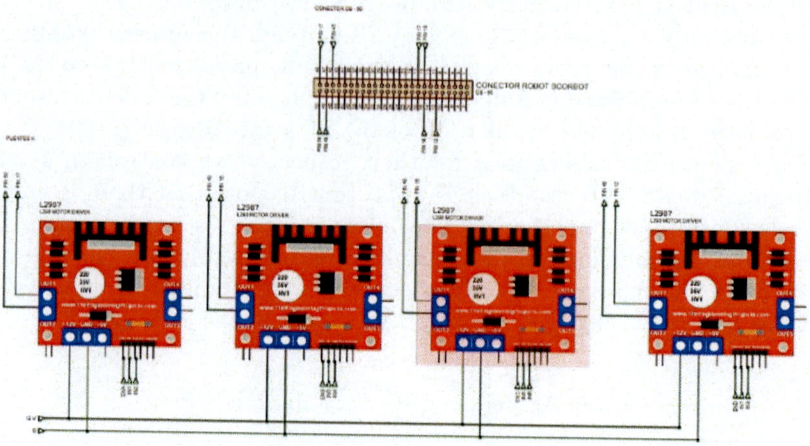

Fig. 11. Connection of the electronic jumper circuit H.

The servo motors of the robotic arm are controlled through message flows using the MQTT protocol in which use is made of a lowcost computer, the Raspberry Pi 3 B+, which uses the Raspbian Linux operating system. Each servo is controlled by a control signal programmed in Python language.

The control and displacement of the Scorbot robotic arm, shown in Fig. 12, allows the user to manipulate each of its joints by means of the electroencephalographic signals that send pulses through the GIPO ports of the Raspberry PI that performs the monitoring and control system. The conditions described by the monitoring and control system are sent through the MQTT protocol and evaluated to activate each of the functions, allowing the correct movement of the Scorbot.

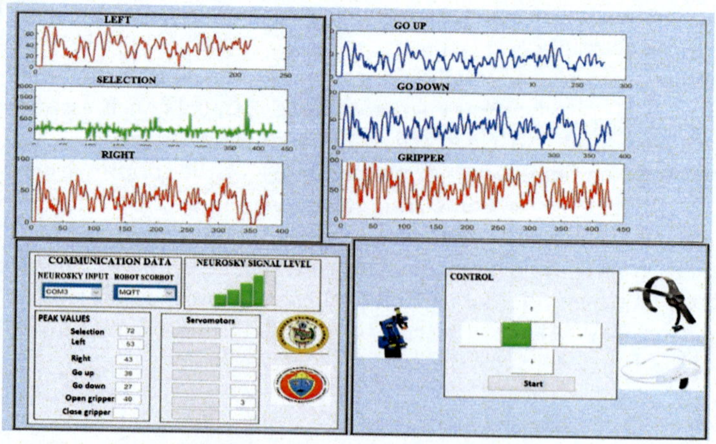

Fig. 12. System implementation.

Simulation of the Proposed System in the UNITY Virtual Environment

To better visualize and understand the operation of the system planned for investigation, the Scorbot Robot was assembled in the virtual environment called Unity. The prototype was developed with this tool, starting in the "Scenes" window, the "explorer" appears, and these elements to build the prototype must be in the same folder. As can be seen in Fig. 13.

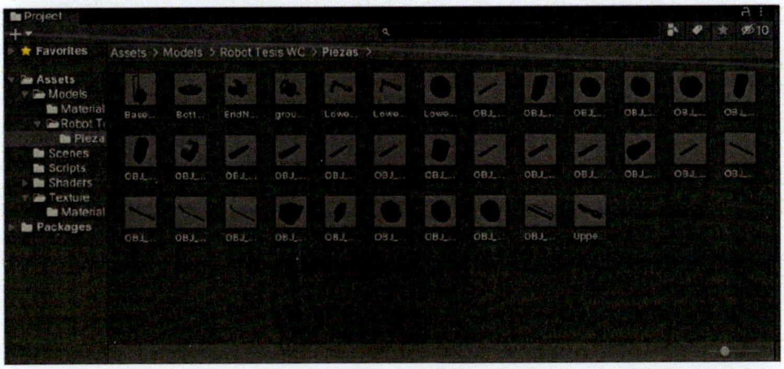

Fig. 13. System implementation.

All the necessary objects were dragged to the "Scenes" window, as we worked with 3D objects, each object was placed in space maintaining its orientation with respect to Unity's three-dimensional axis, as shown in Fig. 14.

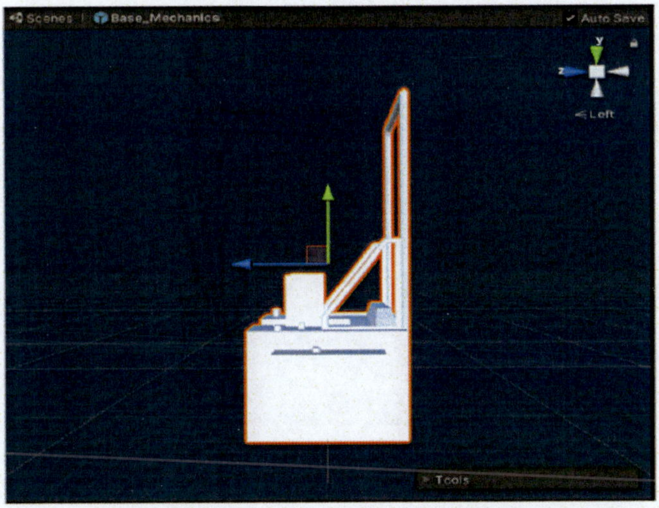

Fig. 14. System implementation.

Each of the pieces were placed keeping the order in order to achieve the assembly of the robotic arm, the "Inspector" window of Unity allows to visualize, add and remove components to the object such as rigid body, collider, mesh, etc. The assembly of the robotic arm was achieved in a precise way by using the "Transform" component. In this way, the entire fixed structure of the robotic arm is built, as shown in Fig. 15.

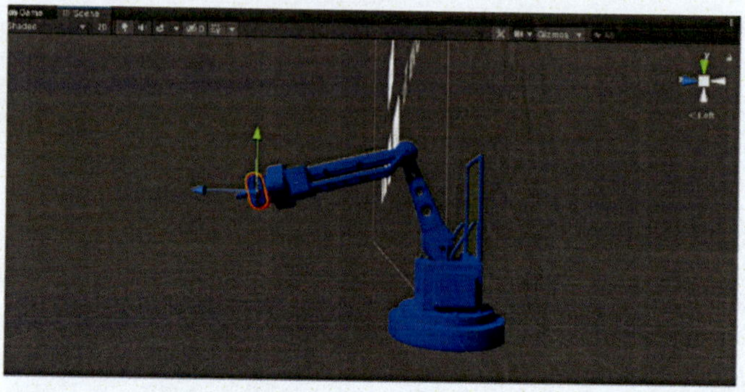

Fig. 15. System implementation.

Unity 3D provides the possibility of communicating through the MQTT protocol, by receiving data from the Matlab graphical interface, and the actions that the physical prototype is performing can be visualized from any computer connected to the system's network. The Unity 3D software was installed on a Windows 10 computer connected to the system network. The 3D model of the arm prototype can be seen in Fig. 16, it was developed with the purpose of visualizing the movements that are executed in the physical robotic arm.

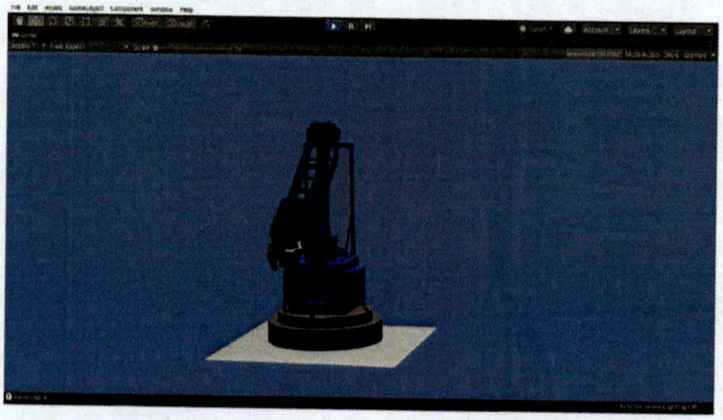

Fig. 16. System implementation.

3 Results

Finally, once the behavior of the Neurosky Mindwave helmet for the identification of commands in the UNITY environment had been evaluated, the Scorbot robotic arm was tested. Figure 17 shows the integration of the system with the Scorbot robotic arm, obtaining a control and monitoring system of the robot through electroencephalographic signals. Communication is achieved through TCP/IP using the MQTT protocol.

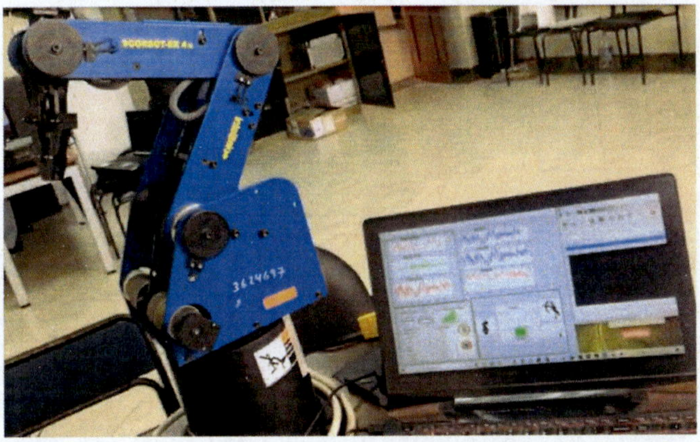

Fig. 17. System implementation.

To better understand the manipulation and monitoring system, the time required to recognize each of the commands sent to the Scorbot Robot was evaluated. The Neurosky Mindwave helmet provides a free library, which allows the acquisition of pure sensor signals. Table 4 shows the tabulation of the data according to the recognition times of the commands for each of the motion conditions.

Table 4. Robotic arm and graphical interface.

Neuroskyn					
Attention 60 uV-80 uV	Attention 80 uV-100 uV	Meditation 80 uV- 100 uV	Meditation 60 uV-80 uV	Attention 40 uV – 60 uV	Meditation 40 uV – 60 uV

6,1	5,6	4,9	5,9	3,9	4,8
6,3	6,6	4,8	5,3	3,8	4,9
6,9	5,7	4,6	7,9	3,7	4,3
5,8	5,8	4,7	6,9	3,7	4,7
7,1	5,9	4,6	5,3	3,8	4,2
5,9	5,3	4,9	5,8	3,8	4,8
7,9	5,6	4,6	5,8	3,8	4,2
6,3	4,9	4,1	5,9	3,2	4,7
5,8	5,3	4,7	5,7	4,1	4,7
5,9	5,7	4,3	5,8	4,2	4,1
7,1	5,7	4,9	5,6	4,9	4,2
7,2	5,9	3,9	5,7	4,7	3,9
7,8	5,9	4,8	5,8	4,8	3,2
7,9	5,7	4,9	5,9	4,8	3,7
7,8	5,3	3,8	5,7	4,1	4,7
7,8	5,3	4,7	5,8	4,9	3,8
7,9	5,7	4,8	5,8	4,1	3,2
7,8	5,4	4,9	5,7	4,7	3,7
7,9	5,7	4,9	5,9	4,3	3,7
7,4	6,0	4,2	4,5	4,7	4,5
7.01	**5.65**	**4.6**	**5.835**	**4.2**	**4.2**

Table 5 shows a summary of the recognition times of the commands by the system for both the Neurosky Mindwave helmet and Unity 3D, for each application the response time has been captured. In Unity the times are lower in relation to the physical robot except for the movement close gripper, this occurs because of the processing of the data and the medium by which it is being transmitted.

Table 5 shows the maximum response time in relation to the motion conditions of the Neurosky Mindwave helmet. It has a maximum response time of 7.01 s and a minimum time of less than 4.20 s, when manipulating the Scorbot Robot. When executing the motion conditions in Unity these response times are lower, when executing these actions, the data start to travel through the MQTT protocol.

4 Analysis

As can be seen in the results obtained, it can be analyzed that the processing time of the electroencephalographic signals taken from the Neurosky helmet and sent to the UNITY virtual environment, presents an average time of 4.25 s, less than the processing time towards the arm, the same one that has an average of 5.25. Although the response times of the arm present a greater latency in approximately 1 s, it is weighed on the quality of sending the information that is achieved with the MQTT protocol, since it can be sent to various devices configured as clients. These results allow the implementation of the MQTT protocol, for sending data to the robotic arm with adequate responses of the movement.

Table 5. Runtimes of the Robot Scorbot and the robot in UNITY 3D.

	Robot Scorbot	Unity 3D
Left	7,01	4,2
Right	5,65	4,3
Go up	4,6	4,2
Go down	5,835	3,9
Open gripper	4,2	3,8
Close gripper	4,2	5,1

5 Conclusions

One of the most relevant points for the correct integration of the proposed system is to have an adequate order in terms of the BCI (Brain Computer Interface) system. The stages for the correct integration are divided into acquisition, conditioning, signal processing and execution of the applications considering the EEG signals. The use of the MQTT protocol in connection with the 802.11n standard establishes a secure and stable connection with the elements of the Scorbot robotic arm. On the other hand, the use of a Rasberry Pi 3 B+ card, as a controller of the Scorbot robotic arm, allows the interaction of different platforms through the implemented MQTT communication protocol. The correct processing of the EEG electroencephalographic signals, captured by the Neurosky Mindwave helmet, reduces the identification time of the movement actions presented in the control and monitoring system for the SCORBOT. The results presented in this investigation allow identifying that the lowest times are obtained in the Unity tool, compared to the execution times obtained with the Scorbot robotic arm, with an average relative error of 0.27, so the system is suitable for its physical implementation.

Future Work. This investigation can become a fundamental pillar for new developments, for the control of different devices such as a wheelchair applying the same principle. This research can become a fundamental pillar for new developments, for the control of different devices such as electric chairs, industrial robotic arms, intelligent vehicles, etc. In the same way, the developed system is a starting point for the evaluation of different communication protocols integrated in IoT devices that can be integrated in controlled environments to facilitate the control of people who have some type of disability.

References

1. Brain-computer interfaces for communication and control. Clin. Neurophysiol. **113**(6), 767–791 (2002). https://doi.org/10.1016/S1388-2457(02)00057-3
2. Test-retest reliability of a single-channel, wireless EEG system. Int. J. Psychophysiol. **106**, 87–96 (2016). https://doi.org/10.1016/j.ijpsycho.2016.06.006
3. Abuzaher, M., Al-Azzeh, J.: Mind-wave wheelchair system (2021)
4. Balon, B., Simić, M.: Using raspberry Pi computers in education. In: 2019 42nd International Convention on Information and Communication Technology, Electronics and Microelectronics (MIPRO), pp. 671–676. IEEE (2019)
5. Delgado Flores, J.C., et al.: Diseño y construcción de un robot cartesiano automatizado para un sistema de hilatura: Transporte de un filamento polimérico (2020)
6. Doherty, J., Hendren, L., Radpour, S.: Kind analysis for MATLAB. In: Proceedings of the 2011 ACM International Conference on Object Oriented Programming Systems Languages and Applications (OOPSLA 2011), pp. 99–118. Association for Computing Machinery (2011). https://doi.org/10.1145/2048066.2048077
7. Dolińska, I., Jakubowski, M., Masiukiewicz, A.: Interference comparison in Wi-Fi 2.4 GHz and 5 GHz bands. In: 2017 International Conference on Information and Digital Technologies (IDT), pp. 106–112. IEEE (2017)
8. Franklin Salazar, L., et al.: Teleoperation and remote monitoring of a SCORBOT ER-4U robotic arm in an academic environment. Int. J. Innov. Technol. Exploring Eng. **8**(6), 40–45 (2019)
9. Garcia, M.V., Irisarri, E., Perez, F., Estevez, E., Orive, D., Marcos, M.: Plant floor communications integration using a low cost CPPS architecture. In: 2016 IEEE 21ST International Conference on Emerging Technologies and Factory Automation (ETFA). IEEE International Conference on Emerging Technologies and Factory Automation-ETFA (2016)
10. Irisarri, E., Garcia, M.V., Perez, F., Estevez, E., Marcos, M.: A model-based approach for process monitoring in oil production industry. In: 2016 IEEE 21ST International Conference on Emerging Technologies and Factory Automation (ETFA). IEEE International Conference on Emerging Technologies and Factory Automation-ETFA (2016)
11. Kodali, R.K., Valdas, A.: MQTT based monitoring system for urban farmers using ESP32 and raspberry Pi. In: 2018 Second International Conference on Green Computing and Internet of Things (ICGCIoT), pp. 395–398. IEEE (2018)
12. Kosiński, J., Szklanny, K., Wieczorkowska, A., Wichrowski, M.: An analysis of game-related emotions using Emotiv EPOC. In: 2018 Federated Conference on Computer Science and Information Systems (FedCSIS), pp. 913–917. IEEE (2018)
13. Marshall, D., Coyle, D., Wilson, S., Callaghan, M.: Games, gameplay, and BCI: the state of the art. IEEE Trans. Comput. Intell. AI Games **5**(2), 82–99 (2013). https://doi.org/10.1109/TCIAIG.2013.2263555
14. Misal, S.R., Prajwal, S., Niveditha, H., Vinayaka, H., Veena, S.: Indoor positioning system (IPS) using ESP32, MQTT and bluetooth. In: 2020 Fourth International Conference on Computing Methodologies and Communication (ICCMC), pp. 79–82. IEEE (2020)
15. Mishra, B., Kertesz, A.: The use of MQTT in M2M and IoT systems: a survey. IEEE Access **8**, 201071–201086 (2020)
16. Nam, D.P., Nguyen, H.Q., Tran, P.N., Yen, T.T.H.: Adaptive dynamic programing based optimal control for a robot manipulator. Int. J. Power Electron. Drive Syst. **11**(3), 1123 (2020)

17. Permana, K., Wijaya, S., Prajitno, P.: Controlled wheelchair based on brain computer interface using Neurosky Mindwave Mobile 2. In: AIP Conference Proceedings, vol. 2168, p. 020022. AIP Publishing LLC (2019)
18. Pham, Q.K., Vo, T.V., Tran, P.T.: On the implementation of a low-cost mind-voice-and-gesture-controlled humanoid robotic arm using leap motion and neurosky sensor. J. Electr. Eng. Technol. **17**(1), 665–683 (2022)
19. Quincozes, S., Emilio, T., Kazienko, J.: MQTT protocol: fundamentals, tools and future directions. IEEE Lat. Am. Trans. **17**(09), 1439–1448 (2019)
20. Rieiro, H., et al.: Validation of electroencephalographic recordings obtained with a consumer-grade, single dry electrode, low-cost device: a comparative study. Sensors **19**(12), 2808 (2019)
21. Rodríguez-Flores, J., Herrera, V.I., Morocho-Caiza, A., Merino, C.: Fractal control design with anti-windup effect for optimal operation of a power flyback source. In: Nummenmaa, J., Pérez-González, F., Domenech-Lega, B., Vaunat, J., Oscar Fernández-Peña, F. (eds.) CSEI 2019. AISC, vol. 1078, pp. 308–328. Springer, Cham (2020). https://doi.org/10.1007/978-3-030-33614-1_21
22. Rusanu, O.A., Cristea, L., Luculescu, M.C.: LabVIEW and Android BCI chat app controlled by voluntary eye-blinks using neurosky mindwave mobile EEG headset. In: 2020 International Conference on e-Health and Bioengineering (EHB), pp. 1–4. IEEE (2020)
23. Sadio, O., Ngom, I., Lishou, C.: Lightweight security scheme for MQTT/MQTT-SN protocol. In: 2019 Sixth International Conference on Internet of Things: Systems, Management and Security (IOTSMS), pp. 119–123. IEEE (2019)
24. Sandilya, K.S., et al.: Interpretation of EEG thought signals for the control and movement using MATLAB. In: 2019 International Conference on Intelligent Computing and Control Systems (ICCS), pp. 297–301. IEEE (2019)
25. Simmonds, B.: Teardown of the TP-link Kasa Smart Wi-Fi plug KP115 (2021)
26. Siswoyo, A., Arief, Z., Sulistijono, I.A.: Application of artificial neural networks in modeling direction wheelchairs using neurosky mindset mobile (EEG) device. Emitter Int. J. Eng. Technol. **5**(1), 170–191 (2017)
27. Strmiska, M., Koudelkova, Z.: Analysis of performance metrics using Emotiv EPOC+. In: MATEC Web of Conferences. EDP Sciences (2018)

Interactive Device for Carpal Tunnel Rehabilitation

Juan Freire[1], Paulina Ayala[1(✉)], and Marcelo-V. Garcia[1,2]

[1] FISEI, Universidad Tecnica de Ambato, UTA, Ambato 180206, Ecuador
{jfreire0899,ep.ayala,mv.garcia}@uta.edu.ec
[2] DISA, University of the Basque Country, EHU/UPV, 48013 Bilbao, Spain
mgarcia294@ehu.es

Abstract. The project proposes the development of an interactive electronic device for the rehabilitation of patients with Carpal Tunnel Syndrome, due to the fact that the rehabilitation process for this type of injury is performed with the intervention of a specialist, this device assists in the rehabilitation stages providing the patient with a different, efficient and progressive method. The device is based on the design of an optimal therapeutic glove that allows performing the necessary movements in the therapy process, through the use of wireless technology with data reception and transmission modules, facilitating the interaction with the patient and the remote control system to direct an object with wrist movements. It consists of a dynamic track that stimulates the sensory and visual senses of the patient during the interaction, through which the robot moves being designed with motion sensors, signaling, lights, and sounds, alerting the arrival of reference points in order to continue with the recovery stage. This device supports the work of the physiotherapist in the rehabilitation processes of their patients, reducing manual intervention, time, and fatigue, in the same way, that helps to stimulate the patient during the rehabilitation stage, recovering the mobility of the wrist quickly and effectively.

Keywords: Interactive · Carpal Tunnel Syndrome · rehabilitation · therapeutic · wireless technology · dynamic track

1 Introduction

In the field of physiotherapy, there are many cases where patients need passive and active rehabilitation of the wrist for certain pathologies and traumatisms, many cases due to forced work activities, sports injuries, neurological accidents or due to the appearance of diseases such as osteoporosis that impede the normal movement of the joints, as well as cognitive or behavioral abilities, about 1.7 million Ecuadorians, suffer from joint injuries of which 80% require neuropsychological care to return to their normal life condition, this means that out of

Supported by Universidad Tecnica de Ambato.

M. V. Garcia and C. Gordón-Gallegos (Eds.): CSEI 2022, LNNS 678, pp. 412–427, 2023.
https://doi.org/10.1007/978-3-031-30592-4_28

every 100 people 10 are affected, also according to the World Health Organization (WHO) more than 15% of the world population has neurological disorders, affecting people between 25 to 64 years of age among them 44% is female and 56% male [6,10].

Currently, for the recovery of patients with Carpal Tunnel Syndrome, the passive-assisted rehabilitation method is used, which activates the motor function of the wrist manually by the physiotherapist, with the aim of restoring its function, another method used is kinesitherapy, a movement treatment used to stimulate and relax the affected tissues, a fundamental technique to prevent muscle atrophy and accelerate recovery with 70% effectiveness compared to other types of therapies [8,13,15]. Each method has a conventional or routine process, so many patients show little interest in being rehabilitated, giving way to long periods of inactivity and therefore a longer recovery time, resulting in severe muscle weakness, joint stiffness, fatigue, limitations in mobility and ability to perform daily activities and other cognitive deficiencies. It is essential that the patient performs therapeutic exercises since this technique contribute to early mobilization and the initiation of activity of the patient through simple exercises, which if not performed, his physical condition will decrease and he will lose muscle strength causing Post-Exercise Fatigue Syndrome (PE Fatigue Syndrome) [3].

There is a high demand for rehabilitation, as well as a limited staff of professionals in the industry, also in Ecuador, the research is focused only on cases of severe trauma, with the use of robotic equipment which increases the cost of treatment and little research on passive physiotherapy, Therefore, in order to contribute with physiatrists in the fulfillment of routines and exercises applied to assisted rehabilitation in patients with carpal tunnel syndrome, it is important to design interactive auxiliary devices that are fundamental to improve the rehabilitation process [1,12].

The project is a passive wrist rehabilitation device, the design is based on a therapeutic glove that allows the patient to perform movements prescribed by a therapist, the glove will determine and census the correct angles to be performed, using wireless technology that will establish the communication link for the transmission and reception of data [2]. Additionally, it will have an interactive track with electronic components, as a support device in the work of the physio-therapist-therapist, the developed device will contribute to stimulating the patient in the process of rehabilitation routines and exercises, improving the movement of the wrist quickly and effectively, and significantly reducing recovery time.

2 Related Researches

Every day new ways, methods, technologies, and paths are discovered to solve everyday problems and make easier the activities that occur daily in people, activities in which the limbs of the body are involved such as the upper and lower, which are in constant physical wear by the execution of actions such as holding or manipulating, being prone to injury, muscle trauma or arrest of movement that are currently treated by specialists and supported in the management of devices that allow an effective solution. For this reason, different researches on this subject have been developed and are presented below:

A study developed a right and left-hand splint capable of measuring the force at the distal level of the fingers of the hand and defining the limit of force to be performed by the user, the device has a dynamic display with LEDs that allow measuring the level of force intensity performed. The graphic interface allows visualizing the data captured in real-time, the maximum force during the test, the average force, and the count that indicates the number of times it was over-passed, the results of this project were based on three areas: the functionality of the product that guaranteed the necessary conditions to provide comfort, versatility, and usefulness to the patient. They defined a functional design of control in the force limits to be performed by the patient, helping him with information about the force applied during the rehabilitation process, and finally in the graphical interface allowing the patient and specialist to visualize real-time data of the progress and thus ensure a better method of support in rehabilitation therapies [10].

Another proposed study developed a mechatronic device for passive wrist rehabilitation with three degrees of operation: flexion-extension, abduction adduction of the wrist, and pronation supination of the forearm, which uses mechanical design software to create the parts of the prototype in the mechanical part, I use a programming code that helps control the engines that include the device, as results when applying the device created ensures that patients perform the exercises executed correctly and safely for use without supervision and without causing any damage, Also making it portable to perform at any time or place, with an application that allows the patient to run or operate the tasks that have the device wirelessly that facilitates the interaction between the user and the machine and constantly improves the regeneration of wrist mobility [17].

The design of a rehabilitation device for the wrist joint from the approach of Concurrent Engineering is a study of rehabilitation in wrist joints developing an autonomous and electronic device, aimed at patients with tendinitis, synovitis, and rheumatoid arthritis, aided in the therapy processes, based on equipment used for CAD rehabilitation. By means of software they evaluated the acquisition and registration of movement data as well as the execution of therapeutic exercises, contributing in the physiotherapist area and in the functional recovery of joints present in the Venezuelan population, this device was able to register and execute routines that help to improve wrist joint injuries, demonstrating that with the use of the device the patient presents less discomfort during the rehabilitation process and providing great help to the physiatrist doctor because he will be able to control and supervise more patients at the same time [16].

Studies of the articular complex of the wrist: Anatophysiological and biomechanical aspects, characteristics, classification, and treatment of the distal radius fracture developed by Carlos Eduardo Medina Gonzalez, Mikhail Benet Rodriguez, Fernando Marco Martinez, yielded results of analysis focused on the scope of radial fracture of the wrist, They analyzed the complex structure of the articular set of the wrist with the parts that constitute it, for example, its multiple joints and metacarpal bone systems allowing to deepen and explain how the stability of the set of ligaments that allow the movement in the main angles present in the wrist such as flexion, extension, abduction, and pronation is maintained.

They explain in a convincing way the affection that causes the immobility of the wrist by external forces, in the same way, they study the types of treatments that play a very important role, with this investigation they contribute in great part to the study of this type of distal radial fracture, contributing with updated methods for the handling of these injuries in patients for the professional in this area [2,7,9].

In another study, the design and drive of a wrist rehabilitator device with the implementation of the structure, composition, and construction of the semi-automatic device drive became functional with two degrees of freedom useful in the active movement of the wrist in its recovery phase, allowing to reach and generate circumduction movement and acquisition with data sample according to the movements generated for rehabilitation, based on the biomechanics of operation, achieving to reproduce the active movement of the wrist joint.

For this purpose, they use actuators in the previously designed structure and the drive programming using a micro controller for data acquisition displayed on an interface, checking the correct operation of the device helps in the rehabilitation of the wrist and the correct assistance of the device in the movements required by the patient in the process of therapy, which is of vital importance be-cause thanks to this device results with the interaction between the mechanism and the patient supporting in the realization and maintaining the active movement of the wrist joint were thrown [20].

3 Theoretical Framework

The most relevant concepts for the development of the problems presented are presented below.

3.1 Carpal Tunnel Syndrome

Referred to as a frequent injury and condition of the wrist in which there is numbness and lack of mobility in the hand and arm in the area of the median nerve, the median nerve is considered one of the main nerves of the hand that make up a set of nerve roots passing through the carpal tunnel, when movements are performed causes pain and discomfort, that's when we talk about carpal tunnel, affecting the synovial membrane that gives lubrication and ease of movement to the tendons, which becomes inflamed or narrowed causing: tingling, weakness, or muscle damage in the hand and fingers. It should be treated immediately by a physiotherapist because it is an injury that worsens over time and without timely treatment may require surgery [4].

According to the World Health Organization, neurological diseases affect about one billion people, naming this disease as the nervous system disorders, or neurological diseases that are responsible for one third of the years that a person has of life within a society [5]. With timely treatment, surgery to repair the lesion can be avoided. For treatment, the use of therapeutic gloves can be considered, which help to place the splint in the correct position and help to perform daytime activities without the danger of incorrect flexion.

Another alternative is to perform various rehabilitation exercises, which help to move and position the ligaments and bones properly, as it can aggravate the injury by keeping the wrist in the same position all the time, for this the therapeutic physician should provide a routine of exercises according to improve the injury present in the carpal tunnel.

3.2 Angles of Movement in Reading Based on the Programming

The condition is called Carpal Tunnel Syndrome, which involves movements such as flexion, extension, radial abduction, and ulnar adduction for rehabilitation [18]. Each movement comprises a range of functions that must be fulfilled, therefore, and through the use of the programming platform, an easy-to-read language is established for the ideal operation, represented in Table 1.

Table 1. Types of movements and reading in programming

Type of movements	Angles of Movements	Range of movements $x = 0$–100 $y = 100$–200
Rest	0°	$45 < x < 55$; $145 < y < 155$
Wrist flexion	80°–90°	$y < 140$
Extension-wrist	70°–90°	$y > 160$
Abduction-wrist (radial)	15°	$x < 40$
Adduction-wrist (ulnar)	30°–45°	$x > 60$

4 Case Study

The treatment of people with Carpal Tunnel Syndrome is performed by the use of a manually assisted therapeutic glove, which is placed on the patient prior to his exercise session, and includes a massage and stimulation of the ligaments by a specialist to help the patient to mobilize his wrist at the required angles and the proposed exercises, according to this, some characteristics were determined for the design and development of the device. Establishing the requirements of the equipment to be developed, a functional diagram was elaborated describing the stages of operation and the components that make it up.

The block diagram in Fig. 1 establishes different stages, where the gesture recognition is achieved thanks to an electronic device placed in the hand in an appropriate manner, when the device is activated, it is calibrated by the

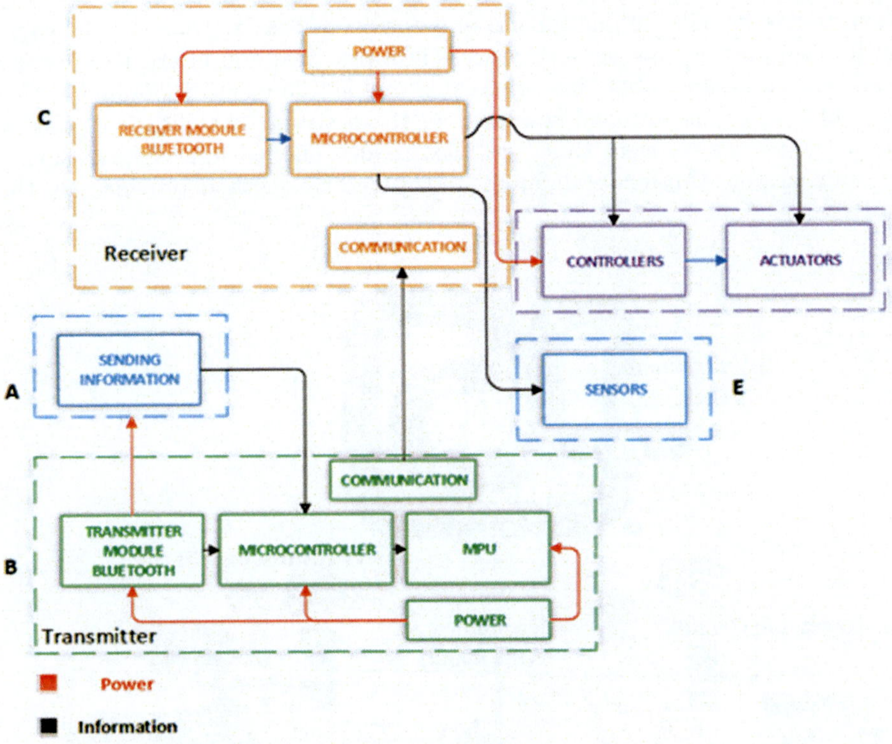

Fig. 1. Device diagram

movement of the wrist. Also involved are modules such as a transmitter, microcontroller, MPU, transmission communication and power supply, responsible for sending the collected data and transmitting them wirelessly to the next stage of reception, in this stage the information processing and control of rotation and speed of movement, which is sent to the receiver is performed. It involves the receiver module, microcontroller, receiving communication and power supply, where the informative commands are received, processed and sent to the controller and actuator. The controller and actuator create links that allow interaction with the components and interpret the instructions received by the microcontroller, understanding the actions performed by the device according to the movements and gesture control performed by the patient during each procedure, thus visualizing the interaction in a practical way and in real time by physically evaluating the movement control of the device performed by the patient and the progress with the use of sensors that indicate control points, rotation and arrival.

Through the analysis and the case study about wrist injuries such as Carpal Tunnel Syndrome, as well as the most used treatments for the passive rehabilitation process, it is determined that the rehabilitation device should help the patient to regain agility of movement, improve psymotor control and wrist coordination, supporting the recovery process in a practical and interactive way by means of a gesticular robot that stimulates the development of movement, this device also helps the medical specialist in the assisted rehabilitation process, in order to reduce recovery time, and also reduce manual intervention, saving time and fatigue, through a different and attractive method provided by the prototype.

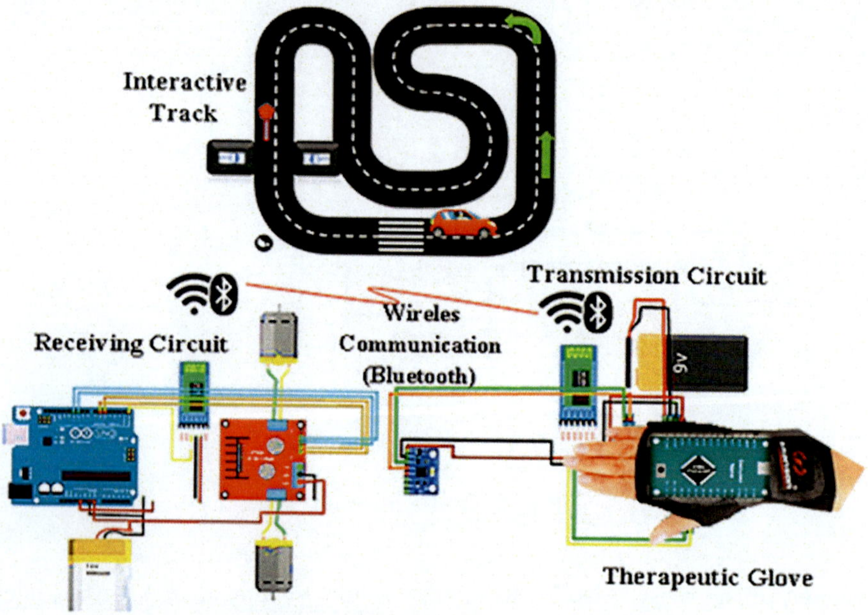

Fig. 2. Device schematic design

4.1 Hardware Architecture

In the general scheme of the device shown in Fig. 2, you can see each element that makes up the design, the first part is the glove made for therapeutic purposes, its design is structured especially to treat wrist injuries such as Carpal Tunnel Syndrome, its shape adapts to all types of hand, consists of a padded coating and a rigid platinum sheet that holds the wrist, It uses a programmable microcontroller suitable to census movements of the wrist, where the exact angles to be performed for the control of the device is determined, linked to the first bluetooth who communicates the data emitted wirelessly to the second bluetooth linking the communication of the second structure that makes up the robot. The microcontroller sends commands to the actuators that interpret and perform the patient's wrist movements that are configured in such a way that they generate an action in the robot as specified below:

– Flexing the robot moves forward.
– Extension the robot moves backward
– Abduction the robot turns to the left
– Adduction the robot turns to the right

Each movement goes according to the design of the track to follow, the device works self-powered, in this case, the use of rechargeable Lipo batteries provides the functionality to the microcontrollers activating the gestural robot, improving the operating space and control for the patient.

According to the established route to follow in the interaction track, the track consists of sensors that alert the patient to different points where he can rest or correct his trajectory and continue, thus developing the mobility and agility skills of the atrophied joint and working on psychomotor control. Figure 3 shows a representation of the hardware architecture between the interaction elements, microcontrollers, gloves, actuators, and robot.

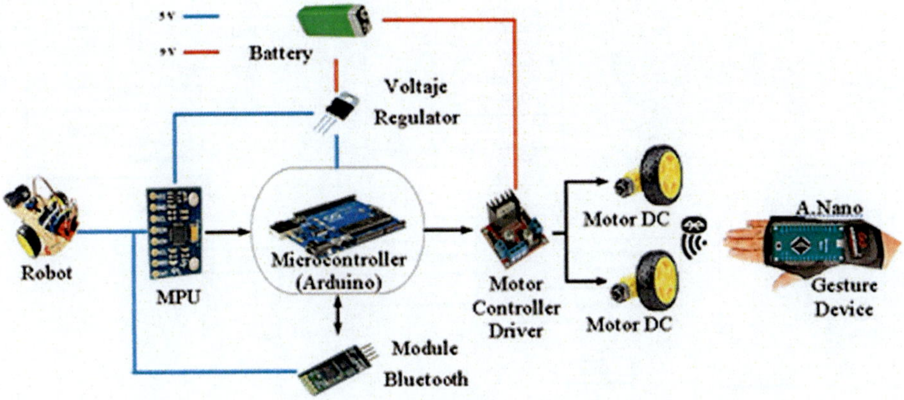

Fig. 3. Hardware architecture between the interaction elements

4.2 Arquitecture Software

The software architecture consists of a level developed in the Arduino IDE, a platform for coding instructions in C++. Developed in an integrated environment of high-level programming language commands and compatible with Arduino incorporating tools for complicated, loading and fast execution in its connection, so the code is based on the detection and interaction of the patient's gestures with the device, specifying the angles of movement to be detected as the actions are performed in their order, the necessary instructions allow the interaction between the device and the patient [8].

4.3 Main Gestural Robot

The functional class diagram of the gesticular robot in Fig. 4 shows how each class is represented with the name, method, and attribute, in this way it can be seen how the Module class has a composition relationship with the Robot class, i.e. this class is the container of the whole process, and in turn, the composition relationship between communication and control can be made, which allows with this the inheritance relationship between the classes of movement of the robot to its respective actuators providing that the Robot control perform the movements inherited by each class.

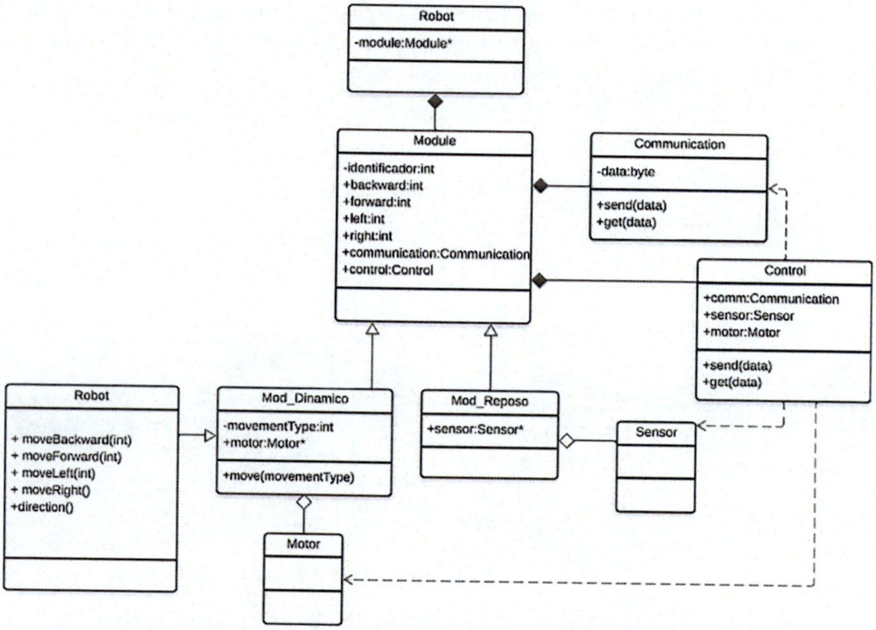

Fig. 4. Main class diagram Gestural Robot

4.4 Connection

The main class Robot contains the programming coding of the transmission module and the functional microcontroller for motion management. In the Module class is the programming part for the movement identifier, in the same way for the communication and control link, then the Communication class is the link between the technology used for wireless communication, sending, and obtaining information data, which proceeds to the Control class where it is attributed the communication received and delivered data for the domain of sensors and motors associated with the communication class so that in the class Mod.

Dynamic defines the instruction actions of the robot inherited by the class of communication and control module, finally, the class Robot controls the actions given by the Dynamic Mod. eh instructions to move forward, backward, or rotate, as well as to address according to the data received.

The class diagram of the connection is represented in Fig. 4, made by means of wireless communication, the main classes are the Bluetooth transmitter and Bluetooth emitter modules since without these classes there is no interaction between the parts for the operation of the device, responsible for connecting and communicating the entire system between modules, robot, and patient (Fig. 5).

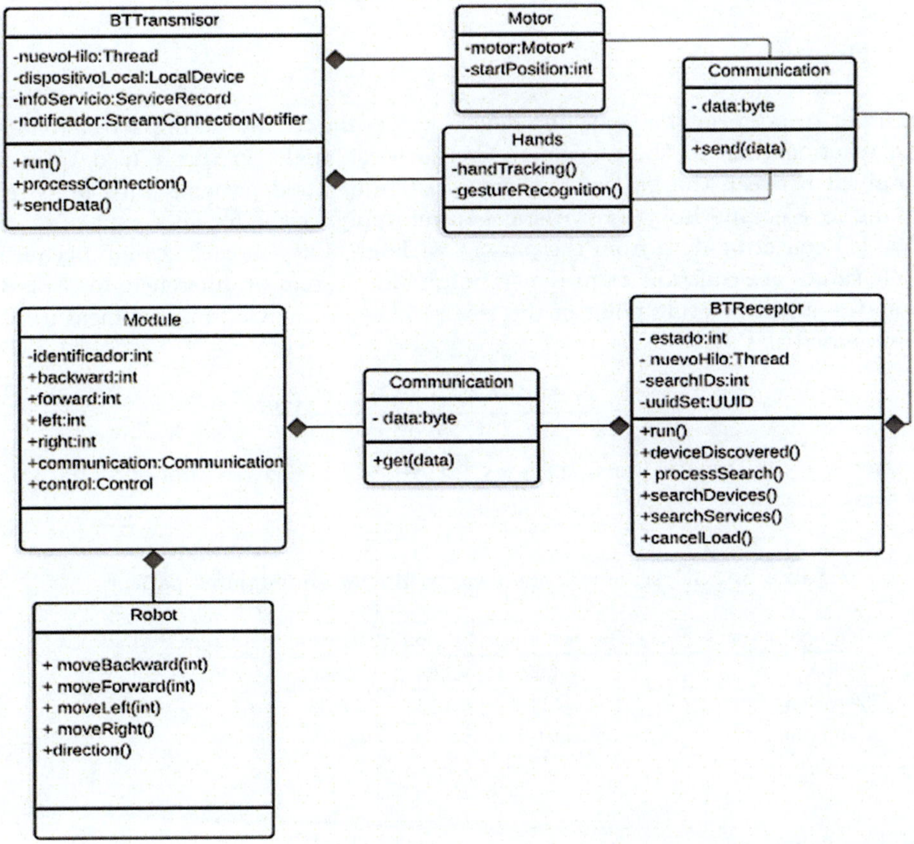

Fig. 5. Connection class diagram

The BTTransmitter class is linked to the main Bluetooth module located in the control glove section to initiate the connection process and send information to other components. In the Hands class the communication is done through the BTTransmitter class to implement the properties of gesture recognition and hand tracking made by the patient then in the Communication class is the interaction part received through the association of the Hands class, it receives and sends the collected data. The BTReceptor class performs search and linking processes with another connection device communicating the data obtained by joining the Modules class in which the instructions received through the communication layer are defined to perform addressing and control actions of the Robot class that has the attributes of action such as the instructions received to execute direction: forward, backward, left, right.

5 Results

In order to perform the tests, the device was evaluated by taking as reference: the reaction time to the movement of the wrist angle, in this way data was obtained between the patient without wrist injury and patients with injuries, taking as a comparison the statistical mean applying the formula expressed in Eqs. (1) collecting data from the patient without injury in each session because their values are constant or present a minimum margin of difference, measured from the initial starting point of the track to the first deviation, this measure is represented as [11].

$$\bar{X} = \frac{x_1 + x_2 + x_3 + x_4 + ... + x_n}{n} \tag{1}$$

where: $\bar{X}$ = Arithmetic mean, $x_1 + x_2 + x_3 + x_4 + ... + x_n$ = sum of all data, n = total data.

Table 2. Analysis of reaction time results for an uninjured patient.

Uninjured patients Exercises	Session 1	Session 2	Session 3	Average Statistics $\bar{X}$
Flexion	≈3,62	≈3,60	≈3,60	3,61
Extension	≈4,58	≈4,55	≈3,66	3,7
Abduction	≈3,19	≈3,20	≈3,18	3,19

The values measured in each movement of the patient without injury are essential since they represent the normal measurements of a healthy person in each session and allow the statistical mean of the measured values to be considered as a basis for comparison with the values taken from people with injury.

According to Table 2, it is observed that in the patient without injury a constant interval was maintained between each exercise performed with the device evaluated during the rehabilitation phase, while in Table 3 the reaction time of

Table 3. Comparison between a patient without injury and patients with injury with respect to movement reaction time using the device.

CriterionAssessable		Flexion	Extension	Adduction	Abduction
Patient Uninjured	$\bar{X}$	3,61	4,56	3,7	3,19
Patient N°1	Session 1	6,11	7,77	7,20	7,00
Patient N°1	Session 3	4,41	4,58	4,52	4,32
Patient N°2	Session 1	6,12	6,13	6,07	6,57
Patient N°2	Session 3	4,1	4,6	4,2	4,28
Patient N°3	Session 1	6,76	6,19	6,49	6,89
Patient N°3	Session 3	4,21	4,66	4,22	4,33

joint mobility and agility in the three remaining patients taken in the first session presents a delayed reaction, but during the 3 sessions, a significant improvement in mobility and reaction response is observed as visualized in Fig. 6.

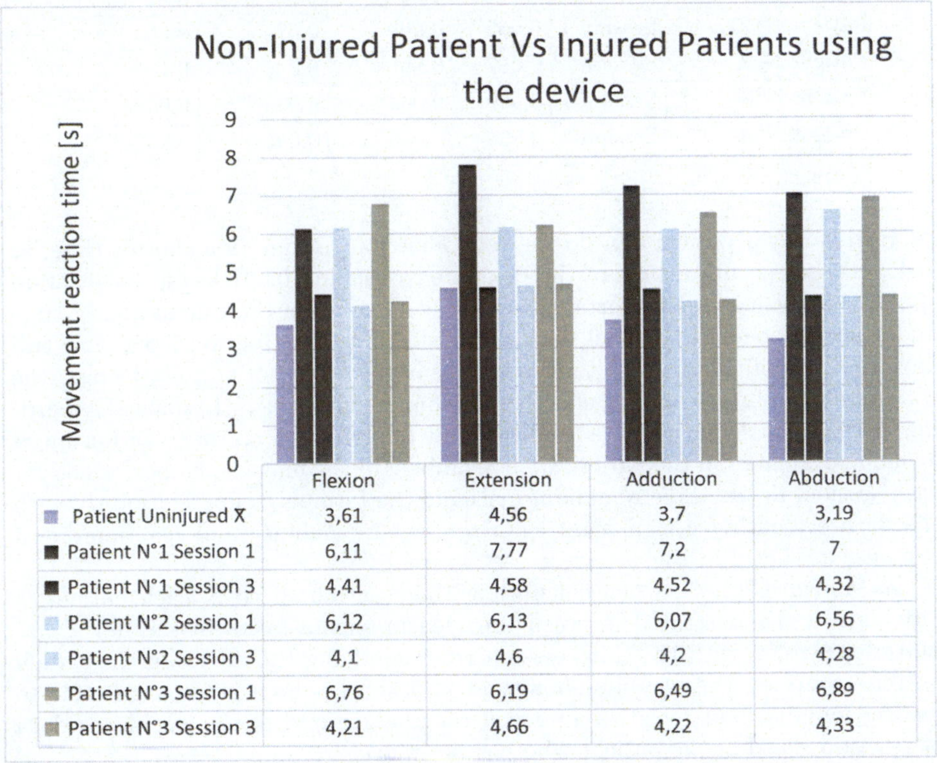

Fig. 6. Connection class diagram

In this way and based on these factors it is possible to calculate the reaction speed of each patient that goes between the variation of displacement traveled from an initial point of the track towards a stretch, related to the decrease of the recovery time, the speed formula expressed in the following Eq. (2) is applied for this purpose [14,19].

$$V = \Delta x/\Delta t = (x_f - x_i)/(t_f - t_i) \tag{2}$$

where: V = Reaction velocity of each patient Δx = Change in initial test position Δt = change in sampling time.

Table 4. Comparison between a patient without injury and a patient with injury with respect to movement reaction speed.

CriterionAssessable		Flexion	Extension	Adduction	Abduction
Patient Uninjured	$\bar{V}$	0,11	0,08	0,11	0,12
Patient N°1	Session 1	0,065	0,05	0,06	0,056
Patient N°1	Session 3	0,09	0,08	0,08	0,09
Patient N°2	Session 1	0,065	0,065	0,07	0,06
Patient N°2	Session 3	0,097	0,09	0,09	0,1
Patient N°3	Session 1	0,059	0,064	0,062	0,058
Patient N°3	Session 3	0,095	0,086	0,095	0,092

In this way a patient can do the rehabilitation routine in a shorter time, as is 4.1 [s] the execution time of the flexion exercise and 0.097 [m/s] the reaction speed, these values are important because it is based on the calculation of time and reaction speed, improving movement skills, reduces between one and two days the rehabilitation stage that normally needs a patient to fully recover the condition of an injury making effective the implementation of the device. According to Table 4, calculation data is obtained by applying the normal reaction speed formula, evaluated in the patient without injury in the movements performed for each session, in the three remaining patients the formula is applied jointly with the data evaluating a significant improvement in mobility and reaction speed response as shown in Fig. 6.

As shown in Fig. 7, when calculating the reaction speed of each patient, a difference is demonstrated between the measurements taken in the first rehabilitation session with the third session when applying the device in their daily exercise routines, these values show that in a short time a patient was able to recover mobility and speed to a normal range compared to the calculated data of reaction speed of the patient who has no injury.

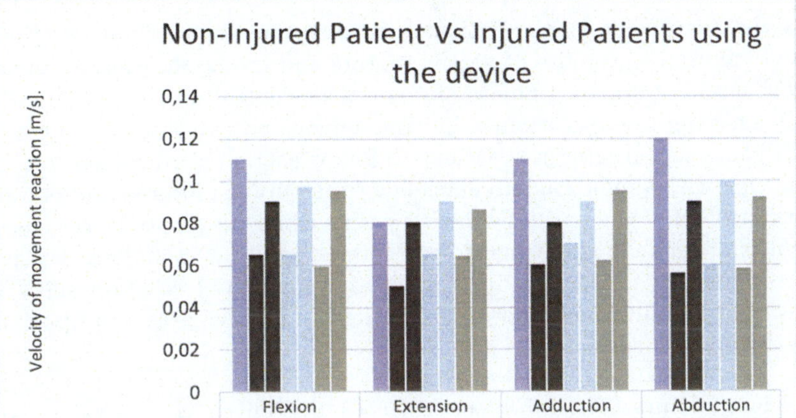

	Flexion	Extension	Adduction	Abduction
Patient Uninjured ∇	0,11	0,08	0,11	0,12
Patient N°1 Session 1	0,065	0,05	0,06	0,056
Patient N°1 Session 3	0,09	0,08	0,08	0,09
Patient N°2 Session 1	0,065	0,065	0,07	0,06
Patient N°2 Session 3	0,097	0,09	0,09	0,1
Patient N°3 Session 1	0,059	0,064	0,062	0,058
Patient N°3 Session 3	0,095	0,086	0,095	0,092

Fig. 7. Connection class diagram

6 Conclusions

The implementation of the interactive electronic device for rehabilitation of patients with carpal tunnel syndrome contributed to improving the rehabilitation process of the patient in the recovery stage, reducing the time in which it is necessary to recover mobility, since the project allowed the patient to resort to the routines in a constant and voluntary way, thanks to the fact that the device offers a new therapeutic method, different and attractive as a motivating factor to the fulfillment of adequate exercises.

The device allows the patient to develop the psychomotor control and agility of movement lost due to the wrist injury, through the use of the therapeutic glove by means of a gesticular sensor that detects the movement gestures transmitting them to a gesticular robot, with the use of actuators and controllers programmed in a way that regulates the speed and sensitivity, establishes the trajectory and direction control through an interactive track composed of lights, sensors, signaling and sounds that stimulate the sensory and visual senses of the patient in the process of interaction with the device when he/she is in the rehabilitation exercises programmed for movements such as flexion-extension/abductionadduction.

The average reaction speed of a person without wrist injury is 0.11 m/s, with this information, it was possible to locate in a higher reference area the values compared to the injured person, which helped the device to adapt and detect the reactive movements of the patient with programming methods of speed and sensitivity as well as the use of geared motors and the application of knowledge in PWM modulation, to regulate the speed of control and use of the device, without affecting the operation, achieving agility and reactive movement in the affected joint, encouraging the patient to follow each prescribed routine.

Verifying with validation and efficiency tests between injured and noninjured patients shows that the patient's recovery time using the device is less compared to traditional rehabilitation, due to the interaction that stimulates the patient to adhere to each procedure, improving movement and reaction skills which reduces the rehabilitation time to the patient and serving as a support to the physio-therapeutic specialist.

Acknowledgements. This work was financed by University Tecnica de Ambato (UTA)and their Research and Development Department (DIDE) under project PFI-SEI31.

References

1. Estimación de la distribución de la velocidad de conducción utilizando la técnica de colisión para evaluar el síndrome del túnel carpiano. In: Conferencia Internacional de 2006 de la Sociedad de Ingeniería en Medicina y Biología del IEEE. https://doi.org/10.1109/IEMBS.2006.259663
2. Telerehabilitación con el guante rutgers master ii después de una cirugía de liberación del túnel carpiano: prueba de concepto. https://doi.org/10.1109/TNSRE.2007.891393
3. Andaluz, V.H., Bustamante, D.C., Silva, S.M., Escudero, V.M., Salazar, P.J.: Carpal tunnel syndrome rehabilitation through force feedback. IEEE Lat. Am. Trans. **16**(9), 2467–2472 (2018). https://doi.org/10.1109/TLA.2018.8789570
4. Arita, S., Hashshizume, H., Honda, M.: A new approach to clarify the fuzziness of medical diagnosis by diagnostic layers - a diagnostic system for carpal tunnel syndrome with two layers of diagnostic filters using clinical indicators. In: 2014 Joint 7th International Conference on Soft Computing and Intelligent Systems (SCIS) and 15th International Symposium on Advanced Intelligent Systems (ISIS), pp. 845–850 (2014). https://doi.org/10.1109/SCIS-ISIS.2014.7044902
5. Babski, K., Crumpton, L.: The use of ergonomic measures to prevent and control the occurrence of cumulative trauma disorders: carpal tunnel syndrome and cubital tunnel syndrome. In: Proceedings of the 1997 16 Southern Biomedical Engineering Conference, pp. 407–410 (1997). https://doi.org/10.1109/SBEC.1997.583326
6. Bandaru, R.S., et al.: Speckle tracking of tendon displacement in the carpal tunnel: improved quantification using singular value decomposition. IEEE J. Biomed. Health Inform. **23**(2), 817–824 (2019). https://doi.org/10.1109/JBHI.2018.2822548
7. Bologna, J.K., Garcia, C.A., Ortiz, A., Ayala, P.X., Garcia, M.V.: An augmented reality platform for training in the industrial context. IFAC-PapersOnLine **53**(3), 197–202 (2020). https://doi.org/10.1016/j.ifacol.2020.11.032
8. Caiza, G., Monta, C., Ayala, P., Caceres, J., Garcia, C.A., Garcia, M.V.: Low-cost fuzzy control for poultry heating systems. In: Yang, X.-S., Sherratt, S., Dey, N., Joshi, A. (eds.) Proceedings of Sixth International Congress on Information and

Communication Technology. LNNS, vol. 236, pp. 109–118. Springer, Singapore (2022). https://doi.org/10.1007/978-981-16-2380-6_9

9. Gualpa, T., Ayala, P., Cáceres, J., Llango, E., Garcia, M.: Smart IoT watering platform based on orchestration: a case study. In: Garcia, M.V., Fernández-Peña, F., Gordón-Gallegos, C. (eds.) CSEI 2021. LNNS, vol. 433, pp. 191–204. Springer, Cham (2022). https://doi.org/10.1007/978-3-030-97719-1_11
10. Jurado, I.C., Rodríguez Vargas, U., Penaloza, C.: AI-VR platform for hand rehabilitation. In: 2020 IEEE International Conference on Systems, Man, and Cybernetics (SMC), pp. 726–730 (2020). https://doi.org/10.1109/SMC42975.2020.9283413
11. Kouchakinejad, F., Mashinchi, M., Khorram, E.: On the linear optimization on a system of arithmetic mean-min fuzzy relation inequalities. In: 2017 5th Iranian Joint Congress on Fuzzy and Intelligent Systems (CFIS), pp. 65–68 (2017). https://doi.org/10.1109/CFIS.2017.8003659
12. Lin, Y.H., Hsieh, M.Y., Lin, C.C., Wang, S.H.: Assessment of the carpal tunnel syndrome by high-frequency ultrasound images of the wrist: A feasibility study. In: 2010 International Computer Symposium (ICS 2010), pp. 923–926 (2010). https://doi.org/10.1109/COMPSYM.2010.5685378
13. Lukacs, P., Pietrikova, E.: Wrist rehabilitation in carpal tunnel syndrome by gaming using EMG controller. In: 2020 18th International Conference on Emerging eLearning Technologies and Applications (ICETA), pp. 412–417 (2020). https://doi.org/10.1109/ICETA51985.2020.9379203
14. Naihui, S., Zhaohui, R., Yongfu, W., Bangchun, W.: A comparison study of PD control of joint velocity in robot arms. In: 2007 Chinese Control Conference, pp. 557–560 (2007). https://doi.org/10.1109/CHICC.2006.4346812
15. Naranjo, J.E., Ayala, P.X., Altamirano, S., Brito, G., Garcia, M.V.: Intelligent oil field approach using virtual reality and mobile anthropomorphic robots. In: De Paolis, L.T., Bourdot, P. (eds.) AVR 2018. LNCS, vol. 10851, pp. 467–478. Springer, Cham (2018). https://doi.org/10.1007/978-3-319-95282-6_34
16. Nathan, P.A., Meadows, K.D., Keniston, R.C.: Rehabilitation of carpal tunnel surgery patients using a short surgical incision and an early program of physical therapy. J. Hand Surg. **18**(6), 1044–1050 (1993). https://doi.org/10.1016/0363-5023(93)90401-N. https://www.sciencedirect.com/science/article/pii/036350239390401N
17. Shah, R., Li, Z.M.: Three-dimensional carpal arch morphology using robot-assisted ultrasonography. IEEE Trans. Biomed. Eng. **69**(2), 894–898 (2022). https://doi.org/10.1109/TBME.2021.3108720
18. Silişteanu, C.S., Crăciun, D.M., David, M.: The importance of the sensor devices in the recovery of the patients with the carpal tunnel syndrome. In: 2016 International Conference and Exposition on Electrical and Power Engineering (EPE), pp. 426–430 (2016). https://doi.org/10.1109/ICEPE.2016.7781376
19. Wang, Y., Piao, L., Li, B.: Study on sensitive mechanism of micro jet angular velocity sensor based on fluid-solid coupling. In: 2019 IEEE 4th Advanced Information Technology, Electronic and Automation Control Conference (IAEAC), vol. 1, pp. 2110–2113 (2019). https://doi.org/10.1109/IAEAC47372.2019.8997585
20. Watanabe, T., Sugiura, Y., Miyata, N., Fujita, K., Nimura, A., Sugimoto, M.: DanceDanceThumb: tablet app for rehabilitation for carpal tunnel syndrome. In: Munekata, N., Kunita, I., Hoshino, J. (eds.) ICEC 2017. LNCS, vol. 10507, pp. 473–476. Springer, Cham (2017). https://doi.org/10.1007/978-3-319-66715-7_66

Electronic Biosignal Monitoring System for the Prevention of Respiratory Diseases by Applying Artificial Intelligence

Santiago Gómez(✉), Juan-P. Pallo, Santiago Manzano, Marco Jurado, and Dennis Chicaiza

Universidad Técnica de Ambato, 180103 Ambato, Ecuador
{sgomez9698,juanppallo,victorsmanzano,marcoajurado, dv.chicaiza}@uta.edu.ec

Abstract. The innovation of health services concerning Telemedicine allows the incorporation of protocols to help in the prevention of respiratory diseases. The project develops and implements an electronic vital sign monitoring system for the prevention of respiratory diseases through the application of artificial intelligence (AI). The data set (dataset) in preparation for the AI algorithm is collected from the Pan American Health Organization (PAHO), Mexican Institute of Social Security, data collected by the private practice Mediprof in the city of Salcedo, and data provided by researchers from the State University of Guayaquil, being labeled as Ambulatory, medical check-up, and hospitalization. The application uses an AI algorithm for supervised learning and classification of Support Vector Machines. The electronic system consists of three stages: 1. Sensing layer using sensors (SPO2, BPM, Temperature), 2. Application layer for data acquisition using a Raspberry Pi Zero with a local database and 3. Cloud service layer through a web interface for monitoring the health status of patients and sending notifications to email. The test group for data validation was 20 people ranging from 20 to 70 years old, between men and women, obtaining device reliability of 85% applying the AI algorithm, and the sensor acquisition data are within the values allowed by the OPS.

Keywords: Artificial intelligence · Supervised learning · Vital signs · respiratory diseases

1 Introduction

Respiratory diseases (SARS COV2, Influenza, HIN1, etc.), have become an unprecedented public health crisis, leaving 582 million people infected and approximately 6 million dead worldwide, according to statistics as of August 5, 2022. [2] Currently, there is a need for a health treatment system using technological tools and artificial intelligence to prevent these diseases. [7] IoT (Internet of things) devices have improved the control and monitoring methodologies of

M. V. Garcia and C. Gordón-Gallegos (Eds.): CSEI 2022, LNNS 678, pp. 428–444, 2023.
https://doi.org/10.1007/978-3-031-30592-4_29

the modern individual, obtaining large amounts of information about a person to process this data efficiently. [11] The project focuses on the implementation of an electronic bracelet that monitors a patient's body temperature, heart rate (BPM), and blood oxygen level (SPO2), through the MAX30100 and MLX90614 sensors, which are connected to a Raspberry Pi Zero W microcomputer through the I2C interface. The diagnosis of respiratory diseases is focused on patients with COVID-19 based on symptomatology, by classifying the risk level, so supervised learning classification algorithms are used, with a dataset or dataset with cases labeled in 3 classes: Outpatient, Medical Review, and Inpatient. The algorithms or models used for prediction are Support Vector Machine, Decision Trees, and Naive Bayes Multinomial. The evaluation of each of these is done with the metrics of "Recall", "Precision" and "F1-Score", determining that the Support Vector Machine algorithm performs the classification more efficiently than the other algorithms. All the information is stored in a local and remote database, which is presented in a WEB interface so that the treating physician can be informed of the patients who are being diagnosed. This is done using a cloud storage service and sending the data from a Python script via HTTP requests. In addition, he can be alerted via email about the patient's health status. This background is the reason for the development of the present research whose topic refers to the development of an electronic vital signs monitoring system for the prevention of respiratory diseases by applying artificial intelligence (AI). Section 2 of the paper details a compendium of related work on the use of telemedicine in the context of respiratory diseases. Section 3 presents the case study, and in Sect. 4 the implementation of the proposed solution. Section 5 discusses the results of the system implementation and finally, Sect. 6 discusses the conclusions and future work.

2 State of the Art

After reviewing different databases and repositories at the national and international level, similar works related to the topic of the study were found, which served as a guide for the development of this research, among these are mentioned: At the University of Cartagena, in 2020 a study was conducted that aimed to develop an IoT system for monitoring heart rate variability in usability tests. The IoT system collects heart rate variability periodically, as well as the mental stress index throughout the user experience [4]. We evaluated the effectiveness of data mining algorithms such as J48 and K-Means, which provide a clearer picture of mental stress behavior as a function of time [10].

In [9], a desktop application is developed that allows clinically instrumented monitoring of psychophysiological variables in multiple patients. The interface was validated by processing signals acquired from a healthy volunteer. The volunteer's information was correctly stored and the features of the acquired biosignals were successfully extracted. An artificial intelligence diagnostic system to aid the accurate diagnosis of Covid-19 was developed in 2021; in the epidemic area and two non-epidemic areas in China. Finally, with the clinical data and AI system, an accurate clinical prognosis is made, which can help physicians consider early, appropriate clinical management and allocate resources [15].

The design of a remote biomedical signal monitoring framework combining Internet of Things (IoT), 5G communication, and artificial intelligence techniques is presented by [13] in 2021. Obtaining as results that the biomedical signals were transmitted through the 5G network to the cloud server where the GRU-AE deep learning model is deployed. It was highlighted that the proposed GRU-AE model can analyze multidimensional biomedical signals in time series. In [12], an electronic vital sign monitoring system and social distancing alerts for respiratory disease prevention are developed using IoT architecture for data management. This system consisted of a portable bracelet for data acquisition, a server that stores the information to display it in a web application, and an assistant in the Telegram application for queries and alerts with a smartphone. The system helped to avoid crowds that generate contagions of these diseases and monitor the vital signs of users to receive timely medical attention if necessary. A home-centric telemedicine system with cloud management and IoT services was developed in 2021. The user's heart rate and SpO2 levels are measured by a low-cost wristband and sent via Bluetooth to an Android TV app. The system was tested through a survey of 40 healthcare professionals to evaluate the system, obtaining high acceptance as a useful system for the medical sector [1].

3 Case Study

3.1 Electronic Device Architecture

This work implements an electronic biosignal monitoring system for respiratory disease prevention by applying artificial intelligence through risk classification using machine learning. The architecture of this device is presented in Fig. 1, which consists of 3 layers: The sensing layer, application layer, and cloud service layer.

In the sensing layer, a wearable bracelet is used for the acquisition of vital signs where the MLX90614 body temperature sensor and the Raspberry Pi Zero are installed. The MAX30100 pulse oximetry sensor is also used, which is connected to the bracelet via cables because it will obtain the data from the patient's index finger. The sensors used work with an input voltage of 3.3 V and communicate through the I2C protocol. The vital signs data is sent to the Raspberry Pi to be displayed on a 1.3-in. OLED screen, which communicates via SPI protocol and works with a 3.3 V power supply. The application layer consists of two elements, the Raspberry Pi and a 1.3-in. OLED display. The first element performs the entire procedure of storing vital signs and processing them along with other data for the diagnosis of respiratory diseases using an artificial intelligence algorithm. In addition, it is connected to WiFi and Bluetooth, both in the 2.4 GHz band. The OLED screen shows the interface that is created by programming in a Python script, which consists of a menu where you can choose several options among which is the diagnosis. When this option is selected using the screen keyboard, the sensors are activated to take the vital signs data. Then the patient's gender, whether or not he/she has been vaccinated, the risk factors, and the symptoms they present when using the electronic device are entered, all

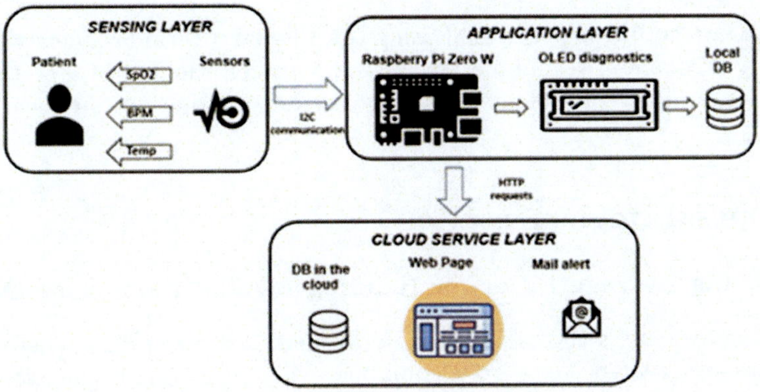

Fig. 1. Biosignal monitoring electronics device architecture.

these data are sent to the artificial intelligence algorithm to be analyzed and finally, the prediction is displayed on the screen. Regarding the vaccine, the type is not considered since the information in the dataset does not include the level of importance of the vaccine type for the classification. All the information is stored in a MySQL database and sent to the cloud server via HTTP requests. In the cloud service layer, all the data is stored in the Raspberry Pi to be displayed in a web interface so that the doctor in charge can be informed of the health status of his patients and take the necessary actions for their well-being. In addition, the physician can be alerted by an e-mail notification if the diagnosis recommends a medical check-up or hospitalization of the patient. This is done through a simple mail transfer protocol (SMTP).

3.2 Important Points in the Algorithm Selection Process

Obtaining the Dataset. The choice of the artificial intelligence algorithm starts from a dataset according to the intended application of the algorithm.

Data Balancing. The data balancing is performed by using the technique of oversampling or "Over Sampling" because the classes do not contain the same number of cases, thus preventing the classification from being biased, i.e. having preference in the classification of the majority class [5,10].

Partitioning. The partitioning of the data is performed by dividing the dataset into a training set and a test set, usually using a ratio of 70:30 respectively [8]. In the dataset there are 678 cases with their respective classes, this amount justifies the use of a 90:10 ratio because the more data in the training set, the more the model will learn. Conversely, if there were a larger amount of data, a lower ratio would be considered for the training set. This also avoids the occurrence of overfitting, which would lead to a simple model with little predictive value [16].

Coding. Data coding is performed using the TF-IDF (Term Frequency - Inverse Document Frequency) technique, because it considers the use of a certain word in a text as an important element in the classification due to its relevance [3].

4 Proposal Implementation

4.1 Moving Average Filter for Reading MAX30100 Sensor Data

The moving average is the calculation performed on a set of data where a series of averages are created, i.e. a list of numbers where each one is the average of a subset of the original data, an example is shown in Fig. 2.

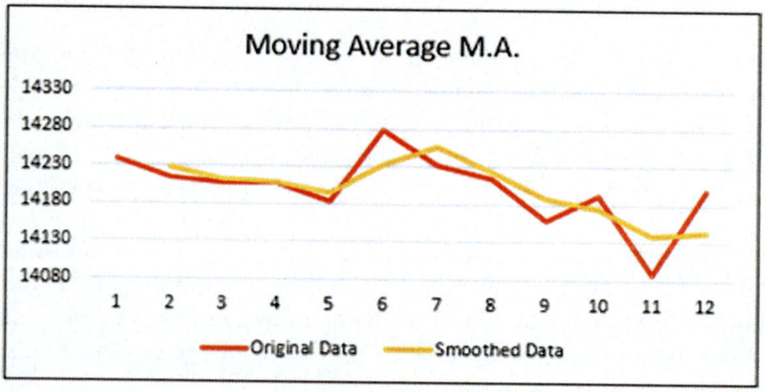

Fig. 2. Data smoothing by application of a moving average filter with a period equal to 2.

Since the MAX30100 sensor reading presents noise, a moving average filter is applied to smooth the peaks. Figure 3a shows the trend of the original heart rate data, where the noise present in the data can be observed. Figure 3b shows the smoothed BPM graph showing a stable trend in the data reading. This process is performed for both heart rate and blood oxygen level readings as shown in Fig. 4a and Fig. 4b.

4.2 MLinear Equation for Fitting Sensor Data

The pulse oximetry and body temperature sensors are adjusted by obtaining a linear equation based on a dispersion graph so that the reading of the sensors tends to approximate the reading of the medical equipment. Therefore, a comparison is made between these two devices. Tests were carried out on 10 people

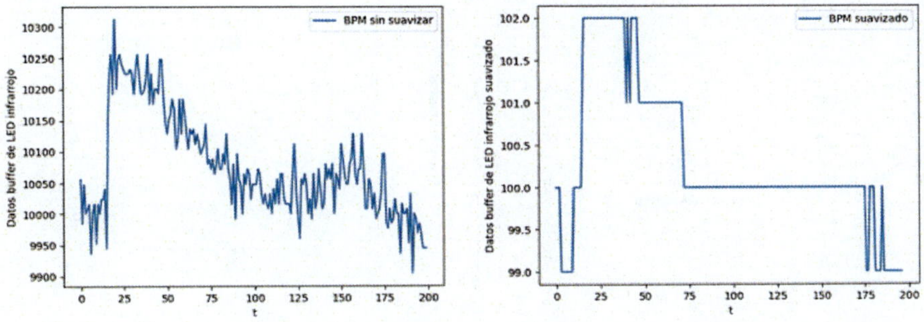

(a) (b)

Fig. 3. Smoothing of BPM data: a) unsmoothed BPM, b) smoothed BPM.

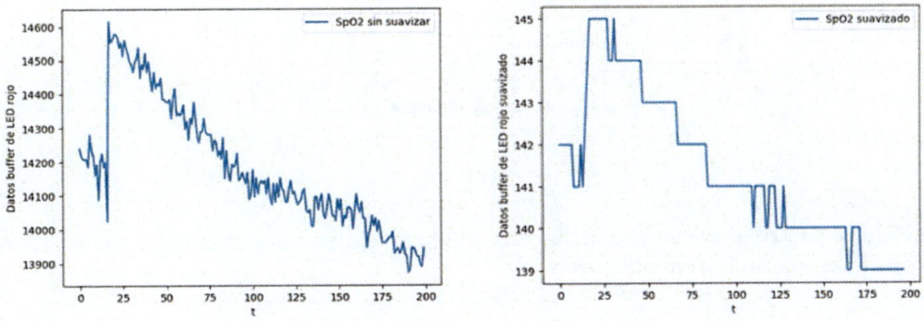

(a) (b)

Fig. 4. Smoothing of SpO2 data: a) unsmoothed SpO2, b) smoothed SpO2.

between 10 and 65 years of age, obtaining the linear equations shown in Fig. 5, which are entered into the data reading code of each sensor to finally adjust them.

4.3 Selection of the Artificial Intelligence Algorithm

Preparation of the Data Set (dataset). The data preparation stage begins with a dataset containing the labeled information, i.e. with its classes, in this case "Outpatient", "Medical Review" and "Hospitalization". The available dataset has four columns: gender, vaccine, symptoms and intensity, these are transformed to data that serve for the classification, in the case of gender it can be observed in Table 1 that there are initials F and M that correspond to Female and Male respectively, therefore, these initials are transformed to their original adjective.

Similarly, in the symptom's column, texts such as "loss of taste and smell", "no taste" or "just had no taste" are observed. These data are treated so that words with the same meaning, but with different syntax are generalized to a single word, where the text "no taste" has been changed to "loss of taste".

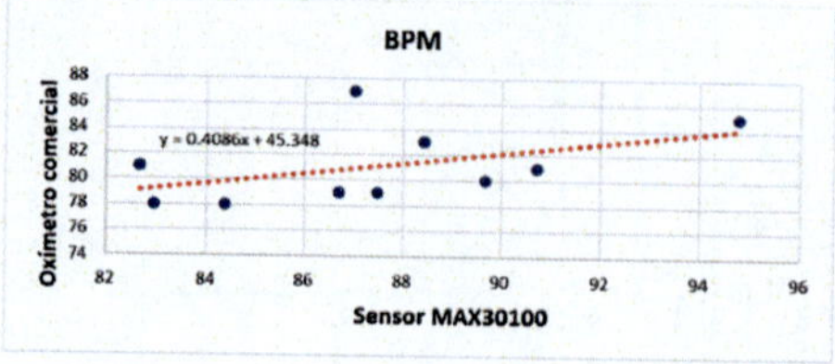

(a)

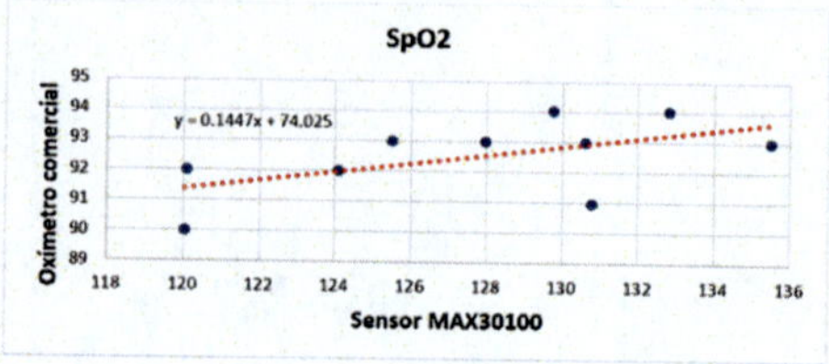

(b)

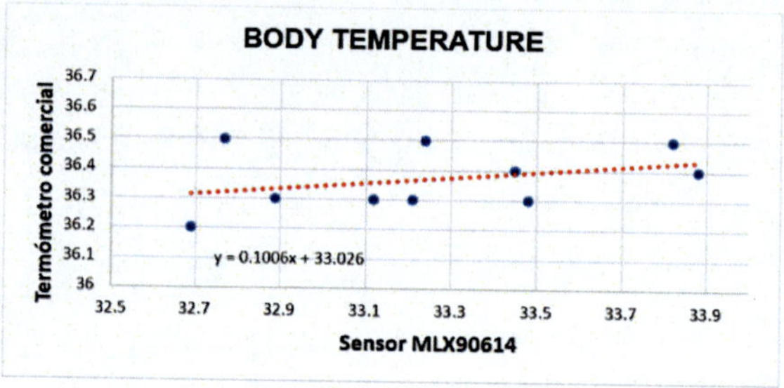

(c)

Fig. 5. Linear trend between sensors and medical equipment: a) Heart rate, b) Blood oxygen level, c) Body temperature.

Table 1. Part of the dataset content to be used.

Gender	Vaccine	Symptom	Intensity
F	Pfizer	loss of taste and smell	Outpatient
M	Sinovac	mild back pain and fever	Outpatient
M	I had no vaccines	chest pain	Outpatient
F	I had no vaccines	headache, dry cough and fever, had no taste, no smell	Outpatient
F	Pfizer	Tasteless	Outpatient
F	I had no vaccines	just had no taste	Outpatient
M	I had no vaccines	total loss of smell	Outpatient
F	I had no vaccines	fever, loss of smell and taste	Outpatient
F	Sinovac	loss of taste and smell, headache, fatigue, fever, fever	Medical revision

Once the text has been transformed, it is also necessary to combine the first 3 columns so that the Machine Learning algorithm can consider all the information. Finally, the dataset is as shown in Table 2.

Table 2. Processed data.

Symptom	Intensity
female, loss of taste, loss of smell	Outpatient
male, back pain, fever	Outpatient
male, no vaccine, chest pain	Outpatient
female, loss of taste	Outpatient
female, no vaccine, loss of taste	Outpatient

Once the data have been processed using Excel, they are scanned in Python using the "Pandas" library to find out the types of data they contain and whether there are any blank cells that generate errors in the vectorization process. Figure 6a shows that the dataset contains two columns, symptoms and intensity, and Fig. 6b shows that there are no null or empty data.

```
datos.info()
<class 'pandas.core.frame.DataFrame'>
RangeIndex: 430 entries, 0 to 429
Data Columns (total 2 columns):
 #    Column       Non-Null  Count     Dtype
----  ----------   --------------------  --------
 0    sintomas     430 non-null         object
 1    intensidad   430 non-null         object
dtypes: object(2)
Memory usage: 6.8+ KB
```

(a)

```
datos_null = datos.isnull().any().sum()
datos_null
```

(b)

Fig. 6. Data exploration: a) Type of data in the two columns, b) Number of null cases.

Data Balancing. Figure 7a shows the number of cases that exist for each category; of these, the one with the most cases is the "mild" category. Therefore, the original distribution of the dataset is modified using the "RandomOverSampler" method to replicate the cases that contain the minority category as shown in Fig. 7b, as mentioned in Sect. 3.2.

Data Training. The training is done with the "train_test_split" method. This method receives as arguments a set of data, the sample size for testing and a random state, this last argument is recommended to use an integer so that, when executing the code several times, new values are not generated for the training and test dataset which can affect the performance of the final model.

- Data set: balanced data.
- Sample size for testing:0.10
- Random state: 42

```
datos.value_counts(['intensidad'])
intensidad
ambulatorio          226
revision_medica      184
hospitalizacion       20
dtype: int64
```

(a)

```
intensidad
ambulatorio          226
hospitalización      226
revision_medica      226
```

(b)

Fig. 7. Data balancing. a) Number of original cases of each class, b) Balanced classes.

It is worth mentioning that due to the amount of data available, it is recommended to use a 90:10 ratio for training and testing, respectively, so that the model can predict efficiently.

Data Encoding. Because the models use numerical data, the need arises to convert the examples, which are texts, into numerical vectors. This is done by using the sklearn module, "TfidVectorizer", which is responsible for defining the relevance of a word in the text as defined in Sect. 3.2. The coding is done to the symptom column for both training and testing because this is going to be the one used in the artificial intelligence algorithm to perform the classification.

Construction and Evaluation of the Algorithm. Once the coding process has been carried out, it remains to test the artificial intelligence algorithms so that, through the evaluation methods, the best one is chosen. The following algorithms will be used.

- Support Vector Machine.
- Decisions Tree.
- Naive Bayes Multinomial

Evaluation of the algorithms. The evaluation of the algorithms is performed using the metrics of "Recall (TPR)", "Precision (P)" and "F1-Score" by means of the equations: 1, 2 and 3. The first equation is the ratio between the True Positives (T.P.) and the sum of these with the False Negatives (F.N.). This indicates how often the model can correctly predict a given class, the closer it is to 1 the better the result. The second equation is the ratio between the True Positives and the sum of these with the False Positives (F.P.). This indicates how often it is correct if the prediction is positive, the closer it is to 1 the better the model. The third equation is the harmonic mean of the pre-accuracy and recall. This indicates how accurate the classifier is as it strikes a balance between accuracy and recall. Its score reaches the best value at 1 and the worst at 0 [6].

$$TPR = TP/(TP + FN) \tag{1}$$

$$Precision = TP/(TP + FP) \tag{2}$$

$$F1 - Score = 2TP/(2TP + FP + FN) \quad (3)$$

These values are calculated starting from the confusion matrix of each algorithm, which, since there are 3 classes for the classification, the matrix obtained will be 3×3, these are shown in the Fig. 8.

```
[[ 18   5   1]
 [  2  24   0]
 [  0   0  18]]
```

(a)

```
[[ 18   4   2]
 [  6  18   2]
 [  0   0  18]]
```

(b)

```
[[ 6   1  17]
 [ 1  12  13]
 [ 0   0  18]]
```

(c)

Fig. 8. Confusion matrices: a) SVM, b) Decisions tree, c) Naive Bayes.

Cross Validation of k Iterations. The k-iteration cross-validation method, also known as K-fold cross-validation, is used, in which the data is partitioned into K subsets. In each iteration, one of the subsets is used for the testing table and the rest for the training table. This method was used to estimate the performance of the algorithms in performing the classifications. In this case, five sets are used, which means that one set of the total data is used for testing and the remaining four for training.

In each iteration, 5 precision and error results were obtained corresponding to the 3 algorithms used for classification. Subsequently, the arithmetic mean of these iterations is calculated for each algorithm, observing that the SVM algorithm for classification is the one with the highest precision, obtaining 84.26% of correct classifications. This can be seen in Fig. 9, which shows the accuracy percentage of each iteration and the arithmetic mean in the legend. In the same way, Fig. 10 shows the error percentage of each algorithm and its respective arithmetic mean, with the error percentage equal to 15.74% for SVM, which is lower than the others.

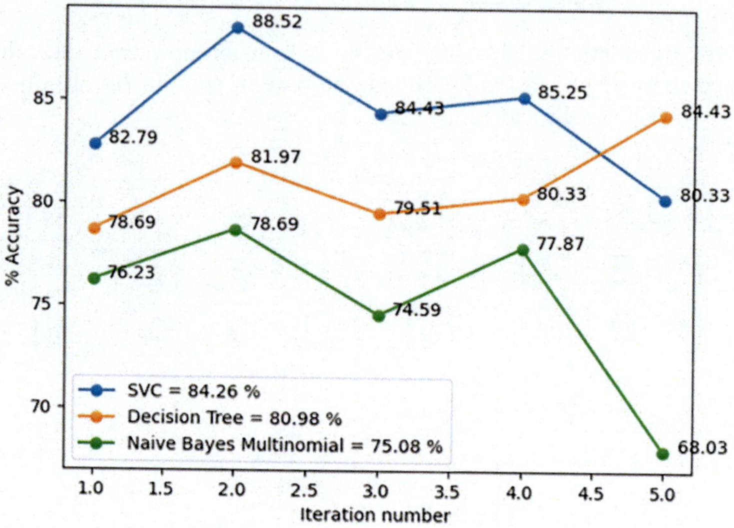

Fig. 9. Accuracy of the models.

Selection of the Algorithm. According to the analysis carried out in Fig. 9 and Fig. 10 and the classification reports presented in Table 3, it can be determined that the algorithm that has the most efficiency in classifying the risk of COVID-19 is the Support Vector Machine algorithm. because the values for "Recall" and "Precision" are higher with respect to the other algorithms, which means that it handles the classification of these classes well. Similarly, through F1-Score, the algorithm classifies the "Hospitalization" class better.

Algorithm Storage. Once the algorithm is chosen, the "pickle" library in Python is used to store the trained data based on the final model [14], this is done by serializing objects where they are converted to byte strings to later write them. in a file with extension ". pickle". This process is carried out to be able to use the file independently in another script without the need to carry out all the steps for choosing the final algorithm, reducing the consumption of Raspberry Pi Zero resources, and increasing the speed of diagnosis. It is worth mentioning that both the final model and the vectorized training and test data must be serialized.

The block diagram in Fig. 11 shows the data entry process for the COVID-19 risk classification.

Finally, the prototype presented in Fig. 12 shows the collection of body temperature data through the sensor installed in the bracelet and the pulse oximetry data from the index finger through the sensor that communicates by cable to the device.

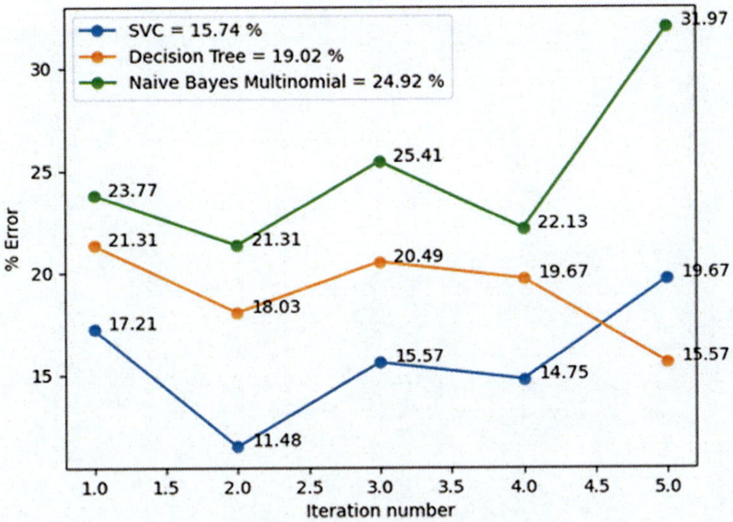

Fig. 10. Models error.

Table 3. Performance of classification algorithms (SVM, Decision Tree and Naive Bayes).

VP	VN	FP	FN	Class	P	TPR	F1
				SVC			
18	42	2	6	Ambulatory	0.9	0.75	0.82
24	37	5	2	Medical check	0.83	0.92	0.87
18	49	1	0	Hospitalization	0.95	1	0.97
				Decision Tree			
18	38	6	6	Ambulatory	0.75	0.75	0.75
18	38	4	4	Medical check	0.82	0.69	0.75
18	46	4	0	Hospitalization	0.82	1.00	0.90
				Naive Bayes Multinomial			
18	20	30	0	Ambulatory	0.86	0.25	0.39
12	41	1	14	Medical check	0.92	0.46	0.62
6	43	1	18	Hospitalization	0.38	1.00	0.55

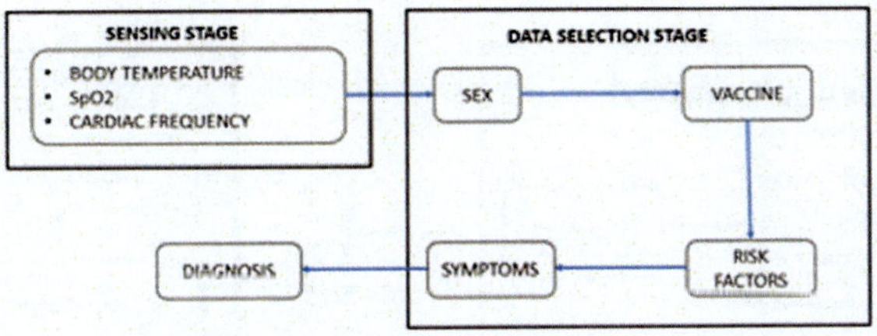

Fig. 11. Block diagram for the diagnosis of COVID-19.

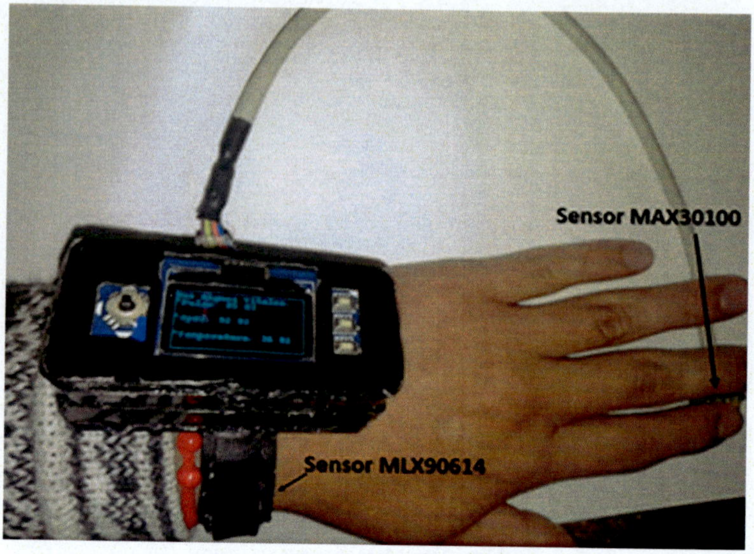

Fig. 12. Electronic system in operation.

4.4 Cloud Storage

The sending of data is done from a Python script using the "request" library, which allows sending HTTP requests in a simple way, from this library the post () method is used to send the data to the web server URL as is shown in Fig. 13. The data is received in a script called "insertar.php" which carries out the process of sending it to the database in the cloud so that it can later be displayed in the web interface, the same as shown in Fig. 14.

Sending Alerts to the Mail. Sending email alerts is done by using the SMTP protocol (Simple Mail Transfer Protocol) that allows a client to communicate with a server to send an email to one or more recipients. This is done from a Python script. The security in the connection to the Internet is done by means of the "ssl" library which secures the sending of data both on the client and server side. Figure 15 shows the alert received by e-mail, with the subject "COVID-19 risk alert" and the patient identification and algorithm classification as the body.

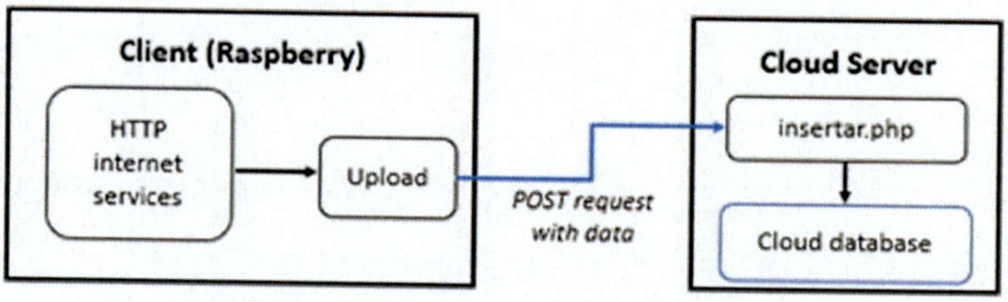

Fig. 13. Method "request.post" using Python.

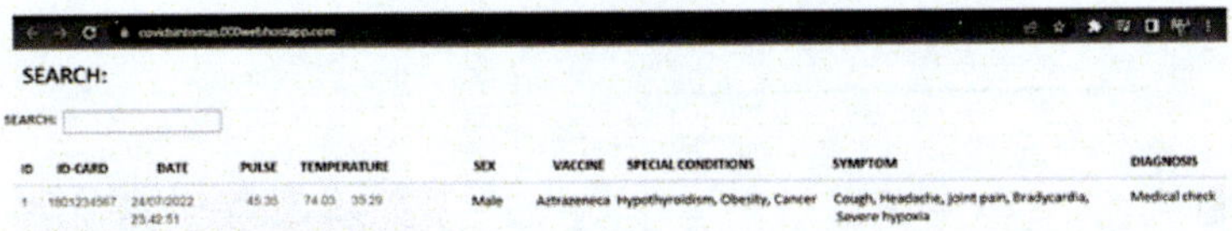

Fig. 14. Method "request.post" using Python.

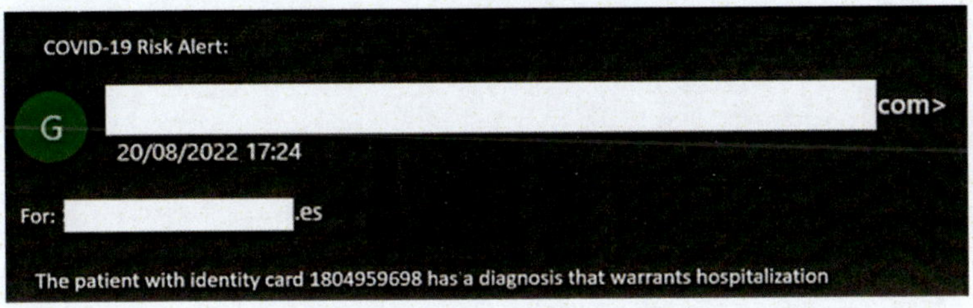

Fig. 15. Alert received by e-mail.

5 Results

5.1 Validation of the Vital Signs Obtained by the Prototype

The validation of the data obtained by the sensors installed in the prototype is done by analyzing the Blant Altman diagram for each vital sign. Figure 16 shows a bias due to the fact that the average line is below zero, this line is located at –2.8 which indicates that the prototype measures 2.8 units more heart rate than the medical equipment.

Figure 16a and 16c show that their average lines are close to 0, indicating that the values of SPO2 and body temperature measurements do not differ much from the medical equipment. Similarly, the values are close to the confidence interval.

5.2 COVID-19 Risk Classification Tests

The prototype is tested on a group of 20 people with suspected COVID-19 to compare with a physician's technical analysis, obtaining 3 incorrect and 17 correct classifications as shown in Fig. 17.

Figure 18a shows the number of classes predicted by the algorithm, where the Medical Review class has the largest number, but this does not mean that all the predictions are correct. This is shown in Fig. 18b, which indicates 1 error and 6 hits for the Ambulatory class, 2 errors and 8 hits for the Medical Review class and 3 hits with 0 errors for the Medical Emergency class. Thus, it could be deduced that, according to Table 1, the algorithm classifies better the Medical Emergency class as indicated in the F1-Score evaluation metric.

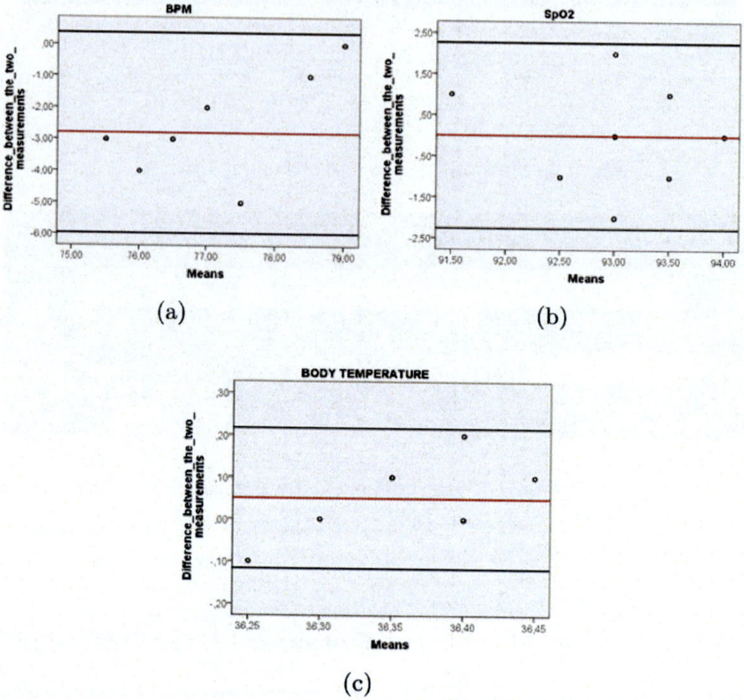

Fig. 16. Bland Altman graphs: a) Heart rate, b) Blood oxygen level, c) Body temperature.

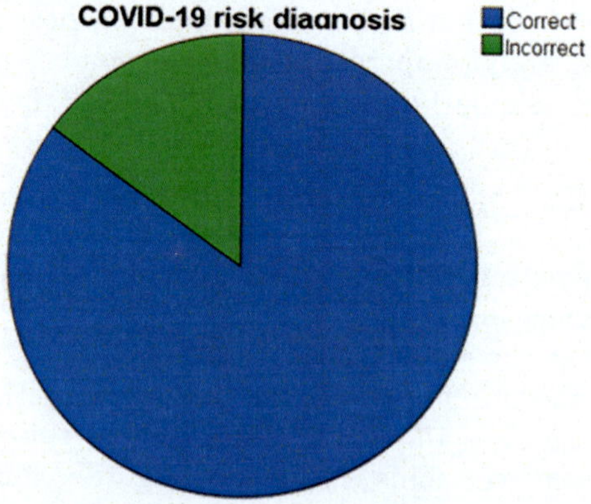

Fig. 17. Global number of hits in COVID-19 risk classification.

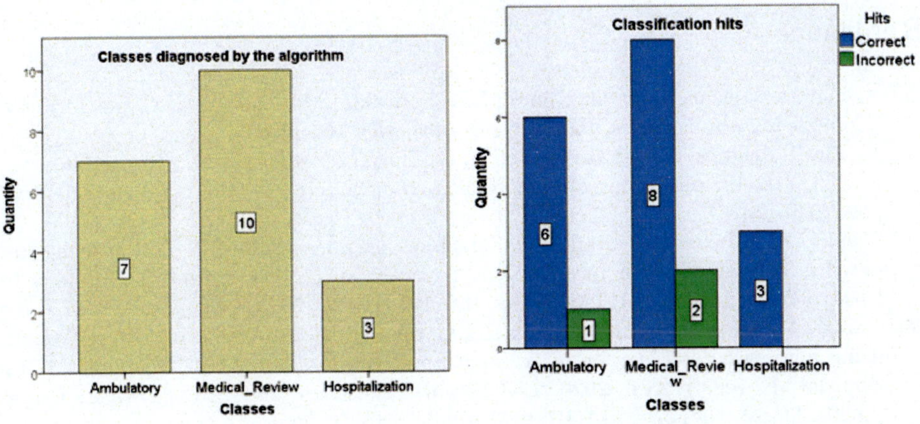

(a) (b)

Fig. 18. COVID-19 risk classification: a) Classes diagnosed by the algorithm. b) Number of successes for each class.

6 Conclusions

The Vector Support Machine model had higher efficiency than the other models. A "Recall" and "Precision" of 0.75 and 0.9 respectively were obtained for the "Ambulatory" class, for the "Medical Review" class values equal to 0.92 and 0.83 and for the "Hospitalization" class 1 and 0.95 which concludes that the model can efficiently handle the 3 classes. The class that the algorithm can predict the most is the "Hospitalization" class because the F1 Score metric has a value equal to 0.97. The vital signs obtained by the prototype had a reliability of 94.98% for heart rate, 98.96% for blood oxygen level and 99.78% for body temperature, thus determining that the values are close to those obtained by medical equipment. The electronic prevention system for respiratory diseases was implemented, focusing on patients with COVID-19 based on the symptomatology applying artificial intelligence, facilitating the interaction between the doctor-patient avoiding the spread of the virus because all data are displayed on a web server so that the treating physician can know the patient's health status and in case of risk can receive an alert to your email.

Acknowledgments. The authors thank the Technical University of Ambato (UTA) and the Research and Development Department (DIDE) for their support in carrying out the project "Electronic biosignal monitoring system for the prevention of respiratory diseases by applying artificial intelligence".

References

1. A. Garces-Salazar, S.M., C. Núñez, J.P.P., Jurado, M.: Low-cost IoT platform for telemedicine applications. RISTI **04**(NE43), 153–166 (2021)
2. Águila Rodríguez, N., Duany Badell, L.E., Herrera Fragoso, L.R., Terry Calderón, M.: COVID-19: perfil clínico-epidemiológico en municipio cumanayagua. cienfuegos, 2020 (2020)
3. Alexan, A., Alexan, A., Ştefan, O.: Machine learning activity detection using ml. net. In: 2020 IEEE 26th International Symposium for Design and Technology in Electronic Packaging (SIITME), pp. 188–191. IEEE (2020)
4. Caiza, G., Bologna, J.K., Garcia, C.A., Garcia, M.V.: Industrial training platform using augmented reality for instrumentation commissioning. In: De Paolis, L.T., Bourdot, P. (eds.) AVR 2020. LNCS, vol. 12243, pp. 268–283. Springer, Cham (2020). https://doi.org/10.1007/978-3-030-58468-9_20
5. Caiza, G., Saeterosa, M., Onate, W., Garcia, M.V.: Fog computing at industrial level, architecture, latency, energy, and security: a review. HELIYON **6**(4) (2020). https://doi.org/10.1016/j.heliyon.2020.e03706
6. Chanchí, G.E., Ospina, M., Perez Medina, J.L.: IoT system to monitoring heart rate variability in usability tests. Espacios **41**, 84–97 (07 2020)
7. Echeverria, B., et al.: A reflection on telemedicine. Revista de Chile **149**, 928–933 (2021)
8. Jackson, Q., Landgrebe, D.: An adaptive classifier design for high-dimensional data analysis with a limited training data set. IEEE Trans. Geosci. Remote Sens. **39**(12), 2664–2679 (2001). https://doi.org/10.1109/36.975001
9. Moreno, J.: ApplicatColombiaion of automatic recording and reporting of physiological parameters for psychophysiological traceability in people post COVID-19. Technical report, Colombia: Bogota (2021)
10. Osorio, J.K.H.: Unbalanced data classification methodology based on submester methods. Technical report, Colombia: Pereira (2019)
11. Placencia, F., Manzano, S., Pallo, J.P., Jurado, M., Chicaiza, D.: Embedded device for blood pressure monitoring. In: 2017 CHILEAN Conference on Electrical, Electronics Engineering, Information and Communication Technologies (CHILECON), pp. 1–5 (2017). https://doi.org/10.1109/CHILECON.2017.8229505
12. Steven Pérez, J.P.P., Santiago Manzano, M.J., Chicaiza, D.: Sistema de telemedicina iot de monitoreo de signos vitales y alertas de distanciamiento físico para enfermedades respiratorias. RISTI, Revista Ibérica de Sistemas e Tecnologias de Informação **04**(NE49), 560–572 (2022)
13. T. Zhongyun, H.H., Chonghuan, X., Kaidi, Z.: Exploring an efficient remote biomedical signal monitoring framework for personal health in the COVID-19 pandemic. Int. J. Environ. Res. Public Health 1–23 (2021)
14. Torres, J.: Python deep learning: practical introduction with Keras and Tensorflow 2. Technical report, España: Barcelona (2020)
15. Zhang, K., et al.: Clinically applicable ai system for accurate diagnosis, quantitative measurements, and prognosis of COVID-19 pneumonia using computed tomography. Cell **181**(6), 1423–1433 (2020)
16. Zhao, G., Liu, Y., Zhang, W., Wang, Y.: TFIDF based feature words extraction and topic modeling for short text. In: Proceedings of the 2018 2nd International Conference on Management Engineering, Software Engineering and Service Sciences, pp. 188–191 (2018)

Industrial Engineering

Application of CFD for the Redesign of a Positive Pressure Mechanical Flow Generation Device in COVID-19 Treatment

Franklin Salazar[1](✉), Diego Núñez[2], Lizette Leiva[2], Kevin Mamarandi[2], and Lisbeth Vargas[2]

[1] Universidad Autónoma de Madrid, Madrid, Spain
fw.salazar@uta.edu.ec

[2] Universidad Técnica de Ambato, Ambato 180103, Ecuador
{df.nunez,le.leiva,kmamarandi7997,lvargas3525}@uta.edu.ec

Abstract. The wide and rapid spread of the covid 19 pandemic has become a global and national health problem. The aggressiveness of the disease produced by the COVID 19 pandemic has caused many countries in Europe and the world to saturate their healthcare systems. Thus, to mitigate the problem of the lack of equipment in health centers and hospitals due to the large number of patients, the CPAC device called "UCLV Ventura CPAP" was developed in the United Kingdom, which is the starting point of this research. This paper proposes a redesign of the device using computational fluid dynamics (CFD) to improve its performance of the device. The redesigned device was made in Polyacetal and was tested by validating the proposed (CFD) model. The tests obtained from the system allowed observation that the device behaves reliably and can work under the required conditions, delivering the right amount of oxygen to the patient.

Keywords: COVID 19 · CFD · CPAP · model · simulation

1 Introduction

The great technological development in the field of engineering makes it possible to predict the behavior of different systems, objects or devices before they are manufactured. The integration of high-level mathematical models in computer tools with user-friendly environments has made simulation programs increasingly indispensable for the final development of a system. One of the new trends at present is the simulation of fluid behavior, which is obtained with the use of computational fluid dynamics, known by its acronym Computational Fluid Dynamics (CFD). Computational fluid dynamics is based on identifying an object, processing it and validating the model to analyze the behavior of the fluid inside or outside of it [1]. The most important point of the CFD simulation stage is

M. V. Garcia and C. Gordón-Gallegos (Eds.): CSEI 2022, LNNS 678, pp. 447–464, 2023.
https://doi.org/10.1007/978-3-031-30592-4_30

to define the object properly, with this starting point the fluid dynamics can be analyzed since the regime in which it moves is known. The dynamics of the fluid inside or outside the materials can be laminar or tubular, for which, the appropriate model for the analysis of this must be determined. Once the object domain is identified, the geometrical construction of the analysis device must be carried out. These data obtained are starting parameters for the validation of the model [9].

The main applications of CFD have been carried out in the industrial area, however, in recent years its application has turned to other fields such as medicine. Thus, Vyaire, a company that focuses on the development of treatment and monitoring devices for respiratory conditions with the help of CFD, has been able to identify that according to the physiology and morphology of each patient, they have a "unique breathing profile" [1,3].

The research carried out by the major medicine industries has made it possible to better cope with the enormous burden on the healthcare system at the peak of the pandemic. The high demand for intensive care units (ICU) led to a shortage of mechanical ventilation systems, so the use of CPAP devices has been presented as a way to convert them into non-invasive mechanical ventilators [4,10]. The principle of operation of these devices is based on the bistable generation of a flow inside a chamber designed for the generation of selfexcited oscillating flow. The new challenge to combat the disease caused by COVID 19, focuses on the delivery of a controlled air flow (flow rate) that delivers an adequate and necessary percentage of oxygen, depending on the patient's condition [3].

One of the main problems of the use of CPAP devices occurring in patients is nasal obstruction related to sleep apnea as presented in [3]. To analyze the presence of this problem, a CFD analysis was performed to determine the airflow, velocity and accuracy loss in patients with and without nasal obstruction. The analysis is performed by creating three-dimensional models that simulate the nasal and nasopharyngeal cavity with the use of a CPAP. Data such as nasal flow rate, pressure and flow rate in the machine allowed validation of the proposed model. In the area of CFD research in medicine, it is necessary to validate the model in a more precise way, since variables that may or may not affect the simulation must be included [6]. The simulations performed indicate that an important aspect in patients is the speed at which air enters the lungs and the percentage of oxygen required by the patient, which must be monitored and adequate according to the patient's condition [12].

On the other hand, the sound generated by CPAP devices, due to the delivery of oxygen, has generated discomfort or rejection of these devices in several patients. Thus, in [13], a computational aeroacoustics test is presented to analyze and reduce the aerodynamic and acoustic behavior inside the device. The high level of turbulence generated inside the device is the cause of the noise generation, which averages 30 dB inside and up to 38 dB outside. The proper design of the device can cause noise to be in a broadband frequency to improve the patient's perception of the noise generated by the device.

The rapid deterioration of patients with COVID has led countries such as the United Kingdom and others belonging to the European Community to recommend intubation and early ventilation to keep patients stable and prevent their death. On the other hand, several countries such as China have proposed the use of CPAP to save the ventilation capacity and resources of an ICU. Thus, in [7], presented at the University Collegue London Hospital, proposed the design of rapidly developing industrial devices based on the Medic-Aid mechanical flow generator developed in 1992. With the release of the Medic-Aid patent in 2019, reengineering work was developed to provide a reliable system like the 1992 system, but with the technological advances of the time. The development of the device was covered in record time with only 4 days to test it. As a result of the use of the device, an improvement in the breathing of patients with COVID 19 was achieved, so that the regulatory agencies of the United Kingdom approved the use of this device. One of the main conditions for the correct use of CPAP devices is the delivery of adequate oxygen for each patient since in several cases this varies in a personal way as mentioned in previous paragraphs the maximum percentage of oxygen required by COVID 19 patients is 70%. Therefore, having a device that allows the oxygen concentration to be varied up to 90% is of utmost importance for its application as an alternative treatment for COVID 19 patients [11].

Based on the studies analyzed, it can be seen that CPAP devices are presented as an alternative for the use of recovery in patients with COVID, so that the proper design and validation with CFD of these devices can reduce costs by increasing the efficiency and effectiveness of its operation. This study presents the design and development of low cost and high efficiency CPAP generation using CFD as an alternative for the recovery of patients with COVID 19 [11].

2 Methods and Materials

The starting point for the development of the device was the CPAP device developed by the University College London Hospital in the United Kingdom (Fig. 1).

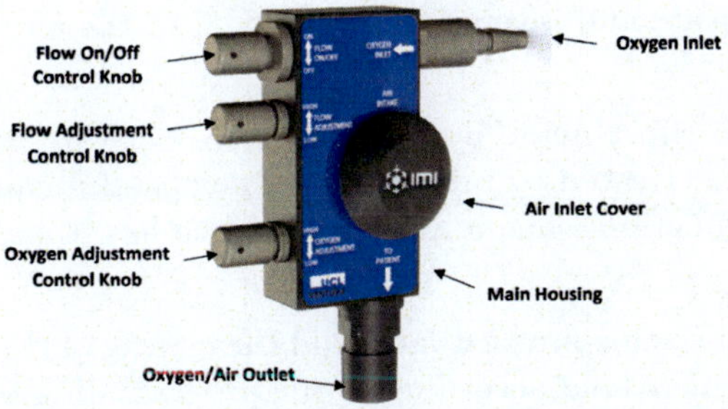

Fig. 1. Main parts of Ventura UCL flow generator

Based on the proposed system, a reengineering is carried out to improve the device circuit, allowing the delivery of large oxygen flows, as well as its contractions, in an adequate way for the recovery of patients with COVID 19 [5]. For the development of this, general mechanical ventilation concepts such as [7,8,11] should be considered:

Flow
It is the quantity of a fluid that enters or leaves a certain space in an interval of time, this is produced thanks to a pressure gradient, for respiration it is used with liters/second [7,8].

Resistance
It is the opposition exerted by the fluids when moving, this is related to the length of the tube, in mechanical ventilation it can be calculated with the pressure gradient divided for the flow, thus obtaining values with units of cm H2O/(l/s) [7,8].

Work
It is the force exerted to obtain a displacement. The unit of measurement is cm H2O $\times$ liters, equivalent to 0.1 J [7,8].

Volume
It is possible to have an instrumental dead volume, which is generated between the generator, tube, filters, and humidifiers; the second volume is the compressible volume, which is trapped in the instrument tubing, and which was not taken into account for gas exchange. To calculate the compressible volume, the peak inspiratory pressure minus the positive end expiratory pressure must be subtracted [7,8].

Inspiratory flow
To obtain this flow, the ventilator must generate a positive pressure, all this energy being produced during the passive expiratory phase. The total pressure necessary to ventilate the patient is defined by the resistive properties, as well as by the pressure gradient, distensibility and volume [7,8].

Expiratory flow
Derived from the pulmonary elastic properties, this flow is directly proportional to the alveolar pressure and inversely proportional to the airway resistance [8].

Activation
It is a signal of the effort made by the patient to activate breathing; through sensitivity it can be activated by pressure or by flow, and it can also be activated by a programmed function when a determined time has elapsed [7,8].

Inhalation
With the opening of the inspiratory valve and the closing of the expiratory valve, an amount of gas is delivered according to the programming that is directed for the patient, in some cases this is controlled with the conditions that the patient has and in others every certain time so that there is no flow deficit [8].

Cycling
It is the change from inspiration to expiration when the parameters of pressure, time, volume, or inspiratory flow decrease have been reached, which was preset [8].

Exhalation
After closing the inspiratory valve, the work of the expiratory valve begins, which has a working time until the machine has a new impulse in the sensitivity of the inspiratory valve; according to studies it was determined that the inspiratory time lasts about 0.8 s and an expiratory time lasts 1.6 s, so it is called that it has a ratio of 1:2 [8].

UCL Venturi CPAP Internal Operation
The device has a mechanical operation, i.e., all actions are performed manually. The valves that make up the device are the most important since they allow regulating both the flow of oxygen that passes through the CPAP, as well as the concentration of oxygen delivered to the patient. The CPAP is a mixing chamber that allows oxygen and air to mix in an adequate way, allowing the flow rate and the amount of oxygen delivered to the patient to be ideal [2]. As can be seen in Fig. 2, once the connection is made the oxygen begins to flow through the dispositive valve B allows the oxygen to pass into the mixing cavity and into the oxygen adjustment chamber. In chamber C an oxygen concentration can be obtained since at this point three fluids are mixed, the oxygen inlet of valve B, the air inlet, and the oxygen valves of chamber C to finally send the fluid to the patient, Fig. 2.

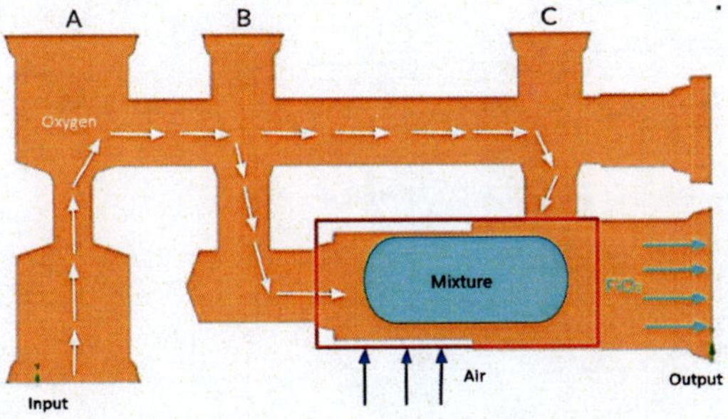

Fig. 2. Fluid flow in Ventura CPAP UCL operation

Analysis and Identification of Zones for the Redesign of the UCL Ventura CPAP Device

To identify the areas that can be optimized with a redesign, the behavior of the UCL Ventura CPAP device is analyzed using the ANSYS software, which is a tool for the analysis of computational fluid dynamics. The software allows the generation of reports to evaluate the different conditions to which the object under study is subjected [14]. Once the base model is obtained, we proceed to the variation of the model to obtain a better response of the device and to propose the appropriate guidelines for its manufacturing. The simulation of the base case allows obtaining adequate data for the redesign of the device to meet the requirements for its use as an alternative for the treatment of patients with COVID 2019. The model analyzed is when you have the opening variations of each of the valves, which allows to observe the fluid dynamics of oxygen and air from the inlet to the delivery of the patient's duct, Fig. 3.

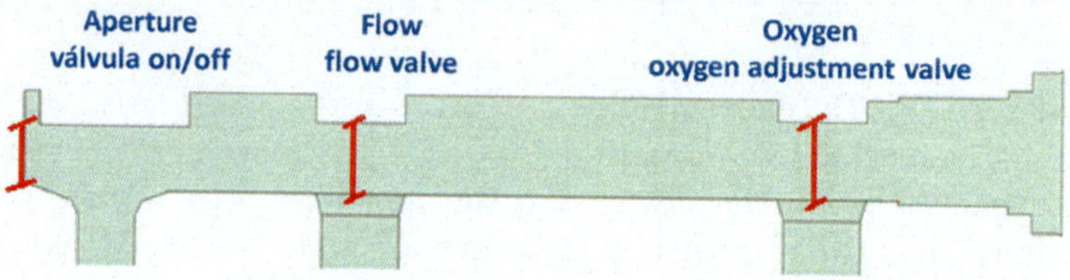

Fig. 3. Valve distribution

To identify the areas that can be intervened and redesigned, an adequate analysis of the process within the device must be carried out. The flow diagram in Fig. 4 shows the process to be carried out to obtain processing and analysis results.

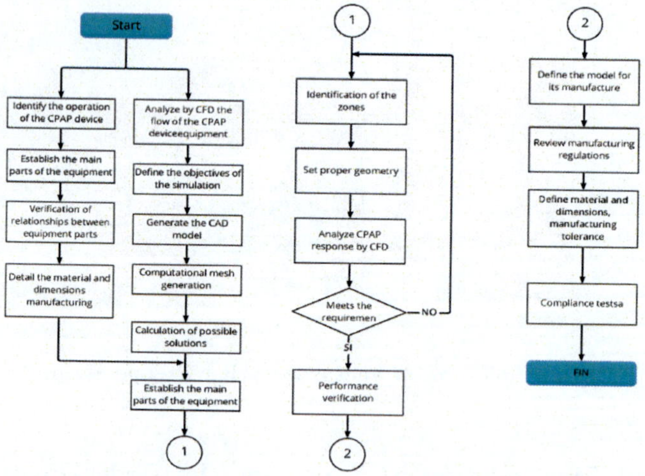

Fig. 4. CPAP device redesign flowchart

Table 1 shows the different components that make up the UCL Ventura CPAP device, which consists of 7 parts [14].

Table 1. UCL Ventura CPAP device details.

N°	Items	Material	Dimensions (mm)
1	on/off Valve	316 Stainless Steel	26x54
2	Flow adjustment valve	316 Stainless Steel	16x54
3	Oxygen Flow Adjustment Valve	316 Stainless Steel	16x54
4	Patient discharge	Tecaform AH MT Black	28x90
5	Oxygen supply	316 Stainless Steel	20x75
6	Air intake plug	Tecaform AH MT Black Stainless Steel	50
7	Body	Tecaform AH MT Black	155x55x30

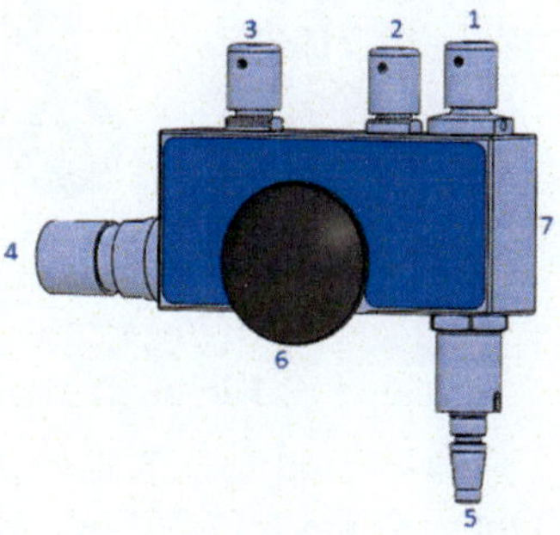

Based on the detail of the geometry of the original model presented in Table 1, we proceed to develop the simplified 2D model in the ANSYS software to avoid the computational consumption of the 3D study, Fig. 5.

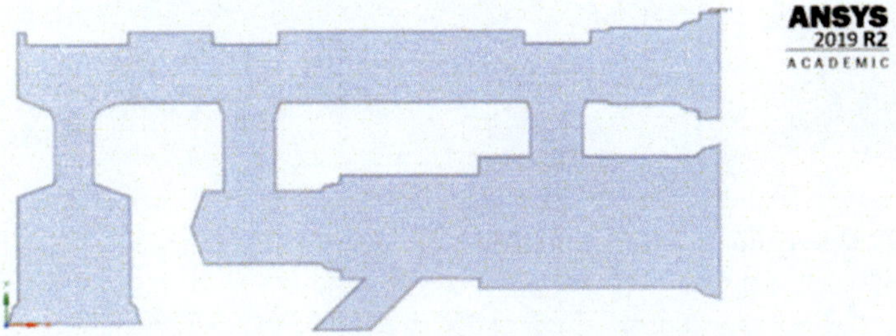

Fig. 5. Simplified model in 2D

Mesh Configuration

Once the two-dimensional model of the device has been made, we proceed to define the details of the mesh, a parameter that allows defining the cavities through which the fluid circulates. For the simulation, two zones have been identified for the configuration of the mesh, zone one is modeled with the "Inflation" technique, which resembles the thickening of the section. This configuration creates a total of 20 equal elements with a size of 0.44 mm. Zone two is configured with the "Face Sinzing" technique with an element size of 0.22 m, which is configured with the help of the software and considering that it is the mixing zone of the fluids to have greater accuracy, Fig. 6.

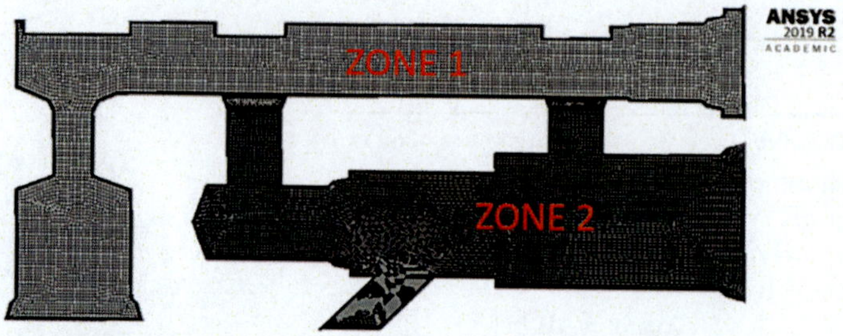

Fig. 6. Determination of computational meshes

Once the mesh has been generated, the boundary zones are defined so that the software can adapt the model to the desired solution. For the case of simulation, A corresponds to the oxygen input, B to the Air input, C the output of the mixture and D to the domain wall.

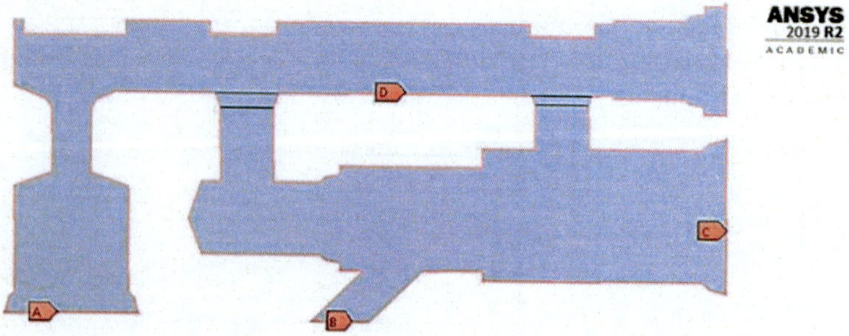

Fig. 7. Determination of geometric locations for boundary configuration

As can be seen in Fig. 7, the inputs and outputs are determined, as well as the limits such as the walls of the model, to perform the simulation. Once the model is built, the processing options must be defined in the ANSYS FLUENT software, the solution will be based on the pressure and the analysis of the simulation is performed in a transient state and evaluating in a 2D, planar space. The equation for the solution of the system is STANDARD since it is the one that best fits a k-epsilon energy model. Since the mixing of oxygen and air takes place in zone two, the SPECIES model is enabled, which works with a volumetric ratio [3]. This allows to set the fraction of air present in the cavity which is set to 21%. To determine the inlet velocity of the oxygen, the inlet diameter of the oxygen, which is 21 mm, is considered. With the help of Eq. (1), the oxygen inlet area is determined, and the oxygen inlet velocity can be calculated.

$$A = \frac{\pi}{4}(\phi)^2 \tag{1}$$

$$A = \frac{\pi}{4}(21)^2$$
$$A = 346.35\,\text{mm}^2$$

where:

$\phi = is\ the\ oxygen\ inlet\ diameter\ (mm)$
$A = Oxygen\ inlet\ area\ in\ (mm^2)$

Once the oxygen area has been determined, the oxygen delivery rate can be calculated using Eq. (2). The oxygen flow rate for the calculation is 15 LPM, which is the maximum value of oxygen delivery in hospitals.

$$Q = A \times v \tag{2}$$

$$v = \frac{Q}{A} \tag{3}$$

$$v = \frac{15\,\frac{\text{L}}{\text{min}} \times \frac{1\,\text{min}}{60\,\text{s}} \times \frac{1\,\text{m}^3}{1000\,\text{L}}}{3.64 \times 10^{-4}\,\text{m}^2}$$
$$v = 0.72\,\frac{\text{m}}{\text{s}}$$

The parameterization of the oxygen delivery values is carried out to carry out the simulation and observe the behavior of the fluid in the CPAP device evaluated, according to the data presented by the manufacturers, it works at a pressure of 4 bar, that is, 40,000 Pa (Fig. 8).

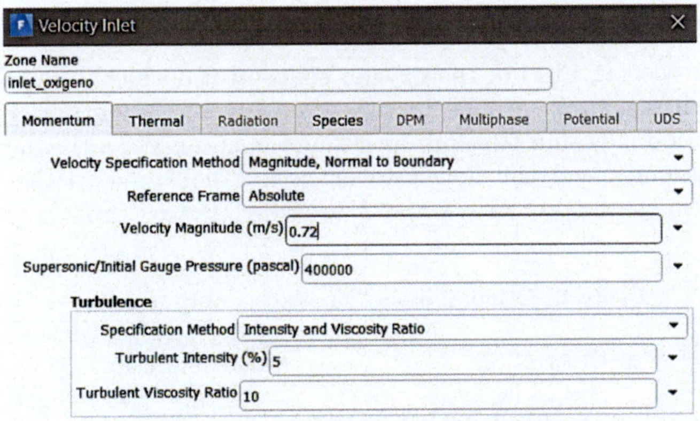

Fig. 8. Oxygen initial conditions setting

After configuring the different parameters, we proceed to perform the simulation for the analysis of the results, with the purpose of identifying the areas that present lower efficiency and can be redesigned to improve oxygen delivery based on the parameters established in the hospital centers. Figure 9 shows the simulation results obtained for both speed and oxygen concentration.

(a) Average speed 1.34.

(b) Average oxygen concentration of 40%. .

Fig. 9. Simulation results of the original model.

As can be seen in Fig. 9b, the oxygen concentration has an average of 40%, which is within the range of 35% to 80% of oxygen at the output of the device, so the simulation is within the appropriate parameters. With the data obtained, the output velocity and the flow rate of oxygen delivered to the patient are calculated with the help of Eqs. 1 and 2 described above, Table 2.

Table 2. Original model simulation output data.

Description	Value obtained
Output Area	$4.66 \times 10^{-4}\,\text{m}^2$
Oxygen Outlet Flow Rate	37.46 LPM

To optimize the design, three simulation stages were carried out, as shown in Table 3, with each of the simulations the area and the outlet flow rate of the different proposed designs were obtained. In the first stage, the space of zone 2 was reduced, as well as the diameter and length of the hypodermic tube, which was parameterized with its characteristic values, and finally, the fluid mixing chamber was reduced. In the second stage, porous filters were included to verify how the model behaves, the oxygen inlets were designed together with the valve to analyze the entire oxygen path as well as the outlet. In the third stage, the length of the oxygen inlet was shortened, and the size of the mixing chamber was increased to analyze the interaction of the fluids; the parameter configuration of stage two was maintained.

Redesigning CPAP Equipment
Based on the speed and oxygen concentration simulations carried out, the redesign of the device is proposed, reducing its size without altering its correct operation. The new proposed dimensions are 84.36 mm in the upper part with a reduction of 24.12 mm from the original, the dimension what occupies of the oxygen mixing chamber with the air and of the hypodermic tube is reduced by 6.87 mm being the final dimension of 44.87 mm, the air inlet was reduced by 0.9 mm and we have a dimension of 14.10 mm; finally the empty space was reduced to 32.94 mm thus having a reduction of 7.04 mm. The comparison of the dimensions of the original and proposed system is presented in Fig. 10.

In the proposed redesigned model, although the space occupied by the mixing chamber and hypodermic tube is reduced, the size of the mixing chamber is increased, taking advantage of space that was unused within the original model, allowing more space for the fluid mixture. In addition, the outlet insert was internally represented to analyze the transport of fluids from the inlet to the outlet of the equipment. The geometric model in two dimensions, as well as the mesh configuration for the redesigned model, is shown in Fig. 11.

Table 3. Simulation data from the three stages.

ETAPA	Geometry	Simulation results	Output values obtained
One		Average speed: 5.58 m/s  Average oxygen concentration: 51% 	Area: $3.71x10^{-5}m^2$ Oxygen flow rate 12.42 LPM
Two	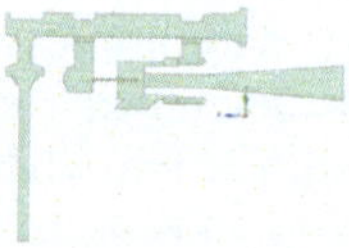	Average speed: 5.67 m/s  Average oxygen concentration: 81% 	Area: $1.90x10^{-4}m^2$ Oxygen flow rate 64.63 LPM
Three		Average speed: 5.82 m/s  Average oxygen concentration: 81% 	Area: $1.90x10^{-4}m^2$ Oxygen flow rate 66.34 LPM

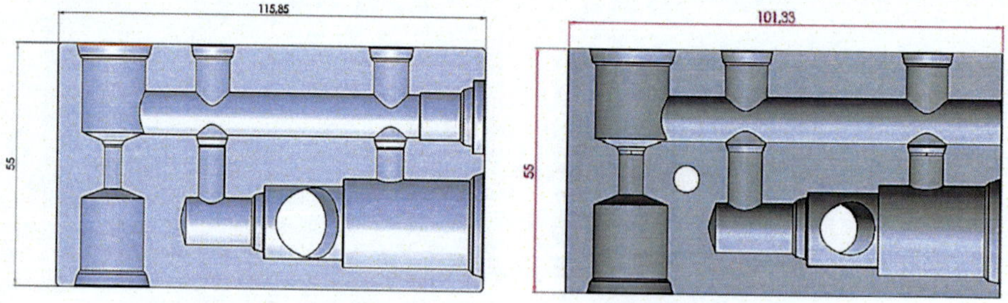

(a) Original dimensions. (b) Redesign dimensioning.

Fig. 10. Overall comparison of the dimensions of the original system vs. the redesigned system.

Once the simulation meshes have been configured, in which the two determined zones can be seen, the boundary conditions shown in Fig. 12 are determined.

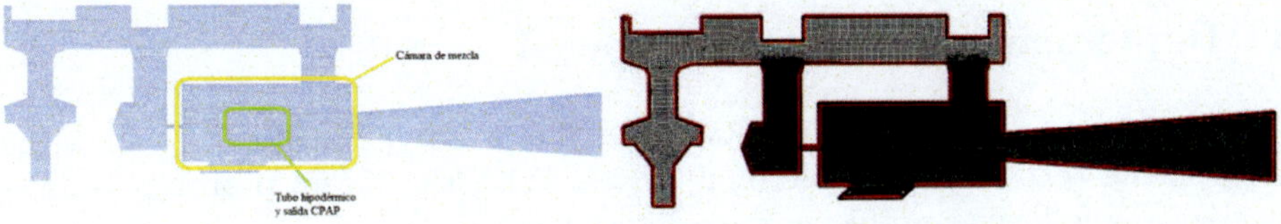

Fig. 11. Two-dimensional geometrical model and mesh configuration for the simulation

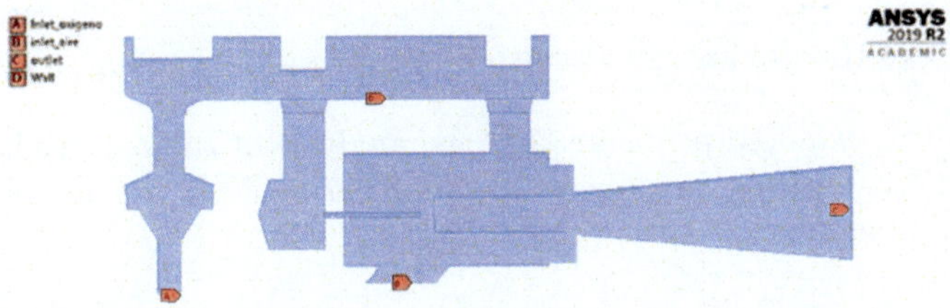

Fig. 12. Setting the boundary conditions of the redesigned model

The results obtained from the simulation performed are presented in Table 4, the results of the simulations were performed with the six variants of the valve positions.

Table 4. Results obtained from the redesigned device.

Geometry	Simulation results	Output values obtained
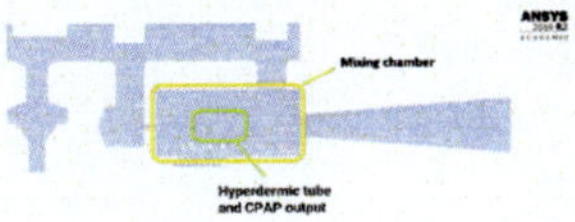	Average speed: 4.86 m/s 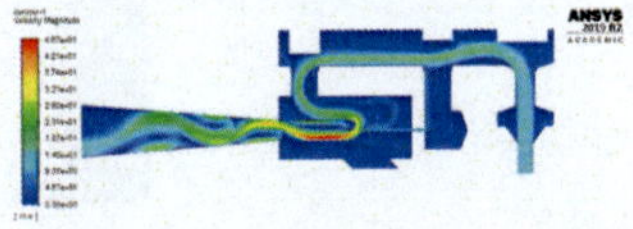 Average oxygen concentration: 86% 	Area: 1.90x10^{-4}m^2 Oxygen flow rate 55.40 LPM

As can be seen in the table of results, it has been possible to obtain an oxygen concentration at the exit of the patient of approximately 86% with an air intake velocity of 1.5 m/s. Having verified the improvement of the air concentration with the redesigned device, we proceeded with the construction.

3 Implementation of the Proposal

For the correct validation of the device, a Nylon prototype was built to allow the adjustments to be made in the final device. With the Nylon prototype implemented, the following problems were determined:

- Problems in the body threads, resulting in fluid leakage.
- Clamping of the patient's exit valve without threads, creating instability of the clamping.
- Poor finish of the device cap, resulting in inadequate sealing of the device.

Based on the detected problems and the analysis of the material, the CPAP device was manufactured in polyacetal. The parts of the device elements are presented in Table 5.

Table 5. Parts redesigned for the construction of the proposed CPAP.

Description	Photography
CPAP body	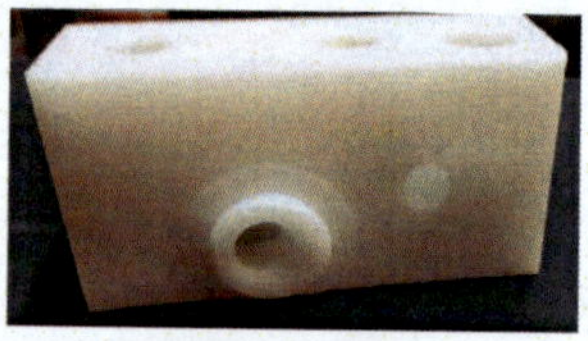
Patient discharge	
Plug	

With the elements designed, proceed to the assembly of the device and to the performance tests by connecting the flow meter and the oxygen concentration meter, Fig. 13.

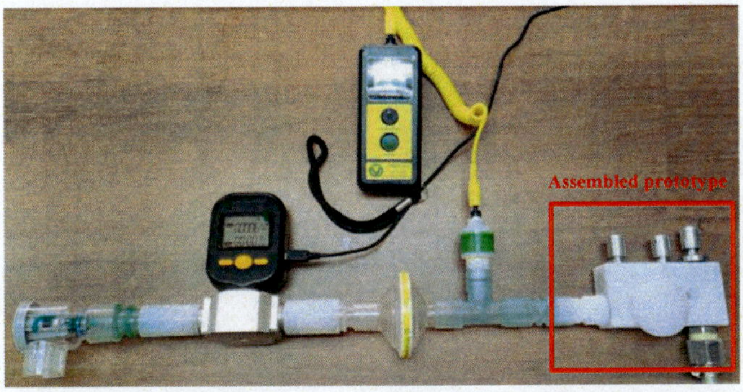

Fig. 13. Redesigned device mounted on the test stand

From the evaluation of the test bench the results presented in Table 6 are obtained.

Table 6. Results obtained with the proposed redesigned CPAP prototype.

Input data based on valve position			Resulted		
Aperture [mm]	Flow [mm]	Oxygen [mm]	Flow [m³/s]	Flow [l/min]	FiO_2 [%]
4.4	2	2	1.94 E-4	11.66	87.9
		4	1.91 E-4	11.43	95.6
		6	1.86 E-4	11.15	95.6
	4	2	2.67 E-4	16.02	74.3
		4	2.44 E-4	14.65	78.9
		6	2.43 E-4	14.58	79.1
	6	2	3.42 E-4	20.49	71.6
		4	2.86 E-4	17.14	76.1
		6	2.84 E-4	17.03	76.5

The tests performed on the original device are presented in Fig. 14. It can be observed that the operating range of the device is between 38% to 95% of FiO_2 while in flow rates values from 5 LPM to 60 LPM are recorded.

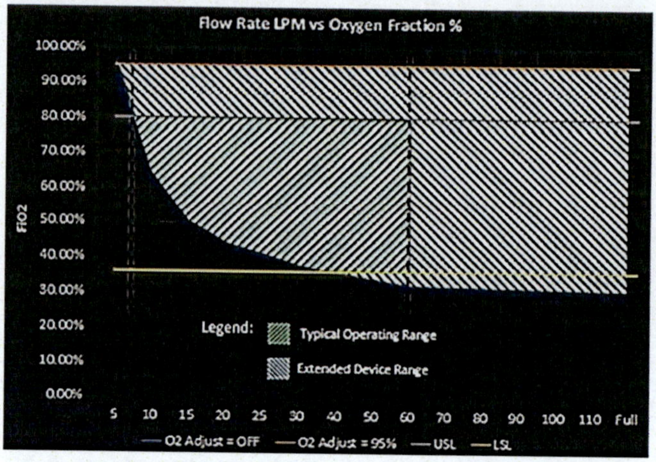

Fig. 14. Results for the original device

The results obtained with the redesigned CPAP device connected to the test bench presents a FiO_2 range of 71.6% to 95.6% with a recorded flow rate value ranging from 11.15 to 20.49 LPM, Fig. 15.

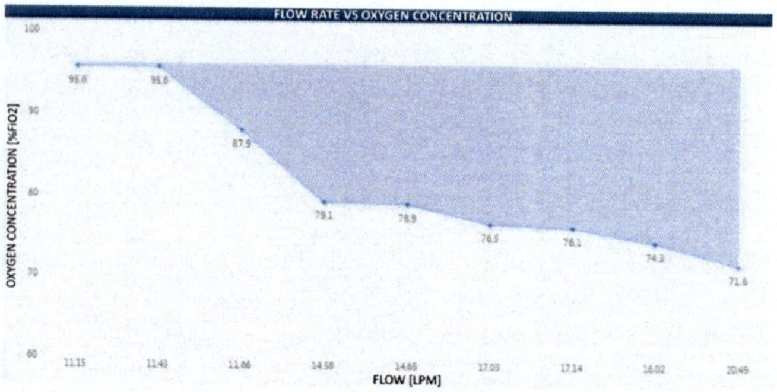

Fig. 15. Results obtained with device redesign

The error obtained from the simulated device with respect to the device implemented on the test bench presents a relative error of 5.83%. This means that the implemented device complies with the error range that exists between design and simulation. From the results presented it can be observed that the proposed device works in a higher FiO_2 zone, raising the FiO_2 percentage from approximately 45% with 20 LPM to 71% with the redesigned prototype.

The positive pressure flow generator, CPAP, is presented as one of the many solutions to the main problem caused by COVID-19, which is the inability to breathe, allowing it to function as a non-invasive mechanical respirator through certain adaptations. The redesigned device allows obtaining oxygen percentage between 70% and 90% of the necessary values required by a COVID-19 patient. Compared to the original model, which has a performance of less than 40%, because it can cause a disproportionate expenditure of oxygen. The CPAP devices are not a replacement for endotracheal intubation, just as the UCL-CPAP is not, since these devices are totally new, health personnel do not have the necessary training to use them, and it is even thought that they could not work correctly.

It should always be kept in mind that these devices are a support option for respiratory support, preventing patients from needing an intensive care unit.

4 Conclusions

For the proper functioning of the CPAP device, the values of oxygen concentration of oxygen to be delivered to the patient should be from 35% FiO_2 to 85% FiO_2, having as maximum value 95% FiO_2; and an equivalent flow rate between 7.5 L/min and 60 L/min with a maximum of 180 L/min. The analysis by means of computational fluid dynamics CFD allows to analyze the mixture between air and oxygen that exists inside the device used for treatment of patients with COVID 19. The redesign in the original CPAP UCL Ventura device was achieved with a reduction in the mixing chamber from 21.6 mm to 16.7 mm and likewise the overall size of the CPAP was reduced from 115.8 mm × 55 mm to 101.3 mm × 55 mm, these changes were made based on computational fluid dynamics ensuring proper operation.

The prototype made of polyacetal proved to be reliable and the results of the performance tests showed that the device will be able to work under the conditions required by the patients. The results of the physical device after the different stages of design and simulation showed that it is possible to obtain an acceptable result starting from a 3D model converted to 2D, as long as the correct decisions are made in the design and also in the simulations, guaranteeing that the 2D geometry resembles the 3D geometry in functionality.

Acknowledgments. The authors would like to thank the Investigation and Development Directorate, DIDE, of the Technical University of Ambato for its special help with the development of this proposal, thanks to the financing of the project "Aplicación de la ingeniería inversa para el diseño, construcción y ensamble de un dispositivo de respiración CPAP para pacientes con COVID-19 en el contexto de la provincia de Tungurahua".

References

1. Broadbent, M.: Covid-19 demand shock and preparedness response (2021)
2. Cruz Gavilan, Y., Valdés Hernández, P.A., Laffita Leyva, A., Gómez Águila, M.V., Chuairey, C.M.: Dinámica de fluido computacional: Revisión y análisis de las aplicaciones en la ingeniería. Revista Ciencias Técnicas Agropecuarias **29**(4) (2020)

3. Dillon, T., et al.: Computational modeling of a low-cost fluidic oscillator for use in an educational respiratory simulator. Adv. Nanobiomed Res. **1**(12) (2021)
4. Garcia, M.V., Irisarri, E., Perez, F.: Vertical integration in factories using OPC-UA and IEC-61499. In: ENFOQUE UTE, vol. 8, pp. 287–299 (2017)
5. Gutiérrez Muñoz, F.: Ventilación mecánica. Acta médica peruana **28**(2), 87–104 (2011)
6. Irisarri, E., Garcia, M.V., Perez, F., Estevez, E., Marcos, M.: A model-based approach for process monitoring in oil production industry. In: 2016 IEEE 21ST International Conference on Emerging Technologies and Factory Automation (ETFA) (2016)
7. Kempley, S., Reeves, J., Wilkins, A.: High oxygen flow rates with the UCL Ventura CPAP device. Lancet Respir. Med. **9**(4), e35 (2021)
8. Munson, B.R., Young, D.F., Okiishi, T.H.: Fundamentals of fluid mechanics. Oceanogr. Lit. Rev. **10**(42), 831 (1995)
9. Ramírez, J., Colmenares, J., González, F., Sarcos, A.: Determinación de la vida útil de una válvula de globo mediante técnicas de dinámica de fluidos computacional
10. Reyes, J., et al.: A study on modeling and simulation of automobile painting process based on flexsim. In: Nummenmaa, J., Pérez-González, F., Domenech-Lega, B., Vaunat, J., Oscar Fernández-Peña, F. (eds.) CSEI 2019. AISC, vol. 1078, pp. 260–273. Springer, Cham (2020). https://doi.org/10.1007/978-3-030-33614-1_18
11. Singer, M., Shipley, R., Baker, T., Cowell, A., Brealey, D., Lomas, D.: The UCL Ventura CPAP device for Covid-19. Lancet Respir. Med. **8**(11), 1076–1078 (2020)
12. Suzuki, M., Tanuma, T.: The effect of nasal and oral breathing on airway collapsibility in patients with obstructive sleep apnea: computational fluid dynamics analyses. PLoS ONE **15**(4), e0231262 (2020)
13. Vuong, T.X., Al-Jumaily, A.M.: An investigation into the noise generation of CPAP devices: experimental and numerical study. Appl. Acoust. **114**, 281–293 (2016). https://doi.org/10.1016/j.apacoust.2016.08.005
14. Woge, O.G., Morán, C.O.G., Chau, A.L.: Introducción al método del elemento finito: Solidworks y matlab. Ideas en Ciencias de la Ingeniería **1**(1), 27–47 (2020)

Hydraulic System A, B, and Standby Operational Test in the Boeing 737-500 Aircraft Flight Simulator

Brandon Calazans(✉), Rodrigo Bautista, Gabriel Inca, and Andrés Arévalo

Universidad de las Fuerzas Armadas ESPE, Sangolquí, Ecuador
{jbcalasanz,rcbautista,gsinca,eaarevalo1}@espe.edu.ec

Abstract. The development of this project is oriented to perform an operational test of the operation of the hydraulic system A, B, and Standby in the flight simulator of the Boeing 737-500 aircraft according to the maintenance manual ATA 29-15-00, making an improvement in the simulator systems in order to be applied as a fundamental complementary tool for education and training. The methodology applied in the study of this project is based on research with a confirmatory quantitative approach that seeks to prove and affirm the benefit and contribution of flight simulators in their optimal conditions for the learning of students in training to aircraft maintenance technicians. The development carried out in relation to the theoretical perspective of the study of this project was based on the review of the academic documentation applicable to flight simulators and their efficiency. In conclusion, the analysis and study of the development of this technical project ends with the application of the operational test for the hydraulic system A, B, Standby in the Boeing 737-500 flight simulator determining and guaranteeing that the operation of its system is in the same conditions required to those of its real aircraft model that is being simulated for the use and practical teaching of the personnel involved in the aeronautical field.

Keywords: simulator · aircraft · operation · system · maintenance

1 Introduction

The hydraulic system is a means of providing power or displacement to various components of an aircraft. This system can vary in its complexity depending on the type of aircraft in which it is located and its main purpose in the same Fig. 1 indicates the system in general. There are hydraulic systems for light aircraft that their configuration will be simple, but hydraulic systems can also be considered for heavy or transport type aircraft where the hydraulic process is involved in the braking system, landing gear, flaps, and flight control surfaces [1]. In recent years the aeronautical industry has evolved rapidly, this requires that both crew and maintenance personnel have high-level training in complex systems in order to maintain operational safety, this equipment allow to represent the operation of a system in certain operating conditions or failure and, since the cost that would entail the actual operation of an aircraft is high, flight simulators have become

M. V. Garcia and C. Gordón-Gallegos (Eds.): CSEI 2022, LNNS 678, pp. 465–477, 2023.
https://doi.org/10.1007/978-3-031-30592-4_31

a safe and effective way to train aeronautical personnel. The systems simulated in real environments by means of training devices correspond to pressurization and air conditioning systems [2], COM/NAV [3], electrical system [4] and engine propulsion system start-up operations [5], among others.

The hydraulic system of the aircraft is essential for the operation of the primary and secondary flight control systems, landing gear, automatic flight control [6, 7]. The safety of the aircraft depends to a large extent on the reliability of the system; therefore, theoretical and practical training must allow the technical personnel to become familiar with the system before reaching the professional field [8]. According to the needs of aeronautical maintenance, this article presents the simulation of the hydraulic system of the Boeing 737 - 500 aircraft, using a SimVimCockpit/HCSCI, configurator integrating the Arduino hardware with the X-PLANE 11 flight simulator, to perform the performance tests of the mentioned system according to the procedures of the Boeing manufacturer's manual - ATA 29, in order to provide line and base maintenance personnel of air maintenance companies with a realistic environment to evaluate the effectiveness of the training, the Fig. 1, shows a hydraulic schematic on board the aircraft [9].

Fig. 1. The function of the hydraulic system is Supply and distribution of the hydraulic fluid in the transport category aircraft.

1.1 Hydraulic System of - Boeing 737 - 500 Aircraft

The hydraulic system of a transport category aircraft such as the Boeing 737 - 500 is made up of three separate and independent systems that provide the energy or power for the movement of the aircraft's flight controls, landing gear, and its thrust reverse system.

The three hydraulic systems shown in the Fig. 2., are: A, B, and Standby, mainly, systems A and B are the ones that work full time in a flight and provide the necessary force distribution through their pressurized fluid for the different technical situations of the aircraft. These two systems, in conclusion, are responsible for keeping all hydraulic components powered. The standby system comes into operation specifically only when a high fluid demand is required [9].

Some additional features of the systems mentioned by Lufthansa and LAN in their training manual are:

- System A and B have one engine pump (EDP) and one electric motor pump (EMDP) each.

- The Standby system has only one electric motor pump (EMDP).
- All three hydraulic systems use a type IV hydraulic fluid - BMS 3-11.

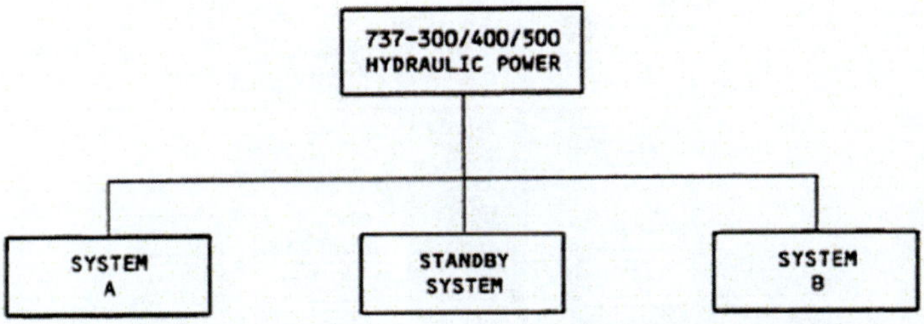

Fig. 2. The hydraulic system of the Boeing 737-300/400/500 aircraft has three systems, which are called System A, System B and Standby System. Hydraulic system of a Boeing 737-500 transport category aircraft.

1.2 Distribution of the Hydraulic System

As detailed in the introduction of this hydraulic system of the Boeing aircraft, it was determined that the aircraft has three independent systems for the distribution of its fluid: system A, system B, and Standby. B, and Standby. The Fig. 3 show the function of the three systems of keeping the whole system powered, guaranteeing the reliability in its performance [8].

The following are two graphs of the power distribution of the hydraulic system and the places through which its fluid can pass within it.

1.3 Hydraulic System Control Panel on B 737 - 500 Aircraft Flight Deck

A flight deck such as that of the Boeing 737-300-500 aircraft has been designed so that it can be efficiently controlled by two pilots in command of the aircraft. Also, its construction and design take into account factors that may arise, such as: the heavy workload for its pilots or situations such as finding the responsibility of the aircraft having to be piloted by a single professional on certain occasions. Due to these observations, the aircraft is configured so that several of its subsystems have their own monitoring and operation autonomy. This reduces the workload for pilots or aeronautical personnel and helps them to perform better and direct their attention to higher priority tasks or areas [1].

The design and construction of Boeing 737 model aircraft and its series from 200 to 500 have been manufactured in such a way that its pilots or aeronautical personnel can once again become familiar with a cabin design that is mostly standard while maintaining the majority of its main configurations so that there is no need to undergo lengthy technical training in these series or in future Boeing 737 model series [10].

The Fig. 4 is a graphic representation of the Boeing 737-500 aircraft flight deck and control cabin and the location of its hydraulic system control panels.

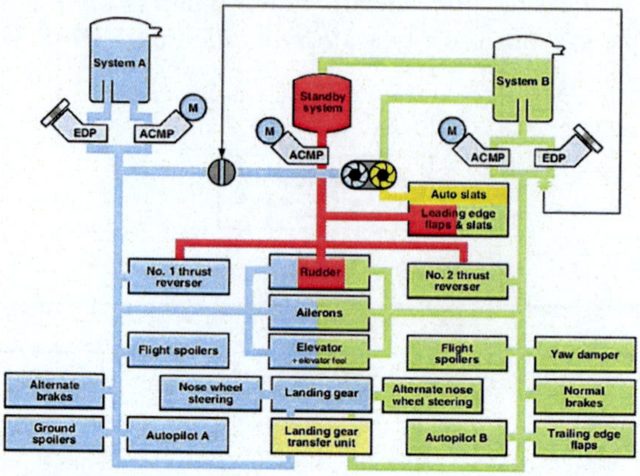

Fig. 3. Both A and B hydraulic systems have an engine-driven pump (EDP) and an ACMP (Alternating Current Motor Pump). The system A engine - driven pump is installed on the number 1 engine and the system B engine-driven pump is installed on the number 2 engine. The AC (Alternating Current) pumps are controlled by a switch on the flight deck. The hydraulic case drain fluid that lubricates and cools the pumps return to the reservoir through a heat exchanger.

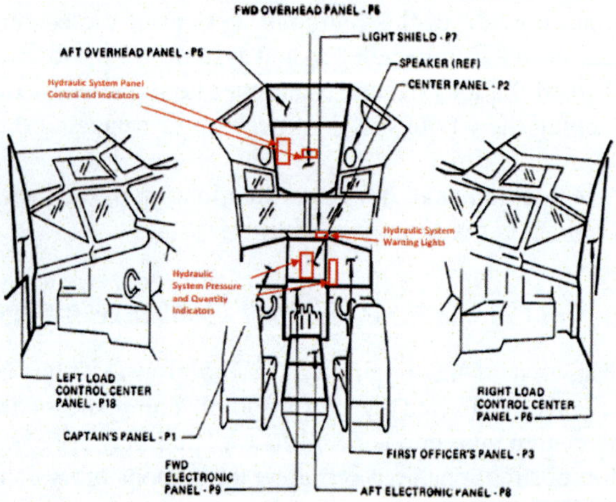

Fig. 4. Boeing 737 - 500 aircraft control cabin and hydraulic system panels, P1 Captain's Panel, P2 Center Panel, P3 First Officer's Panel, P5 Overhead Panel and Aft Overhead Panel, P6 Right Load Control Center Panel, P7 Light Shield, P8 Aft Electronic Panel, P9 FWD Electronic Panel, P18, Left Load Control Center Panel

The process carried out for the fulfillment of the final scope of the present project, which determines to guarantee an operational check in the hydraulic system A, B, and Standby in the flight simulator of the Boeing 737 - 500 aircraft, is detailed below [11].

2 Structural Reconditioning Process of the B 737 - 500 Flight Deck

The structural process was carried out in this project in order to repair, modify, and paint the external structure of the flight deck, thus improving its presentation and finishing conditions in which this cockpit will house all the flight simulation equipment and components, guaranteeing a safe place for their performance Fig. 5.

Fig. 5. Boeing 737 - 500 aircraft cabin installed in a suitable place and painted preserving the condition of a real aircraft.

This process can be divided into the following stages: Structural repairs, this process of structural repair of the Boeing flight deck consisted of completing the missing structure of the lower edges of its cockpit, adjusting and adapting the nose of the aircraft. In the pickling process, a pickling process was performed on the Boeing aircraft's flight deck structure where the old paint was removed from its surface, which consisted of three layers that had been part of its previous paint treatment.

Prior to the painting process, a visual check of its structure was performed to identify any type of damage, crack or crack to avoid possible damage to the internal equipment of the simulator. Painting, this process consisted of a coat of primer to protect the material from any threat against corrosion, another coat of industrial anticorrosive paint, and a coat of industrial synthetic paint as the final finish of its structure.

2.1 Internal Reconditioning Process of the Boeing 737 - 500 Flight Deck

Diagram of the internal process carried out for the assembly of the hydraulic system is shown in the Fig. 6, to implement the panels in the simulation cockpit, the control modules were configured directly so that the behavior of the elements is similar to that shown in the virtual simulator, this process was performed using the SimVimCockpit/HCSCI configurator. The SimVimCockpit/HCSCI is a proprietary interface for X-PLANE 11, allows to configure the hydraulic panel of the cockpit of an aircraft providing a variety of options for inputs and outputs, the system works with the HCSCI add-on, responsible for communicating the simulator data configured with the hardware controller [12].

The HCSCI configurator has maps and images that include the elements and control devices of the different cockpit panels which allowed to easily identify switches, lights and elements associated with the system and automatically assign them to the pins on the Arduino board. Once the configuration was established, the physical components

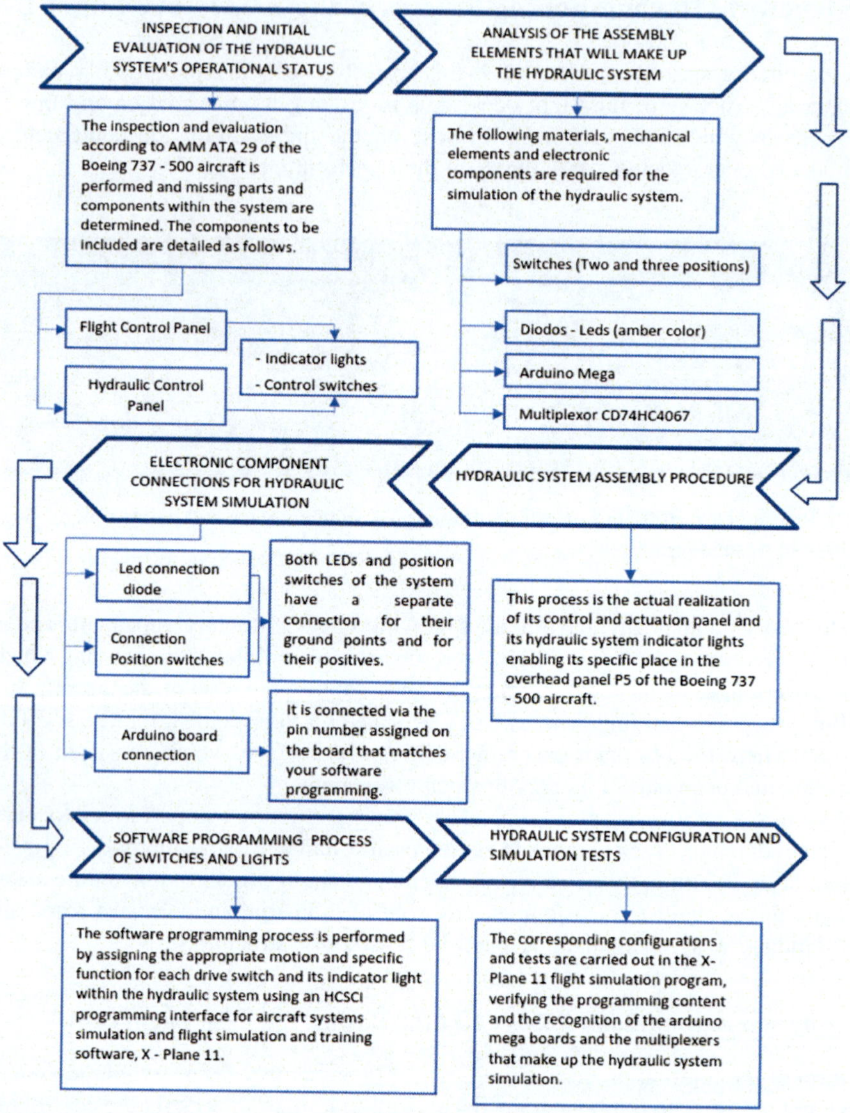

Fig. 6. The diagram shows the process of initial inspection, evaluation of the hydraulic system and analysis of the assembly of the elements.

(switches, lights and elements associated with the electrical system) were connected. Each element is individually connected so that in case of failure it can be easily replaced [13].

2.2 Initial Inspection and Evaluation of the Hydraulic System's Operational Status

For the inspection and evaluation process of the hydraulic system, the specific parameters corresponding to the operational status in which the system must be found for its correct functioning and operation are taken into account, as detailed in the AMM, Fig. 7, specifically in ATA 29 regarding Systems A, B, and Standby of the Boeing 737-500 aircraft [8, 9].

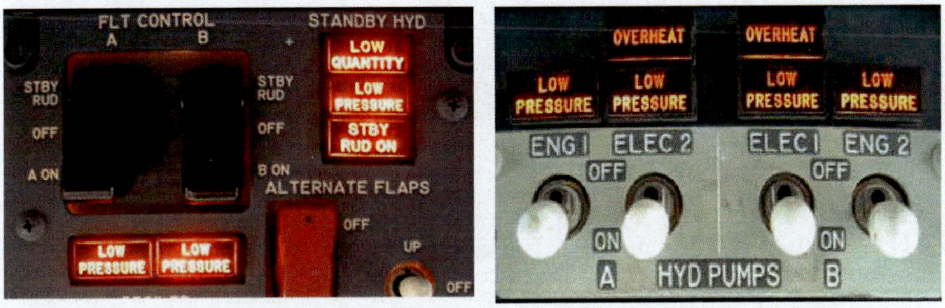

Fig. 7. Inspection and evaluation of the elements in hydraulic system panels.

2.3 Elements of Aircraft Hydraulic System

The Fig. 8, shows all the elements that compose the hydraulic system for its simulation within the parameters established by the Boeing 737 - 500 aircraft manual is composed of Switches such as:

- Drive for each pump of systems A and B.
- Engine pump.
- Electric pump.
- Drive for the flight control panel (spoilers, yaw damper, flaps, and rudder).

Simulated lights are represented by high intensity (amber) LEDs for the indicator lights:

- Low pressure.
- Over temperature indication.

Fig. 8. Hydraulic system assembly in the flight deck.

3 Hydraulic System Assembly Procedure

It is carried out taking as reference the location, dimensions and design characteristics of the hydraulic system panels of a real Boeing 737 - 500 aircraft. All its assembly, operation and functioning has been developed in such a way that pilots, aeronautical mechanics and any other personnel related to the aviation field in general can perform a training and familiarization of the flight deck with the same characteristics as those of the real model of the aircraft being characterized [14].

3.1 Connections of the Electronic Components for the Simulation of the Hydraulic System

The connection of electronic components in the hydraulic system simulation was determined and developed under the following analysis parameters and procedures:

This type of connections was made in parallel separately, i.e., there is a parallel circuit only of lights where they share their ground or negative in common, and a parallel circuit only of switches where in the same way they share their ground or negative in common, Fig. 9 shown the simulated circuit [15].

The Arduino mega in Fig. 10, allows making the connections and sending signals in the drive and simulation of the aircraft systems, Fig. 10. Where we used part of its signal pins, 24 pins of inputs and outputs (10 outputs for LEDs and 14 inputs for switches) of the 54 pins with which this Arduino has in this type of digital connections [16].

It must be taken into account the space for the 2 inputs to ground (negative) of both the LED circuit and the switches.

Once the number of pins needed for the system connections has been identified, the circuits are closed and their positive and ground wiring is done separately by performing a micro soldering process [17].

As a result of this process, 24 wires should be obtained for the 24 digital connection pins of the Arduino mega board.

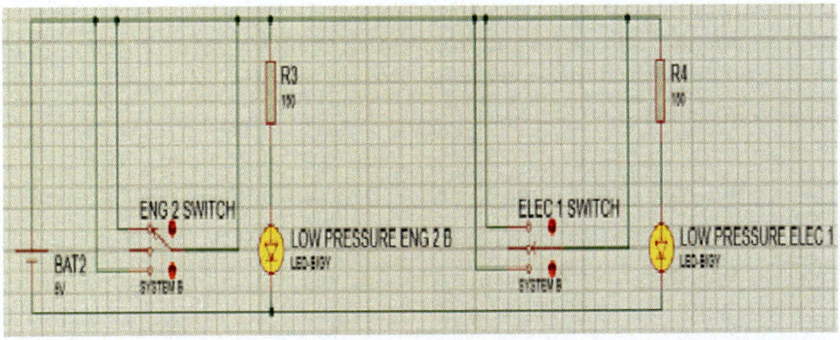

Fig. 9. Simulated circuit of the hydraulic system.

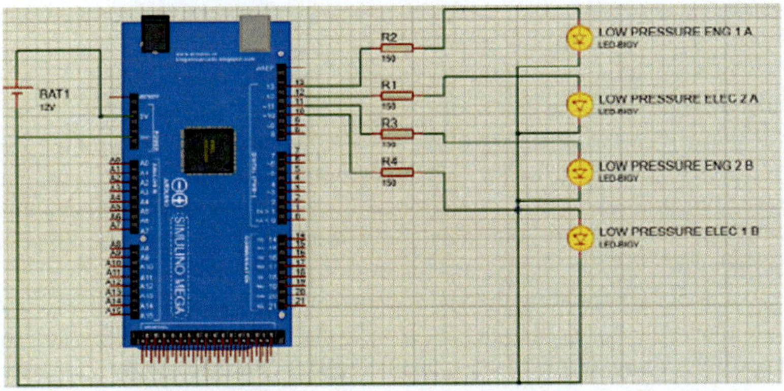

Fig. 10. Simulated circuit of system switches with Arduino mega.

It should be emphasized that one of the most important details in this whole connection procedure is to remember the pin number assigned within the Arduino board for each wire that comes from a led or switch with a specific action to be fulfilled, Fig. 11.

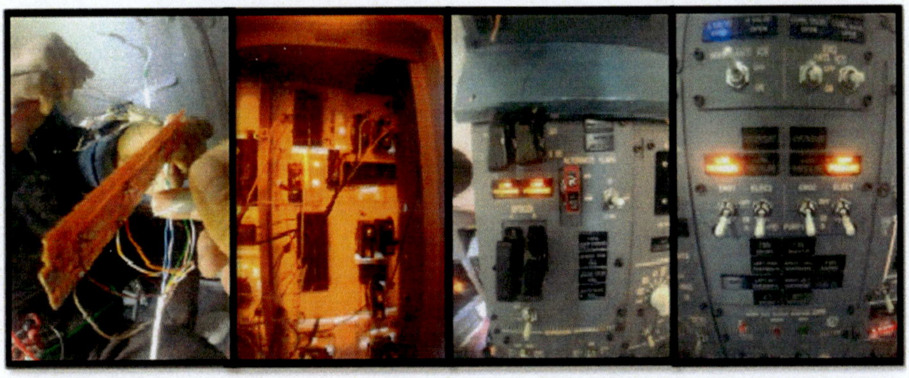

Fig. 11. Connection of the electronic components of the hydraulic system.

Finally, in this way, the software programming will be able to assign its function based on this pin number connection on the Arduino board and ensure that it fulfills and performs its specific task.

3.2 Switch and Light Programming and Software Configuration Process

Prior to the programming process, a continuity check of the circuit connections of both LEDs and switches must be performed, for all the wires and points that have undergone a micro soldering process and that are connected directly to their assigned pin inside the Arduino mega board.

The programming process of lights and switches was done through the programming interface called HCSCI [18], which is an online configurator, which has images and maps of the design that represent all aircraft systems with all the parameters that must meet each element of cabin, with this interface can be configured quickly and easily any panel of the cabin that includes control elements such as switches, lights, push buttons, potentiometers and others, Fig. 12.

In the aforementioned interface we proceeded to perform the coding for the turning on of leds and the activation of switches of the hydraulic system through the communication protocol with the motherboard (Arduino Mega2560), this type of communication allows to receive control data that are sent to the motherboard to convert them into commands that can simulate the operation of a specific device [19, 21–23].

The process of downloading the HCSCI interface is performed and we proceed to install it on the computer and configure it with the X-PLANE 11 program, this user interface gives this simulator the very realistic flight experience so that the performance of various systems during the flight can be predicted. Once the HCSCI programming

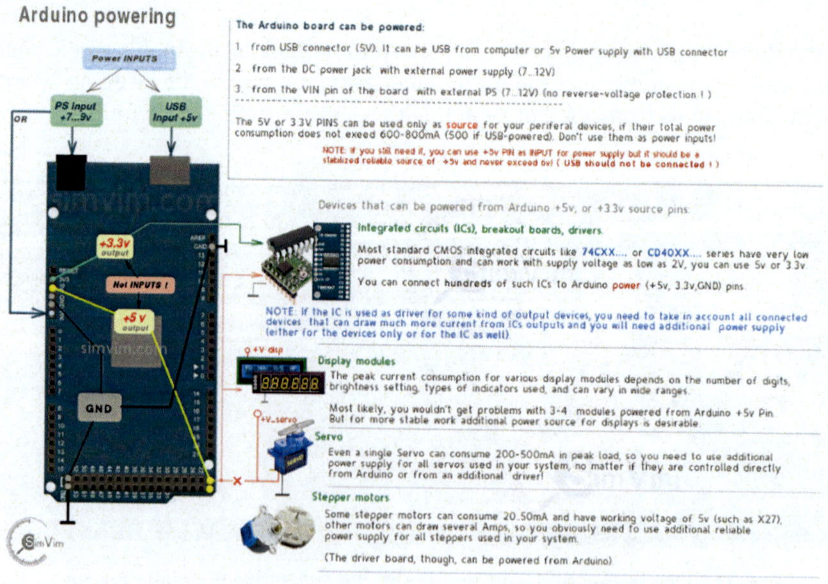

Fig. 12. Selection of switches and indicator lights of the hydraulic system.

interface is downloaded to the computer, you return to its official web page for the configuration process and assignment of switch actuation and light indications corresponding to your system [20].

The communication between the elements of the real panel with those that are virtualized is given by means of the programming contained in the file "Data. Cfg" file previously configured in HCSCI, this file provides a communication channel between the elements connected to the Arduino board with the data reference file that has the X-PLANE 11 program [24], which evaluates the conditions in which each state of the real switches is and compares them with the position of the virtualized switches, the HCSCI data are sent in the form of position bits (0 = off, 1 = on), thus changing the position of the switches as they are manipulated, Fig. 13.

Fig. 13. Execution of the X-PLANE 11 program

Finally, a check of the programming and configuration that was assigned through the HCSCI official web interface is performed by entering the X-PLANE 11 program, these operational tests are subject to the tasks that all lights and switches must reflect in

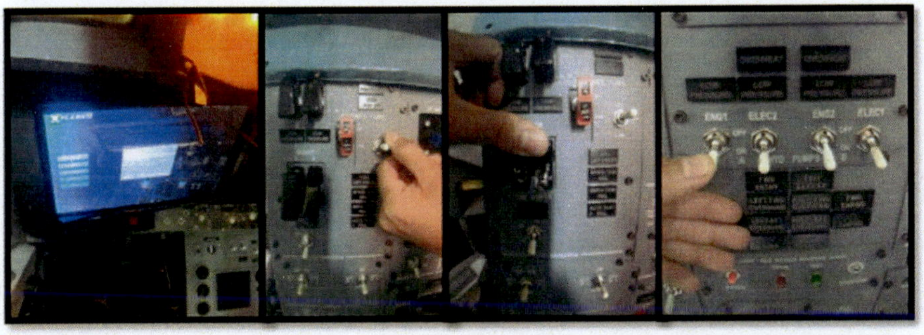

Fig. 14. Programming and software configuration process of switches and indicator lights in the hydraulic system.

an original way to the real parameters of an aircraft, taking into account the guidelines established by the AMM- ATA 29 maintenance manual [8], Fig. 14.

4 Conclusions

The tests in the simulator of the hydraulic system of the Boeing 737-500 aircraft were carried out in compliance with the procedure of the maintenance manual in which the aircraft was started to use the mechanical pumps installed in each engine; also, the operation of the electric pumps was checked, in which each one delivers a value ranging between 3000 P.S.I. The inspection allowed determining the regulatory parameters of the status and performance of a hydraulic system in accordance with the technical documentation in which it complies with features such as: three-position drive switches for engine pumps and electric engine pumps of its A, B, and Standby systems, visual indicator lights for low pressure and low quantity for warnings of the same, among others.

The architecture of this system in the training cabin complied with the operation verification requirements, which allows training aeronautical maintenance professionals through repetition and practice of the system operation procedures, as well as improving theoretical competencies and increasing expertise in problem solving. Future work consists of model refinement and flexible testing of other systems involving constant speed devices, including the generation control unit, and other critical aircraft control systems whose training will effectively contribute to aircraft maintenance.

References

1. F. A. A. U. S. Department of Transportation, Aviation Maintenance Technician Handbook - Airframe Volume 1, Aviat. Maint. Tech. Handb. - Airframe, vol. 1, p. 588 (2018)
2. Jarrin, D., Inca, G., Bautista, R., Arévalo, E.: Implementation of the pressurization and air conditioning system, through technical information from the Ata 21 maintenance manual, in the simulator of the Boeing 737–800 aircraft belonging to the university of the armed forces-Espe. Polo del Conoc. **7**(4), 1938–1955 (2022). https://doi.org/10.23857/pc.v7i4.3930
3. Moreira, X., Inca, G., Arévalo, E., Bautista, R.: Com/Nav system of the flight simulator of the Boeing 737–800 aircraft belonging to the university of the armed forces Espe. Dominio las Ciencias **8**(1), 497–516 (2022). https://doi.org/10.23857/dc.v8i1.2585
4. Ruiz, L., Inca, G., Bautista, R., Arévalo, E.: Simulation of a Boeing 737-500 Aircraft Electrical System, pp. 3–7 (2022)
5. Espín, D., Bautista, R., Inca, G., Arévalo, E.: Starting of the Cfm56-7b engine in the flight simulator of the Boeing 737–800 aircraft belonging to the university of the armed forces ESPE. Dominio las Ciencias **8**(2), 519–544 (2022). https://doi.org/10.23857/dc.v8i1.2585
6. Allerton, D.J.: The impact of flight simulation in aerospace. Aeronaut. J. **114**(1162), 747–756 (2010). https://doi.org/10.1017/S0001924000004231
7. Yongjun, Q., Xiaoming, B., Xiaofang, X., Yongsheng, L.: Visual simulation of aircraft carrier assistant landing system. In: Proceedings of the 2008 Pacific-Asia Workshop on Computational Intelligence in Industrial Application PACIIA 2008, vol. 2, no. Lcl, pp. 714–717 (2008). https://doi.org/10.1109/PACIIA.2008.57
8. Boeing, Aircraft Maintenance Manual, no. June, pp. 1–24 (2019)

9. LAN, L.T.T.: Training Manual ATA 29 Hydraulic Power, vol. I, p. 40 (2020)
10. Wyatt, D., Tooley, M.: Aircraft Electrical and Electronic Systems (2009). https://doi.org/10.4324/9780080942971
11. Wyatt, D.: Aircraft Flight Instruments and Guidance Systems (2014). https://doi.org/10.4324/9781315858975
12. RealSimControl, Cockpit Input/Output control Interface (2020). https://realsimcontrol.com/index.html
13. Zhi, C., Gao, Y., Li, X.: A distributed system for aircraft hydraulic system simulation test based on LXI, AUTOTESTCON (Proceedings), no. September, pp. 185–188 (2008). https://doi.org/10.1109/AUTEST.2008.4662609
14. Tooley, M.: Aircraft Digital Electronic and Computer Systems Principles, Operations and Maintenance, pp. 1–172 (2007)
15. Lawhorn, D., Rallabandi, V., Ionel, D.M.: A network graph technique for the design of electric aircraft power systems. In: 2020 IEEE Transportation Electrification Conference and Expo, ITEC 2020, pp. 809–813 (2020). https://doi.org/10.1109/ITEC48692.2020.9161492
16. Tan, C.F., Chen, W., Van Den Boomen, G., Rauterberg, M.: Application of automation for low cost aircraft cabin simulator. In: ICCAS 2010 - International Conference on Control, Automation and Systems, pp. 44–47 (2010). https://doi.org/10.1109/iccas.2010.5669948
17. Athalye, P., Maksimović, D., Erickson, R.: High-performance front-end converter for avionics applications. IEEE Trans. Aerosp. Electron. Syst. **39**(2), 462–470 (2003). https://doi.org/10.1109/TAES.2003.1207258
18. Sharma, A., Capoor, G.K., Chattopadhyay, A.B.: Advanced aircraft electrical systems to enable an All-Electric aircraft. In: Electrical Systems for Aircraft, Railway and Ship Propulsion, ESARS, vol. 2015-May, p. 5 (2015). https://doi.org/10.1109/ESARS.2015.7101423
19. Wang, D., Yang, S., Wang, L., Liu, W.: Hardware-in-the-loop simulation for aircraft electric power system. In: 2016 International Conference on Electrical Systems for Aircraft, Railway, Ship Propulsion and Road Vehicles International Transportation Electrification Conference (ESARS-ITEC 2016), pp. 1–5 (2016). https://doi.org/10.1109/ESARS-ITEC.2016.7841331
20. Jirgl, M., Boril, J., Jalovecky, R.: The identification possibilities of the measured parameters of an aircraft model and pilot behavior model on the flight simulator. In: ICMT 2015 - International Conference on Military Technologies 2015, no. 1 (2015). https://doi.org/10.1109/MILTECHS.2015.7153726
21. Bravo, J.M., Santos, V.: Backtesting recurrent neural networks with gated recurrent unit: probing with Chilean mortality data. In: Garcia, M.V., Fernández-Peña, F., Gordón-Gallegos, C. (eds.) Advances and Applications in Computer Science, Electronics, and Industrial Engineering. CSEI 2021. LNNS, vol. 433, pp. 159–174. Springer, Cham (2022). https://doi.org/10.1007/978-3-030-97719-1_9
22. Garcia, C.A., Naranjo, J.E., Garcia, M.V.: Analysis of AMQP for industrial internet of things based on low-cost automation. In: Iano, Y., Arthur, R., Saotome, O., Kemper, G., Padilha França, R. (eds.) BTSym 2019. SIST, vol. 201, pp. 235–244. Springer, Cham (2021). https://doi.org/10.1007/978-3-030-57548-9_22
23. Naranjo, J., Sanchez, D., Robalino-Lopez, A., Robalino-Lopez, P., Alarcon-Ortiz, A., Garcia, M.: A scoping review on virtual reality-based industrial training. Appl. Sci. (Switzerland) **10**(22), 1–31 (2020). https://doi.org/10.3390/app10228224
24. X.-P. XPlane 11, pp. 22–24 (2022). https://www.x-plane.com/

Check for updates

Immersive Technology-Based Guidance Module for Induction Motor Diagnosing

Jose-E. Naranjo[1(✉)], Gustavo Caiza[2], Veronica Gallo-C.[3], Santiago Alvarez-T.[3], Wilson-O. Lopez[4], and Marcelo-V. Garcia[1,5]

[1] Universidad Tecnica de Ambato, UTA, Ambato 180103, Ecuador
{je.naranjor,mv.garcia}@uta.edu.ec

[2] Universidad Politecnica Salesiana, UPS, Rumichaca y Moran Valverde av., Quito 170146, Ecuador
gcaiza@ups.edu.ec

[3] Instituto Superior Tecnológico María Natalia Vaca, ISTMNV, Bolivariana Av. and El Condor Av., Ambato 180205, Ecuador
{vgallo,salvarez}@institutos.gob.ec

[4] Universidad Tecnologica Indoamerica, UTI, Ambato 180103, Ecuador
wilsonlopez@indoamerica.edu.ec

[5] University of Basque Country, UPV/EHU, Bilbao 48013, Spain
mgarcia294@ehu.eus

Abstract. One of the main objectives of virtual reality (VR) is to simplify human resource interactions with equipment once a process is in operation, especially in the case of techniques that are related and whose devices communicate with each other. VR allows operators to become familiar with the production environment and the process in a tangible way during training, making them aware of the context where they will be working, both the physical space and the contingencies that may arise. This paper presents an immersive training module for detecting and correcting faults in induction motors. Two groups were considered for testing the effectiveness of the developed system: a control group and an experimental group. The tests showed that with this technology, the teaching-learning process time is optimized by more than 50%. In addition, with a p-value lower than the selected alpha level, it was confirmed that the use of VR, in this case, is more efficient than the use of conventional training methodologies. Finally, the usability of the proposal was evaluated, concluding that, although the system has several areas for improvement, it is beneficial for the industry.

Keywords: Virtual Reality · Induction Motor troubleshooting · Training · SUS

1 Introduction

The learning of human resources must advance together with technology. They are essential to optimize the productive chain of a company or industry. At

M. V. Garcia and C. Gordón-Gallegos (Eds.): CSEI 2022, LNNS 678, pp. 478–493, 2023.
https://doi.org/10.1007/978-3-031-30592-4_32

present, it is new to incorporate training with virtual mechanisms; this means reducing costs and guaranteeing the learning of the information imparted to the personnel. The central commitment of virtual training is to generate fast and easy-to-use solutions for human resources [4,8]. This is an extension of previous work presented by authors in [3].

Industrial revolution 4.0 promotes this type of technology, thus allowing the entry of information and efficient resource management. Most industries have been forced to enter this new era in order to obtain more efficient production techniques. The digital tools within this fourth industrial revolution are VR and augmented reality (AR), used for staff training.

The primary beneficiaries of this technology are the goods industries since it allows the automated management of their processes. By having such a wide field of utilisation, it has been seen that these technologies have applications from education, military and medicine to the maintenance of machinery at an industrial level [1].

The health emergency and the economic crisis that the world is experiencing make technology a decisive factor in overcoming these difficulties. At an industrial level, there are also severe effects since employers, ensuring the health of their employees, seek to avoid physical contact and investigate new ways of performing automated work. Virtual applications and platforms are used to create products for the industrial area, which is a way to counteract the damages caused by the pandemic [21].

It is evident that with the application of virtual procedures in the industrial environment, changes are forged in the working frequency and interpersonal relationships [18]. A technological alternative to the present and future is VR since it avoids the industrial environment and maintains the integrity of people. It is an available option in all processes of industrial production [22].

Similarly, the continuous progress of technology allows the application of virtual fields in all industries and even in people's daily lives. It is a tool that helps to avoid accidents since it allows to obtain a virtual didactic automated handling. Likewise, since many companies have adopted this model to carry out training in all the industrial sector processes, it is a preponderant factor. One of the main advantages adopted with the application of VR is better performance in industrial projects. Besides, it is a factor that seeks continuous improvement and competitive industrial advantage over other organizations [6,22].

As described in previous paragraphs, this article presents the development of a VR system for training in the detection of failures of three-phase motors with squirrel-cage rotor, as well as its comparison with a conventional training system. The system has been developed using the Unity 3D graphics engine and Blender. The HMD and HTC VIVE controls have also been integrated.

This article is divided into eight sections, including the introduction. Section 2, presents a brief description of works similar to the research proposal, in Sect. 3, the objectives of the study are delimited. In Sect. 4, the development and usefulness of the interfaces are described, while in Sect. 5, the methodology used for the usability evaluation is shown. Section 6, defines the selected sample, and the usability questions asked. In Sect. 7, the obtained results are discussed; and, in Sect. 8, the conclusions and future works are displayed.

2 Background Knowledge

VR technology is increasingly used in industrial processes. The fourth industrial revolution is essential since it allows optimizing the training of human resources. This generates efficiency in the time and cost that the industries budget. For this reason, several researchers have made VR simulators to optimize the productive chain of corporations. Some works are cited below.

Lacko [10], in his study, shows the use of VR to provide health and safety training for industries. He focuses on solving possible failures through automated work that avoids accidents and economic losses. The practice was carried out in two groups of workers, group A was trained with traditional apprenticeships, and group B used VR. At the end of the simulation, a questionnaire of 20 questions was applied to all participating workers. Group A obtained a percentage of correct answers of 87%, and in group B, 97% of the answers were positive. When applying the questionnaire after one month of training, the correct answers of group A decreased to 68%, while group B obtained 87% effectiveness.

Shen [20], research similarly exposes a comparison between conventional training and VR. He described that the technology applied in the industry is a factor that saves costs. The study focused on a two-hand gearbox assembly task. The four participants in the simulation indicated that the use of VR generates learning in less time, requires less effort when it comes to learning about new topics and obtaining better performance in practice according to what was taught in the experiment.

Implementing a VR system on practices for welding is aimed at beginners who want to learn about the subject. This is how Chibani [5], explains the main processes that the SPIDAR-WELDER system has to help the student manipulate the welding tools and generate essential techniques for the participants. At the end of the study, VR's main advantages were to avoid possible accidents in beginners, generate didactic learning on how to operate the machines, and carry out several pieces of training that ensure a critical and optimal performance in workers.

The main objective of Lustosa [12], is to implement the fourth industrial revolution in corporations. The system was based on a graphical mechanism and forklift control interfaces, using a joystick. The effect caused by the simulator in the eleven workers with an age range between 22 and 26 years was positive since, after undergoing a quantitative test, most of them answered that the training is easy to understand and execute. The test was carried out with a score of 1000 possible points, to which the users obtained a result of 827 in module 1 and 850 in module 2.

Randeniya [17], divided 6 participants, 3 through traditional training and 3 with VR. The users who experimented with regular learning did so through PowerPoint, Slides, Videos, among others. The cognitive and psychomotor skills of the participants were evaluated, as well as the main results of time, intuition, completion of work and precision. The 3 participants with VR training scored 9.09, and the three users with conventional training averaged 6.31, which shows that VR is better than traditional training.

Nurkertamanda [16], develops the issue of complications and errors that can occur in lathe motors. The objective of the work is to find a solution to the failures through a VR system made in Unity 3D and Blender. The system consists of two essential menus; i) an introduction where the parts and the functions of the lathe are shown, and ii) A menu divided into five user training options. Overall, users answered 336 questions, 24 each. Three hundred thirty questions were fully agreed. The system is efficient since it shows that users obtain better performance and knowledge thanks to this type of learning.

3 Case Study

In the industrial rhythm in which we find ourselves today, highly qualified operators are the ones who make the difference when measuring the productivity of a process or production line. For this reason, Industry 4.0 has become vitally important, providing new and efficient methodologies for human resources training.

Among these tools is VR, which is defined as the simulation of reality utilizing a computer. VR's advantages are focused on optimizing resources such as time or money and preventing workers from seeing their physical integrity exposed unnecessarily.

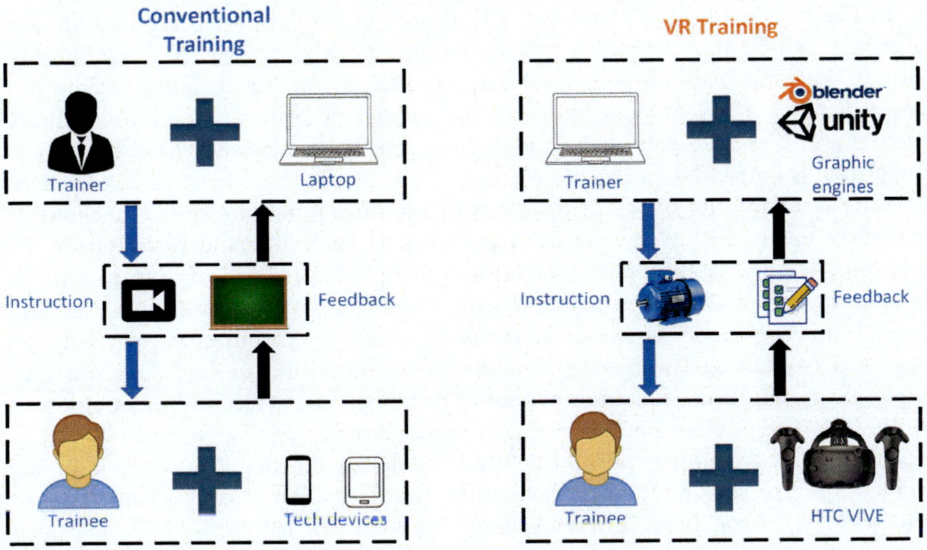

Fig. 1. Conventional and VR training

This study compares two forms of training in terms of fault detection and correction in three-phase electric motors. The first is based on the teaching adopted as usual within the health crisis that we are daring: videos, slides, and virtual classrooms through video calling software. The second uses an immersive VR application, within which interactive content and sensors are used that generate a sense of complete immersion for the user. See Fig. 1.

Because there is a wide range of possibilities in terms of failures within this type of motors, those that have been taken into account for this study, as they are the most common and because of their easy detection, are described below:

- **Broken Bars:** It causes a decrease in the regular performance of the induction motor. The failure begins with the fracture of a bar, and if it is not solved in time, it triggers the breakage of more rotor bars. Therefore it is necessary to give continuous maintenance to these machines [13].
- **Inter-turn short citcuits:** are caused by damage to the coating that insulates the coil windings. This can affect the entire coil and even adjacent coils. Mechanical stress, high current peaks, or thermal impacts are some of the causes of the insulation. When this failure occurs, the machine has vibration problems, causing a major breakdown in its magnetic field [7].

4 System Description

The graphical environment limits the virtual space where the user will develop and learn to identify the possible damages of an electric motor. Here, the Unity 3D graphics engine and Blender design software was used. Besides, the interaction of the interfaces with the HMD and HTC VIVE controls was achieved.

In Fig. 2, it is observed that the VR system is made up of three modules, i) Overview, which allows acquiring technical notions about three-phase induction motors. This has been implemented through a video. ii) Motor faults, where the user will be able to learn how to recognize an error due to broken bars or short circuits and how to repair each one of them. iii) Test, necessary to evaluate the knowledge acquired by the apprentices.

When starting the system, the user will find himself immersed in a mechanical workshop in order that his sense of presence is complete and feels within an environment that facilitates industrial learning. Here, a menu of options will be displayed where he will be able to choose between starting the training or exiting the system. See Fig. 3. If the user agrees to start the training, a video will be displayed explaining the basic principles of an induction motor, its parts and its applications. When finishing watching the video, a button will be enabled to continue to the next scene. This can be appreciated in Fig. 4.

In Fig. 5, the running motor's nominal data and the signal of a phase in the oscilloscope are shown. It also states that the first error the user will have to deal with is broken bars. The characteristic sound of this error has also been simulated to familiarize the user with their surroundings and with the engine. In this interface, a button will take the apprentice to see, on a board, the difference in signal in each phase of a motor without fault and one with fault.

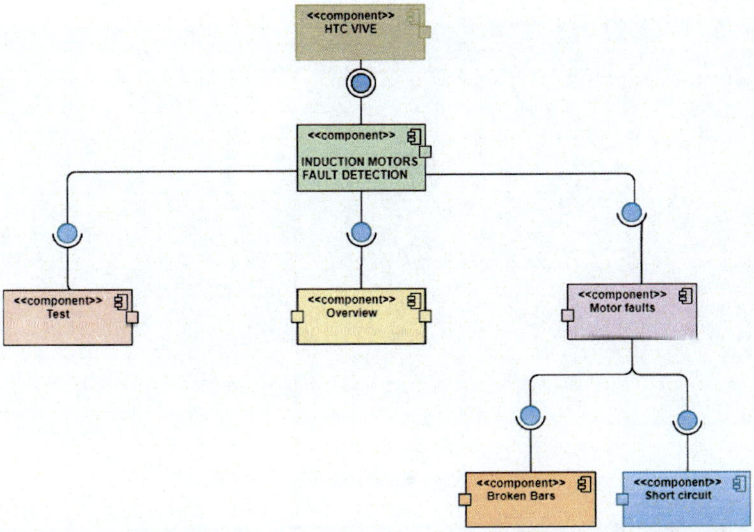

Fig. 2. System component diagram

Fig. 3. Options menu

In this scene, there is auditory feedback, which explains that the amplitude in the frequency range of 75 to 110 Hz decreases to lower values of –40 dBV compared to a standard signal. On the other hand, it is also taught that there is an increase in amplitude in the frequency range from 100 to 110 Hz. Finally, it is established that another characteristic of this type of failure is the creation of lateral peaks near 70, 40 and 50 Hz with oscillating amplitudes around –10 dbV. See Fig. 6.

Once the user has captured this information, the disassembled motor will be displayed, i.e. without its casing and stator. A red swinging arrow will iden-

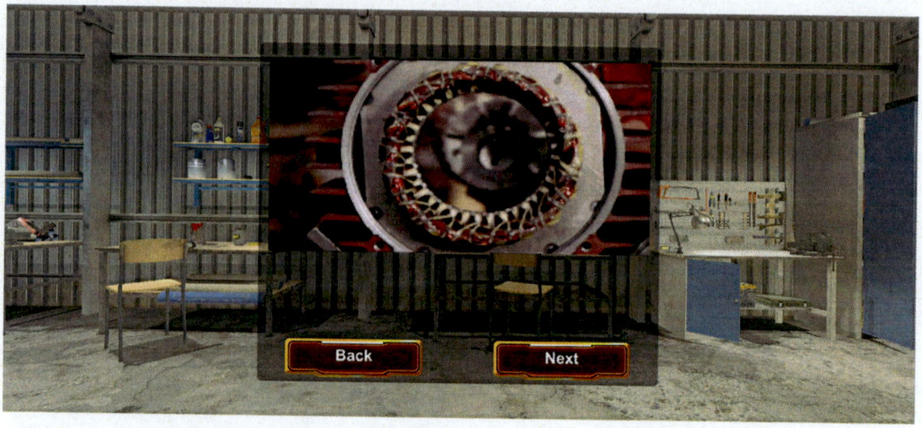

Fig. 4. Tutorial

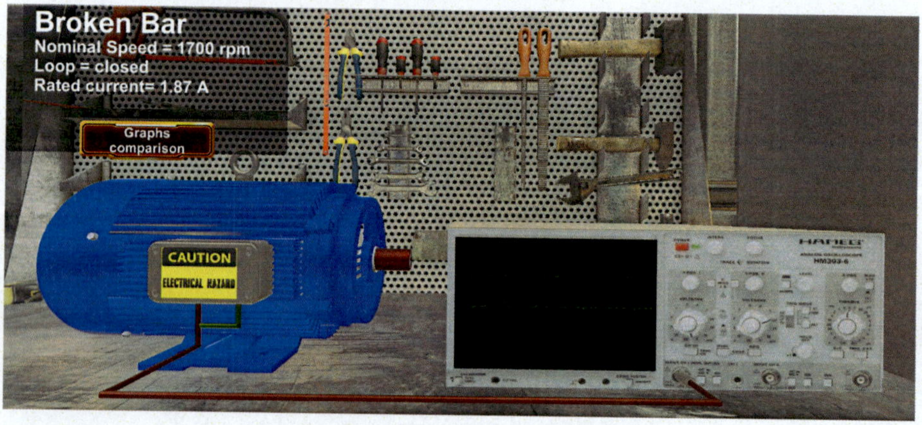

Fig. 5. Broken Bars fault

tify physical damage to be visible and draw the trainee's attention. The above described can be seen in Fig. 7. After this, the user will be trained on how to solve it. Here, the chosen method is brazing since it is the most common way to solve small rotor bars' small failures. When the repair process is finished, the damage disappears, faithfully simulating the actual repair process, having the option to continue with the subsequent failure. See Fig. 8.

The scenes developed for the inter-turn short circuits failure are similar. The nominal motor data displayed on the screen differs so that this fault's actual conditions can be created. Also, in Fig. 9 it can be seen that when the user goes on to compare the graphs, he will be able to learn that, in phase 3, there is a relatively small increase of 0.53 dBv in the frequency 300 Hz. In phase 1, the spectral signal moves 11 dBv above the signal without failures. Finally, in the

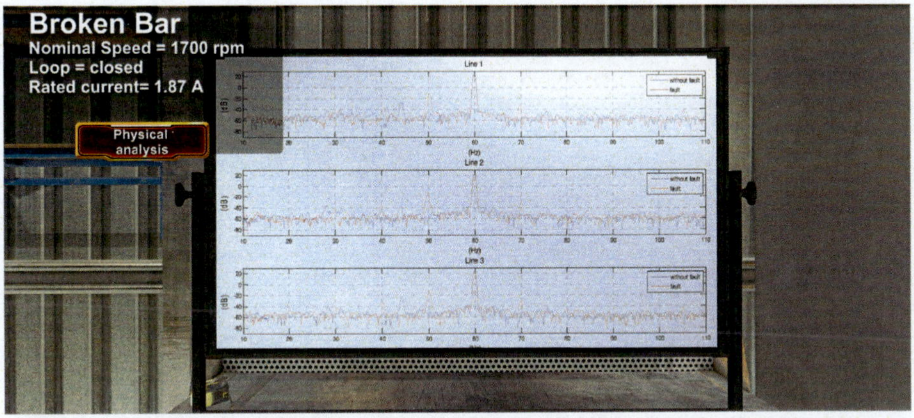

Fig. 6. Broken Bars: graphs comparison

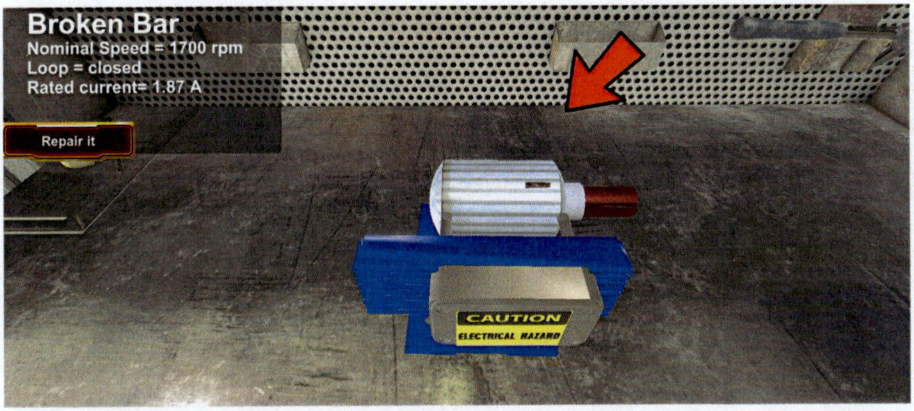

Fig. 7. Broken Bars fault: Physical damage

three phases, it can be seen that 180 Hz, the signal spectrum increases 23 dBv above the noise level [2].

Finally, to evaluate the knowledge acquired by each user, a questionnaire of ten multiple-choice questions has been implemented. Four have been directed to understanding the broken bars fault, four to the intern-turn short circuit and two to the general understanding of three-phase electric motors. Each question has three options, of which only one corresponds to the appropriate result. See Fig. 10. To guarantee the answers' reliability, each time the user selects an answer, it will be stored, and there will be no way to modify it. After this, he will automatically be immersed in the next question. In Fig. 11, it can be understood that when this evaluation has finished, a screen with the appropriate feedback will be displayed, i.e., username, score obtained, and the time spent.

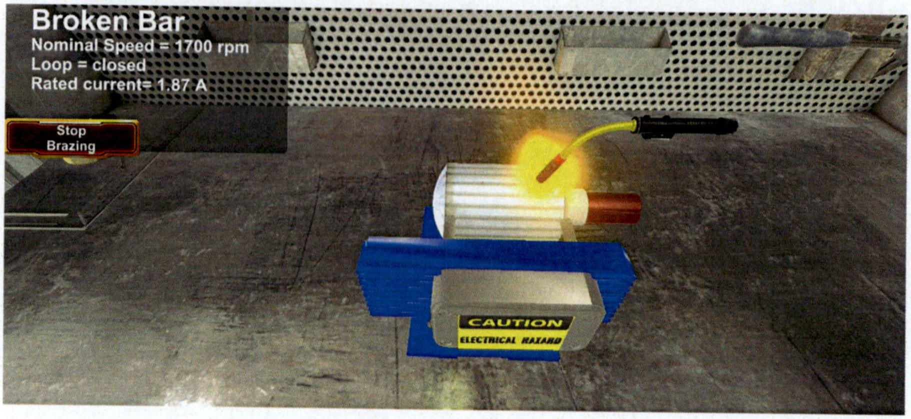

Fig. 8. Broken Bars fault: Brazing process

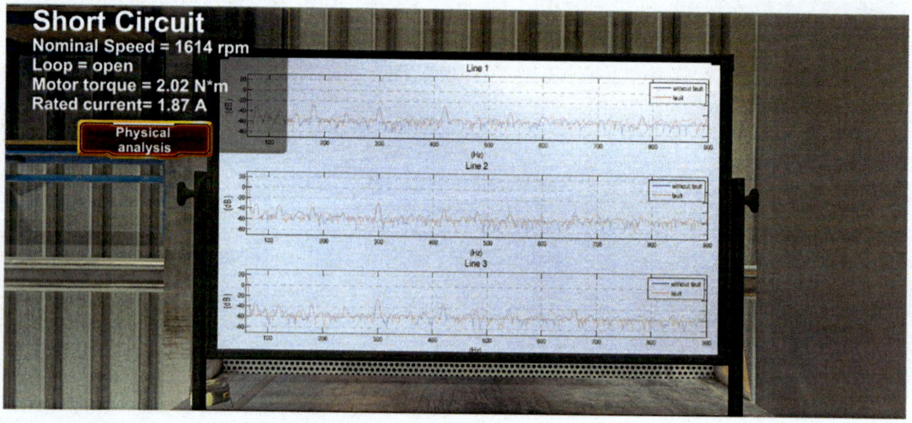

Fig. 9. Short circuit: graphs comparison

5 SUS

It is a procedure that helps to measure the usability of technological systems. It was developed in 1986 by John Brooke and was based on ten statements where users respond through a satisfaction scale about each virtual operator's functionality. It is a test that allows users to determine virtual platforms purpose and level of service [11].

A rating from 0 to 100 determines how efficient the developed systems are. Those that obtain qualifications greater than 80 are excellent products, which do not have much to correct, and their qualification is A. Systems that are in a range between 70 and 80 are considered acceptable and have the designation B. Systems that are in the range of 50 and 70 are considered poor and have a C rating. Lastly, products that obtain a score below 50 are considered to have

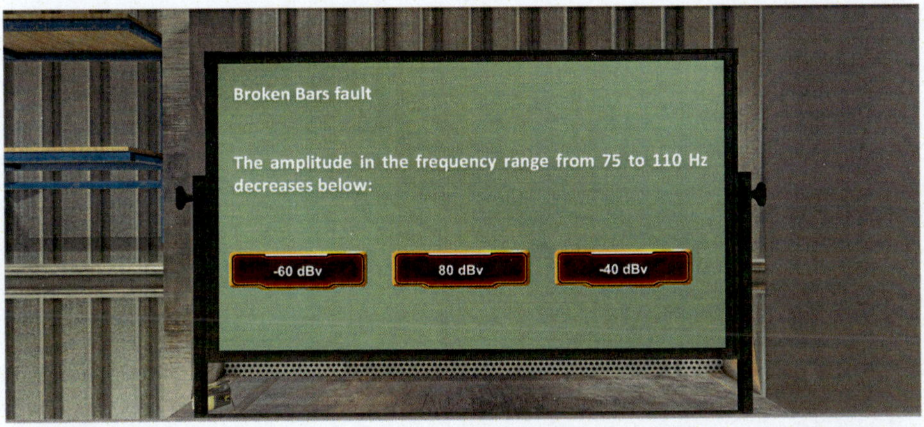

Fig. 10. Test

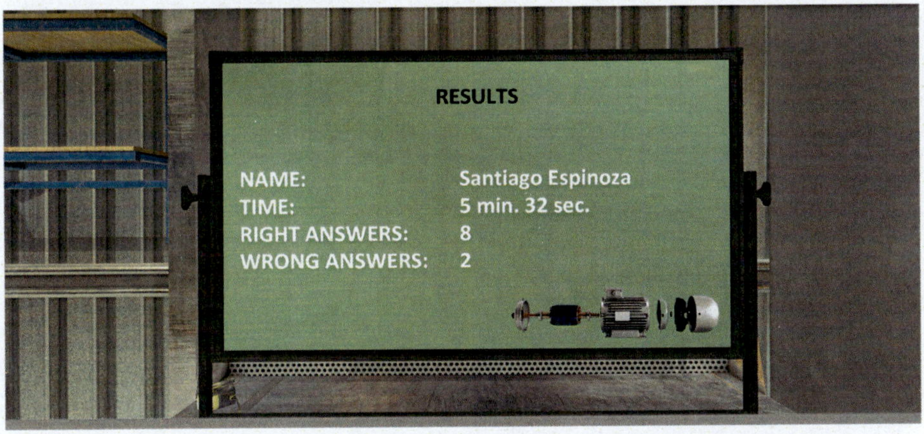

Fig. 11. Test feedback

very little functionality and are valued with a D rating. The general guideline is shown on Table 1 [14].

The formula used is shown in Eq. 1, where X represents the answer to each question. Each of the even-numbered questions is scored by subtracting the user's score from 5, that is (5 − X). Instead, subtracting one from the user-generated score (X − 1) is performed for odd-numbered questions. To determine the final score, the numbers of the positive and negative questions are added and multiplied by 2.5, thus generating the final result of the usability of a VR system [19].

$$S = 2.5 * [\sum_{i=1}^{5}(X_{2i-1} - 1 + 5 - X_{2i})] \tag{1}$$

Table 1. SUS scores description.

SUS Score	Grade	Adjective Rating
>80	A	Excellent
>70–80	B	Good
50–70	C	Poor
<50	D	Awful

6 Experiment Setup

To fulfil this research's objective, an experiment was designed with the participation of students and professionals from the city of Ambato-Ecuador with basic knowledge about electronics and automation. 30 people were chosen to take part in this experiment, and they were separated into two groups of 15 (a control group and an experimental group). 2 questions were included in this research:

- Q1: Does the use of digital systems like VR allow the optimization of training times?
- Q2: Is the VR system usable for induction motor fault detection and correction training?

As in other studies, the evaluation of acceptance was carried out through a survey following the guidelines presented by SUS [9,15]. Although there are other methods for measuring usability, SUS has been widely used in the evaluation of digital systems and VR systems [11].

The questionnaire's responses were established following the Likert scale: 1-Strongly disagree, 2-Disagree, 3-Neutral, 4-Agree, 5-Strongly agree. The following items were included in the survey:

1. I think that I would like to use this VR system frequently.
2. I found the VR system unnecessarily complex.
3. I thought the VR system was easy to use.
4. I think that I would need the support of a technical person to be able to use this VR system.
5. I found the various functions in this VR system were well integrated.
6. I thought there was too much inconsistency in this VR system.
7. I would imagine that most people would learn to use this VR system very quickly.
8. I found the VR system very cumbersome to use.
9. I felt very confident using the VR system.
10. I needed to learn a lot of things before I could get going with this VR system.

7 Results

7.1 Test

The conventional training process's main results indicate that in question 1, 33.33% correct answers and 66.67% wrong answers were obtained by the participants. In question 2, 53.33% of the people answered correctly and 46.67% wrong. Question 3 was successful at 53.33% while 46.67% failed. The participants in question 4 reached a higher number of correct answers, having only 33.33% wrong answers. In question 5, 80% correct answers were obtained. The result of question 6 was 73.33% correct answers and the remainder incorrect. The value of correct answers was 46.67% in question 7. Success in questions 8 and 9 was 53.33%. Finally, in question 10, 40% of the people got it right, and 60% failed.

Using VR methodology, more significant learning is obtained since it is reflected in the number of correct answers for each question. In question 1, 80% favorable responses were obtained, and only 20% unfavorable responses. Question 2 shows that 73.33% of the participants answered correctly and only 26.67% incorrect. Of the 15 people evaluated, 80% were successful in answering questions 3, and 4, and 20% failed. Question 5 had the highest number of hits, reaching 93.33%. Question 6 was answered 86.67% correctly. Questions 7 and 8 obtained a 73.33% success in answering. Finally, 80% of the participants got questions 9 and 10 correct.

According to the average of the two groups' qualifications, it reflects a favorable result for the VR methodology, since 80% of the participants clearly understood the different ways to repair an induction motor's errors. On the other hand, 55.33% of the people trained with conventional methods generated essential knowledge for the maintenance of the mechanisms. All this can be seen in Fig. 12.

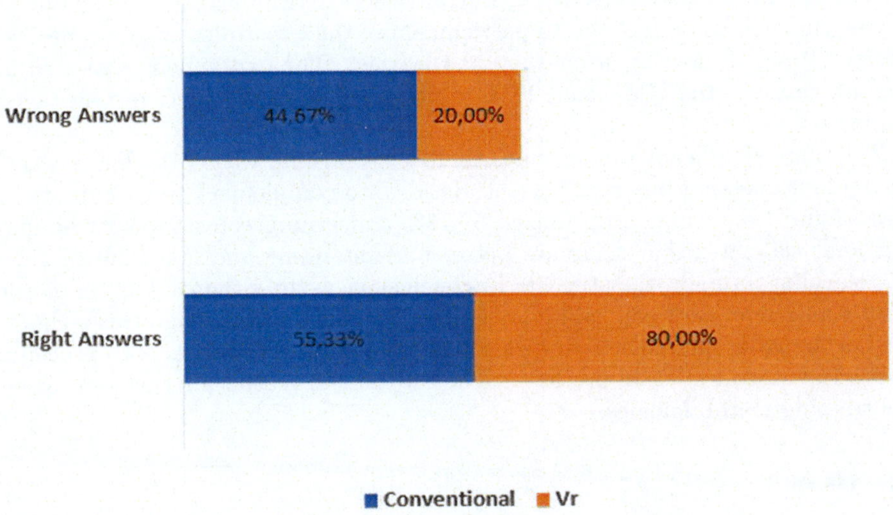

Fig. 12. Scores percentage comparison

To statistically verify the efficiency of VR training, the T-student tool has been used, and it has been based on two hypotheses: i) H_0 There is no significant difference between conventional training and VR training; y ii) H_1 There is significant difference between conventional training and VR training. The confidence level selected corresponds to $\alpha = 5\% = 0{,}05$.

On the other hand, the Shapiro-Wilk test identifies the normality of the samples' observations smaller than 30 data. The significance values of conventional training and VR were 0.908 and 0.934, i.e., more significant than the p-value 0.05. These statistical data affirm that the series are normally distributed. To determine the variances' equality, the Levene test is performed, where a significance value equal to 0.346 was obtained.

With the assumptions of normality and equality of the verified variances, the significance value (0.000003) was calculated. Being less than 0.05, this value allowed rejecting H0 and accepting H1, concluding that VR training is more effective than conventional training, providing solid knowledge, information retention, and correct understanding of the processes shown.

7.2 Training Time

To determine the time resource's optimization, each of the two induction motor maintenance training was timed. The conventional instruction was carried out with the control group, which was trained in a constant time of 60 min. During this time, the professional's presentation, the content of the module studied, and the evaluation were considered. In contrast, VR training was carried out at varying times since each of the 15 participants in the experimental group needed different times to develop what they had learned. The experiment consisted of learning through the HMD and HTC controls. The total experience averaged 25 min 36 s.

There are significant time savings thanks to learning with VR, thus answering the first question posed in Sect. 6. Besides, to complement the information, it was found that time is optimized 57.73% concerning conventional training. This analysis states that there are better understanding and less training time with learning through Industry 4.0. In conclusion, we find that induction motor maintenance operators can gain more experience and learn through this technology. On the other hand, the optimization of resources in industrial entrepreneurs is essential since the costs are lower and can be the reason for the creation of new training methodologies.

7.3 Usability

One of the primary purposes of the study was to verify the use that VR training can generate. This section indicates the result of the second question written in Sect. 6. For this reason, SUS was concerned to determine how acceptable to the public is the solution of errors in induction motors by VR. The experimental group responded using a Likert scale to the ten statements made by the procedure. The average obtained was 73.33 out of 100 possible. According to the methodology, it is concluded that it is a program that is friendly, easy to use, and intuitive since it is within the range of 70 and 80 points.

In Fig. 13, the participants' qualification is indicated after being subjected to the methodology proposed by John Brooke. After performing the arithmetic operations, it was obtained that the participants find an appropriate use in the system in all areas concerning an efficient VR system. According to the SUS procedure, training to solve induction motors' errors is acceptable for 12 people since it is in the range of 70 to 80 points. On the other hand, three people obtained grades between 65 and 70, so they find the application with minor usability.

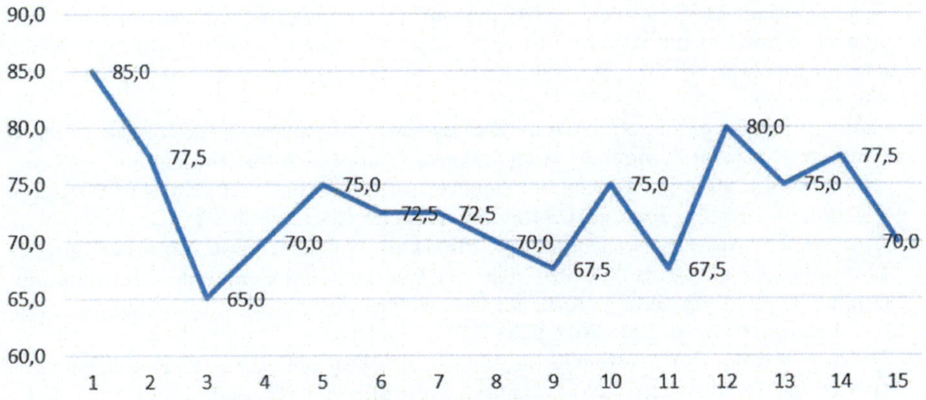

Fig. 13. SUS scores

8 Conclusions and Future Research

An immersive proposal for industrial training in detecting faults in three-phase electric induction motors has been evaluated. Using technologies such as VR allows the operator to interact directly with the process without risking his physical integrity or that of the equipment. There is still room for improvement in the proposed features, such as feedback, immersion, and fidelity. However, it has been proven that this system is suitable for use within the industrial environment previously described.

For future work, an ergonomic study of the system is proposed to know the operators' appropriate exposure time to the VR equipment. Finally, it is intended to enlarge the sample and form a third experimental group to which another teaching methodology different from the conventional and VR will be applied.

References

1. Baygin, M., Karakose, M., Akin, E.: ITHET 2016. In: 2016 15th International Conference on Information Technology Based Higher Education and Training (ITHET) : 8–10 September 2016, Istanbul, Turkey. Institut of Electrical and Electronics Engineering (2016)
2. Caiza, G., Garcia, C.A., Naranjo, J.E., Garcia, M.V.: Assessment of engineering techniques for failures simulation in induction motors using numerical tool. In: Iano, Y., Arthur, R., Saotome, O., Kemper, G., Padilha França, R. (eds.) BTSym 2019. SIST, vol. 201, pp. 307–319. Springer, Cham (2021). https://doi.org/10.1007/978-3-030-57548-9_28
3. Caiza, G., Riofrio-Morales, M., Veronica Gallo, C., Santiago Alvarez, T., Lopez, W.O., Garcia, M.V.: Virtual reality system for training in the detection and solution of failures in induction motors. In: 33rd European Modeling and Simulation Symposium, EMSS 2021, pp. 199–207 (2021). https://doi.org/10.46354/i3m.2021.emss.027
4. Cervera, M., Grandon, N., Rivera, M., Besoain, F.: Improving the selection of IQF raspberries in processing lines: a virtual reality approach for training and selecting personnel. In: 2018 IEEE Biennial Congress of Argentina, ARGENCON 2018, pp. 1–7 (2019). https://doi.org/10.1109/ARGENCON.2018.8646167
5. Chibani, D., Achour, N., Daoudi, A.: SPIDAR-Welder a haptic interface for virtual welding training. In: CCSSP 2020–1st International Conference on Communications, Control Systems and Signal Processing, pp. 293–297 (2020). https://doi.org/10.1109/CCSSP49278.2020.9151477
6. Davila Delgado, J.M., Oyedele, L., Beach, T., Demian, P.: Augmented and virtual reality in construction: drivers and limitations for industry adoption. J. Constr. Eng. Manag. **146**(7), 04020079 (2020). https://doi.org/10.1061/(asce)co.1943-7862.0001844
7. Fireteanu, V., Constantin, A.I., Leconte, V., Lombard, P.: Analysis of the evolution of stator short-circuit and rotor bar breakage faults in a squirrel-cage induction motor. In: SPEEDAM 2018 - Proceedings: International Symposium on Power Electronics, Electrical Drives, Automation and Motion, pp. 190–195 (2018). https://doi.org/10.1109/SPEEDAM.2018.8445226
8. He, Q., Cheng, X., Cheng, Z.: A VR-based complex equipment maintenance training system. In: Proceedings - 2019 Chinese Automation Congress, CAC 2019, pp. 1741–1746 (2019). https://doi.org/10.1109/CAC48633.2019.8996496
9. Kim, S.U., Lee, K., Cho, J.H., Koo, K.C., Kim, S.B.: Toward an evaluation model of user experiences on virtual reality indoor bikes. Eur. Sci. J. **7881**(June), 1857–7881 (2017)
10. Lacko, J.: Health safety training for industry in virtual reality. In: Proceedings of the 30th International Conference on Cybernetics and Informatics, K and I 2020, pp. 1–5 (2020). https://doi.org/10.1109/KI48306.2020.9039854
11. Lewis, J.R., Sauro, J.: The factor structure of the system usability scale. In: Kurosu, M. (ed.) HCD 2009. LNCS, vol. 5619, pp. 94–103. Springer, Heidelberg (2009). https://doi.org/10.1007/978-3-642-02806-9_12
12. Lustosa, E.B.S., De MacEdo, D.V., Rodrigues, M.A.F.: Virtual simulator for forklift training. In: Proceedings - 2018 20th Symposium on Virtual and Augmented Reality, SVR 2018, pp. 18–26 (2018). https://doi.org/10.1109/SVR.2018.00016

13. Maloma, E., Muteba, M., Nicolae, D.V.: Effect of rotor bar shape on the performance of three phase induction motors with broken rotor bars. In: Proceedings - 2017 International Conference on Optimization of Electrical and Electronic Equipment, OPTIM 2017 and 2017 Intl Aegean Conference on Electrical Machines and Power Electronics, ACEMP 2017, pp. 364–369 (2017). https://doi.org/10.1109/OPTIM.2017.7974997
14. Meldrum, D., Glennon, A., Herdman, S., Murray, D., McConn-Walsh, R.: Virtual reality rehabilitation of balance: assessment of the usability of the Nintendo Wii ® fit plus. Disabil. Rehabil. Assist. Technol. **7**(3), 205–210 (2012). https://doi.org/10.3109/17483107.2011.616922
15. Naranjo, J., Urrutia Urrutia, F., Garcia, M., Gallardo-Cardenas, F., Franklin, T., Lozada-Martinez, E.: User experience evaluation of an interactive virtual reality-based system for upper limb rehabilitation. In: 2019 6th International Conference on eDemocracy and eGovernment, ICEDEG 2019 (2019). https://doi.org/10.1109/ICEDEG.2019.8734389
16. Nurkertamanda, D., Saptadi, S., Widharto, Y., Maliansari, A.N.: Feasibility evaluation: virtual laboratory application based on virtual reality for lathe engine training simulation. In: 2019 6th International Conference on Frontiers of Industrial Engineering, ICFIE 2019, pp. 6–10 (2019). https://doi.org/10.1109/ICFIE.2019.8907772
17. Randeniya, N., Ranjha, S., Kulkarni, A., Lu, G.: Virtual reality based maintenance training effectiveness measures - a novel approach for rail industry. In: IEEE International Symposium on Industrial Electronics 2019-June, pp. 1605–1610 (2019). https://doi.org/10.1109/ISIE.2019.8781351
18. Rozo-García, F.: Revisión de las tecnologías presentes en la industria 4.0. Revista UIS Ingenierías **19**(2), 177–191 (2020). https://doi.org/10.18273/revuin.v19n2-2020019
19. Sharfina, Z., Santoso, H.B.: An Indonesian adaptation of the system usability scale (SUS). In: 2016 International Conference on Advanced Computer Science and Information Systems, ICACSIS 2016, pp. 145–148 (2017). https://doi.org/10.1109/ICACSIS.2016.7872776
20. Shen, S., Chen, H.T., Leong, T.W.: Training transfer of bimanual assembly tasks in cost-differentiated virtual reality systems. In: 26th IEEE Conference on Virtual Reality and 3D User Interfaces, VR 2019 - Proceedings, pp. 1152–1153 (2019). https://doi.org/10.1109/VR.2019.8797917
21. Urbina Pérez, M.G., Mecalco Reyes, J., Muñoz Herrera, C.A., Cruz Silva, S.P.: Las fortalezas y debilidades de los ambientes virtuales/digitales en crisis sanitarias; caso de estudio COVID-19 en México, no. November, pp. 1–7 (2021). https://doi.org/10.4995/inn2020.2020.11853
22. Wang, C.: Research on panoramic image processing technology based on virtual reality technology. In: Proceedings - 2019 International Conference on Virtual Reality and Intelligent Systems, ICVRIS 2019, pp. 55–58 (2019). https://doi.org/10.1109/ICVRIS.2019.00022

Statistical Model for Production Planning in a Vehicle Assembler Applying Lean Manufacturing

José-L. Gavidia and Christian-J. Mariño(✉)

Universidad Tecnica de Ambato, UTA, Ambato 180103, Ecuador
{jl.gavidia,christianjmarino}@uta.edu.ec

Abstract. The main objective of the present study is to design a statistical model for the planning of production in a vehicle assembly plant, using the Lean Manufacturing methodology. Initially, the identification of the different assembly processes that are executed in ten workstations in an entity taken as a case study. Applying an instrument to collect current information (Check List) for the selected 5s Lean Manufacturing tools, it was obtained: (Select 11.5%, Order 13.4%, Clean 13.6%, Standardize 11.5%, Self-discipline 11.5%, Lean Structure 19.2%, Production Planning 19.2%). The data of the investigative work high-light the need to provide a statistical tool that guides efficiently and sustainably in the short, medium, and long term the planning of production in the assembly area of the company Ciauto Cía. Ltda. Based on the historical Production versus Quality and by means of a Simple Linear Regression applying the statistical software R-Studio, a statistical model was established which indicates that for a quality objective of ninety percent (90%) it is obtained a production of one hundred forty-six (146) units approximately.

Keywords: Assembly · Planning · Production · Quality · Statistical Model · Tool

1 Introduction

The objective of the research work focused on designing a Statistical Model for production planning, in a vehicle assembly company, through the Lean Manufacturing methodology, based on a qualitative and quantitative research approach, with an inductive and deductive method, making use of bibliographical and documentary research [14,15].

For the development of this research work, where it is intended to come up with a statistical model for production planning, as a relevant part, it begins by obtaining the data, for this we apply a Checklists and tables. In addition, the application of the R-Studio statistical software, with which it is possible to obtain a simple linear regression model, relating two variables; Dependent Variable: (Production), Independent Variable: (Quality). Vehicle assembly companies in Ecuador, such as Aymesa, General Motors, Ciauto, carry out production

M. V. Garcia and C. Gordón-Gallegos (Eds.): CSEI 2022, LNNS 678, pp. 494–507, 2023.
https://doi.org/10.1007/978-3-031-30592-4_33

planning based on local market demand, plus the export plan, this activity is carried out in consensus with all the areas involved. Of the assembly plant and the Coach Manager (dealer), where there is no statistical or mathematical model that can forecast the planned production based on the quality objective set by the plant [6,13].

In view of the above, it can be pointed out that an adequate statistical model for production planning has as its fundamental objective, to provide, through a simplified, mathematically formalized way, of approximating reality on the quantities of raw material, machinery, and labor. Works in advance, based on production capacity and projected sales in a given period of time [4,8,9].

Reviewing the state of the art, we can find similar research, such as the Mathematical model for production planning in the leather sector in the parish of Quisapincha, where the study was to develop a mathematical model that meets the needs of economic improvement of the leather sector in the parish of Quisapincha Ambato Ecuador, in order to properly plan production through the use of a quantitative tool that helps support decision-making and improve administrative management, as a result they achieved maximize your profits [3,12].

The main limitation of this exploratory work is the number of sample cases collected, which could be expanded if the other areas that make up the company are considered, taken as a case study.

2 Materials and Methods

2.1 Materials

For the development of this research work, the following materials are used:

— Current production plan.
— Instrument for data collection (Check List)
— R-Studio Statistical Software
— Computer
— Printing machine.
— Field NotebookLetreros de identificación

2.2 Methodology

The investigation begins with the study of the state of the art, and with the documentary analysis of the current production process of the plant, using a Check List, which is applied in the assembly area in seven (7) Lean tools. Manufacturing taken for the study (Select, Order, Clean, Standardize, Discipline, Lean Structure, Production Planning), entering the data collected in the R statistical software, the Simple Linear Regression model is obtained [2,11].

5S's – Applied Lean Manufacturing Tool
The 5S's method, named after the first letter of the Japanese name designated in five stages, is a Japanese management technique based on five simple

principles. Whose characters begin with the letter "S" and correspond as shown in Table 1 [16].

Table 1. The five Japanese words whose characters start with "S".

N.º	Strategy	Peculiarities
1	Seiri (proper arrangement)	It consists of clearly distinguishing between necessary and unnecessary items and discarding the latter
2	Seiton (order)	It is based on always keeping the necessary items in the right place, so that anyone can find them immediately
3	Seiso (cleaning)	The factory must be kept washed and clean
4	Seiketsu (purity or extremely clean)	This is the condition when we keep the first three S. Standardize
5	Shitsuke (discipline)	It consists of making a habit of maintaining established procedures. (Hirano, 2001)

The benefits of applying and managing the 5's tools in the industry are many, among the most representative we can mention the indicated in Table 2.

Table 2. The main benefits of the 5's.

N.º	Strategy	Characteristics
1	Zero waste	Lower costs and higher capacity
2	Zero damage	Improvement in security
3	Zero breakdowns	Better maintenance
4	Zero defects	Highest quality
5	Zero change of tools	Facilitating the diversification of production
6	Zero delays	Reliability in deliveries
7	Zero complaints	Increasing reliability and trust

Research Type

[1] in their work Research Methodology, argue that all research work is based on two main approaches: the quantitative approach and the qualitative approach, which together form a third approach: The mixed approach.

The qualitative approach was precisely required for the issuance of a diagnosis on items linked to the research variables: Select, Order, Clean, Standardize, Discipline, Lean Manufacturing Tools and Production Planning, allowing in turn give the corresponding conclusions and recommendations of the analysis carried out. The quantitative approach allowed the determination of the percentage of each one of the items linked to the research variables, for the evaluation and diagnosis of the Ciauto Cía. Ltda., taken as a case study.

When using the approaches, the aim is to achieve accuracy in data measurements to support knowledge and through statistical analysis to test the previously established hypothesis to faithfully establish behavior patterns in the study population.

Variables
Independent variable: Lean Manufacturing tools.
Dependent variable: Production panning.
Observation unit: Vehicle assembler.

2.3 Methods

Deductive Method
It is necessary to analyze the problem of study in a global way, to establish specific criteria, these allow to determine the relationship between production planning and the Lean Manufacturing tools established by the researcher, to reach the particularity of the planning of the assembler vehicular.

Analytical Method
This research process consists of determining in depth the information and analysis criteria that will help determine the action criteria to establish the statistical model, and in this way understand the behavior of the applicability of Lean tools Manufacturing.

Correlational Method
In this investigative process, the interaction of the study variables is analyzed, in this way the relationship of the phenomena that are not susceptible to manipulation can be specified and in this way establish the criteria for acceptance of the relationship and approval of the investigative hypotheses, allowing later to formulate a proposal for a solution to the problem under investigation.

Instrument for Data Collection
The instrument is provided by a diagnostic Check List technique, taking into account the integrated evaluation ISO 9001:2015. Where: C (compliance), C/P (partial compliance), NC (non-compliance) for the assembly process. See Table 3.

Subsequently, an analysis was developed for each of the questions that make up the diagnostic Check List, the proportion of the results is taken into account with respect to all the points evaluated, for subsequent verification, and their percentage, as shown in Tables 4 and 5 respectively.

Table 3. Diagnostic Checklist.

N.º	ÍTEMS	C	C/P	NC	TOTAL
1	**SELECT**				
1.1	Are there useless things that can disturb the work environment?	9	25	14	48
1.2	Is there any type of tool, screw, spare part, tool or similar in the work environment?	18	17	13	48
1.3	Are frequently used objects arranged in their location and correctly identified in the work environment?	13	18	17	48
1.4	Are all the cleaning elements: rags, brooms, gloves, products in their location and correctly identified?	19	17	12	48
1.5	Is there unused machinery in the work environment?	28	12	8	48
1.6	Are unnecessary items identified as such?	15	26	7	48
	Total	102	115	71	288
2	**ORGANIZE**				
	Total	129	106	101	336
3	**CLEAN UP**				
	Total	109	120	111	340
4	**STANDARDIZE**				
	Total	75	108	105	288
5	**SELF-DISCIPLINE**				
	Total	43	124	121	288

Table 4. Situational diagnosis contingency matrix

N.º	ITEMS	C	C/P	NC	TOTAL
1	Select	102	115	71	288
2	Organize	129	106	101	336
3	Clean up	109	120	111	340
4	Standardize	75	108	105	288
5	Self discipline	43	124	121	288
6	Lean structure	163	219	98	480
7	Production planning	153	218	109	480
	TOTAL	774	1010	716	2500

Analysis and Interpretation

The use of the Checklist technique allowed to show that; of the 48 elements taken as deferential points to make the diagnosis of the good use of Lean Manufacturing in the assembly area, 40.4% show partial compliance (CP) in the use of Lean

Table 5. Contingency matrix of the situational diagnosis in percentages

N.º	ÍTEMS	C (%)	C/P(%)	NC(%)	TOTAL(%)
1	Select	4.1	4.6	2.8	11.5
2	Organize	5.2	4.2	4.0	13.4
3	Clean up	4.4	4.8	4.4	13.6
4	Standardize	3.0	4.3	4.2	11.5
5	Self discipline	1.7	5.0	4.8	11.5
6	Lean structure	6.5	8.8	3.9	19.2
7	Production planning	6.1	8.7	4.4	19.2
	TOTAL	31.0	40.4	28.6	100.0

tools, 31.0% of those evaluated highlight a total compliance (C), and finally 28.6% of these indicate that there is a non-compliance (NC) of the tools. In this way, an easy reading is obtained on what should be the causes of the problem that must be attacked. See Fig. 1.

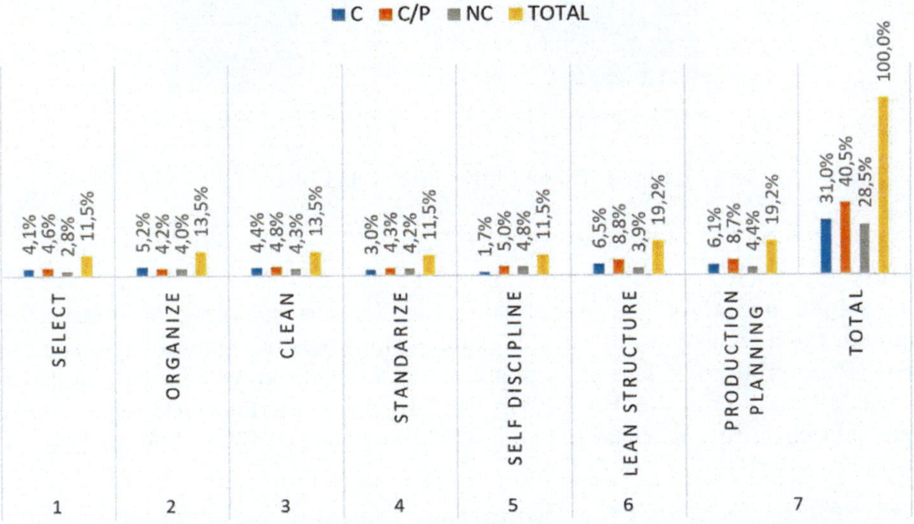

Fig. 1. Results of the situational diagnosis to the assembly area - Checklist.

Analysis of Planning Vs Actual Production

The data present the records of planned and produced units, respectively. This show that in the months of January and February the production is stable and guarantees a correct analysis of the data, considering that it is a preliminary analysis of the production histories and they follow a trend to the current data.

Figure 2 present the variation of planned and produced units, respectively, in which it can be clearly evidenced that 82% compliance with production planning is reached versus actual production, being an aspect to be considered transcendental to propose the design of an adequate statistical model for production planning, which corrects each deficiency detected in the various analyzes carried out.

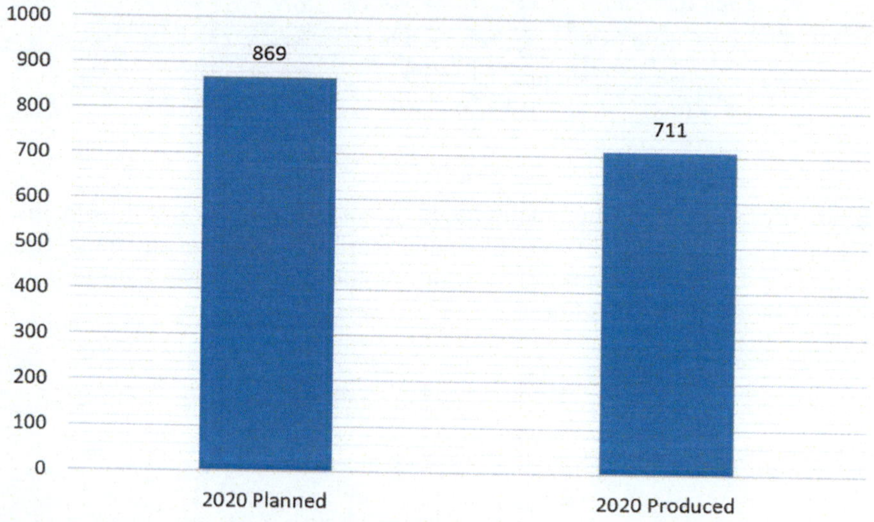

Fig. 2. Comparative production plan.

Study Area

The determination of the statistical model for the planning of production through the management of Lean Manufacturing tools, is done in the assembly area of the productive line of compact vehicles, made up of ten work stations called: E1, E2, E3, E4, E5, E6, E7, E8, E9, E10, distributed in the three assembly lines that constitute it, these being: TRIM Line, Chassis Line, Release Line.

2.4 Statistical Model for Production Planning in a Vehicle Assembly Plant, Using the Lean Manufacturing Methodology

Resolution with the Linear Programming Method for the Production Process

Once each of the data that is part of the production with its respective variables is known, the linear programming model begins to be assembled within the statistical software R - Studio.

Simple Linear Regression Model
This is most often used because it is much simpler to model. See Eq. 1.

$$mx + b = y \tag{1}$$

In addition, it depends on the objectives of the study, for example, (finding the relation-ship of the variables), the higher the Quality, the higher the Production, it is a directly proportional relationship. See Eq. 2.

$$yi = B0 + B1xi + ei; i = 1, 2, 3, ...n \tag{2}$$

where ei is an unobservable error. Regression means going back, going back to the origin, where the independent variable "x" is called a regressor because it somehow contributes to giving rise to the dependent variable "y".

$$E[yi] = E[B0 + B1Xi + ei] = B0 + B1.Xi \tag{3}$$

$$V[yi] = V[B0 + B1xi + ei] = V[ei] = Sigma^2 \tag{4}$$

Therefore "e" and "y" are random variables, also B0 and B1 are unknown constants also called parameters. Where B0 is the intersection of the line on the "y" axis and B1 is the slope.

3 Results

3.1 Determination of the Statistical Model in R

We entered the data in R-Study for the M4 model. In this way it has been possible to confirm that the Lean Manufacturing Tools affect each other, that is, non-compliance with one affects the other. With a good management of the Lean Tools, the planned Production is fulfilled, we observe that these have a directly proportional relationship with the Quality of the Product, for a better implementation and management of the Lean methodology, higher quality. Now, we can clearly see that between Quality and Production there is a directly proportional relationship, the higher the Quality, the higher the Production [7,10]. See Table 6.

If we have a quantitative variable "x" y that is associated with the response of interest, also quantitative "y", then the simplest Statistical Model that we can postulate is:

$$yi = B0 + B1 * xi + ei; i = 1, 2, 3 \ldots\ldots\ldots 12 \tag{5}$$

This is a Simple Linear Regression Model, simple because we only have one independent variable "x", also called a regressor. Linear because the relationship is linear in the Bs, where B0 and B1 are unknown parameters in the R that we must estimate from the data, and ei is an unobservable error that we assume is present in each observation. All the mentioned above, can be seen in Figs. 3, 4, and 5.

Table 6. Production and Quality of the M4 Model – 2019

Month	Production (units)	Quality (%)
January	150	92
February	115	78
March	115	80
April	136	82
May	138	84
June	120	81
July	144	88
August	144	90
September	115	79
October	138	87
November	115	78
December	120	82

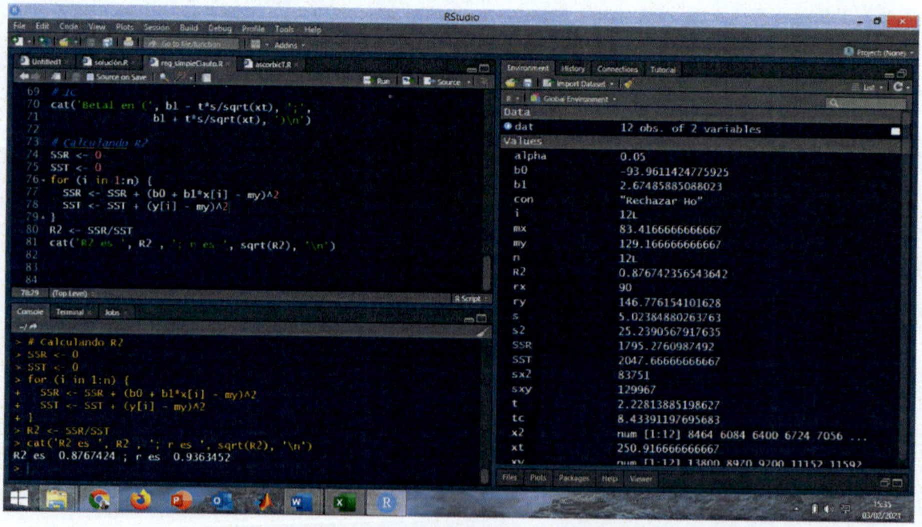

Fig. 3. Results in R-Studio Simple Linear Regression Model.

Given the results obtained in R, with B0 = −93.96 and B1 = 2.67 with an Alpha = 0.05; with two degrees of freedom, i = 12 L, so that the Simple Linear Regression Model is:

$$yi = -93.96 + 2.67xi + ei \tag{6}$$

$$i = 1, 2, 3 \ldots\ldots\ldots.12 \tag{7}$$

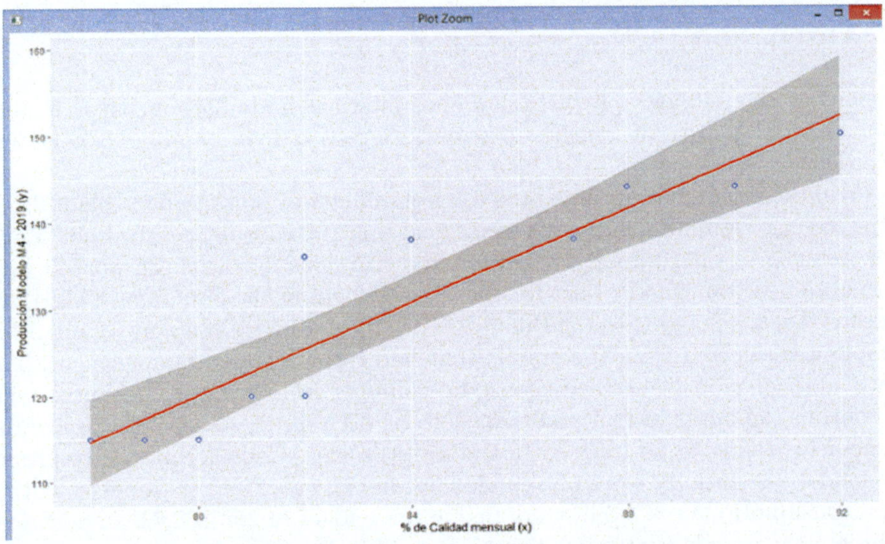

Fig. 4. Simple Linear Regression Model M4

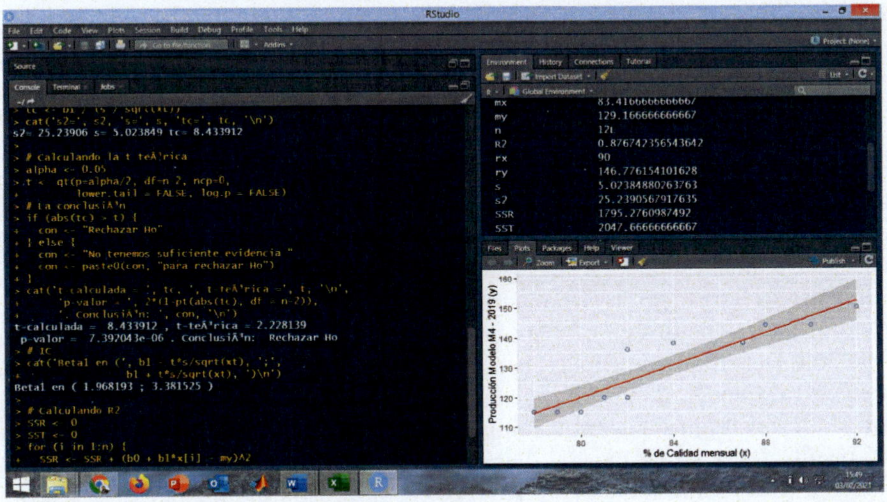

Fig. 5. R-Studio results

The pvalue data = 7.392043e−06 less than alpha, indicates that H0 should be rejected in favor of Ha, in addition, the value of R2 = 87.67% indicates that the model explains the variation in the production of the M4 model with this percentage.

4 Discussion

Reviewing the bibliography, studies on statistical and mathematical models applied in the area of textiles (leather), forestry, food, among others, are recorded.

[5] in the research that aimed to model thinning in Pinus caribaea plantations using linear programming as a mathematical tool and considering the budget and fuel assigned to the company as economic restrictions. For this, we worked with data obtained from the Project for the Management of the San Juan y Martínez Forestry Unit, belonging to the Pinar del Río Agroforestry Company - Cuba. The sample was selected from the stands that had recommended management. The mathematical model obtained plans the thinning in a total of 30 months with the optimal amounts of fuel and budget to be used for it, so as to maximize the number of stands to be thinned in the forestry unit. Considering the results of the model, the activity would be carried out in two years and six months, with a total consumption of 973.2 liters of fuel and a budget of 298,708.74. In addition, it orders the stands that must receive the silvicultural treatment, showing the type of thinning and its order, so that forestry activity is optimized.

[4] in their work that studied the supply chain taking as a reference the tire production sector in Ecuador, through the use of a Mixed Integer Linear Programming model. They concluded that: for the design of a statistical model for the supply chain; the determination of the amount of raw materials that must be delivered, as well as which suppliers must supply the plant, based on the demand for each product, was transcendental. Additionally, it should be taken into account that for the design of a model, different scenarios should be considered; In an optimistic scenario, the expansion of consumption areas is suggested due to the fact that several distribution centers converge in some sectors. The conservative model indicates that the distribution centers are optimal, while the pessimistic scenario requires the installation of more centers. of distribution. On the other hand, the design of the network is carried out, where the application of the Sole Distributor approach can be observed, since each distribution center covers certain areas, in this way it is possible to cover the demand of the company in all consumption areas, avoiding duplication of efforts and resources. The final result of the study can be described as the development of a mathematical model that allows identifying the best composition of the logistics network, analyzing different scenarios, so that these models can be adapted to different segments of the industry.

[12], in his research carried out with the aim of developing a mathematical model that is adjusted to the production planning of the leather sector in the Quisapincha parish of the Ambato canton. It concludes that: the investigation began because inadequate production planning was evidenced, later information was collected regarding the study variables with the use of scientific articles and books on production planning variables and mathematical models; with their respective fundamental categories. In addition, in order to collect information regarding the sector, surveys of current and potential customers were used. Subsequently, a correlation and multiple regression analysis was carried out. Through

the Anova analysis, it was possible to identify the variables that had the highest correlation with each other: i) control, ii) planning, iii) forecast and iv) production process; obtaining a Cronbach's alpha of 0.770; which helped to conclude that the reliability of the instrument used is good, with a tendency to excellent. Finally, the proposal was established to provide the sector with a mathematical model using the Excel Solver statistical tool to identify the resources needed per week and the optimal amount to produce in order to maximize profits.

[17], in the research that aimed to analyze the impact on continuous improvement and optimization of a production system through the implementation of Lean Manufacturing tools, as well as the changes that are generated in different companies through an instrument; concludes that: this is achieved using different research methods and techniques, such as documentary review of different literature, documentary analysis and data collection. Within the results, tables, and figures are obtained that show the efficiency of this tool, which proves its validity through success stories where it was implemented, as well as relevant information that could be used as a basis in companies that have not opted. For your application. Additionally, they state that within the contributions obtained from carrying out this research project is to easily identify the reasons, advantages, and applications of Lean Manufacturing in the different companies, in order to make the decision to establish this technique. Some limitations are also considered, such as the time required to obtain results, since at least two years are needed.

As can be seen in the results presented in the previous point and compared with similar research, companies that have implemented Lean Manufacturing tools as a work philosophy have achieved significant reductions in production costs, inventories, purchase, quality costs and lead time applying mathematical or statistical models.

This research presents a statistical model for production planning through the Lean Manufacturing methodology (Select, Order, Clean, Standardize, Discipline, Lean Structure, Production Planning), applying R-Studio, a Regression model is obtained Simple Linear.

5 Conclusions

With good management and application of Lean Manufacturing tools, better quality can be obtained in all operational areas.

For this particular case of the assembly company, by improving quality, production increased, for a goal of 90% quality, a production of approximately 147 units can be obtained.

Taking into account the level of quality generated by the plant, production can be planned, which can be determined with the Statistical Model of Linear Programming.

References

1. Bradshaw, M., Wood, L., Williamson, S.: Applying qualitative and quantitative research: a social impact assessment of a fishery. Appl. Geogr. **21**(1), 69–85 (2001)
2. Cabrera-Abad, K., Pinos-Úrgiles, P., Jara-Diaz, O., Duque-Córdova, L., Escobar-Segovia, K.: Ergonomic working conditions in workers under the modality of "homeoffice" due to a COVID-19 pandemic, in a bottling company in Ecuador. In: Garcia, M.V., Fernández-Peña, F., Gordón-Gallegos, C. (eds.) CSEI 2021. Lecture Notes in Networks and Systems, vol. 433, pp. 41–56. Springer, Cham (2022)
3. Caiza, G., Torres, F.I., Garcia, M.V.: Identification of patterns in the involvement of novice software developers in software testing processes. In: 2021 IEEE International Conference on Artificial Intelligence and Computer Applications. ICAICA 2021, pp. 378–382 (2021). https://doi.org/10.1109/ICAICA52286.2021.9498137
4. Campoverde, J.A., Romero, C.A., Naula, F.B., Loyola, D.M., Coronel, K.T., Jimenez, J.A.: Aplicación de un modelo matemático para el diseño de la cadena de suministro en el sector de neumáticos en Ecuador. Espacios 40 (2019)
5. Edua, B.R.C., Mendoza, Z.A.: Modelación del raleo mediante el uso de la Programación Lineal en plantaciones de Pinus caribaea Morelet de la Empresa Agroforestal Pinar del RÃo, Cuba. Arnaldoa **25**, 597–614 (2018). https://doi.org/10.22497/arnaldoa.252.25215
6. Cruzado, J.A.G., Chancusig-Chisag, J.C., Lozada, J.V.M., Acuña, A.J.A.: Proposal of a systemic model for integration of strategic planning in corporate level with balanced scorecard. In: Garcia, M.V., Fernández-Peña, F., Gordón-Gallegos, C. (eds.) CSEI 2021. Lecture Notes in Networks and Systems, vol. 433, pp. 71–88. Springer, Cham (2022). https://doi.org/10.1007/978-3-030-97719-1_4
7. Harrison, S., Bansee, J., Chowdary, B.V.: Improvement of service operations at a tyre servicing centre: application of lean manufacturing and lean digitisation. Adv. Eng. Res. **49**, 159–181 (2022)
8. López, S., Lema, F., Rosero, C., Sánchez, C., López, J., Tigre, F.: Management by integrated processes with biosafety parameters. case study SMEs manufacturing rest footwear in the province of tungurahua. In: Garcia, M.V., Fernández-Peña, F., Gordón-Gallegos, C. (eds.) Advances and Applications in Computer Science, Electronics, and Industrial Engineering CSEI 2021. Lecture Notes in Networks and Systems, vol. 433, pp. 107–123. Springer, Cham (2022). https://doi.org/10.1007/978-3-030-97719-1_6
9. Morales, L., Silva, D., Moreno, V., Collantes, S.: Symptomatology of musculoskeletal pain related to repetitive movements. Preliminary study "post-harvest in floriculture companies". In: Nummenmaa, J., Pérez-González, F., Domenech-Lega, B., Vaunat, J., Oscar Fernández-Peña, F. (eds.) CSEI 2019. AISC, vol. 1078, pp. 329–340. Springer, Cham (2020). https://doi.org/10.1007/978-3-030-33614-1_22
10. Motahari, R., Alavifar, Z., Zareh Andaryan, A., Chipulu, M., Saberi, M.: A multi-objective linear programming model for scheduling part families and designing a group layout in cellular manufacturing systems. Comput. Oper. Res. **151**, 106090 (2023)
11. Ojo-Gonzalez, K., Prosper-Heredia, R., Dominguez-Quintero, L., Vargas-Lombardo, M.: A model DevOps framework for SaaS in the cloud. In: García, M.V., Fernández-Peña, F., Gordón-Gallegos, C. (eds.) Advances and Applications in Computer Science, Electronics and Industrial Engineering. AISC, vol. 1307, pp. 37–51. Springer, Singapore (2021). https://doi.org/10.1007/978-981-33-4565-2_3

12. Ortiz, J., Teneda, W.: Modelo Matemático Para La Planificación De La Producción Del Sector Cuero En La Parroquia De Quisapincha. Ph.D. thesis, Universidad Tecnica de Ambato (2017)
13. Riofrio-Morales, M., Garcia, M.V.: Training virtual reality-based system for detection and simulation of motors failures. J. Phys.: Conf. Ser. **1983**(1) (2021). https://doi.org/10.1088/1742-6596/1983/1/012099
14. Comas Rodríguez, R., Simón-Cuevas, A., Sánchez Fleitas, N., García Lorenzo, M.M.: An ontology-based data management model applied to a real information system. In: Nummenmaa, J., Pérez-González, F., Domenech-Lega, B., Vaunat, J., Oscar Fernández-Peña, F. (eds.) CSEI 2019. AISC, vol. 1078, pp. 36–50. Springer, Cham (2020). https://doi.org/10.1007/978-3-030-33614-1_3
15. Salazar-Moya, A., Garcia, M.V.: Lot streaming in different types of production processes: a Prisma systematic review. Designs **5**(4), 67 (2021). https://doi.org/10.3390/designs5040067
16. Tomašević, I., Slović, D.: Lenses of lean in non-repetitive manufacturing: systematic literature review. In: Mihić, M., Jednak, S., Savić, G. (eds.) Sustainable Business Management and Digital Transformation: Challenges and Opportunities in the Post-COVID Era SymOrg 2022. Lecture Notes in Networks and Systems, vol. 562, pp. 490–508. Springer, Cham (2023). https://doi.org/10.1007/978-3-031-18645-5_31
17. Vargas-Hernández, J.G., Muratalla-Bautista, G., Jiménez Castillo, M.T.: Sistemas de producción competitivos mediante la implementación de la herramienta lean manufacturing. Ciencias Administrativas **11**, 81–95 (2018)

KPI'S Model Focused on the Evaluation of the Inventory Management of a Textile Company. A Case Study

Paul Moya-Carranza[1], Carlos Sánchez-Rosero[1], Freddy Lema[1], Christian Mariño[1], Jessica López[1], and César Rosero-Mantilla[1,2](✉)

[1] Universidad Técnica de Ambato, UTA, 180103 Ambato, Ecuador
{pmoya6516,carlohsanchez,fr.lema,christianjmarino, jp.lopez,cesararosero}@uta.edu.ec

[2] Escuela Politécnica Nacional, EPN, 170525 Quito, Ecuador
cesar.rosero@epn.edu.ec

Abstract. Small and medium-sized textile enterprises work to satisfy a changing market, which makes efficient inventory management difficult; therefore, several control methodologies have been developed to help to make this task easier. Though, only few researchers have sought to evaluate the performance of inventory management considering the multiple existing control methodologies and that none of them can be adapted one hundred percent to the characteristics of a company. In this paper, a model of key development indica-tors for the evaluation of the inventory management of a textile company dedicated to the manufacture of leisure footwear is carried out. For its development, the Fink methodology was used for the bibliographic research, with which 16 KPI's were obtained, 7 of which were selected through the application of weighting factors where the needs and characteristics of the company were considered. The model was applied to evaluate the company's performance in the last year, obtaining as a result that its inventory management is deficient, since no indicator obtained favourable values in accordance with the acceptable ranges of each one. However, thanks to the implementation of this model, the company knows its current situation, understands the aspects in which it must improve and has a tool that will allow it to verify its evolution if improvement actions are taken.

Keywords: SME's · textile company · KPI's · inventory · inventory management

1 Introduction

Over time, companies have determined that one of the most significant problems in their management is inventory control, and it is a problem that they have been dealing with forever and to which they have not yet found a solution that

M. V. Garcia and C. Gordón-Gallegos (Eds.): CSEI 2022, LNNS 678, pp. 508–519, 2023.
https://doi.org/10.1007/978-3-031-30592-4_34

is one hundred percent effective and applicable to all companies, because each company is totally unique and different [8]. But even so, inventory is necessary, since it is the factor that allows a company to: guarantee the availability of a product, satisfy the preferences and requirements of its customers, respond to market variations, or prevent any type of inconvenience related to its supply, its production line, its deliveries, among others [3,17].

However, those who suffer most from these consequences are small and medium-sized enterprises (SME's). Large companies have the economic and knowledge capacity to respond to this context, and thousands of companies have already developed methodologies based on high inventory turns and wide product ranges, which have brought excellent results [11,15]. Unfortunately, SME's cannot assimilate this behaviour due to their limited capacity in production acquisition processes and the high risks of maintaining production lines [14]; in addition, the globalized economy is also a huge disadvantage for these companies, most of which are trying to reach a local market [1].

It is for all these reasons that the application of inventory management systems has become so important at all levels of companies, because now, more than a tool, they are considered as a competitive strategy and a factor that determines the profitability and productivity of the company [4,8,16]. Several control methodologies have been developed so that every company can apply them in its management, which, although they are useful and generate great benefits [6,7], do not guarantee one hundred percent satisfactory results, since none of them can be completely adapted to the characteristics, needs and socioeconomic context of a company [10,13].

In view of this reality, there are multiple investigations focused on the textile industry, which have attempted to develop new methodologies or modify some existing ones to better respond to the reality of each company. All of them are based on covering as many variables as possible, from the most common and important ones such as: waiting times, demand, inventory levels, safety stocks [2,12]; to specific variables such as: distribution costs, supply chain, demand forecasting, service restrictions, among others [9]. But what guarantees the success of these methodologies, and of inventory management in general, is their continuous monitoring to ensure that they have a positive impact [9].

Nowadays, Inventory management is fundamental in companies to ensure their competitiveness in the market and there are several applicable control methodologies, so it is likely that every company has implemented one or is in the process of applying it; and therefore, it is reasonable that several research are focused on collaborating with it. Nevertheless, very few research have ventured into evaluating the performance of inventory management, before and after applying any methodology. This would allow to know the initial performance of a company and to evaluate it constantly to see if the improvement actions implemented are being effective or not; and even to evaluate the context in which a methodology is more effective than another.

That is why the purpose of this research is to pro-pose a KPI's model that allows the evaluation of the performance of the inventory management of a textile company, as a first step to the development of a tool that records its evolution over time and provides accurate and relevant information for decision making by the company, when it comes to make changes.

2 Methodology

As a first step, it was proposed to determine the focus of the bibliographic re-search by applying Fink's methodology [5], thus defining that the research would be focused on inventory management, the information most commonly used in control methodologies and the evaluation of their performance through key performance indicators (KPI's). For the search, only journal or conference articles in the English language with a publication date equal to or greater than 2015 and which, of course, are related to the areas of knowledge related to the subject matter addressed were considered, obtaining as a result 24 documents that support this research. From the information collected, 16 indicators were extracted, which are among the most relevant and necessary for the recognition of the performance of inventory management and for the supply of inventory control methodologies applicable to SMEs. However, to achieve these, it is necessary to know the demand, the inventory level, and the storage costs; data that can be obtained from the accounting records of the company's activities. Given the characteristics of the company and its needs, the indicators were evaluated using the weighted factor method [12,18]. Seven were selected for the KPI model, which were those that obtained a total score greater than or equal to 7.0 after having been evaluated under four factors: the relationship with inventory control (F1), the relevance for the company (F2), its contribution to an inventory management methodology (F3) and its feasibility of implementation (F4). These indicators are detailed in Table 1.

Table 1. Proposed indicators for the model.

Name	Formula	
Duration of inventory	$\frac{Final\ inventory}{Average\ sales} * 30$	(1)
Inventory turnover	$\frac{Cost\ of\ goods\ used}{Average\ inventory\ of\ goods}$	(2)
Days of supply	$\frac{1}{Inventoryturnover} * 30$	(3)
Storage cost per unit	$\frac{Storage\ costs}{Number\ of\ units\ in\ storage}$	(4)
Storage cost per square meter	$\frac{Warehouse\ operating\ cost}{Warehouse\ area} * 100$	(5)
Obsolescence	$\frac{Number\ of\ times\ a\ product\ enters}{Rotation}$	(6)
Covering rate	$\frac{Average\ stock}{Average\ demand}$	(7)

3 Results and Discussion

3.1 KPI's Model Focused on Inventory Management

The implementation of a KPI's model, focused on the evaluation of inventory management performance, is a useful tool to know how efficient inventory management is and whether it is within the appropriate parameters according to each of the indicators. It also allows to have a record with which to compare any type of modification or improvement implementation.

For the application of the model in the company, some relevant considerations were considered regarding its operation and the operating conditions of the last year. Among the operating considerations are that the company works on the basis of previous purchase orders, so it does not record finished product inventory; its production line and product delivery dates are planned in advance and take into account previous orders, which means that the company does not handle in-process product inventory, since all material exits from a process are planned to match their entry into the subsequent process until their respective dispatch; The company only manages raw material inventory corresponding to fabrics; and finally, the company does not have any inventory management methodology, since it only manages it empirically.

Regarding the considerations related to operating conditions in the last year, it is worth mentioning that the study was conducted for 2020, a year in which the company recorded irregular movements due to the sanitary emergency caused by COVID-19. Although the company recorded a regular level of demand and even a little higher in certain months because people had to take refuge in their homes and chose to avoid entering with normal footwear, which triggered a high demand for this rest footwear intended for indoors.

Taking these considerations into account, the following are the first indicators extracted from the company's accounting records:

Demand. This indicator was derived from monthly sales, which in this case corresponds to the number of pairs produced. On average their demand is around 28,200 pairs per month or 338,400 pairs per year.

Inventory Level. The company records the monthly inventory level of raw material, which in monetary value is approximately $ 431,428.10 per month, although its value increases over time. In this analysis, the company divides its raw material into fashion fabrics and standard fabrics, the latter subdivided into high, medium, and low rotation fabrics. By analysing the Kardex cards, a tool used by the company, the current inventory levels of each fabric are broken down according to their category as shown in Fig. 1.

Storage Costs. This is made up of three types of costs: management, infrastructure and operational. The details of this information can be found in Fig. 2.

Based on the first 3 indicators, the 7 selected for the model can be calculated, the results of which are shown below:

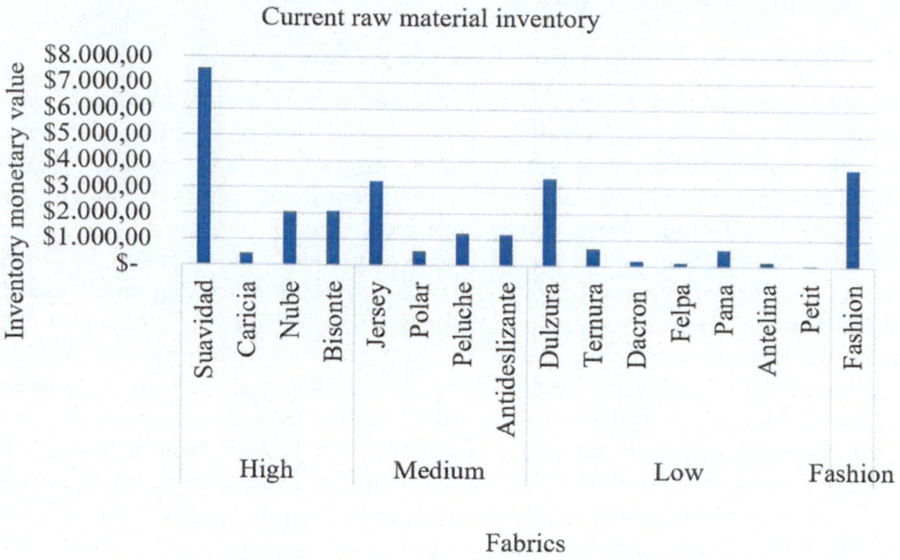

Fig. 1. Current inventory level record in meters of fabric.

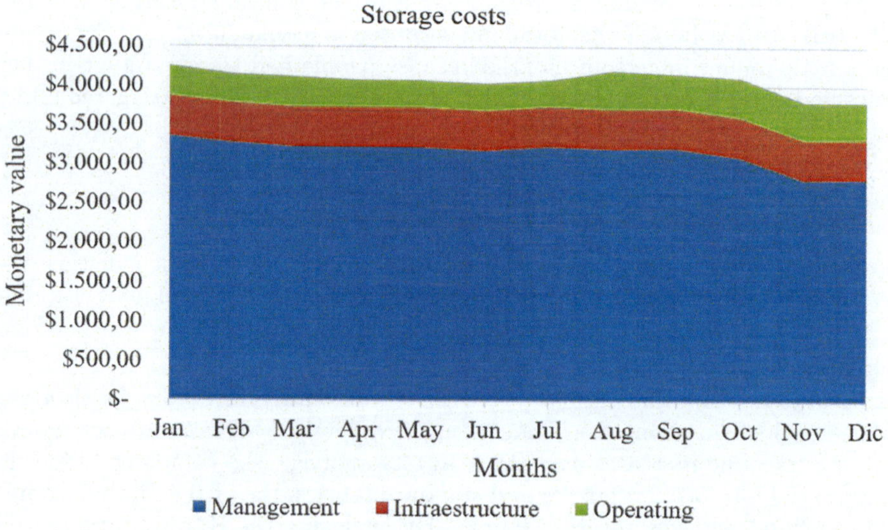

Fig. 2. Record of storage costs.

Duration of Inventory. For the calculation of this indicator, an analysis of the raw material was considered in general and for each of the types of fabrics; the results are shown in Fig. 3.

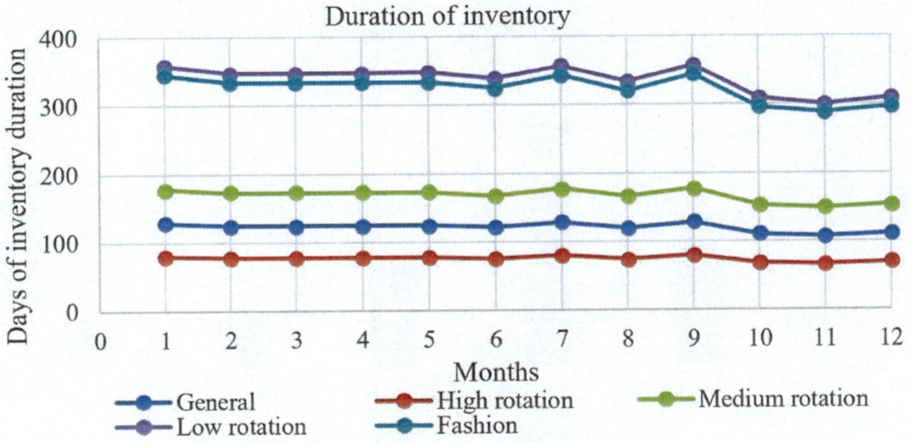

Fig. 3. Duration of inventory in days.

As a result, it is found that the inventory in general remains in storage for an average of 121 days, equivalent to approximately 4 months; which is a very high value for this indicator, as the optimum would be 1 month, taking into account that the production capacity is 200 pairs per hour, its production orders are 100 pairs on average and that the waiting time for the delivery of raw materials is 15 days.

Inventory Turnover. For the calculation of this indicator, the same considerations were taken as for the previous indicator, the results of which are shown in Fig. 4. The results of this indicator show that the company has an overall inventory turnover of 2,973 times, which means that its stock was fully used and has been replenished in approximately 4 months (12 months/2,973), a value that is very low in relation to the annual demand of approximately 340000 pairs; but it is consistent with the results of the inventory duration indicator.

Days of Supply. This indicator is the inverse of inventory turnover, so the exact same considerations were taken for its calculation. The results indicate that the stock stored by the company would run out in approximately 150 days, which would be equivalent to approximately 5 months, a fairly high value considering that the waiting time for restocking is only 15 days. In the individual analysis of the types of fabric, no favourable results were obtained either, as the values range from 93 days in the best of cases, corresponding to high rotation fabrics, to 420 days, in the case of low rotation fabrics. The graph of the results can be seen in Fig. 5.

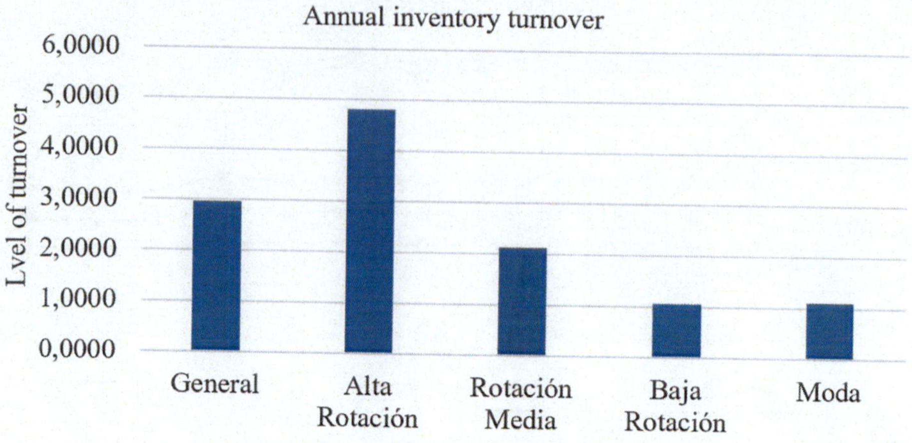

Fig. 4. Annual level of inventory turnover.

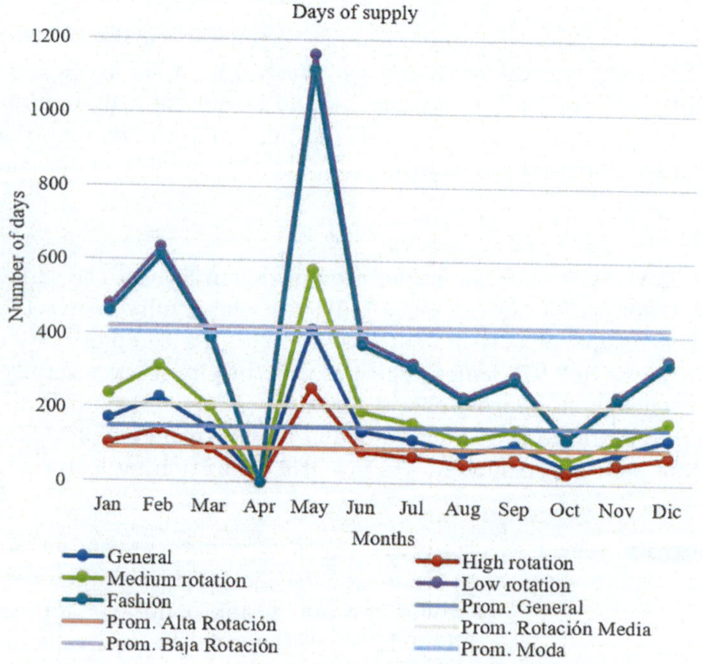

Fig. 5. Days of supply.

Storage Cost per Unit. For its calculation, the storage costs obtained previously and the monthly inven-tory level in meters for the last year were considered. Their results can be seen in Fig. 6.

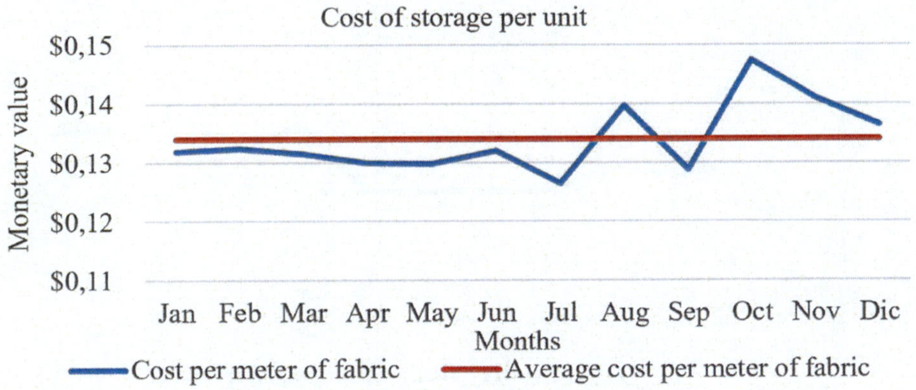

Fig. 6. Cost of storage per meter of fabric.

As a result, it costs the company an average of $0.13 per month for each metre of fabric stored. And considering that it is presented in rolls of 50m each, it costs the company an average of $6.70 per month for each roll of fabric.

Storage Cost per Square Meter. For its calculation, formula 11 was applied, considering the operating costs, broken down by storage costs, and the 204.07 m2 that the company has set aside for warehousing. The results are shown in Fig. 7.

The result of this indicator shows that it costs the company an average of $1.99 per month for each square meter of floor space available for inventory storage.

Obsolescence. This indicator was analysed for the general inventory and for each of the types of fabric. The results obtained can be seen in Fig. 8.

Due to the fact the results of this indicator are above the optimal range (from 0.5 to 1.5) the company's inventory has a high obsolescence rate, which means that a large part of the raw materials can no longer be used to produce recreational footwear, either due to physical deterioration, insufficient material or because they are no longer to the liking of customers.

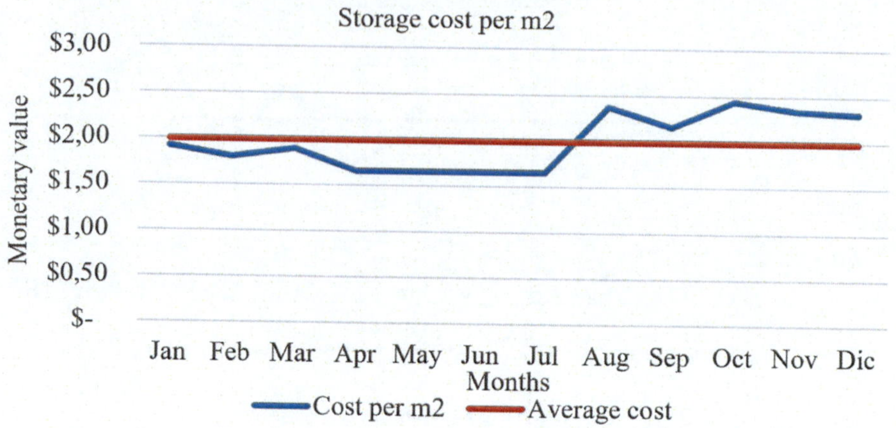

Fig. 7. Storage cost per square meter.

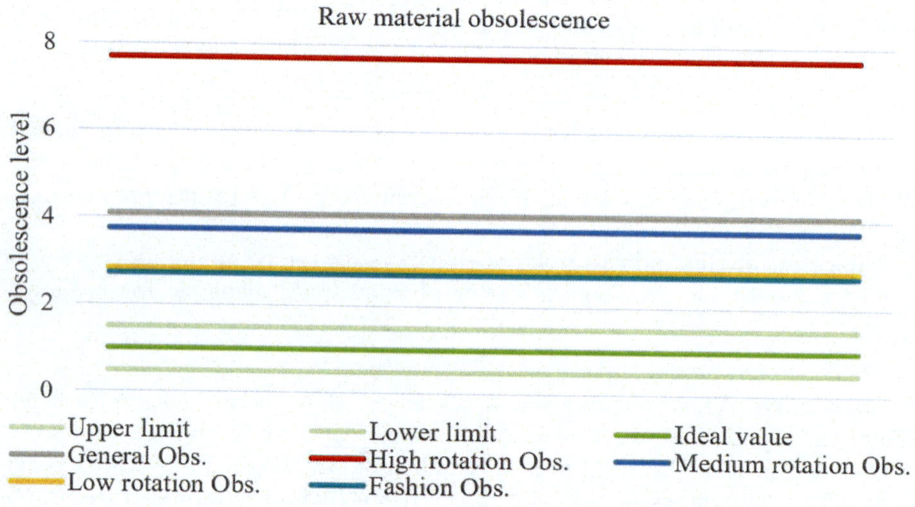

Fig. 8. Obsolescence level.

Covering Rate. Like other indicators, this analyses the inventory in general and specifically for each type of fabric. Its results are shown in Fig. 9.

According to the results obtained, there is such an amount of inventory that it could cover the production of approximately 4 months, which is an extremely long time. The ideal values that would be expected for this case study would be between 15 days, which is the time it takes the company to restock the product, and a month, which could be considered acceptable.

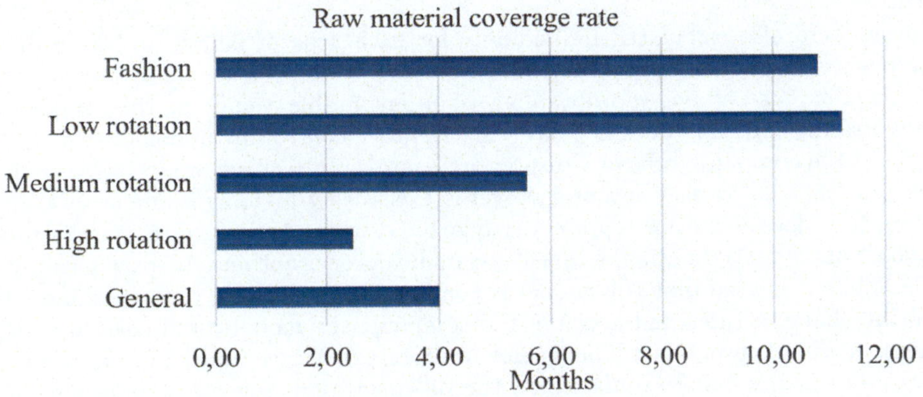

Fig. 9. Covering rate.

4 Conclusions

Textile SMEs are the most problematic when it comes to inventory, as due to the nature of their market they are constantly confronted with changing factors such as fashion, trends, seasons, and consumer preferences. Therefore, several re-search and control methodologies have been developed that are applicable to these types of companies. However, there is very little research focused on the evaluation of a company's inventory management and the performance of the methodologies applied, as not all companies will need or have the same effectiveness with the same methodology.

In view of this, this document proposes a model for the continuous evaluation of the performance of inventory management based on the calculation of key development indicators. This model is applied to a textile company dedicated to the manufacture of slippers and, in response to its characteristics and needs, 7 indicators are determined for the KPI model as such, which are calculated from the information gathered from the company.

For the application of the model, it was considered that the company only generates raw material inventory thanks to its production planning and its policy of working under purchase orders. This led to the analysis being focused purely on the storage of raw material, which comprises only fabric and which is classified, according to its purchase and consumption behaviour, into standard fabrics: high, medium, and low rotation; and fashion fabrics.

With these considerations, it was obtained that the company has a deficient inventory management, as for none of the indicators it obtained favourable results. Their results are that they have an inventory that on average remains in stock for 4 months, has a turnover of 2.97 times a year, is capable of supplying production for 150 days in case it is not replenished, represents a storage cost of \$0.13 per month per metre of fabric or \$6.70 per month per roll of fabric, generates a storage cost per square metre of \$1.99 per month, has a level of obsolescence of 4.10 and has a coverage rate of 4 months. Although some indicators were also calculated individually for each type of fabric, no favourable values were obtained in any case.

In this case study, the obtaining of unfavourable values in the model is because the company does not have an inventory control methodology, this has caused the company to have a disproportionate supply of raw materials, as it is evident that its average inventory level is excessively high, as its monthly consumption does not even represent a quarter of this. And since its purchasing behaviour is not a function of inventory, its inventory continues to grow steadily.

While it is true that this model is composed of indicators that are adapted to the characteristics and needs of the company, the model could cover a wide number of indicators that would allow a more exhaustive evaluation according to the company being evaluated. In the future, various inventory management methodologies applied could be evaluated and an analysis could be made of those whose application gives the most favourable results in small and medium-sized textile companies. For future research, the opportunity remains open to apply the KPIs obtained to micro, small and medium-sized textile companies in the sector to strengthen the model with a greater amount of information.

Acknowledgments. The authors thank the Technical University of Ambato for the support provided during the execution of this work and the VLIR-UOS for the funding within the framework of the research project called "Incorporating sustainability concepts into the management models of Micro, Small and Medium Textile Companies (SUMA)".

References

1. A me-based rough approximation approach for multi-period and multi-product fashion assortment planning problem with substitution. Expert Syst. Appl. **84**, 127–142 (2017). https://doi.org/10.1016/j.eswa.2017.04.051
2. Inventory control model using exponential smoothing control chart. J. Phys. Conf. Ser. **1217**(1), 012095 (2019). https://doi.org/10.1088/1742-6596/1217/1/012095
3. Chinello, E., Lee Herbert-Hansen, Z.N., Khalid, W.: Assessment of the impact of inventory optimization drivers in a multi-echelon supply chain: case of a toy manufacturer. Comput. Ind. Eng. **141**, 106232 (2020). https://doi.org/10.1016/j.cie.2019.106232
4. Espinosa, R.V., Soto, M., Garcia, M.V., Naranjo, J.E.: Challenges of implementing cleaner production strategies in the food and beverage industry: literature review. In: García, M.V., Fernández-Peña, F., Gordón-Gallegos, C. (eds.) Advances and Applications in Computer Science, Electronics and Industrial Engineering. AISC, vol. 1307, pp. 121–133. Springer, Singapore (2021). https://doi.org/10.1007/978-981-33-4565-2_8
5. Fink, A.: Conducting Research Literature Reviews. From the Internet to Paper, 3rd ed. SAGE, Publications Inc., Thousand Oaks (2010)
6. Garcia, C.A., Montalvo-Lopez, W., Garcia, M.V.: Human-robot collaboration based on cyber-physical production system and MQTT. Procedia Manuf. **42**, 315–321 (2020). https://doi.org/10.1016/j.promfg.2020.02.088

7. Gómez, C.: Proyecto De Factibilidad Para La Creación De La Finca Agroturística Gomcomp Ubicada En San Jorge, Cantón Santa Clara, Provincia De Pastaza, Auspiciada Por Gad-Municipal De Santa Clara, 2017. Ph.D. thesis (2018)
8. Islam, M.T., Azeem, A., Jabir, M., Paul, A., Paul, S.K.: An inventory model for a three-stage supply chain with random capacities considering disruptions and supplier reliability. Ann. Oper. Res. **315**(2), 1703–1728 (2022). https://doi.org/10.1007/s10479-020-03639-z
9. Li, G., Kang, Y.C., Xu, G.: Optimal decision for fashion supply chains with service level constraint and controllable lead time. Math. Probl. Eng. **2015**, 1–9 (2015). https://doi.org/10.1155/2015/758494
10. Luis, J., Valenzuela, M.: Programa de Pronóstico e Inventario para la Elaboración de un Plan de Producción Programa de Pronóstico e Inventario para la Elaboración de un Plan de Producción Jorge Luis Maurer Valenzuela Roberto Andrés Oviedo Espín. Universidad San Francisco De Quito (2016)
11. Ma, J., Xie, W.: Application and research on the complex dynamics behaviour under price-discount promotion policy in inventory control system. Int. J. Logist. Res. Appl. **19**(6), 520–536 (2016). https://doi.org/10.1080/13675567.2016.1168378
12. Mamdouh, M., Abido, M.A., Hamouz, Z.: Weighting factor selection techniques for predictive torque control of induction motor drives: a comparison study. Arab. J. Sci. Eng. **43**(2), 433–445 (2018). https://doi.org/10.1007/s13369-017-2842-2, http://link.springer.com/10.1007/s13369-017-2842-2
13. Montenegro, J., Ayala, P., Ibarra, A.: Predictive dynamic matrix control (DMC) for ball and plate system used in a Stewart robot. In: Garcia, M.V., Fernández-Peña, F., Gordón-Gallegos, C. (eds.) Advances and Applications in Computer Science, Electronics, and Industrial Engineering. CSEI 2021. LNNS, vol. 433, pp. 237–253. Springer, Cham (2022). https://doi.org/10.1007/978-3-030-97719-1_14
14. Muchaendepi, W., Mbohwa, C., Hamandishe, T., Kanyepe, J.: Inventory management and performance of SMEs in the manufacturing sector of Harare. Procedia Manuf. **33**, 454–461 (2019). https://doi.org/10.1016/j.promfg.2019.04.056
15. Nayak, R., Singh, A., Padhye, R., Wang, L.: RFID in textile and clothing manufacturing: technology and challenges. Fash. Text. **2**(1), 1–16 (2015). https://doi.org/10.1186/s40691-015-0034-9
16. Park, J., Nam, K.: Group recommender system for store product placement. Data Min. Knowl. Discov. **33**(1), 204–229 (2018). https://doi.org/10.1007/s10618-018-0600-z
17. Sun, Y., Wissel, J., Jackson, P.L.: Separation results for multi-product inventory hedging problems. Ann. Oper. Res. **237**(1-2), 143–159 (2016). https://doi.org/10.1007/s10479-013-1473-6
18. Wang, R., Zhao, H., Wu, Y., Wang, Y., Feng, X., Liu, M.: An industrial facility layout design method considering energy saving based on surplus rectangle fill algorithm. Energy **158**, 1038–1051 (2018). https://doi.org/10.1016/j.energy.2018.06.105

Facility Layout Proposal for a Tannery, Evaluated by the Simulation Software-Flexsim

Fabricio Calapiña, Daysi Ortiz(✉), Alex Pazmiño, and Israel Naranjo

Technical University of Ambato, Ambato, Ecuador
{ocalapina0007,dm.ortiz,apazmino4147,ie.naranjo}@uta.edu.ec

Abstract. The present paper shows the development of Facility Layout Design for the Tenería San José Cía. Ltda. on a new building using decision making tools, facility layout and manufacturing simulation. The layout alternatives were proposed and analyzed with the Systematic Layout Planning (PLC) method for being later assessed through the process simulator Flexsim, resulting a distribution that generates a saving of 35,50% in the distance walked and a reduction of 60,13% in materials transportation costs, the average level of work in process decreases to 2892 leather bands, finally the utilization of the bottleneck machine shows an increase of 1,11% while production capacity increases by 1.62%.

Keywords: Facility layout · SLP · simulation in Flexsim

1 Introduction

Currently the business presents difficult challenges and to remain in the market must necessarily increase their efficiency, quality and delivery speed [1], the goal of a production facility is to manufacture products with both minimum costs and maximum benefits [2] a feasible path to achieve this goal is the optimization of the facilities design [3], earlier this theme had been ignored, but due to the increased competition, the manufacturing industries have been forced to implement potential strategies focused on the plant distribution design in a manner that will optimize the production process and the organizational effectiveness [4], the layout design is one of the important factors that has a significant impact on productivity, profitability and performance of any industry no matter the category and size of the plant [5,6], by optimizing the facility design, the noticeable improvements are efficiency in the use of spaces, reduction of time transport and increase in the performance of assets [3,7].

The company Tenería San José Cía. Ltda. currently located in Ambato City, Ecuador, it is a plant that processes bovine skins intended mostly for footwear, clothing and leather goods; this is involved by several problems attributed to the facilities design, to mention a few: the absence of marked areas, transit and mobility routes congested with work-in-process, spaces that are not evenly distributed, excessively long distances between areas especially in the sections of

M. V. Garcia and C. Gordón-Gallegos (Eds.): CSEI 2022, LNNS 678, pp. 520–531, 2023.
https://doi.org/10.1007/978-3-031-30592-4_35

conditioning and finishes; beyond these dilemmas, the building is built in a residential area, forcing it in the future to seek a new site for its operation; the present research proposes to carry out a study focused on the physical distribution of the company's facilities projected in a new building as a long-term strategy maintaining technological and human resources.

This study focuses on Fendi, Nubuck, Vitello and Nuvola products that represent 81.18% of the sales of the third quarter of 2019; the skins to be transformed into finished leather go through five working sections, ribera, tanning, retanning, conditioning and finishing; the ribera stage aims to remove the excess of hair, fats and bait; the tanning stage converts the skin into leather by the action of chemical processes with chrome; the retanning stage provides basic characteristics to the leather such as caliber, strength and color; the conditioning stage prepares the leather belts for the final processes through mechanical processes such as drying and stretching; the finishing stage provides pigments, surface finishing by means of painting machines, presses and planers.

The Systematic Layout Planning (SLP) is a procedure to set the layout of the place of work that is based on the logical relationship of its activities, it is achieved by placing the areas with high frequency close to each other [8,9], this allows to generate design alternatives aligned with the objectives and corporate strategies of the institution [10], previous research on SLP developed by Da Silva shows that this method is still valid and effective for making changes in the design of both new and existing facilities, reducing travel distances, material movement costs and production time [11].

The simulator processes Flexsim is a technological tool that is used to give a solution to problems of manufacturing aimed at the optimization of processes [12], with SLP method alternatives for improvement are generated with a qualitative approach, while with Flexsim quantitative data is obtained without the need to physically implement, which serve later as the basis for the selection of the design proposal [13] in the research presented by Zeng, Flexsim software is used to evaluate the impact before and after the improvements implemented in the facilities design [14].

2 Methodology

The procedure used to determine the facilities design for the new production plant of the company Tenería San José consists of three steps, analysis of the current situation, design of facilities based on the SLP method, evaluation, and selection of alternatives of using Flexsim software. The simulation program was used as an analytical tool to inform the decision making in the selection supported by quantitative data [15].

2.1 Analysis of the Current Situation

Observations and preliminary studies are necessary to identify and analyze the current conditions of the plant facilities design; process diagrams were used to describe in detail the sequence of activities, the type of operation performed, the time used for each of these and the distances between workstations [16]; the development of the study of times and movements was used to obtain the standard times used later for the generation of random variables used as input information in the simulation models, as well as to determine the theoretical production capacity.

2.2 Design of Distribution Proposals Based on SLP Method

With SLP method also known as the Muther method [17], the flow of material is expected to be faster and with the lowest cost, carrying a logical order of activities. The new design was determined based on the following stages [15].

Stage I: Surface Requirement Estimation. At this stage the recommended area for the construction of the new production plant building was calculated; this area is divided into two, the production area occupied by machinery and waiting sites, the auxiliary area intended for warehouses, toilets, nursing, among others. The production area is calculated accord-ing to Guerchet's criteria while the auxiliary area is evaluated together with the management through checklists.

Stage II: Determination of the Place Where the New Facility Will Be Built. At this point, the optimal location for plant relocation was determined, in accordance with the management this should be located in an industrial zone, the Ambato Industrial Park and the Santa Rosa Industrial Park are the proposed alternatives; for the selection we used the Analytical Process Hierarchical (AHP) and the software Expert Choice, which were evaluated under the criteria detailed in Fig. 1.

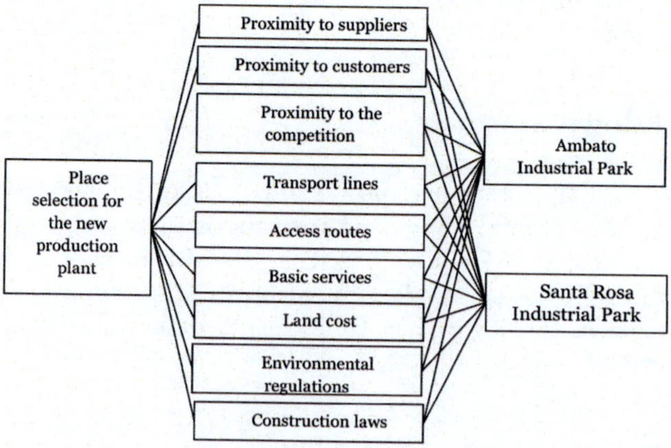

Fig. 1. Facility location

Stage III: Determine the Facilities Design of the Company. The design proposals were based on the level of relationship between the activities of the production process and the limitations attributed in the current plant distribution identified in the analysis stage; the degree of relationship was determined with the following criteria: process sequence, type of work and safety, type of effluent, hygiene and aesthetics, sharing of resources and information.

Stage IV: Obtaining Results from Proposals. To obtain the results, simulation models were created that represent the current and proposed scenario of the plant facilities design through the Flexsim 2019 student version software, in these the behavior of the variables shown in Table 1 was analyzed, the cost attributed to the material transport equipment (operators) was based on the unified basic salary 2020 in Ecuador.

Table 1. Variables to experience.

Variable	Meaning	Equation
Total distance travelled	Back and forth distance by transport operators	$DR = \sum_{i=0}^{n} RcO_i + \sum_{i=0}^{n} RsO_i$ (1)
Materials transportation cost	Cost attributed to transport equipment (operators)	$CT = \left(\sum_{i=0}^{n} TRcO_i + \sum_{i=0}^{n} TRsO_i \right) * CO_i$ (2)
Production capacity	Units produced by the model	-
Use of machinery	Percentage of use of machinery	$Um_i = \frac{Tsetip_i + Tprocessing_i}{28800\ seg * simulated\ days} * 100\%$ (3)
Work in process	Unfinished units held in queues and temporary storage	-

2.3 Design of Distribution Proposals Based on SLP Method

At this stage, the indicators obtained in the simulation that helped as analytical support for the selection of the alternative with the best score were evaluated.

3 Results

3.1 Assessment of the Current Situation

The manufacturing process for obtaining the finished leather has two stages, the first is the wet section that involves the processes of soaking, fleshing, tanning, retanning and drying; the second stage considered the dry section embeds the conditioning and finishing processes, these finishing processes are different for each product.

The time required for processing 2 batches of 150 skins in the wet area is 104.97 h with a covered distance of 11,987.30 m. In the dry area, to the processing of 100 bands of Fendi leather is used 13,17 h with 403,08 m in distance, for the Nubuck type is necessary 2,19 h and 161,27 m in distance, the Vitello type is used 11,71 h and 404,46 m in distance, the Nuvola type requires a time of 5.24 h and 248,31 m in distance.

The process that sets the work pace is vacuum drying with a standard cycle time 1.55 min/belt, the theoretical production capacity is 309 leather belts daily.

3.2 Area Requirement

The estimated production area is 2,040.64 m^2 which entails an increase of 196.08 m^2 in relation to the current installation, which is mostly destined to the ribera area. The resulting auxiliary surface is 1,430.86 m^2 where it presents an increase of 75.30 m^2 compared to the current installation due to the increase of space in the warehouse of raw material, finally the total basic surface for the construction of the building is 3,471.50 m^2.

3.3 Facilities Location

Ambato Industrial Park (PIA) is the best place for the construction of the new production plant with 56.40% acceptance with an inconsistency index equal to zero shown in Fig. 2, the soil occupation coefficient (COS) of this place is 60% which the actual size of the land amounts to 5,785 m^2.

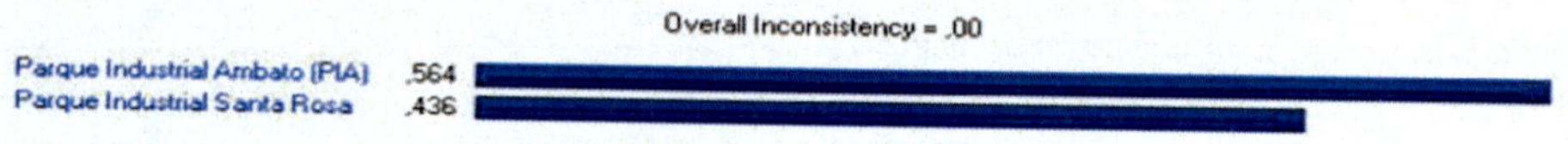

Fig. 2. Overall Inconsistency

There is a reduction in space of 28.70% compared to the area of the current land, this is mainly due to the elimination of unused spaces to those that were used by maquila.

3.4 Facilities Layout Proposal

The objective of the proposals is to shorten the transport distances, for this there are two alternatives to improve the layout that meet the level of relationship between operations (see Fig. 3); in the current distribution there are several practical limitations to consider in the new production plant, these have a favorable impact in terms of the flow of materials, development of complementary activities and effluent discharge, these are listed below:

- Place the stripping and splitting machines near an access door to relieve solid effluent discharge activities.
- Place the sanding and polishing machines on an adjacent wall to the maneuver yard so that the dust deposit resulting from these processes does not interfere with other operations.
- Install a skin transfer mechanism between the ribera drums and the fleshing machine.

The surface requirement for each department is shown in Table 2.

Proposal 1 is shown in Fig. 4, the departments as warehouses of inputs and chemicals are at the entrance of the facility, the flow of material resembles an "U" that goes from left to right, from top to bottom, the offices are on the top floor above the finishing area.

Proposal 2 is shown in Fig. 5, in which the auxiliary departments are placed in the lower right, the flow of material resembles an "U" that goes from right left from top to bottom, in the same way the offices are located on the top floor above the finishing and packaging area.

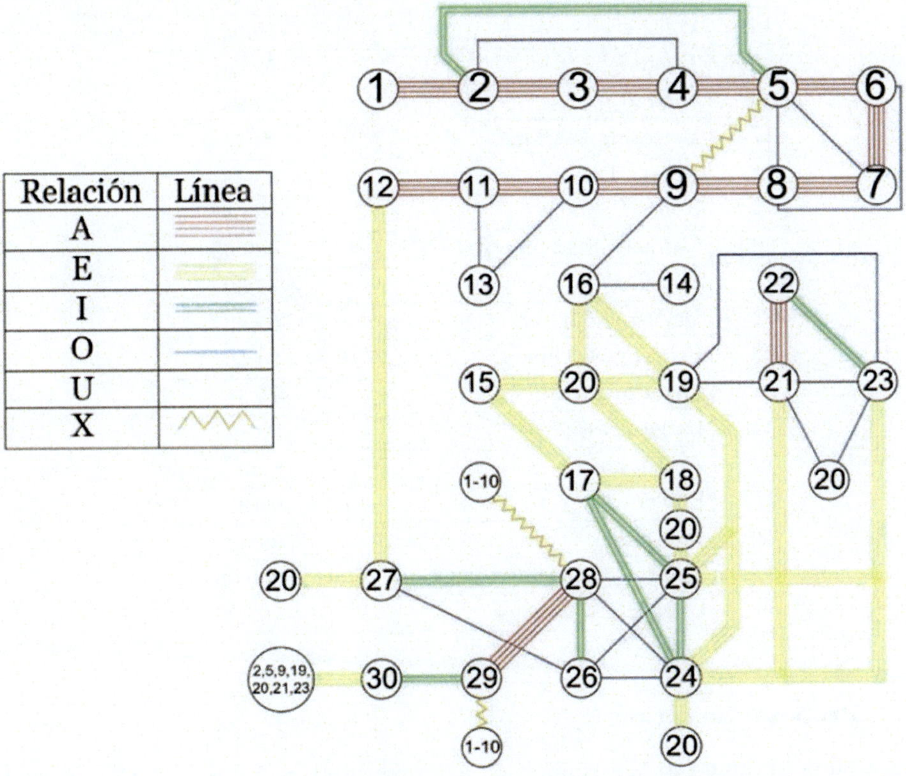

Fig. 3. Relationship diagram

Table 2. Space required for each department.

Code	Department	Surface [m^2]
1	Raw Material Warehouse	173.60
2	Ribera drums	78.22
3	Debarker	46.93
4	Divider	107.77
5	Tanning drums	71.02
6	Drainer	64.76
7	Classified area WB	69.40
8	Reducer	21.52
9	Retanning drums	53.28
10	Decoiler	20.68
11	Vacuum and Air Drying	107.53
12	Mollisa	34.38
13	Wetting	27.13
14	Polisher	19.35
15	Toggling	99.83
16	Sander	72.95
17	Battering Machine	34.31
18	Rotary Press	23.48
19	Roller Pigment	169.32
20	Carousel Pigmentator	489.65
21	Painting	32.39
22	Drying cabin	52.40
23	Printing Machine	24.50
24	Medium Press	15.31
25	Large Press	20.82
26	Acetinator	6.53
27	Planer	9.19
28	Finishing	88.58
29	Finished materials Warehouse, offices	390.00
30	Chemical warehouse	152.90

3.5 Proposal Selection

The results obtained in the simulation are presented in Table 3, the simulation time is 3 months.

The two proposals compared to the current model present an improvement in all measured variables, proposal 1 reduces 72,075.17 m/month representing

Table 3. Space required for each department.

Variable	Current	Proposal 1	Proposal 2
Distance travelled [m/month]	203,024.73	130,949.56	138,498.96
Material transportation cost [$/month]	320.92	125.95	137.15
Production Capacity [bands/day]	309.00	314.00	311.00
WIP [average bands]	3,030.00	2,892.00	2,906.00
Utilization rate [%] (Bottleneck Machine)	91.33	92.44	91.94

Fig. 4. Proposal 1

a decrease of 35.50%. Proposal 2 reduces 64,525.77 m/month representing a reduction of 31.78%. The materials transportation cost in proposition 1 reduces $192.97/month representing a reduction of 60.13% while proposition 2 reduces $183.77/month representing a reduction of 57.26%. The level of work in progress in proposal 1 is reduced by an average of 138 leather bands, in proposal 2 by an average of 124 leather bands.

The production capacity in proposal 1 increases 5 leather bands per day representing an increase of 1.62% while proposal 2 increases 2 leather bands per day representing an increase of 0.64%, finally the use of the vacuum drying machine considered the bottleneck increases its use by 1.11% which is equivalent to a time of 1 h 46 min 33.6 s per month in proposal 1, while in proposal 2 it

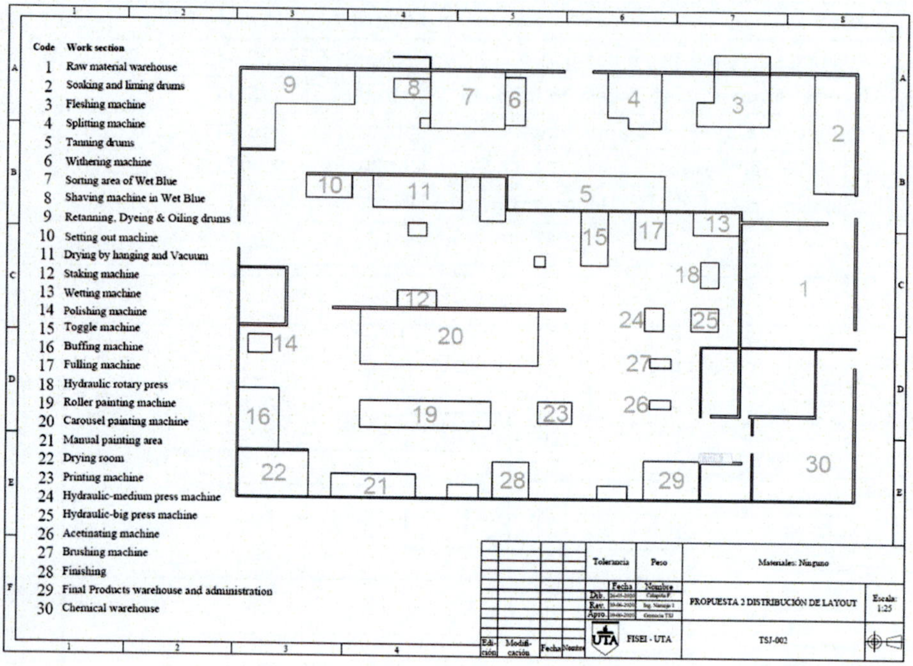

Fig. 5. Proposal 2

Fig. 6. Simulation Model - Ribera área.

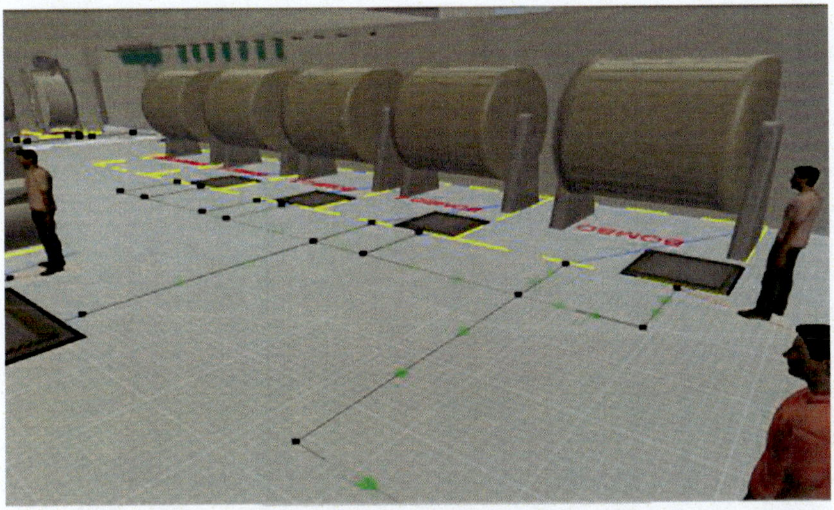

Fig. 7. Simulation Model - Tanning área.

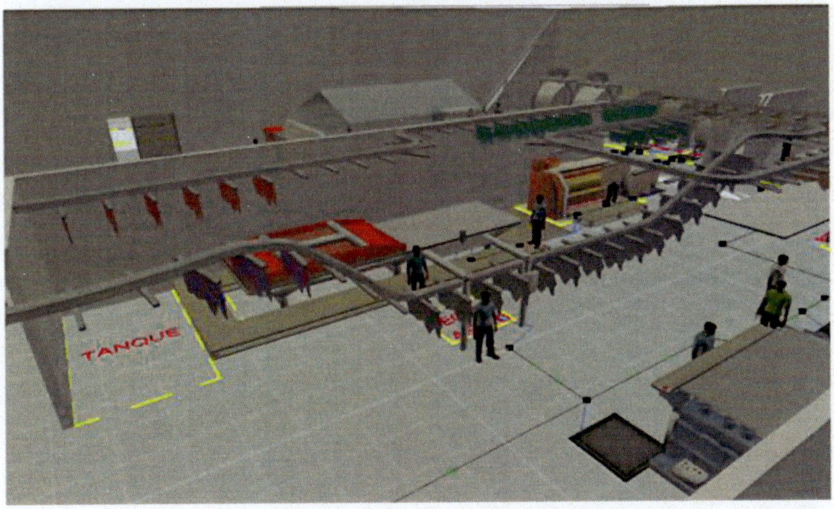

Fig. 8. Simulation Model - Conditioning área.

increases by 0.61% which is equivalent to a time of 0 h 58 min 33.6 s. Figure 6 to 9 shows views of the proposed model 1 workspace in the simulation environment (Figs. 7, 8).

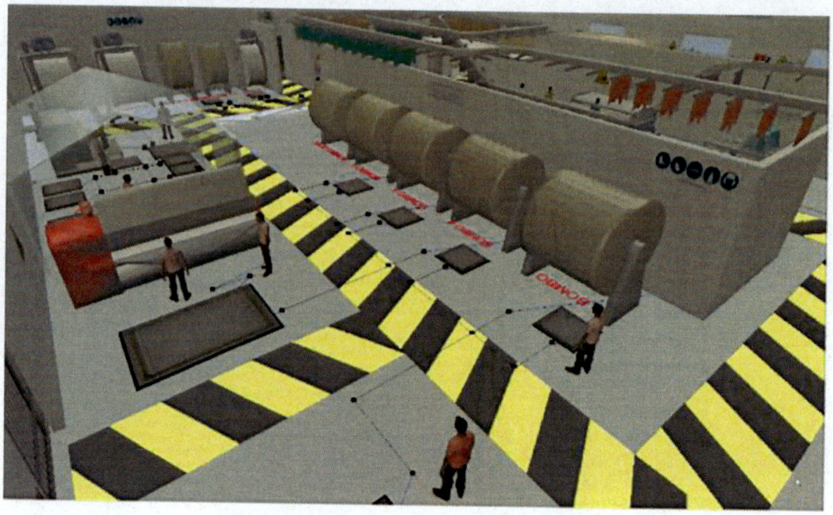

Fig. 9. Simulation Model - General view.

4 Conclusions

This study shows the use of multicriteria decision tools, facility distribution and manufacturing simulation, carrying out a complete analysis of the design and industrial plant organization with the sustenance that each stage of analysis requires. The application of the SLP method generates two improvement alternatives which are evaluated with Flexsim software giving as a result that the first alternative offers better benefits compared to the current model and the alternative 2; the distance traveled between departments is reduced in a 35,50%, materials transportation cost is reduced in 60,13%, the level of work-in-progress low, on average, 138 bands of leather, the bottleneck machine increases its use in a 1,11% and the capacity production increases in a 1.62%.

References

1. Suhardi, B., Juwita, E., Astuti, R.D.: Facility layout improvement in sewing department with systematic layout planning and ergonomics approach. Cogent Eng. **6**(1), 1597412 (2019)
2. Potadar, O.V., Kadam, G.S.: Development of facility layout for medium-scale industry using systematic layout planning. In: Vasudevan, H., Kottur, V.K.N., Raina, A.A. (eds.) Proceedings of International Conference on Intelligent Manufacturing and Automation. LNME, pp. 473–483. Springer, Singapore (2019). https://doi.org/10.1007/978-981-13-2490-1_43
3. Naranje, V., Reddy, P.V., Sharma, B.K.: Optimization of factory layout design using simulation tool. In: 2019 IEEE 6th International Conference on Industrial Engineering and Applications (ICIEA), pp. 193–197. IEEE (2019)
4. Chakraborty, B., Das, S.: Development of plant layout for improving organizational effectiveness by hybridizing GT, TOPSIS and SLP. In: Venkata Rao, R., Taler, J.

(eds.) Advanced Engineering Optimization Through Intelligent Techniques. AISC, vol. 949, pp. 515–525. Springer, Singapore (2020). https://doi.org/10.1007/978-981-13-8196-6_45

5. Arunyanart, S., Pruekthaisong, S.: Selection of multi-criteria plant layout design by combining ahp and DEA methodologies. In: MATEC Web of Conferences, vol. 192, p. 01033. EDP Sciences (2018)
6. Escobar-Naranjo, J., Caiza, G., Garcia, C.A., Ayala, P., Garcia, M.V.: Applications of artificial intelligence techniques for trajectories optimization in robotics mobile platforms. In: 4th International Conference on Industry 4.0 and Smart Manufacturing Procedia Computer Science, vol. 217, pp.543–551 (2023)
7. Gualpa, T., Ayala, P., Cáceres, J., Llango, E., Garcia, M.: Smart IoT watering platform based on orchestration: a case study. In: Garcia, M.V., Fernández-Peña, F., Gordón-Gallegos, C. (eds.) CSEI 2021. LNNS, vol. 433, pp. 191–204. Springer, Cham (2022). https://doi.org/10.1007/978-3-030-97719-1_11
8. Swati Jain and Tarun Kumar Yadav: Systematic layout planning: a review of improvement in approach to pulse processing mills. Int. Res. J. Eng. Technol. (IRJET) **4**(5), 503–7 (2017)
9. Naranjo, J.E., Caiza, G., Velastegui, R., Castro, M., Alarcon-Ortiz, A., Garcia, M.V.: A scoping review of pipeline maintenance methodologies based on industry 4.0. Sustainability (Switzerland) **14**(24), 16723 (2022)
10. Rochman, R.A., Wahyuningsih, S., Ramelan, A.H., Hanif, Q.A.: Preparation of nitrogen and sulphur co-doped reduced graphene oxide (rGO-NS) using N and S heteroatom of thiourea. In: IOP Conference Series: Materials Science and Engineering, vol. 509, pp. 012119. IOP Publishing (2019)
11. da SILVA, J.M.N., de Araújo VIEIRA, E.M., TORRES, M.G.L., de Matos COSTA, A.N., SANTOS, L.C.: Planejamento sistemático do layout: aplicação em uma indústria de panelas esmaltadas. Revista ESPACIOS— Vol. 36 (N° 09) Año 2015 (2015)
12. Cuesta, S., Siguenza-Guzman, L., Llivisaca, J.: Optimization of assembly processes based on lean manufacturing tools. case studies: television and printed circuit boards (PCB) assemblers. In: Botto-Tobar, M., Zambrano Vizuete, M., Torres-Carrión, P., Montes León, S., Pizarro Vásquez, G., Durakovic, B. (eds.) ICAT 2019. CCIS, vol. 1195, pp. 443–454. Springer, Cham (2020). https://doi.org/10.1007/978-3-030-42531-9_35
13. LI, Q., Haidong, L.: Workshop facility layout optimization design based on SLP and Flexsim. Int. J. Simul. Syst. Sci. Technol. **17** (2016)
14. Zeng, Q., Dong, X., Cao, W.: Research on furniture workshop layout optimization based on SHA and SLP. In: Chien, C.-F., Qi, E., Dou, R. (eds.) IE&EM 2019, pp. 134–144. Springer, Singapore (2020). https://doi.org/10.1007/978-981-15-4530-6_14
15. Suhardini, D., Septiani, W., Fauziah, S.: Design and simulation plant layout using systematic layout planning. In: IOP Conference Series: Materials Science and Engineering, vol. 277, pp. 012051. IOP Publishing (2017)
16. Jaramillo, J.E., López, Y.M.: La lúdica como apoyo al proceso de enseñanza-aprendizaje en la universidad de san buenaventura-medellín: Temáticas, diagramas del proceso y kanban. INVESTIGACIÓN FORMATIVA EN INGENIERÍA, 296 (2017)
17. Nieble, B., Freivalds, A.: ingeniería industrial. Métodos, estándares y diseño del trabajo, 12ª Edición McGraw-Hill Interamericana (2009)

A FlexSim-Based Approach to Efficient Layout Planning for a Tire Company

Juan-F. Reyes[1], M.-Xavier Lopez[2], Edwin-O. Portero[3], Wilson-O. Lopez[4], Marcelo-V. Garcia[2,5], and Jose-E. Naranjo[2(✉)]

[1] Unidad Educativa Vicente Anda Aguirre, Pedro Vicente Maldonado 171201, Ecuador
juanf.reyes@educacion.gob.ec
[2] Universidad Tecnica de Ambato, UTA, Ambato 180103, Ecuador
{mx.lopez,mv.garcia,je.naranjor}@uta.edu.ec
[3] Electrosis ETS CIA. LTDA, Ambato 180155, Ecuador
eportero@electrosis.com.ec
[4] Universidad Tecnologica Indoamerica, UTI, Ambato 180103, Ecuador
wilsonlopez@indoamerica.edu.ec
[5] University of Basque Country, UPV/EHU, Alameda Urquijo, 48013 Bilbao, Spain
mgarcia294@ehu.eus

Abstract. Plant layout is an issue that is not considered at the industrial level, so the physical construction of the facilities is done empirically and with designs adjusted to existing physical spaces. This research focuses on the analysis and design of industrial facilities. Several methodologies were used, considering the human factor, distances, and spaces. An approach was chosen that guarantees the reduction of transportation, changing the physical layout of the facilities, reducing the obstruction, and directly influencing the generation of unnecessary and unproductive time. FlexSim software was used for the simulation of the current marketing process flow and the proposal. The following results were observed, an increase in productivity of 400% and a need to reduce from five workers to one. In addition, using the plant layout software for production and operations management, quantitative methods POM-QM, and new work areas were established, which decreased transitions and transportation distance by 35.67%.

Keywords: Layout · industry · operations · productivity · FlexSim

1 Introduction

Facility layout is an issue that plays a fundamental role in the operation of a company. The impact it has on system performance and efficiency is well known [17]. Problems such as design become very complex issues since they depend mainly on finding the most optimal layout that seeks to optimize resources [21].

M. V. Garcia and C. Gordón-Gallegos (Eds.): CSEI 2022, LNNS 678, pp. 532–542, 2023.
https://doi.org/10.1007/978-3-031-30592-4_36

The designs of facilities are varied and generate a working flow, mainly stimulating the development of activities. However, they can be an aggravating constraint to the correct development of the work since they directly impact service times and operating costs [15].

In several investigations, logistics is generally taken as the only factor to be evaluated, without considering the handling of materials between facilities, which is predominant. The widely dominant factor attributed to the operational part is people. Under this precept, the physical characteristics of the workers are vital indicators to limit the whole system in the operations [9].

In previous studies such as [5], the need for the effort factor performed by workers as part of an analysis from the occupational safety and health point of view is denoted. Mainly, it is projected that the design of installations must enact physical safety and accessibility of the interested parties to avoid accidents and reduce performance capabilities in compliance with legal frameworks in each country.

Globally, a plant design known as optimal is one of the operational strategies capable of significantly reducing costs and increasing productivity [4,10]. In addition, it denotes an increase in the flow of work in the operation route traced with its activities, decreasing the damage to the integrity of the workers due to the physical risk to which they are exposed and maintaining their capabilities at the same level throughout the workday. The FLP is one of the classic methods for problems focused on planning, which consists of allocating the different resources of the company, such as machines, products, workers, work areas, sanitary services, customer service, warehouses, canteens, water dispensers, and offices. Solving plant layout problems based on FLP leads to a strategic decision that dramatically impacts operating expenses, meeting delivery promises, and productivity in general. [1,7,20,22].

The systematic layout planning method [SLP], which is the solution to the FLP, allows obtaining step-by-step the plant layout of a company from the arrivals and activities to evaluate effectiveness. The result is a new plant layout with a better flow of the marketing process and the correct use of the facilities, which also helps to increase the productivity of their workers while preserving their physical capabilities [3,8,11,14,19].

This research has five sections, including the introduction. The second section is state-of-the-art, where related works are discussed. The third is the case study, where the previous information of the company in which the study is made is presented, the fourth section explains the results, and in the last section, the conclusions are presented.

2 State of the Art

[20] focuses on considering the dimensions of the dynamic departments to increase the facility design's efficiency. As a consequence of this analysis, the mixed integer programming model was developed, and the addition of two new objective functions, consisting of minimizing the area distributed by the departments, generated a decrease in the operating cost of $45883.

The integrated FLP resolution model proposed by [6], ndicates in the first instance that their approach relates to various designs through mathematical model programming. The aim is to minimize the total cost, especially with the reduction of transportation between all facilities. Through the opinion of several experts, a quantitative qualification was assigned to each alternative, and it was possible to generate some designs, initially only with the transportation cost. Then they were related to the rankings made based on quantitative and qualitative criteria, which showed that those designs that are less important considering the transportation cost are discriminated. However, there are better results with other factors such as information flows, equipment, plant personnel, and links with supervision.

Another method to apply FLP consists of the following steps, where Sahragard and Bashiri [16] indicate a structure of how to obtain good results:

1. Solving the problem using a software construction algorithm (ALDEP) based on successive selection, regions' location, and proximity requirements also present a limitation for the maximum number of installations.
2. Describe the dimensions of the facilities, which must have areas intended for their use and specific elements such as corridors, and stairways, among others.
3. Assign preferences in locations between departments from essential (first letters of the alphabet) to undesirable (last letters of the alphabet).
4. Establish a format for the number of distributors to be counted.

The ALDEP algorithm allows the construction of an optimal design by itself, which provides knowledge of the relationships of parts in the process related to operations activities.

Oscar Calapiña [2] indicates that the optimal plant layout is obtained through virtual tools, especially for selecting a design option with a 3-month simulation time using FlexSim software, resulting in a 60.75% reduction in material transportation costs, which indicates a decrease in operations time.

[13] presents an improvement in the distribution of a meat facility, specifically in the separating area, with a relational methodology of activities, proposing the design of the spaces used by the area through AutoCAD software. He also evaluated with FlexSim software to simulate the process and observed the tonnage reduction over the distance traveled in kilometers by the workers by almost 10%, which allowed an increase in general productivity of 0.12%.

In the study proposed by [12], a plant distribution design is presented for the separation process of the company Alimentos Cárnicos S.A.S. to reduce the transportation that the operators perform internally in the facilities. The results obtained yielded data that indicate a benefit of reducing the load and the distance of approximately 10% less than the current method of operation. In addition, the conclusions were structured based on the simulation carried out with FlexSim software, which allowed them to evaluate variables such as workload, frequency of activities per task, and distances showing a behavior of the model close to reality in time and leaving a tool for continuous improvement to be evaluated in the coming years.

3 Study Case

3.1 Process Analysis

Tire marketing companies in Ecuador have an empirical physical distribution that does not meet customer service needs nor comply with international norms or standards suitable for operations of this purpose. So the use of the experience of administrative and operational staff has shaped their designs and their respective validation, not thinking about the customer or being competitive but looking for interior comfort, sometimes even imposing customs on the consumer. One of the prominent tire marketing companies is described in Fig. 1, and the flowchart of the activities carried out in the commercial process of tires is appreciated.

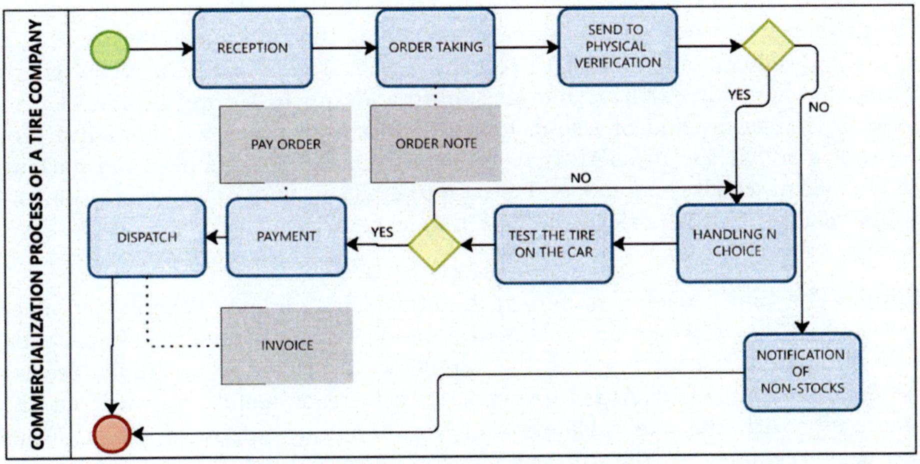

Fig. 1. Company information flowchart.

The commercial process begins with the reception, where a salesperson attends to customers. Here, the customer's needs and product options are stipulated. Then, the selected product is transported from the company's warehouse to the parking lot, where it is tested in the vehicles. If one of the options is adequate, the client proceeds to the invoicing area where, after receiving the order form, he/she pays the value stipulated for the product purchased and receives the invoice or sales note; once the salesperson verifies this document, the dispatch of the product is notified.

If the process does not comply with those mentioned above and the product required by the customer is not in stock, the customer will be notified, and the need will be managed by estimating a date or specifying the waiting time for the arrival of this input.

Warehouse Routes. The drawings correspond to the physical space available to the company in its new facilities. The route during the commercial process is shown in Fig. 2 where the distance with yellow lines that the salesperson moves can be seen. In addition, the product transfers while it is in the company can be appreciated.

The commercial process also maintains strategic allies such as the tire installation, balancing, and alignment services that are part of the cashier department and that, in moments, are part of the queue treatment in the invoicing service. It is also necessary to consider the use of personnel not linked to commercial functions but rather to administrative functions that do not add value to the service provided by the company.

Distance. The initial operating data for this company is the distance traveled. The distances traveled are made up of a corridor that measures 26.4 m; also, a corridor of 4.36 m can be seen to reach the rear stands of 8.19 m. The length of the warehouse is 8.095 m, and this route adds up to a total of 47.045 m to perform the verification or search for tires. In the distance for the delivery of the tire is added the measure that extends from the reception to the parking lot of 2.535 m which generates an increase to 49.58 m, under this description the commercial process is performed.

Time. The time traveled (minutes) is detailed in Table 1. Where the values that each of the activities are indicated, in addition, the standards of operation were stipulated considering the due supplements using the technique recommended and applied by Alfredo Neira [18], who recommends carrying out the time study with the due calculation of the minimum number of observations and the assignment of the due supplements corresponding to the commercial activities and personal needs.

4 Results and Discussion

4.1 Layout

The research sought to produce probable scenarios and simulated behaviors using the mobility of areas and labor to reduce time, which is an exhaustible resource for human beings and represents various types of resources. In the proposed model of plant distribution, necessary parameters are estimated, such as physical effort, safety, and health at work of the salespeople, as well as the physical design of the facilities, which are the fundamental axes of development and increase of productivity and other quality indicators. The elements such as customers served and orders filled are also preponderant factors for optimal work and the increased business benefits of positioning, competitiveness, and monetary gains.

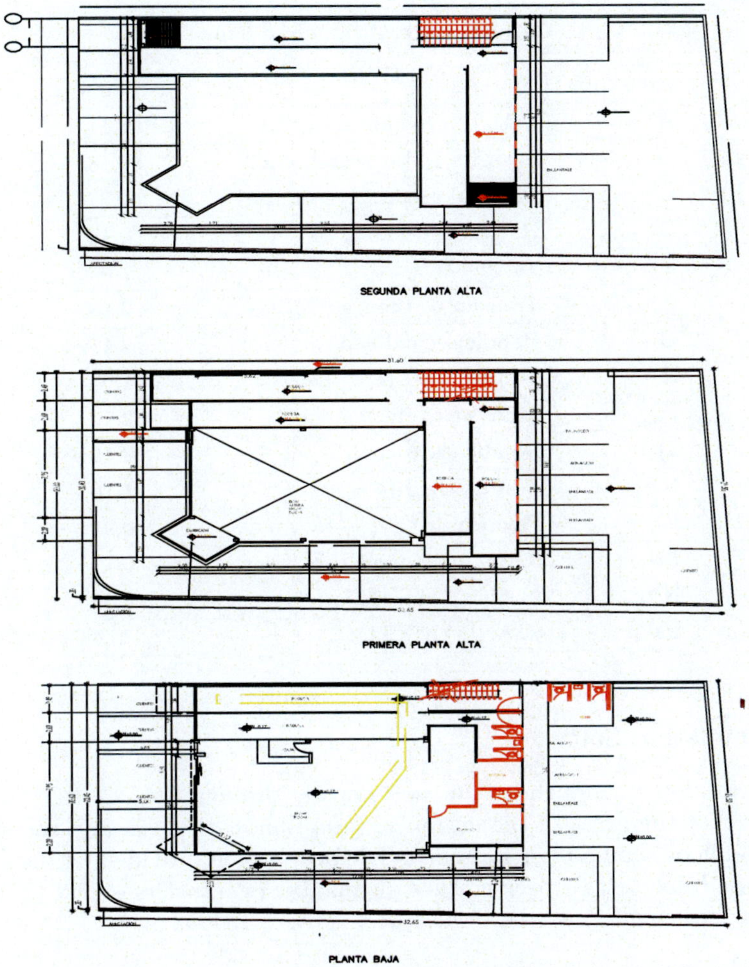

Fig. 2. Transportation drawings.

The advantages shown by the proposal are mainly based on the fact that it does not directly affect the company's functional processes, and it does not seek to increase operating costs due to momentary changes for experimentation or, even worse, to cause discomfort in the work of the interested parties. The company improves its operating time by 64.29%, representing better competitive returns and greater use of labor, reducing downtime and improving the quality of service, making it more agile. In addition, there is an increase of 3,345 tires per worker.

Table 1. Current Standard Time

Operations	Activities	Time
A)	Take the customer's order	1,11
B)	Transfer to the warehouse	2,18
C)	Physical verification of stock	1,07
D)	Handle first choice	2,14
E)	Testing the tire	1,09
F)	Transfer to the warehouse	2,18
G)	Handle second choice	2,17
H)	Testing the tire	1,09
I)	Transfer to the warehouse	2,16
J)	Handle third choice	2,15
K)	Testing the tire	1,10
L)	Payment	0,92
M)	Receiving payment for selected tires	0,73
N)	Invoicing and dispatch	0,35
ST	STANDARD TIME	20,44

4.2 Simulation Software

The distribution recommended for the company through the POM-QM software indicates that the areas of the entrance, reception, cash desk, and the stairs to the second floor should be maintained. The best option provided by the software is shown in Table 2. It is specified that the number of transfers is 85064 compared to the current conditions where the transfers were 164052.

Table 2. Simulation of proposed layout in POM-QM.

Department	Room
Dept 1	Room 1
Dept 2	Room 2
Dept 3	Room 3
Dept 4	Room 5
Dept 5	Room 6
Dept 6	Room 8
Dept 7	Room 7
Dept 8	Room 4
Total movement	85064
Total cost	0

With the use of FlexSim software, modifications were structured to the layout recommended by the POM-QM software that allows the layout of areas seeking the shortest distance traveled, recommending a transfer of the warehouse from the third floor to the first floor, which reduces by 41.15% the load per distance, and the proposed model is shown in Fig. 3.

The layout of the departments corresponds to the use of the physical space on the second floor. Each department was differentiated with the process elements from the beginning: order taking, waiting room, warehouse stock verification, and invoicing. This shows a decrease in the number of spaces used for the commercial process.

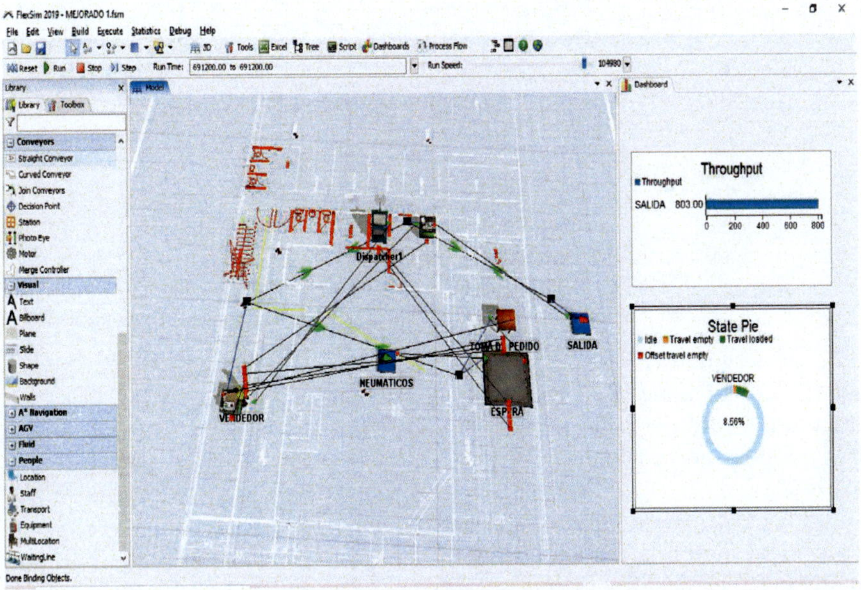

Fig. 3. Proposed layout simulation

Through a time analysis, the optimization of the process improves in such a way that a comparison between the current standards and the proposed ones shows a significant reduction of 7.30 min in the entire commercial process, as shown in Table 3.

Estimates of previous operating standards are necessary since, without them, it is impossible to make a comparison of continuous improvement, and, in turn, their subsequent revision must be periodic to adjust to international levels and comply with current labor regulations.

Operations B), D), F), G), I), J) are the ones that considerably reduce their operation time in such a way that they optimize the sales assistants' work time and reduce their physical load. Mainly on the issue of transfers between areas that generate fatigue and decrease in their performance.

Table 3. Current Standard Time

Operations	Activities	Time
A)	Take the customer's order	1,11
B)	Transfer to the warehouse	0,95
C)	Physical verification of stock	1,07
D)	Handle first choice	0,94
E)	Testing the tire	1,09
F)	Transfer to the warehouse	0,95
G)	Handle second choice	0,95
H)	Testing the tire	1,09
I)	Transfer to the warehouse	0,94
J)	Handle third choice	0,94
K)	Testing the tire	1,10
L)	Payment	0,92
M)	Receiving payment for selected tires	0,73
N)	Invoicing and dispatch	0,35
ST	STANDARD TIME	13,14

The reduction of the standard time is 35.71 percent per worker. If it is transferred to the calculation in the working day of 8 h per day, an average of 24 customers per day per salesperson is obtained in the current method; however, with the reduction in the operating time, the capacity to serve 37 customers is increased.

In addition to the benefits of reduced transportation and unnecessary time, the beneficial factor is the reduced time of exposure to interaction within the company between customers and workers, which reduces the biological risk associated with Covid-19.

Productivity, for this case, is defined as the relationship between customers served and the time invested per worker in the development of commercial activities. The current productivity is 3.345 tires per worker per hour, while productivity with the improved model is 4.18 tires per worker per hour. This indicator represents not only greater efficiency in the use of resources but also better utilization of labor and even contributes to strategic decision-making for personnel leveling and a reduction in operating costs.

Moving the number of salespeople and using a methodology that allows the same working conditions in the simulation shows an increase of 30% in time occupation. It indicates an imbalance in work for the salespeople and also an oversizing of the labor force. It is suggested that for the simulation, a training plan for the labor force in good manufacturing practices should be previously structured in order to increase at least 10% of the learning curve and also to add a master production plan that allows the leveling of inputs and keep the service stock practically working in a first in - first out system.

5 Conclusions and Ongoing Work

The proposed plant distribution model allows the best layout of the company's space and increases productivity by referencing time and relating it to sales based on historical data.

The analysis of the productivity indicators allows to carry out strategic level processes driven by the company's top management, structure new scopes in the business objectives, and manage the corporate image better, so it should be a periodically updated indicator.

The proposal in the study not only has an impact on the reduction of internal transportation but also on the feeling of quality of service to the client, given that he/she must wait less time to carry out the transaction and the wait time to comply with the commercial process is considerably reduced.

The simulation of the business process with different software has no impact on the company's operations; however, the suggested adjustments can generate a highly beneficial or detrimental impact depending on the case study.

In future work, it is expected to carry out measurement projects to follow up on the new design and improve it when it requires a new reorganization. In addition, studies should be conducted in the area of human resources for the selection of personnel and thus have the human element required for the job, and finally, a six sigma study of continuous improvement is proposed to maximize the benefits of the new design.

References

1. Barojas, E., Diaz, C., Juárez, V., Medina, J., Márquez, S.: Distribución en planta: análisis y diseño. Memorias del Congreso Internacional de Investigación Academia Journals Oaxaca 2019 **11**, 116–121, June 2019
2. Calapiña Caguana, O.F.: Distribución de instalaciones en la Planta de Producción de la Empresa Tenería San José Cía. Ltda. B.S. thesis, Universidad Técnica de Ambato. Facultad de Ingeniería en Sistemas ... (2020)
3. Carhuamaca Shapiama, J.C., Zelada Morales, J.J.: Distribución de planta para mejorar la productividad de la empresa extintores apad srl, san juan de lurigancho, 2020 (2021)
4. Cartagena, P.D., Naranjo, J.E., Saltos, L.F., Garcia, C.A., Garcia, M.V.: Multifunctional exoskeletal orthosis for hand rehabilitation based on virtual reality. Adv. Intell. Syst. Comput. **884**, 209–221 (2019). https://doi.org/10.1007/978-3-030-02828-2_16
5. Delgado Bernal, M.R., Zenteno Benítez, M., et al.: Diseño de un plan de higiene, seguridad industrial y salud ocupacional en la Empresa Wiled Paper SRL. Ph.D. thesis (2018)
6. Duero, W., Solís, J.: Propuesta de diseño de distribución en planta para el sistema productivo de una empresa metalmecánica contra pedido. caso: empresa fb ingeniería y estructuras (2020)
7. Garcia, C.A., Naranjo, J.E., Campana, L.A., Castro, M., Beltran, C., Garcia, M.V.: Flexible robotic teleoperation architecture under IEC 61499 standard for oil gas process. In: IEEE International Conference on Emerging Technologies and Factory Automation, ETFA 2018-September, pp. 1269–1272 (2018). https://doi.org/10.1109/ETFA.2018.8502520

8. Guzman, D., Brito, G., Naranjo, J.E., Garcia, C.A., Saltos, L.F., Garcia, M.V.: Virtual assistance environment for deaf people based on an electronic gauntlet. In: 2018 IEEE 3rd Ecuador Technical Chapters Meeting, ETCM 2018 (2018). https://doi.org/10.1109/ETCM.2018.8580281
9. Katiraee, N., Battini, D., Battaia, O., Calzavara, M.: Human diversity factors in production system modelling and design: state of the art and future researches. IFAC-PapersOnLine **52**, 2544–2549 (2019). https://doi.org/10.1016/j.ifacol.2019.11.589
10. Kikolski, M., Ko, C.H.: Facility layout design - review of current research directions. Eng. Manag. Prod. Serv. **10**, 70–79 (9 2018). https://doi.org/10.2478/emj-2018-0018
11. Naranje, V., Reddy, P.V., Sharma, B.K.: Optimization of factory layout design using simulation tool, pp. 193–197. IEEE, April 2019. https://doi.org/10.1109/IEA.2019.8715162
12. Orozco, H.P., Truque, J.C., Pemberthy, L.P., Sinisterra, H.A.: Propuesta para un diseño de distribución en planta en el área de separado para la empresa de alimentos cárnicos sas, evaluada mediante una herramienta de simulación-flexsim. Publicaciones e Investigación **12**(2), 83–93 (2018)
13. Orozco, H.P., Dorado, J.A.S.: Propuesta de un diseño de distribución de un centro logístico frutícola para la zona franca agroindustrial en la región patía: caso colombia. Publicaciones e Investigación **15**(1) (2021)
14. Pan, C., Yu, S., Du, X.: Optimization of warehouse layout based on genetic algorithm and simulation technique, pp. 3632–3635. IEEE, June 2018. https://doi.org/10.1109/CCDC.2018.8407753
15. Payán, E.B., Díaz, C.O., Rivera, V.J., Domínguez, S.M., Cervantes, M.J.M.: Artículo: Distribución en planta, análisis y diseño (2019)
16. Phanden, R.K., Demir, H.I., Gupta, R.D.: Application of genetic algorithm and variable neighborhood search to solve the facility layout planning problem in job shop production system, pp. 270–274. IEEE, March 2018. https://doi.org/10.1109/ICITM.2018.8333959
17. Pierreval, H.: Integrated simulation optimization for layout problems, pp. 1486–1490. IEEE, December 2018. https://doi.org/10.1109/IEEM.2018.8607591
18. Reyes, J.F., Barragan, M.A., Sanchez, P., Garcia, M.V.: Planificación del diseño de las instalaciones de la industria de comercialización de neumáticos. Ibérica de Sistemas e Tecnologias de Informação **42**, 616–629 (2021)
19. Sahragard, M., Bashiri, M.: Layout design for large scale problems with a hybridized clustering based heuristic method, pp. 1610–1614. IEEE, December 2016. https://doi.org/10.1109/IEEM.2016.7798149
20. Sharma, P., Singhal, S.: A review of objectives and solution approaches for facility layout problems. Int. J. Ind. Syst. Eng. **24**, 469 (2016). https://doi.org/10.1504/IJISE.2016.080289
21. Solano, J., Moreno, J., Ramos, F.: Ant colony optimization algorithm for facility layout problem. Visión electrónica **10**, 6 (2016)
22. Zhao, D., Yang, J., Zhou, H.: Layout design of warehouse based on systematic layout planning and GA-ACO algorithm, pp. 7101–7104. IEEE, November 2020. https://doi.org/10.1109/CAC51589.2020.9327612

Distribution of Facilities to Improve the Raw Material Storage System

Washington Calderón, Daysi Ortiz(✉), Alex Pazmiño, and Israel Naranjo

Technical University of Ambato, Ambato, Tungurahua, Ecuador
{wcalderon6628,dm.ortiz,apazmino4147,ie.naranjo}@uta.edu.ec

Abstract. The problem arises from the need to have a distribution of facilities that provides adequate spaces, properly distributed areas and a storage system according to the requirements of a company, since currently, the distribution of facilities is not adequate. For the development of the research, the methodology of logistics diagnosis of warehouses and distribution centers (IPISI) was used to identify the current conditions of the warehouse, obtaining as a result deficiencies in the quality of the floor, non-existence of loading-unloading docks, limitations in maneuvering spaces, inadequate distance between aisles and lack of performance indicators, triggering problems in the processes of reception, storage and dispatch of raw materials. To complement the study, a categorization was made by product family according to the level of consumption and an analysis based on the criticality of the stored raw materials, identifying the most critical materials through the use of tools such as ABC, BOM and criticality matrix. On the other hand, the PRISMA methodology (Preferred Reporting Items for Systematic reviews and Meta-Analyses) was used to obtain application criteria and distribution techniques that allowed the design of the new distribution of the warehouse facilities with which a reduction of about 3,000 m is expected and 66% in material handling costs, reducing internal movements and transports through the application of a heuristic location allocation method.

Keywords: Storage systems · IPISI · categorization systems · heuristic allocation method · storage principles. · distribution techniques

1 Introduction

Storage systems are a key element in the logistics management of warehouses in companies, as they help with the development of internal operations, improving performance, organization and increasing the level of response within the storage centers. They are a method that consists of placing goods, products or raw materials together in physical structures, promoting optimization through the use of available space [1]. Problems in storage systems within companies have existed for a long time, due to poor inventory management and poor planning

M. V. Garcia and C. Gordón-Gallegos (Eds.): CSEI 2022, LNNS 678, pp. 543–564, 2023.
https://doi.org/10.1007/978-3-031-30592-4_37

and distribution of storage space [2]. To solve these problems, warehouse management must be improved through the design or restructuring of the distribution of the facilities, because having a correct distribution positively influences the organizational environment of the company, generating great benefits in staff productivity, dispatch quality and safety and occupational health indicators, staff availability increases, the execution of internal activities is improved, transportation and times between activities decrease, increasing the performance of the activities [3].

The distribution of facilities consists of the layout of the physical space determining the location of the departments, groups or work stations and material storage points within the facilities, in this sense, its purpose is to organize these elements guaranteeing the continuous flow of products or work [4–6]. The distribution of facilities must ensure an adequate internal warehouse management, since a correct distribution favors the securing, conservation and handling of goods or products [7,8]; the design should be based on six guidelines called basic storage principles: 1. Maximizing space. 2. Make use of the best charging unit. 3. Minimize movements. 4. Control movements and location. 5. Promote a safe place. 6. Easy stock control [9]. Compliance with these principles ensures greater usability, better cost-effectiveness of space use, allows for smooth placement, arrangement and removal activities, and that stored products are properly preserved.

Also, to have distribution techniques that complement the proposed principles such as: 1. Assigning locations through the application of heuristic method. 2. Adapt spaces for each activity (reception, storage, retrieval and dispatch) and 3. The proper equipment for the handling and transportation of materials; with this, times between processes are reduced, distances traveled are shortened, transports and movements are limited, storage costs are optimized, safety and occupational health are increased, guaranteeing an effective flow of materials [10–12].

Large industries are constantly innovating their processes to remain solvent and flexible in the market. The implementation of new technologies is common and the immersion in Industry 4.0 is a reality, currently storage centers have devices capable of reducing human errors, through the use of technology applications such as: barcodes, smart cards, smart tablets, radio frequency identification (RFID), real-time location and localization systems, integrated AGV, truck loading and unloading automation, lancer-based retrieval systems and the implementation of conveyor robots creating an intelligent warehouse logistics management system and a cyber-physical system (CPS) recreating virtual copies of real environments and interacting with virtual worlds, turning them into warehouses 4.0 warehouses, as in the case of Amazon, which has made it possible to eliminate waste in internal logistics, optimize the level of response in the preparation of orders, increase warehouse capacity without increasing its physical capacity, reduce costs and reduce the number of failures, among other benefits [13].

The objective of this article is to pose a proposal for the improvement of the storage system in a warehouse of raw materials of a footwear company in the city of Ambato - Ecuador, using techniques of distribution of facilities, storage principles and findings in various case studies identified through a literature review supported by PRISMA methodology. Unlike published studies in which the effective management of storage systems is analyzed independently, reducing costs in material handling, space optimization, warehouse automation, racking system design, and so on, the present study seeks to unify this set of approaches and develop a global analysis of the facilities, starting from a situational analysis through the IPISI methodology, continuing with a classification of the raw materials handled in the warehouse, sectorization of the physical space and location of the materials using storage principles and distribution techniques of facilities to finally propose a proposal that is validated through a process simulation software.

2 Methodology

2.1 The PRISMA Methodology

This methodology is used for the execution of the research work, since it allows obtaining information from the systematic literature review, based on guidelines such as: (i) research questions, (ii) document search, (iii) document selection and (iv) information extraction [14,15].

Research Questions. The research is aimed at investigating aspects related to the research topic, with emphasis on the following points of view: distribution of facilities (VP1), storage systems (VP2), storage principles (VP3), distribution techniques and allocation of locations (VP4), and innovations in large warehouse-gas plants (VP5 - VP6) in order to answer the following questions, Table 1 details the research questions and their motivation.

Table 1. Research questions.

N°	Research questions	Motivation
RQ1	How does the distribution of facilities improve the storage system in raw material warehouses?	Identify the influence of the distribution of facilities in the raw material storage systems
RQ2	What storage principles and distribution techniques are applied in ware-houses?	To know the principles of storage of raw materials used in warehouses worldwide
RQ3	What are the benefits of applying storage principles and distribution techniques in warehouses?	Improve storage systems and the distribution of items within a raw materials warehouse
RQ4	What innovations are large raw material warehouses making to improve their storage systems?	Knowing the actions of large companies regarding the storage of their products and assessing them for application at the local level

Document Search. Aimed at examining information of interest to the study, logical connectors are used to simplify the search process by focusing objectively on certain documents related to the topic. The bibliographic search is made up of scientific publications from 2011 to 2022, because with the passing of time new methods have been implemented that comply with the requirements of the industrial sector seeking to establish themselves in strategic places, understanding strategic place as the geographical location that favors the economic activity of the company in a certain location. For this reason, in the basic search, terms were used in accordance with the aforementioned aspects. For VP1 ("Distribution" & "facilities" & "warehouses"); VP2 ("System" & "storage" & "warehouses"), VP3 ("Principles" & "storage" & "warehouses"), VP4 ("Techniques" & "distribution" & "warehouses"), VP5 ("Innovations" & "warehouses" & "raw materials") and VP6 ("Innovations" & "Systems" & "storage") were selected, as well as the articles that had related terms in the titles and in the case study abstracts. The terms and the combinations of these allowed to reach optimal results. A total of 310 articles were obtained, 101 from Web of Science, 10 from Scielo, 84 from Taylor and Francis and 115 from Redalyc.

Document Selection. Activity aimed at discerning information by excluding those documents that do not provide useful information, focusing on certain documents whose information is conducive to the research. The selection process of the articles was classified based on the inclusion and exclusion criteria detailed in Table 2.

Table 2. Criteria for inclusion and exclusion of documents.

N°	Inclusion	Exclusion
C1	Related articles	Duplicate articles
C2	Articles in English and Spanish	Conferences, books, reports, web page
C3	Articles related to the distribution of warehouse facilities	Research done in areas other than industry
C4	Articles related to the raw materials storage system	Articles not related to the topic
C5	Articles related to the principles of raw material storage	Articles not related to the topic

Information Extraction. It once the documents of interest have been identified, the information is extracted according to subject, importance, context and synthesis. The extraction of documents is as shown in Fig. 1.

2.2 The IPISI Methodology

In turn, the methodology of logistics diagnosis of warehouses and distribution centers IPISI is used, determining the initial conditions of the distribution of

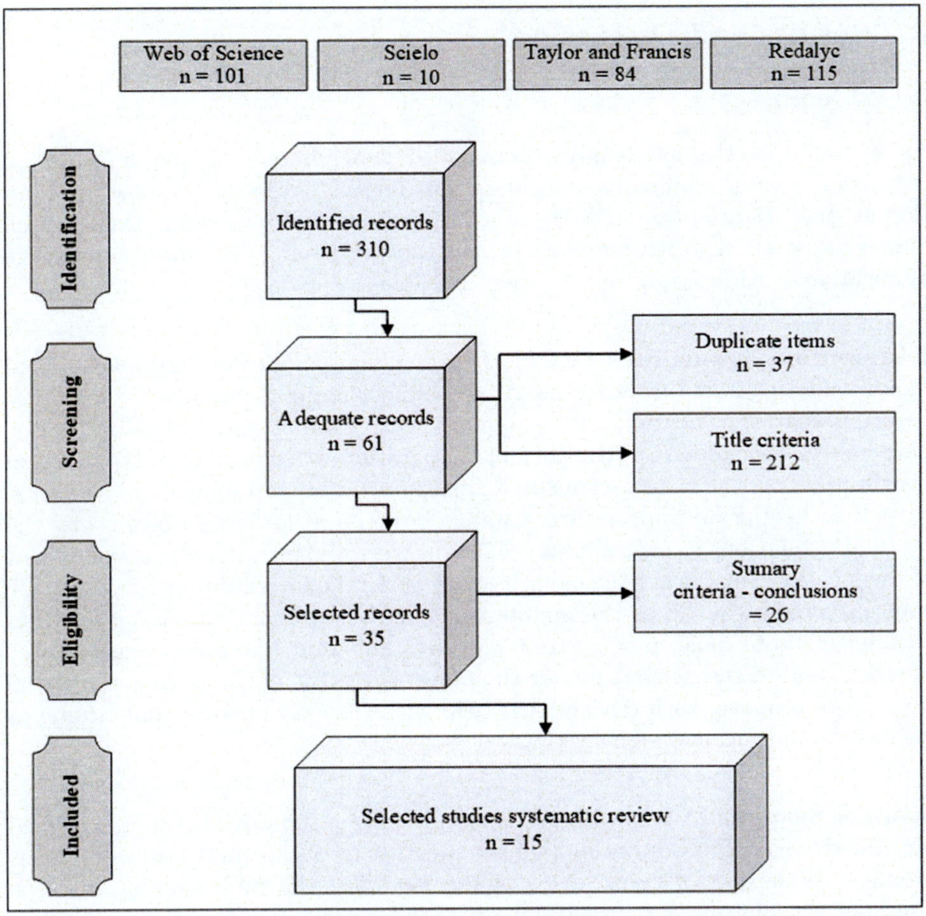

Fig. 1. Information extraction.

facilities of the companies, the methodology contains 5 phases for the analysis of: Infrastructure and areas, Processes, Inventories, Work safety, Indicators [16], as well as empirical methods related to visual inspections and the application of interviews for information gathering. The IPISI methodology with which warehouses and distribution centers are analyzed will allow to establish comprehensive strategies to improve its productivity. It can be comparative (analyzes), descriptive (exposes) or normative (evaluates). It is not the same to look for alternatives to a specific problem than to look for problems or root cause of these.

3 Results and Discussion

3.1 Findings

As a result of the systematic literature review, based on the information extracted, several applicative considerations for the design of the new distribution of facilities are presented, the analysis is focused on answering the research questions posed (i) with emphasizing on the proposed (VPi) viewpoints. The applicative considerations are detailed in Table 3.

Distribution of Facilities (VP1). It presents an applicative approach aimed at consolidating and analyzing design and dimensioning techniques for facilities, ensuring that the locations are equipped with loading and unloading docks, have spaces for storage, retrieval and preparation of orders, that the internal distribution has the necessary aisles for transportation and internal movements, as well as having an appropriate distance between aisles as established by the internal regulations in each country [17]. Likewise, the infrastructural conditions of the location must grant the capacity to develop the internal processes in a safe way, ensuring at all times the maintenance of the products, providing facilities that have stable roofs, protect from sun rays and rain, the floors must be free of cracks and openings that hinder the movements, the entrance and exit doors must work properly with devices that help with the easy opening and closing as well as the appropriate signage [18].

Storage Systems (VP2). Based on establishing a structural system designed for the storage of products in large quantities in a safe and stable manner, oriented to the use of cubic space of the facilities, the bibliographic research considers the joint use of physical structures called racks and shelves because they allow a better control of inventories, a correct access and recovery of products, the storage is by product family in fixed positions, there is a better stability of the shelves because they are based on uniform distributions, that is to say the products of greater but and/or use are positioned in the lower parts providing stability to the structures, besides proposing indicators as the percentage of use to complement in analysis of the storage systems [19,20,24].

Storage Principles (VP3). Aimed at improving the operation within the storage centers through the application and use of the principles of storage, the research proposes to promote the unitization of loads by making use of the best unit of grouping of loads as it allows greater transport of products in less time and distance, ensure the proper places for the storage of products by improving the use of cubic space, promote activities and tasks of safe allocation for storage by providing procedures and personal protection equipment to employees and ensure the implementation of machinery equipment and tools for storage, retrieval, handling and transport of loads to optimize the total transport within the location, this makes the warehouses in flexible centers with a fast level of response and agile stock control [21,22,25].

Table 3. Application considerations - PRISMA.

VP	Approach	Application consideration
VP1	Station design and sizing	- Loading and unloading docks [17] - Storage areas [17] - Order picking and retrieval areas [17] - Distance between aisles [17] - Number of aisles [18] - Location assignment policies [18]
VP2	Structural storage system	- Implementation of racks and shelving [19] - Percentage of system utilization [20] - Improved inventory control [20] - Racks stability [20] - Distribution uniformed [20]
VP3	Efficient operation of the storage center	- Use of the best unit load [21] - Assigning storage tasks [21] - Minimize movements [22] - Ensuring favorable storage locations [22] - Equipment of machinery and tools for storage and recovery operations [22]
VP4	Assignment of locations and dispatching techniques	- Heuristic bin assignment methods [11] - Control of movements and locations [10]
VP5 - VP6	Innovations Large companies - Industry 4.0 (Warehouses 4.0)	- Use of mobile picking robots [13] - Barcodes [13] - Smart cards and tablets [13] - Radios frequency identification [23] - Cyber-physical systems [23]

Distribution and Location Allocation Techniques (VP4). Focused on determining optimally the best location for internal items (stored products), for the distribution and allocation of locations adaptive designs are presented, association rules can be used for the calculation of supports relating the rotation of products also considering the distance between the location of the product, the reception and dispatch, similarly heuristic strategies are used based on various aspects such as: Iterated local search, hill climbers characterized by attributes, heuristics related to the principle of Tabu search for the processing and preparation of orders, these applicative parameters drive a reduction in the distances traveled and transports, minimization of times in the activities of recovery and preparation of orders presenting an improvement in the flexibility of operations [10–12].

Innovations in Large Warehouse (VP5-VP6). Related to the improvement and optimization of the logistics process of warehousing based on the innovations of industry 4.0, provide useful technologies to compete globally seeking to convert them into warehouses 4.0 Among the main technological applications to be used are: bar codes, smart cards, integrated tablets, radio frequency identification (RFID), real-time location and localization systems, integrated AGV, recovery systems based on lanceras, implementation of mobile robots transporters, cyber physical systems improving response times, utilization of facilities, inventory control and reducing errors [13,23,26].

3.2 Problem Description and Criticality Analysis

Applying the IPISI warehouse and distribution center logistics diagnostic methodology, the following results were obtained for each of the analysis phases (Table 4).

Table 4. Diagnosis of current conditions IPISI methodology.

PHASE 1: INFRSTRUCTURE AND AREAS	
Focused on aspects related to: floor and ground quality, loading and unloading docks, facility layout, maneuvering spaces and distance between aisles	
Quality of floor and ground	There are flaws, cracks and openings as a result of wear and tear and deterioration of the facilities
Loading and unloading docks	There are no loading and unloading docks because the doors are at ground level
Layout of facilities	There are two warehouses, the facilities cover a total area of 868 m^2 (31 × 28) and a total volumetric area of 6.510 m^3 (31 × 28 × 7.5)
Maneuvering spaces	It does not have the necessary space for vehicle handling operations
Distance between aisles	The distance on certain routes is less than 0,50 m, making it impassable and hindering product storage, recovery, transportation and dispatch operations
PHASE 2: PROCESSES	
Internal processes are related to: reception, storage and dispatch of raw materials	
Receipt of raw materials, Storage of raw materials, Dispatching of raw materials	The processes present: physical overexertion and exhaustion of personnel, lack of knowledge of raw material locations, lack of product mapping, lack of process standardization, lack of OHS control, long process times, lack of machines, materials and equipment for handling and transporting raw materials, lack of storage systems, chaotic storage, excess of materials, poor access to required materials, limited distance between aisles, lack of loading and unloading docks and facilities with contamination (dust)
PHASE 3: INVENTORIES	
The company has more than 405 different items, around 8353 cardboard packages containing soles, insoles, laces, activators, cleaners, among others	
PHASE 4: OCCUPATIONAL HEALTH AND SAFETY	
Lack and control of OHS, instability in the vertical storage of packages, permanent storage height exceeds 1.5 m from the floor, the few internal racks are made of wood and do not have the necessary safety guarantees, and workers do not have the necessary PPE	
PHASE 5: PERFORMANCE INDICATORS - KPI'S	
There are currently no performance indicators for the raw material storage logistics process	

ABC Analysis and Bill of Material (BOM). ABC analysis is used to classify the products of greatest demand and importance within organizations, segmenting their items, inventories or raw materials based on the profitability they represent for the company, taking into account the percentages for each of the categories. Table 5 shows the type of product according to the determined category corresponding to the ABC analysis of products from the quantity produced in pairs of shoes in the 2021 period and based on the accumulated share. Table 6 details the BOM of the first product, the materials of which will be taken to a criticality analysis.

Table 5. ABC analysis.

Product Family	Sale $	Participation	Cumulative Participation	Product type
Industrial Safety	5200	27,23%	27,23%	A
Trekking	4580	23,98%	51,20%	A
Casual	2940	15,39%	66,60%	A
Tactical boot	2500	13,09%	79,69%	A
Urban	1600	8,38%	88,06%	B
Sport	1300	6,81%	94,87%	B
Child	980	5,13%	100,00%	C
TOTAL	19100			

Table 6. BOM - Industrial Security.

N°	Material	Amount	Unities
1	Water-repellent Leather	10,30	Decimeters
2	Leather base	14,50	Decimeters
3	Cordura	0,09	Meter
4	Textile lining	0,06	Meter
5	Foam 10 mm	0,025	Meter
6	Sponge 2 cm	0,02	Meter
7	Plastic eyelets	8,00	Unities
8	Plastic washers	8,00	Unities
9	Plastic cord	8,00	Unities
10	Cord	1,00	Pair
11	hread	8,00	Grams
12	Reata	0,10	Meter
13	Tag	1,00	Unities
14	Template and assembly recount	0,02	Meter
15	Arming buttress	2,00	Unities
16	Composite/polycarbonate tip	1,00	Pair
17	Nitrile rubber sole/non-slip	1,00	Pair
18	Industrial insoles	1,00	Pair
19	Glues	0,13	Liters

Criticality Analysis. The criticality analysis is used to determine the level of importance and incidence of the raw material for the final product, since criticality is presented as a potential threat to the organizations. Criticality is determined from the application of a matrix shown in Table 7, which associates the multiplication of the probability of having disturbances in the supply of raw materials and the consequences caused.

Table 7. Criticality matrix

CRITICALITY MATRIX						
		Consequence				
		Minimum	Minor	Moderate	Major	Maximum
Probability		1	2	4	8	16
Very high	5	5	10	20	40	80
High	4	4	8	16	32	64
Medium	3	3	6	12	24	48
Low	2	2	4	8	16	32
Very low	1	1	2	4	8	16

The criticality matrix consists of the probability of occurrence and the consequence generated in case the event occurs, presenting different considerations and scores, it has four weighted indicators within a numerical interval, from 1 to 4 indicates an acceptable critical raw material, from 5 to 12 indicates a tolerable critical raw material, from 16 to 24 indicates a high critical raw material, and from 32 to 80 indicates a very critical raw material.

To determine the level of criticality, a characteristic color is used for quick identification according to Table 8. The criticality matrix shows 4 items with a high level of criticality, 11 with a tolerable level of criticality and 13 with an acceptable level of criticality. Table 9 shows the criticality level obtained by applying the criticality matrix for each of the stored raw materials.

Table 8. Level of criticality

Level of criticality	
Acceptable Criticality	
Tolerable Criticality	
High Criticality	
Extreme Criticality	

Mapping and Sizing of Current Facilities. Warehouse 1 stores 5.976 cartons. The arrangement of raw materials in the warehouse isn't adequate, since the distance between aisles is below the distances suggested by national regulations, aisle AB before the ramp has a distance of 1,30 m, and then the distance is reduced to 0,80 m, aisles BC and CD have a distance of 0,50 m, aisle DF has a distance of 1,17 m, aisle E at the back has a distance of 1,00 m, but it is impassable, as shown in Fig. 2.

Warehouse 2 stores 433 packages of insoles, 1.944 packages of soles and 855 rolls of fabrics; it also stores chemical products and plastic and steel tips, but

Table 9. Level of criticality for each prime material

Prime Material	Probably	Consequence	Level of criticality
Sole	Low	Greater	High Criticality
Insoles	Low	Greater	High Criticality
Glues	Low	Greater	High Criticality
Fabric	Low	Greater	High Criticality
Washer	Low	Moderate	Tolerable Criticality
Eyelets	Low	Moderate	Tolerable Criticality
Cleansing cream	Low	Moderate	Tolerable Criticality
Steel tips	Low	Moderate	Tolerable Criticality
Plastic tips	Low	Moderate	Tolerable Criticality
Thread	Medium	Moderate	Tolerable Criticality
Bravo 800 Plates	Low	Moderate	Tolerable Criticality
Artecol	Low	Moderate	Tolerable Criticality
SAR Glue	Low	Moderate	Tolerable Criticality
AM-11 Yellow Glue	Low	Moderate	Tolerable Criticality
Foam	Medium	Moderate	Tolerable Criticality
Cord	Low	Minor	Acceptable Criticality
Hook Pulleys	Low	Minor	Acceptable Criticality
Reata	Very Low	Minimal	Acceptable Criticality
Pin	Very Low	Minimal	Acceptable Criticality

since it does not have an adequate storage area, they are stored in the office. The conditions are similar to warehouse 1, adding that in this warehouse not only packages are stored, but also rolls of fabrics and chemical products related to glues, as can be seen the aisles are not well defined, aisles KJ and LK have a distance of 1,62 m and 1,60 m respectively, aisle M has a distance of 0,63 m, aisle N has a distance of 0,62 m, aisle Ñ has a distance of 1,14 m, these aisles are not well defined, 14 m, these corridors are located in the back part from the ramp that joins the two locations, the intermediate corridor RM has a distance of 0,82 m, in the front part there are no specific corridors, however they have been considered, the corridor G has a distance of 0,54 m, the corridor GI 1,56 m, the corridor GP has a distance of 0,67 m and finally the access corridor to the office GQ has a distance of 1,1 m, as shown in Fig. 2.

Current Distances Traveled. The distances traveled are obtained by measuring the displacements for each of the processes presented. For the processes of reception and storage of raw materials, 100 loads are considered and for the dispatch process, 18 loads are considered (any item required for production) (Table 10).

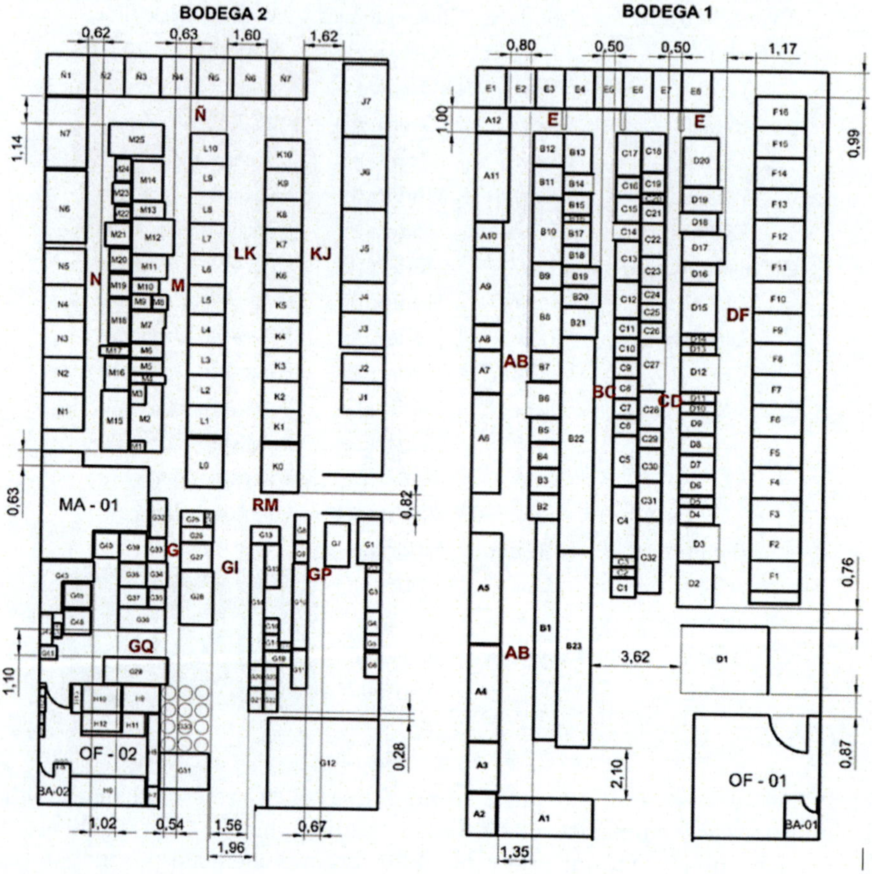

Fig. 2. Mapping and sizing - warehouse 1 and 2.

Table 10. Current distances traveled.

Process	Distance traveled	Observation of the process
Reception	1.419,82 m	Manual operations. 200 transports are used. Time employed of 3 h 28 min and 36 s
Storage	2.009,02 m	Manual and trolley operations. 80 transporters are used. Time employed of 9 h 41 min and 25 s
Dispatch	1113,59 m	Manual operations. 36 transports are used. Time employed of 8 h 34 min and 08 s

3.3 Proposed Distribution

The proposed distribution is based on the adaptation of certain research-related application considerations, in addition to the use of the current regulations in Ecuador, since they provide the necessary information to develop the new distribution of the company's raw material warehouse facilities.

National Regulations. "Reglamento de Seguridad y Salud de los Trabajadores y Mejoramiento del Medio Ambiente de Trabajo - Decreto Ejecutivo 2393". This regulation mentions the criteria related to structural safety, distance between aisles, handling and storage of materials, circulation of vehicles, carts or carts, fire prevention, handling of hazardous materials, signaling, among others, which are used for the preparation of the proposal [27].

Matrix of Application Considerations. In order to analyze the applicative considerations of the basic principles of facility distribution and storage, it is necessary to work on them together, for which Table 11 has been used.

Applied Heuristics. A heuristic allocation method is used to determine the best location for the stored raw materials, the method consists of: 1. Rank all positions from low to high di. 2. Rank all SKUs from high to low ni, and 3. Assign next highest SKU (ni) to next lowest location (di). Equation 1 seeks to minimize the total labor time (distance) to store and retrieve:

$$Min\,Z = \sum_{i} (d_i * n_i) \tag{1}$$

where:

di = distance for pallet location i from receiving to shipment location
ni = average number of times location i is visited per year

For this purpose: 1. The distances from point 0 to each of the new storage points of the new distribution are measured and ordered from smallest to largest. 2. SKUs will be classified according to a categorization system (ABC) based on the criticality and consumption level of each of the raw materials stored. 3. Category A materials are assigned to the locations closest to the points, maintaining the loop by product families and not diversifying raw materials, in order to have better inventory control.

By applying this method, it is possible to reduce distances, positioning those raw materials with the highest turnover and consumption close to the loading and unloading points, minimizing movements and costs.

Table 11. Application Consideration - Principles of storage and distribution [9,28]

Principles of distribution	Principles of storage	Application considerations
Joint integration	Maximize space, use of the best unit load, minimize movements	- Use of racks to maximize space - Use of pallets to make use of the best unit load - Implement a forklift for transportation and minimize movements
Minimum distance traveled	Using the best loading unit, minimize and control movements and location	- To have materials that delimit the number of total transports - Unitization allows the movement of a greater quantity of materials at the same time, minimizing movements and controlling the location of stored products
Material flow	Maximize space, use of the best unit load, minimize and control movements and location, easy stock control	- To have a uniform, signposted distribution, with aisles in accordance with internal activities, allowing materials to circulate in a timely manner without interruptions - To have a facility that uses volumetric space optimally by increasing storage spaces - The effective allocation of locations allows for a reduction in movements - Acquire computerized equipment to facilitate stock control by monitoring movements and locations
Volume occupied	Maximize space, use of the best unit load, minimize and control movements and location, easy stock control	- Implementation of racks - Unitization of loads - Use of the forklift helps to fully utilize the occupied volume within the location
Safety and satisfaction	Promote a safe place	- A layout that allows workers to carry out their daily activities without risk. The necessary signage, fire-fighting equipment and security guards
Flexibility	Maximize space, use the best unit load, minimize and control movements and location, promote a safe place, easy stock control	- The application of all the previous considerations allows us to respond quickly to market requirements, since compliance with all storage and distribution principles makes it possible to have a location with the capacity to react quickly and constantly over time

Proposed Layout. Figures 3, 4, 5 and 6 show the new distribution of facilities, warehouse 1 has 2 main aisles (AB - CD) along the location and 1 PS aisle that interconnects the aisles mentioned and warehouse 2 has 2 main aisles (FE - HG)

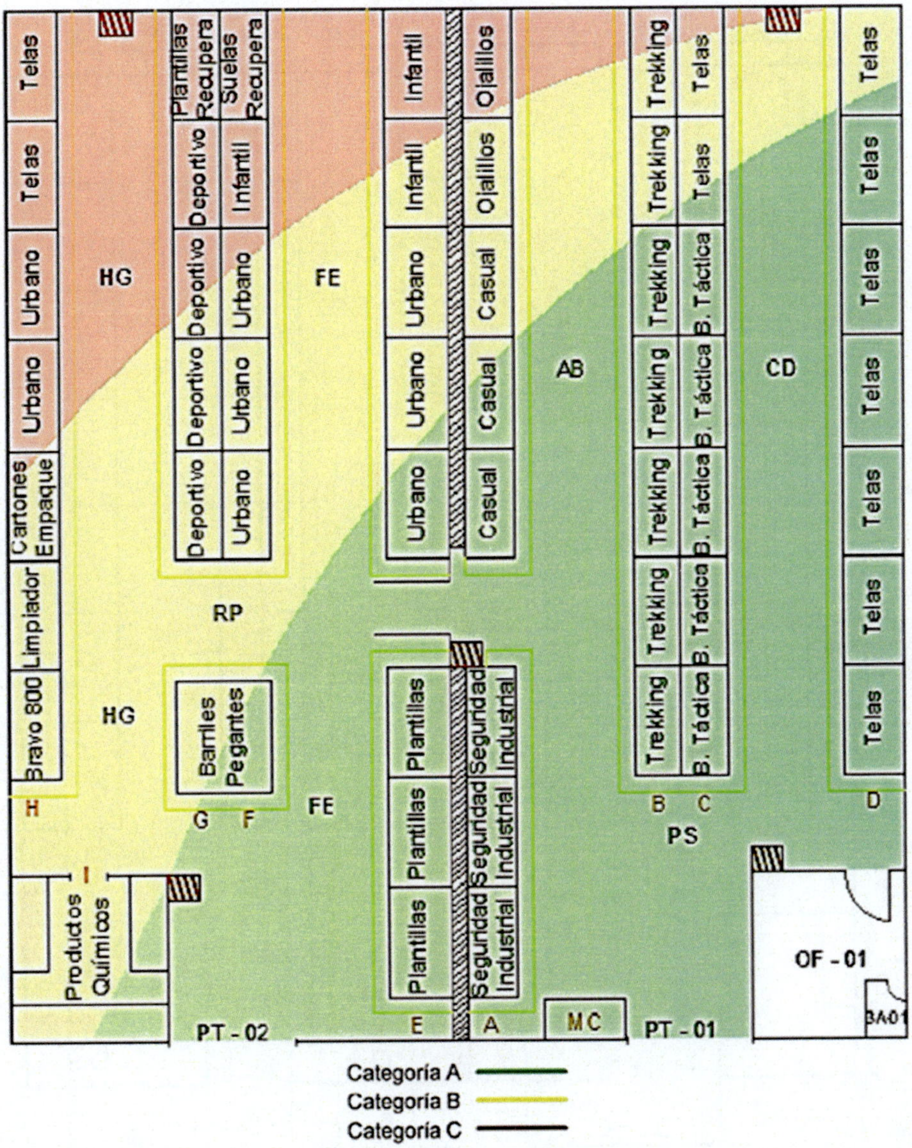

Fig. 3. Assignment of locations.

along the location and 1 RP aisle that connects the aisles mentioned and allows access to warehouse 1, overall, the new distribution has 4 main aisles and 2 cross aisles.

The proposed distribution of facilities will have 54 selective racks, distributed by columns A, B, C, D, E, F, G and H, the lengths of the racks are constant for all columns but at certain points space is added depending on the materials to

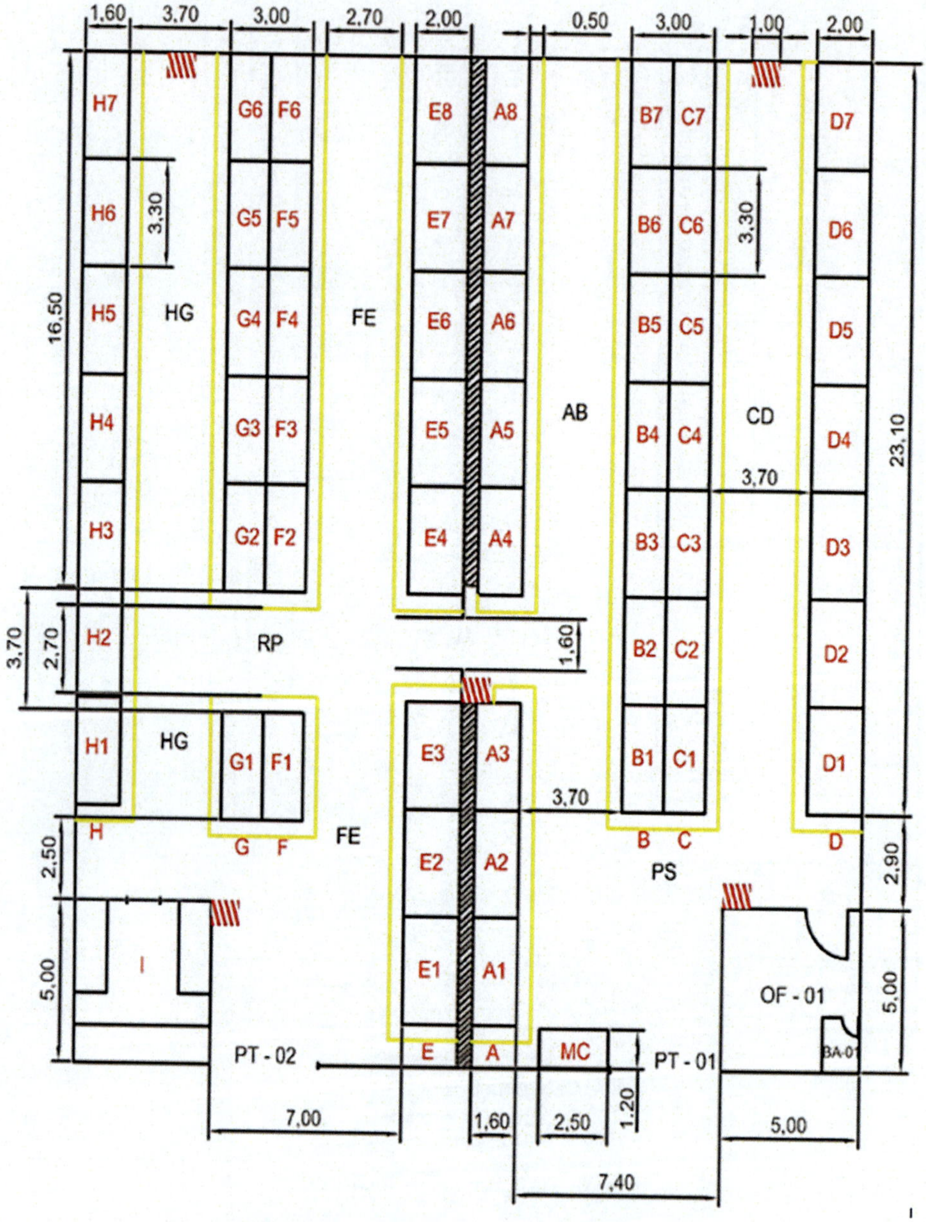

Fig. 4. Proposed distribution of facilities.

be stored, columns A and H have a width of 1,60 m (1,50 m of the rack ± 0,10 m) columns B, C, F and G have a width of 1,50 m, and columns D and E have a width of 2,00 m (1,50 m of the rack ± 0,50 m), the height of the racks is 4,5 m and 3,30 m long, in sectors G1 and F1 racks are not used because these spaces

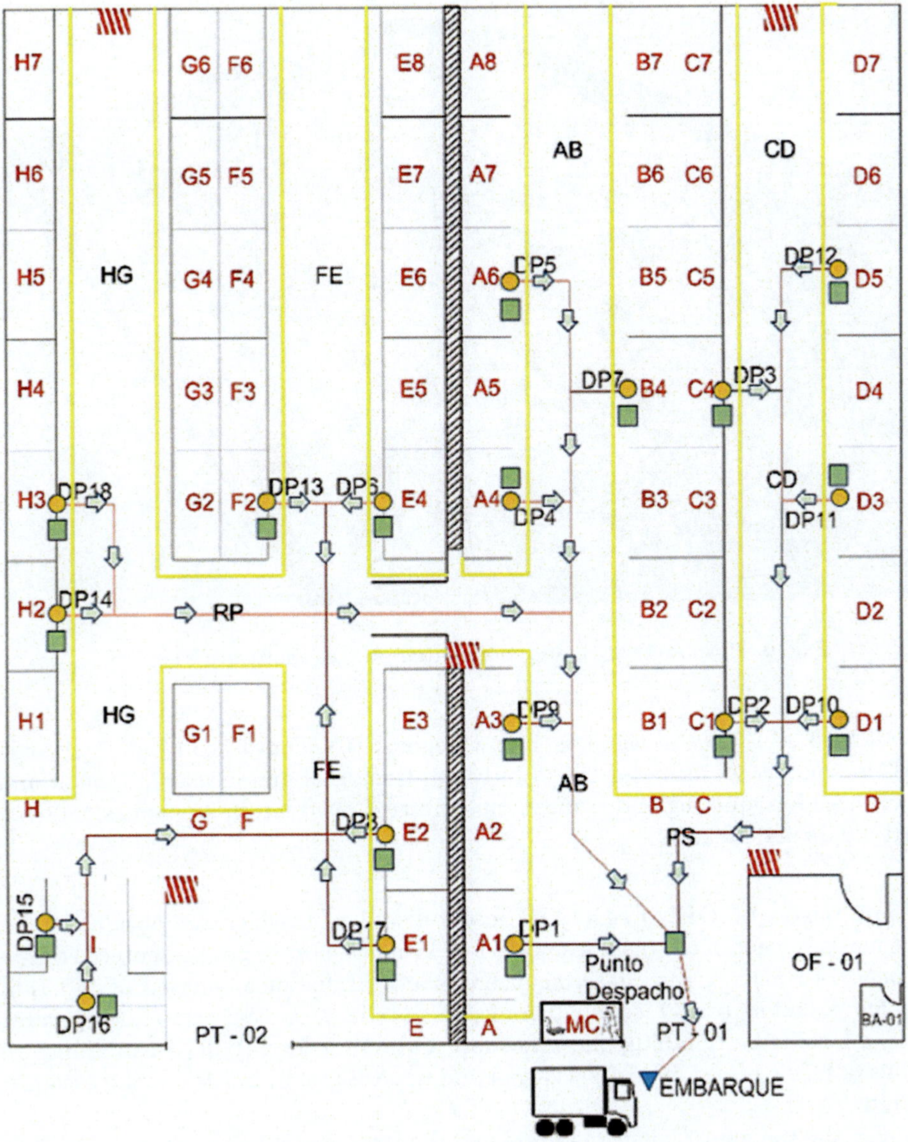

Fig. 5. Proposed raw material dispatch process.

are intended for storing large tanks of glue. There are 4 main aisles AB - CD - FE - HG with a width of 3,70 m each and a length of 31 m for the aisles AB - FE and 26 m for the aisles CD - HG. There are also 2 transversal aisles PS - RP, the PS aisle has a width of 2,90 m and a length of 12,4 m, the RP aisle that connects the two warehouses internally has an initial width of 3,70 m up to the ramp and

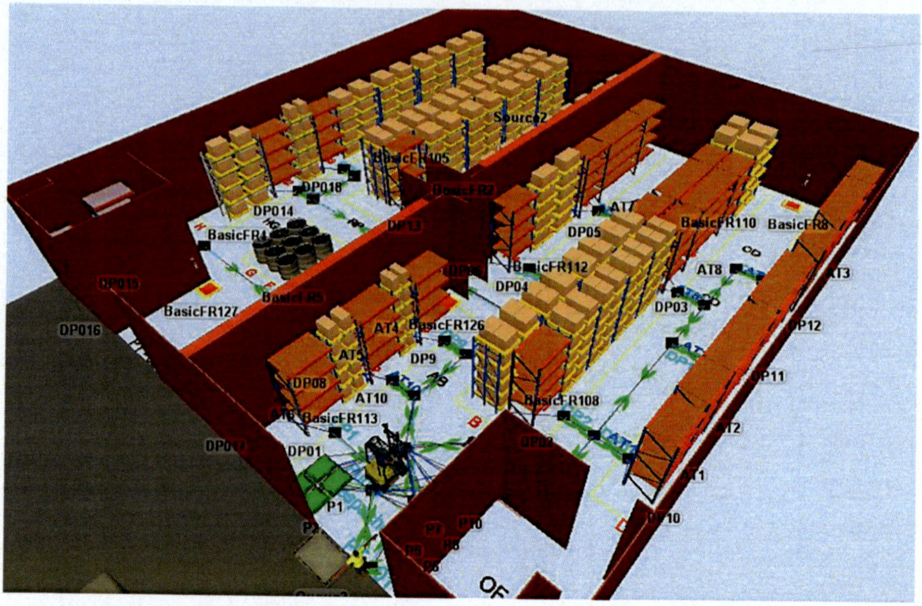

Fig. 6. Proposed raw material dispatch process. Software FlexSim.

is reduced to 1,60 m to the attached warehouse, its length is 14.7 m. A specific MC location is set aside for the orderly positioning of materials, equipment and tools used for handling and transporting internal raw materials, such as: ladders, metal cars and forklifts.

Comparison of Distances. The actual distances traveled are obtained from the measurement of the displacements for each of the processes presented. For the process of receiving raw materials there is a total distance traveled of 140,31 m for an unloading of 100 loads, 20 transports are used, because now the operator uses a forklift for handling and transport and a unit load (pallet) containing 10 (rolls of fabric or cardboard packages) and deposits them in each of the assigned points.

For the raw material storage process, the total distance traveled is 423,45 m for a permanent storage of 100 loads, 20 transports are used, the operator takes the entire unit load (pallet) containing 10 (rolls of fabric or cardboard packages) and deposits them in the permanent storage points (racks).

For the process of dispatching raw materials there is a total distance traveled of 987 m for a delivery of 18 loads (any item that requires production process) 36 transports are used because the operator takes the item in the stored place and deposits it at the dispatch point established for shipment, this process is done until the entire shipment is completed (Table 12).

Table 12. Analysis of results - Comparison of total distances traveled.

Process	Current			Proposal		
	N° Loads	N° of transports	Distance traveled (meters)	N° Loads	N° of transports	Distance traveled (meters)
Reception	100	200	1,419.82	10	20	140.31
Storage	100	80	2,009.02	10	20	423.45
Dispatch	18	36	1,113.59	18	36	978.00

The time taken for the proposed distribution is determined through an evaluation based on the activities, operations and transports, emphasizing the distances traveled for the two distributions and according to the activities that represent a significant reduction (Table 13).

Table 13. Analysis of the results - Comparison of process times.

Process	Current		Proposal	
	N° Distance (meters)	Time (h:min:s)	Distance (meters)	Time (h:min:s)
Reception	1,419.82	3:28:36	140.31	0:47:29
Storage	2,009.02	9:41:25	423.45	4:48:08
Dispatch	1,113.59	8:34:08	978.00	6:56:17

The comparison of the current vs. proposed cost is based on the analysis of material handling costs according to the loads handled and distances traveled, the cost determined is for each time the process is carried out (Table 14).

Table 14. Analysis of the results - Comparison of process times.

Process	Current			Proposal		
	Distance traveled (meters)	Cost × meter ($*meter)	Total ($)	Distance traveled (meters)	Cost × meter ($*meter)	Total ($)
Reception	1,419.82	0.043	61.05	140.31	0.043	6.03
Storage	2,009.02	0.043	86.39	423.45	0.043	18.21
Dispatch	1,113.59	0.043	47.88	978.00	0.043	42.05

4 Conclusions

The application of the IPISI methodology made it possible to diagnose the current conditions of the raw material warehouses of the case study company, which revealed an inadequate distribution of equipment and materials, inappropriate floor surfaces, lack of loading and unloading docks, limited maneuvering spaces, inadequate distance between aisles, lack of control in aspects related to labor and occupational safety, and the lack of performance indicators, causing problems in the processes of receiving, storing and dispatching raw materials.

In the ABC analysis, the products were categorized according to their level of consumption, identifying within category A the product families: Industrial Safety, Trekking, Casual, and Tactical Boot; subsequently the critical level of the raw materials stored in the facilities was determined, being: Soles, Insoles, Glues and Fabrics the most important since their non-existence will represent restrictions in production and delays in delivery times.

Based on the PRISMA methodology, the most relevant information was obtained regarding the distribution of facilities, storage systems, storage principles, distribution techniques, allocation and dispatch and innovation, which served as a reference for the design of the new distribution of facilities for the raw materials warehouse, and also through the application of the simple heuristic method, a more efficient distribution was proposed for each of the SKUs handled in the warehouse.

With the implementation of the new distribution of facilities, a reduction in the distances traveled is expected, in this sense, the distance traveled for the reception process decreases by 1.279,51 m, the distance traveled for the storage process decreases by 1.585,57 m, the distance traveled for the dispatch process decreases by 135,59 m, thanks to the reduction in the number of total transports for each process, also reducing the cost of material handling by 66%. The use of process simulation software has become an essential tool for the assessment of possible scenarios, in this study the proposal has been analyzed both graphically and results using the FlexSim software allowing to visualize and test changes in logistics and material handling operations.

References

1. Rueda, J.C.: Mejoramiento de los procesos de gestión de inventarios, almacenamiento y planeación de requerimiento de materias primas para la empresa calzado tiger pathfinder, con base en el software erp accasoft. Universidad Industrial De Santander, Bucaramanga, Santander (2015)
2. Mateo, P.S.: La importancia de las materias primas en la economía española. evolución histórica de precios y expectativas de futuro. 40Publicación cuatrimestral de la Caja de Ahorros y MP de Zaragoza, Aragón y Rioja (2009)
3. Rospigliosi Iparraguire, D.M.: Rediseño de almacén y su impacto en la gestión de almacenamiento de una empresa minera (2019)
4. Naranjo, M., Narvàez, R.: Distribución de las instalaciones en una fábrica de servicios. Caribeña de Ciencias Sociales, (mayo) (2018)
5. García, M.V., Irisarri, E., Pérez, F., Estévez, E., Marcos, M.: Automation architecture based on cyber physical systems for flexible manufacturing within oil&gas industry; [arquitectura de automatización basada en sistemas ciberfísicos para la fabricación flexible en la industria de petróleo y gas]. RIAI - Revista Iberoamericana de Automatica e Informatica Industrial **15**(2), 156–166 (2018)

6. García, C.A., Castellanos, E.X., García, M.V.: UML-based cyber-physical production systems on low-cost devices under IEC-61499. Machines **6**(2) (2018)
7. Serrano, M.J.E.: Logística de almacenamiento 2. Ediciones paraninfo, SA (2019)
8. García, C.A., et al.: Fuzzy control implementation in low cost CPPS devices. In: IEEE International Conference on Multisensor Fusion and Integration for Intelligent Systems, pp. 162–167 (2017)
9. Flamarique, S.: Manual de gestión de almacenes. Marge Books (2019)
10. Chiang, D.M.-H., Lin, C.-P., Chen, M.-C.: The adaptive approach for storage assignment by mining data of warehouse management system for distribution centres. Enterp. Inf. Syst. **5**(2), 219–234 (2011)
11. Henn, S., Schmid, V.: Metaheuristics for order batching and sequencing in manual order picking systems. Comput. Ind. Eng. **66**(2), 338–351 (2013)
12. Henn, S., Wäscher, G.: Tabu search heuristics for the order batching problem in manual order picking systems. Eur. J. Oper. Res. **222**(3), 484–494 (2012)
13. Zoubek, M., Poór, P., Broum, T., Šimon, M.: Methodology proposal for storage rationalization by implementing principles of industry 4.0. in a technology-driven warehouse. Trans. FAMENA **44**(4), 75–98 (2020)
14. Molins, F., Serrano, M.A.: The neural bases of loss aversion in economic contexts: a systematic review according to the PRISMA guidelines. Revista de Neurologia **68**(2), 47–58 (2019)
15. Page, M.J., et al.: The PRISMA 2020 statement: an updated guideline for reporting systematic reviews. Syst. Rev. **10**(1), 1–11 (2021)
16. Granados, J.A.V.: Metodología de diagnóstico logístico de almacenes y centros de distribución. Realidad y Reflexión, pp. 93–105 (2019)
17. Amorim-Lopes, M., Guimarães, L., Alves, J., Almada-Lobo, B.: Improving picking performance at a large retailer warehouse by combining probabilistic simulation, optimization, and discrete-event simulation. Int. Trans. Oper. Res. **28**(2), 687–715 (2021)
18. Posada, J.G.A.: Aspectos a considerar para una buena gestión en los almacenes de las empresas (centros de distribución, cedis). J. Econ. Financ. Adm. Sci. **16**(30), 83–96 (2011)
19. Saderova, J., Poplawski, L., Balog, M., Jr., Michalkova, S., Cvoliga, M.: Layout design options for warehouse management. Pol. J. Manag. Stud. **22**(2), 443 (2020)
20. Yang, D., Yaohua, W., Ma, W.: Optimization of storage location assignment in automated warehouse. Microprocess. Microsyst. **80**, 103356 (2021)
21. Leng, J., et al.: Digital twin-driven joint optimisation of packing and storage assignment in large-scale automated high-rise warehouse product-service system. Int. J. Comput. Integr. Manuf. **34**(7–8), 783–800 (2021)
22. Saderova, J., Rosova, A., Sofranko, M., Kacmary, P.: Example of warehouse system design based on the principle of logistics. Sustainability **13**(8), 4492 (2021)
23. Mirzaei, M., Zaerpour, N., de Koster, R.B.M.: How to benefit from order data: correlated dispersed storage assignment in robotic warehouses. Int. J. Prod. Res. **60**(2), 549–568 (2022)
24. Chen, Z., Li, X., Gupta, J.N.D.: A bi-directional flow-rack automated storage and retrieval system for unit-load warehouses. Int. J. Prod. Res. **53**(14), 4176–4188 (2015)
25. Cardona-Tunubala, J.L., Orejuela-Cabrera, J.P., Rojas-Trejos, C.A.: Gestión de inventario y almacenamiento de materias primas en el sector de alimentos concentrados. Revista eia **15**(30), 195–208 (2018)
26. Zadorozhnyi, Z.-M., Muravskyi, V., Pochynok, N., Hrytsyshyn, A.: Innovation management and automated accounting in the chaotic storage logistics (2020)

27. del Trabajo, S.G.R.: Decreto ejecutivo 2393 reglamento de seguridad y salud de los trabajadores y mejoramiento del medio ambiente de trabajo (1998)
28. Mejia, H., Wilches, M.J., Galofre, M., Montenegro, Y.: Aplicación de metodologías de distribución de plantas para la configuración de un centro de distribución. Scientia et technica **3**(49), 63–68 (2011)

Plant Distribution Based on a Resilient Approach in Textile SMEs

Franklin Tigre[1](✉), Estefanía Llerena[1], Carlos Sánchez[1], César Rosero[1,2], and Freddy Lema[1]

[1] Faculty of Systems, Electronics and Industrial Engineering, Technical University of Ambato, Ambato, Ecuador
{fg.tigre,ellerena4044,carloshsanchez,cesararosero,fr.lema}@uta.edu.ec, cesar.rosero@epn.edu.ec

[2] National Polytechnic School, Quito, Ecuador

Abstract. Flexibility, rapid response to changes and the ability to respond to external factors are strategies that organizations must develop to ensure their continuity in the market and face adverse situations. The research project presents a proposal for the distribution of facilities for the company dedicated to the manufacture of rest footwear, considering resilient aspects focused on the global pandemic caused by the COVID-19 virus. Through software programmed by the research group "ResilTEX" a management system is developed to solve problems of distribution of facilities, the parameters that are considered to define in the study are: distances traveled, cost of transfer of raw material, the application of the methodology of systematic design planning (SLP) and resilient indicators. The development of the project consists of distinct stages, beginning with an analysis of the current situation of the production process, the diagnosis of the facilities, the process flow diagram, and a time study. Resilient aspects are defined that are considered for the elaboration of the distribution of facilities applying the SLP method, where the proximity between work areas is evaluated; In this way, the "ResilTEX" software is used, it proposes two alternatives, the best one is alternative 2 with a cost of $97.54 with respect to material handling.

Keywords: Systematic Layout Planning · pandemic · distribution of facilities · resilience · indicators

1 Introduction

The design of manufacturing systems is an especially important topic. In conventional systems, the design rarely changes after its initial implementation. However, as market demands change more frequently, design configurations must be able to reconfigure the layout of resources to accommodate new production requirements, minimizing material handling and relocation costs and maximizing savings. in material flow and inventory costs [1].

Facility design issues have a significant impact on the productivity and efficiency of manufacturing systems [2] Companies use Six Sigma, Lean Manufacturing, and Facility Layout Optimization to efficiently operate production and

M. V. Garcia and C. Gordón-Gallegos (Eds.): CSEI 2022, LNNS 678, pp. 565–585, 2023.
https://doi.org/10.1007/978-3-031-30592-4_38

service systems [3]. In this time of change, caused by the health emergency due to COVID-19, production problems that affect companies and denote the fragility of their supply chains stand out. Evidence of this situation can be seen in small and medium-sized enterprises (SMEs) due to different factors: economic resources, little adaptability to new restrictions, and inflexible or null responses to external factors [4].

The spread of the SARS-CoV-2 virus has caused a worldwide humanitarian and economic crisis creating substantial detrimental impacts on the manufacturing industry [5]. Many countries have had to implement new policies to ensure the well-being of people, forcing confinements and cessation of non-essential activities, affecting the economy [6]. However, some companies close their economic activities, while others made changes to their organizational structures and production systems, causing considerable changes in the face of disruptive events. The sudden change forced companies to have a prominent level of adaptation and short response times, many of the changes made were made based on classic engineering methodologies or business administration proposals. Large and well-organized companies managed to reduce the impact of the pandemic, but this was not the case for many small and medium-sized companies that could not adapt to the new conditions and restrictions [3].

The context of the industry has changed according to two aspects: flexibility and resilience [7]. The facility design problem (FLP) is defined as the problem that seeks to organize the layout of a factory's facilities to develop its activities efficiently without violating any constraints [2]. For this reason, focusing on flexibility through the FLP will improve the performance of the industry in the current context of the health emergency. The latter can be defined as a continuous and adaptive process that allows organizations to face various serious problems and strategic challenges by improving their capacity to respond and reinvent themselves to achieve organizational innovation [8]. These two aspects together guarantee the continuity of production by reducing negative external impacts [9]Resilience is considered as the ability to withstand a shock (or a series of shocks) and recover the performance of a company, the recent outbreak of coronavirus COVID-19 It clearly shows the need for this new perspective [10].

The objective of this article is to propose the elements to consider in the distribution of the Plant, analyzing variables and resilient business factors most frequently applied in the industry. Finally, the relationship between two themes resilience and SLP is analyzed. The problem of the distribution of facilities is widely studied, however, the new conditions and restrictions of resilience and industrial hygiene are still under study. In the reviewed literature, resilience is considered from different points of view, including from the philosophical perspective, thus increasing the complexity of various theoretical models. While post-pandemic industrial hygiene activities are being evaluated in most companies, because the control agencies in the country gave general recommendations for the prevention of contagion, given that when a new virus is present, there is no sufficient information to mitigate the risks of contagion. In this case study, a rest footwear industry is analyzed, its industrial activities, the relationship between processes, to propose a distribution of facilities with a resilient approach.

2 Methodology

It begins by considering a theoretical approach to resilience and technical approaches to the facility distribution solution. Next, a theoretical approximation of the plant distribution problem is sought to analyze how to link resilient indicators and other restrictions that modify organizational behavior. In general, design problems deal with the optimization of facility space and locations (for example: machines, and departments) and are aimed at optimizing system performance within the facility space. The most widely used technique for facility design is SLP with the following approaches: exact, heuristic, metaheuristic, and evolutionary [2].

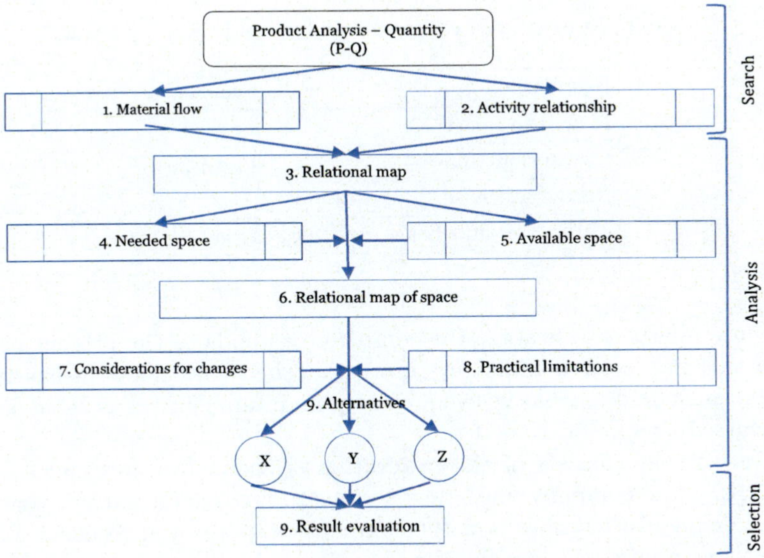

Fig. 1. SLP methodology

Research to determine a methodology to solve the distribution problem in a systematic way began in the 50s of the last centuries. However, Muther in 1961 was the first to find a method, ordered systematic distribution planning (SLP), applicable to the resolution of the problem regardless of its nature [11,12]. SLP interacts with variables such as: quantity of products to be transported, production processes, movement of people, transportation costs, as well as special restrictions on the shared use of certain equipment [13–15]. The relationship between these variables can be assumed for several reasons: proximity, connection (maintain flows), object integration (maintain the relationship with contour and spatial integration) and configuration (spatial area designation), an example of the classic model is found in Fig. 1, where: P = product, Q = quantity, R = restrictions, S = special relationships.

Based on this model, it is proposed to modify the analysis section, in order to add evaluations from the resilient point of view. According to the literature studied, a starting point is through models that contain industrial safety criteria. To start with the plant distribution model, the resilience phases and the industrial safety criteria are defined, to then define the indicators to be incorporated into the modified SLP. The practical use of resilience requires an understanding of how the four skills fit together and thus depend on each other, hence the importance of thinking about the built system as a whole and providing effective concepts and methods to manage overall performance. The dependencies are show in Table 1, where they have been classified into 3 levels related to security [16,17].

Table 1. Approaches between Resilience and Security levels

Resilience	Security
Response phase	Level 1
Prevention Phase	Level 2
Recovery Phase	
Mitigation and Learning Phase	Level 3

Response phase, in this stage the company responds to the present negative event, in this way several indicators are established that allow measuring the response it must said a negative event or help maintain control as the disruptive event occurs [16,18].

Recovery phase consists of the restoration of social routines and economic activities [19]. The company must take actions to face this negative event, indicators are defined to measure how the company prepares and recovers.

Prevention phase, the company learns from the negative event, and the defined indicators can be a key component for it to be prepared for a new negative disruptive event [20].

Mitigation and learning phase, indicators are defined that help to visualize how the system obtained experiences of the disruptive event, and how the effect that originated in a certain term can be measured to create a greater strategic development [20].

With the approaches analyzed, the indicators that will be used to link them with the SLP are defined. For the model in question, the pandemic caused by covid-19 will be used as a disruptive event, and a search is made in the literature for possible indicators.

Resilience can be measured with objective and subjective indicators, the former can deal with a descriptive statistical analysis based on historical data, while the latter can use human understanding justified in dimensions and scale measurements, resilience is a broad concept and complex that is considered a key component of sustainability [21].

The indicators in security systems are based on management models such as the International Standard Organization (ISO), which proposes guidelines for establishing indicators, in the same way as resilience indicators, they can be objective and subjective. However, in terms of industrial safety and health at work, the organizations themselves can implement their indicators considering the particularities of their system. It must be considered that objective indicators base their robustness on the amount of data collected, while subjective indicators are widely used, since they capture the human understanding of the concept of resilience, they are criticized for their level of precision [22]. It must be considered that the objective indicators do not allow the results obtained by different studies to be compared, since the conditions and restrictions are different [21].

Once the criteria or phases of resilience have been established, it is necessary to define indicators in each of these phases to be considered with the application of the SLP methodology, in this way a literary review was carried out to define and establish indicators that are focused on business resilience; with the help of different experts in the area, "Incorporating Sustainability concepts to management models of textile Micro, Small and Medium Enterprises (SUMA)" and "Resilient plant distribution model for SMEs with a focus on productivity and occupational safety (ResilTEX)", Table 2 presents information on the proposed indicators.

Table 2. Proposed indicators for each resilient phase for the disruptive event

resilient phase	Indicators
Response	Expenses for accidents (EA), incidents (EI) and occupational diseases (OD)
	% Applicability of corrective actions against Covid-19 cases
	% Of distancing compliance in each plant area
	% Implementation of a security management system
	% Perception of safety of co-workers
	% Of evaluation/risk management/risk matrices raised
	% Of legal compliance of the security system regarding the location of machines
Recovery/Prevention	% Applicability of preventive actions
	% Compliance with planned measures
	% Reduction in the rate of mechanical risks
	% Of preventive plans for catastrophic security events
	% Of infections given in the company by Covid -19
Mitigation and learning	% Of actions implemented after a disruptive event - Covid-19
	% System flexibility in the face of catastrophic events

In phase one, the indicators measure how the company reacts to the disruptive event (pandemic), in phase two it analyzes the actions that could anticipate the disruptive event, and in phase three they evaluate the actions carried out by the company that allow learning of this event.

Proposal for the Modified SLP

By having various considerations, all of them of a subjective nature and due to the complexity related to obtaining information, from the traditional approach starting with the SLP and modifying the analysis stage, Fig. 1, to later optimize the process through the method quadratic assignment adjusted to the case study. The SLP consists of several steps and levels, however, in the case study, it is proposed to modify steps 6, 7 and 8 of the classic models, the methodology to be followed is as follows:

Search level

P-Q Analysis. - Identify the relationship between the products and the amount of raw material required, to establish the importance between the location of the material and the assigned process.

1. **The flow of Materials**. - Evaluate the current flow of the system considering the distribution and available space, if possible, assess the traffic between sections, the distance matrix or the cost evaluation matrix. Having the layout and its structural limitations is essential at this point, to identify possible relationships of interest or desirable distances.
2. **Relationship between activities. -** Identify how the activities are linked or not in the process, the importance of the proximity between the areas is evaluated.
3. **Process Relationship Diagram. -** Represent graphically how the plant is currently distributed and the relationship between the areas and the surface they occupy.

4–5. Space Requirements vs. Available Space. - Determine the distances, considering structural obstacles, the feasibility of a plant distribution will be established in the current conditions of the system, considering the production processes using the scale layout.

Analysis level

6. **Modified Space Relations Diagram** Table 3 the different proximity evaluators are described to establish the relationship that exists between the different sections of the plant, representing each one with a code.

As it is a qualitative evaluation, the proximity values are assigned subjectively based on the knowledge and experience of the personnel who work in the company, Table 4 so scales were used to assign the weight or importance of each proximity reason. These subjective criteria provide a first approximation of a solution, which must be analyzed and evaluated under other specific criteria depending on each system, Fig. 2.

Table 3. Proximity Evaluators

Code	Proximity evaluators
A	Absolutely necessary
E	specifically important
I	Important
O	normal proximity
U	Not necessary
X	Undesirable

Table 4. Relationship of indicators with the proximity ratio

Code	proximity ratio
1	Safe flow of materials and people
2	staff security
3	Avoid an unsafe act
4	Prevention

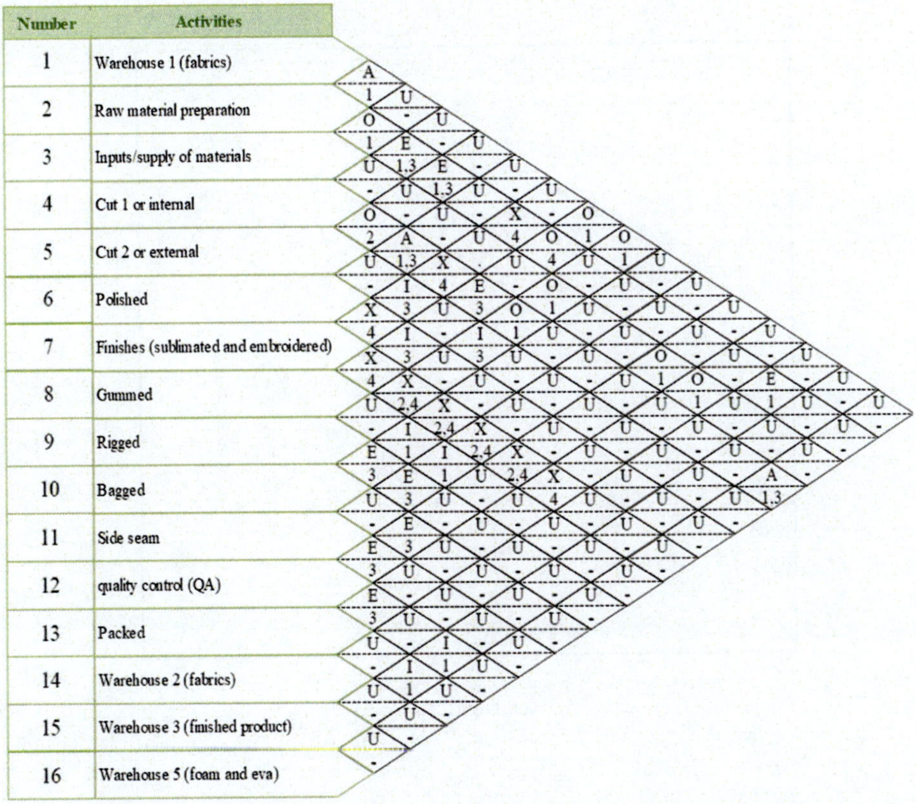

Fig. 2. Relationship between activities of the company CM Original

Table 5 indicates the proximity line code that is used to carry out the connection between the different activities according to the proximity evaluators that were previously presented in Table 4; In addition, the number of relationships that exist in the case study is show Fig. 3.

Table 5. Proximity lines code

Code	proximity evaluators	line code	Color	Number
A	Absolutely necessary		Green	3
E	specifically important		Blue	9
I	Important		Yellow	7
O	normal proximity		Orange	9
U	Not necessary	-	-	-
X	Undesirable		Black	9

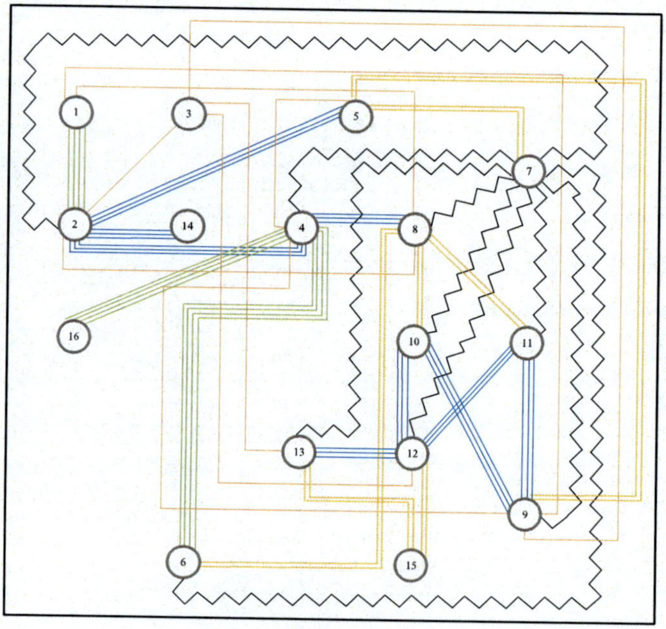

Fig. 3. Relationship diagram

3 Result and Discussions

This section details the information collected from the case study company, which contributed to the development of the project and the use of the software programmed by the participants of the "ResilTEX" research project. This software was designed for the management of plant distribution problems, first the general information must be registered, then the data referring to the product being analyzed is entered, finally, the best installation distribution proposal is selected.

Stage 1: Product management. - The products to be analyzed are placed in the plant layout. For the case study, the demand for the slingback slipper was analyzed, Table 6 show the ABC analysis, based on historical data according to the monthly demand that the product had in the last year 2020.

Table 6. ABC analysis of the products of the case study (Summary)

Zone	Range	No. of products	% Of investment
A	0–90%	5	77.81%
B	81–95%	4	15.68%
C	96–100%	12	6.51%

In the case study, the necessary information on the processes was collected as show Fig. 4.

Stage 2: Management of areas or work center. - In this stage the work areas are registered, the areas of the entire industrial plant were considered. It should be noted that for the analysis carried out later, only the production areas were considered, that is, the administrative and design areas are placed as fixed areas within the software. A matrix from - to is obtained, in which the number of trips between work areas is placed, in this case there are 31 trips in total as show Fig. 5.

Stage 3: Management of machinery or areas. - The layout of the company's areas is represented, starting from the center of each work area taking the coordinates in X, and, these data are entered in the software, it is also necessary to know the width and height of the areas placed in the stage 2. Due to this, Fig. 6 contains said information that was taken from the current distribution of the company, placing the point (0,0) in the lower left part of the current layout as a reference.

Figure 7 shows the matrix of the centers of gravity of each area, and Fig. 8 shows the current distribution of the company.

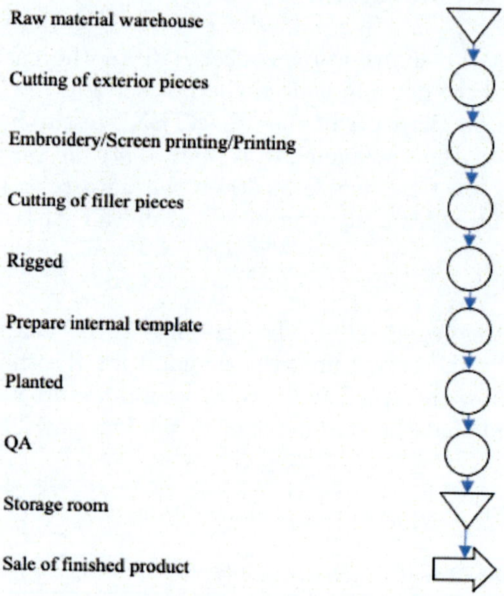

Fig. 4. Flow of the operation for manufacturing footwear

Stage 2 - Management of areas or work centers

Registration of areas or work centers

	Warehouse 1	Raw material	Supply of materials	Warehouse 2	Cutting 1	Cutting 2	Gumming	Finishing	Side seam	Trimming	Bagging	Quality Control	Warehouse 3	Polishing	Packaging	Warehouse 5	Design area	Administrative area
Warehouse 1	0	0	0	0	0	0	0	0	0	0	0	0	0	0	0	0	0	0
Raw material preparation	0	0	0	0	1	2	0	0	0	0	0	0	0	0	0	0	0	0
Supply of materials	0	0	0	0	0	0	0	0	0	0	0	0	0	0	0	0	0	0
Warehouse 2	0	0	0	0	0	0	0	0	0	0	0	0	0	0	0	0	0	0
Cutting 1	0	0	0	0	0	0	4	0	0	3	0	0	0	1	0	0	0	0
Cutting 2	0	0	0	0	0	0	0	3	0	4	0	0	0	0	0	0	0	0
Gumming	0	0	0	0	0	0	0	0	0	0	1	0	0	0	0	0	0	0
Finishing	0	0	0	0	0	0	0	0	0	3	0	0	0	0	0	0	0	0
Side seam	0	0	0	0	0	0	0	0	0	0	0	0	0	0	0	0	0	0
Trimming	0	0	0	0	0	0	0	0	0	0	2	0	0	0	0	0	0	0
Bagging	0	0	0	0	0	0	0	0	0	0	0	1	0	0	0	0	0	0
Quality Control	0	0	0	0	0	0	0	0	0	0	0	0	1	0	0	0	0	0
Warehouse 3	0	0	0	0	0	0	0	0	0	0	0	0	0	0	1	0	0	0
Polishing	0	0	0	0	1	0	0	0	0	0	0	0	0	0	0	0	0	0
Packaging	0	0	0	0	0	0	0	0	0	0	0	0	0	0	0	0	0	0
Warehouse 5	0	0	0	0	3	0	0	0	0	0	0	0	0	0	0	0	0	0
Design area	0	0	0	0	0	0	0	0	0	0	0	0	0	0	0	0	0	0
Administrative area	0	0	0	0	0	0	0	0	0	0	0	0	0	0	0	0	0	0

Fig. 5. Matrix from-to number of trips

Stage 4: Distance Between Components. - The research project "SUMA" provides the cost of transfer per minute found in Table 7.

For this stage, it is necessary to know the cost of transporting the material, for which, based on the cost in Table 7, the cost of transport per meter is calculated for one pair and for a production lot of 143 pairs. This requires the rectilinear distances traveled by the operator, taken from the current distribution of the

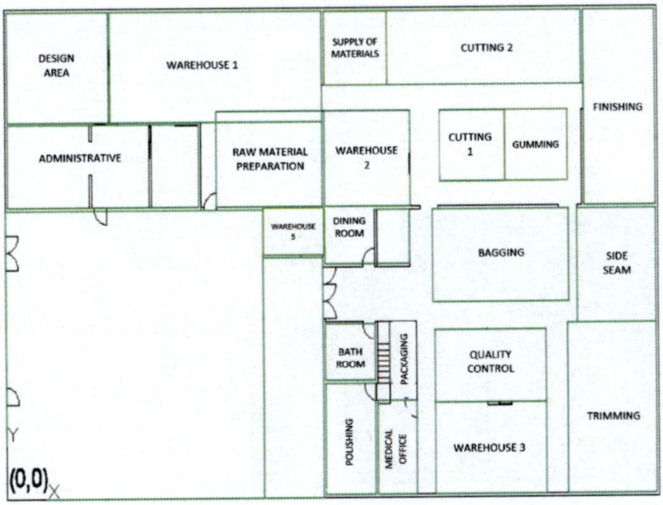

Fig. 6. Reference for centers of gravity

Stage 3: Machinery management
Center of gravity matrix

	Vertex 1		Vertex 2		Vertex 3		Vertex 4		Center of gravity		Dimensions	
Name	X1	Y1	X2	Y1	X2	Y2	X1	Y2	Xm	Ym	Width	Height
Warehouse 1	7.655	19.605	22.965	19.605	22.965	25.335	7.655	25.335	15.310	22.470	15.310	5.730
Raw material preparation	15.265	15.180	22.955	15.180	22.955	20.160	15.265	20.160	19.110	17.670	7.690	4.980
Supply of materials	23.120	21.490	27.280	21.490	27.280	25.330	23.120	25.330	25.200	23.410	4.160	3.840
Warehouse 2	23.125	15.180	29.395	15.180	29.395	20.160	23.125	20.160	26.260	17.670	6.270	4.980
Cutting 1	30.525	1.425	35.275	1.425	35.275	5.135	30.525	5.135	32.900	3.280	4.750	3.710
Cutting 2	27.285	21.490	41.755	21.490	41.755	25.330	27.285	25.330	34.520	23.410	14.470	3.840
Gumming	35.275	1.425	39.045	1.425	39.045	5.135	35.275	5.135	37.160	3.280	3.770	3.710
Finishing	41.920	15.185	47.100	15.185	47.100	25.335	41.920	25.335	44.510	20.260	5.180	10.150
Side seam	31.140	15.190	40.380	15.190	40.380	18.850	31.140	18.850	35.760	17.020	9.240	3.660
Trimming	40.735	6.140	47.105	6.140	47.105	15.020	40.735	15.020	43.920	10.580	6.370	8.880
Bagging	31.030	10.160	40.790	10.160	40.790	15.020	31.030	15.020	35.910	12.590	9.760	4.860
Quality Control	31.150	5.895	39.150	5.895	39.150	9.685	31.150	9.685	35.150	7.790	8.000	3.790
Warehouse 3	39.100	0.160	47.100	0.160	47.100	4.920	39.100	4.920	43.100	2.540	8.000	4.760
Polishing	23.125	0.160	26.755	0.160	26.755	5.920	23.125	5.920	24.940	3.040	3.630	5.760
Packaging	27.815	5.925	29.785	5.925	29.785	9.175	27.815	9.175	28.800	7.550	1.970	3.250
Warehouse 5	18.610	12.475	21.410	12.475	21.410	15.025	18.610	15.025	20.010	13.750	2.800	2.550
Design area	0.165	19.605	7.495	19.605	7.495	25.335	0.165	25.335	3.830	22.470	7.330	5.730
Administrative area	0.165	15.180	13.995	15.180	13.995	19.440	0.165	19.440	7.080	17.310	13.830	4.260

Fig. 7. Center of gravity matrix

Table 7. Cost provided by the "SUMA" project.

Model	Cost per minute of transfer in cents	Unit of measurement
slingback	$0.076342	Cost per minute to move 1 pair

company. In this way, Table 8 shows the total distance, the total time and the meters covered in one minute.

Next, the cost of transfer by meter is presented; For a pair of slippers, the calculation is made by dividing the cost by the distance traveled in one minute, finally obtaining the cost of 0.25 usd for the production batch of 143 pairs. This calculated cost is placed in the software and the distances between work

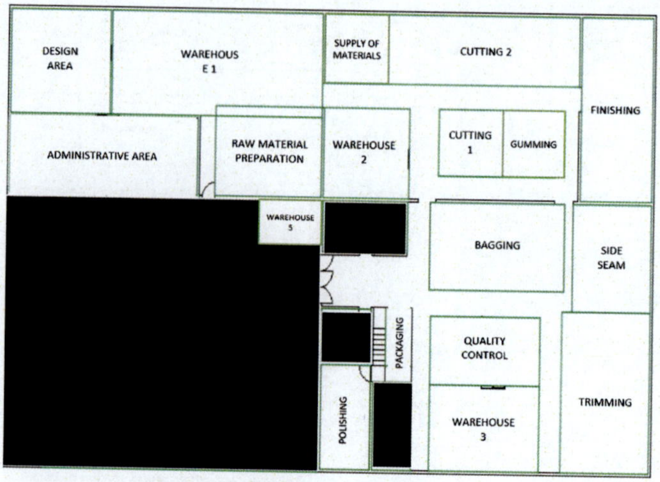

Fig. 8. Current distribution in the ResilTEX software

Table 8. Distance traveled in one minute

Since	Until	Distance (m)	time(s)
Warehouse 1	Raw material preparation	7.75	fifteen
Raw material preparation	Cutting area 1	15.56	28
Raw material preparation	Cutting Area 2	21.21	38
Cutting Area 2	Finishing area	4.34	fifteen
Warehouse 5	Cutting area 1	25.90	43
Cutting area 1	Polishing Area	17.61	fifteen
Polishing Area	Cutting area 1	17.61	fifteen
Cutting Area 1	gumming area	9.02	8
gumming area	bagging area	12.71	25
Finishing area (capelladas)	parking area	17.82	twenty
Cutting area 2 (template strips)	parking area	18.49	25
Cutting area 1 (liners)	parking area	15.90	10
parking area	bagging area	13.61	16
bagging area	Quality control area	12.94	12
Quality control area	Warehouse 3	2.68	5
Warehouse 3	packing area	7.89	12
TOTAL		221.04	302
Time (min)			5.03
Distance traveled in one minute (m)			43.92

areas are automatically calculated, which are from center of gravity to center of gravity, that is, they are Euclidean distances; Fig. 9 shows the distances between components.

Stage 4 - Distance between components

0.25

Automatic distance calculation | Apply default cost (1$)

	Warehous e 1	Raw material preparatio n	Supply of materials	Warehous e 2	Cutting 1	Cutting 2	Gumming	Finishing	Side seam	Trimming	Bagging	Quality Control	Warehous e 3	Polishing	Packaging	Warehous e 5	Design area	Administrative area
Warehouse 1	0	2.150	2.708	3.938	5.698	5.038	6.765	7.853	9.850	11.620	7.620	8.865	9.938	7.265	7.103	3.355	2.870	3.348
Raw material preparation	2.150	0	2.958	1.788	3.863	5.288	4.930	6.998	7.700	9.470	5.470	6.715	7.788	5.115	4.953	1.205	5.020	3.098
Supply of materials	2.708	2.958	0	1.700	3.460	2.330	4.528	5.615	7.613	9.383	5.383	6.628	7.700	5.158	4.865	3.713	5.578	6.055
Warehouse 2	3.938	1.788	1.700	0	2.075	3.500	3.143	5.210	5.913	7.683	3.683	4.928	6.000	3.988	3.165	2.543	6.808	4.885
Cutting 1	5.698	3.863	3.460	2.075	0	1.425	1.068	3.135	4.153	5.923	1.923	3.168	4.240	6.063	3.970	4.618	8.568	6.960
Cutting 2	5.038	5.288	2.330	3.500	1.425	0	2.198	3.285	5.283	7.053	3.053	4.298	5.370	7.488	5.395	6.043	7.908	8.385
Gumming	6.765	4.930	4.528	3.143	1.068	2.198	0	2.068	3.085	4.855	2.000	3.635	4.708	7.130	5.038	5.685	9.635	8.028
Finishing	7.853	6.998	5.615	5.210	3.135	3.285	2.068	0	2.118	4.063	4.068	5.703	6.775	9.198	7.105	7.753	10.723	10.095
Side seam	9.850	7.700	7.613	5.913	4.153	5.283	3.085	2.118	0	1.945	2.230	3.585	4.658	7.080	4.988	6.495	12.720	10.618
Trimming	11.620	9.470	9.383	7.683	5.923	7.053	4.855	4.063	1.945	0	4.000	2.755	2.713	5.135	4.518	8.265	14.490	12.388
Bagging	7.620	5.470	5.383	3.683	1.923	3.053	2.000	4.068	2.230	4.000	0	1.635	2.708	5.130	3.038	4.265	10.490	8.388
Quality Control	8.865	6.715	6.628	4.928	3.168	4.298	3.635	5.703	3.585	2.755	1.635	0	1.073	3.495	1.763	5.510	11.735	9.633
Warehouse 3	9.938	7.788	7.700	6.000	4.240	5.370	4.708	6.775	4.658	2.713	2.708	1.073	0	2.673	2.835	6.583	12.808	10.705
Polishing	7.265	5.115	5.158	3.988	6.063	7.488	7.130	9.198	7.080	5.135	5.130	3.495	2.673	0	2.093	3.910	10.135	8.033
Packaging	7.103	4.953	4.865	3.165	3.970	5.395	5.038	7.105	4.988	4.518	3.038	1.763	2.835	2.093	0	3.748	9.973	7.870
Warehouse 5	3.355	1.205	3.713	2.543	4.618	6.043	5.685	7.753	6.495	8.265	4.265	5.510	6.583	3.910	3.748	0	6.225	4.123
Design area	2.870	5.020	5.578	6.808	8.568	7.908	9.635	10.723	12.720	14.490	10.490	11.735	12.808	10.135	9.973	6.225	0	2.103
Administrative area	3.348	3.098	6.055	4.885	6.960	8.385	8.028	10.095	10.618	12.388	8.388	9.633	10.705	8.033	7.870	4.123	2.103	0

Fig. 9. Distance between components.

Stage 5: Practical limitations. - It is based on the limitations that could be generated when moving an area from one place to another, or limitations to move the machinery. For the case study, there are no practical limitations, this is omitted in the software, as shown in Fig. 10.

Stage 5 - Practical constraints

Id	Name	Description
19	No practical constraints	No practical constraints were raised in this case study.

Fig. 10. Practical limitations

Stage 6: Evaluation of indicators. - It reflects the indicators considered to be evaluated in the disruptive event, an assessment obtained through surveys carried out on the company's personnel, these indicators raised. The criteria were given by the "ResilTEX " research group and the score of the indicator recorded by the company contrasted with its reality. In Fig. 11 there is a summary of the indicators with the global value established for each of these, for the value of expenses and flexibility of the system in the event of catastrophic events, the company does not have a specific value, for which a value of 0 was assigned.

Stage 7: Relationship of indicators with proximity ratio. - The information corresponding to the distribution of facilities was taken by applying SLP stage 1, so the 4 proximity reasons for this analysis are:

Stage 6 - Evaluation of indicators

Code	Name	Descripción	Unit	Value
I01	Expenditures	Accidents (AE), incidents (IE) and occupational illnesses (OE) Expenses	Money	0.0
I02	Corrective corrective actions	% Applicability of corrective actions for Covid-19 cases	Percentage	100.0
I03	Distancing	% Compliance with distancing in each plant area	Percentage	90.0
I04	Safety management Management	% Implementation of a safety management system	Percentage	82.0
I05	Perception of Security	% Perception of safety by co-workers % Risk assessment / management / enrollment	Percentage	97.7
I06	Risk assessment Risk assessment	% Risk assessment / risk management / risk matrixes raised	Percentage	100.0
I07	Legal compliance	% legal compliance of the safety system with respect to location of machinery	Percentage	100.0
I08	Preventive Preventive actions	% applicability of preventive actions	Percentage	100.0
I09	Planned Planned measures	% Compliance with planned measures	Percentage	84.0
I10	Mechanical Risks	% Reduction of mechanical risk rate % Preventive plans for safety events	Percentage	15.0
I11	Catastrophic catastrophic events	% preventive plans for catastrophic safety events % # contagions occurring in the plant	Percentage	86.0
I12	COVID Contagions	% # of contagions given in the company by Covid - 19	Percentage	29.0
I13	Post-contagion actions	% actions implemented after a disruptive event - Covid 19	Percentage	90.0
I14	Flexibility of the system	% flexibility of the system in the face of catastrophic events	Percentage	0.0

Fig. 11. Global value of indicators

1. Safe flow of materials and people
2. Staff security
3. Avoid an unsafe act
4. Prevention

In Fig. 12 the results of proximity are show of case of study.

Stage 8: Relationship diagram. - In this stage, the evaluators and proximity ratio are placed, this can be seen in Fig. 13, in addition, the relationship diagram is placed. Figure 14 shows the analysis that was carried out of the distribution of facilities applying the modified SLP.

Stage 9: Diagram of spatial relationships - alternative graphic distribution. - The ResilTEX software allows to place a proposed base distribution for the company, in this way it is proposed to move some work areas according to the criteria presented in the relationship diagram, Fig. 15. In Fig. 16 you can see the proposed distribution 1 of the company, detailing the centers of gravity, height and width of the areas that will be redistributed within the production plant.

Stage 7 - Relation of indicators with proximity ratios

Code	Proximity ratio	Indicator	Value
1	Safe flow of materials and people	Distancing	90
1	Safe flow of materials and people	System flexibility	0
1	Safe flow of materials and people	Legal compliance	100
2	Personnel safety	Distancing	90
2	Personnel safety	Post-contagion actions	90
2	Personnel safety	Security management	82
2	Personnel safety	Perception of security	97.73
2	Personnel Safety	Planned measures	84
2	Personnel safety	Catastrophic events	86
3	Preventing an unsafe act	Corrective actions	100
3	Preventing an unsafe act	COVID Contagions	29
3	Preventing an unsafe act	Mechanical risks	15
3	Preventing an unsafe act	Risk assessment	100
4	Prevention	Preventive actions	100
4	Prevention	Post-contagion actions	90
4	Prevention	Risk assessment	100
4	Prevention	Planned measures	84

Fig. 12. Relationship of indicators with proximity ratio

Stage 8 - Relationship diagram

Reference information

Proximity evaluators

Code	Name
A	Absolutely necessary
E	Specifically important
I	Important
O	Normal proximity
U	Not necessary
X	Not desirable

Proximity reasons

Code	Name
1	Safe flow of materials and people
2	Personnel safety
3	Preventing an unsafe act
4	Prevention

Fig. 13. Referential information-relationship diagram

Stage 10: Matrix of alternative distances. - From stage 9, the distances for the proposed distribution are obtained, in the same way the cost of moving material presented in stage 4 is used: distance between components. Figure 17 shows the matrix of alternative distances.

Cost calculation for the current and proposed distribution. - Once the plant distribution problem management stages have been carried out, the algorithm is executed. In addition, the complexity of the plant distribution problem is calculated, that is, the closer the value is to 0, the more complex the problem resolution will be. In this case, the software provides a value of 57,847 as shown in Fig. 18.

The software provides a cost based on all the factors entered, such as the travel matrix, distances, resilient indicators, and relationship diagram; performs a weighting with all the data entered. This is how the initial cost according to

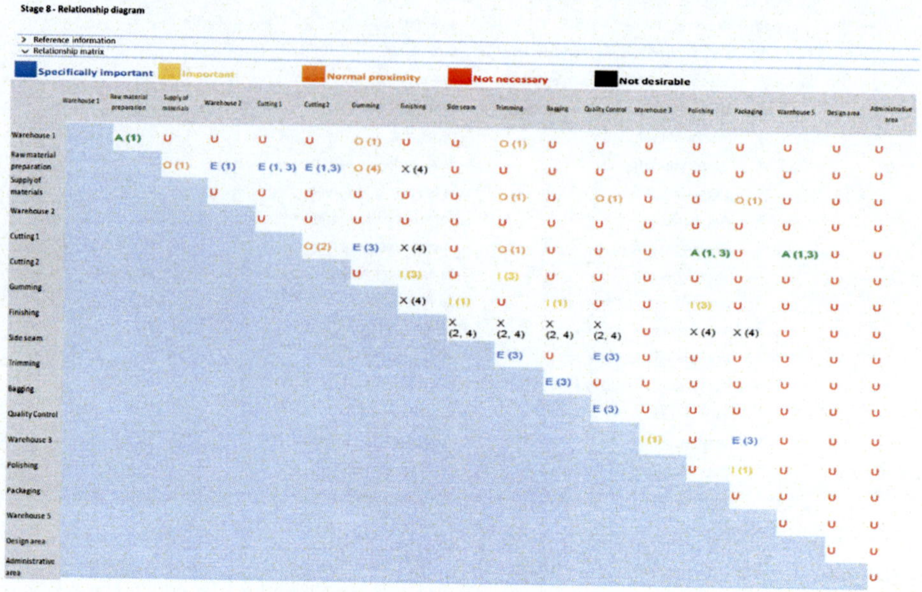

Fig. 14. Relationship diagram

Stage 9: Spatial Relationship Diagram - Alternative Graphical Layout
Alternative center of gravity matrix

Name	Vertex 1 X1	Vertex 1 Y1	Vertex 2 X2	Vertex 2 Y1	Vertex 3 X2	Vertex 3 Y2	Vertex 4 X1	Vertex 4 Y2	Center of gravity Xm	Center of gravity Ym	Dimensions Width	Dimensions Height
Warehouse 1	7.655	19.605	22.965	19.605	22.965	25.335	7.655	25.335	15.31	22.47	15.31	5.73
Raw material preparation	15.265	15.18	22.955	15.18	22.955	20.16	15.265	20.16	19.11	17.67	7.69	4.98
Supply of materials	23.12	21.49	27.28	21.49	27.28	25.33	23.12	25.33	25.2	23.41	4.16	3.84
Warehouse 2	23.125	15.18	29.395	15.18	29.395	20.16	23.125	20.16	26.26	17.67	6.27	4.98
Cutting 1	31.555	16.445	36.305	16.445	36.305	20.155	31.555	20.155	33.93	18.3	4.75	3.71
Cutting 2	27.285	21.49	41.755	21.49	41.755	25.33	27.285	25.33	34.52	23.41	14.47	3.84
Gumming	36.345	16.615	40.055	16.615	40.055	19.985	36.345	19.985	38.2	18.3	3.71	3.37
Finishing	41.92	15.185	47.1	15.185	47.1	25.335	41.92	25.335	44.51	20.26	5.18	10.15
Side seam	41.44	9.045	47.1	9.045	47.1	15.015	41.44	15.015	44.27	12.03	5.66	5.97
Trimming	40.735	0.16	47.105	0.16	47.105	9.04	40.735	9.04	43.92	4.6	6.37	8.88
Bagging	31.03	10.16	40.79	10.16	40.79	15.02	31.03	15.02	35.91	12.59	9.76	4.86
Quality Control	31.13	4.935	39.13	4.935	39.13	8.725	31.13	8.725	35.13	6.83	8	3.79
Warehouse 3	31.13	0.16	39.13	0.16	39.13	4.92	31.13	4.92	35.13	2.54	8	4.76
Polishing	23.125	0.16	26.755	0.16	26.755	5.92	23.125	5.92	24.94	3.04	3.63	5.76
Packaging	27.815	5.925	29.785	5.925	29.785	9.175	27.815	9.175	28.8	7.55	1.97	3.25
Warehouse 5	18.61	12.475	21.41	12.475	21.41	15.025	18.61	15.025	20.01	13.75	2.8	2.55
Design area	0.165	19.605	7.495	19.605	7.495	25.335	0.165	25.335	3.83	22.47	7.33	5.73
Administrative area	0.165	15.18	13.995	15.18	13.995	19.44	0.165	19.44	7.08	17.31	13.83	4.26

Fig. 15. Centers of gravity for proposal 1

the current distribution of the company is $95.58 for handling material, considering the factors of resilience and relationship diagram, meeting the demand established above.

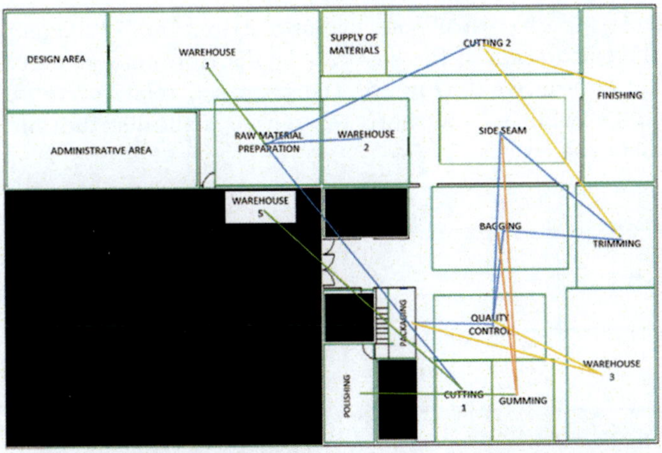

Fig. 16. Proposed distribution 1, "ResilTEX" software

Stage 10 - Distance between components

Cost of moving material: 0.25

Calculate alternative distances

	Warehouse 1	Raw material preparation	Supply of materials	Warehouse 2	Cutting 1	Cutting 2	Gumming	Finishing	Side seam	Trimming	Bagging	Quality Control	Warehouse 3	Polishing	Packaging	Warehouse 5	Design area	Administrative area
Warehouse 1	0	2.150	2.708	3.938	9.195	5.038	10.260	7.853	6.475	10.125	7.620	8.630	11.930	7.265	7.103	3.355	2.870	3.348
Raw material preparation	2.150	0	2.958	1.788	7.045	5.288	8.110	6.998	4.325	7.975	5.470	6.480	9.780	5.115	4.953	1.205	5.020	3.098
Supply of materials	2.708	2.958	0	1.700	6.958	2.330	8.023	5.615	4.238	7.888	5.383	6.393	9.693	5.158	4.865	3.713	5.578	6.055
Warehouse 2	3.938	1.788	1.700	0	5.258	3.500	6.323	5.210	2.538	6.188	3.683	4.693	7.993	3.988	3.165	2.543	6.808	4.885
Cutting 1	9.195	7.045	6.958	5.258	0	5.438	1.065	7.148	4.150	4.580	3.080	1.690	2.735	2.050	2.093	5.840	12.065	9.963
Cutting 2	5.038	5.288	2.330	3.500	5.438	0	5.693	3.285	1.908	5.558	3.053	4.063	7.363	7.488	5.395	6.043	7.908	8.385
Gumming	10.260	8.110	8.023	6.323	1.065	5.693	0	6.083	3.785	3.515	2.640	1.630	1.670	3.115	3.158	6.905	13.130	11.028
Finishing	7.853	6.998	5.615	5.210	7.148	3.285	6.083	0	2.998	2.568	4.068	5.458	4.783	9.198	7.105	7.753	10.723	10.095
Side seam	6.475	4.325	4.238	2.538	4.150	1.908	3.785	2.998	0	3.650	1.145	2.460	5.455	6.200	4.108	4.755	9.345	7.243
Trimming	10.125	7.975	7.888	6.188	4.580	5.558	3.515	2.568	3.650	0	2.505	2.890	2.215	6.630	4.538	6.770	12.995	10.893
Bagging	7.620	5.470	5.383	3.683	3.080	3.053	2.640	4.068	1.145	2.505	0	1.390	4.310	5.130	3.038	4.265	10.490	8.388
Quality Control	8.630	6.480	6.393	4.693	1.690	4.063	1.630	5.458	2.460	2.890	1.390	0	3.300	3.740	1.648	5.275	11.500	9.398
Warehouse 3	11.930	9.780	9.693	7.993	2.735	7.363	1.670	4.783	5.455	2.215	4.310	3.300	0	4.665	4.828	8.575	14.800	12.698
Polishing	7.265	5.115	5.158	3.988	2.050	7.488	3.115	9.198	6.200	6.630	5.130	3.740	4.665	0	2.093	3.910	10.135	8.033
Packaging	7.103	4.953	4.865	3.165	2.093	5.395	3.158	7.105	4.108	4.538	3.038	1.648	4.828	2.093	0	3.748	9.973	7.870
Warehouse 5	3.355	1.205	3.713	2.543	5.840	6.043	6.905	7.753	4.755	6.770	4.265	5.275	8.575	3.910	3.748	0	6.225	4.123
Design area	2.870	5.020	5.578	6.808	12.065	7.908	13.130	10.723	9.345	12.995	10.490	11.500	14.800	10.135	9.973	6.225	0	2.103
Administrative area	3.348	3.098	6.055	4.885	9.963	8.385	11.028	10.095	7.243	10.893	8.388	9.398	12.698	8.033	7.870	4.123	2.103	0

Fig. 17. Alternative distances provided by the software

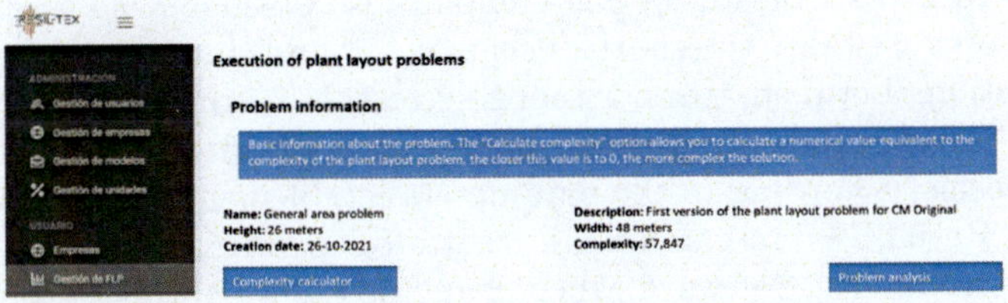

Fig. 18. System complexity value

After calculating the cost of the current distribution of the company, it continues with the calculation of the cost for the distribution for the two proposals presented. For alternative 1 (Fig. 16) the proposed relative cost is 133.62 for material handling, while for alternative 2 another plant distribution is presented in Fig. 19, where the cost is $97.54.

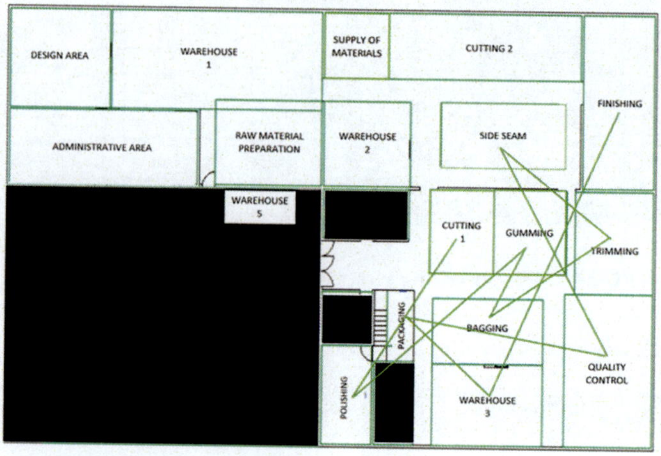

Fig. 19. Proposed distribution 2, "ResilTEX" software

Below is a summary in Table 9 with the cost of the current distribution and the cost of the proposed distribution 1 and 2

Table 9. Summary of the cost of current and proposed distribution

Costs calculated by the software		
Distribution Current	Distribution Proposal 1	Distribution proposal 2
$95.58	$133.62	$97.54

The software tries to optimize the production plant, which is why the cost provided for each distribution of the company is considered as a relative cost, because the software takes the shortest sequence it finds in the environment, performing iterations between areas. According to the proximity evaluators established in the SLP relationship diagram table, being alternative 1 the cheapest of the two proposals, according to the parameters established in the software developed in the "ResilTEX" research project. It all depends on how the weightings have been entered in the Software, when changing the weighting of some parameter of the model, the values could change. Alternative two considers the adequate spaces so that a workstation can function according to the disruptive event but based on the modified SLP model that seeks to have a better distribution according to the established criteria, it seeks the optimization of space, therefore In the initial situation, the biosafety parameters were not considered, possibly one of the reasons why a lower value is obtained when considering the proximity of the jobs.

In the alternative for distances and sequence of the process, it is proposed to move the cutting stations 1, gluing, side seam, stacking and warehouse 3, in order to improve the production process, but when entering the values in the software the cost is high with respect to the initial situation, for this reason alternative two is prepared, in which it would change the location of cut 1, gumming, side seam, lining, quality control and bagging, obtaining a better cost than alternative one. It must be considered that the company handles several categories of models around 21 models, therefore, managers would have to decide which alternative would be more convenient for the company based on the models it produces and the demand for each of them. This study would have the possibility of being extended to more models in order to analyze the most economical distribution that meets the resilient indicators proposed in the model.

4 Conclusions

The COVID-19 pandemic has affected production in small and medium-sized enterprises (SMEs); however, most have had the ability to adapt to change. The development of this case study proposed 14 indicators in total, 7 for the response phase, 5 for the Recovery/Prevention phase and 2 for the Mitigation and Learning phase, where from these indicators the company and has a compliance of 74.90% of resilience on the negative event COVID-19. These were evaluated in a critical, transparent, and reproducible way, allowing the objective of the study to be met. Based on this information, by applying the SLP methodology, it was applied to the distribution of facilities, including resilience criteria. The pandemic favored the innovation of organizations. With the application of the SLP methodology, a relationship table between work areas was elaborated, which was the fundamental basis for working in the "ResilTEX" software that performs an optimization process and finds minimum distances between areas, and also manages: the product with the highest demand of the company, the areas or work centers establishing the number of trips, distances between components with a transfer cost of $0.25 per meter and the relationship of resilient indicators with respect to the proximity ratio; these data were entered for a production lot of 143 pairs. Through the "ResilTEX" software, a cost of $95.58 per month for handling materials is obtained for the current distribution, taking into consideration factors of resilience, in addition, alternative proposals are analyzed for the company CM Original, which are evaluated through the SLP, carrying out a weighting with the different factors entered into the software as a matrix of trips, distances and global value of the resilient indicators; Thus, the software executes the calculation for the proposed distribution 1 with a cost of $133.62 per month and for the proposed distribution 2 with a monthly cost of $97.54, being the proposed distribution 2 the best according to the parameters established in the software developed in the research project.

Acknowledgments. The authors thank the Ecuadorian Corporation for the Development of Research and Academia - CEDIA for the financial support provided to this research, development and innovation work through its CEPRA program, especially CEPRA XV-2021-01- ResilTEX. In addition, this research was partially funded by the University of Cuenca, the Technical University of Ambato, and the University of Azuay.

References

1. Maganha, I., Silva, C., Ferreira, L.M.D.F.: The layout design in reconfigurable manufacturing systems: a literature review. Int. J. Adv. Manuf. Technol. **105**(1), 683–700 (2019)
2. Salmani, M.H., Eshghi, K., Neghabi, H.: A bi-objective MIP model for facility layout problem in uncertain environment. Int. J. Adv. Manuf. Technol. **81**(9), 1563–1575 (2015)
3. Hosseini-Nasab, H., Fereidouni, S., Ghomi, S.M.T.F., Fakhrzad, M.B.: Classification of facility layout problems: a review study. Int. J. Adv. Manuf. Technol. **94**(1), 957–977 (2018)
4. Flores-Siguenza, P., Siguenza-Guzman, L., Lema, F., Tigre, F., Vanegas, P., Aviles-González, J.: A systematic literature review of facility layout problems and resilience factors in the industry. In: Botto-Tobar, M., Montes León, S., Torres-Carrión, P., Zambrano Vizuete, M., Durakovic, B. (eds.) ICAT 2021. CCIS, vol. 1535, pp. 252–264. Springer, Cham (2022). https://doi.org/10.1007/978-3-031-03884-6_19
5. Kamarthi, S., Li, W.: Technology enablers for manufacturing resilience in the COVID-19 and post–COVID-19 era. Smart Sustain. Manuf. Syst. **4**(3), 20200064 (2020)
6. Cai, M., Luo, J.: Influence of COVID-19 on manufacturing industry and corresponding countermeasures from supply chain perspective. J. Shanghai Jiaotong Univ. (Sci.) **25**(4), 409–416 (2020)
7. Sharma, P., Singhal, S.: Design and evaluation of layout alternatives to enhance the performance of industry. Opsearch **53**(4), 741–760 (2016)
8. Herbane, B.: Rethinking organizational resilience and strategic renewal in SMEs. Entrepre. Reg. Dev. **31**(5–6), 476–495 (2019)
9. Bemthuis, R., Iacob, M.-E., Havinga, P.: A design of the resilient enterprise: a reference architecture for emergent behaviors control. Sensors **20**(22), 6672 (2020)
10. Ivanov, D., Dolgui, A.: Viability of intertwined supply networks: extending the supply chain resilience angles towards survivability: a position paper motivated by COVID-19 outbreak. Int. J. Prod. Res. **58**(10), 2904–2915 (2020)
11. Mauricio, D., Leyva, M., Bacalla, J.S.: Una taxonomía del problema de distribución de planta por procesos y sus métodos de solución. Ind. Data **16**, 132–143 (2013)
12. Van Donk, D.P., Gaalman, G.: Food safety and hygiene: systematic layout planning of food processes. Chem. Eng. Res. Des. **82**(11), 1485–1493 (2004)
13. Van Donk, D.P., Gaalman, G.: Food safety and hygiene: systematic layout planning of food processes. Chem. Eng. Res. Des. **82**(11), 1485–1493 (2004)
14. Wiyaratn, W., Watanapa, A.: Improvement plant layout using systematic layout planning (SLP) for increased productivity. Int. J. Ind. Manuf. Eng. **4**(12), 1382–1386 (2010)

15. Caiza, G., Riofrio-Morales, M., Gallo, C.V., Alvarez, T.S., Lopez, W.O., Garcia, M.V.: Virtual reality system for training in the detection and solution of failures in induction motors. In: 33rd European Modeling and Simulation Symposium, EMSS 2021, pp. 199–207 (2021)
16. Pęciłło, M.: Identification of gaps in safety management systems from the resilience engineering perspective in upper and lower-tier enterprises. Saf. Sci. **130**, 104851 (2020)
17. Hollnagel, E.: Resilience engineering and the built environment. Build. Res. Inf. **42**(2), 221–228 (2014)
18. Escobar-Naranjo, J., Caiza, G., Garcia, C.A., Ayala, P., Garcia, M.V.: Applications of artificial intelligence techniques for trajectories optimization in robotics mobile platforms. Procedia Comput. Sci. **217**, 543–551 (2023)
19. Gisbert, R.S.: Propuesta de una Herramienta y Metodología para el Análisis y Mejora de la Resiliencia Empresarial (2017). Accessed 22 Sept 2022
20. Ranasinghe, U., Jefferies, M., Davis, P., Pillay, M.: Resilience engineering indicators and safety management: a systematic review. Saf. Health Work **11**(2), 127–135 (2020)
21. Assarkhaniki, Z., Rajabifard, A., Sabri, S.: The conceptualisation of resilience dimensions and comprehensive quantification of the associated indicators: a systematic approach. Int. J. Disaster Risk Reduct. **51**, 101840 (2020)
22. Cai, H., Lam, N.S.N., Qiang, Y., Zou, L., Correll, R.M., Mihunov, V.: A synthesis of disaster resilience measurement methods and indices. Int. J. Disaster Risk Reduct. **31**, 844–855 (2018)

Diagnosis of Digital Maturity of SMEs in the Province of Imbabura - Ecuador

Irving Reascos(✉), Diego Trejo, Katerin Benavides, Bryan Aldás, and Bryan Quilo

Universidad Técnica del Norte, Ibarra 100150, Ecuador
{imreascos,djtrejo}@utn.edu.ec
http://www.utn.edu.ec/

Abstract. Digital transformation is the realignment of technology, business models, and processes to create value for customers and employees in a changing digital economy. The problem addressed in this article is that we currently have a limited understanding of how SMEs in the province face the modernization of their businesses. This research aims to understand how SMEs adapt to digital transformation and identify the individual and group maturity level of three types of business sectors (retail, textile, and industrial). We conducted this research following a mixed approach. Qualitative to understand the adaptation process and quantitative to identify SMEs' maturity level in Imbabura. The qualitative results obtained are the following: Identification of strengths, opportunities, weaknesses, and threats of the SMEs; In addition, we identified digital transformation activities in five dimensions: strategy, organizational culture, technology, processes, and environment. On a quantitative level, we determined the level of digital maturity of each company, as well as the sectoral group to which it belongs. We concluded that SMEs adapt to new changes empirically and without a strategic plan; business processes and customer experience are the most challenging dimensions.

Keywords: Digital maturity · digital transformation · Industry 4.0 · SME

1 Introduction

Information Technology (IT) applications are unavoidable constituents of enterprises, including Small and Medium Enterprises (SMEs). They participate in executing enterprise functions anytime and anywhere; those functions involve information processing. Information is unavoidable in human and social action, so IT applications have extensive roles in any enterprise.

The digital economy is becoming increasingly complex due to empowered and informed customers, more aggressive competitors, and new products and services with very short life cycles. The technological revolution we are experiencing in

Supported by Universidad Técnica del Norte.

M. V. Garcia and C. Gordón-Gallegos (Eds.): CSEI 2022, LNNS 678, pp. 586–601, 2023.
https://doi.org/10.1007/978-3-031-30592-4_39

all aspects of society has dramatically affected the business environment, forcing companies (regardless of their size or market) to change their conventional business models for ones that adapt to a digital economy. Failure to make these changes can even lead to the closure of companies [17].

It is no longer enough to produce faster, cheaper, and higher quality than competitors, defending the competitive advantage. The industry needs to introduce new types of innovative and *"digital"* production strategies to maintain the current competitive advantage in the long term [11].

Despite the above, the adoption of IT and its use for business transformation remains particularly low among SMEs, even for relevant technologies, such as cloud computing [1,7].

Digital transformation integrates emerging technologies such as the Internet of Things (IoT), Big Data, cloud computing, and Blockchain, among the main ones [3,10,15]. These can intervene in some or all areas of a company to improve efficiency, improve customer service, manage risk, and discover new opportunities for revenue generation [8,13]. However, digital transformation is also a cultural issue that each manager or entrepreneur must address since they must know how the customer reaches their company to define strategies [12].

Technologies based on digital transformation are acting as a disruptive force that is reshaping companies' operations and organizational structure, giving rise to open and distributed innovation processes. It also allows low-cost access to global markets and knowledge networks [5]. In these projects, the strategy must determine which technologies should be implemented, under what objectives, and what results are expected from the adopted systems. In addition, the company must train human talent to make the most of the tools they acquire to optimize resources and time [4,6].

The problem addressed in this article is that we currently have limited knowledge of how SMEs in the province face the modernization of their businesses. In this study, we try to understand how SMEs are modernizing; we also want to identify the level of maturity of these companies in terms of the use of IT.

For this, We use a hybrid research approach to understand how SMEs in the province of Imbabura face their digital transformation. On the one hand, we conducted interviews with SME managers, and on the other hand, we used a questionnaire to measure these companies' digital maturity levels.

Next, the article presents the following sections: research design, results obtained at a qualitative and quantitative level, conclusions, and recommendations.

2 Research Design

We conducted this research using a mixed approach: qualitative (interviews) and quantitative (questionnaires), as shown in Fig. 1. For the interviews, we used as a reference the methodology to carry out Case Studies proposed by Yin [16]. To determine an SME's maturity level regarding digital transformation, we used a questionnaire submitted by the Junta de Andalucía - Spain [2]. The time frame in which we collected data was from January to July 2021

2.1 Field Study

The first phase of the study consists of conducting a field study, which we divided into four stages: planning, data collection, data analysis, and presentation of results.

Planning. This stage comprises the following activities: Definition of the unit of analysis, preparation of data collection instruments, and planning of interviews with business-people in the province of Imbabura.

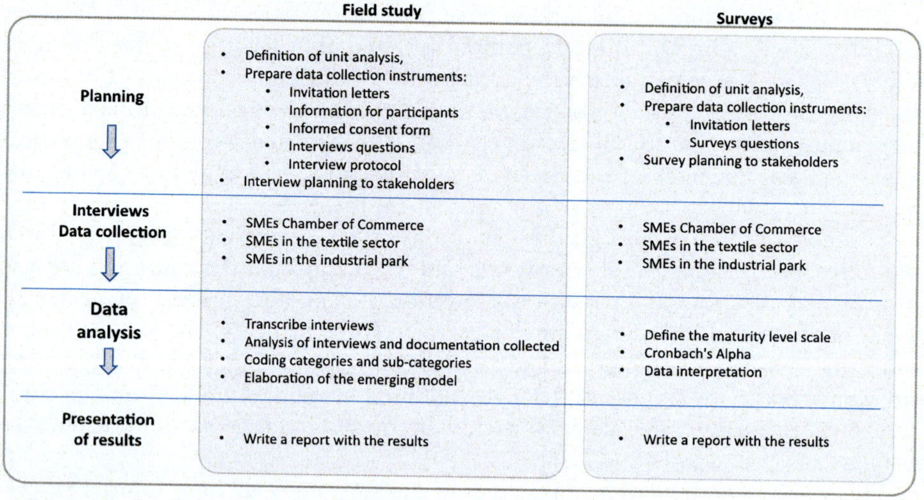

Fig. 1. Research design, using a mixed approach.

The unit of analysis with which the study was carried out is: *"How SMEs are carrying out their digital transformation."* The research questions associated with the unit of analysis are: What problems do SMEs face in Digital Transformation? What strategies are SMEs implementing to achieve their Digital Transformation? How are they adapting to the technological change that is emerging? And what is their level of digital maturity?

The next activity consists of preparing the instruments for data collection. The tools created were an invitation letter, information for participants, an informed consent form, questions for the interview, and interview protocol. Invitation letters were delivered via email (primary) or personally (secondary) to contact those interested in collaborating with the research. In addition, we define the modality of the interview and the survey (face-to-face or virtual).

We use membership lists of three business associations to invite participants to the study: Chamber of Commerce and Production, Industrial Park, and Association of Textiles. We sent invitations to approximately 100 SMEs, of which 36 entrepreneurs agreed to participate in the interviews.

Data Collection. Once the entrepreneurs agreed to participate in this study, we conducted the interviews. These were carried out in person or virtually (using the Zoom platform). The interviews were recorded with the prior authorization of the interviewee and lasted an average of forty minutes.

Audio or video recordings allow us to support the investigation. Subsequently, we conduct an in-depth analysis of the research topic from the managers, chefs, or executives who manage or direct the SMEs.

Analysis of Data. Before data analysis, it is necessary to transcribe the interviews in a clear verbatim format. Next, we use the MaxQDA software to perform the qualitative analysis of each interview, following the recommendations of Kuckartz [9] and the considering the research questions.

The result of the qualitative analysis is a set of codes and sub-codes; the elaboration of the code system is based on the combination of the deductive and inductive methods. Finally, we reviewed the coding obtained to unify the analysis criteria.

It is important to emphasize that during the data analysis phase, the research focus must be maintained around the unit of analysis so that the research questions raised are answered.

Results Presentation. Once we obtained the codes and sub-codes of the interviews, we proceeded to interpret and report on the investigation. For its interpretation, we consider the primary codes obtained in the data analysis: strengths, weaknesses, opportunities, and threats (SWOT), digital transformation, processes, organizational culture, infrastructure and technology, marketing, leadership, customer experience, strategy, and environment.

2.2 Survey to Measure the Digital Maturity of SMEs

The study's second phase consists of conducting a survey; we divide this phase into four stages: planning, data collection, data analysis, and presentation of results.

Planning. This stage comprises the following activities: definition of the unit of analysis, preparation of the data collection instruments, and planning of the surveys to be carried out on business-people in the province of Imbabura.

The unit of analysis with which the study was carried out is: *"What is the level of digital maturity of SMEs"*. The next activity consists of preparing the instruments for data collection (letters of invitation and questionnaires). The proposed questionnaire was adapted from the digital maturity model proposed by the Junta de Andalucía - Spain [2]. This adapted model consists of five dimensions: i) digital strategy, ii) customer experience, iii) digital culture, iv) infrastructure and technology, and v) processes.

Digital Strategy. Digital transformation must be part of the company's objectives, strategic planning, and organizational culture and values.

Customer Experience. Determines the level at which the company uses digital media in its customer relationship.

Digital Culture. The company's human team must be trained for digitization and be flexible to adapt and take advantage of these changes.

Infrastructure and Technology. It is vital to have the necessary technologies and infrastructures to face the organization's digital transformation challenge.

Processes. Technological tools can be used for digitization. The internal operations of the SME value chain are likely to be improved.

Data Collection. The questionnaire was applied to SMEs in these economic sectors for data collection. It should be noted that no sampling technique was used for this information gathering since the invitation was sent in a general way, and the data collected was from the companies that responded positively to our call.

Initially, we applied the questionnaire to the SMEs that participated in the interviews. Subsequently, we applied the questionnaire to other SMEs.

Analysis of Data. For data collection and analysis, an Excel matrix was used, which was designed to determine the maturity index according to the scale proposed by [14], which is shown in Fig. 2. Next, we explain the descriptors by maturity level.

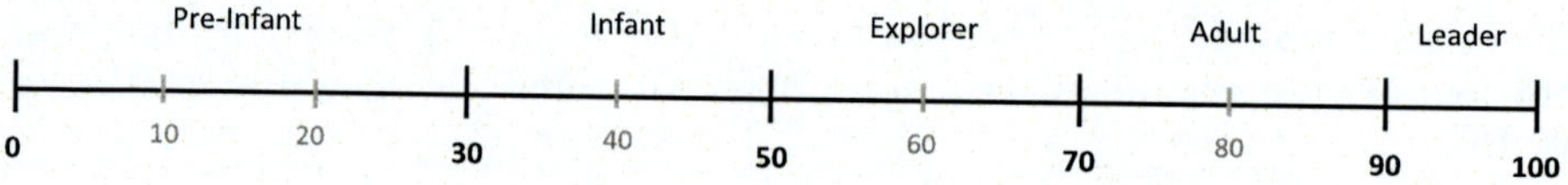

Fig. 2. Maturity level in the range of [0, 100], adapted from [14]

Pre-infant (0–30). There is a shallow maturity level with many weaknesses, including problems in complying with legal provisions, a chaotic system; first steps towards digital transformation; few technological tools.

Infant (31–50). Low maturity level, poorly developed, with several evident weaknesses and few technological tools; it acquires awareness of the importance of digital transformation.

Explorer (51–70). Like the average, enough to survive but not to stand out. Sufficient technology to keep the company competitive; projects and initiatives to develop advanced capabilities.

Adult (71–90). Good level of maturity compared to the average and direct competition, achieving certain prestige in the market; the technology used is relatively high about direct competition; Development of projects and initiatives optimized together with the company.

Leader (91–100). High or exceptional. There is knowledge of state of the art in international contexts, and the company strives to be the best; Use of cutting-edge technology; is a multiplier of their practices to others.

Results Presentation. After conducting the surveys by sector, the data were entered into an Excel matrix (one per company). This matrix was prepared to weigh the responses and present the results for each dimension and their respective total value. These individual results were delivered privately to each of the SMEs. Subsequently, graphs were generated by sector to visualize the level of maturity of the SMEs.

3 Results

3.1 Qualitative Results (Interviews)

SWOT Analysis. Based on the interviews, we prepared a strengths, weaknesses, opportunities, and threats (SWOT) analysis of the SMEs in the province of Imbabura. In the following, we present this analysis.

Strengths. The main power mentioned by SMEs is their infrastructure (commercial premises, offices, machinery, and innovation); they also say customer service, quality products, experience, customer recognition, and a variety of products or services. Innovation is a strength that is more relevant in companies in the textile sector.

Weaknesses. The main weakness we identified in the SMEs is their adaptation to IT; there is still distrust or fear of using them. Other weaknesses are lack of trained personnel in areas specific to the line of business, human talent with limited knowledge in the use of IT, lack of organization (control - processes), low profitability margins, weak internal communication, little customer follow-up, difficulty in accessing credit and employee conformism. In addition, the pandemic caused a lack of liquidity and a shortage of raw materials.

Opportunities. Most SMEs see growth as an opportunity; however, during the pandemic, the main challenge was to stay in the market. It is also essential to consider that the pandemic forced companies to reinvent themselves; faced with a new scenario, they had to innovate and adopt new ways of reaching their customers. Another opportunity mentioned by entrepreneurs is the acquisition and implementation of new machinery and technology to produce and market their products.

Threats. The main danger identified by the entrepreneurs is market competition, essentially unfair competition (companies that do not pay taxes, smuggling, altered products, etc.). In addition, they mention the scarce control of authorities towards irregular businesses, the country's weak economy, proximity to the border, use of digital platforms, computer security, and delinquency. We should also mention that the pandemic was their main threat when it started.

Strategy. Few companies have a strategic plan; most SMEs in the province do not have a formalized (documented) program, and some say that their strategic plan is not up to date. SMEs have goals for what they want to do in the future, but they do not necessarily work to make them concrete; the plans they mention are to grow, increase physical infrastructure, and adopt IT.

Customer Experience. One of the ways to reach the customer is through marketing, which is why SMEs are trying to gain a presence in digital media through websites, social networks, and electronic messaging (Whatsapp). In addition, in some cases, they maintain traditional advertising, only for special seasons (Christmas) or high seasons. Also, this type of company places great value on personal references, which is why they take great care in their service.

Regarding sales channels, this type of company prioritizes physical stores. Lately, these companies have promoted their products on social networks and take orders electronically (Whatsapp). Only one company mentioned that it has e-commerce and a small number of companies consulted make use of Marketplace available in the country.

The way SMEs interact with a customer is basically by word of mouth; some companies have enabled Whatsapp channels to receive feedback from customers. Some companies also review comments on social networks to know the opinion of users. However, one company commented that they do not have a presence on social networks since the comments received are detrimental to the company and do not conform to the truth.

Regarding the means of payment, these are primarily cash and bank transfer. Other means of payment are checks, banking services, and credit card use. Many SMEs mention that they do not receive payments by credit card due to the high transaction costs of the credit card provider.

Organizational Culture. In general, the working environment in SMEs is good; there is trust and camaraderie among employees; managers encourage employee confidence, teamwork, and collaboration among workers.

Communication within SMEs is horizontal and direct (verbal) since employees and managers work together. They also mention that regular meetings are held weekly, biweekly, or monthly to analyze the company's performance, problems, and proposals for improvement.

The cultural level of SME personnel is low; the labor supply in these sectors is basically at the operational level (machine operator, sales, warehouse, reception, etc.), which is why personnel rarely have formal education. Professional jobs are found in this type of company, mainly administrative accounting.

Some SMEs are already preparing for change through employee training, adoption of information technologies (websites, electronic invoicing, e-commerce), creation of digital catalogs, search for new markets, incorporation of new machinery, and changes in their mentality of employees. However, training in these companies is scarce, and many entrepreneurs and employees do it independently. Training on technologies and information systems and their importance in current times is not frequent.

Technology. Regarding technological infrastructure, SMEs in this sector rely mainly on inventory, invoicing, and accounting systems. In recent years, SMEs have adopted electronic invoicing as government regulation. In addition, in recent times, they have been trying to use social networks to benefit the business. However, these companies are not yet considering the use of e-commerce.

Most interviewees stated that technology brings positive changes for the company, but its implementation still generates doubts. One SME interviewed mentioned that it had difficulties implementing a new marketing system because the market is not yet ready for these changes.

Processes. In most SMEs, processes are not formalized or documented. SMEs in these sectors have informal processes with defined people in charge, but everything is done empirically. An additional disadvantage is that employees perform several activities simultaneously; there is no specialization.

As mentioned in the Technology section, many companies have automated inventory, invoicing, and accounting processes, depending on the company's size. It is worth noting that in these companies, the software package (COTS) implanted imposes the process on the company.

SMEs that offer financial services have established processes as government regulations require them.

Environment. Many entrepreneurs distrust the central government and also mention that they do not know whether local, provincial, and national governments provide incentives to SMEs. They also say that there is no support for SMEs and that the only thing the government is concerned about is collecting taxes. In addition, it is clear from the interviews that business-people do not expect anything from the government and therefore try to develop independently. On the other hand, they also mention that doing business with the government is a high risk since it does not always pay on time, and for an SME, this is very dangerous due to the small amount of capital it has.

3.2 Quantitative Results (Surveys)

This section shows the results obtained from the surveys conducted in three economic sectors of the province. As mentioned above, we prepared the questionnaire to measure the maturity index of SMEs in five dimensions: strategy, customer experience, digital culture, technological infrastructure, and processes; in Fig. 3, we show a summary of the definitions of each dimension.

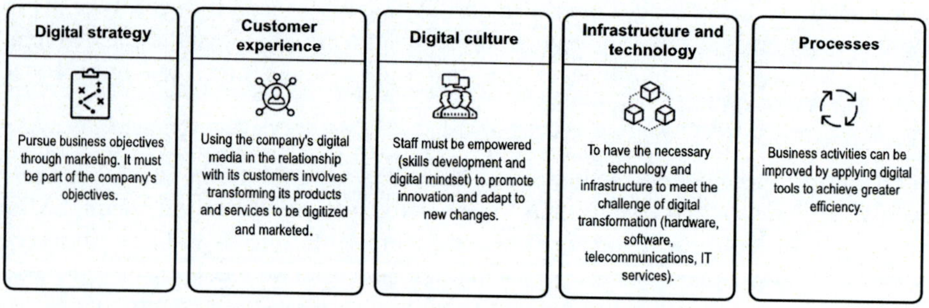

Fig. 3. Dimensions of digital maturity, adapted from [2]

Each dimension was evaluated out of twenty points, obtaining 100 points. We determined the company's maturity level from the evaluation obtained in each size. For example, if a company scored 60 points, we decided that the company was at an Explorer maturity level, as shown in Fig. 4. Also, Fig. 4 explains the meaning of each maturity level.

Once we determined the dimensions and level of maturity to be evaluated, we conducted 168 surveys of SMEs in three economic sectors grouped as follows: chamber of commerce (110), textile manufacturing (18), and industrial park (40).

The next step was to analyze this data in general by economic sector. Table 1 shows the average for each dimension and the total points obtained by the SMEs: i) The chamber of commerce, on average, has a maturity level of 48.15; that is, the average of these companies is at an infant maturity level. ii) The textile sector has an average maturity level of 51.09, which indicates that the average of these companies is at an explorer maturity level. iii) The industrial park has an average maturity level of 40.10; this indicates that the average of these companies is at an infantile maturity level.

Concerning the dimensions, we can mention that the best results have been obtained in the dimensions of technological infrastructure (12.06) and strategy

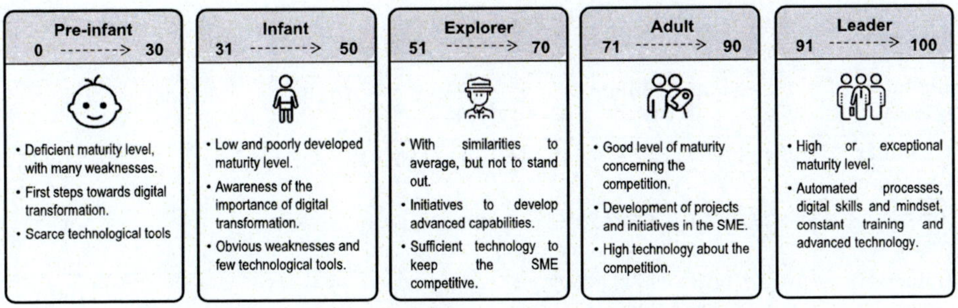

Fig. 4. Levels of digital maturity, adapted from [14]

Table 1. Results of the applied maturity model.

Economic sector	NSMEs	Strategy	Customer experience	Digital Culture	Technological infrastructure	Processes	Total
Chamber commerce	110	11,92	8,06	7,75	13,59	6,83	48,15
Textile sector	18	11,76	9,59	9,92	11,03	8,79	51,09
Industrial park	40	10,45	4,94	7,50	11,56	5,65	40,10
Total	168	11,38	7,53	8,39	12,06	7,09	

(11.38); this is because entrepreneurs are seeking to cover their needs for infrastructure and technology; in addition, they are always seeking new strategies to improve the company. Poor results have been found in the dimensions processes (7,09), customer experience (7.53), and digital culture (8.39); this indicates that companies still lack established procedures, that digital culture is low, and that customer experience is still not considered as a requirement to improve customer relations and get feedback about their products or services.

From the data, we determined the number of SMEs (and their percentage) by maturity level. Table 2 and Fig. 5 show these results.

We also identified that the SMEs in the textile sector are more developed than the SMEs in the industrial park because the textile sector companies have more competition than those in the industrial park. In addition, the concern for the customer experience and process dimensions is more significant in the textile sector.

Figure 5 shows the Chamber of Commerce's SMEs at each maturity level: 14.55% of the companies are at the Pre-infant level, 40.00% at the Infant level, 34.55% at the Explorer level, and 10.91% at the Adult level.

Figure 5 also shows the percentage of SMEs in the textile sector at each maturity level: 22.22% of the companies are at the Pre-infant level, 22.22% at the Infant level, and 33.330% at the Explorer level, and 22.22% at the Adult level.

Table 2. Number of SMEs by the level of maturity.

Economic sector	Pre-infant	Infant	Explorer	Adult	Leader	Total
Chamber commerce	16 (14,55%)	44 (40,00%)	38 (34,55%)	12 (10,91%)	–	110 (100%)
Textile sector	4 (22,22%)	4 (22,22%)	6 (33,33%)	4 (22,22%)	–	18 (100%)
Industrial park	10 (25,00%)	21 (52,50%)	8 (20,00%)	1 (2,50%)	–	40 (100%)

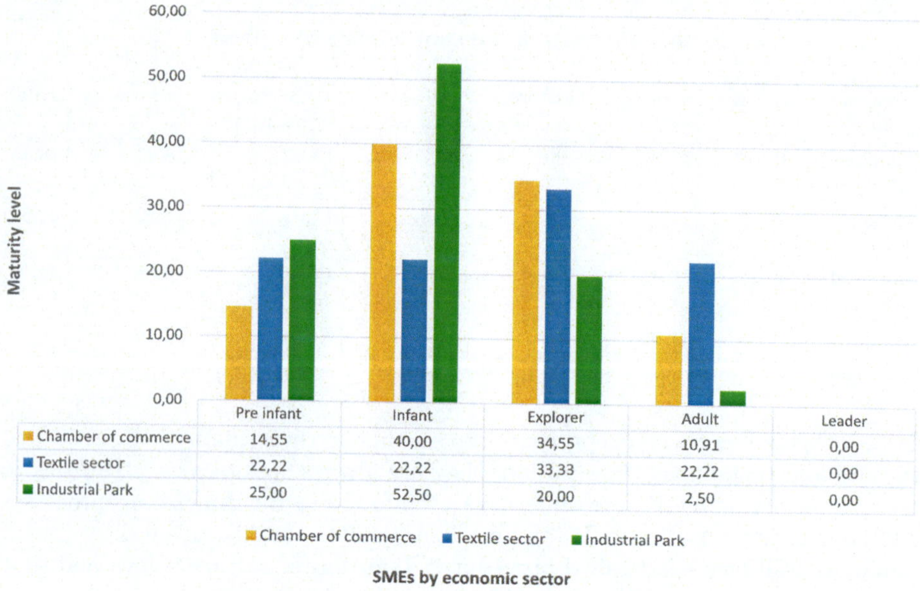

Fig. 5. Maturity level of SMEs by economic sector

Also, Fig. 5 shows the SMEs in the industrial park by maturity level: 25.00% of the companies are at the Pre-infant level, 52.50% at the Infant level, 20.00% at the Explorer level, and 2.50% at the Adult level.

Finally, from Fig. 5, we can see that no SMEs in the province of Imbabura are in the Leader dimension.

Maturity Level by Sector: Chamber of Commerce. Figure 6 shows the digital maturity level of one hundred and ten companies in the Chamber of Commerce. The company with the best results obtained an index of 83.4 (Adult), the company with low results acquired an index of 18.0 (Pre-infant), and the average of the companies has an index of 47.98 (Infant).

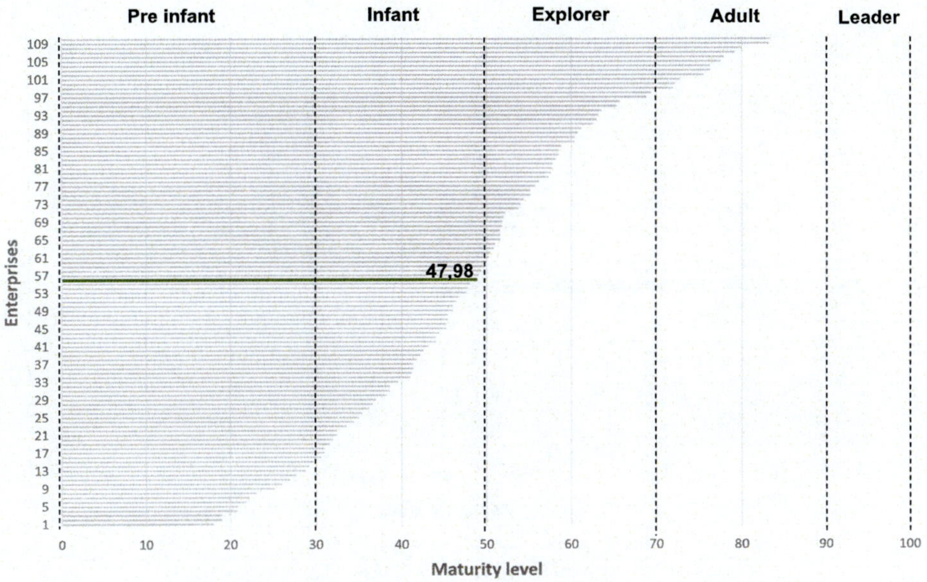

Fig. 6. Digital Maturity Level - SMEs Chamber of Commerce.

The activity of SMEs in this sector is focused on the marketing of products. The aspects to be improved in this sector are: At a strategic level, identify the advantages provided by IT in addition to conceiving the expansion of areas coverage; Regarding the customer experience, expand payment methods as well as enable channels to obtain customer feedback; In digital culture, train workers in the use of IT; In Infrastructure, implement electronic commerce at the individual or group level; At the process level, describe the optimal processes for sales management.

Maturity Level by Sector: Textile. Figure 7 shows the digital maturity level of eighteen companies in the textile sector in a general way. The company with the best results obtained an index of 79.6 (Adult), the company with low results acquired an index of 18.66 (Pre-infant), and the average of the companies has an index of 51.01 (Infant).

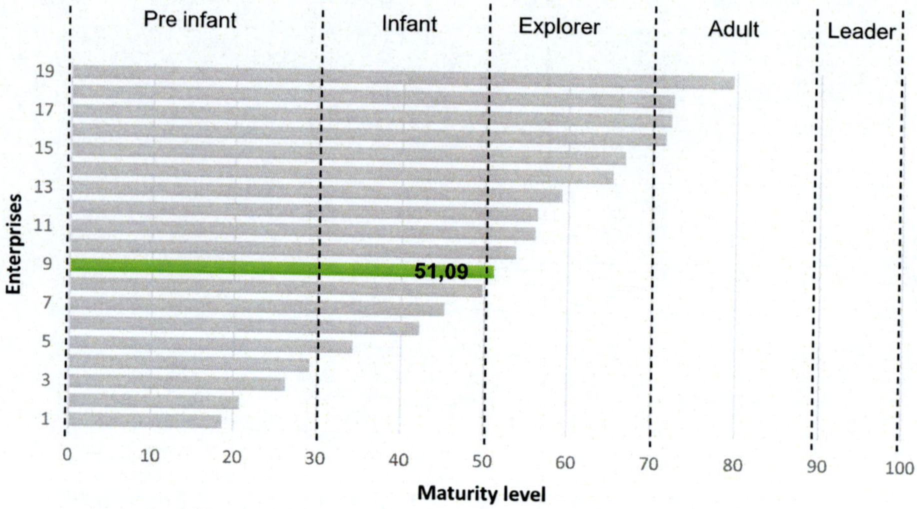

Fig. 7. Digital Maturity Level - SMEs Textile Sector.

The activity of SMEs in this sector is focused on the manufacture of clothing. Although this sector is better positioned compared to the other two, the aspects of being improved are: At a strategic level, identify the advantages that ICTs provide in the manufacturing and commercialization processes; In customer experience, expand payment methods as well as enable channels to obtain feedback regarding garments made; In digital culture, monitor fashion trends; In Infrastructure, implement electronic commerce at the individual or group level; At the process level, describe the optimal processes for the preparation and marketing of products.

Maturity Level by Sector: Industrial Park. Figure 8 shows the digital maturity level of forty companies in the industrial park. The enterprise with the best results obtained an index of 71.2 (Adult), the company with low results acquired an index of 13.0 (Pre-infant), and the average of the enterprises had an index of 40.10 (Infant).

The SMEs in this sector are different; for example, there are SMEs dedicated to automotive maintenance, industrial mechanics, commerce, construction, and clothing, among the main ones. These SMEs are the ones that have the most significant difficulties regarding digital maturity, which is why the following is

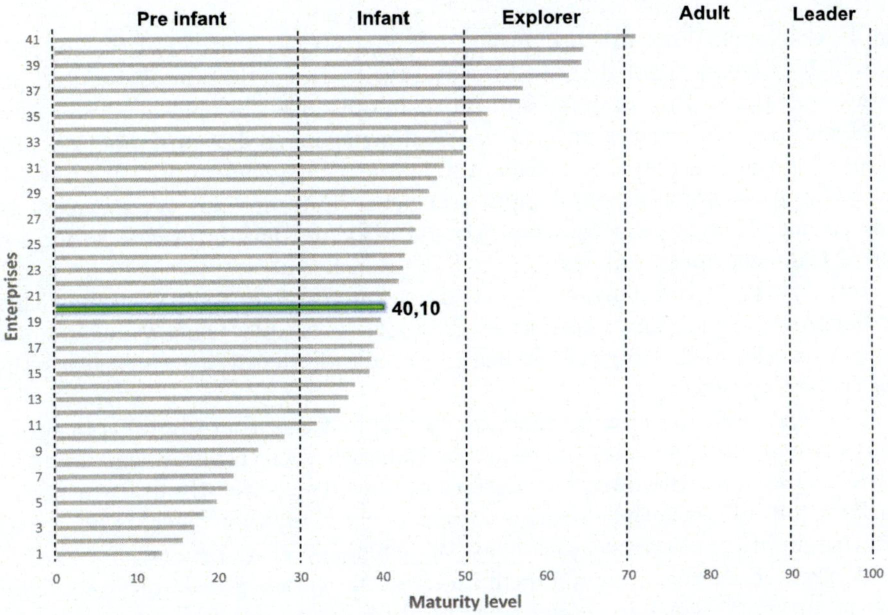

Fig. 8. Digital Maturity Level - SMEs Industrial Park.

proposed: At a strategic level, identify the advantages that ICTs provide in management, manufacturing, and marketing processes; In customer experience, have an internet presence, expand payment methods, enable channels to obtain customer feedback; In digital culture, train workers in the use of ICTs, in addition to identifying the advantages provided by these new tools; In Infrastructure, implement electronic commerce at the individual or group level; At the process level, describe the optimal processes for manufacturing and marketing your products.

4 Conclusions and Recommendations

Because we conducted this research during the pandemic, we found that the priority of most companies was to stay in the market and gradually reactivate as COVID comes under control.

The digitization of SMEs is being done empirically, according to the initiative of the company leader. We found that SMEs do not have a strategic plan that allows them to move towards digital transformation systematically. We could also mention that the leaders of the companies do not have an urgency to digitize their companies; they still do not adequately dimension the advantages and benefits that the digitization of the SME could bring.

The dimensions in which SMEs have difficulties are processes and customer experience. In the processes dimension, these are carried out empirically by employees; operations are not formalized, optimized, and, even worse, automated. It is recommended that SMEs start with the definition of essential processes to organize the company and, in turn, optimize resources.

Regarding the customer experience dimension, SMEs are not yet fully adapted to the new digital culture in the market. They lack alternative payment methods (preferably cash and bank transfers); they are not yet aware of the importance of having a presence on the Internet and lack strategies to improve and obtain customer feedback.

In this type of company, entrepreneurs are unaware of tools that allow them to diagnose the company's current situation. For example, using a questionnaire to measure the SME's digital maturity level allows identifying the aspects that should be improved.

There is confusion among entrepreneurs; they think that digital transformation is only the use of social networks and websites; these tools are part of digital marketing but not the integral transformation of the SME. Digital transformation is a cultural issue that involves the entire company to achieve its objectives, and the adoption of technologies helps the efficiency of its processes.

Workers in micro, small and medium-sized enterprises should receive training in contemporary issues to reduce uncertainty in using information systems and technology within the SME. The manager should encourage adopting a learning culture to develop new skills and competencies of workers. Also, this research identifies a total disarticulation between SMEs and local, regional, and national governments. In this sense, we can mention that the digital transformation of SMEs is a challenge in which the private sector, the public sector, and academia must work in coordination.

References

1. Alarcon, O.A., Velastegui, H.J., Garcia, M.V., Gallo, C.V., Espejo, V, P., Naranjo, J.E.: Monitoring and control system approach for native threatened species. In: Hallo, M., Molina, M., Valdivieso, L. (eds.) 2019 International Conference on Information Systems and Software Technologies (ICI2ST), pp. 85–91 (2019). https://doi.org/10.1109/ICI2ST.2019.00019
2. Andalucía, J.: Test de autodiagnóstico: lleva la digitalización de tu negocio al siguiente nivel - Empresa Digital (2021). https://www.programaempresadigital.es/
3. Caicedo, F., Moreno, H., Moreta, J.: Air to air communication system for collaborative drone work in rural areas. In: Garcia, M.V., Fernandez-Peña, F., Gordon-Gallegos, C. (eds.) Advances and Applications in Computer Science, Electronics, and Industrial Engineering CSEI 2021. Lecture Notes in Networks and Systems, vol. 433, pp. 307–321. Springer, Cham (2022). https://doi.org/10.1007/978-3-030-97719-1_18
4. Caiza, G., Calapaqui, C., Regalado, F., Saltos, L.F., Garcia, C.A., Garcia, M.V.: Upper limb rehabilitation with virtual environments. In: De Paolis, L.T., Bourdot, P. (eds.) AVR 2019. LNCS, vol. 11613, pp. 330–343. Springer, Cham (2019). https://doi.org/10.1007/978-3-030-25965-5_24

5. Crupi, A., et al.: The digital transformation of SMEs – a new knowledge broker called the digital innovation hub. J. Knowl. Manage. **24**(6), 1263–1288. publisher: Emerald Publishing Limited (2020). https://doi.org/10.1108/JKM-11-2019-0623
6. Ekos: Transformación digital (2020). http://dp.hpublication.com/publication/46855167/
7. Eller, R., Alford, P., Kallmünzer, A., Peters, M.: Antecedents, consequences, and challenges of small and medium-sized enterprise digitalization. J. Bus. Res. **112**, 119–127 (2020). https://doi.org/10.1016/j.jbusres.2020.03.004, https://www.sciencedirect.com/science/article/pii/S0148296320301600
8. Jonathan, G.M.: Digital transformation in the public sector: identifying critical success factors. In: Themistocleous, M., Papadaki, M. (eds.) EMCIS 2019. LNBIP, vol. 381, pp. 223–235. Springer, Cham (2020). https://doi.org/10.1007/978-3-030-44322-1_17
9. Kuckartz, U.: Qualitative Text Analysis. a Guide to Methods, Practice and Using Software. First edition edn. (2014). https://us.sagepub.com/en-us/sam/qualitative-text-analysis/book240393
10. Manzano, S., Granizo, C., Velasteguí, H., Guilcapi, J., Benalcazar, F., Gordón, C.: Enabling electronic system for emergency alerts. In: Nummenmaa, J., Pérez-González, F., Domenech-Lega, B., Vaunat, J., Oscar Fernández-Peña, F. (eds.) CSEI 2019. AISC, vol. 1078, pp. 184–198. Springer, Cham (2020). https://doi.org/10.1007/978-3-030-33614-1_13
11. Matt, D.T., Rauch, E.: SME 4.0: the role of small- and medium-sized enterprises in the digital transformation. In: Matt, D.T., Modrák, V., Zsifkovits, H. (eds.) Industry 4.0 for SMEs, pp. 3–36. Springer, Cham (2020). https://doi.org/10.1007/978-3-030-25425-4_1
12. Olarte, E.E.: Factores de la cultura organizacional que deben ser gestionados : un proceso transformación digital percibidos por un grupo de altos directivos de una empresa de servicios de tecnologías de la información (2020). http://repository.eafit.edu.co/handle/10784/16008, accepted: 2020-04-13T23:46:00Z Publisher: Maestría en Administración - MBA
13. Sucozhañay, G., Cabrera, F., Sucozhañay, D., Guaman, R., Siguenza-Guzman, L., Vanegas, P.: Toward a sustainability balanced scorecard for managing corporate social responsibility: a conceptual model. In: García, M.V., Fernández-Peña, F., Gordón-Gallegos, C. (eds.) Advances and Applications in Computer Science, Electronics and Industrial Engineering. AISC, vol. 1307, pp. 279–298. Springer, Singapore (2021). https://doi.org/10.1007/978-981-33-4565-2_18
14. Vivares, J.A.: Modelo de madurez para valorar el sistema de producción y formular la estrategia de manufactura. Technical report (2017)
15. Wolf, M., Semm, A., Erfurth, C.: Digital transformation in companies – challenges and success factors. In: Hodoň, M., Eichler, G., Erfurth, C., Fahrnberger, G. (eds.) I4CS 2018. CCIS, vol. 863, pp. 178–193. Springer, Cham (2018). https://doi.org/10.1007/978-3-319-93408-2_13
16. Yin, R.: Case Study Research. Design and Methods, 5th edn. SAGE Publications, California (2014)
17. Zorro, E.: Modelos de madurez digital en pymes – Caso de estudio de una pyme de telecomunicaciones de Colombia (2020). https://repositorio.unal.edu.co/handle/unal/77419. accepted: 2020-04-16T16:39:01Z

Model Production Based on Industry 5.0 Pillars for Textile SMEs

Fabiola Reino-Cherrez[1](✉), Julio Mosquera-Gutierres[2], Franklin Tigre-Ortega[3], Mario Peña[4], Patricio Córdova[3], Dolores Sucozhañay[1], and Israel Naranjo[3]

[1] University of Cuenca, Cuenca, Ecuador
{fabiola.reinoc,dolores.sucozhanay}@ucuenca.edu.ec
[2] Centre for Statistics, University of Azuay, Cuenca, Ecuador
juliomosquera@uazuay.edu.ec
[3] Faculty of Systems, Electronics and Industrial Engineering, Technical University of Ambato, Ambato, Ecuador
{fg.tigre,edgarpcordovac,ie.naranjo}@uta.edu.ec
[4] Department of Applied Chemistry and Systems of Production, Faculty of Chemical Sciences, University of Cuenca, Cuenca, Ecuador
mario.penao@ucuenca.edu.ec

Abstract. One of the most significant sectors in the global economy is the textile industry, which has been influenced by events such as globalization, technological development, global warming, and the recent COVID-19. It has caused important changes in the way of producing and marketing. Therefore, this article proposes the construction of a theoretical textile production model that incorporates sustainability practices, resilience and focuses on the human being. These concepts are part of the Industry 5.0 (I5.0) philosophy. Through a systematic review of the literature, 58 articles published between 2012 and 2022 were analyzed. The analysis was done using Atlas. ti software, with the application of three questions (who, what, and where) that were answered. In addition, a new complementary literature review was carried out to deepen the research. According to the literature reviewed, still do not have a culture of applying the elements of I5.0 in production models for SMEs. Therefore, in the research, it was not easy to understand the intrinsic variables of I5.0 that SMEs have applied in the textile area. The results show that there are significant deficiencies in the literature on sustainable, resilient production models with a focus on the human being in the textile sector, thus, a theoretical production model is proposed. This model presents the core dimensions of a theoretical model of textile production called "ProS-5.0", considering innovative approaches based on the concepts of I5.0. These are presented in a holistic way from the point of view of production models, processes, subprocesses, and activities. Likewise, the benefits and challenges of "ProS-5.0" are established. Practical implications and directions for future research are also mentioned.

Keywords: Industry 5.0 · production models · textile · Small and Medium Enterprises (SMEs)

M. V. Garcia and C. Gordón-Gallegos (Eds.): CSEI 2022, LNNS 678, pp. 602–624, 2023.
https://doi.org/10.1007/978-3-031-30592-4_40

1 Introduction

Textile SMEs organizations are more vulnerable to external and internal chaotic events. Such as the global pandemic of COVID-19 that caused re-adjustments and re-configurations in production processes in record time. This with the purpose of manufacturing biosecurity implements that contribute to sanitary services. Likewise, with technological advances in designs, models and innovations in machinery and production techniques. This demands lower costs, production times and increases complexity. Therefore, the textile sector must respond on multiple fronts simultaneously while working to protect the safety of its workers, safeguard operational viability and manage risks.

In this context, both global production and society and its economy, in general, in recent years have begun to propose a new industry: Industry 5.0 (I5.0) and the enormous revolution it entails. This new industry focuses on people and the development of solutions, supported by exponential technologies, that respond to their needs and benefit them through resilience, sustainability and human-centeredness. So, the activities are not necessarily directed to progress, but to the benefit and convenience of people [40], through the labor very necessary and incisive in the results of the processes, directly impacting the quality of the final product [44].

I5.0 brings with it different approaches such as: sustainability, resilience and human-centeredness. Which when integrated into production models, contribute to high flexibility, efficiency and quality. In addition, they contribute to smart production through the use of devices, equipment and materials, seeking customer satisfaction [68]. Integrating software and modern technologies in textile manufacturing, facilitates decision making in complex processes [16]. Thus, the textile sector attracts worldwide attention due to its high volatility in production [61]. Similarly, achieving sustainability is one of the most important challenges for organizations, evidenced by widespread economic difficulties, social inequality and a deteriorating natural environment [1].

In this sense, production models should be built with a sustainable ethical vision; with a cleaner production and social focus, where the employer, employee and consumer are concerned about the environment [2]. The positive impacts expected with sustainable practices range from cost reduction, increased productivity, improved image to reduced environmental and occupational risks [18]. On the other hand, the dynamics of the textile sector has been strongly influenced by events in the international arena. In which, different trends have been developed around competitiveness as a fundamental element for the globalization process, among which the implementation of procedures and activities related to sustainable development stand out. Mainly, in the productive process for the textile sector, where it is observed that in this sector the subject is handled in an incipient way, especially in developing countries [22].

Therefore, the objective of this work is to build a proposal for a theoretical production model with a focus on I5.0, based on the literature review of the concepts of sustainability, resilience and its contribution to human care [4], for SMEs belonging to the textile sector. Such organizations should adopt these approaches as an alternative transition to more sustainable, resilient and human-centered operations [3]. In this sense, this article is initially organized with the literature review establishing a theoretical framework of the models with the use of the qualitative tool Atlas.ti, to answer three questions. This serves to obtain theoretical information on the main characteristics, elements and barriers between the main production models. Section 2 presents the analysis of the main characteristics of the models. The third section details the methodology, and the fourth section presents the results of the research on the elements, characteristics, applications, models, approaches, discussion and limitations that were also part of the study. Finally, the fifth section formulates conclusions and some recommendations for future research.

2 Literature Review

2.1 Textile Production Systems

Textile fashion companies are constantly exploring new sustainable solutions to reduce their social and environmental impact and increase economic benefits. Therefore, they integrate environmental analysis into their production systems by evaluating the levels of wastewater pollution and carbon footprint, with the purpose of creating cleaner textiles [60]. In addition, they seek to optimize the number of batches produced and the order size determined by means of a meta-heuristic approach. In [70] an important production system was raised in which an improved systems-oriented design framework based on life cycle assessment (LCA) was built, applied in polyester-cotton manufacturing that includes the general principles of Eco-design. In [28] evaluated current best practice scenarios for environmental impact reduction by implementing circular design strategies in the production of bio-based fashion materials. In [34] an integrated conceptual system showing drivers, ex ante and ex post barriers, and circularity practices is established. In [8] a practical study is presented under the circular approach, on the use of upcycling design and production methods. Which consists of reusing waste to create products with a higher added value than traditionally recycled products in the mass production of garments in order to redirect the surplus material to the production of new garments. These production systems focus on the sustainable performance of organizations, but lack a resilience approach and are not human-centered.

2.2 Production Models

The impacts of World War II in Japan restricted and re-established production patterns. To adapt to the post-war scenario and financial inability to acquire new machinery. Therefore, Japanese industry had to work with existing production equipment more efficiently. In addition, re-evaluate the use of resources, production methods and change the paradigms of manufacturing production. Thus, new production models were developed to increase productivity using existing resources. Antunes [5], discusses in detail the main current theories in a central or derived way. These theories refer to the future of Production Models.

Among these theories are Postfordist, Toyotism, Lean Production, New Production Concepts, Manufacturing Systems, New Industrial Relations and Industrial Governance. Some of these theories do not have the production process at their functional center, although they do take it into account, and are more interested in a global vision of the forms of economic growth that organizations can achieve. The just-in-time (JIT) production philosophy together with automation laid the foundations for the Toyota Production System (TPS) in 1945. TPS improved the relentless waste disposal and helped Toyota to overcome the Japanese economic crisis [25]. In an attempt to replicate the success achieved by Toyota, the automotive industry adopted TPS as a production model and then forged a new manufacturing philosophy, later termed "lean production" [9].

The lean production principles (waste eradication, flow, value, value stream and zero defects) crossed the boundaries of manufacturing. These principles permeated various production systems and industries, such as software development [51], even reaching business models [35]. At the same time, manufacturing formulated mathematical explanations of production management. Studies of production process variables-a process is a repetitive collection of interrelated tasks aimed at achieving a specific goal-provided quantitative meaning to production management practices in manufacturing. The equations could now represent queuing systems [33], lead time , machine maintenance [29], Kanban, inventory management [17], throughput, cycle time, and work in process [5]. The set of equations allowed a deeper understanding of current practices and the results of their interaction.

The fourth industrial revolution (Industry 4.0) generated a shift in industrial development towards a new perspective that brings together software engineering and production systems, as a fundamental pillar. The textile sector, for example, improved processes achieving several goals by applying these technologies in their fields and have improved the economy of many countries in which industry 4.0 was deliberately implemented, managing to generate higher revenues, reduce costs, increase productivity, reduce response times to customers and above all provide better human quality [41]. Now, model design is an optimal way, if not the only possible way, to adopt these theories and update the structure of production systems and manufacturing processes. The main focus of the design should allow to reduce risks, decrease unproductive times, improve quality, intensify the control of technological and technical compliance, reduce the degree of influence of uncertainty, multiple factors, inconsistency, distribution, multiple criteria, temporal diversity and many other factors, which ultimately facilitates obtaining objective results with high accuracy and efficiency [47].

Production models are classified into theoretical, mathematical and structural. First, theoretical models refer to the way organizations produce in: closely related dimensions, stages or phases, showing in some cases the barriers that companies must face for their application [39,54,55,67]. Secondly, mathematical models clearly show variables, parameters, objective function and constraints, whose main function is to model the production process from the input of raw materials to the preparation and dispatch of finished products [50,63,65]. Mathematical models are most often recorded as a computer program. They include the basic characteristics of a system and may, in certain cases, exhibit a connection to the environment [36].

Thirdly, structural models are usually shown in graphical schemes with a deep and properly grounded theoretical basis and include mainly: a production process management system, a monitoring system or information management system and the monitoring of productive resources [66,67]. Therefore, a simulation model can be used, which is an excellent tool to verify the operation of the processes allowing a clear visualization of the selected assumptions. The use of simulation models facilitates a preliminary analysis of how the processes develop and a verification of the suggested changes according to the approved conditions. Applied simulation makes it possible to study specific scenarios and the development of processes without interfering in their course, as well as to examine the chosen solutions without the risk of testing specific assumptions [36].

2.3 Production Models with Focus on Industry 5.0

Theoretical, mathematical and structural production models consider tools whose purpose is to evaluate sustainability in the aspects of: design, materials, manufacturing, type of consumers, durability of garments, transportation in order to use or eliminate during the life cycle of the product [39]. Which, constitutes the right solution for textile manufacturers pursuing sustainable development strategies with transition to circular economy. Which improves the resource efficiency of materials, processes and business operations in general. As is the case of some companies that manufacture their products from recycled materials, avoiding the use of virgin raw materials [67]. The commitment to sustainability and social responsibility can be seen in human talent management policies; in the hiring of workers from nearby areas and with this, that the working day is reduced.

Among the theoretical models that work based on dimensions, stages or phases closely related, showing in some cases the barriers that companies must face for its implementation, approaches such as: circular economy [56,67], industrial textile waste management [54] and commercial especially in the field of sustainable fashion [48] are handled. Textile companies belonging to a sector that has adverse environmental, economic and social impacts, are adopting sustainability practices, this can be observed in the manufacture of sportswear, footwear and even lingerie [39]. This forces them in a certain way to develop sustainable products and services that consider these criteria throughout the product life cycle.

In this case, models can work with applied indicators to capture the innovations brought by the Circular Economy [55] and focus on sustainable production based on four ambitions: (1) phase out substances of concern and the release of microfibers (the new textile economy must ensure that the material input is safe and healthy to enable recycling and avoid negative impacts during the production, use and after-use phases); (2) transform the way clothing is designed, sold, and worn to break free from its increasingly disposable nature; (3) radically improve recycling by transforming the design, collection, and reprocessing of garments; and (4) make effective use of resources and move to renewable inputs [67]. This finding may be of interest and should be characterized by (i) high labor intensity and (ii) highly mature production processes [59]. Resilience practices are introduced at one of four levels of innovation: 1. product level, 2. social service product level, 3. systems level and 4. resilient, adaptive and circular level [39].

Mathematical models focus on solving production scheduling problems with explicit consideration of pollution reduction objectives that are fundamental to the long-term goal of sustainable manufacturing [52]. According to [19], the models are evaluated based on sustainability elements in six dimensions-Environment, Economy, Society, Functionality, Manufacturability, and Reusability-to provide a new comprehensive framework for sustainability assessment in a closed-loop life cycle system. Stochastic multi-objective mixed-integer optimization stochastic models can be employed to ensure production efficiency under uncertainty and meet the requirements of sustainable development.

The efficiency of the production system is ensured through objective functions that optimize order backlog, machine uptime and customer satisfaction [14]. Regarding resilience practices the goal is to integrate industrial IoT for flexible management and optimization of cyber-physical systems in factories. Factory assets such as machines and advanced smart tool control system will be integrated within this IoT network, enabling near real-time remote process monitoring, optimization, control and resilience [10]. On the other hand, it is crucial that the human dimension is also included in an overall factory model. The relevance of considering human behavior is crucial from a manufacturing optimization point of view, however, this area is at a very early stage of development. In particular, the validation of models, the collection of data from real workers to provide model inputs and the comparison of simulation results with manufacturing systems, as well as sensitivity analysis, are significant fields of innovation where more research is needed, as, it will allow to improve factory design both from a performance and resilience point of view [45].

Regarding structural models the trend is towards sustainability, circularity and resource efficiency of materials, processes and general business operations. Also, new consumer business models based on sharing productive resources and final products [67]. The sustainability approach is shared with the theoretical models. On the other hand, resilience integration focuses on quantifying the integration of various operations, such as the integrated control system and its maintainability. The main factors contributing to resilience are divided into two main categories: reliability factors and recoverability factors [58]. To focus on human can develop a human production management system by using multi-agent technology that fully reflects the initiative, intelligence and coordinated value for production management and decision analysis work, improving work efficiency, achieving humanization management of business requirements and strongly boosting IT management [66].

2.4 Elements of Production Models with a Focus on Industry 5.0

Similarly, it is essential to understand the elements that constitute the theoretical, mathematical and structural production models. These are focused on guiding organizations in the direction of: circular economy [55,67,69], sustainability [39], textile waste management [54], and sustainable fashion [48]. The components of the theoretical production model from the sustainability approach are: environmental, social, economic, aesthetic and cultural [39]. Other evaluation elements of this group are segmented by components in six dimensions: environment, economy, society, functionality, manufacturability and reusability [50], and focus on environmental measurement specifically on waste management [19].

On the other hand, the components of circular economy theorists' models are: reduction of raw materials and remanufacturing [55,67], applied in procurement and sourcing processes, established from the perspective of sustainable fashion. Whereas, the mathematical production pattern focuses on establishing mixed integer linear programming [50], which includes the optimization element of sustainable development from the stochastic approach [65] with the use of industrial IoT tools for flexible management and optimization. In addition, in this model from resilience and human-centered approaches, decisions to improve work efficiency, humanization management of business requirements and strongly drives IT management are analyzed [66]. Finally, structural models are characterized by graphical schemes with a deep theoretical basis and shaped by environmental, social, resilient and human-centered elements. These, through the use of substances and the release of microfibers, facilitate the work of recycling and seek to avoid negative impacts during the production, use and disposal phases. In addition, by transforming the design, sale and use of textiles, they examine alternatives to radically improve recycling through the reprocessing of garments, and the effective use of renewable resources and inputs [67].

2.5 Production Models Applications and Impacts

Over the years, textile companies have been evolving their production models seeking to improve their production processes to provide better products to their customers and to find the best ways to conserve the environment. The traditional linear production model applied in the fashion industry, based on extract, produce and waste, brought with it serious negative environmental, economic and social consequences [54]. Several studies have sought to better understand the textile sector and propose various methods to adjust their production models to provide the best results to companies. Thus, for example, a model has been created for the analysis of industrial textile waste flows and the development of scenarios for reasonable waste management, allowing the generation of preventive solutions through sustainability indicators that represent the effect of the scenarios for environmental, economic and social aspects [52].

Similarly, cleaner production models have been generated to manage waste in textile companies, since they have limited capacity for this management [56]. Thus, the transition towards the implementation of circular economy models begins, which implies substantial changes [19]. From another approach, studies have been conducted to ensure the efficiency of production under conditions of uncertainty and meet the requirements of sustainable development, improving in many aspects compared to manufacturing practices under the same production conditions as those normally used. This type of model also ensures the production manager's discretion on whether or not to adopt production options such as backorders, overtime and employment of temporary workers, generating a baseline for managers to adjust the production system in different situations [65]. From a strategic point of view, the profitability of each manufacturing plant depends on its production plan to support customer demand, thus models have emerged that allow profit maximization over the planning horizon, which consists of a finite number of periods [67].

2.6 Barriers in the Implementation of Textile Production Models

Textile manufacturing industries face different obstacles to produce and implement production models that include practices focused on sustainability, resilience and human-centeredness. Among the main barriers to implementing production models found in the literature, five categories stand out: (1) management and decision-making, (2) human resources, (3) rules and regulations, (4) costs, and (5) external constraints. First, management decision-making is characterized by a lack of clear, standardized information in quantitative and qualitative measurement of objectives; lack of information and feedback, which influences discrepancies in data and performance indices. According to [10] in this performance group are seven indicators: (1) initial investment cost, (2) environmental costs (externalities), (3) governance issues of liabilities and infrastructure ownership, (4) limited implementation of new sustainable business models, (5) lack of information sharing among stakeholders, (6) limited reverse logistics capacity, and (7) lack of measurement.

Second, human resources refers to the absence of staff eco-literacy. So, it is important to know the steps of productive processing and updating [53]. Among the most representative constraints are: obstacles to the adoption of circular economy business practices, influence of green policies depends on the sector in which they operate, lack of technical skills prevents SMEs from taking advantage of green economy opportunities and lack of support from the supply and demand network and'green supply chain' initiatives require the involvement of external stakeholders (such as'green procurement') generally have a low level of adoption.

Third, non-compliance with rules and regulations on certain standards, such as waste collection and recycling, can lead to sustainability problems in ensuring quality, efficiency, traceability and measurement of production. This complicates the prevention of negative environmental impacts, complicates the measurement of the (long-term) benefits of projects [62].

Fourth, the cost of investment in new technologies, certification processes and employee training is an impediment to the implementation of production models for SMEs. Those who do not have sufficient resources to invest in integration in this group are the high investment for the adoption of technology, for human resources training, the high cost of sustainable raw materials, the initial cost of production of the new model and government investments for the sector [23,37].

Fifth, among the external obstacles, the COVID-19 pandemic stands out, which interrupted the supply chain of the world textile market. This outbreak has a global impact due to travel restrictions, which affects the shortage of raw materials. Through a case study on a ready-made garment (RMG) industry in Bangladesh, which showed that supply chain disruption due to COVID-19 would increase the cost of production. This contributes to the textile industry being forced to make changes in long-term projection [15].

3 Research Methodology

This article is comprised of two sections, the first part by the systematic literature review and the second by the construction of a proposal for a theoretical production model. Initially, a qualitative research approach of interpretative nature was applied, through the literature review on textile production models that apply sustainability, resilience and human-centered approaches. In the first part of the systematic literature review, the methodology proposed by Fink [20] was followed to ensure the quality and consistency of the work in terms of textile production models. The method was composed of seven steps: 1) selection of research questions, 2) definition of database sources, 3) selection of search terms, 4) application of practical selection criteria, 5) application of methodological criteria, 6) literature review, and 7) synthesis of results.

In Step 1, the PICO (Population, Intervention, Comparison and Outcome) strategy [38] was used to formulate the research questions, resulting in three questions correctly: a) What production models in textile companies have incorporated sustainability, resilience and human-centeredness practices? b) What elements are considered in a textile production model that incorporates sustainability, resilience and human-centeredness practices? and c) What have been the applications and impacts of the production models in the textile companies? In steps 2 and 3, the databases selected for this literature review were: Scopus, Web of Science, Scielo and Latindex Catalogue. The keywords and search strings used to answer the research questions were: "Production models" AND "textile"; ("Sustainability" OR "Resilience" OR "Empowerment") AND "Production models" AND "textile"; ("Production techniques" OR "Production models") AND "textile"; "Elements" AND "Production models" AND ("Sustainability" OR "Resilience" OR "Empowerment").

Steps 4 and 5 allowed filtering the search in different databases to ensure that the articles retrieved in the present study were current and relevant. The inclusion criteria used were as follows. 1) articles in Spanish, English and Portuguese; 2) scientific journal and conference articles; 3) articles published between 2012 and 2022; 4) Subject areas: Environmental Sciences, Engineering, Computer Science, Mathematics, Economics, Business and Management, and Social Sciences. In addition, the exclusion criteria used were 1) irrelevant articles; 2) duplicate articles; and 3) articles with academic publication remarks.

After passing the exclusion criteria, the articles sought to answer the following questions: Is the research design of this study internally and externally valid? Are the data sources used in the study reliable and valid? Are the analytical methods appropriate, given the characteristics and quality of the study data? Were the results measured in practical and statistical terms? If the articles found do not answer one or more questions, their methodological quality decreases. In compliance with Step 6, the Atlas. Ti software was used to review the retrieved articles. This software is a qualitative tool that allowed, through codes, to highlight and classify the information according to its characteristics. A set of 81 codes, described in detail in Annex A, was used in this study. Finally, in Step 7, the results were synthesized. Figure 1 summarizes the number of articles resulting from the review and application of the methodology. As can be seen, in a base of 6,647 research studies, a total sample of 58 studies were identified, which referred specifically to textile production models. While others were oriented to the textile part with topics related to: sustainability, resilience, empowerment and human-centered. However, it was subsequently found that there were deficiencies in the literature on: textile production systems, sustainability, resilience and human-centeredness.

The second methodological part consisted of the construction of a theoretical production model proposal, which required a new complementary literature review, a survey of thirteen experts in the textile sector to determine the production processes and the characteristics that integrate the production model proposal. This article consists of the main findings of the literature review, followed by the results of the research through a content analysis and a brief discussion. Finally, it details the proposal of a theoretical textile production model that considers Industry 5.0 approaches: sustainability, resilience and human-centeredness and summarizes the main conclusions.

4 Results

4.1 Main Findings

Sustainability Approach in the Textile Sector. The concept of sustainable development was popularized in 1987 with its publication in the Brundtland Report, Our Common Future [31], under the auspices of the United Nations, where it is defined as "development that meets the needs of the present without compromising the ability of future generations to meet their own needs". The literature on its definition has focused mainly on its environmental dimension,

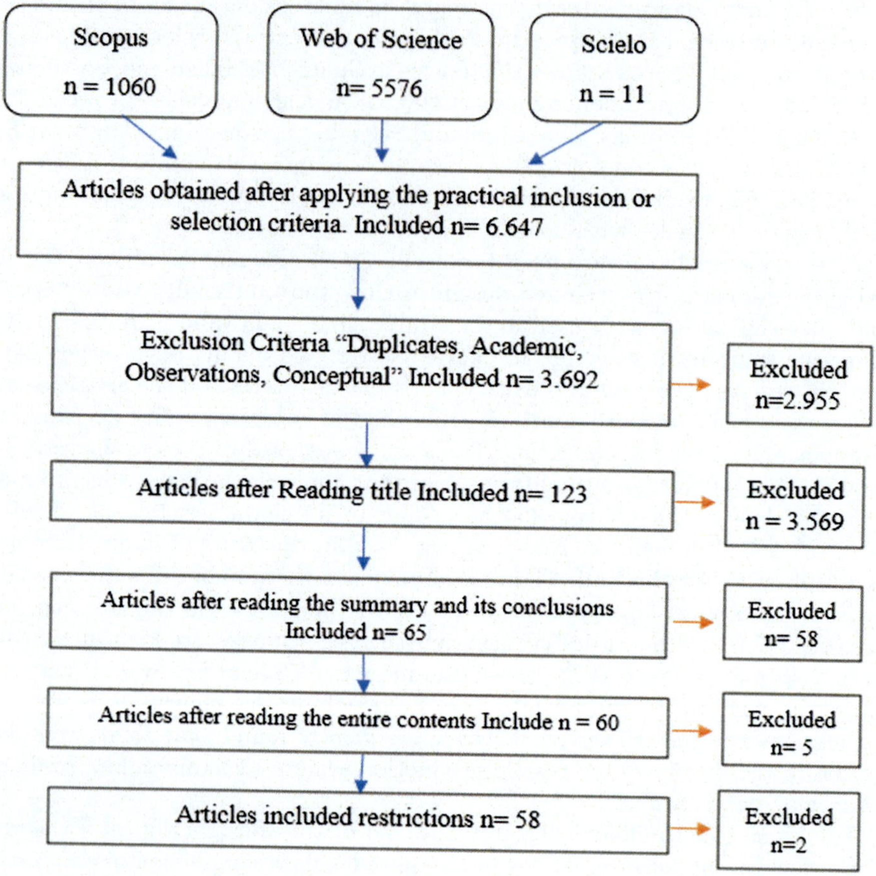

Fig. 1. Steps involved in a research literature review [20].

followed by the economic, and to the detriment of its social perspective [42]. The application of sustainable techniques has simultaneously prioritized environmental, social and economic benefits to improve the performance of product design, material selection, manufacturing processes, etc., throughout the entire life cycle [19]. Sustainability focuses on respecting the environment surrounding industrial activities, be they with the environment, people and the economic part that represents a profitability for companies. Large companies periodically publish reports through different media on the development of their sustainability activities; similarly, several textile SMEs are adopting sustainability practices [39] being forced to develop sustainable products and services that consider environmental, economic and social criteria.

The sustainability approach should aim to maximize total profit, minimize emissions in the production process (water use, energy conservation, and wastewater pollution reduction) [49], minimize labor force turnover, minimize order backlogs, maximize machine uptime, and maximize customer satisfaction [65]. This is achieved by (1) using more sustainable textile fibers, (2) zero waste when using all fabric for product details (3) using renewable energy sources, which reduce environmental impact and allow for periodic measurements, (4) optimal management of textile garments discarded and sent back to the company to make a new article and (5) sending waste to recycling companies, is important in a model for industrial textile waste management [24,54,56].

The textile industry needs to adopt sustainability actions such as: remanufacture, recondition, reuse, recycle, reduce, repair, recover, redesign and refurbish [12,13]. Thus, the probability of success of a company improves with the implementation of sustainability and its application in all its activities, considering not only economic variables, but also environmental and social ones [21]. By addressing concepts such as cleaner production, eco-efficiency, eco-design and product development, an improvement proposal appropriate to the real needs and limitations of each organization is sought [32].

Resilience Approach in the Textile Sector. According to [6], resilience refers to flexibility, inclusiveness, restructuring and survival. It is described as "how various types of systems respond to unexpected shocks". Other authors define it as a dynamic process that can be developed or inhibited due to the interaction between protective and risk factors [27]. While, in [64] calls resilience as the ability to recover from extremes of trauma and stress. According to [64], resilience constitutes the ability and capacity to be robust under conditions of enormous stress and change. Whereas, corporate resilience is the organization's ability to cope with uncertainty and learn from past failures (citation 2), including crisis planning and providing organizations with the ability to survive and thrive in the face of adverse conditions, including adaptation and learning. On the other hand, according to [27], resilience structure is linked to entrepreneurship in two aspects, firstly, to issues of aptitude, tenacity, sustainability or self-efficacy. Secondly, cognitive, behavioral entrepreneurial characteristics, ways of adapting to new conditions and the ability to provide long-term sustainability through innovation [30].

On the other hand, resilience in the textile sector, according to [64] is made up of three essential elements: companies, relationships and institutions. Firms pave the way forward with their strategic responses. Relationships or clustering form the interactional network without which the model does not work and finally institutions refer to sets of organizations including associations and regional governments that support the sector, providing it with representativeness and legitimacy. This theoretical framework was used in the analysis of the main changes in the processes, products and markets of the Spanish textile industry trying to capture one of the specific responses of the fashion industry. The findings indicated that textile companies more actively involved in their immediate context generated and obtained more external economies, which could benefit from territory-based industrial policies.

Human Being Approach in the Textile Sector. According to [26], decision support models for production have largely neglected human factors in the past. For this reason, industry requires the development of more realistic planning models for production problems that incorporate human factors. Although in many circumstances Industry 4.0 is still in the process of implementation, the promise of Industry 5.0 presents us with the possibility of empowering and repositioning the human being at the center of economic processes, production and with control and command over them, all of course linked to the extraordinary development of Industry 4.0 enabling technologies, especially digital technologies, robotics and, of course, artificial intelligence [68].

Thus, Industry 5.0 seeks extreme connectivity between humans and machines, where production is limited to the supply of products tailored to customer needs. The methods have been created by various researches and surveys. Production and process management under such conditions necessitate the use of new technologies to maximize system efficiency and productivity. Entering the fifth industrial revolution, companies worldwide have adopted new technologies to successfully meet the quality and value-added demands of customers. With the use of new transparent technologies with intelligent components, time saving, real-time product monitoring, information availability, new software possibilities, more meaningful data and information security [63].

In this context, it is crucial that the human dimension is also included in manufacturing models. For human capital is now, much more than before, a critical resource for the competitiveness of a company and for the improvement of the industry. The relevance of considering human behavior is crucial from the point of view of manufacturing optimization, however, this area is at a very early stage of development. In particular, validation of models, collection of data from real workers to provide model inputs and comparison of simulation results with manufacturing systems, as well as sensitivity analysis, are significant areas of innovation where more research is needed, as, it will enable improvement of factory design from both a performance and resilience point of view [10]. Therefore, it is the responsibility of this sector to promote clothing, adding value to the product that aims to dress, protect and bring elegance to the final consumer. However, despite the availability of modern machinery, clothing is currently a productive sector where labor is highly necessary and incisive in the results of the process. This means that the manual skill of the operators directly affects the quality of the final product; that is, the greater the presence of the human factor, the greater the chances of generating failures and defects, so finding ways to combine human labor with new technological resources is extremely important to improve the quality of production within the textile apparel sector [44].

4.2 Model Development

Meredith [46] defines a conceptual model as a set of definitions used to represent or describe a process. It can be a description, a taxonomy or an inductive reflection. Based on the exhaustive review of the literature on production models, a new proposal for a theoretical textile production model called Pro.S.5-0 was created. It is based on the article by the author Jinfu Wang [66], entitled "Multi-Agent based Production Management Decision System Modelling for the Textile Enterprise". In which, a textile production process management system design is proposed, which strengthens the interactivity functions of both the user and the system. In addition, the system meets the requirements of production management, intelligent decision, customized production and heterogeneous integration between the different components of Industry 5.0.

The understanding of how all organizational processes work in textile companies forms the basis for the approach of a production model focused on the characteristics of Industry 5.0. This model is made up of three general areas shown in Fig. 2. The first part corresponds to the support area through Information Management, whose main activity is to support the other two sections through the collection and processing of information on the levels or performance of Industry 5.0. In the second part is the production process management system divided into two types of areas, internal (production processes) and external (supply management, facility distribution and logistics). Only for this model proposal, the internal processes will be analyzed in depth, since they can be measured and controlled within an organization. The manufacturing section was established on the basis of all the processes, sub-processes and activities that are part of textile manufacturing. Three general subprocesses were established, which are: planning and design; manufacturing and packaging. Figure 2 also shows, in an oval shape, the constant primary activity in the three subprocesses of control and monitoring. This activity is constant and is chained to all the subprocesses: before starting with the planning and design of the product batches, a stock monitoring is carried out, based on this, the purchasing management is performed, then the raw materials arrive, which are checked to ensure that they meet the requested quality standards.

Next, the cut is made and transferred to the manufacturing subprocess, where the operators monitor that the cuts are accurate. Then, before the third packaging sub-process, the last and most rigorous review is carried out. The last and most rigorous check is made to ensure that there are no threads coming out of the garments, that the size labels and washing instructions are correctly made, and that there are no stains. In addition, this model ensures that the monitoring and quality control activity also focuses on reviewing compliance with sustainability, resilience and human-centeredness standards. Finally, the third part breaks down the three components of the Industry 5.0 approach: sustainability, resilience, human empowerment and the elements that comprise it.

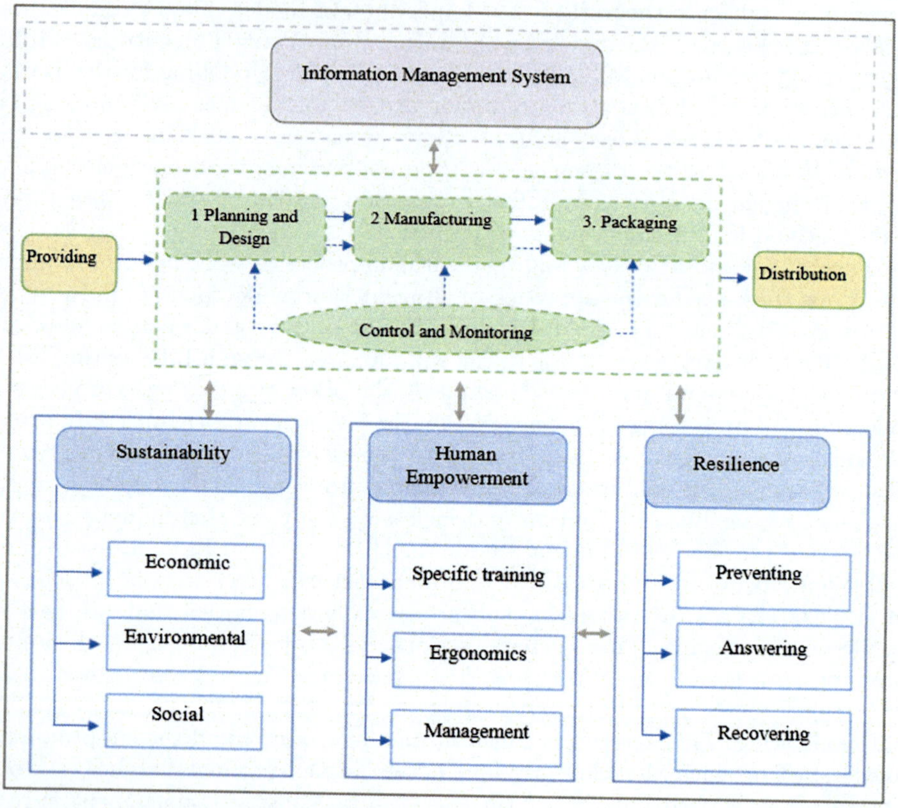

Fig. 2. Proposed Model of Textile Production with a Focus on Industry 5.0.

Sustainability Approach. The probability of a company's success improves with the implementation of sustainability and its application in all processes, by considering not only economic variables, but also environmental and social variables [21]. The elements that are applied should provide results that support the linkages between sustainability commitment and collaborative capabilities both within and between companies; through the collaborative capabilities and performance of social, environmental and economic sustainability components [43].

Social Dimension. The social category addresses practices that seek to promote the development of workers, including issues such as the inclusion of apprentices, training or introduction offered to workers at the start of their work and the existence of additional education programs for employee development [21]. Thus, the categories proposed by Flores et al. [21] for the social dimension are: employment relationship, freedom of association and collective bargaining, occupational health and safety, fair wages, social security, working hours, forced labor, child labor, equal opportunities, worker development, behavior in the event of dismissal, sexual harassment and community. These categories seek to generate development for the worker through the creation of performance evaluation plans and programs in which employees are increasingly involved in activities designed to improve the sustainability of consumption and production [11]. Similarly, the relationship and work environment in accordance with current regulations, contributing with fair wages and working hours, social benefits and equal opportunities, thus creating a human management system to improve work efficiency and at the same time promoting safety. Social issues in the workplace are seen as a symbol of integration for productivity. It may not only be able to reduce potential risks and injuries, but also influence to improve customer satisfaction and employee performance [59].

Environmental Dimension. The environmental category is based on the efficient use of the planet's natural resources and maintaining them for future generations, fully protecting the systems and providing a high quality of life for people [43]. According to Flores et al. [21] the components of the environmental part are: environmental management, energy, water, materials, waste, emissions, effluents and environmental noise, which seek to carry out measurements in the production processes. The aim is to minimize the use of resources, making their use more effective and focusing on energy conservation through periodic measurements of production processes, reduction of materials and pollution. Through the optimal management of discarded textile waste, proposing a priori and above all a reduction of consumed resources by applying, for example, various options for reuse or repair based on the collaborative economy [56]. This category also aims at long-term pollution reduction and sustainable manufacturing [52] supported by compliance with environmental laws and participation in external environmental initiatives related to the company's production activity. A number of recent findings show that environmental sustainability has attracted the attention of textile companies, not only in pollution reduction and energy savings, but also in the implementation of new technologies such as renewables, which generates cost savings through renewable energy, resource management, recycling of materials and efficiency in waste management [59].

Economic Dimension. Economic indicators aim to evaluate the organization from two economic dimensions, wealth generation and economic stability. Wealth generation addresses the general objective of economic activities to optimize the utility for society towards an increase in wealth. Economic stability aims at a self-sustainable economic system over time. In this category the economic indicators are made up of: Profitability, Productivity, Business diversity, Long-term investment and Consumer satisfaction. Profitability, refers to the creation of value or added value by the company, which is mainly based on the current market situation; Productivity, understood as production in relation to employee participation and human development; Business diversity, analyzed as the rate of diversification of the organization and the complexity of the organization's products; Long-term investments, which determine the long-term success and survival in the market and include dependencies on different types of capital and; Consumer satisfaction, which represents the usefulness of a product from the consumer's perspective [7].

Resilience Approach. Resilience is the ability to adapt to adverse situations, unforeseen events and setbacks in organizations. This seeks to confront changes or unexpected events in the globalized market due to different situations of uncertainty and crises in different environments, mainly natural and political emergencies [66]. In addition, to have the ability to anticipate, respond and adapt to changes, sudden interruptions in order to survive and prosper. The concept of resilience from the perspective of Industry 5.0, is not yet well established [30] and may differ in key elements between different authors and publications. However, in a theoretical review by Flores et al. [21], three categories of resilience are proposed: preventing, answering, and recovering. The preventing category seeks to anticipate unexpected events, in this group are the assets and resources and infrastructure. Assets and resources seek savings in tangible goods and intangible financial assets. The mastery of physical resources facilitates the overcoming of problems due to crisis effects and organizational performance.

Financial assets are intended to provide the organization with capital as a buffer against negative impacts due to crises or as a reserve for critical times. The presence of preventive plans contributes to anticipation and a way to face production interruption problems due to lack of materials. Similarly, preventive measures in the manufacturing and storage infrastructure seek to prevent internal and external events. This allows knowing the physical characteristics of the establishment, as well as the advantages of its location. The availability of facilities is also sought to meet preventive storage requirements for: type, size and weight of the material to be stored. In these available manufacturing facilities, the location of machines, work equipment, material and personnel flow, storage, workshops, recreational areas, among others, are evaluated. With the purpose of preventing internal accidents in the production plant.

Moreover, the answering that organizations have to face unexpected events in terms of flexibility and dynamic competitiveness to learn and adapt to the changing environment. This includes organizational strategic alliances that can be developed and controlled internally through inter-organizational networks such as: mergers and strategic alliances. These contribute to the transfer and exchange of intangible resources (knowledge and complementary relationships). In addition, they evaluate the organization's flexibility to compete in the market in the face of external events. With the purpose of evaluating the robustness, complexity of the institutions and the capacity to recover the functionality after the occurrence of a disruptive event and adaptation to them.

Furthermore, the recovering emphasizes the behavior and leadership of organizations during crises. Thus, organizational learning and culture contribute to organizational decision making through employee responsibility and ownership. This contributes to knowledge sharing and leads to well-being through the alignment of corporate and organizational values. Through the perception of the sense of leadership the organization evaluates the workforce on the level of leadership exercised by the employer during and after a crisis, through an environment of active communication and the preservation of calm under pressure. Also through the satisfaction of the measures taken by the organization with respect to trust, acceptance and awareness of belonging to the organization through the commitment of workers to be part of something, feel identified with the values and objectives of the company and with the team of which it is part.

Human Empowerment Approach. Another focus of Industry 5.0, in the production model proposed, is to place the human being at the center of the production processes, establishing a close collaboration between highly trained personnel and the use of technology to create unique or individualized products. Such as workstation designs and specific ergonomic features that enhance the synergy between human and machine. This allows machines or robots to do the heavy, repetitive and dangerous tasks, while the individual contributes his or her talent and creativity to carry out more complex functions. According to Flores et al. [21], the categories of the human-centered approach are the following three: specific training, ergonomics, and management.

Specific training corresponds to programs aimed at production personnel to improve their skills, the performance of their functions and the management of the technology used in their workplace. In addition, we seek to determine the level of new knowledge obtained and the level of worker satisfaction with the programs received. The ergonomics component refers to physical and cognitive ergonomics. Physical ergonomics focuses on incorporating anthropometric criteria in work spaces in organizations. Such as industrial design focused on the human being, for example in machinery adapted according to gender, age, human structure, the activity performed, etc.

Cognitive ergonomics evaluates if the company has risk assessment plans related to aspects of memory, concentration, attention, perception, etc., in the operators linked to the productive activity. The management corresponds to plans and policies at the managerial level focused on promoting and applying human empowerment strategies, in which it is evaluated whether the company has human empowerment policies that strengthen the capabilities, confidence, vision and protagonism of the human being hand in hand with technology [57]. It also analyzes the presence of implementation plans of technological innovation tools for workers in which it is intended that operators have technological innovation tools such as 3D glasses, exoskeletons, digital twins in order to enhance the skills in the execution of their activities.

5 Conclusion

The proposal to create the conceptual production model called "ProS-5.0" arises from the lack of literature on the application of sustainable, resilient and human-centered production models in the textile sector. Based on the systematic theoretical review of 58 articles evaluated in a critical, transparent and reproducible manner, and the guidelines are established. This paper takes the initiative to consider the information management system as a powerful tool for support, feedback and evaluation of the levels or performance of Industry 5.0 approaches. From the knowledge of the textile case studies, all the sub-processes that integrate the production system are established in a general way, starting with the supply of primary materials, planning, design, manufacturing, packaging and ending with the distribution of finished products. Additionally, the Control and Monitoring activity is the constant within this flow and fulfills the objective of supervision and compliance with quality standards, approaches and their categories that integrate the organization.

The main contribution of this research is to present the elements that integrate the model with a focus on Industry 5.0: sustainability, resilience and focus on the human being. From the perspective of sustainability, the environmental, social and economic components are explored in depth and practical analysis strategies are proposed to anticipate, respond, evaluate performance and correct deviations. As with the human-centered component, the aim is for workers to perform their activities better with specific training in the area in which they work. In addition, the aim is to prevent injuries or accidents through ergonomics, by means of management plans and policies that integrate the human being as the central axis of production. Finally, resilience with the components of prevention, response and recovery provide important insights into production management, as the proposed framework offers a useful tool for decision makers.

This paper can be extended in several other directions, including product inspection strategies, determination of levels or performance of sustainability, resilience and human-centered approaches. Also, as design of an Industry 5.0. Standards compliance inspection scheme for a production system. In addition, anticipate the difficulties that arise when implementing this model.

Acknowledgments. The authors would like to thank to Corporación Ecuatoriana para el Desarrollo de la Investigación y Academia - CEDIA for the financial support given to the present research, development, and innovation work through its CEPRA program, especially for the CEPRA XV-2022-01-Pro.S.-5.0. In addition, this research was partially funded by the University of Cuenca, Technical University of Ambato and University of Azuay.

Annexes

Annex A

The list of codes used in Atlas.ti for the systematic literature review can be found online at https://imagineresearch.org/txmdl5-annex-a/.

References

1. Abreu, L.I., Cavalieri, A.V., Schlatter, P., Vinuesa, R., Henningson, D.S.: Resolvent modelling of near-wall coherent structures in turbulent channel flow. Int. J. Heat Fluid Flow **85**, 108662 (2020)

2. Akundi, A., Euresti, D., Luna, S., Ankobiah, W., Lopes, A., Edinbarough, I.: State of industry 5.0-analysis and identification of current research trends. Appl. Syst. Innov. **5**(1), 27 (2022)
3. Anatan, L., et al.: A review on MSMEs resilience: Strategies and policies post COVID-19 pandemic. Rev. Integr. Bus. Econ. Res. **10**, 386–394 (2021)
4. Andrieu, N., et al.: Co-designing climate-smart farming systems with local stakeholders: a methodological framework for achieving large-scale change. Front. Sustain. Food Syst. **3**, 37 (2019)
5. Antunes, R., Gonzalez, V.: A production model for construction: a theoretical framework. Buildings **5**(1), 209–228 (2015)
6. Appel, A., Hardaker, S.: Strategies in times of pandemic crisis-retailers and regional resilience in Würzburg, Germany. Sustainability **13**(5), 2643 (2021)
7. Audretsch, D.B., Link, A.N.: Entrepreneurship and knowledge spillovers from the public sector. Int. Entrepreneurship Manage. J. **15**(1), 195–208 (2019)
8. Aus, R., Moora, H., Vihma, M., Unt, R., Kiisa, M., Kapur, S.: Designing for circular fashion: integrating upcycling into conventional garment manufacturing processes. Fashion Text. **8**(1), 1–18 (2021). https://doi.org/10.1186/s40691-021-00262-9
9. Banker, R.D., Potter, G., Schroeder, R.G.: An empirical analysis of manufacturing overhead cost drivers. J. Account. Econ. **19**(1), 115–137 (1995)
10. Bécue, A., Maia, E., Feeken, L., Borchers, P., Praça, I.: A new concept of digital twin supporting optimization and resilience of factories of the future. Appl. Sci. **10**(13), 4482 (2020)
11. Bengtsson, M., Alfredsson, E., Cohen, M., Lorek, S., Schroeder, P.: Transforming systems of consumption and production for achieving the sustainable development goals: moving beyond efficiency. Sustain. Sci. **13**(6), 1533–1547 (2018)
12. Billy Ardiansyah, G., Tjahjadi, B., Soewarno, N.: Measuring customer profitability through time-driven activity-based costing: a case study at hotel x Jogjakarta. In: SHS Web of Conferences, vol. 34 (2017)
13. Caiza, G., Llamuca, E.S., Garcia, C.A., Gallardo-Cardenas, F., Lanas, D., Garcia, M.V.: Industrial shop-floor integration based on AMQP protocol in an IoT environment. In: 2019 IEEE 4th Ecuador Technical Chapters Meeting, ETCM 2019 (2019). https://doi.org/10.1109/ETCM48019.2019.9014858
14. Cao, C., Li, C., Yang, Q., Zhang, F.: Multi-objective optimization model of emergency organization allocation for sustainable disaster supply chain. Sustainability **9**(11), 2103 (2017)
15. Chakraborty, S., Biswas, M.C.: Impact of COVID-19 on the textile, apparel and fashion manufacturing industry supply chain: case study on a ready-made garment manufacturing industry. J. Supply Chain Manage. Logist. Procure. **3**(2), 181–199 (2020)
16. Demir, K.A., Döven, G., Sezen, B.: Industry 5.0 and human-robot co-working. Procedia Comput. Sci. **158**, 688–695 (2019)
17. Diamantidis, A.C., Papadopoulos, C.T., Heavey, C.: Approximate analysis of serial flow lines with multiple parallel-machine stations. IIE Trans. **39**(4), 361–375 (2007)
18. Dias, J.G.: Environmental sustainability measurement in the travel & tourism competitiveness index: an empirical analysis of its reliability. Ecol. Ind. **73**, 589–596 (2017)
19. Feng, C., Mai, Y.: Sustainability assessment of products based on fuzzy multi-criteria decision analysis. Int. J. Adv. Manuf. Technol. **85**(1), 695–710 (2016)
20. Fink, A.: Conducting Research Literature Reviews: From the Internet to Paper. Sage Publications (2019)

21. Flores-Sigüenza, P., Marmolejo-Saucedo, J.A., Niembro-Garcia, J., Lopez-Sanchez, V.M.: A systematic literature review of quantitative models for sustainable supply chain management. Math. Biosci. Eng. **18**(3), 2206–2229 (2021)
22. Freire-Aradas, A., et al.: Tracking age-correlated DNA methylation markers in the young. Forensic Sci. Int. Genet. **36**, 50–59 (2018)
23. Garcia, C.A., et al.: CPPS on low cost devices for batch process under IEC-61499 and ISA-88. In: Proceedings - 2017 IEEE 15th International Conference on Industrial Informatics, INDIN 2017, pp. 855–860 (2017). https://doi.org/10.1109/INDIN.2017.8104884
24. Garcia, M.V., Irisarri, E., Perez, F., Marcos, M., Estevez, E.: Engineering tool to develop CPPS based on IEC-61499 and OPC UA for oil&gas process. In: IEEE International Workshop on Factory Communication Systems - Proceedings, WFCS (2017). https://doi.org/10.1109/WFCS.2017.7991969
25. Ghinato, P.: Sistema toyota de produção: mais do que simplesmente just-in-time. Production **5**, 169–189 (1995)
26. Grosse, E.H., Calzavara, M., Glock, C.H., Sgarbossa, F.: Incorporating human factors into decision support models for production and logistics: current state of research. IFAC-PapersOnLine **50**(1), 6900–6905 (2017)
27. Hamedi, H., Mehdiabadi, A.: Entrepreneurship resilience and Iranian organizations: application of the fuzzy DANP technique. Asia Pac. J. Innov. Entrepreneurship **14**, 231–247 (2020)
28. Hildebrandt, J., Thrän, D., Bezama, A.: The circularity of potential bio-textile production routes: comparing life cycle impacts of bio-based materials used within the manufacturing of selected leather substitutes. J. Clean. Prod. **287**, 125470 (2021)
29. Hopp, W.J., Wu, S.C.: Machine maintenance with multiple maintenance actions. IIE Trans. **22**(3), 226–233 (1990)
30. Houston, J.B.: Community resilience and communication: dynamic interconnections between and among individuals, families, and organizations. J. Appl. Commun. Res. **46**(1), 19–22 (2018)
31. Imperatives, S.: Report of the world commission on environment and development: our common future. Accessed Feb **10**, 1–300 (1987)
32. Islam, M., Ingham, V., Hicks, J.: Flooding in Bangladesh: an in-depth analysis of community disaster resilience applying the 'production model'. Int. J. Interdisc. Environ. Stud. **11**(4), 59–74 (2016)
33. Ivanov, A., Jaff, T.: Manufacturing lead time reduction and its effect on internal supply chain. In: Campana, G., Howlett, R.J., Setchi, R., Cimatti, B. (eds.) SDM 2017. SIST, vol. 68, pp. 398–407. Springer, Cham (2017). https://doi.org/10.1007/978-3-319-57078-5_38
34. Jia, F., Yin, S., Chen, L., Chen, X.: The circular economy in the textile and apparel industry: a systematic literature review. J. Clean. Prod. **259**, 120728 (2020)
35. Kabeer, S.J., Nayebi, M., Ruhe, G., Carlson, C., Chew, F.: Predicting the vector impact of change-an industrial case study at brightsquid. In: 2017 ACM/IEEE International Symposium on Empirical Software Engineering and Measurement (ESEM), pp. 131–140. IEEE (2017)
36. Kikolski, M.: Study of production scenarios with the use of simulation models. Procedia Eng. **182**, 321–328 (2017)
37. Kirchherr, J., et al.: Barriers to the circular economy: evidence from the European Union (EU). Ecol. Econ. **150**, 264–272 (2018)

38. Kloda, L.A., Boruff, J.T., Cavalcante, A.S.: A comparison of patient, intervention, comparison, outcome (PICO) to a new, alternative clinical question framework for search skills, search results, and self-efficacy: a randomized controlled trial. J. Med. Libr. Assoc.: JMLA **108**(2), 185 (2020)
39. Kozlowski, A., Bardecki, M., Searcy, C.: Tools for sustainable fashion design: an analysis of their fitness for purpose. Sustainability **11**(13), 3581 (2019)
40. Kurniawan, A., Komara, B.D., Setiawan, H.C.B.: Preparation and challenges of industry 5.0 for small and medium enterprises in Indonesia. Muhammadiyah Int. J. Econ. Bus. **2**(2), 155–160 (2019)
41. Leal, L.R.G., de Alba, L.O., Conrado, A.R.: Desarrollo, tendencias, aplicaciones y herramientas de la industria 4.0 en el sector textil. Boletín Innov. Logíst. Oper. **2**(1), 98–101 (2020)
42. López, I., Arriaga, A., Pardo Buendía, M.: La dimensión social del concepto de desarrollo sostenible: ¿ la eterna olvidada? (2018)
43. Luzzini, D., Brandon-Jones, E., Brandon-Jones, A., Spina, G.: From sustainability commitment to performance: the role of intra-and inter-firm collaborative capabilities in the upstream supply chain. Int. J. Prod. Econ. **165**, 51–63 (2015)
44. Maestri, G., Bessa, G.C., Oliveira, F.R., Steffens, F.: Revoluções tecnológicas e a relação com o setor têxtil: perspectivas baseadas em indústria 3.5, indústria 4.0 e indústria 5.0. Rev. Eletrôn. Perspect. Ciência Tecnol. **13** (2021). ISSN 1984-5693
45. Masi, D., Kumar, V., Garza-Reyes, J.A., Godsell, J.: Towards a more circular economy: exploring the awareness, practices, and barriers from a focal firm perspective. Prod. Plann. Control **29**(6), 539–550 (2018)
46. Meredith, J.: Theory building through conceptual methods. Int. J. Oper. Prod. Manage. **13**(5), 3–11 (1993)
47. Moeuf, A., Pellerin, R., Lamouri, S., Tamayo-Giraldo, S., Barbaray, R.: The industrial management of SMEs in the era of industry 4.0. Int. J. Prod. Res. **56**(3), 1118–1136 (2018)
48. Muthu, S.S., Li, Y., Hu, J., Mok, P.Y.: Quantification of environmental impact and ecological sustainability for textile fibres. Ecol. Ind. **13**(1), 66–74 (2012)
49. Naqvi, S.A., Arshad, M., Farooq, A., Nadeem, F.: Implementation of sustainable practices in textile processing mills of Lahore, Pakistan. Pol. J. Environ. Stud. **29**(1) (2020)
50. Oliveira, E.A., Costa, C.B.B., Ravagnani, M.A.D.S.S.: < b> an optimization model for production planning in the drying sector of an industrial laundry. Acta Scientiarum. Technol. **39**(1), 69–77 (2017)
51. Poppendieck, M., Poppendieck, T.: Lean Software Development: An Agile Toolkit. Addison-Wesley (2003)
52. Rapsikevičienė, J., Gurauskienė, I., Jučienė, A.: Model of industrial textile waste management. Environ. Res. Eng. Manag. **75**(1), 43–55 (2019)
53. Rizos, V., Behrens, A., Kafyeke, T., Hirschnitz-Garbers, M., Ioannou, A.: The circular economy: barriers and opportunities for SMEs. CEPS Working Documents (2015)
54. Rossi, E., Bertassini, A.C., dos Santos Ferreira, C., do Amaral, W.A.N., Ometto, A.R.: Circular economy indicators for organizations considering sustainability and business models: plastic, textile and electro-electronic cases. J. Clean. Prod. **247**, 119137 (2020)
55. Ruan, X., Ding, N., Yang, J.: Dual-level eco-efficiency analysis in sustainable management: an empirical study on textile manufacturers and supply chains in china. J. Environ. Manage. **304**, 114227 (2022)

56. Rumanti, A.A., Sunaryo, I., Wiratmadja, I.I., Irianto, D.: Cleaner production through open innovation in Indonesian batik small and medium enterprises (SME). TQM J. (2020)
57. Sabadie, J.A.: Technological innovation, human capital and social change for sustainability. Lessons learnt from the industrial technologies theme of the EU's research framework programme. Sci. Total Environ. **481**, 668–673 (2014)
58. Sarwar, A., Khan, F., James, L., Abimbola, M.: Integrated offshore power operation resilience assessment using object oriented Bayesian network. Ocean Eng. **167**, 257–266 (2018)
59. Savino, M.M., Shafiq, M.: An extensive study to assess the sustainability drivers of production performances using a resource-based view and contingency analysis. J. Clean. Prod. **204**, 744–752 (2018)
60. Tayyab, M., Jemai, J., Lim, H., Sarkar, B.: A sustainable development framework for a cleaner multi-item multi-stage textile production system with a process improvement initiative. J. Clean. Prod. **246**, 119055 (2020)
61. Tayyab, M., Sarkar, B.: An interactive fuzzy programming approach for a sustainable supplier selection under textile supply chain management. Comput. Industr. Eng. **155**, 107164 (2021)
62. Tura, N., Hanski, J., Ahola, T., Ståhle, M., Piiparinen, S., Valkokari, P.: Unlocking circular business: a framework of barriers and drivers. J. Clean. Prod. **212**, 90–98 (2019)
63. Valencia-Rodríguez, O., Olivar-Tost, G., Redondo, J.M.: Modeling a productive system incorporating elements of business sustainability. Dyna **85**(207), 113–122 (2018)
64. Verdone, F., Cantarero, S., Puig, F.: Capturing the resilience of the textile companies as a specific response of the fashion industry. In: Rienda, L., Ruiz-Fernández, L., Drylie Carey, L., García-Medina, I. (eds.) Firms in the Fashion Industry, pp. 141–161. Springer, Cham (2021). https://doi.org/10.1007/978-3-030-76255-1_9
65. Wang, C.N., Nhieu, N.L., Tran, T.T.T.: Stochastic Chebyshev goal programming mixed integer linear model for sustainable global production planning. Mathematics **9**(5), 483 (2021)
66. Wang, J.: Multi-agent based production management decision system modelling for the textile enterprise. J. Glob. Inf. Manage. (JGIM) **25**(4), 1–15 (2017)
67. Warwas, I., Podgórniak-Krzykacz, A., Przywojska, J., Kozar, Ł.: Going green and socially responsible-textile industry in transition to sustainability and a circular economy. Fibres Text. Eastern Europe (2021)
68. Yavari, F., Pilevari, N.: Industry revolutions development from industry 1.0 to industry 5.0 in manufacturing. J. Industr. Strategic Manage. **5**(2), 44–63 (2020)
69. Zhang, R., Carboni, J.: The sustainable project management utilization in French wineries while adapting to climate change. Eur. J. Sustain. Dev. **10**(4), 85 (2021)
70. Zhang, Y., et al.: Improved design for textile production process based on life cycle assessment. Clean Technol. Environ. Policy **20**(6), 1355–1365 (2018). https://doi.org/10.1007/s10098-018-1572-9

Theory of Restrictions for the Improvement of Production Capacity in Textile SMEs

Ana Sánchez-Zapata, Edith Tubón-Núñez(✉), Sandra Carrillo-Ríos, and Franklin Tigre-Ortega

Faculty of Systems, Electronics and Industrial, Technical University of Ambato, Ambato, Ecuador
{asanchez8672,ee.tubon,sandralcarrillor,fg.tigre}@uta.edu.ec

Abstract. The research deals with the development of the Theory of Restrictions for the improvement of the productive capacity within the textile company, through the collection of information corresponding to the inventories and sales records of the products offered in the sports line, the current situation of the company and the ABC analysis was performed. In addition, through a time study, the operational problems of the organization were identified to apply the DBR algorithm. The data to compare the production capacity was taken from the simulation in FlexSim, which generated an improvement percentage of 86.48% within the cutting process, while in the manufacturing process the improvement was 58.29% and in the packing no improvement was obtained. In the current and proposed process, the effective transfer rate is 9.00 dollars and operating expenses are reduced by 5.03%, which is verified in the financial indicator, where the net profit of the process increased by 4.30%.

Keywords: Theory of restrictions · ABC analysis · time study · DBR algorithm · transfer rate · operating expenses · net income

1 Introduction

Manufacturing companies associated with textile products have had a growth in the technification of production processes worldwide. On many occasions, productivity problems come from inadequate stock management of warehouse products, which implies a higher financial expense and comes to reduce working capital [6]. In addition, some constraints may arise during the manufacturing process that affect the yield, inventory, and operating costs of this process [12].

The application of TOC (theory of constraints), is used to identify, analyze, and eliminate the constraints that restrict the value adding process of a company [12], and establish a continuous flow between the operations of the productive process [4], which influences the increase of capacity and reduction of processing costs [10]. The TOC is evaluated by means of the following indicators: a) the effective transfer rate based on the speed at which the company generates money through product sales or inventories, b) operating expenses related to machinery, labor, premises, among others, excluding those related to raw materials for

M. V. Garcia and C. Gordón-Gallegos (Eds.): CSEI 2022, LNNS 678, pp. 625–638, 2023.
https://doi.org/10.1007/978-3-031-30592-4_41

inventory production and c) net profit which is the effective economic benefit obtained by the company [1].

In TOC, system bottlenecks are critical to improve productivity and profitability of the production system. The purpose is to develop practical steps to identify bottlenecks in production systems characterized by uniform flow [12]. Raising the bottleneck capacity, eliminating defects and the use of those resources that do not limit production and exceed capacity, are the main practices applied to improve competitiveness and performance in companies [5,8,17]. Likewise, following the steps of the TOC methodology, a model can be established to simulate production processes [18].

The research begins with the ABC analysis, based on the current demand data [8,16]. Subsequently, the study of times and movements of the production process and the calculation of the current productivity in units per hour were carried out, then, the adequate model that adapts to the needs of the textile company was determined, focusing on planning, work organization and communication between the preset areas. All these actions were aimed at establishing a point of improvement of the production process to achieve economic benefits. In addition, the bottlenecks that limit production were identified to know exactly the amount of material, inputs, and products within the company [7,11,15].

2 Methodology and Methods

2.1 Methodology

DBR (Drum-Buffer-Rope) Algorithm Derived from TOC

Using the DBR algorithm based on the Theory of Constraints Logistic application approach, the following steps were executed [9,14]:

(a) Identify the system constraint, in this case the bottleneck.
(b) Establish the respective strategy to take advantage of the bottleneck in terms of productivity.
(c) Direct the entire workflow according to the bottleneck, lifting and removing the constraint.
(d) Start the process again to find a new bottleneck.

2.2 Methods

The applied research seeks the generation of knowledge to achieve the main purpose of the work, focused on improving the production flow through the TOC; which allowed the optimization of the production flow, control of raw material inventories and use of existing resources in the company. Supported by the field research, the principles, and algorithms for the application of the theory of constraints are based.

The case study company has four workers, one in the cutting area, two in the clothing area and one in the warehouse area; inexperienced personnel in terms of production planning and raw material requirements procedures; a situation that limits the availability of objective and detailed information.

Through direct observation during the visits to the company, the current situation in the production and inventory management areas was determined; this allowed contextualizing the proposed solution that adapts to the reality of the study unit. In addition, interviews were conducted with those responsible for the production area, to collect relevant data on the production flow and the main problems generated in the production process and delivery to the end customer (supply chain). Using Microsoft Excel, the data was coded, tabulated, represented, and statistically analyzed, obtaining information that reflects the company's reality.

3 Results

The analyzed study unit handles a wide range of products of the sports line such as: sweaters, pants, T-shirts, pants; made with quality raw material. With historical sales data for the years 2018 and 2019, using the variables of unit price of the product and average annual sales, the ABC analysis was applied, based on Eq. 1 ($ valorization) and 2 (% consumption) [2].

$$\$Valorization = \text{unit price} \times \text{annual average} \tag{1}$$

$$\%\text{ of consumption} = \frac{Valorization * 100\%}{\sum Valorization} \tag{2}$$

Once the percentage of consumption of each product is generated, it is ordered from highest to lowest and the percentage of accumulated consumption is calculated to establish the zone to which each product belongs, see Eq. 3 [2].

$$\%\text{accumulated consumption} = \%\text{accumulated consumption}_{i-1} + \%\text{accumulated consumption}_{i} \tag{3}$$

Product SDG_CH_36 (Fig. 1) represents 80% of the annual consumption with a valuation of $ 795.00 and a consumption percentage of 3.29%. This item has the greatest impact on the company and is in zone A of the ABC analysis.

Fig. 1. Product SDG_CH_36

Flow diagrams were developed with information collected through interviews and on-site visits, applied for the survey of processes, detailing the areas and activities involved in the production process [11].

For the time study, the General Electric criterion was applied [9]. The performance factor and supplements determined that the TS (Standard Time) used in the cutting, tailoring, and packing process is 20.07 min, 46.57 min and 2.48 min, respectively.

By means of the analytical cursogram and Eq. 4, the productive time used for the cutting, making, and packing process was determined to be 95.26%, 8.74% and 71.41%, respectively. Actions that allowed identifying the activities that affect the productive time of the process [3,13].

$$Productivity = \frac{\text{operation time}}{\text{Total time (operation, inspection, waiting, transport and storage))}} \quad (4)$$

According to the data obtained in the time study, the TS of the activities performed by the operators in the different machines (cutter, overlock, straight or coating machine) was detailed. Four phases were identified for the cutting process, fifteen phases in the making-up process and two phases in the packing process; with a total of 21 phases that group the times of the repetitive activities during each process.

3.1 Algoritmo DBR Derivado de TOC

Step 1: Identify the Process Constraining Activity

The activities that limit the current process: Phase 10 - Tailoring process (performed by operator 3):

(a) prepare machine, sew zipper
(b) stitch zipper and collar

They have an operation TS of 11.88 min for one unit.

Step 2: Exploiting the Determined Constraint

As a measure to exploit the restriction in the action that limits the process, personnel were added to the activities to reduce operating time and increase production.

In addition, it was observed that there are ANV related to the cutting process, so it was recommended to standardize the molds of the pieces that make up the garment; this will not increase the time in the cutting process since it is an external activity that will be performed only once, until the molds have the correct dimensions; allowing the generation of pieces according to the dimensions of the design that fit correctly without the need to cut the excess fabric in the manufacturing process.

Finally, it is observed that the TS of the proposed process decreased from 45.95 min to 42.58 min; the cutting process maintained the TS of 19.70 min as well as the packing process with a TS of 2.10 min, without considering the transport TS; generating that the TS per unit goes from 67.72 min (1.13 h) to 64.35 min (1.07 h) for the product. In the same way, the TPf (average phase time) of the proposed process decreased from 19.70 min to 9.83 min in the cutting process and from 41 min to 23.43 min in the packing process, but the TPf of 2.10 min was maintained in the packing process (Fig. 2).

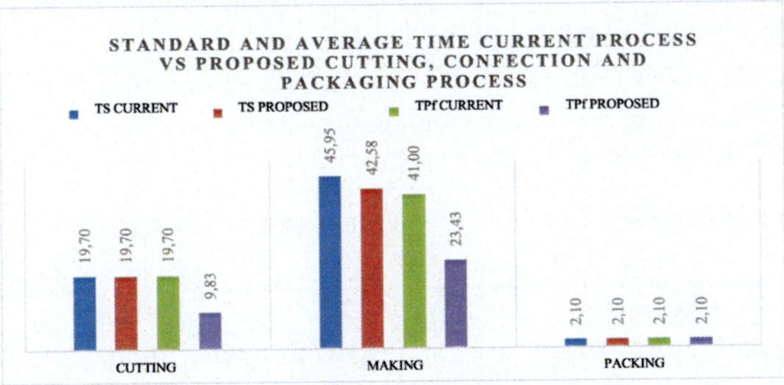

Fig. 2. Standard and average time of the actual vs. proposed cutting, making, and packing process of code SDG_CH_36

Step 3: Subordination of activities according to the bottleneck

According to the TOC methodology, the production rate known as drum is the activity that limits the process. Therefore, by equalizing the resources in the phases of the process and the operation time of the workers, a continuous flow is sought. By means of the simulation, phases 1, 2, 3 and 4 of the cutting process and phases 2, 5, 7, 9, 10, 11, 12, 13, 14 and 15 of the making-up process were subordinated, dividing their activities for two operators avoiding the inventory in process and ensuring a continuous flow of raw material between the phases, then proceeding to eliminate the transport of the cutter, from the cutting table to the finished cuts table and the transport of operator 2, from the overlock machine 1 to the finished cuts table (See Fig. 3).

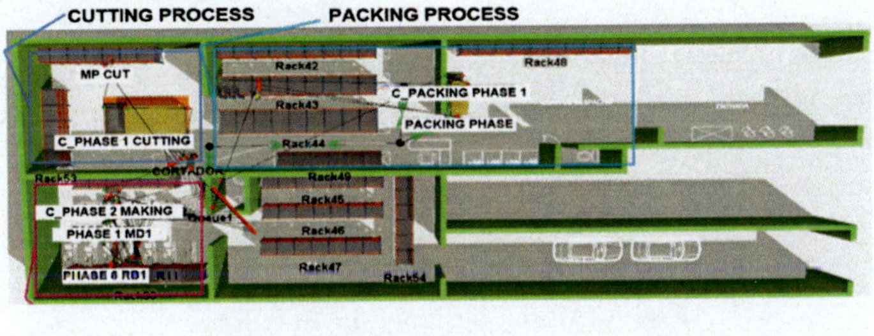

Fig. 3. Model designed in FlexSim

Step 4: Lifting of the System Constraint

By means of the simulation, it was possible to increase the bottleneck, thus reducing the product manufacturing time. The results presented in Table 1 show the productivity (simulation time in seconds) of the current process compared to the proposed process for cutting, making, and packing (See Fig. 4). By means of the dashboard called production, the performance of the workers involved in the current process compared to the proposed method of cutting, making and packing was specified as shown in Table 1.

Table 1. Productivity of the current vs. proposed cutting, making, and packing process of the SDG_CH_36 code.

CURRENT PROCESS		PROPOSED METHOD	
	Time of simulation (s)		Time of simulation (s)
CUT	2229,42	CUT	1195,51
CONFECTION	161954,69	CONFECTION	102103,51
PACKING	169443,69	PACKING	109592,55

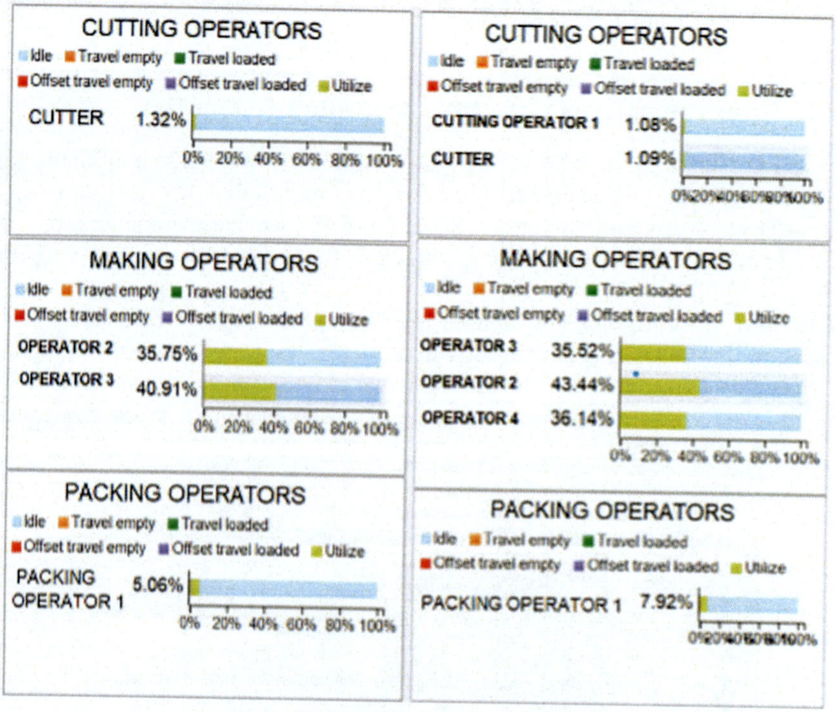

Fig. 4. Model designed in FlexSim

Step 5: Return to Step 1

With the application of the TOC methodology, the activities that limit the process are eliminated to ensure the non-existence of events that alter the continuous flow of the process, for which the following activities were carried out:

(a) Control of the supply of inputs
(b) Weekly production planning
(c) Hiring of personnel with experience in cutting and making.
(d) Training of personnel

4 Discussion

The simulation of the current and proposed process was carried out using FlexSim software, to determine how long it will take to produce an estimated demand of 50 units, data considered by the organization. The current process involved 4 operators, with an accumulated time of 169443.69 s (5.88 days), which is divided into 3 processes detailed in Table 2.

Table 2. Time of the simulation of the actual cutting, making, and packing process.

Proposed Process	Time		
	Accumulated (s)	Process (s)	Process (days)
Cut	2229,42	2229,42	0,08
Confection	161954,69	159725,27	5,55
Packing	169443,69	7489,00	0,26
Total		169443,69	5,88

The proposed process involved 5 operators, with an accumulated time of 109592.55 s (3.80 days), which is divided into 3 processes, as detailed in Table 3.

Table 3. Time of the simulation of the proposed cutting, making, and packing process.

Proposed Process	Time		
	Accumulated (s)	Process (s)	Process (days)
Cut	1195,51	1195,51	0,04
Confection	102103,51		3,50
Packing	109592,55	7489,04	0,26
Total		109592,55	3,80

It was observed that the time required to process the batch of 50 units of the current vs. proposed process decreased from 5.88 days to 3.80 days.

Based on the results obtained from the simulation, the percentage improvement is calculated using Eqs. 5 [4]. The percentage improvement in the cutting, making, and packing process was 86.48%, 58.29% and 0%, respectively.

$$\% \text{ Improve} = \frac{\text{process compare} - \text{process reference}}{\text{process reference}} * 100 \tag{5}$$

The Cp (production capacity) [8] of the cutting process is limited by the linear cutting machine of 20.32 cm (8 in.), which allows to perform a single cut optimizing the molding and cutting time. In addition, the table allows to lay a maximum of 2.30 m of fabric, and in the machine enter 96 layers equivalent to 48 layers of violet fabric corresponding to 4 rolls considering the type of fabric as well as its stiffness, composition and weight, the material on average has 65 m and a width of 1.55 m according to Fig. 5. When molding the white pieces as shown in Fig. 6, it is determined that for 2 sweaters 1.72 m and 1.5 m wide of white fabric are required, using only 232.16 m equivalent to 3.57 rolls of fabric, allowing to cut up to 192 sweaters each time the process is executed.

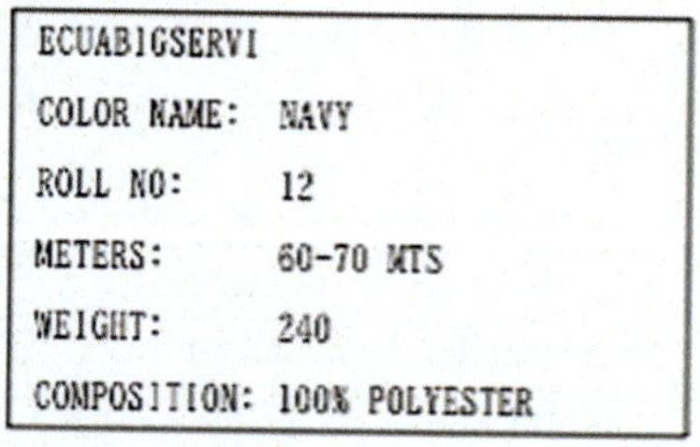
ECUABIGSERVI
COLOR NAME: NAVY
ROLL NO: 12
METERS: 60-70 MTS
WEIGHT: 240
COMPOSITION: 100% POLYESTER

Fig. 5. Specifications of the biotope fabric

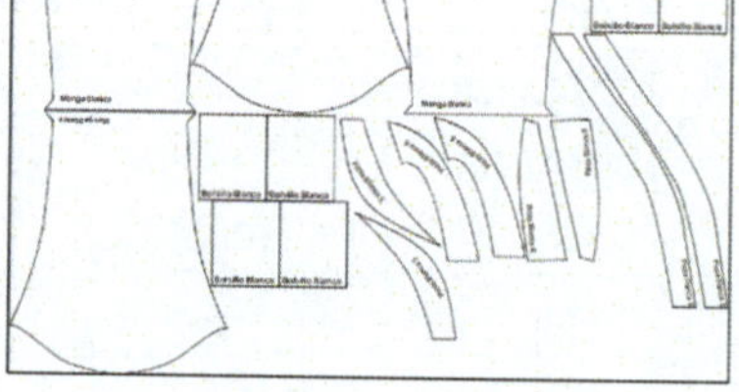

Fig. 6. Mold on the white fabric

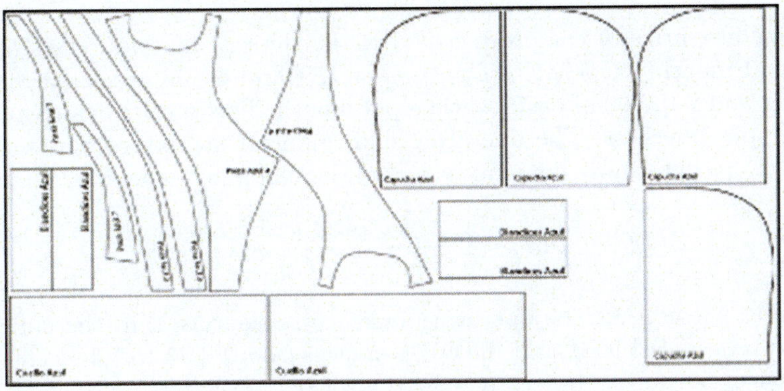

Fig. 7. Molding on blue fabric

The cuts of the blue pieces are complements of the white pieces, so less meters of fabric are needed for their production, that is, the Cp of the machine can produce these units since it uses a smaller amount of material than the white parts, so the Cp of the cutting process is determined by the cuts of the white pieces. The Fig. 7 shows that for 2 sweaters, 1.5 m wide and 0.78 m high of blue fabric is required, using only 187.04 m equivalent to 2.88 rolls of fabric. The Cp of the current vs. proposed process of making up and packing based on the simulation, in the proposed process allowed to increase the Cp from 80.74 units/hour to 150.56 units/hour in the cutting process, from 1.13 units/hour to 1.78 units/hour (9 units/day to 14 units/day) in the making up process, but in the packing process it remained at 24.04 units/hour (192 units/day), as shown in Fig. 8.

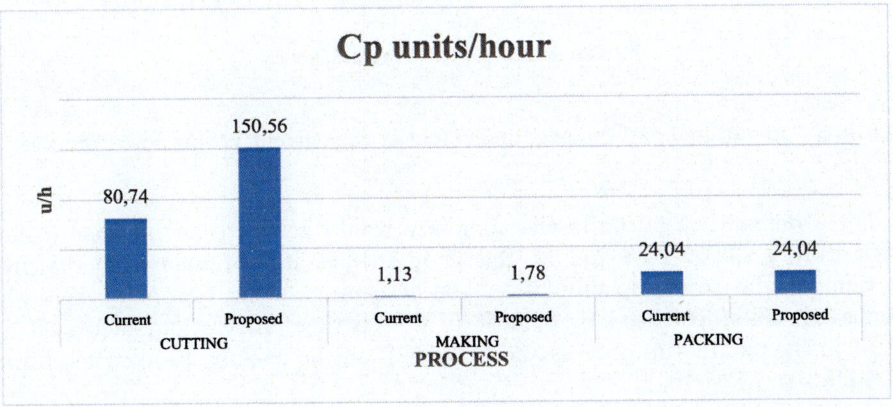

Fig. 8. Production capacity of current vs. proposed cutting, making, and packing process

The selling price of the product is \$ 15.00, the raw material costs (MP) are \$ 6.00 and monthly $OperatingExpensive(OE)$ are: \$ 400 of salary, \$ 44.60 of insurance and \$ 100.00 of basic services per month. This results in a cost of \$ 2.27 per hour per operator. The operating and financial indicators were calculated using Eq. 6 to determine how much money the company generates.

$$OE(\$/unit) = \frac{2,27 * \#\text{production operator}}{Cp} \tag{6}$$

Figure 9 shows that the operating indicators, such as T in the current and proposed process is \$ 9.00, and the OE decreases from \$ 4.15 to \$ 3.94 considering a demand of 50 units, that is, a reduction of 0.21 ctvs per product (5.03%) is generated, which is verified in the financial indicator, where the net profit of the process increases 4.30%, that is, from \$ 4.85 to \$ 5.06 per product.

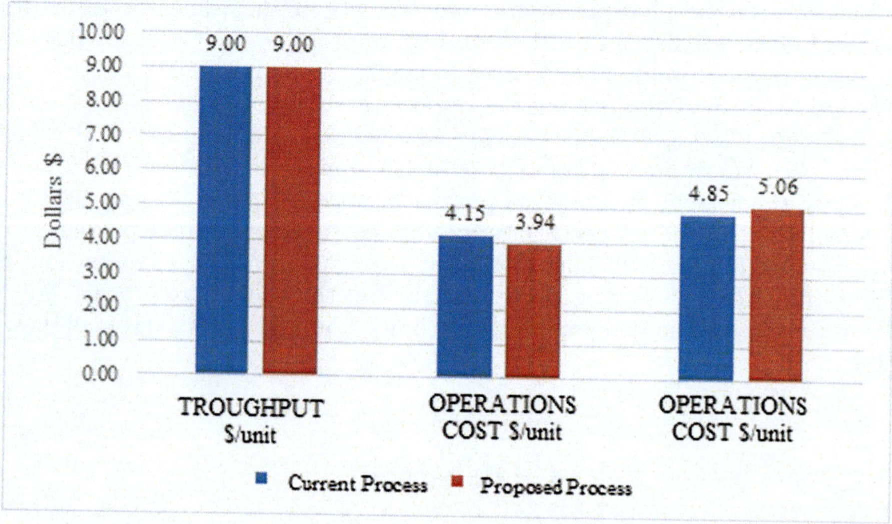

Fig. 9. Operational and financial indicators of the current vs. proposed process

For a demand of 50 units the company would generate an increase in net profit from \$ 242.52 to \$ 252.96. Based on the results obtained from the production Dashboard of the simulation, the percentage utilization of workers was calculated. This decreased by 21.66% in the cutting process and increased by 0.10% in the making-up process and by 35.87% in the packing process, as shown in Table 3 and Fig. 10.

Table 4. Average utilization and percentage improvement of workers in the cutting, making, and packing process current vs. proposed

Operators	Usage		% Improvement
	% Current Process	% Proposed Process	
Cut	1,32	1,09	21,66
Confection	38,33	38,37	0,10
Packing	5,06	7,89	35,87
Total		109592,55	3,80

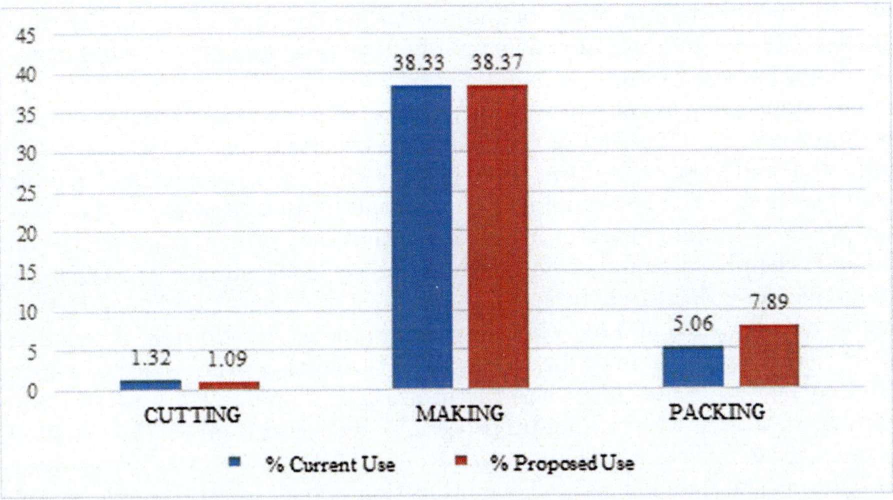

Fig. 10. Percentage improvement in operator utilization of current vs. proposed process

In addition, it is recommended to implement the digital patternmaking system to eliminate the time of the shaping activity from the cutting process (Table 4).

5 Conclusions

Through the application of the steps of the Theory of Constraints (TOC) methodology, it was possible to identify the activities that lead the production rate of the making process (phase 10) performed by operator 3. In addition, based on the step established in the TOC methodology that proposes to respect the determined restriction, a simulation of the production area was performed by means of the FlexSim software to subordinate and raise the restriction of the making process system, thus dividing its activities for two operators avoiding the inventory in process and ensuring a continuous flow of the same.

By means of the simulation, an improvement percentage of 86.48% was achieved in the cutting process, increasing its production capacity from 80.74

units/hour to 150.56 units/hour in a batch of 50 units, with respect to the confection process with an improvement of 58.29% going from 9 units/day to 14 units/day, however, in the packing process no improvement percentage was obtained, being its production capacity of 192 units/day. Likewise, the percentage of utilization of workers in the current vs. proposed process decreased by 21.66% in the cutting process but increased by 0.10% in the making-up process and by 35.87% in the packing process. In addition, the operating and financial indicators were calculated, as the throughput in the current and proposed process was $ 9.00, and operating expenses decreased from $ 4.15 to $ 3.94, that is, a reduction of 5.03% per product was generated, which is verified in the financial indicator, where the net profit of the process increased 4.30%, that is, from $ 4.85 to $ 5.06 per product, in a demand of 50 units the net profit would increase from $ 242.52 to $ 252.96.

The simulation of the current and proposed process was carried out with FlexSim software, a demand of 50 units was estimated, the improvement of the production and operation of the company was given by means of the DBR algorithm known as TOC methodology. The indicators used to measure the process were productivity and operator utilization within each process, through the data table and the production dashboard programmed in the simulation. Where, the productive time decreased from 5.55 days to 3.50 days in the manufacturing process but remained at 7284 s in the packing process. In addition, it was determined that the utilization of the operators in the cutting process decreased from 1.32% to 1.08% for the cutter and the utilization of operator 4 was reduced by 1.09%. Like-wise, in the making-up process, it decreased from 35.75% to 35.52% for operator 2 (overlock machine) and from 40.91% to 43.44% for operator 3 (straight machine) and generated the utilization of operator 4 (straight machine and overlock machine) by 36.14%. Finally, in the packing process, the utilization of operator 1 increased from 5.06% to 7.89%.

Several techniques were used to establish current productivity, such as the study of times, valuation, and supplements, in addition to analytical charts for the cutting, making, and packing processes belonging to code SDG_CH_36. The company Edy Sánchez Sport has a wide range of products for sale, one of them being the sports line, each product contributing in different ways to the company's income. Through the application of the ABC analysis, the most representative garments of the company were established with a total of 24 products that are within zone A of the analysis, representing 80% of the demand, resulting in code SDG_CH_36, which was the object of study for the realization of the project.

References

1. Diseño, implementación y análisis de una metodología para aplicar TOC a empresas metalmecánicas con restricciones físicas internas – caso de aplicación: Colombia (2016). https://www.revistaespacios.com/a16v37n31/16373130.html. Accessed 23 Sept 2022

2. Acosta, R.M., Resendiz, A.L., Lozano, C.I.L.: Análisis de la cadena de suministro por clasificación ABC: el caso de una empresa mexicana. RAN: Revista Academia Negocios **4**(2), 83–94 (2019). https://dialnet.unirioja.es/servlet/articulo?codigo=6750256
3. Andrade, A.M., A. Del Río, C., Alvear, D.L., Andrade, A.M., A. Del Río, C., Alvear, D.L.: A study on time and motion to increase the efficiency of a shoe manufacturing company. Información tecnológica **30**(3), 83–94 (2019). https://doi.org/10.4067/S0718-07642019000300083
4. Baron Chuman, M.A.: La teoría de restricciones como estrategia para incrementar el nivel de servicio en la Empresa Tablenorte S.A.C. Ph.D. thesis, Universidad Católica Santo Toribio de Mogrovejo (2020). https://tesis.usat.edu.pe/handle/20.500.12423/2898?locale=es
5. Barona-Lopez, L.I., Valdivieso-Caraguay, A.L., Benalcazar, M.E., Aguas, X., Zea, J.A.: Feature evaluation of EMG signals for hand gesture recognition based on mutual information, fuzzy entropy and RES index. In: García, M.V., Fernández-Peña, F., Gordón-Gallegos, C. (eds.) Advances and Applications in Computer Science, Electronics and Industrial Engineering. AISC, vol. 1307, pp. 101–119. Springer, Singapore (2021). https://doi.org/10.1007/978-981-33-4565-2_7
6. Bombón, M., Jordán, A., Jordán, J.: Teoría de restricciones como herramienta de desarrollo estratégico productivo del sector textil. 593DPCEIT **4**(5), 52–66 (2019). https://doi.org/10.33386/593dp.2019.5.116
7. Caiza, G., Riofrio-Morales, M., Robalino-Lopez, A., Toscano, O.R., Garcia, M.V., Naranjo, J.E.: An immersive training approach for induction motor fault detection and troubleshooting. In: De Paolis, L.T., Arpaia, P., Bourdot, P. (eds.) AVR 2021. LNCS, vol. 12980, pp. 499–510. Springer, Cham (2021). https://doi.org/10.1007/978-3-030-87595-4_36
8. Cajigas, M., Ramirez, E., Ramirez, D.: Capacidad de producción y sostenibilidad en empresas nuevas. Revista ESPACIOS **40**(43) (2019). https://www.revistaespacios.com/a19v40n43/19404315.html
9. Chase, R.B.: Administración de operaciones : producción y cadena de suministros. - 13. edición. McGraw-Hill (2014)
10. Díaz, F.R.G., Allaica, J.C.M., Sevilla, D.I.R., Rodríguez, B.G.G.: La teoría de restricciones (TOC) y su incidencia en los costos de producción. Caso empresa MIVIRN de Riobamba-Ecuador. 1. **3**(3.1), 285–306 (2020). https://doi.org/10.33262/concienciadigital.v3i3.1.1395
11. Evans, J.R., Lindsay, W.M.: Managing for Quality and Performance Excellence. South-Western College Publication (2007)
12. Farouk, H.H.A.: The application of theory of constraints in a production planning process with reference to its application in ABB company
13. Gómez, I.G., Aguilar, J.B.: Administración de operaciones. Editorial UIDE (2020). https://ulibros.com/administracion-de-operaciones-baomh.html
14. Marín, W.M., Gutiérrez, E.V.G.: Desarrollo e implementación de un modelo de teoría de restricciones para sincronizar las operaciones en la cadena de suministro. Revista EIA **10**(19), 67–77 (2013). https://dialnet.unirioja.es/servlet/articulo?codigo=4529770
15. Méndez, F., Calupiña, L., García, M.V., Caiza, G.: Detection of incipient faults in three-phase motors through analysis of stator currents. In: Salgado Guerrero, J.P., Chicaiza Espinosa, J., Cerrada Lozada, M., Berrezueta-Guzman, S. (eds.) TICEC 2021. CCIS, vol. 1456, pp. 250–263. Springer, Cham (2021). https://doi.org/10.1007/978-3-030-89941-7_18

16. Morales, J., Morales, V., Bayona, S., Morales, M.: Formative assessment model using an analytical rubric for written tasks. In: Garcia, M.V., Fernández-Peña, F., Gordón-Gallegos, C. (eds.) CSEI 2021. LNNS, vol. 433, pp. 143–157. Springer, Cham (2022). https://doi.org/10.1007/978-3-030-97719-1_8
17. Pusdá-Chulde, M.R., Salazar-Fierro, F.A., Sandoval-Pillajo, L., Herrera-Granda, E.P., García-Santillán, I.D., De Giusti, A.: Image analysis based on heterogeneous architectures for precision agriculture: a systematic literature review. In: Nummenmaa, J., Pérez-González, F., Domenech-Lega, B., Vaunat, J., Oscar Fernández-Peña, F. (eds.) CSEI 2019. AISC, vol. 1078, pp. 51–70. Springer, Cham (2020). https://doi.org/10.1007/978-3-030-33614-1_4
18. Rubio Tapia, X.E.: Teoría de Restricciones en el área de producción en industrias textiles para el incremento de la productividad. Master's thesis, Universidad Técnica de Ambato. Facultad de Ingeniería en Sistemas, Electrónica e Industrial. Maestría en Producción y Operaciones Industriales (2021). https://repositorio.uta.edu.ec/handle/123456789/33285

Information and Communication Technologies Adoption Model for SMEs. Case Studies

José Santisteban[1], Vicente Morales[2](✉), Juan-Carlos Chancusig[3], and Marjorie Morales[4]

[1] Universidad Privada Norbert Wiener, Av. Arequipa 440 con Jr. Larrabure y Unanuc 110. Urb., Santa Beatriz, Peru
jose.santisteban@uwiener.edu.pe

[2] Universidad Técnica de Ambato, Los Chasquis y Río Payamino, 180103 Ambato, Ecuador
jvmorales99@uta.edu.ec

[3] Universidad Técnica de Cotopaxi, Av Simón Rodríguez, 050108 Latacunga, Ecuador
juan.chancusig@utc.edu.ec

[4] LaGuardia Community College, 31-10 Thomson Avenue, Queens, NY 11101, USA
mmorales0011@uta.edu.ec

Abstract. External support is critical to the success of ICT adoption in SMEs, as these businesses are lagging due to a general lack of IT literacy, skills, and training resources. The initial aim of this study is to identify those factors that influence the adoption of ICT in SMEs in emerging countries. According to the review of the literature, there is no specific model that integrates all the fac-tors that influence the adoption of ICT in SMEs, so a model that integrates the most relevant factors that influence the adoption of ICT is proposed. This study proposes a model to identify the factors that influence the adoption of ICT in SMEs in emerging countries. The proposed model was validated through case studies, for which the 10 steps outlined in the Methodological Guide were applied. The findings obtained made it possible to highlight that the factors related to caring for the environment, and the experience and knowledge of ICT of employees exert a strong positive influence on employers when adopting ICT, and resistance to change exerts a strong negative influence.

Keywords: ICT Models · Adoption Factors · ICT Adoption · SMEs

1 Introduction

1.1 A Subsection Sample

Small and medium-sized enterprises (SMEs) represent the main source of employment in both developed and developing countries, and Information and

M. V. Garcia and C. Gordón-Gallegos (Eds.): CSEI 2022, LNNS 678, pp. 639–661, 2023.
https://doi.org/10.1007/978-3-031-30592-4_42

Communication Technologies (ICT) benefit this type of company in the ability to generate new products, businesses, and the possibility of reaching new markets. ICTs represent an area of opportunity for SMEs. The challenge is that these companies will necessarily have to strategically adopt and incorporate ICTs into their organization [48]. For an adequate adoption of ICT in SMEs, the authors [48] argue that the key factor is to determine the appropriate technology, economically and culturally, in relation to the company's available resources. According to [35, ?], in the case of ICT, there is a certain consensus that if SMEs do not adopt new technologies in their buying and selling strategies, they will very possibly not be able to survive in the new environment.

In the case of developing countries, the adoption of applications based on computing and the Internet by SMEs constitute the main source of sustained growth in their economies. According to [19,45], the access and use of ICTs have substantially improved the efficiency of SME businesses, because the use of the Internet has opened new opportunities in all fields. New product development, marketing, purchasing, distribution, and customer service have been improved, regardless of geographic distance.

On many occasions, the adoption of ICT in SMEs finds inhibiting factors or barriers, mainly because they do not know what technologies can be good for their business, it also happens that current information technologies (IT) do not adapt to their needs [24,45]. Regarding the use of the Internet, some SMEs find barriers in the availability of PCs, there are SMEs that have difficulty accessing the Internet, lack of knowledge of the service, complexity in its use, concern about security and cost [11,21]. Many owners of SMEs state that they do not see its utility, that it is difficult to use, that the acquisition price is high and due to the existence of a maintenance cost, they do not adopt ICTs in their businesses [10,37].

This paper is organized in 5 sections. Section 2 reviews the models that have incorporated the adoption of ICTs in SMEs in developing countries. Section 3 shows the Methodological Guide for the implementation of the proposed model. Section 4 shows the model proposed to identify the factors that influence the adoption of ICTs. Section 5 presents case studies to identify the factors influencing ICT adoption. Finally, Sect. 6 shows the conclusions of the investigation.

2 Literature Review

In the work of [40] they indicate that among the theories or models of ICT adoption, the most used are the technology acceptance model (TAM), the unified theory of the acceptance and use of technology (UTAUT) and the theory of planned behavior (TPB) [18] in this case as individual ICT adoption models, and the theory of diffusion of innovation (DOI) and the model of technology, organization and environment (TOE), such as theories of ICT adoption at the organizational level.

In developing countries, some of the challenges when adopting ICTs are legal and regulatory issues, weak adoption strategies, lack of research and development, excessive dependence on foreign technology, and weak implementation processes of ICTs. ICT [16]. In the study by [50], it was found that technology-savvy SME owners tend to have a more positive attitude towards ICT adoption. In the study [47] they concluded that the management (the direction) plays a fundamental role in the adoption of ICT in SMEs. According to [46], it was found that the characteristics of the industry sector and the company are factors that contribute to the adoption and use of ICT by SMEs.

The younger workforce is more enthusiastic about adopting ICT in their businesses. An employee with a good education tends to create the necessary flexibility to adopt ICT [36]. Owners/managers of SMEs who are knowledgeable about ICT tend to have a positive attitude towards innovation and are likely to adopt ICT in their business [38]. [1] state that perceived benefits is the main enabler for ICT adoption in the context of Indian SMEs. According to [49] they concluded that the adoption of ICT in Malaysian SMEs is highly influenced by technological factors.

In the work of [12], the literature on the determining factors that stimulate the adoption and use of ICT in SMEs is analyzed in depth, and the determining factors are classified into 5 groups: individual, organizational, environmental, technological, economic. In the study by [20] the authors developed a model of factors related to the ICT adoption process in SMEs. The model categorizes the factors into internal (characteristics of the owners/managers, organizational characteristics, company resources and IT user) and external (IT products in the market, external competitive pressure, suppliers and external consultants of IT and government support).

[2] investigated to examine the determinants of ICT adoption in SMEs in the state of Malays in Malaysia using the Theory of Reasoned Action (TRA) to measure the main factors that influence the adoption. The model of this research shows that the adoption of ICT depends on subjective norms and the attitude of executives.

The study by [2] identified factors such as lack of knowledge, lack of skill, lack of realization of benefits, and infrastructure problems as factors that impede adoption in SMEs. The unique characteristics of SMEs can be highlighted as one of the main reasons that differentiate when adopting ICT in SMEs with respect to large organizations. [15] identified the lack of ICT skills and knowledge on the part of SME employees as one of the main inhibitors when it comes to adopting ICT in their businesses. Ignorance about technology and limited resources are barriers to Internet adoption. The study on the impact of ICT on SMEs found that the unique nature of SMEs has been shown to be an important consideration for ICT adoption [5].

The results of the study [32] in SMEs show that the lack of technical skills, the lack of IT knowledge and the high cost of IT applications are important barriers to the adoption of electronic commerce. Furthermore, the authors found that, despite the exponential growth of ICT, the rate of adoption in these companies has remained relatively low. In the study by [25] investigated the adoption of ICT in SMEs in Nigeria and found that one of the main factors impeding the diffusion and intensive use of ICT is their poor physical infrastructure.

According to [7], one of the barriers why SMEs cannot use ICTs is that these types of companies still think that ICTs cannot provide benefits. Although the price is getting cheaper and more affordable for most SMEs, they are still perceived as an expensive asset. In [29] it is indicated that SMEs in developing countries, their limited resources and poor infrastructure are some of the factors that contributed to the slow adoption of ICTs. [1] states that the lack of strategic direction and the high cost of ICT are the most important barriers to ICT adoption in Indian SMEs. The Confederation of [9] study shows that ICT penetration in SMEs in India is very low, and this is attributed to very few employees of an SME acquiring specialized ICT skills in their training.

A study was carried out to arrive at several potential determinants influencing the adoption of ICT by SMEs in Zimbabwe [33]. The study revealed both internal and external barriers to ICT adoption. The author developed a model of barriers to the adoption of ICT in SMEs Zimbabwe, this model groups the internal barriers in characteristics of the owner/manager, characteristics of the organization and the cost and return on investment and the external barriers in Infrastructure, Social, Cultural and Political. Factors influencing ICT adoption in SMEs in Lagos State in Nigeria are analyzed in [25]. The results indicate that cost is the main barrier in the adoption of ICT in SMEs. The research model for this study is made up of five variables: cost, company size, infrastructure, government support, and management support. The main objective of the study [27] was to identify and evaluate the factors that are facilitating or impeding the adoption of ICT in SMEs in the Indian automotive sector, whose results show that the level of adoption in SMEs is low. This study reveals that perceived benefits and perceived competitive pressure are enabling factors for adoption. However, lack of financial capacity, small-scale operations, and a lack of IT manpower within the company inhibit adoption.

The study by [41] identified international barriers to ICT adoption in small businesses in Iran. The author developed a model of barriers to adoption based on two levels; the barriers at the company level and the barriers at the individual level, in turn the company level is made up of the technological, organizational and environmental barriers (taken from the TOE Model) and the individual level is made up of the factors lack facility of perceived use and perceived lack of usefulness [14].

In his study [51] shows that the adoption of ICT in SMEs is relatively low, and that the main reason for adopting ICT is the benefits to increase business competitiveness and performance. The factors that impede adoption are cost, scarcity of human resources and complexity of ICT. In addition, a research model is proposed that is composed of four factors that determine the adoption of ICT: individual characteristics, technology characteristics, business context and environmental context.

Finally, SMEs, being a special type of company, are generally affected by the presence of some factors, including government support with special programs and pressure from competitors as key factors for the adoption of ICT [20]. Another factor studied was the characteristics of the owner, since the business leader must have passion, determination, and motivation to achieve the success of his business [2,20]. However, there are also factors that prevent small companies from adopting ICTs in their businesses, such as the high cost of acquisition [23,39], the ease of use by the entrepreneurial leader, and the lack of adequate infrastructure for this type of company [39].

3 ICT Adoption Factors Model

This section presents the proposal that allows identifying the factors of ICT adoption in small and medium-sized enterprises (SMEs) of emerging countries. In addition, the meta-analysis of the adoption factors, the motivation, the model map, a description of each of its components, the hypotheses and finally a Methodological Guide for the implementation of the model is shown.

3.1 Factor Analysis

Table 1 shows the influence of the 14 factors identified in previous research. The symbols ++, +, - -, - are used, which are explained below.

Notation:

- ◇ ++ indica que el factor influye de manera positiva y ha sido validado experimentalmente. De ahora en adelante serán llamados factores muy positivos.
- ◇ + indica que el factor influye de manera positiva, pero no ha sido validado experimentalmente. De ahora en adelante serán llamados factores positivos.
- ◇ - - indica que el factor influye de manera negativa y ha sido validado experimentalmente. De ahora en adelante serán llamados factores muy negativos.
- ◇ - indica que el factor influye de manera negativa, pero no ha sido validado experimentalmente. De ahora en adelante serán llamados factores negativos.
- ◇ Sin símbolo, indica que el factor no ha sido considerado por los autores en sus investigaciones.

Analysis:

- ◇ There is no model that integrates all the ICT adoption factors found in the literature review for SMEs.
- ◇ There are common factors identified (these are those factors that have been cited in different investigations that deal with the issue of ICT adoption), among these factors we can mention the pressure of competitors, the support of the government, the characteristics of the owner/manager, organizational characteristics, ICT costs, perceived benefits of ICT, and lack of infrastructure
- ◇ The government support factor is a doubtful factor, this is since in the investigations of [20] and [25] it was found that it had a positive influence. But in the investigations of [12] and [33] it was found that it had a negative influence.
- ◇ There are very positive factors (+++), among them we can mention: the innovation needs of the clients, the characteristics of the owner/manager, the perceived benefits of the use of ICT and the changes in the business environment

Table 1. Influence of factors in the adoption of ICT [Author]

No.	Factors	[20]	[23]	[25]	[27]	[12]	[33]	[2]	[39]
1	Government support	+		++		-	- -		
2	Customers' innovation requirements				++	+			
3	Pressure from competitors	+			++	+			
4	Experience and ICT knowledge of employees					+			
5	Owner/Manager Characteristics	+	++	++	+	++	++		
6	Organizational characteristics	-		- -	- -	-	- -		
7	ICT products available on the market	-					-		
8	External ICT consultants and providers	+							
9	Lack of perceived usefulness and ease of use		-				-		- -
10	ICT cost			- -		-	- -		- -
11	Lack of IT infrastructure	-		- -	- -	-	- -		- -
12	Perceived benefits of ICT use				++		++	++	
13	Changes in the business environment				++				
14	Social and culture						- -		

◇ There are very negative factors (–), among them we can mention the organizational characteristics, the lack of perception of usefulness and ease of use, the costs of ICTs, the lack of IT infrastructure, social and cultural factors.

3.2 Proposed Model

According to the review of the literature on ICT adoption factors, the solutions found correspond to SMEs that do not belong to the reality of Latin America, and there is currently no model that integrates all the adoption factors. In addition, no adoption model has considered the factors: prior experimentation with ICTs, environmental factors, and resistance to change. This research will propose to design a model that integrates all the ICT adoption factors found in previous research, plus the three new factors mentioned.

Different studies have tried to identify those factors that influence the ICT adoption process in SMEs, finding divergent results in the literature. Figure 1 shows the proposed model of adoption factors. The proposed improvement consists of the identification of 3 new factors that influence the adoption of ICTs, these new factors are prior experimentation with ICTs, resistance to change and environmental factors.

The proposed model of adoption factors shown in Fig. 1 has been classified first into facilitating and inhibiting factors and then into internal and external factors. Among the facilitating factors, the following have been identified: ICT experience and knowledge of the employees, prior experimentation with ICT, pressure from competitors, external ICT consultants and providers, and environmental factors. Within the inhibiting factors have been identified: resistance to change and ICT products available on the market.

Previous ICT Experimentation. According to [7], experimentation is the degree to which an innovation can be used on a trial basis before adoption occurs. [42] found that the experimentation of an innovation, according to the perception of the members of an organization, is positively related to the rate of adoption. [3] suggested that experimentation has become an important feature of innovation, as it provides a means for future adopters to reduce their uncertainties regarding technologies or products that they do not know about. Therefore, [17] conclude that the experimentation and complexity of electronic commerce influence its adoption by small businesses. This agrees with [13] who studied Internet-based ICT adoption in Malaysian SMEs, where they found that experimentation was significantly associated with Internet adoption. In the study by [8], experimentation was found to exert a positive and direct influence on the adoption of e-business in SMEs.

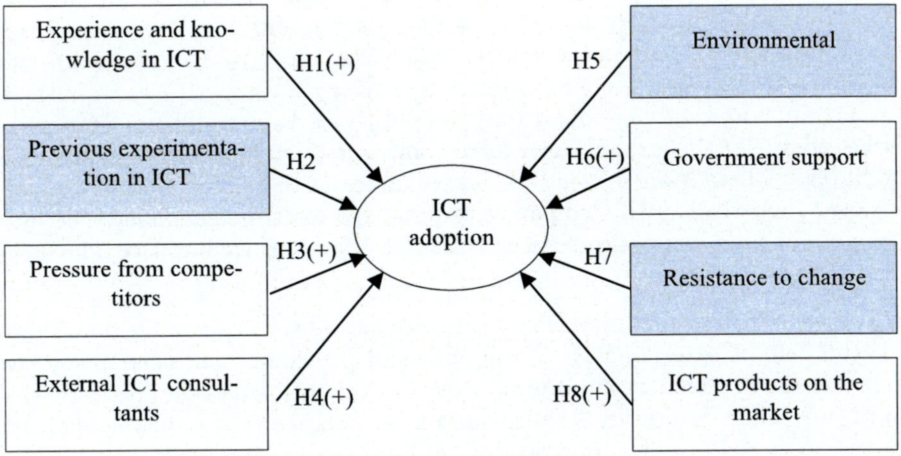

Fig. 1. Proposed Model

ICT Products Available in the Market. The ICT adoption process in SMEs also depends on the characteristics of the ICT products on the market, that is, on the set of factors that include type, compatibility with company processes, ease of use, popularity of ICT, quality of software available in the market and ICT costs [43]. Since systems and IT are considered the main enablers of superior business performance, the quality of ICT products available on the market (for example, their reliability and usefulness) could be a determining factor when deciding on adoption of ICT among SMEs. Several studies on the adoption of ICT in SMEs found that the adoption and use of ICT is significantly affected by the compatibility of technological products.

Government Support. Significant positive relationships could be found between ICT adoption and government support. Due to their size and lack of resources, SMEs tend to rely more on external resources and support from other companies [44]. A study of small businesses in six European countries indicates that while governments have tried to help SMEs adopt ICT through increased public spending on technology projects, there are nevertheless builtin barriers to adoption in the mechanisms of government agencies designed to help these businesses [16]. These adoption barriers are due to the difference between what is really required from SMEs and what is established by the government [44]. Despite the above results, government support has been found very helpful, recent studies, particularly in developing countries, have shown that ICT adoption in SMEs has been significantly improved through policies and government initiatives.

Competitor Pressure. A company can feel pressure when it sees that more competitors in the industry are adopting technologies and therefore the company decides to adopt to remain competitive [13]. The study by [13] also confirmed that SMEs are interested in adopting the technology if competitors and business partners are adopting. [34] found that competitive pressure and expectations of market trends in the industry can force companies to adopt ERP, e-commerce, and Suplain Chain Management (SCM) technologies to gain competitive advantage over their rivals [31]. Companies could adopt electronic commerce because of the use of their competitors, in order not to lose their competitive advantage [42].

ICT Consultants and External Providers. There is research that shows that the help of external ICT consultants and providers, and their respective quality, is one of the most important aspects of the ICT adoption process within SMEs [17]. Their professional skills could have positive effects on the adoption process, most SMEs suffer from the lack of hiring external consultants [43]. The study by [13] points out that the lack of internal expertise has seriously hampered the sophistication and evolution of small businesses, therefore they must overcome this problem through seeking help from external sources or developing its own internal computing capabilities. [20] indicates that external consultants and providers are the main sources of professional expertise in ICT implementation within small businesses. In general, the functions carried out by the external experts include project management, encouraging employees to accept new systems, overcoming their fear of new technologies, analysis of business needs, user training and recommending hardware and suitable software [20].

Environmental. According to [6], several of the small companies reflect that before their ICT adoption phase, they apply an evaluation of respect for the environment in all its components within their purchasing process. In his study, respect for the environment influences organizations to adopt and implement technologies in all their business processes. Another study suggests that one of the factors for the adoption of technologies in most cases is the environment, according to the author, the drivers of ICT adoption are factors that do not fall into the economic or social categories, and are the legal, environmental and cognitive factors [50]. Several studies dealing with the issue of ICT adoption classified those factors that influence adoption into economic, demographic, and environmental factors [30].

Experience and Knowledge in ICT of the Employees. In [50] it was found that if workers in small businesses have a very high level of ICT skills, it would be one of the main drivers for small businesses to adopt IT. According to [32], small businesses tend to avoid ICT in their business if it looks so complex to use. This is not surprising, as SMEs always lack skills among workers to use it. The capacities, the knowledge, the ICT skills of the employees, increases the possibility of the use of ICT among SMEs. The presence of employees with ICT knowledge could facilitate the adoption of technology if the owner believes that this technology can only be used by having a specialized staff [26]. The presence of adequate technical and administrative staff with sufficient IT experience has a significant influence on adoption [22]. [28] showed that having technical knowledge is an important driver for ICT adoption in SMEs in developing countries. [41] found that small businesses are not likely to adopt more sophisticated technologies if they are not familiar with basic technologies. [51] stated that the technical ability of the employees is more important for the adoption of advanced technologies such as e-commerce and ERP software.

Resistance to Change. According to [4] it is normal to find resistance to a new technology, since it takes time to get to know it and then be accepted. [42] shows that the adoption of a new technology has few adopters at the beginning and over time the percentage of people who adopt is increasing, in his research he found that some of the participants in his study do not want to switch to new technologies, because they find that the normal way (what they always do) is much better and more profitable for them. On the other hand, [4] he obtained from a series of interviews with 16 Saudi companies, a list of factors that inhibit or discourage retailers from adopting technologies. These inhibitors were identified as the cost of technologies, delivery issues, and resistance to change. [33] identify three types of barriers that prevent the successful adoption of electronic market systems, and these are the lack of description of electronic products, the narrowness of the market (lack of critical mass) and the resistance to the market change (inertia of the old ways of doing business). Finally, in the research of [4], they identified resistance to change as one of the main inhibitors of e-commerce adoption in SMEs.

3.3 Hypothesis of the Proposed Model

Table 2 shows the hypotheses to test the relationship between facilitating and inhibiting factors (technological, environmental, and individual) and the adoption of ICT in SMEs.

Table 2. Influence of factors in the adoption of ICT [Author]

Code	Description
H1	The experience and knowledge in ICT of the employees positively influence the adoption
H2	Previous experimentation with ICT has a positive influence on adoption
H3	Competitive pressure positively influences adoption
H4	External ICT consultants and providers positively influence adoption
H5	Environmental factors positively influence adoption
H6	Government support positively influences adoption
H7	The ICT products available in the market have a negative influence on adoption
H8	Resistance to change negatively influences adoption

4 Methodological Guide for the Application of the Model

Next, a guide for the application of the proposed model is shown to describe each of the steps. This Methodological Guide considers the flowchart presented in Fig. 2. Next, two case studies are examined, for which we will carefully follow each of the 10 steps outlined in the Model Application Guide. Both case studies are aimed at identifying the factors that positively or negatively influence the adoption of ICT in small and mediumsized enterprises (SMEs).

5 Case Studies

5.1 Case Study 1: SMEs in Metropolitan Lima

Previous studies indicate that the mortality rate of small businesses is higher in developing countries than in developed countries, and although SMEs play a very important role in the economy, the terrain where they develop is still very fragile, since low rates of survival and business consolidation of these companies are observed [27].

In Peru, the statistics of the National Directorate of the Ministry of Labor and Employment Promotion (MTPE), referring to SMEs, show that the participation of SMEs in the GDP has been 42% and they have represented 98% of the total number of companies in Peru and 80.96% of the economically active population. In addition, according to the Global Entrepreneurship Monitor study carried out in Peru, it is estimated that 9.2 million Peruvians are involved in an entrepreneurial activity, of which 75% have a period of operations between 0 and 42 months, and the remaining 25% exceeded this period. These figures show the need to help this type of company making use of ICT. In this sense, this research analyzes the factors that influence the adoption of ICTs with the aim of increasing their chances of success, reducing the mortality rates of these companies, and increasing their chances of survival in the long term. Next, the steps of the Methodological Guide described in the previous section are followed.

Description of the Organization's Business. As the object of our study is a group of SMEs from Metropolitan Lima, this step does not apply. It would apply in the case that it was for a single SME.

Description of the Organization's Business. The work team is made up of:

◇ Principal investigator.
◇ Owners of SMEs interested in identifying adoption factors.
◇ Statistical, as support for the processing of survey data.

For our research, no support staff has been required for the surveys, because they have been carried out using an online form (google docs). The construction of these forms and their processing has been assumed by the principal investigator.

Selection of Adoption Factors. Continuing with the development of the Guide, we are going to select all the facilitating and inhibiting factors of adoption (considering the 2 rules defined in the Methodological Guide) considering that the research seeks to validate previous studies and serve as a reference for future research on the adoption of ICT in SMEs, we will select all the facilitating and inhibiting factors, as shown in Table 3.

Table 3. Influence of factors in the adoption of ICT [Author]

Facilitating factors		Inhibitory factors	
A. Internal Individual		**A. Internal Individual**	
1. Experience and knowledge in the use of ICT	✓	7. Resistance to change	✓
A. External Technological		**B. External Technological**	
2. Previous ICT experimentation of the environment	✓	8. ICT products available on the market	
3. Pressure from competitors	✓		
4. External ICT consultants and providers	✓		
5. Environmental	✓		
6. Government support	✓		

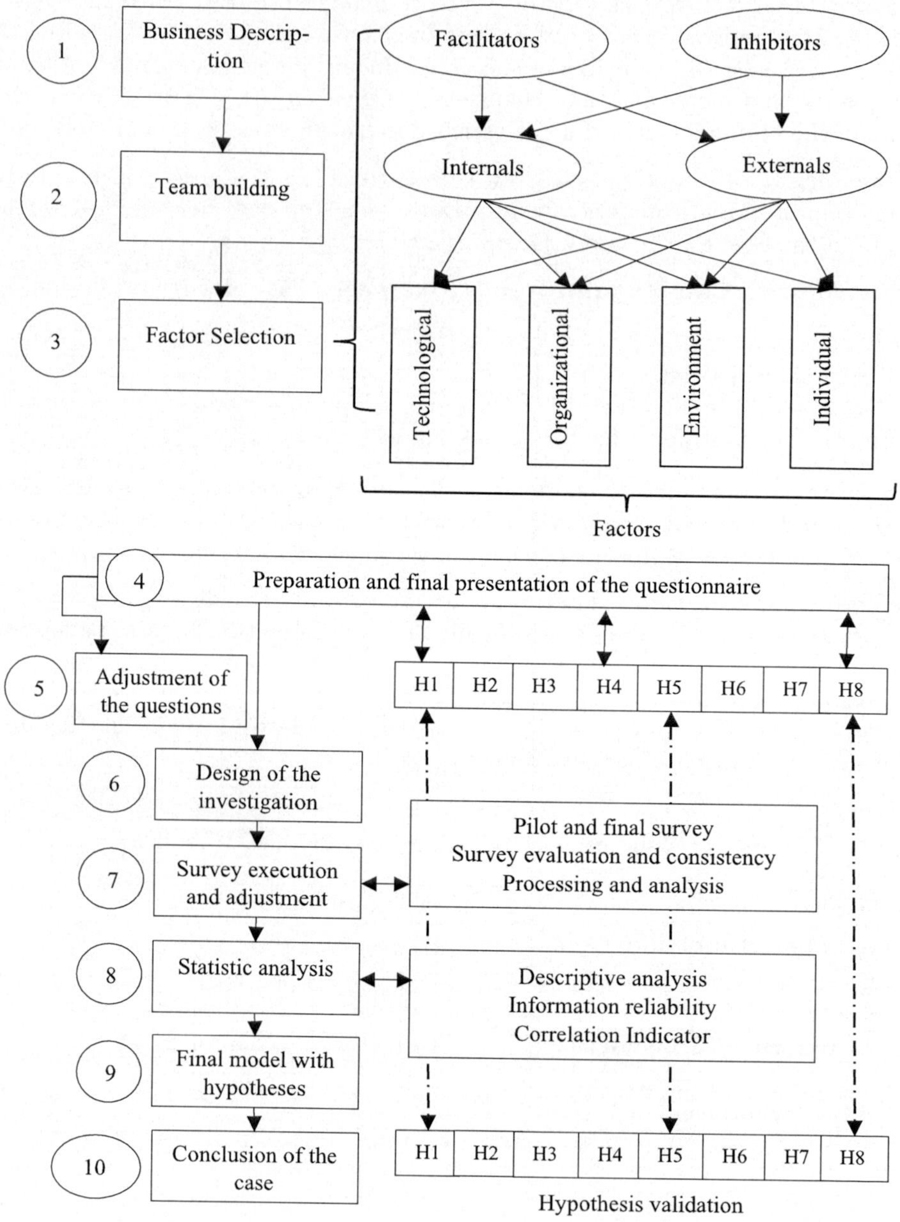

Fig. 2. Flowchart of the Methodological Guide

Preparation and final presentation of the questionnaire. The survey has 3 parts:

- ◇ Part I: Shows the general characteristics of SMEs (10 questions).
- ◇ Part II: Shows the questions related to ICT adoption factors (33 questions). These questions have been asked on a Likert scale, being (1) Totally disagree (2) Disagree (3) Neutral (4) Agree (5) Totally agree
- ◇ Part III: Shows the questions related to ICT Tools commonly used by SMEs (12 questions).

Question Settings. The relevance and wording of the survey questions have been submitted to expert judgment, in which 3 personalities from the business world with work experience in the country's public and private sectors, and in teaching, participated.

Design of the Investigation. To achieve the objectives proposed in the research, the methodological option chosen is the quantitative approach. This allows an analysis of each ICT adoption factor in SMEs, that is, the information can be quantified, analyzed, and classified. Quantitative research uses statistical models to explain data and research.

Design Type. The case study is an investigation with a non-experimental design (implies correlation/causality) of a transversal type, since it is a study where the objective is to know the perception of the CEOs/owners of SMEs regarding a possible adoption of ICTs, without carrying out any type of intentional manipulation, but knowing their behavior and perception in a natural way, without involving any external effector. Likewise, it is cross-sectional because the data collection has been developed for 3 weeks. The present study is basically considered as an exploratory investigation.

Target Population. The first thing that will be done is to define the sampling frame and the unit of analysis, then proceed to delimit the population that will be studied and on which the results are intended to be generalized; For this, the National Registry of Associations of Small and Medium Enterprises (RENAPYME) was taken as a reference (sample framework). Considering the National Registry of Small and Medium Enterprises (RENAPYME) it has been determined:

- ◇ Universe: The 24 SME associations that participate in RENAPYME.
- ◇ Informants: The CEOs or owners of the SMEs who received a letter or email for their participation in the survey.
- ◇ Selection of the type of sample: With the main objective of achieving the greatest coverage of the case study, it was not considered necessary to limit the sample by any type of professional area or other criteria. On the contrary, the surveys were carried out in one stage. Which was addressed to the entire universe of owners/managers/CEO of an SME registered in the RENAPYME, for this a letter was prepared, in which the support of the Vice Ministry of

mype and Industry of the Ministry of Production is requested. (PRODUCE) of Peru for the dissemination of the online survey. For the online survey, a form was prepared in google docs, which was sent to all SME associations registered in RENAPYME so that they could enter the electronic questionnaire, giving them two weeks to respond.

◇ Type of sampling: The sample is not probabilistic; therefore, no sampling has been used in the case study.
◇ Size of the survey: A total of 96 responses were obtained (electronic by google docs). Of which 22 were discarded because they were incomplete. Considering that an electronic form was used, where the tool allows us to validate the data entry, we have obtained 77% of valid responses.

Execution and Adjustment of the Survey. The relevance and wording of the survey questions have been submitted to expert judgment, in which 3 personalities from the business world with work experience in the country's public and private sectors, and in teaching, participated. **Pilot Survey.** In our case study, a pilot survey was carried out to determine some statistical indicators and observe the problems of its application in a real scenario and take note of the defects or problems to overcome them in the final survey. This pilot survey was also applied online. The pilot survey questionnaire was made up of 3 parts. Additionally, the content of the questionnaire was submitted to expert judgment, which led to the exclusion of some items and the adjustment of others. **Final Survey.** Considering the problems found in the pilot survey, the final survey was prepared, which included the following:

◇ Correct writing errors.
◇ Include new questions to measure the ICT adoption factor.
◇ Apply the questionnaire electronically, using the google docs forms. This option corrects the problems of incomplete questionnaires, validates the data entered, and data capture and analysis are much faster.

The new survey was submitted to an expert judgment in which 3 personalities from the business world with experience in teaching and work in the country's public and private sectors participated. After coordination with the Vice Ministry of My-pe and Industry, all SME associations that belong to RENAPYME were invited to participate in answering the questionnaire by entering an electronic link. The final questionnaire is made up of a total of 55 questions divided into 3 parts, as can be seen in Table 4. The sample size is 74 in total, because an electronic form was used, where data entry and unanswered questions were validated, all responses were validated. **Evaluation and Consistency of Surveys.** Once the deadline for responding to the questionnaire had expired, the google docs file was downloaded in Excel file format and with the confidence of using a digital process, 74 questionnaires were given as valid. **Processing and Analysis.** For our validation process we have used the IBM SPPS Version 20.0 software for the statistical reading of the data and MS Excel 2010 to convert SPSS charts taking advantage of its visual presentation capabilities.

Table 4. Parts of the final questionnaire [Author]

Items	parts	No. Questions
General characteristics of SMEs	I	10
Adoption Facilitators	II	20
Adoption Inhibitors		8
Adoption intent		5
Commonly used ICT tools	III	12
Total questions		55

Statistical Analysis

Information Reliability - Cronbach's Alpha and Correlation Indicator - Spearman's Correlation. Table 5 shows the results obtained from the Reliability analysis using Cronbach's Alpha statistic. It should be noted that values greater than 0.60 would indicate that the constructs are defined by the variables involved in the measurement of the same construct. In addition, the results obtained from the Spearman correlations are also presented.

Table 5. Reliability analysis Cronbach's Alpha and Spearman's Correlation - Case 1 [Author]

No.	Factors	Cronbach's Alpha	Spearman correlation
1	Experience and ICT knowledge of employees	0.74	0.60
2	Previous ICT experimentation	0.82	0.78
3	Government support	0.83	0.79
4	Pressure from competitors	0.77	0.68
5	External ICT consultants and providers	0.89	0.80
6	Environmental	0.92	0.91
7	Resistance to change	0.78	−0.75
8	ICT products available on the market	0.93	−0.58

When observing the results of the evaluated adoption factors, they have a very high Cronbach value, which confirms the reliability of our variables and their respective data.

In addition to Spearman's correlation indicators between adoption factors and the intention to adopt ICTs, we can state that:

◇ There is statistical evidence to affirm that there is a correlation (association) between the factors measured and the adoption of ICTs.

◇ It is confirmed that the factors, experience, and knowledge in ICT of employees, previous experimentation with ICT, competitive pressure, consultants and external ICT providers, environmental factors, and government support, positively influence adoption of ICT.
◇ It is confirmed that the factors, resistance to change and ICT products available in the market, negatively influence the adoption of ICT

Final Model of Case 1. Based on the results obtained, the final accepted model is presented in Fig. 3.

Analysis of the Facilitating Factors of Adoption. The experience and knowledge in ICT of the employees (0.602), the previous experimentation of the ICT (0.787), the support of the government (0.799), the pressure of the competitors (0.689), the consultants and external suppliers of ICT (0.809) and the environmental ones (0.916). These are factors that positively influence the adoption of ICT in SMEs. The most important facilitating factor is "environmental" with a coefficient of 0.916.

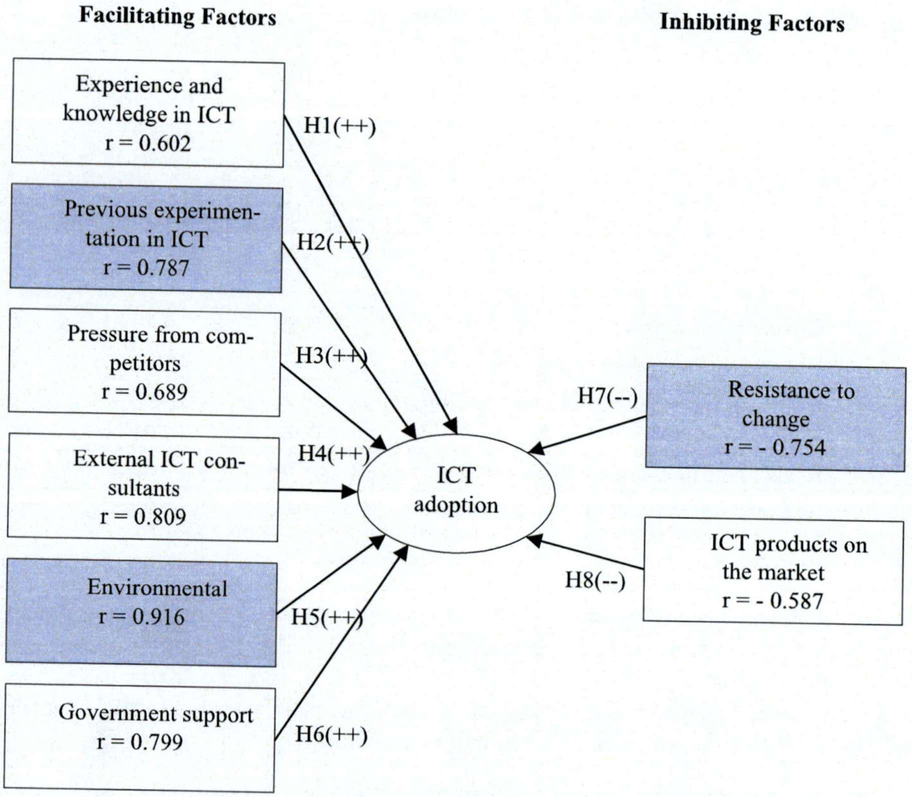

Fig. 3. Final Model - Case 1

Analysis of Factors Inhibiting Adoption. The resistance to change (-0.754) and the ICT products available in the market (0.587). These are factors that negatively influence the adoption of ICT in SMEs. The most important inhibiting factor is 'ICT products available on the market' with a coefficient of 0.916.

Conclusions of the Case. There is statistical evidence to affirm that the adoption of ICT is very well represented by its variables. When performing the analysis of correlations and associations between the factors and the adoption of ICT, it is concluded that all the factors evaluated are necessary to explain the intention to adopt ICT in SMEs, some with a higher incidence and others with a lower incidence.

Within the entire model, the most important factors that facilitate adoption are those related to caring for the environment, followed by consultants and external ICT providers. Likewise, the factors that inhibit adoption are the resistance to change and the ICT products available in the market.

These conclusive results indicate that for SMEs to decide to adopt ICT in their businesses, it is essential that the benefits of its use be communicated in their organizations. In other words, if we want to consider that entrepreneurs intend to adopt ICT in their business, it is necessary that people perceive its tangible benefit in the organization; To do this, it is important to train key people on this issue.

5.2 Case Study 2: SME - Benito Commercial

The results of the statistical analysis for case 2 of a particular SME are shown below. Steps 1 to 7 of the Model Application Guide will be skipped because they are the same as those explained in Case 1.

Statistical Analysis

Information Reliability - Cronbach's Alpha and Correlation Indicator - Spearman's Correlation. Table 6 shows the results of the Reliability analysis using Cronbach's Alpha statistic and the results obtained from the Spearman correlations, for the specific case of the SME "Benito Commercial".

Table 6. Reliability analysis Cronbach's Alpha and Spearman's Correlation - Case 2 [Author]

No.	Factors	Cronbach's Alpha	Spearman correlation
1	Experience and ICT knowledge of employees	0.84	0.96
2	Previous ICT experimentation	0.75	0.50
3	Government support	0.81	0.86
4	Pressure from competitors	0.71	0.79
5	External ICT consultants and providers	0.76	0.56
6	Environmental	0.89	0.94
7	Resistance to change	0.74	−0.78
8	ICT products available on the market	0.84	0.51

When observing the results of the evaluated adoption factors, they have a very high Cronbach value, which confirms the reliability of our variables and their respective data.

Observing the Spearman correlation indicators between the adoption factors and the intention to adopt ICTs, for case 2, we can state that:

- ◇ There is statistical evidence to affirm that there is a correlation (association) between the factors measured and the adoption of ICTs.
- ◇ It is confirmed that the factors: ICT experience and knowledge of employees, prior ICT experimentation, competitive pressure, external ICT consultants and providers, environmental factors, and government support, positively influence adoption of ICT. Unlike case 1, the factor of ICT products available in the market, according to the results of the experimentation, this time is a factor that positively influences the adoption of ICT.
- ◇ It is confirmed that the resistance to change factor has a negative influence on the adoption of ICTs.

Final Model of Case 2. Based on the results obtained, the final model accepted for Case 2 is presented in Fig. 4.

Analysis of the Facilitating Factors of Adoption. ICT experience and knowledge of employees (0.961), prior ICT experimentation (0.503), government support (0.866), pressure from competitors (0.799), external ICT consultants and providers (0.508), environmental (0.946) and ICT products available on the market (0.514). These are factors that positively influence the adoption of ICT in SMEs. The most important facilitating factor is 'Employees' ICT knowledge and experience' with a coefficient of 0.961.

Analysis of Factors Inhibiting Adoption. The resistance to change (-0.782). It is the factor that negatively influences the adoption of ICT in SMEs.

Conclusions of the Case. When performing the analysis of correlations and associations between the factors and the adoption of ICT, we conclude that all the factors evaluated are necessary to explain the intention to adopt ICT in SMEs, some with a higher incidence and others with a lower incidence.

We can conclude that, within the entire model, the most important factors or those that facilitate the adoption of ICTs are the employees' experience and knowledge of ICTs; followed by the factor referring to caring for the environment. Likewise, the factor that, as in the experimentation of case 1, inhibits the adoption of ICTs is resistance to change. These conclusive results indicate that for the Commercial Benito SME to decide to adopt ICTs in its business, it is essential that the benefits of its use be communicated in their organizations. In other words, it is necessary for the owner of that SME to perceive the tangible benefit of the use of ICTs in the organization, for this, it is important to train the key people on this subject.

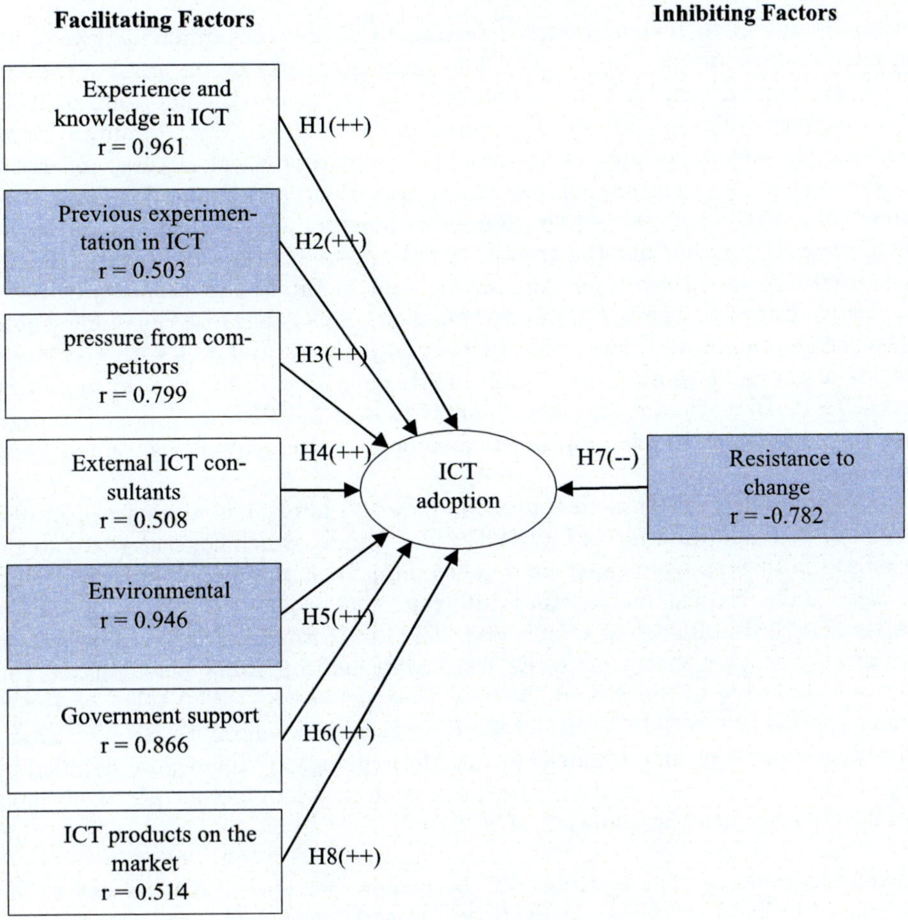

Fig. 4. Final Model - Case 2

6 Conclusions

In Sect. 3, the model of factors that influence the adoption of ICTs was designed. For this achievement, the environmental and individual technological factors that influence the adoption of ICT in SMEs have been identified and, unlike other models, new factors have been included such as care for the environment, prior experimentation with ICT and resistance to change. Models of ICT adoption factors applied to SMEs in different developing countries have been reviewed. The most outstanding among them, what was done by [16], with its proposal of a model of ICT adoption factors. Likewise, the proposal made by [41] is very important, where the authors proposed a model of ICT adoption barriers in SMEs in Iran. Finally, [25] investigated the case of SMEs in Nigeria, proposing an ICT adoption model with five factors. These and many other investigations have allowed us to review models referring to SMEs from different developing countries, either directly in the investigation or indirectly by the state of the art.

Three new factors have been identified that influence the adoption of ICTs that are particular to the reality experienced by SMEs in Metropolitan Lima, such as previous experimentation with ICTs, environmental factors, and resistance to change. According to the experimentation carried out, for Case 1, of these three new proposed factors, the environmental ones turned out to be the facilitating factor that has the greatest incidence when adopting ICTs in SMEs. Consequently, the most important factors that facilitate the adoption of ICTs are those related to caring for the environment. This is because the circle where entrepreneurs operate forces them to be permanently updated with the latest trends in the management of IT tools, a situation that has increased more due to the COVID-19 pandemic, where many CEOs of MYPES were forced to digitize their processes to offer a product and/or services using at least one digital channel.

A Methodological Guide has been prepared to implement the new proposed model of ICT adoption factors in SMEs. The Guide, which consists of 10 steps, details step by step what must be done to implement the model and with them identify those critical factors that influence the process of ICT adoption in SMEs. The Methodological Guide was validated through 2 case studies. Case 1 was applied to a group of SMEs from Metropolitan Lima belonging to the National Registry of Small and Medium Enterprises (RENAPYME), and case 2 was applied to the SME Commercial Benito, also belonging to RENAPYME. The stages and their development of the Methodological Guide were detailed in Sect. 5, where the main findings are presented on what factors positively and negatively influence the adoption of ICTs.

Acknowledgments. The gratitude for the support provided in the research to the Technical University of Ambato, to the Research and Development Directorate (DIDE), and to the Norbert Wiener Private University.

References

1. Ahuja, V., Yang, J., Shankar, R.: Study of ICT adoption for building project management in the Indian construction industry. Autom. Constr. **18**(4), 415–423 (2009). https://doi.org/10.1016/j.autcon.2008.10.009
2. Alam, S.S., Jani, M.F.M., Omar, N.A., Hossain, T., Ahsan, M.N.: Empirical study of theory of reason action (TRA) model for ICT adoption among the Malay based SMEs in Malaysia. Bus. Manag. Strategy **3**, 43–53 (2012). https://doi.org/10.5296/bms.v3i2.2911
3. Alam, S., Noor, M.: ICT adoption in small and medium enterprises: an empirical evidence of service sectors in Malaysia. Int. J. Bus. Manag. **4** (2009). https://doi.org/10.5539/ijbm.v4n2p112
4. AlGhamdi, R., Nguyen, A., Jones, V.: A study of influential factors in the adoption and diffusion of B2C e-commerce. Int. J. Adv. Comput. Sci. Appl. **4** (2013). https://doi.org/10.14569/IJACSA.2013.040113

5. Beheshti, H.: The impact of it on SMEs in the united states. Inf. Manag. Comput. Secur. **12**, 318–327 (2004). https://doi.org/10.1108/09685220410553532
6. Buchalcevova, A., Gála, L.: Green ICT adoption survey focused on ICT lifecycle from the consumer's perspective (SMEs). J. Competitiveness **4**, 109–122 (2012). https://doi.org/10.7441/joc.2012.04.08
7. Busaidi, N., Bhuiyan, A., Zulkifli, N.: The critical review on the adoption of ICTs in the small and medium enterprises (SMEs) in the developing countries. Int. J. Small Medium Enterp. **2**, 33–40 (2019). https://doi.org/10.46281/ijsmes.v2i2.437
8. Caldeira, M., Ward, J.: Using resource-based theory to interpret the successful adoption and use of information systems and technology in manufacturing small and medium-sized enterprises, vol. 12, pp. 1159–1169 (2001). https://doi.org/10.1057/palgrave.ejis.3000454
9. Chatzoglou, P., Fragidis, L., Chatzoudes, D., Symeonidis, S.: Critical success factors for ERP implementation in SMEs. In: 2016 Federated Conference on Computer Science and Information Systems (FedCSIS), pp. 1243–1252 (2016)
10. Chicaiza, F., Garcia, C.A., Castellanos, E.X., Sanchez, C., Rosero, C., Garcia, M.: Agile architecture based on ISA-88 for the design of execution control chart in distributed applications using IEC-61499. ENFOQUE UTE **9**(1), 149–165 (2018). https://doi.org/10.29019/enfoqueute.v9n1.186
11. Cobos, X., Oñate, W., Caiza, G.: Interactive application for learning electric motors using leap motion controller. In: García, M.V., Fernández-Peña, F., Gordón-Gallegos, C. (eds.) Advances and Applications in Computer Science, Electronics and Industrial Engineering. AISC, vol. 1307, pp. 233–248. Springer, Singapore (2021). https://doi.org/10.1007/978-981-33-4565-2_15
12. Consoli, D.: Literature analysis on determinant factors and the impact of ICT in SMEs. Procedia Soc. Behav. Sci. **62**, 93–97 (2012). https://doi.org/10.1016/j.sbspro.2012.09.016
13. Cragg, P., King, M., Hussin, H.: IT alignment and firm performance in small manufacturing firms. J. Strateg. Inf. Syst. **11**(2), 109–132 (2002). https://doi.org/10.1016/S0963-8687(02)00007-0
14. Davis, F., Davis, F.: Perceived usefulness, perceived ease of use, and user acceptance of information technology. MIS Q. **13**, 319 (1989). https://doi.org/10.2307/249008
15. Duan, Y., et al.: Addressing ICTs skill challenges in SMEs: insights from three country investigations. J. Eur. Ind. Training **26** (2002). https://doi.org/10.1108/03090590210451524
16. Eze, S.: Examining information communication technology (ICT) adoption in SMEs: a dynamic capabilities approach. J. Enterp. Inf. Manag. **31** (2018). https://doi.org/10.1108/JEIM-12-2014-0125
17. Eze, S., Olatunji, S., Chinedu-Eze, V., Bello, A.: Key success factors influencing SME managers' information behaviour on emerging ICT (EICT) adoption decision-making in UK SMEs. Bottom Line **31** (2018). https://doi.org/10.1108/BL-02-2018-0008
18. Fishbein, M., Ajzen, I.: Belief, Attitude, Intention, and Behavior: An Introduction to Theory and Research. Addison-Wesley, Reading (1975)
19. Garcia, V.M., Irisarri, E., Perez, F., Marcos, M., Estevez, E.: From ISA 88/95 meta-models to an OPC UA-based development tool for CPPS under IEC 61499. In: 2018 14th IEEE International Workshop on Factory Communication Systems (WFCS 2018) (2018)

20. Ghobakhloo, M., Hong, T., Sabouri, M., Zulkifli, N.: Strategies for successful information technology adoption in small and medium-sized enterprises. Information **3** (2012). https://doi.org/10.3390/info3010036
21. Gilaninia, S., et al.: The role of ICT in performance of small and medium enterprises. Interdisc. J. Contemp. Res. Bus. **3**(9), 7 (2012)
22. Hashim, J.: Information communication technology (ICT) adoption among SME owners in Malaysia. Int. J. Bus. Inf. **2**, 221–240 (2007)
23. Honarmand, A., Manian, A.: International it adoption barriers among smaller firms. Glob. J. Technol. **1** (2013)
24. Icaza, D., Borge-Diez, D.: Energy supply of a hybrid system of biomass and wind turbines of the Pichacay landfill towards an intelligent network for the city of Cuenca-Ecuador. In: Nummenmaa, J., Pérez-González, F., Domenech-Lega, B., Vaunat, J., Oscar Fernández-Peña, F. (eds.) CSEI 2019. AISC, vol. 1078, pp. 287–307. Springer, Cham (2020). https://doi.org/10.1007/978-3-030-33614-1_20
25. Irefin, I.A., Abdul-Azeez, I.A., Tijani, A.A.: An investigative study of the factors affecting the adoption of information and communication technology in small and medium scale enterprises in Nigeria. Aust. J. Bus. Manag. Res. **02**, 01–09 (2012). https://doi.org/10.52283/NSWRCA.AJBMR.20120202A01
26. Jones, P., Beynon-Davies, P., Muir, D.: E-business barriers within the SME sector. J. Syst. Inf. Technol. **7**, 1–25 (2003). https://doi.org/10.1108/13287260380000771
27. Kannabiran, G., Dharmalingam, P.: Enablers and inhibitors of advanced information technologies adoption by SMEs. J. Enterp. Inf. Manag. **25**, 186–209 (2012). https://doi.org/10.1108/17410391211204419
28. Kartiwi, M., Macgregor, R.: Electronic commerce adoption barriers in small to medium-sized enterprises (SMEs) in developed and developing countries: a cross-country comparison. ERA - 2010 **5** (2007). https://doi.org/10.4018/jeco.2007070103
29. Khasawneh, M., Ibrahim, H.: Toward an information and communication technology development in developing countries. Commun. IBIMA **1**, 908–913 (2008)
30. Kraemer, K., Ganley, D., Dewan, S.: Across the digital divide: a cross-country multi-technology analysis of the determinants of IT penetration. J. Assoc. Inf. Syst. **6**(12) (2005). https://doi.org/10.17705/1jais.00071
31. Kshetri, N., Dholakia, R.: Factors impacting the adoption of the internet among SMEs. Small Bus. Econ. **23**, 311–322 (2004). https://doi.org/10.1023/B:SBEJ.0000032036.90353.1f
32. MacGregor, R., Vrazalic, L.: A basic model of electronic commerce adoption barriers. J. Small Bus. Enterp. Dev. **12**, 510–527 (2005). https://doi.org/10.1108/14626000510628199
33. Manuere, F.: Barriers to the adoption of ICT by SMEs in Zimbabwe: an exploratory study in Chinhoyi district (2012)
34. Martin, L., Matlay, H.: "Blanket" approaches to promoting ICT in small firms: some lessons from the DTI ladder adoption model in the UK. Internet Res. **11**, 399–410 (2001). https://doi.org/10.1108/EUM0000000006118
35. Middleton, K., Byus, K.: Information and communications technology adoption and use in small and medium businesses: the influence of Hispanic ethnicity. Manag. Res. Rev. **34**, 98–110 (2011). https://doi.org/10.1108/01409171111096496
36. Mohamad, R., Ismail, N.A.: Electronic commerce adoption in SME: the trend of prior studies. J. Internet Banking Commer. **14**, 1 (2009)
37. Mokaya, S.O.: The adoption of information and communication technology by small enterprises in Thika municipality, Kenya (2012)

38. Nguyen, T.: Information technology adoption in SMEs: an integrated framework. Int. J. Entrepreneurial Behav. Res. **15**, 162–186 (2009). https://doi.org/10.1108/13552550910944566
39. Nurhadi, N.: Model of it adoption for increasing the MSME's competitive advantage. J. Econ. Bus. Accountancy Ventura **16** (2013). https://doi.org/10.14414/jebav.v16i1.122
40. Oliveira, T., Martins, M.R.: Literature review of information technology adoption models at firm level. **14**, 1566–6379 (2011)
41. Owusu, A., Broni, F., Penu, O., Boateng, R.: Exploring the critical success factors for cloud BI adoption among Ghanaian SMEs. In: AMCIS 2020 Proceedings (2020)
42. Perlego, E.M.R.: Diffusion of Innovations 5th edn
43. Premkumar, G.: A meta-analysis of research on information technology implementation in small business. J. Organ. Comput. Electron. Commer. **13**(2), 91–121 (2003). https://doi.org/10.1207/S15327744JOCE1302_2
44. Sarosa, S., Zowghi, D.: Strategy for adopting information technology for SMEs: experience in adopting email within an Indonesian furniture company. Electron. J. Inf. Syst. Eval. **6** (2003)
45. Selamat, Z., Jaffar, N., Kadir, H.: ICT adoption in Malaysian SMEs (2011)
46. Shiels, H., McIvor, R., O'Reilly, D.: Understanding the implications of ICT adoption: insights from SMEs. Logist. Inf. Manag. **16**, 312–326 (2003). https://doi.org/10.1108/09576050310499318
47. Susanty, A., Sari, D.P., Anastasia, D.: Critical success factors for the internet technology adoption by SMEs and its impact for the performance. In: 2016 2nd International Conference on Science in Information Technology (ICSITech), pp. 35–40 (2016). https://doi.org/10.1109/ICSITech.2016.7852604
48. Tan, K., Eze, U., Chong, S.C.: Effects of industry type on ICT adoption among Malaysian SMEs. J. Supply Chain Customer Relat. Manag. 1–12 (2012). https://doi.org/10.5171/2012.113797
49. Thurasamy, R., Mohamad, O., Omar, A., Marimuthu, M.: Technology adoption among small and medium enterprises (SME's): a research agenda. Int. J. Econ. Manag. Eng. **3**(5), 512–515 (2009)
50. Vivas, G.V., et al.: Revisión de modelos para identificar los factores de adopción de TIC en Pymes. Associação Ibérica de Sistemas e Tecnologias de Informacao, pp. 496–511 (2021)
51. Zhang, Y., Sun, J., Yang, Z., Wang, Y.: Critical success factors of green innovation: technology, organization and environment readiness. J. Clean. Prod. **264**, 121701 (2020). https://doi.org/10.1016/j.jclepro.2020.121701

Early Stage Proposal of a Multi-tool Lean Manufacturing Methodology to Improve the Productivity of a Textile Company

Carlos Sánchez-Rosero, Jessica P. Lalaleo, César Rosero-Mantilla, and Jose E. Naranjo(✉)

Universidad Tecnica de Ambato, UTA, Ambato 180103, Ecuador
{carloshsanchez,jlalaleo6664,cesararosero,je.naranjor}@uta.edu.ec

Abstract. When a company does not take advantage of the resources at its disposal, its efficiency and the quality of its products are affected. This is what Lean Manufacturing is all about, a work philosophy applied to project management that aims to reduce waste, understood as anything that does not add value and increase productivity. One of the keys to success is its human dimension since people are the company's most important capital. A lean corporate culture implies collaboration and communication at all company levels. However, this is not always fulfilled, significantly affecting the company's productivity and even stakeholders' physical and mental health. This article gives way to the initial stage of a Lean multi-tool methodology focused on optimizing waste in a textile factory in Ambato-Ecuador. The results show that there are five primary wastes that dramatically affect the company's productivity, for which five lean manufacturing tools have been proposed. Noticing that there are several tools that can be used for the same purpose, a weighting method has been used, obtaining the following outcomes: 5S, TMP (Total Productive Maintenance), Kaizen, Andon, and Jidoka.

Keywords: Lean Manufacturing · Productivity · Textile Company · Initial stage

1 Introduction

Manufacturing industries worldwide have demonstrated great competitiveness over the years, in addition to the constant search for continuous improvement and opportunities to increase productivity through different methods and techniques deduced from quantitative analysis and operational management, thus creating competitive advantages to obtain higher profits [26].

This is how Asian countries are pioneers in the application of lean manufacturing tools in their daily life, in the educational field, and large industries, always showing efficiency in their productivity and being exemplary leaders in the fact that this change of mentality contributes significantly to the improvement of

M. V. Garcia and C. Gordón-Gallegos (Eds.): CSEI 2022, LNNS 678, pp. 662–677, 2023.
https://doi.org/10.1007/978-3-031-30592-4_43

working conditions and the elimination of waste generated in the different companies [3,11,23].

Several successful cases of the application of this Lean methodology have occurred in the United States, where companies such as Nike obtained outstanding results in terms of decreasing the defect rate to 50%, delivery time to 40%, improving their productivity by 20%, and also recognizing the application of the model much faster than in other companies [16,25].

On the other hand, in Ecuador, according to the National Institute of Statistics and Census (INEC), in October 2021, it was recorded that 8.3% of companies were concentrated in the manufacturing sector. Contributing 21% of sales and generating 12.5% of registered jobs, all this in the period from 2012 to 2020, thus denoting the importance of this sector in the country, positioning it in third place in the structure of the distribution of jobs [13,17].

The textile and clothing sector is one of the leading industries in Ecuador, as it is in third place in the manufacturing sector with a 7% contribution to the national manufacturing GDP. It is also the second largest generator of formal employment. According to data provided by the INEC, it has about 158 thousand direct employees, which is why the Corporation Mucho Mejor Ecuador promotes the consumption of textile products to encourage job creation while recovering and improving production and sales levels that were affected by Covid-19 [12].

It is also known that the textile sector is constantly growing, which is why small and medium-sized companies need to innovate, creating new strategies for continuous improvement that generate benefits in increased productivity and improved working conditions, eliminating everything unnecessary in the manufacturing process, allowing adequate compliance with market requirements and responding promptly to the competition generated [22,27].

As can be seen, even though this industry is one of the economic mainstays of Ecuador, in Ambato, there is still a lack of knowledge about good practices to improve productivity, usefulness and efficiency of human resources. Due to this, this article proposes developing a multi-tool methodology, which allows attacking the most common wastes within the textile industry. It is necessary to clarify that this is the first part of this research, so in future works, the results obtained through the application of this methodology will be demonstrated.

This document is divided into seven sections, including the introduction. Section two establishes the context of the research. This section explains why this particular research topic is important and essential to understanding the main aspects of the study. Section three provides details about the textile company to be analyzed and determines the phases of the study. Section four explains the company's baseline. Section five explains the results, section six presents the discussion and finally, the conclusions and future work are detailed in section seven.

2 Background Literature

The objective of [19] is to design and implement a logistics system within a textile production company focused on the automotive sector. This study was based on two non-traditional Lean Manufacturing tools: Mizusumashi and e-Kanban. With these methodologies, they could control the amount of raw material to be handled, the necessary workforce, the safety stock and the raw material replenishment time. The results showed that the amount of stock in the production area could be reduced by up to 50%. In addition, it is emphasized that an internal supply system can be redesigned with the investment of a minimum amount of money.

Competition within the textile industry has increased significantly worldwide in recent years due to the quality, volume and delivery times offered by first-world industries. In the Peruvian industry, delays in delivery and low productivity have led certain companies almost to bankruptcy. Due to this, [4] proposes an integrated model that acts on the management of production and warehouse, using Lean tools such as JIT (Just in Time), Kanban and standardized work. The results show that it has been possible to reduce critical indicators such as defective items or the rate of late delivery, obtaining unprecedented indicators in this country.

The textile sector presents several problems in its production chain, mainly due to the lack of management systems. [2] evaluates a small enterprise textile, where it starts from its baseline to define the deficiencies and factors that affect its competitiveness. Among the findings, poor quality management, high waiting times and lack of clear procedures can be mentioned. Jidoka and SMED (Single Minute Exchange of Die) were used to solve these problems. Incorporating these two methodologies reduced defective products, labour overload and machine set-up time.

On the other hand, [7] establishes that the textile sector has majority participation in the manufacturing industry in Latin America, especially in Peru. Therefore, seeing the inefficiency of these companies in their production and delivery area, he proposes an optimization model using three tools, Just in Time 5S and Lean Warehouse, to reduce the number of pending orders, minimize delivery times and minimize the entry of defective products. Their results, based on a simulation performed in Arena software, are encouraging. The number of non-optimal orders was reduced by 55%, and process efficiency was increased by more than 5%. Finally, the reduction in the cost of purchases was 40.55%.

Meeting order deadlines is an activity that most textile companies worldwide do not do efficiently, generating distrust in their customers and losing their target audience. [5] proposes the Systematic Layout Planning (SLP), which, under a PDCA plan, aims to improve the pace of work and the percentage of activities that do not add value to the production chain. Through a simulation, the results showed a more than 20% reduction in delivery times, thus meeting customer needs and exceeding their expectations.

A root cause analysis was carried out by [24] at a textile company in Pakistan. A triangulation method was used to decide the obstacles to implementing Lean Manufacturing in this sector. Several expert interviews were conducted to understand how human resources refuse to improve. It was found that poor leadership, poor knowledge, poor communication and not seeking help from experts are the significant constraints to implementing Lean tools. However, it was also found that the most widely used tool that has been a resounding success in the textile industry is the 5S. This research serves as a guide so that companies just entering the textile world can control their production from the beginning and have the least amount of waste.

As it can be seen and as far as it has been investigated, no methodology efficiently combines Lean Manufacturing tools to solve productivity problems within a textile company in Ecuador. For this reason, the present work contributes significantly to the Andean-industrial literature since it serves as a guide to identifying waste and seeking efficient solutions through industrial optimization methodologies.

3 Case Study

Lily Toy's is one of the leading companies in the textile manufacturing sector in the city of Ambato, responsible for the manufacture of cushions, stuffed animals, blankets, and slippers, among others, located in Izamba, canton Ambato, province of Tungurahua. Its products are sold through virtual platforms, shopping malls, in the same branch, and in the city of Guayaquil, where it currently has a point of sale.

Lily Toy commits senior management to solve the problems that arise in production, among which the presence of waste that does not add value to the product and generate delays in the process can be highlighted. The company has not obtained good results in terms of productivity due to the lack of a good culture for eliminating waste and improving its processes. The disorder in the facilities is evident. The unproductive times present reflect economic losses and a poor working environment.

This study was carried out in the company's production area mentioned above. For the study, all the workers in the different work departments and senior management were considered the population. Not having a population of more than 100 people, it is not necessary to have a representative sample, so the entire population was selected. It corresponds to 19 employees.

Three phases were taken into account for the development of this research, which have been adapted from [8,14]:

1. Survey of information
   The information collected was reviewed, showing the current status of the production process of the most relevant product and its detailed description. All the operations for the confection of stuffed animals were analyzed, and all the information obtained was interpreted, focusing on the waste generated, in addition to a time study in which the initial conditions of the production process were obtained. ABC analysis, VSM, and graphic procedures were used to carry out this phase.
2. Manufacturing tool selection
   Lean manufacturing tools were identified and selected to improve the company's productivity based on its current situation. The VSM tool, descriptive matrices, and the weighted factors method were used for this phase.

3. Results evaluation
The lean manufacturing tools were analyzed and interpreted for the development of the productivity improvement proposal in the company. Finally, as future work and as part of the final stage of this research project, the current situation and the proposal will be compared through a simulation of the process with the improvements to be implemented. In addition, once approved by the top management, all the selected tools will be implemented progressively.

4 Current Status of the Company

4.1 Layout

The plant distribution or layout is arranging the elements that make up the production system in the physical space so that the production objectives are achieved most appropriately and efficiently possible [20]. The production plant currently has ample space to execute its work, where several products can be found waiting to complete production orders, giving a vision of disorder and waste of space. On the other hand, products are filled daily in the center of the plant in the finishing area, generating insecurity when passing through the workstations. This can be seen in Fig. 1. The production plant works from Monday to Friday from 7:00 am to 3:30 pm, with a lunch break from 1:00 to 1:30 pm, legally complying with the eight working hours. It also works on particular Saturdays to meet the required demand.

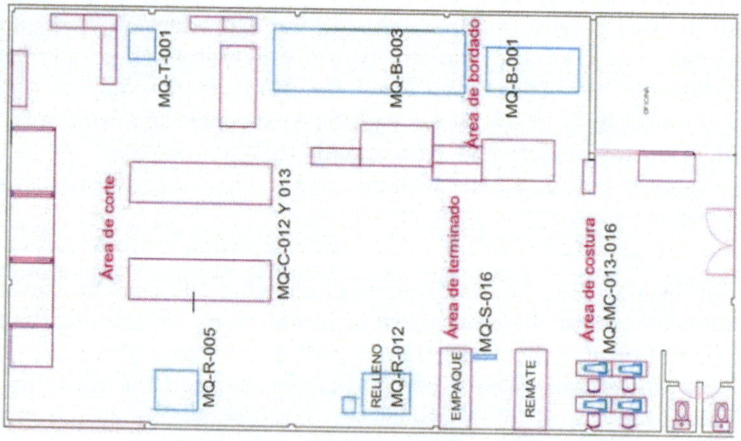

Fig. 1. Production plant layout.

4.2 Production Process

In Fig. 2, it can be seen that the production operating process has four main areas for the production of the five families of products offered, including the cutting area where the fabric measurement is developed, cutting using the mobile cutter, and cutting through the die-cutter, the embroidery, sewing and finishing area where the stuffing, finishing, and packaging of the products are carried out.

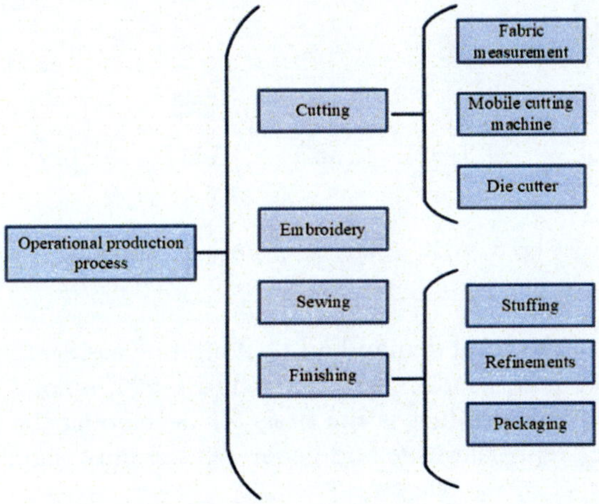

Fig. 2. Production process.

4.3 ABC Analysis

ABC analysis is a system used to design the distribution of inventories in warehouses. This methodology aims to optimize the organization of products in such a way that the most attractive and requested by the public are within a more direct and quicker reach. This reduces search times and increases efficiency [1]. Lily Toy's manufactures a large number of products that are differentiated by their production processes presented in families, including stuffed animals (1), cushions (2), blankets (3), hearts (4), and slippers (5). Through the sales report for the year 2021, the family of products with the highest demand was obtained. As shown in Fig. 3 the product family most in demand by customers is stuffed animals, which will account for the highest percentage of sales in 2021.

After determining that the most demanded family is stuffed animals, another ABC analysis is performed to determine the product with the highest sales in 2021. From this analysis, it can be deduced that 131 products belonging to the stuffed animals family represent 5.05% of the demand, therefore belonging to classification C, 66 products represent 15.30% belonging to classification B, and finally, 78 products of the stuffed animals family manufactured in the company

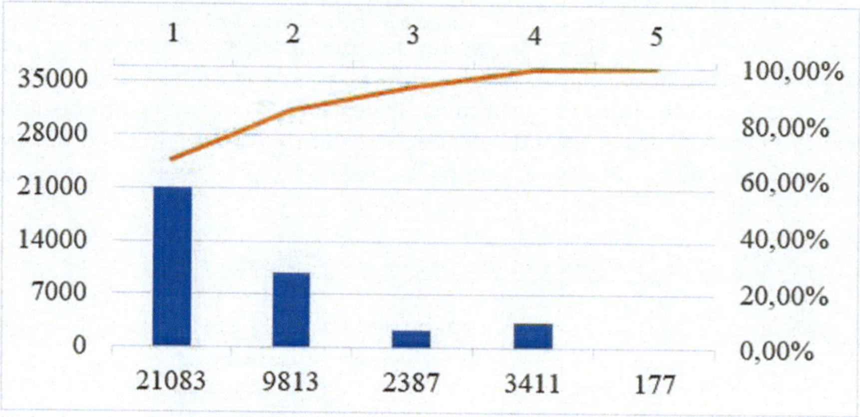

Fig. 3. ABC Analysis of product families.

represent 79.66% of the total demand and belong to classification A, this can be seen in Fig. 4. The main product has the code LT-994, among the 78 products of classification A. This product is the basis for developing the research project since it is the most representative and generates the most significant impact on the company.

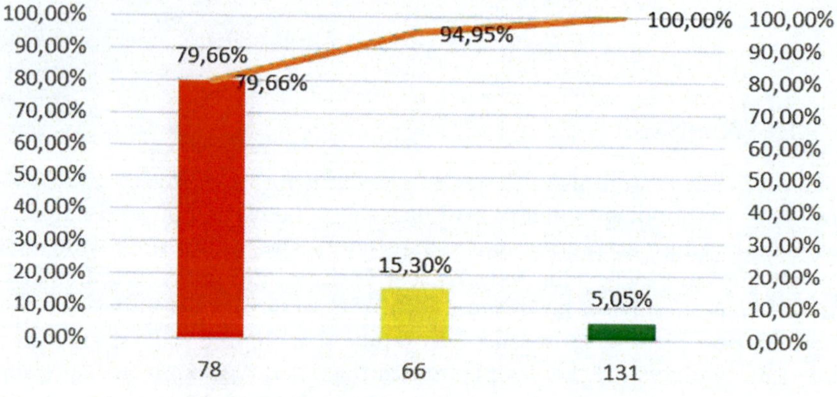

Fig. 4. ABC Analysis of the stuffed animals.

4.4 Time Study

Work measurement is the application of techniques to determine the time a skilled worker spends in carrying out a defined task according to a pre-established standard of performance [10]. The time study was carried out for a production order of 60 units, where the General Electric table was used to know the number of cycles to be taken in each operation. The performance factor and worker supplements were calculated from the Westinghouse system. Table 1, shows a summary of the standard times of each area, as well as its production capacity.

Table 1. Standard time and capacity.

Area	Activities	Standard Time (s)	Capacity (batch/h)
Cutting	Fabric measurement	1689.9	2.13
	Mobile cutting machine	7207.1	0.50
	Die cutter	796.01	4.52
Embroidery	Embroidery	14818.8	0.24
Sewing	Sewing	50800.6	0.14
Finishing	Stuffing	3017.26	1.19
	Refinements	9812.01	0.37
	Packaging	5323.04	0.68
	Total (s)	93464.7	1557.745

4.5 Productivity

Productivity is an indicator that defines how many products or services have been produced for each resource used in their production (labor, time, and capital, among others) within a given period [21]. For the calculation of the current productivity, it is necessary to know the costs of the inputs, which are raw material, labor, inputs, and energy, and the cost of the outputs knowing that a weekly order of 60 units is sold, which in this case is $425.4, according to the data obtained from the sales area. The raw material for completing a production order is $198.84. It is known that the basic salary of the Ecuadorian worker is $425 for 30 working days and rest days and that the employer's contribution is 12.15%, so a worker earns $2.37 per hour of work. The calculation is made with the standard time obtained from the time study for obtaining an order of 60 units and the cost of the man-hour, obtaining a total of $20.7.

Inputs include sleeves, cardboard labels, fabric labels, and yarns for $19.9. According to the company's spreadsheets, an average calculation was made to obtain the cost of electric energy used for an order, obtaining $7.95. The average calculation is $4.80 for each production order, in external sewing, the cost is $21, in external embroidery, it is $37.35, and among the personnel in charge of production control $16.2 for each order, obtaining a total of $79.35. In indirect labor, maintenance of the existing machinery is considered. With this data, it can be established that productivity is equal to 1.30, which, being greater than 1, reflects that the income generated is more significant than its cost.

4.6 Value Stream Mapping

VSM, which stands for Value Stream Mapping, is a tool included in the Lean Manufacturing methodology. It is a graphical representation that allows visualizing, analyzing and improving the production flow. In addition, this representation becomes an excellent aid in improving the capture and analysis of the information produced during the production process. It consists of a flow chart with symbols representing the different work activities and information flows. Thus, each step of the production process is recorded according to whether it adds value or not from the customer's point of view. This way, the company will know which steps can be eliminated because they do not add value [9]. To obtain the initial diagnosis of the company, the VSM is used. This map is made to identify the wastes that are generated during the production process. Therefore, it is used to reduce or eliminate them by applying Lean Manufacturing tools.

The value chain begins when customers place orders to sellers, either by telephone, website or email, and the requested quantity is checked in the warehouse. In the case of non-existence is requested to the production area by a phone call. The production manages with the warehouse manager the shipment of raw material to the cutting area and begins the process with the measurement of fabric, cutting by the mobile cutter, die cutter, embroidery, sewing, stuffing, finishing and packaging, and the necessary numerical data are obtained from the study of times and baseline detailed in previous subsections. This can be verified in Fig. 5.

5 Results

According to the deficiencies found in the company's production area, the different Lean Manufacturing tools used to reduce and eliminate waste after a correct implementation are determined. The following is a description of the waste, the possible causes and the tools that can be used for the Lean Manufacturing implementation plan.

1. **Unnecessary movements**
   These are present in all the work areas since there are several movements in the search for work tools such as scissors, measuring tape, product moulds, and threads. One of its causes is disorder and disorganization under the tables and on the shelves for storing raw materials and cuts made. In addition, obstacles in the aisles constantly hinder the correct circulation of the operators, which is a constant threat to their safety and makes it challenging to transport production orders from one area to another.
2. **Transportation of material or tools**
   Several types of transport generate long distances that operators must travel daily with the materials for each production order made per day. In this case, the total distance travelled in the finishing area is 36.8 m between filling, finishing and packing. This inconvenience should be solved by reducing the transportation of materials in each process.

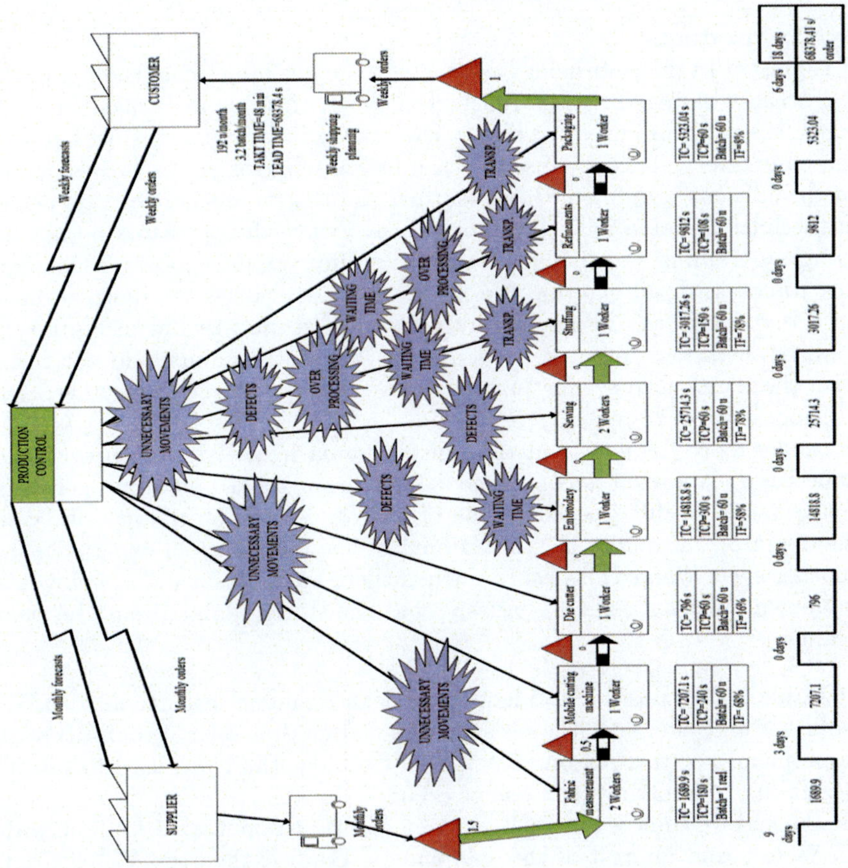

Fig. 5. Current VSM, where TCP is the starting changeover time, TC is the cycle time and TF is the running time of the machines.

3. **Waiting time**
   It frequently happens in the embroidery area because when the automation is misused, the operators are at the service of the machine, i.e. they wait until the embroidery is mechanized, walking from one side to the other, without taking advantage of the time to have the necessary implements ready to continue with the embroidery. On the other hand, waiting has a great weight in the finishing area, in the stuffing process since on many occasions the bad planning of the maintenance of the machine causes unforeseen stops, so the product must wait without contributing any value until the problem in the machinery is solved. Finally, the main inconvenience of waiting in this process is the product quality from previous positions. The most common is the sewing since when stuffing and finishing off, it is possible to observe the faults of the stuffed animals in their different parts or even the existence of holes.

4. **Faulty products**
   It is present in the embroidery area, where the machine signal is often ignored when the winding thread is finished or when the thread breaks. In several cases, the patch is cut very close to the sewing line causing visible failures in the embroidery. This is caused by the lack of control in the process and by the insufficient training of the operators. Another area where defects occur is in the finishing area in the stuffing process, since the operators do not have sufficient training and therefore stuff the stuffed animals poorly. This may be due to excess down entering the machine, which causes the down to have a hard texture, or to not moving properly while stuffing or not using the right tool. In the sewing area, it occurs when the stuffed animals do not comply with the established sewing patterns, have holes or the stitches burst.
5. **Unnecessary processes**
   It occurs more frequently in the finishing area in the stuffing process since unnecessary approval is constantly sought from other operators, generating delays in the stuffing process. One of its causes may be the insecurity that the operator has due to lack of training in his job or lack of experience in it, another event where this occurs is when there are defects, and in many cases, the down must be removed entirely, and the stuffed animal must be stuffed again.

In Table 2, it is possible to identify the lean manufacturing tools focused on the solution of waste, and it is necessary to clarify that several tools attack the same problem. For this reason, a weighted factor method was used to establish the feasibility of applying each one of them.

In the case of unnecessary movements, the best application tool, according to the factors and interest of the company, is the 5 s, then the Visual Control or Andon, the TPM and finally the Heijunka. For waiting waste, the tool with the best score is the TPM. For the waste generated by defective products, the Kaizen and Jidoka tools are used. In the case of transportation waste, the 5 s is the implementation tool. Finally, for waste due to overprocessing or unnecessary processes, the Jidoka and Kaizen tools are used.

6 Discussion

Latin America is home to textile-producing nations emerging as potential growth markets, driven by domestic demand, economic growth and purchasing power. In the textile field, Peru is the largest exporter of clothing in South America, thanks to well-differentiated raw materials such as the fibres of its camelids, alpacas, llamas and vicuñas, and its Pima and Tanguis pieces of cotton. For this reason, it is the country that has the largest amount of research on the optimization of resources, such is the case of [6,15,18].

Despite the Covid-19 pandemic, forecasts in Latin America in this health crisis were encouraging, estimating that spending on clothing and footwear will grow by 7.2%, reaching 221 billion dollars by 2021. This has generated much

Table 2. Lean Manufacturing Tools.

L.M Tools	Unnecesary Movements	Transp.	W. Time	Unnecessary Processes	Defects
5S	x	x	x	x	x
SMED			x		
Kanban			x	x	x
TPM	x	x	x		x
Heijunka	x	x			
Poka Yoke			x	x	x
Andon	x		x		
JIT					
Kaizen				x	x
Jidoka				x	x

interest from companies that manage foreign brands. In order to capture new markets and maintain the ones it has achieved, Latin America will have to face substantial challenges, such as market dynamics, where factors such as innovation and sustainability will have to be maintained to be more competitive.

Another challenge has to do with technological trends. Technical advances must always go hand in hand with the production of resources since keeping up with the latest developments is an indispensable factor for survival in the textile world. In this sense, since Latin America is a region with little machinery production, it will have to select, transfer and adapt the best technology that fits the environment in which it is located.

Diversification in printing, yarns and textiles will play an essential role in strengthening the industry, both locally and regionally, as long as local, regional and international brands crystallize continuous development and innovation.

Lean manufacturing is a business strategy focused on increasing efficiency and reducing waste in a company on an ongoing basis. This is particularly true in the digital age, when new technologies can make companies more agile. The last thing a company, especially in the textile industry, wants to do is waste its limited resources. When done right, it can significantly affect the bottom line. Most senior managers in Latin America are not thoroughly familiar with this culture and are wary of the benefits to be gained. However, many companies have already implemented tools in their lines of business that provide them with an optimized and low-cost operating environment.

Process optimization is an ongoing exercise that pays to provide teams with the tools to remain agile and competitive. Among the principles of lean manufacturing, we can mention: i) Define value: lean manufacturing should seek to understand the value customers assign to the company's products. Knowing how much they are willing to pay helps to establish a production budget. ii) Mapping the value flow: The product's complete life cycle, from its conception to its final delivery, must be covered. By mapping the flow, it will be possible to identify where value is added. Then, the steps or actions that do not add value should be eliminated, classifying these elements as waste.

iii) Create an optimal flow: The flow of value should flow steadily. Any bottleneck generates waste. It must be ensured that all the processes of the product life cycle are optimized and that they flow uniformly from one process to another. iv) Establish a pull: Inventory is one of the most significant potential sources of waste. Limit inventory and work-in-process items to only those required by the customer. Seek just-in-time delivery in the exact quantities needed to eliminate waste.

As proposed in this work, an efficient combination of Lean tools and respect for its principles can allow the Ecuadorian textile industry to access great benefits. i) Manufacture defect-free products: If the sold items have defects, they could cost the extra money needed to make corrections. Lean manufacturing focuses on designing and manufacturing a perfect product. Making a product free of defects provides the opportunity to devise a thorough manufacturing process. ii) Not over-processing: Over-processing means the company spends more on making a product than it is worth to the consumer. Lean manufacturing seeks to manufacture products without unnecessary add-ons. The main objective is to reduce the production process to only what is essential and required.

iii) Reduce inactivity: Lean manufacturing seeks to reduce the waiting time between two or more processes. This usually happens when one operation has to wait for the other to continue working. Applying lean manufacturing saves time as well as money. In this case, by finding ways to reduce the time it takes for a process to complete a job or by assigning additional tasks to idle employees.

iv) Reduce unnecessary movements: It is crucial to find ways to make the movements of employees, machinery and products as efficient as possible. Being lean means making a complicated movement of someone or something more straightforward and eliminating all movements that have no utility. Applying lean manufacturing ranges from better organizing tools and products to changing the layout of a workstation to reduce walking.

7 Conclusions and Future Work

This work has presented a proposal to combine several Lean Manufacturing tools to improve the productivity of a textile company in Ambato-Ecuador. The application of Lean Manufacturing in an industrial company implies eradicating and suppressing everything that hinders and makes its production processes more expensive, achieving a more competitive final result due to the reduction in costs that it causes.

This is produced by performing a total minimization of resources, eliminating those parts that are unproductive or have little performance (waste). However, it is not necessary to implement all Lean Manufacturing tools. Each of them has specific characteristics and purposes. Therefore, each company must choose those that best suit its production process. These tools can be implemented in isolation and gradually, focusing on the medium and long term. As more tools are implemented, the company's productivity and profits significantly increase as they build on each other.

In future work, it is proposed to reach the last phase of this study, i.e., to implement the previously selected tools and carry out a follow-up-monitoring process, in order to observe the progress of each one of them as well as the individual and global impact they have on the company's productivity.

References

1. Annie Rose Nirmala, D., Kannan, V., Thanalakshmi, M., Joe Patrick Gnanaraj, S., Appadurai, M.: Inventory management and control system using ABC and VED analysis. Mater. Today Proc. 922–925 (2022). https://doi.org/10.1016/j.matpr.2021.10.315
2. Arica-Hernandez, M.A., Llagas-Llontop, S.E., Khaburzaniya, I.: Implementation of lean manufacturing principles to increase productivity in SMEs in the manufacturing sector of clothing. In: 2022 The 3rd International Conference on Industrial Engineering and Industrial Management, pp. 162–166. ACM, New York, NY, USA, January 2022. https://doi.org/10.1145/3524338.3524363
3. Buer, S.V., Strandhagen, J.O., Chan, F.T.S.: The link between industry 4.0 and lean manufacturing: mapping current research and establishing a research agenda. Int. J. Prod. Res. (8), 2924–2940 (2018). https://doi.org/10.1080/00207543.2018.1442945
4. Canales-Jeri, L., Rondinel-Oviedo, V., Flores-Perez, A., Collao-Diaz, M.: Lean model applying JIT, Kanban, and Standardized work to increase the productivity and management in a textile SME. In: 2022 The 3rd International Conference on Industrial Engineering and Industrial Management, pp. 79–84. ACM, New York, NY, USA, January 2022. https://doi.org/10.1145/3524338.3524351
5. Carhuayano-Paucar, C., Caceres-Cipriano, J., Viacava-Campos, G.: Case study: a medium-sized baby clothing manufacturing company, a production model based on standardization and SLP. In: 2022 8th International Conference on Information Management (ICIM), pp. 191–195. IEEE, March 2022. https://doi.org/10.1109/ICIM56520.2022.00042
6. Cespedes-Pino, R., Hurtado-Laguna, J., Macassi-Jaurequi, I., Raymundo-Ibañez, C., Dominguez, F.: LEAN production management model based on organizational culture to improve cutting process efficiency in a textile and clothing SME in Peru. In: IOP Conference Series: Materials Science and Engineering, no. 1, p. 012004 (2020). https://doi.org/10.1088/1757-899X/796/1/012004
7. Coronel-Vasquez, J., Huamani-Lara, D., Flores-Perez, A., Collao-Diaz, M., Quiroz-Flores, J.: Logistics management model to reduce non-conforming orders through lean warehouse and JIT: a case of study in textile SMEs in Peru. In: 2022 The 9th International Conference on Industrial Engineering and Applications (Europe), pp. 19–24. ACM, New York, NY, USA, January 2022. https://doi.org/10.1145/3523132.3523136
8. Deshmukh, M., Gangele, A., Gope, D.K., Dewangan, S.: Study and implementation of lean manufacturing strategies: a literature review. Mater. Today Proc. 1489–1495 (2022). https://doi.org/10.1016/j.matpr.2022.02.155

9. Dinesh, S.N., Shalini, M., Vijay, M., Vijey Mohan, R.C., Saminathan, R., Subbiah, R.: Improving the productivity in carton manufacturing industry using stream mapping (VSM). Mater. Today Proc. (2022). https://doi.org/10.1016/j.matpr.2022.05.015. ISSN: 22147853
10. Dong, Q., Lu, J., Kan, S.: A study on man-hour calculation model for multi-station and multi-fixture machining center. In: Jin, D., Lin, S. (eds.) Advances in Electronic Commerce, Web Application and Communication. AISC, vol. 149, pp. 403–411. Springer, Berlin, Heidelberg (2012). https://doi.org/10.1007/978-3-642-28658-2_62
11. Garcia, C.A., Naranjo, J.E., Alvarez-M., E., Garcia, M.V.: Training virtual environment for teaching simulation and control of pneumatic systems. In: De Paolis, L.T., Bourdot, P. (eds.) AVR 2019. LNCS, vol. 11613, pp. 91–104. Springer, Cham (2019). https://doi.org/10.1007/978-3-030-25965-5_8
12. Ibujés Villacís, J.M., Benavides Pazmiño, M.A.: Contribución de la tecnología a la productividad de las pymes de la industria textil en Ecuador. Cuadernos de Economía (115), 140–150 (2018). https://doi.org/10.1016/j.cesjef.2017.05.002
13. Instituto Nacional de Estadisticas y Censos: Boletin Tecnico DIEE (2020)
14. Kumar, N., Shahzeb Hasan, S., Srivastava, K., Akhtar, R., Kumar Yadav, R., Choubey, V.K.: Lean manufacturing techniques and its implementation: a review. Mater. Today Proc. 1188–1192 (2022). https://doi.org/10.1016/j.matpr.2022.03.481
15. León-Guizado, S., Castro-Hucharo, A., Chavez-Soriano, P., Raymundo, C.: Production model under lean manufacturing and change awareness approaches to reduce order delays at small and medium-sized enterprises from the clothing sector in Peru. In: Iano, Y., Arthur, R., Saotome, O., Kemper, G., Padilha França, R. (eds.) BTSym 2019. SIST, vol. 201, pp. 391–400. Springer, Cham (2021). https://doi.org/10.1007/978-3-030-57548-9_36
16. Llamuca, E.S., Garcia, C.A., Naranjo, J.E., Garcia, M.V.: Cyber-physical production systems for industrial shop-floor integration based on AMQP. In: Proceedings - 2019 International Conference on Information Systems and Software Technologies, ICI2ST 2019, pp. 48–54 (2019). https://doi.org/10.1109/ICI2ST.2019.00014
17. Llamuca, J.D., Garcia, C.A., Naranjo, J.E., Rosero, C., Alvarez-M, E., Garcia, M.V.: Integrating ISA-95 and IEC-61499 for distributed control system monitoring. Adv. Intell. Syst. Comput. **1099**, 66–80 (2020). https://doi.org/10.1007/978-3-030-35740-5_5
18. Martinez-Condor, B., Mamani-Motta, F., Macassi-Jaurequi, I., Raymundo-Ibañez, C., Perez, M.: Lean production model aligned with organizational culture to reduce order fulfillment issues in micro- and small-sized textile businesses in Peru. In: IOP Conference Series: Materials Science and Engineering, no. 1, p. 012016 (2020). https://doi.org/10.1088/1757-899X/796/1/012016
19. Matos, A.R., Dias, J.P., Lima, R.M., Sousa, R.M., Carvalho, M.S.: Redesign of the internal logistics system of a textile supplier for the automotive industry, pp. 49–60 (2022). https://doi.org/10.1007/978-3-031-09360-9_5
20. Mouli, C.S.B.C., Kulkarni, S.D., Deepak, S.: Productivity improvement of a small scale industry by the application of an effective plant layout and weld-fixture design. Mater. Today Proc. 367–372 (2022). https://doi.org/10.1016/j.matpr.2021.09.060
21. Nieto, M.J., Rodríguez, A., Hernández, V.: International sourcing and the productivity of SMEs in transition countries: formal and informal region effects and the communist footprint. J. Bus. Res. 347–359 (2022). https://doi.org/10.1016/j.jbusres.2022.02.069

22. Orejuela-Escobar, L.M., Landázuri, A.C., Goodell, B.: Second generation biorefining in Ecuador: circular bioeconomy, zero waste technology, environment and sustainable development: the nexus. J. Bioresour. Bioprod. (2), 83–107 (2021). https://doi.org/10.1016/j.jobab.2021.01.004
23. Palange, A., Dhatrak, P.: Lean manufacturing a vital tool to enhance productivity in manufacturing. Mater. Today Proc. 729–736 (2021). https://doi.org/10.1016/j.matpr.2020.12.193
24. ur Rehman, M.Z., et al.: Enablers and barriers for adoption of lean system: a mix method approach. Int. J. Product. Qual. Manag. (2), 220 (2022). https://doi.org/10.1504/IJPQM.2022.121311
25. da Silva, R.C., da Silva, T.M., dos Santos, F.A., Robles Junior, A.: Estudo comparativo entre a Adidas e a Nike sob o enfoque do ambiente Lean: manufacturing e accounting. CAFI - Contabilidade, Atuária, Finanças & Informação (1), 68–85 (2019). https://doi.org/10.23925/cafi.v2i1.40912
26. Sun, Y., Li, L., Shi, H., Chong, D.: The transformation and upgrade of China's manufacturing industry in Industry 4.0 era. Syst. Res. Behav. Sci. (4), 734–740 (2020). https://doi.org/10.1002/sres.2714
27. Vayas-Ortega, G., Larzabal-Fernandez, A., Rivera, M.G., Perez, A., Gonzalez, A.: Characterization of corporate social responsibility in the textile industry of jeans: Pelileo-Ecuador. A case study. Adv. Sci. Lett. (12), 9291–9295 (2018). https://doi.org/10.1166/asl.2018.12257

Author Index

M. V. Garcia and C. Gordón-Gallegos (Eds.): CSEI 2022, LNNS 678, pp. 679–680, 2023.
https://doi.org/10.1007/978-3-031-30592-4